electrical trade practices

electrical trade practices

CENGAGE

Electrical Trade Practices
1st Edition
Ralph Berry
Phill Chadwick

Publishing manager: Dorothy Chiu
Senior publishing editor: Sophie Kalienicki
Developmental editor: Tharaha Richards
Project editor: Michaela Skelly
Art direction, cover & text design: Danielle Maccarone
Editor: Anne Mulvaney
Proofreader: James Anderson
Indexer: Julie King
Permissions/Photo researcher: Wendy Duncan
Reprint: Magda Koralewska
Typesetter: MPS Limited

Acknowledgements
Cengage Learning Australia wishes to acknowledge the contribution of TAFE NSW South Western Sydney Institute in the preparation of this book.

For product information and technology assistance,
in Australia call **1300 790 853**;
in New Zealand call **0800 449 725**

For permission to use material from this text or product, please email **aust.permissions@cengage.com**

National Library of Australia Cataloguing-in-Publication Data
Creator: Berry, Ralph James, author.
Title: *Electrical Trade Practices* / Ralph Berry; Phillip Chadwick.
ISBN: 9780170236164 (paperback)
Subjects: Electricity--Study and teaching (Higher)
Other Creators/Contributors: Chadwick, Phillip, author.
Dewey Number: 621.307

Cengage Learning Australia
Level 7, 80 Dorcas Street
South Melbourne, Victoria Australia 3205

Cengage Learning New Zealand
Unit 4B Rosedale Office Park
331 Rosedale Road, Albany, North Shore 0632, NZ

For learning solutions, visit **cengage.com.au**

Printed in China by China Translation & Printing Services.
4 5 6 7 8 21 20 19 18

Brief contents

Contents

Preface

This book has been written to help apprentice electricians in the electrocommunications and energy utilities areas study the practical aspects of the Certificate III in Electrotechnology Electrician course (National course code UEE30811).

The text and illustrations follow the layout of the required knowledge and skills set out in each competency of the UEE11 training package.

We have not set out to create a comprehensive information source but a study guide with margin notes to emphasise safety issues and help locate information during revision.

Web references are provided where additional information can be sought. The book is generally written to the student where possible so they feel more connected with the subject.

About the authors

Ralph Berry

Ralph spent 17 years working as an electrician in mining, construction and heavy transport before starting work as a TAFE teacher with the Electrical Trades School in Launceston in Tasmania.

During 31 years of teaching Ralph was involved in many projects including training for power station electricians in Samoa and more recently the Electroflex and e-Learning for Hydro Generation Technician Trainees projects for the Flexible Learning Framework. Ralph also assisted with the development and organising of online materials to help apprentices study from home and in their workplaces.

Phillip Chadwick

Phillip started his career as an apprentice electrical fitter mechanic in western Sydney's electrical supply industry. Work included the installation and maintenance of low voltage district substations, connection of new services and consumer metering equipment. His electrical experience extends into domestic, commercial and industrial contracting as well as the entertainment industry. He joined TAFE NSW in 1990 and has teaching positions at North Sydney, Granville and Miller colleges. His teaching experience includes development of student resources, classroom and workplace delivery and assessment.

Acknowledgements

Ralph Berry

I would like to thank Idris Reese from TasTAFE at Launceston who helped me locate the equipment to photograph, and to Andrew Brown and Joe Kurtukoff with whom I have shared the teaching of electrical practice over many years. I would especially like to thank my wife Maureen for spending countless evenings without her husband while I wrestled with the text for each chapter.

Finally I would like to thank the team at Cengage Learning who have made this publication possible, in particular Sophie Kaliniecki for inviting me to be a co-author of this book, and Tharaha Richards for helping me to set out the text and materials, and to the contributors listed below.

Phillip Chadwick

Writing a book of this nature has been an incredible experience and I feel very privileged for the opportunity to do so. I would like to express my thanks to Peter Phillips who has played a vital role as a mentor; without his assistance this work would have not been possible.

To all of the electrical teachers at Miller College I am very grateful for your support and encouragement during this process. I would also like to acknowledge my appreciation to Frank Cahill; his guidance over many years has been central to my development as a teacher of electrical trades.

I share Ralph's sentiments for the team at Cengage Learning, I thank Sophie Kaliniecki for giving me the opportunity, and Tharaha Richards for keeping me on track. Most of all I would like to thank my family and partner Julie for putting up with me and patiently supporting me.

The publishers and the authors would like to thank the following people for contributing, reviewing the text and providing valuable feedback.

- Kris Heel, Industry, formerly TAFE Queensland
- Aldo Vergan, Endeavour Energy
- Keith Heavens, SkillsTech Australia
- Joe Pagnoccolo, Box Hill Institute
- Michael Henderson, College of Electrical Training

A special thanks go to the Photographer, Ed Hawkins.

Resources guide

As you read this text you will find features in every chapter to enhance your study of electrical trade practices, which will help you understand how your studies are applied in the real world.

Chapter outline

The **chapter outline** lists the key topics covered in the chapter.

An **introductory paragraph** provides the context of the chapter to give you a clear sense of what the chapter will cover.

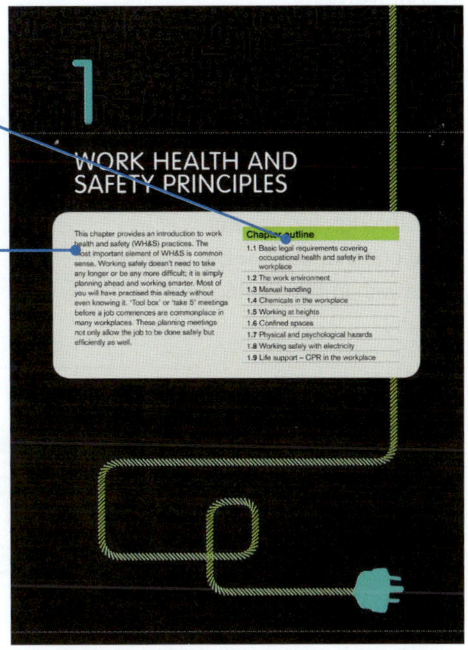
1

WORK HEALTH AND SAFETY PRINCIPLES

This chapter provides an introduction to work health and safety (WH&S) practices. The most important element of WH&S is common sense. Working safely doesn't need to take any longer or be any more difficult; it is simply planning ahead and working smarter. Most of you will have practised this already without even knowing it. 'Tool box' or 'take 5' meetings before a job commences are commonplace in many workplaces. These planning meetings not only allow the job to be done safely but efficiently as well.

Chapter outline

1.1 Basic legal requirements covering occupational health and safety in the workplace
1.2 The work environment
1.3 Manual handling
1.4 Chemicals in the workplace
1.5 Working at heights
1.6 Confined spaces
1.7 Physical and psychological hazards
1.8 Working safely with electricity
1.9 Life support – CPR in the workplace

Safety note

Safety note boxes run throughout the chapters to give you important information on safe work practices.

> Adjustable spanners can slip off the nut or bolt head unexpectedly if you apply too much torque.

Sustainability

Sustainability boxes highlight the key energy sustainability issues relevant to the electrical industry.

> Drilled holes should be cleaned with a hand pump or vacuum so the dust does not cause pollution.

Information

Information boxes provide additional useful information about the concepts being discussed.

> In the AS/NZS 3000:2007 index, a wiring enclosure can also be a called a conduit or a duct; a wiring system can be a cable or TPS or building wire in conduit; a bend can also be known as a change of direction.

Key concepts

The **key concepts** mentioned in the chapters are emphasised in the margin for your quick, easy reference, and to help you in your revisions.

Compliance with the electrical safety act

Examples

Examples throughout the book show how you can work through the various situations that you will be faced with in the electrical industry.

EXAMPLE 9.3

Heat energy transfer

Determine the heat loss from a room through a glass window 4.0 m × 2.00 m × 5 mm thick. The inner room temperature is 24°C and the outside temperature is 12°C, given that the **k** for glass is 0.84 j/s.m.°C.

$$\frac{\Delta Q}{\Delta t} = k \times A \times \frac{T1 - T2}{l}$$

$$= 0.84 \times 8 \times \frac{(24 - 12)}{0.005}$$

$$= 1612.8 \text{ j/s}$$

$$= 1612.8 \text{ watts}$$

If this situation was continued for 10 hours, the amount of electrical energy wasted in commercial units would be 16 kilowatt hours (Kwh). Check your latest electricity account to see what this would cost.

Summing up

Each key learning area is summarised in a clear, succinct and easy-to-follow manner. The key learning points are all included in **summing up** to assist your studies.

SUMMING UP...

- A flat TPS cable comprises one or more cable cores surrounded by a sheath laid up in flat formation.
- TPS cables can provide double insulation when required.
- Flat TPS is the most commonly used wiring system.
- Systems are colour coded. Green, yellow and green/yellow are prohibited.
- Sizes of multicore cables are limited due to installation difficulties with heavy cables.
- Cables must be supported by clips to prevent disturbance or strain.
- Additional protection is required in high mechanical risk situations.
- Two configurations used in lighting circuits are loop at the light and loop at the switch.
- An earth terminal must be provided at all lighting points.

End-of chapter materials

At the end of each chapter, you will find several additional learning tools to help you revise.

Review questions

Review questions are provided for you to revise the content in the chapter. A solutions section provides an answer for all of these questions at the back of the book.

REVIEW EXERCISES

Answer these questions and check your answers at the back of the book.

Review questions 4.1

1 What is an ISO standard and how do these standards apply to Australian/New Zealand standards?
2 Name the standard for general technical drawing.
3 List two uses for a thick dashed line in technical drawings.

Review questions 4.2

1 Give two examples of pure metals.
2 Briefly define toughness.

Review questions 4.3

1 When marking out, how can scribed lines on metal be made to stand out more clearly?
2 Describe how you would locate the centre of a piece of round stock using jenny calipers.

Review questions 4.4

1 Which type of vice is best suited to hold steel conduit?
2 What direction should the teeth of a hacksaw blade point in relation to the handle of the hacksaw frame?

Review questions 4.5

1 What are the functions of the flutes in a twist drill?
2 How is a drill sharpened for drilling brass?

Review questions 4.6

1 State the three types of thread cutting taps, and where each type is used.
2 State the purpose of a die nut.

Review questions 4.7

1 What is the typical weight of an engineer's hammer suitable for general use?

Online resources

You can follow up additional **online resources** to help you extend your knowledge and understanding of the key points through wider reading.

ONLINE RESOURCES

Sample safety forms
WorkSafe Victoria
http://www.worksafe.vic.gov.au > Forms and Publications > Safety and Prevention
Asbestos Awareness
http://www.asbestosawareness.com.au

Codes of practice: Managing electrical risks at the workplace, Confined spaces, etc.
WorkCover NSW
http://www.workcover.nsw.gov.au > Forms and Publications > Work, Health and Safety > Codes of practice

Documents related to HV work, including access permits
TransGrid
http://www.transgrid.com.au > Safety, health & environment > Health and safety > Power system safety rules

For the instructor

Cengage Learning is pleased to provide you with a selection of online supplements to help you prepare for your classes. These resources have been specifically developed to complement *Electrical Trade Practices*, first edition, and are available from **http://login.cengage.com**

Competency mapping grid

The **mapping grid** is mapped to the essential knowledge and associated skills needed to complete the UEE30811 Certificate III in Electrotechnology Electrician, as endorsed by the National Quality Council on 17 April 2014. Please check for details on the latest version by going to TGA (http://training.gov.au), the official database on Vocational Education and Training in Australia.

Instructor's manual

The **instructor's manual** provides you with additional review exercises complete with answers to help you test your students.

PowerPoint presentations

These **PowerPoint** presentations have been prepared to cover the main points of the chapters in *Electrical Trade Practices*, first edition, and can be edited to suit your course. Use these slides to enhance your lectures and reinforce the key principles of the course.

Testbank

A **test bank** of questions covering each chapters learning objectives and key topics, has been prepared for your use. These are available in a Word file format and can be downloaded directly into your own Learning Management System, and can be customised to meet your students' learning requirements.

Artwork

Chapter by chapter **artwork** files, which feature all of the images, diagrams and tables from the text, can be used by you in a variety of ways, including lecture presentations, student handouts and supplementary material in course management systems.

1

WORK HEALTH AND SAFETY PRINCIPLES

This chapter provides an introduction to work health and safety (WH&S) practices. The most important element of WH&S is common sense. Working safely doesn't need to take any longer or be any more difficult; it is simply planning ahead and working smarter. Most of you will have practised this already without even knowing it. 'Tool box' or 'take 5' meetings before a job commences are commonplace in many workplaces. These planning meetings not only allow the job to be done safely but efficiently as well.

Chapter outline

1.1 Basic legal requirements covering occupational health and safety in the workplace

The underlying principles of WH&S

A six-point approach has been devised to help implement effective work health and safety systems. The six points, as seen in Figure 1.1, are:

1. Develop an OWHS policy and related programs.
2. Set up a consultation mechanism with employees.
3. Establish a training strategy.
4. Establish a hazard identification and workplace assessment process.
5. Develop and implement risk control.
6. Promote, maintain and improve these strategies.

Source: 'It's better to be safe than sorry', Adelaide Business Hub, http://adelaidebusinesshub.com.au

FIGURE 1.1 The six elements of WH&S risk management

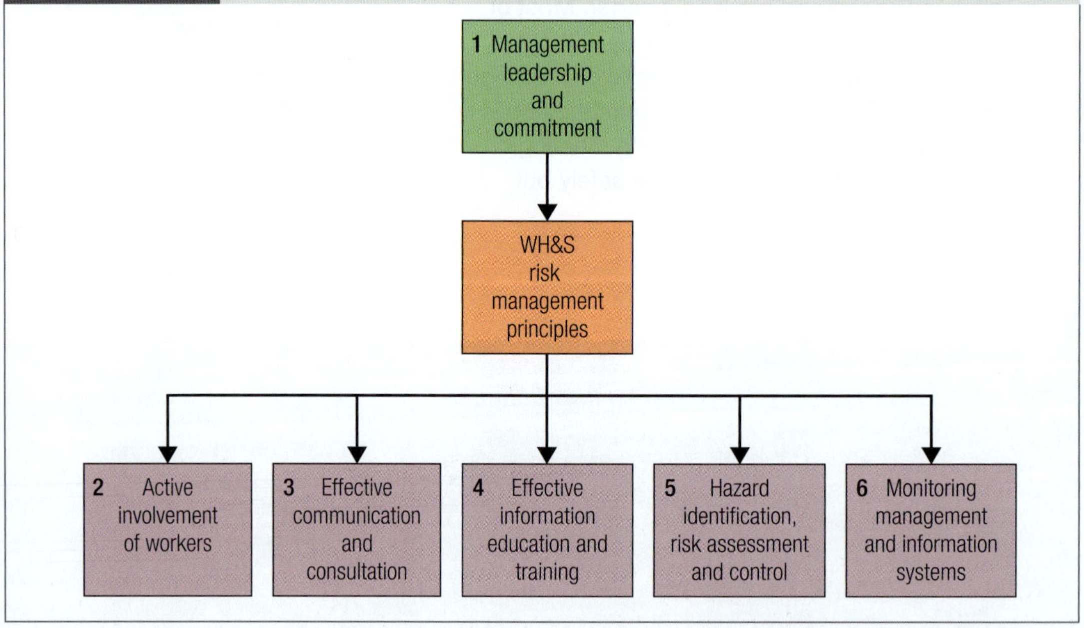

General aims and objectives of WH&S legislation

The Commonwealth and most of the state/territory governments have agreed to harmonise their work health and safety laws, including regulations and codes of practice, so they are similar in each jurisdiction. The *Model Work Health and Safety Act 2010* is the template that state and territory WH&S laws are to be based on. This act is administered by Safe Work Australia, the body also responsible for developing an action plan in regard to workers' compensation.

There are a number of federal and state government agencies that develop, implement and enforce WH&S legislation throughout Australia. Some of these regulatory bodies are also responsible for the administration of workers' compensation.

The general aims are focused on:

- reducing high incidence/severity risks
- improving the ability of business operators and workers to manage occupational safety and health effectively

- preventing occupational related diseases
- eliminating hazards at the design stage
- strengthening the ability of government to influence workplace safety and health outcomes
- increasing the competitiveness of state and territory economies through productive, healthy and safe workplaces
- promoting productive, healthy and safe workplaces for workers and employers.

Responsibilities, rights and obligations

As noted in Safe Work Australia's 'Explanatory Memorandum – Model Work Health And Safety Bill' (http://www.safeworkaustralia.gov.au), both employers and employees are required to do what is reasonably practicable in ensuring health and safety, which includes taking into account:

- the likelihood of the hazard or the risk concerned occurring
- the degree of harm that might result from the hazard or the risk
- what the person concerned knows, or ought reasonably to know, about the hazard or the risk
- ways of eliminating or minimising the risk
- the availability and suitability of ways to eliminate or minimise the risk
- assessing the extent of the risk and the available ways of eliminating or minimising the risk, and then weighing this up against the cost.

Employers

Employers are referred to as persons conducting business or undertakings (PCBUs) and have both WH&S rights and responsibilities. Under WH&S legislation company directors are termed 'officers' of the PCBU. A self-employed person is also considered a PCBU. A PCBU must do what is *reasonably practicable* to make the workplace safe (see Table 1.1).

TABLE 1.1 PCBU's primary duty of care

Duty	Description
Health and safety	Of workers and subcontractors
Health and safety	Of other persons, i.e. customers and visitors
Safe work environment	Provision and maintenance
Safe plant and structures	Provision and maintenance
Safe systems of work	Provision and maintenance; safety includes the design stage of processes
Safe use, handling and storage	Of plant, structures and substances
Adequate facilities	For the welfare at work of workers including ensuring access to those facilities, i.e. site sheds, meal rooms, toilets and showers
Information, training, instruction or supervision	To protect all persons from risks to their health and safety
Monitoring	The health of workers and the conditions at the workplace
PCBU owned accommodation	Must be maintained so that the worker occupying the premises is not exposed to risks to health and safety.

Source: Model Work Health and Safety Bill, http://web1.chost.com.au/~rmfconsu/wp-content/uploads/2011/07/ModelWHSBill_RevisedDraft11May2010.doc

As outlined in the Model Work Health and Safety Bill, a self-employed person must ensure, as far as is reasonably practicable, their own health and safety while at work.

Employees

A worker is defined as a person who carries out work in any capacity for a PCBU. This includes an employee, a contractor or subcontractor, an apprentice or trainee, and a student gaining work

Your WH&S responsibilities

experience. A PCBU is also a worker if he/she is an individual who carries out work in that business or undertaking.

As defined by the NSW *Work Health and Safety Act 2011*, s. 28, responsibilities of workers include:

- taking reasonable care for his or her own health and safety
- taking reasonable care that his or her acts or omissions do not adversely affect the health and safety of other persons
- complying, so far as the worker is reasonably able, with any reasonable instruction that is given by the PCBU to allow the person to comply with the WH&S Act
- cooperating with any reasonable policy or procedure of the PCBU relating to health or safety at the workplace that has been notified to workers.

Major functions of safety committees and representatives

Health and safety committees

As described by SafeWork SA (http://www.safeworkaustralia.gov.au/sites/swa/ > about > publications > Model Codes of Practice > Work Health and Safety Consultation, Co-operation and Co-ordination), a health and safety committee brings together workers and management to assist in the development and review of health and safety policies and procedures for the workplace. A PCBU must establish a health and safety committee within two months after being requested to do so by five or more workers, or by a health and safety representative (HSR), at the workplace. Health and safety committees must meet at least every three months. HSRs may choose to be members of the health and safety committee. In total, at least half of the members of the committee must be workers who are not nominated by management. The major functions of a health and safety committee include:

- having regular, structured discussions about health and safety matters
- encouraging a cooperative approach to health and safety
- bringing together a group of worker and business representatives to discuss and develop ways of improving systems for managing health and safety at the workplace
- encouraging the development and retention of corporate knowledge on health and safety matters for the workplace.

Health and safety representatives (HSR)

Safe Work Australia's *Worker Representation and Participation Guide* (Safe Work Australia, p. 4; http://www.safeworkaustralia.gov.au) outlines that a worker may ask the PCBU for the election of a health and safety representative to represent them on work health and safety matters. Work groups are formed to enable workers to elect HSRs to represent them on health and safety matters. The process requires the PCBU and workers to negotiate and agree on the formation of work groups. To establish a work group, the PCBU must:

- take all reasonable steps to commence negotiations with the workers within 14 days after a worker makes the request
- negotiate with a workers' representative (such as a union official) if a worker makes a request to do so
- notify the workers of the outcome of the negotiations and of any work groups determined by the agreement.

The WH&S Act sets out specific powers and functions that a HSR can perform in the interests of the workers they represent. These powers are listed in Table 1.2.

TABLE 1.2 The functions of HSRs

Function	Task
Represent workers	in their work group in relation to work health and safety matters.
Monitor	the measures taken by the PCBU to comply with the WH&S Act in relation to their work group members.
Investigate	complaints from work group members about work health and safety.
Inquire	into anything that appears to be a risk to the health or safety of work group members, arising from the conduct of the business or undertaking.

Source: Adapted from Safe Work Australia, *Worker Representation and Participation Guide*, p. 13, http://www.safeworkaustralia.gov.au

Table 1.3 shows some of the ways a HSR may exercise their powers.

TABLE 1.3 The powers of HSRs

Power	Description
Inspect	the workplace at any time after giving notice to the PCBU or at any time following an incident or any situation involving a serious risk.
Attend interviews	between one or more work group members and an inspector or the PCBU.
Request	the establishment of a Health and Safety Committee (HSC).
Direct	a work group member to cease unsafe work.
Issue	a Provisional Improvement Notice (PIN).

Source: Adapted from Safe Work Australia, *Worker Representation and Participations Guide*, p. 13, http://www.safeworkaustralia.gov.au

Powers given to WH&S inspectors

Inspectors must, when entering a workplace, take reasonable steps to notify the PCBU (employer) as well as any HSRs. Inspectors may enter any workplace without prior notice.

The Community and Public Sector Union (CPSU) fact sheet 'Role of inspectors in compliance and enforcement' states that 'An inspector is not required to provide notification if that would defeat the purpose of entry or would cause delay in investigating the suspected contravention' (p. 1).

FIGURE 1.2 Enforcement measures

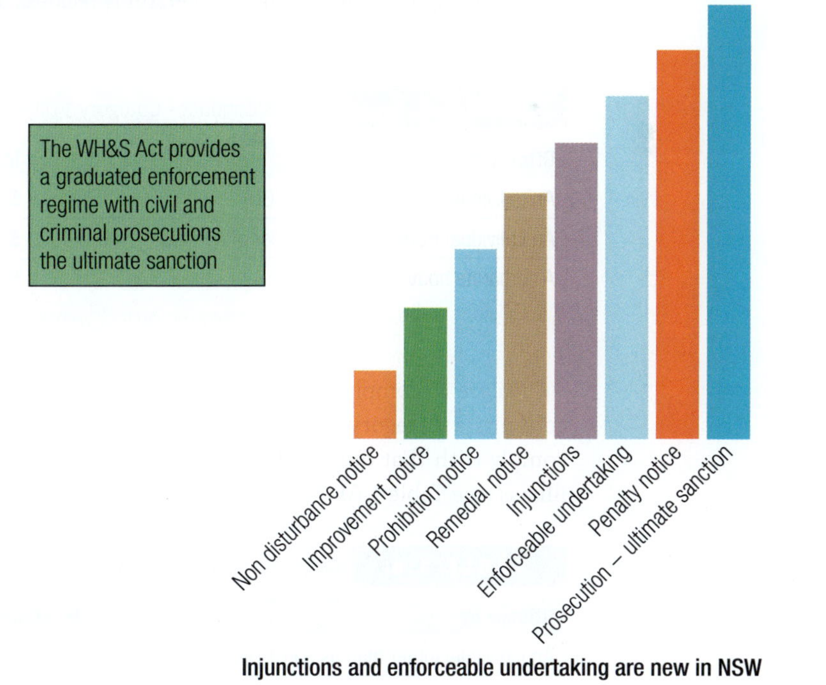

Injunctions and enforceable undertaking are new in NSW

Notifiable incident

The WH&S Act provides that a 'notifiable incident' arising out of work carried out in a workplace is:

- the death of a person
- a serious injury or illness
- or a dangerous incident.

Only the most serious safety incidents are intended to be notifiable, and they trigger requirements to preserve the incident site pending further direction from the regulator (see Figure 1.2 and Table 1.4).

TABLE 1.4 Notices issued by WH&S inspectors

Notice type	Description
Non-disturbance notices	A non-disturbance notice may require a person to preserve a site where a notifiable incident has occurred, or prevent the disturbance of a particular site in other circumstances.
Improvement notices	An inspector can issue an improvement notice if they believe that a person is contravening the model WH&S Act.
Prohibition notices	An inspector can issue a prohibition notice if they believe there is a serious risk to the health or safety of a person at a workplace resulting from an exposure to a hazard. A prohibition notice may also be issued if there is the potential for such an activity to occur.
Remedial action and injunctions	If a person issued with a prohibition notice fails to comply with the notice, the regulator may take remedial action to make the workplace safe.

Source: Adapted from Community and Public Sector Union, 'Role of inspectors in compliance and enforcement: Fact sheet', pp. 3, 4

Penalties

Safe Work Australia's 'Explanatory Memorandum – Model Work Health And Safety Bill' (http://www.safeworkaustralia.gov.au) outlines that prosecutors must prove all matters relating to non-compliance with duties of care, including whether the PCBU failed to do what was 'reasonably practicable' to ensure the health and safety of workers and, in relation to Category 1 offences, whether the reckless conduct was engaged in without reasonable excuse.

This reflects the generally accepted principle that in a criminal prosecution, the onus of proof to the standard of beyond reasonable doubt normally rests on the prosecution.

Reckless conduct is known as a Category 1 offence. A person commits a Category 1 offence if the person has a health and safety duty and engages in conduct that exposes an individual to a risk of death or serious injury or illness and the person is reckless as to the risk to an individual of death or serious injury or illness (see Table 1.5).

TABLE 1.5 Reckless conduct – Category 1 offence

Offence by	Maximum penalty
An individual (other than as a PCBU)	\$300 000 or 5 years imprisonment or both
An individual PCBU or as an officer of a PCBU	\$600 000 or 5 years imprisonment or both
A corporate body	\$3 000 000

Source: NSW *Work Health and Safety Act 2011*, s. 31, Austlii, http://www.austlii.edu.au/au/legis/nsw/consol_act/whasa2011218/s31.html

Failure to comply is known as a Category 2 offence (covered in s. 32 of the NSW Act). A person commits a Category 2 offence if the person has a health and safety duty and the person fails to comply with that duty and the failure exposes an individual to a risk of death or serious injury or illness (see Table 1.6).

TABLE 1.6 Failure to comply – Category 2 offence

Offence by	Maximum penalty
An individual (other than as a PCBU)	\$150 000
An Individual PCBU or as an officer of a PCBU	\$300 000
A corporate body	\$1 500 000

Source: NSW *Work Health and Safety Act 2011*, s. 31, Austlii, http://www.austlii.edu.au/au/legis/nsw/consol_act/whasa2011218/s32.html

1.2 The work environment

Every environment has elements that may pose some risk to the health and safety of workers. Some workplaces such as construction sites are considered to be 'high risk' locations due to the nature of the work being done. Locations such as retail outlets and business offices will pose less risk but still present their own unique hazards.

Typical hazards associated with work environments

The release of **AS/NZS 31000:2009 Risk management – Principles and guidelines** has resulted in a number of changes in the terminology used in risk assessments, and Table 1.8 shows some of them.

TABLE 1.8 Risk assessment terms (old and new)

AS/NZS 4360:2004 – RISK ASSESSMENT TERMS (OLD)	
Term	**Description**
Hazard	Potential for harm, or adverse effect on an employee's health
Risk	Likelihood the hazard will cause injury or ill health
Exposure	When a person comes into contact with a hazard
AS/NZS 31000:2009 – RISK MANAGEMENT TERMS (NEW)	
Risk source	An element which alone or in combination has the intrinsic potential to give rise to risk
Risk management	Effect of uncertainty on objectives. An effect is a deviation from the expected
Event	Change of a particular set of circumstances. An event can sometimes be referred to as an 'incident' or 'accident' An event without *consequences* can also be referred to as a 'near miss', 'incident', 'near hit' or 'close call'

Sources: Adapted from Standards Australia, AS/NZS ISO 31000:2009 Risk management – Principles and guidelines

Risk sources or hazards are classified into five different types, as listed in Table 1.9.

TABLE 1.9 Classification of hazards

Term	Description
Biological	Includes bacteria, viruses, mould, mildew, insects, vermin and animals
Chemical	Includes chemical substances such as acids, alkalis or poisons and those that could lead to fire or explosion, cleaning agents, dusts and fumes from various processes such as welding
Electrical and/or mechanical	Includes electricity, machinery, equipment, pressure vessels, dangerous goods, forklifts, cranes and hoists
Physical	Includes floors, slippery surfaces, stairs, steps, work platforms, ladders, falling objects, manual handling (lifting, pushing, pulling), excessively loud and prolonged noise, vibration, heat and cold, radiation, poor lighting, ventilation, air quality and fire
Psychological	Includes workplace stressors arising from a variety of sources such as bullying and harassment

Source: Australian Government – Comcare, 'OHS10 Identifying Hazards in the Workplace', http://lib.mlaholdings.com.au/mlatechdata/Health & Safety/Hazard Identification & Risk Assessment/Comcare OHS10.htm

Some physical and chemical hazards can lead to fire, explosion and other safety hazards.

Principles of risk assessment/management

Risk management is a systematic process, as shown in Figure 1.3.

FIGURE 1.3 The risk management process AS/NZS 31000:2009

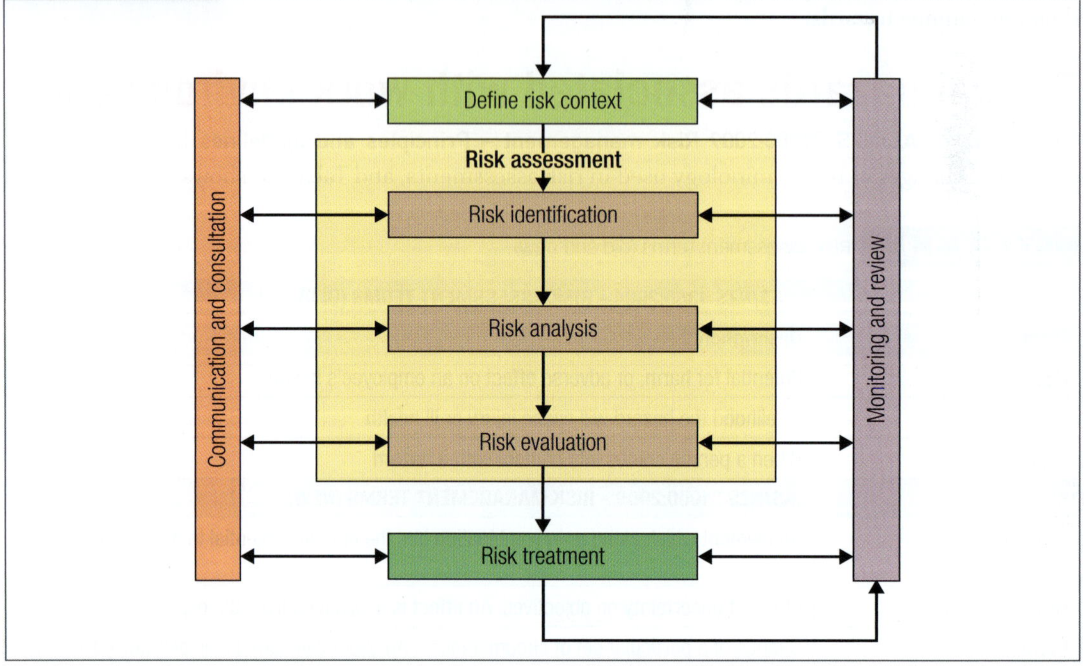

Safe Work Australia's 'How to consult on Work Health and Safety Draft Code of Practice' (http://www.safeworkaustralia.gov.au) describes consultation as

> a two-way process between a PCBU and workers where you:
>
> - talk to each other about health and safety matters
> - listen to each other's concerns and raise your own concerns
> - seek and share views and information
> - as a PCBU, consider what workers say before decisions are made…
>
> …Consultation does not mean telling workers about a health and safety decision or action after it has been taken. (p. 6)

The risk management process requires consultation between all stakeholders; in most cases that will be the employers and employees. That is they must:

1 Define risk criteria – this identifies what work will be included in the risk assessment.
2 Assess the risk – using a tool such as the risk assessment matrix shown in Figure 1.4. A low score indicates a high priority in controlling the risk source.
3 Control the risk using the hierarchy of WH&S control measures.
4 Monitor and review the process. The risk assessment is repeated until the risk is reduced to an acceptable level.

i Basic risk control measures

Hierarchy of WH&S control measures

The systematic approach used to remove or reduce a risk source is called a control measure. In order of importance they are rated in levels:

- *Elimination of the hazard – remove it completely*. For example, if noise is a hazard, remove the source of the noise (Level 1).
- *Substitution*. Where it is not possible to completely eliminate the hazard, minimise the risk by substituting it with a hazard of lower risk. For example, instead of using electric power tools replace them with pneumatic (air powered) tools (Level 1).

FIGURE 1.4 The risk matrix

	Very likely – could happen any time	Likely – could happen some time	Unlikely – could happen but very rarely	Very unlikely – could happen but probably never will
Kill or cause permanent disability or ill health	1	1	2	3
Long-term illness or serious injury	1	2	3	4
Medical attention and several days off work	2	3	4	5
First aid needed	3	4	5	6

- *Isolation*. This can mean either to isolate personnel from the risk or vice versa. For example, use remote control for the opening and closing of high voltage switch gear (Level 2).
- *Engineering*. Apply engineering controls such as putting guards on dangerous machinery, installing automatic cut-off switches, or providing assistive equipment such as a trolley for lifting heavy loads (Level 2).
- *Administration*. Further down the hierarchy is the option of applying administrative controls. Examples include developing workplace procedures, providing specialist training or putting a supervisor in place to monitor the environment (Level 3).
- *Personal protective equipment (PPE)*. Considered only as a last resort, personal protective equipment such as safety glasses will reduce the risk, but not remove the hazard (Level 3).

The hierarchy of control measures is shown in Figure 1.5. The final step in risk management is to establish and maintain systems which provide for regular evaluation and review procedures. The Australian Government's Comcare website says 'Evaluation means examining control measures to ensure risks are removed or reduced and have not caused new hazards presenting unacceptable risk' and further that 'Best practice is embedding work health and safety risk management into daily usage at all levels of a workplace' and in everyday business practice.

FIGURE 1.5 Hierarchy of control measures

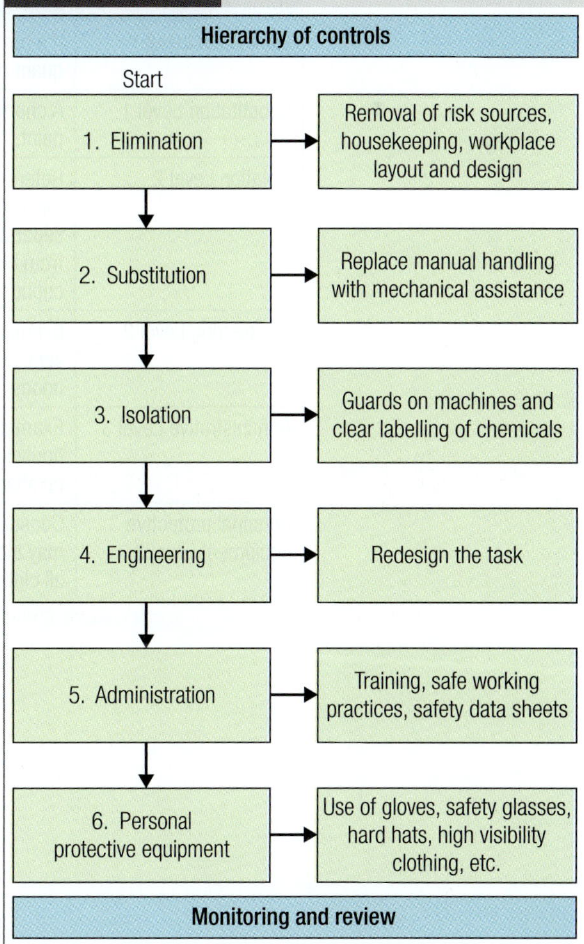

Controlling risks

Tables 1.10 to 1.13 provide examples of how the hierarchy of control can be applied to a particular risk source.

TABLE 1.10 Biological risk sources

Term	Description
Elimination Level 1	Is rarely possible, as the biological risk is normally required for the process.
Substitution Level 1	Control measures can be as simple as replacing reusable items for disposable ones, e.g. using disposable gloves.
Isolation Level 2	Isolating the risk by containing it in a special room or container. This must be planned for during the design phase of the process.
Engineering Level 2	Using engineering controls in a building – ventilation systems and construction seals to create negative-pressure rooms, bio-safety hoods, with specific ventilation systems.
Administrative Level 3	Includes general good housekeeping, disinfection, regular hand washing, worker training, no eating in work areas, no sharing of personal items, limited exposure times.
Personal protective equipment Level 3	Includes latex gloves, correct mask for biohazards and eye protection. Ordinary dust masks do not prevent the spread of viruses and bacteria. Consideration is necessary when choosing the most appropriate mask or respirator for biological hazards.

Source: 'Biological hazards overview', http://www.slideshare.net/search/slideshow?searchfrom=header&q=biological+hazards+overview

TABLE 1.11 Chemical hazards

Term	Description
Elimination Level 1	If a hazardous chemical cannot be completely eliminated it may be possible to reduce the quantity, so if accidental exposure occurs the risk will not be so great.
Substitution Level 1	A chemical swapped for a less hazardous chemical such as a lead-based paint or an acrylic paint.
Isolation Level 2	Refers to the safe use, storage and mixing of chemicals. For example, making sure any form of cyanide does not come into contact with an acid, keeping concentrated nitric acid separate from organics such as solvents and oils, or keeping oxidising agents isolated from combustibles. Isolation can be achieved, for example, by the use of separate storage cupboards.
Engineering Level 2	Examples of controlling biological hazards, using engineering controls in a building, are ventilation systems and construction seals to create negative-pressure rooms, bio-safety hoods, with specific ventilation systems.
Administrative Level 3	Examples of administrative controls for biological hazards include general good housekeeping, disinfection, regular hand washing, worker training, no eating in work areas, no sharing of personal items, limited exposure times.
Personal protective equipment Level 3	Consult MSDS sheets and choose protective equipment to suit the chemical exposure which may include: gloves – butyl, latex or similar – safety glasses or goggles, fully buttoned cover-all clothing and rubber boots.

Source: 'Chemical hazards', http://www.slideshare.net/search/slideshowsearchfrom=header&q=biological1hazards1overview

Electrical and/or mechanical risk sources

The control of electrical and or mechanical hazards is effectively the control of energy. Electric shock can cause ventricular fibrillation, asphyxia, severe muscle contraction and burns. Arcing faults can blind or burn workers and cause fires and explosions.

Mechanical energy presents risks associated with moving machinery such as cuts, loss of body parts and crushing. It can also be present in the form of stored energy in springs and hydraulic reservoirs.

- *Elimination* is achieved through disconnection of the energy source, that is, the machine is turned off, powered down and any stored energy released. Tests are then performed to ensure all energy has been removed.
- *Substitution* of electrically powered hand tools with pneumatically (air) or hydraulic (fluid) powered tools is common practice to control electrical hazards.
- *Isolation*: **AS/NZS 3000:2007 Wiring Rules** clause 1.5.4.1 provides details of basic protection (protection against direct contact). This rule requires most electrical cables to be insulated; this is actually an isolation control measure to reduce the risk of electrical hazards. The insulation prevents conductors coming in contact with each other and controls the electrical hazard to workers. Isolation control measures can be as simple as placing a protective screen around an area in which electrical arc welding is taking place to shield nearby workers from ultra violet rays.

  Electrical/mechanical hazards are typically controlled by:
  - enclosures
  - barriers
  - insulation
  - placing out of reach.
- *Engineering* controls used to reduce the risk of electrical hazards include:
  - residual current devices (RCDs; also known as safety switches)
  - isolation switches
  - circuit protective devices such as circuit breakers and fuses
  - lockout devices
  - stands for extension leads
  - electrical interlocks
  - isolation transformers.

  For mechanical hazards the risk may be controlled by:
  - fitting guards to machines
  - dead man switches
  - safety curtains and interlocks.
- *Administrative* control of electrical and mechanical hazards is achieved by the use of:
  - Standard operating procedures (SOPs) or safe work method statements (SWMS). The documents list the correct way to operate and maintain equipment in a safe manner.
  - Access permits: these define what areas are to be worked in or on and ensure that all energy either electrical or mechanical has been removed or made safe. The permit must be signed off to indicate all workers have completed the task and have left the area before the energy source can be turned back on.
  - Danger tags.

- *Personal protective equipment*: electrical workers have a specialised range of PPE. This equipment may include rubber insulating gloves, hoods, sleeves, matting, blankets, safety glasses, and industrial protective helmets designed to reduce electric shock hazard. Clothing should be of heavy cotton material or similar.

TABLE 1.12 Physical hazards

Term	Description
Elimination Level 1	Can be achieved at the design stage of a process. Good housekeeping will control slips, trips and falls.
Substitution Level 1	Redesign of the process to decrease the risk.
Isolation Level 2	Operators of machinery can be placed in an air-conditioned soundproof booth to reduce heat and noise stress.
Engineering Level 2	Use of lifting devices such as cranes or forklifts instead of manual handling.
Administrative Level 3	Marking of lines on floor. Training workers to lift correctly. High visibility clothing to prevent workers being hit by vehicles.
Personal protective equipment Level 3	Includes earplugs, safety glasses, body harnesses, long-sleeved clothing, etc.

TABLE 1.13 Psychological hazards

Term	Description
Elimination Level 1	Well-planned work processes.
Substitution Level 1	Multi-skilling so that workers may be rotated through a variety of jobs.
Isolation Level 2	Relocation of a person involved in bullying other workers.
Engineering Level 2	A well-maintained workplace, pleasant surroundings and landscaping. Adequate lighting and ventilation.
Administrative Level 3	Policies to encourage meaningful communication between staff. Open lines of communication. Job statements defining the role of each worker.
Personal protective equipment Level 3	Not applicable.

Documentation for a risk assessment

Risk assessments, or 'riskys' as they are known in the trade, can be quite involved, as they need to be individualised for each company, each job and each site. The overall approach is repetitive and once you have completed a few, the process is quite easy. It does not matter whether the risk assessment is for an electrician, a plumber or office worker, the principle is the same. Definition of the risk context is shown by the names of the company, contractor, project, operator and task being performed. The risk source (hazard) is identified, analysed and evaluated by rating the risk 1–6. The risk is treated by the controls listed. Consultation, monitoring and review is achieved through a tool box meeting. Figures 1.6 and 1.7 show examples of 'riskys'.

FIGURE 1.6 Example of a 'risky'

A & A Hire Service

PO BOX 155 GLENORIE NSW

PH: 9652-2391

SAFE WORK METHOD STATEMENT

CONTRACTOR:	PROJECT:
SITE ADDRESS:	PREPARED BY: Place name here Sign : .... Date : ..../.........../..................

TASKS=

EXCAVATOR OPERATION

General excavation

TASK	RISKS IDENTIFIED (Risk class-1=high, 2=med or 3=low)	CONTROLS THAT WILL BE USED	LEGISLATION/ CODES APPLICABLE
Load/unload machine to/off transport	Manual handling 2	Ensure correct posture and position when lifting load ramps in place. Ensure safe access to machine on transport. Secure ramps and any gates before moving machine.	Operator manual Cop_riskassessment_963 WH&S Reg 2001/OH&S ACT 2000. Cop_moving_plant_1310
	Movement hazards 1	Check that all designed hazard controls are operating. Qualified persons only to operate machinery. Buzzers and flashing lights operating as required by conditions.	Cop_moving_plant_1310 WH&S Reg 2001/OH&S ACT 2000. Cop_riskassessment_963
	Roll over 1	Do not exceed designed operational angle of machine. Ensure level and stable surface while loading / unloading machine off transport. Move machine in way designed.	Operator manual Cop_riskassessment_963 WH&S Reg 2001/OH&S ACT 2000. Cop_moving_plant_1310
General excavation <1500 depth	**Limited visibility/Contact hazard/slewing** 1	Exclusion zone if required Use high visibility clothing. Spotter required at all times. Qualified persons only to operate machinery. Buzzers and flashing lights operating as required by conditions. Do not exceed designed limits of reach.	Cop_moving_plant_1310 WH&S Reg 2001/OH&S ACT 2000. Cop_riskassessment_963 Cop_excavation_312
	Stockpiling spoil/ Dislodgement/ Unstable excavation/ collapse 1	Inspect nature of strata upon exposure. Stockpile outside zone of influence. Stockpile material to lower side of excavation. Minimise ground disturbance and vibration. Minimise strata exposure time. Backfill as work progresses. Spotter required at all times. Qualified persons only to operate machinery.	Cop_excavation_312 Cop_moving_plant_1310 OH&S Reg 2001/OH&S ACT 2000. Cop_riskassessment_963

SWMS No: …………. Page 1 of 10 Date: …………/………/…………

Source: Alan Chadwick

FIGURE 1.7 Example of a 'risky'

RESPONSIBILITIES (Persons involved in activity)	QUALIFICATIONS EXPERIENCE	TRAINING REQUIRED
SPOTTER	**As per WH&S REG 2001**	Industry induction as per regulation, WorkCover induction training, Site specific induction, Project specific training.
MACHINE OPERATOR	CERTIFICATE OF COMPETENCY	Industry induction as per regulation, WorkCover induction training, Site specific induction, Project specific training.

CHECKLIST OF ITEMS THAT MAY BE REQUIRED FOR THIS ACTIVITY	
WORKCOVER APPROVALS/ CERTIFICATES	WH&S GREEN CARD, EXPLOSIVE POWER TOOLS, FORKLIFT, MACHINE OPERATOR
CODES OF PRACTICE/ LEGISLATION APPLICABLE	OH&S ACT 2000, WH&S REG, 2001 CODE OF PRACTICE 'MOVING PLANT' CODE OF PRACTICE 'PUMPING CONCRETE' CODE OF PRACTICE 'CONSTRUCTION TESTING CONCRETE PUMPS' CODE OF PRACTICE 'FORMWORK'
PLANT & EQUIPMENT	MANUAL WORK TOOLS, PLANT & MACHINERY
MAINTENANCE CHECKS	PERIODICAL SERVICE, DAILY CHECKS, MONTHLY ELECTRICAL TAGGING
PERSONAL PROTECTIVE EQUIPMENT	HARD HATS, HIGH VISIBILITY VESTS, PROTECTIVE FOOTWEAR, EYE PROTECTION, GLOVES, SUNSCREEN

Record of Tool Box Talk					
Workplace: Supervisor/presenter:			Date:		
Subject: WMS sign off and weekly tool box talk			Duration: 45 mins		
Persons Present:					
Print Name	Signature	Date	Print Name	Signature	Date
Comments and points raised:					
Inspection of site and WH&S work method statement and amendments					
General on-site work issues and work method and safety					
Correct use of safety equipment					
Preparation and use of hazardous materials					
Site induction requirements on safety					
I have read and understood the safe work method statement					

Corrective Action	Action by	Action Complete	
		Sign off	Date

SWMS No: ………… Page 10 of 10 Date: …………/………/…………

Source: Alan Chadwick

Typical workplace safety signs

AS 1319:1994 Safety signs for the occupational environment specifies a number of signs to regulate and control safety related behaviour, warn of hazards and provide emergency information, including fire protection information. These signs are shown in Figure 1.8.

Regulatory signs

Regulatory signs contain instructions where failure to comply with them is an offence. They are:

- *Regulatory Prohibition* signs that indicate an action or activity is not permitted. In other words it is *not* allowed (see Figure 1.9).
- *Regulatory Mandatory* signs that indicate an instruction *must* be carried out. In other words you *must* do this (see Figure 1.10).
- *Regulatory Restriction* signs that place a defined limit on an activity (see Figure 1.11).

FIGURE 1.8 Colour and shape requirements for safety signs

Safety signs		
Sign function	**Symbol**	**Text**
Regulatory prohibition		Black
Regulatory mandatory		White
Regulatory restriction		Black
Hazard warning		Black
Hazard danger	DANGER	Black
Emergency information		White
Fire sign		White

FIGURE 1.9 Regulatory prohibition sign

FIGURE 1.10 Regulatory mandatory sign

FIGURE 1.11 Regulatory restrictive signs

Hazard warning signs

Warning signs contain instructions where failure to comply is not an offence but give a warning of a particular hazard that is likely to be *not* life-threatening (see Figure 1.12).

FIGURE 1.12 Hazard warning signs

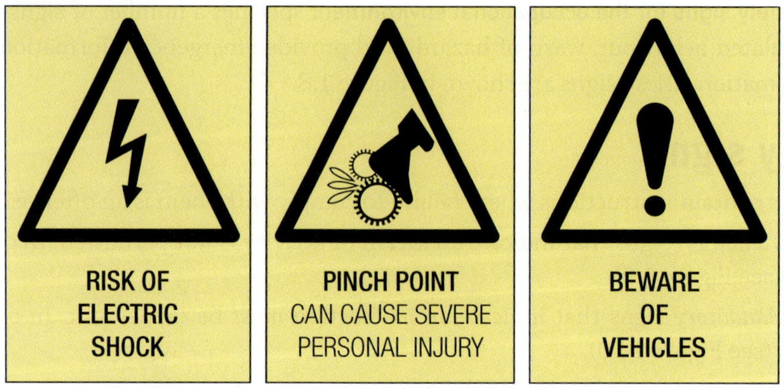

Hazard danger signs

Danger signs contain instructions where failure to comply is not an offence but give a warning of a particular hazard that is likely to be *life-threatening* (see Figure 1.13).

Emergency information signs

Emergency information signs indicate the location of emergency related facilities such as exits, safety equipment or first-aid facilities (see Figure 1.14).

FIGURE 1.13 Hazard danger signs

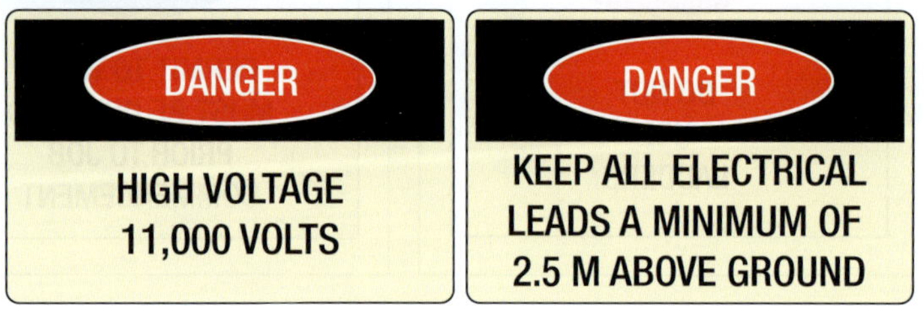

FIGURE 1.14 Emergency information signs

Fire signs

Fire signs show the location of fire alarms and fire-fighting facilities (see Figure 1.15).

Notice signs

Notice signs are for general information only, and are not supported by any legislation (see Figure 1.16).

FIGURE 1.15 Fire signs

FIGURE 1.16 General information signs

Workplace emergencies

PCBUs must ensure that an emergency plan is prepared for the workplace, as well as for workers who may work at multiple sites. **AS 3745:2010 Planning for emergencies in facilities** specifies the requirements for workplace evacuations. Private dwellings are not required to have an evacuation plan.

Safe Work Australia's 'Emergency plans fact sheet' (http://www.safeworkaustralia.gov.au/sites) outlines the types of workplace emergencies to plan for, including fire, explosion, medical emergency, rescues, incidents with hazardous chemicals, bomb threats, armed robberies and natural disasters. External hazards should also be considered in preparing an emergency plan, for example, a chemical storage facility across the road.

An emergency evacuation plan may include information for workers such as:

- emergency contact details for key personnel who have specific roles or responsibilities under the emergency plan, for example, fire wardens, floor wardens and first aid officers
- contact details for local emergency services, for example, police, fire brigade and ambulance
- a description of the mechanisms for alerting people at the workplace to an emergency or possible emergency, for example, siren or bell alarm
- the location and route to an emergency assembly evacuation point (see Figure 1.17)
- arrangements for assisting any hearing, vision or mobility impaired people ('Emergency plans fact sheet', pp. 2, 3).

Safe Work Australia further recommends that training must be provided to employees and short-term contractors or visitors at the workplace to include:

- emergency procedure training in induction courses for new workers
- training for existing workers
- specific training for individuals who have a formal role in an emergency, for example, fire wardens, floor wardens, first aid officers
- practice of emergency evacuation drills ('Emergency plans fact sheet', p. 3).

Once the emergency plan has been developed, it must be trialled within 12 months. All areas of a facility should participate in at least one emergency response drill at least once a year.

FIGURE 1.17 Emergency assembly point sign

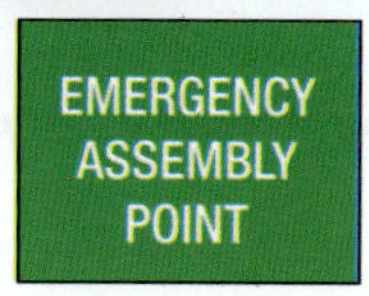

Fire extinguishers for a given type of fire

Fires are identified according to *fire classes*. Each class designates the fuel involved in the fire (see Table 1.14). The classifications allow selection of the extinguisher which is the most effective at putting out that type of fire, as well avoiding unwanted side effects. For example, non-conductive extinguishers are used for electrical fires, to prevent electrocuting the firefighter (Utah Fire Equipment, 'Fire 101: How does it work?', http://www.utahfire.com).

TABLE 1.14 Classification of fires

Class	Combustible material or fuel
A	Ordinary combustibles: solid organic materials such as paper, wood, cloth, rubber and some plastics
B	Flammable liquids: petrol, kerosene, methylated spirits etc.
C	Flammable gases: includes acetylene, ammonia, butane, carbon monoxide, ethylene, hydrogen, methane and propane (LPG)
D	Combustible metals: includes sodium, titanium, magnesium, potassium, uranium, lithium, plutonium and calcium. Magnesium and titanium fires are common
E	Electrical equipment
F	Cooking oil or fat

Fire extinguishers are therefore classified by the class of fire they are most suitable to extinguish. If the wrong class of extinguisher is used on the wrong class of fire the results can be disastrous. It should be noted that none of the fire extinguishers shown in Table 1.15 are suitable to fight 'Class D' combustible metal-fuelled fires.

TABLE 1.15 Application of fire extinguishers

Know Your Extinguisher

YES / NO — TYPE OF EXTINGUISHER Colour scheme - AS 1841.1 (Pre 1997 / Post 1997)	A Wood, Paper & Plastic	B Flammable & Combustible Liquids	C Flammable Gases	E Energised Electrical Equipment	F Cooking Oils & Fats	NOTES: *Limited indicates that the extinguishant is not the agent of choice for the class of fire, but that it will have limited extinguishing capability. Class D fires (involving combustible metal(s) use only special purpose extinguishers and seek expert advice. COMMENT: Refer Appendix A of AS 2444
Powder ABE	YES	YES	YES	YES	NO	Special Powders are available specifically for various types of metal fires. Seek expert advice.
Powder BE	NO	YES	YES	YES	YES	Special Powders are available specifically for various types of metal fires. Seek expert advice.
Carbon Dioxide (CO_2)	LIMITED	LIMITED	NO	YES	NO	Generally not suitable for outdoor fires. Suitable only for small fires.
Water	YES	NO	NO	NO	NO	Dangerous if used on flammable liquid, energized electrical equipment and cooking oil/fat fires.
Foam	YES	YES	NO	NO	LIMITED	Dangerous if used on energized electrical equipment.
Wet Chemical	YES	NO	NO	NO	YES	Dangerous if used on energized electrical equipment.
Vaporising Liquid	YES	LIMITED	LIMITED	YES	NO	Check the characteristics of the specific extinguishant.
Fire Blanket	NO	NO	NO	NO	YES	Use blanket to wrap around a human torch. Ensure you replace the blanket with a new one after use.
Fire Hose Reel	YES	NO	NO	NO	NO	Ensure you maintain a path of egress between you and the nearest exit.

Protecting People & Property

WORMALD® *A Tyco Business*

Source: Courtesy of Wormald

FIGURE 1.18 Fire extinguisher location signs – AS 2444

Location of portable fire extinguishers

AS 2444:2001 Portable fire extinguishers and blankets – Selection and location requires fire extinguishers to be located in a conspicuous and readily accessible position. If possible, extinguishers should be located along normal paths of travel and near exits. There must be a sign showing the location of the fire extinguisher and the sign must be (Figure 1.18) not less than 2 m above floor level and in a position clearly visible for a distance up to 20 m in all directions.

The extinguisher itself is required to be securely mounted using a bracket or cabinet no higher than 1.2 m, and no less than 0.1 m from the floor (Figure 1.19).

FIGURE 1.19 Fire extinguisher mounting

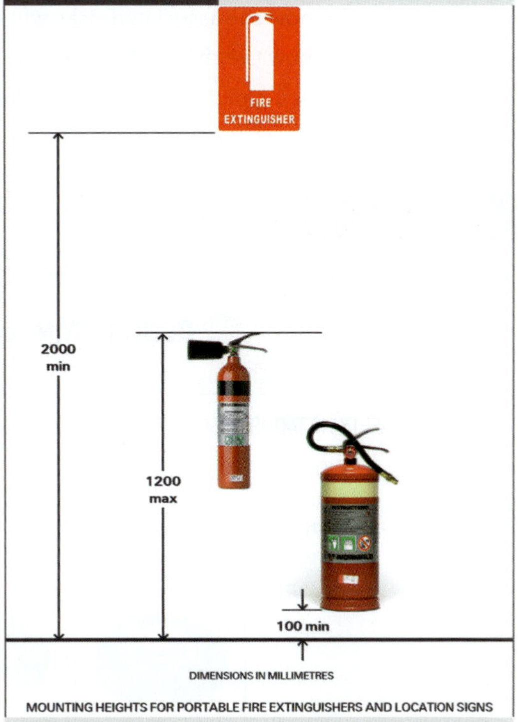

Source: Courtesy of Wormald

Basic process of fighting a fire

Fires start when a flammable and/or a combustible material, in combination with the right mixtures of oxygen (oxidiser) are exposed to a heat source. The fuel and oxygen start a chemical reaction called an *exothermic* reaction. Heat is produced as a by-product of the chemical reaction; if the heat produced is sufficient to continue the chemical reaction, a fire is started. The combination of fuel, oxygen and heat is known as the *fire triangle* (see Figure 1.20).

Removing any one or more elements of the fire triangle will extinguish the fire. Water-based agents will smother the oxygen and cool the heat. Foam and dry powders cover the fuel and cut oxygen to the fire. Carbon dioxide (CO_2) agents replace the oxygen, starving the fire of oxygen. Vaporising liquids actually absorb the oxygen surrounding the fire in much the same way as the older halon agents did, but without the environmental impact on the ozone layer.

When fighting fires, if possible, first turn off the fuel source by closing a valve or tap, or in the case of an electrical fire isolate or turn off the power source. Prepare the extinguisher before you approach the fire, then release the extinguisher aiming for the base of the fire until the fire is completely out. Watch out for reignition.

FIGURE 1.20 The fire triangle

Importance of safe premises – consequences of non-compliance

PCBUs have a duty of care in ensuring the safety of workers and visitors in the workplace. This duty of care also extends to the general security of the workplace. Just as good housekeeping provides both the benefit of controlling hazards and giving the workplace an organised, professional appearance, a secure workplace reduces the risk of theft and malicious damage, as well as the risk to visitors and workers of assault and other types of harassment from outsiders.

i Fighting a fire

Standard work procedures

Standard operating procedures (SOPs) or safe work method statements (SWMS) are an administrative control measure. They provide guidance to workers on the correct, safe way to perform a task and the use of plant and equipment. While the name of these procedures may vary between workplaces, the overall principle remains the same. SOPs must be readily accessible, easy to read and provide:

- the context or scope of the risk, e.g., using a bench grinder
- the risk sources or hazards associated with, in this example, the bench grinder

- relevant mandatory regulation signs giving instruction on the appropriate PPE to use
- a description of how the associated risks will be controlled
- a description of how the risk control measures will be implemented, monitored and reviewed.

Figure 1.21 shows a typical SOP that would be laminated and displayed adjacent to a bench grinder in a workshop.

FIGURE 1.21 Standard operating procedures

BENCH GRINDER – Light material

SAFETY HAZARDS ASSOCIATED WITH THIS EQUIPMENT:

- Dust and grit particles
- Abrasive injury
- Sharp edges and burns
- Electricity
- Torn or loose clothing
- Long hair to be covered
- Gloves NOT to be worn
- Eye injuries

1. Eye protection MUST be worn.
2. Foot and clothing protection MUST be worn.
3. Hair should be covered by wearing a cap.
4. Gloves must NOT be worn when operating machine.
5. Familiarise yourself with and check all equipment operations, controls and emergency stop equipment.
6. Ensure power is OFF
7. Check leads and switches for damage before use.
8. Check grinding wheel and machine for damage.
9. Adjust guides to as close as possible to wheel.
10. Be aware of other people in the immediate vicinity.
11. Check that machine lighting is adequate.
12. Ensure work area and grinding wheel is clear before use.
13. After all checks have been made turn main power on.
14. Keep hands clear of moving components.
15. Do not stand in front of wheel when starting equipment.
16. Move job across the wheel to even wear on grinding wheel.
17. Turn off power supply immediately when finished.
18. Do not overload the motor.
19. Never leave running machine unattended.

REPORT ALL FAULTS IMMEDIATELY TO SUPERVISOR
IF YOU ARE UNSURE ABOUT ANY PROCEDURE – PLEASE ASK

Review No.	01	02	03	04	05	06	07	08	09
Initials:									
Date:									

SUMMING UP...

- Risk sources (hazards) are classified as biological, chemical, electrical and/or mechanical, physical and psychological.
- Risk assessment/management procedure – 1, define risk context; 2, risk identification; 3, risk analysis; 4, risk evaluation; 5, risk treatment. Throughout the process there must be consultation with workers, plus monitoring and review.
- Control measures (risk treatment) include elimination, substitution, isolation, engineering, administration and PPE.
- Workplace safety signs. Prohibited – red circle with red stripe through means not permitted; Mandatory – blue circle background with white print means must do, red circle with black print shows legal limit; Hazard warning – yellow triangle with black border and black print means hazard is not life-threatening; Hazard danger – red circle on black background with white and black print means hazard which is life-threatening.
- Dry powder, CO_2 and vaporising liquid are the only types of fire extinguishers that are suitable to use on electrical fires.
- Standard operating procedures (SOPs) and safe work method statements (SWMS) provide information on the correct way to do a task or use plant and equipment.

1.3 Manual handling

Manual handling is a physical risk. In its *Service Agreement Information Kit for Funded Organisations*, the Victorian Government Department of Human Services (Section 4.7 Occupational Health and Safety; http://www.dhs.vic.gov.au) describes it as covering

> a wide range of activities including carrying, holding, lifting, throwing, pushing and pulling. It includes repetitive tasks such as packing, typing, assembling, cleaning and sorting, using hand tools, and operating machinery and equipment. While not all manual handling tasks are hazardous, manual handling accounts for about a quarter of all workplace injuries.

Manual handling injuries and their effects

Safe Work Australia's 'Code of Practice for Manual Handling' relates that unsafe manual handling can cause a variety of injuries and conditions including:

- muscle sprains and strains
- chronic pain
- hernia
- injuries to muscles, ligaments, intervertebral discs and other structures in the back
- injuries to soft tissues such as nerves, ligaments and tendons in the wrists, arms, shoulders, neck or legs
- musculoskeletal disease (MSD) which includes conditions such as repetitive strain injury (RSI) (p. 5).

The following text is adapted from 'Anatomy of back pain' (adapted from *Spine Universe* by Kamiah A Walker, http://www.spineuniverse.com/conditions/back-pain/anatomy-back-pain). Muscles are strands of tissues that act as the source of power for movement. Ligaments are the strong, flexible bands of fibrous tissue that link the bones together, and tendons connect muscles to bones (Figure 1.22). Blood vessels provide nourishment. These parts all work together to help you move. Muscles can be torn and ligaments can be stretched or snapped.

FIGURE 1.22 The knee showing muscle and ligament

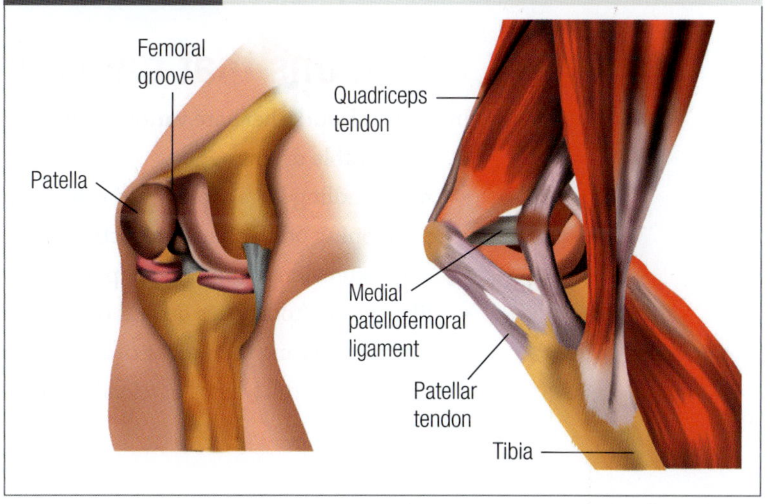

FIGURE 1.23 The human spine

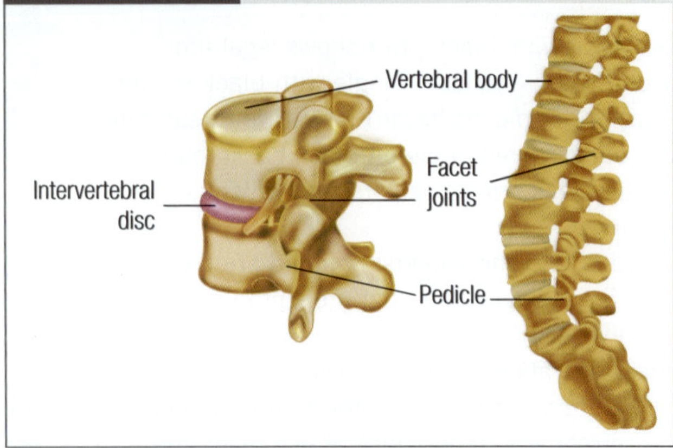

FIGURE 1.24 Back muscle

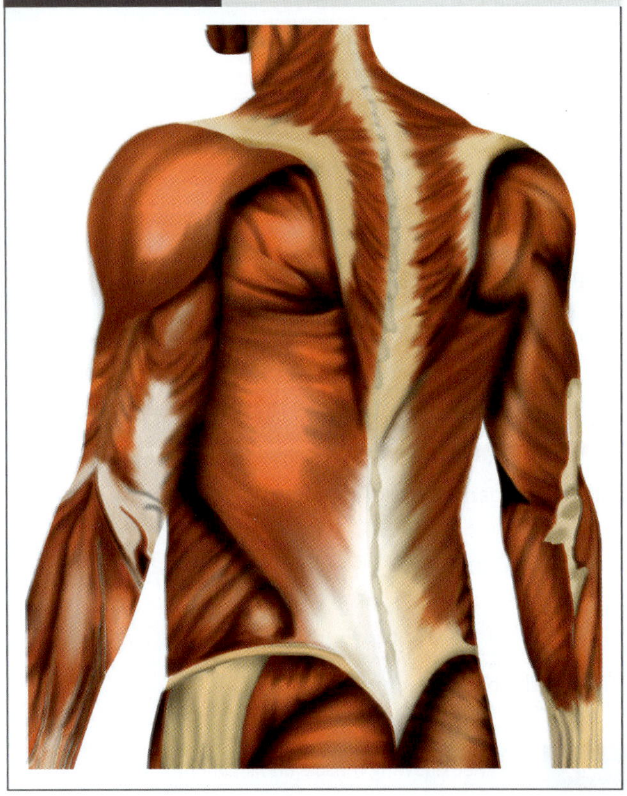

A person's ability to perform manual handling will vary with their age, sex, physical strength and level of fitness. There is no regulated limit to what an individual can lift, as lifting ability will vary from person to person. If that person exceeds their limit, they will risk injury.

Some manual handling injuries cause damage to the back and spine. The adult human spine is made up of 33 bones called vertebrae; the spine also has facet joints, which are on the back of the vertebrae. These joints help facilitate movement and are very important to flexibility. Another vertebral bone structure is the pedicle. These are on either side of the vertebrae (the ridges on the back), and they also form part of the spinal canal.

In between the vertebrae are the intervertebral discs. These act like pads or shock absorbers for the spine as it moves. Each disc is made up of a tyre-like outer band and a gel-like inner substance. Together, the vertebrae and the discs provide a protective tunnel (the spinal canal) to house the spinal cord and spinal nerves. These nerves run down the centre of the vertebrae and exit to various parts of the body, where they help with feeling and movement (see Figure 1.23).

The back also has muscles, ligaments, tendons and blood vessels (Figure 1.24). Back pain may be a result of injury to any or all of these body parts. Injury to the muscles, ligaments or tendons can result in sprains or strains, which are generally not considered serious. However, injury to bones, nerves or blood vessels may be more serious. Also, the outer layers of the intervertebral discs can get tears or cracks, allowing parts of the discs to bulge out. Any of these injuries can cause inflammation and pain.

The spine supports most of a person's bodyweight and is involved in nearly all body movement. Severe spinal injuries can leave the victim paralysed or cause death. Manual handling injuries are rarely that serious but can result in a lifetime of restriction and pain. Manual handling injuries can result in a number of physical and psychological effects that can alter a person's lifestyle, such as:

- loss of physical fitness and sporting ability
- increased fatigue as a result of constant pain
- decreased income due to loss of employment opportunities
- addiction to pain-killing drugs
- depression
- breakdown of personal relationships.

Situations that cause manual handling injuries

In electrical work there are various situations that are a manual handling risk, such as:

- lifting cable drums
- manhandling large cables onto cable trays and racks
- positioning oddly shaped objects such as air-conditioning units and hot water systems into place
- over-extension of upper body parts while working upon ladders or elevated work platforms
- lifting tools/tool boxes on and off vehicles
- underestimating the weight of a load being lifted
- carrying the full weight of a load after a lifting partner drops their end unexpectedly.

Lifting and carrying to prevent manual handling injuries

Young workers make up a disturbingly large number of manual handling injuries. Lack of experience and training and the desire to 'show off' their physical strength to other workers are contributing factors. As with any risk source, manual handling can be managed using various control measures, the aim being to prevent injuries.

For example, a work process can be redesigned to eliminate the need for lifting. The size of heavy loads may be broken down into smaller, easier-to-handle loads. Manual handling can be substituted for a mechanical device. Lifting devices can be used to raise and hold heavy objects such as air-conditioning ducts into position until they are secured into place (Figure 1.25). Mechanical devices such as cranes can be used to lift heavy loads. Cable winches are used to assist to pull heavy cables onto cable trays. Forklifts and trolleys are used to move loads around a work site.

Some manual handling will always be necessary. In these cases, administrative controls of risks require training in correct manual handling procedures, and the correct marking of packaging. Manual handling using the SMART approach is shown in Figure 1.26.

FIGURE 1.25 Lifting hoist

Source: Dreamstime/Uatp1

FIGURE 1.26 SMART lifting procedures

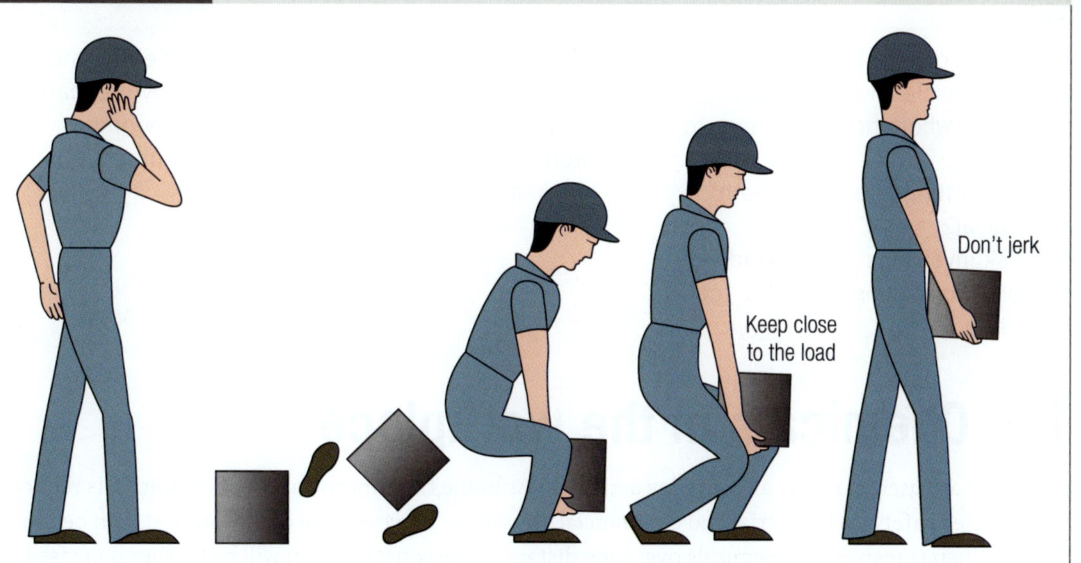

When lifting packages such as cardboard cartons it is impossible to 'size up the load'. To assist, International Standard **ISO R/780:2000 Packaging – Pictorial marking for handling of goods** provides manual handling guidance on packaging, as shown in Figure 1.27.

FIGURE 1.27 ISO R/780 symbols

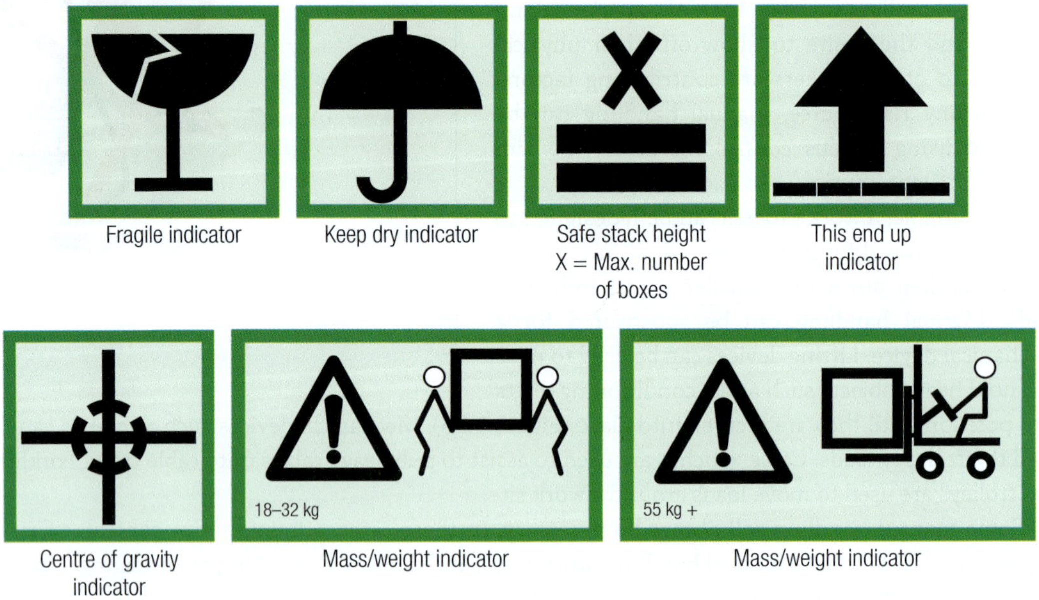

SUMMING UP...

- The most common injuries caused by manual handling are damage to muscle and ligaments.
- More serious injury occurs when there is damage to the back, spine and discs.
- The effects of manual handling injuries are both physical and psychological.
- Potential sources of risk involving manual handling should be identified prior to starting work at a 'tool box' or 'take five' meeting.
- Manual handling risk sources can be eliminated through process design, substituted by mechanical aids, controlled administratively by training workers and the use of ISO R/780 symbols on packaging.

1.4 Chemicals in the workplace

Chemicals are used in nearly all everyday activities. At home the concentration of chemicals is reduced for increased safety. In industrial and commercial situations, however, the concentrations can be much higher. When transporting chemicals over long distances the concentration will be further increased to reduce costs, but will also increase handling risks.

When working with chemicals it is vital to understand what chemicals you are working with. Some chemicals are considered 'caustic', which means they will burn the skin, eyes or lungs. Chemicals can also react with one another, creating unexpected risk. Mixing common cleaning products such as bleach with ammonia can release deadly chloramine gas. These gases are invisible and most aren't noticed by smell, since the products they are combining with already have a strong odour.

Hazardous substances and dangerous goods

A hazardous substance contains ingredients that may be solids, liquids or gases that can harm people, other living organisms, property or the environment in the immediate and long term. The Hazardous Substances Information System (HSIS) at the Safe Work Australia website (http://www.safeworkaustralia.gov.au) provides a database of hazardous substances.

Under the 'Managing Risks of Hazardous Chemicals in the Workplace Code of Practice' (2012) dangerous goods are classified on the basis of immediate or short-term physical or chemical hazards, such as fire, explosion, corrosion and toxicity, that may affect life, health, property or the environment. Details of substances considered to be 'dangerous goods' can be found in the Australian Dangerous Goods Code at the National Transport Commission website (http://www.ntc.gov.au/).

Classification

Hazardous substances are defined by their direct health effects on people and the environment, both immediate and long term. From January 2012, hazardous substances are identified by the Globally Harmonised System (GHS) of Classification and Labelling of Chemicals. Many substances are both dangerous goods and hazardous substances. Where there is an overlap between the two groups, both sets of laws apply. Classification assigns the substance to a hazard category (Physical, Health or Environment) and a subdivision. For example, a substance categorised as Physical has subdivisions such as explosive, gases under pressure, flammable liquid or solids. A substance in the Health category has subdivisions including skin corrosion, eye irritation and carcinogenic.

Dangerous goods and hazardous substances are classified according to different criteria. Dangerous goods are classified on the basis of immediate physical or chemical effects that may impact on people, property or the environment. They are:

- Class 1 Explosive substances and articles
- Class 2 Gases
- Class 3 Flammable liquids
- Class 4.1 Flammable solids, self-reactive substances and solid desensitised explosives
- Class 4.3 Substances which, in contact with water, emit flammable gases
- Class 5.2 Organic peroxides
- Class 6.1 Toxic substances
- Class 6.2 Infectious substances
- Class 7 Radioactive material
- Class 8 Corrosive substances
- Class 9 Miscellaneous dangerous substances and articles.

Requirements for labelling chemicals

Under WH&S regulations, manufacturers and importers of a substance, mixture or article must correctly classify them. PCBUs must ensure that any hazardous chemical at the workplace is correctly labelled, except where the hazardous chemical is a consumer product in household quantities.

Every known chemical substance is given a unique CAS (Chemical Abstracts Service) Registry Number which is globally recognised. For example, sodium chloride (table salt) has a CAS Number of 8028-77-1. Hazardous chemicals may also be given a United Nations (UN) number. You may have noticed the label UN1075 on LPG bottles.

Pictograms provide a pictorial multilingual representation of the type of hazard that can be easily recognised at a glance. Table 1.16 shows the nine hazard symbols adopted by the Globally Harmonised System of classifying and labelling chemicals.

TABLE 1.16 Globally Harmonised System

GLOBALLY HARMONISED SYSTEM LABELLING OF CHEMICALS			
	Symbol	**Pictogram**	**Hazard class**
Physical hazards	Exploding bomb		Explosives; self-reactive substances; organic peroxides
	Flame		Flammable gases, includes aerosols, liquids, solids and self-reactive substances
	Flame over circle		Oxidising gases or liquids
	Gas cylinder		Gases under pressure
	Corrosion		Corrosive to metals

	Symbol	Pictogram	Hazard class
Health hazards	Skull and crossbones		Acute toxicity
Health hazards	Corrosion		Skin corrosion; serious eye damage
Health hazards	Exclamation mark		Health hazards such as skin or eye irritation, narcotic effects and skin sensitisation
Health hazards	Health hazard		Chronic health hazards including carcinogenicity, reproductive toxicity; aspiration hazard
Environmental	Environment		Acute and chronic hazards to the aquatic environment

Source: Worksafe NT, Safework Australia

The Safe Work Australia Globally Harmonised System of Classification and Labelling of Chemicals describes the following labels:

- *Signal words* are used to indicate the relative level of severity of a hazard. The GHS uses 'Danger' and 'Warning' as signal words. 'Danger' is used for a more severe or significant hazard, while 'Warning' is used for the less severe hazards.
- *Hazard statements* describe the nature of a hazard, including the degree of hazard, where appropriate. A unique hazard statement is assigned to each hazard class and category. Examples of hazard statements include 'Flammable liquid and vapour', 'May cause cancer', 'Fatal if inhaled' and 'Causes severe skin burns and eye damage'.
- *Precautionary statements* provide instructions for risk prevention, response, storage and disposal. Examples include 'Use explosion proof electrical equipment', 'Do not breathe vapours', 'Do not get in eyes, on skin or clothing', 'Use only outdoors or in a well ventilated area' and 'Do not attempt to fight fire'.
- *Unmarked containers*. If you find a container that does not have a label or is incorrectly labelled, it must be relabelled 'Caution – do not use: unknown substance'. Such a container should be stored in isolation until its contents can be identified and, if hazardous, the container is appropriately labelled (Safe Work Australia, 'Globally Harmonised System of Classification and Labelling of Chemicals (GHS)', http://www.safeworkaustralia.gov.au).

Labelling commercial quantities of a hazardous chemical is covered by legislation and must include the product identifier (CAS or United Nations number), manufacturer/importer details, a hazard pictogram and information about the hazards ('Draft Labelling of Workplace Hazardous Chemicals Code of Practice', Safe Work Australia).

Dangerous goods labelling provides guidance to transport companies on correct handling, and to emergency services such as police, ambulance and fire brigade on how to respond in the case of an accident. Table 1.17 shows the more commonly used dangerous goods transport pictograms. There are a total of 12 pictograms covering goods such as radioactive, spontaneously combustible and infectious substances.

TABLE 1.17 GHS dangerous goods transport pictograms

Classification	Pictogram	Example
Class 1 Explosive		Gunpowder, gelignite, fireworks, fuses, detonators
Class 2.1 Flammable gas		Acetylene (dissolved), LPG (liquefied petroleum gas), many aerosols
Class 2.2 Non-flammable non-toxic gas		Carbon dioxide, oxygen (compressed), air (refrigerated liquid), nitrogen

Classification	Pictogram	Example
Class 2.3 Toxic gas		Chlorine (gas), methyl bromide, anhydrous ammonia
Class 3 Flammable liquid Flashpoint not greater than 60.5°C		Petrol, kerosene, paint thinners

Safe storage procedures for chemicals

Hazardous chemicals must be stored in quite different ways than ordinary materials. In the case of a fire or structural failure, if chemicals from incompatible groups combine, unpredictable or catastrophic consequences can occur. The most commonly used control measures are to:

- limit the quantity of the hazardous substance stored in one place (administrative)
- segregate incompatible hazardous substances from each other, either enclosing them in lockable steel cupboards or placing at a distance from each other (isolation)
- provide ventilation and/or temperature control of the storage area (engineering).

The Tan Storage System website explains that the area where hazardous substances or dangerous goods are stored should be:

- well ventilated and lit
- separated from ignition sources
- secured from the public
- protected from temperature fluctuations and direct sunlight.

Shelving and cupboards should be:

- constructed of chemically resistant materials
- provide a lip at the front of the shelf.

Source: http://tanstoragesystem.blogspot.com.au/2008/01/storing-of-dg.html

As well, secondary containment (e.g. spill trays) should be considered for liquid dangerous goods stored outside of a chemical storage cabinet. Dedicated areas should be provided for measuring out chemicals (see http://tanstoragesystem.blogspot.com.au/2008/01/storing-of-dg.html).

Packaging groups

Substances have been assigned in the Australian Dangerous Goods Code (ADG) to one of three Packaging Groups according to the degree of danger they present, using the following Roman numerals:

- PG I – high danger
- PG II – medium danger
- PG III – low danger.

Restrictions are placed on incompatible substances such as flammable liquids and oxidisers being stored together or transported in the same vehicle.

Storage areas

Storage areas are categorised by the volume of the hazardous substances stored at that location. They include areas known as:

- minor storage
- retail storage
- intermediate bulk container
- bulk.

Table 1.18 shows how limiting the amount of fuel for a given situation relieves the need for storage cabinets. This shows why most petrol containers for lawn mowers are restricted to 5 litres, so they can be stored within a residence. If a workplace exceeds the limits of minor storage, it will be required to provide storage cabinets. Chemicals from incompatible groups such as acids and bases can be 'segregated' by placing them in different compartments in the cabinet.

TABLE 1.18 Limits for minor storage areas

Liquid	UN number	CAS number	PG	INDOORS		
				House	Shed > 1 m from house	Workshops and factories
Petrol	1203	8006-61-9	II	5 L	100 L	1 L per 2 m^2 or 250 L per 500 m^2 maximum
Diesel	1202	68334-30-5	III	25 L	250 L	1 L per 1 m^2 or 500 L per 500 m^2 maximum

FIGURE 1.28 Australian standard storage cabinet

Source: Ed Hawkins

The goal of minor storage is to keep quantities so small or spread out that the substances will present little risk. And, if they are flammable they will not significantly contribute to a building's fire load, if stored in this way – nor would they hinder the activities of the emergency services.

If quantities can be kept below minor storage limits then the storage requirements are not as stringent (see BCS Plastics, http://www.bcsplastics.com.au). This could mean that storage can take place safely in different locations within the workplace without having to purchase an Australian Standard approved cabinet (Figure 1.28). Larger quantities of chemicals are stored in intermediate bulk containers (IBC) (Figure 1.29).

The most commonly used dangerous good in the electrical trade is liquid petroleum gas (LPG). LPG cylinders should be stored outside, in a well-ventilated area, and be carried and stored upright at all times.

There have been a number of cases where LPG bottles have been left overnight in work vans with the valves not fully closed. Overnight, gas filled the interior of the van, and when the driver used the remote control to open the van door, the central door locking caused a spark which ignited the gas. The resulting explosion destroyed the van and severely injured the driver. Always ensure valves are turned off firmly when not in use, and make sure that pressure relief valves face away from dwellings and combustible materials where possible. Do not store or use petrol, flammable liquids or aerosols near LPG cylinders, as the combined fuel source will make a fire uncontrollable.

NSW Fire Brigades 'Fact Sheet No 43' (http://www.fire.nsw.gov.au) states that when transporting LPG cylinders in a car or work vehicle the total capacity must not exceed 9 kg. The cylinders should be transported in an upright position, secured, preferably in the boot or storage area. No more than two cylinders should be carried in a car at any one time and LPG bottles should not be left in a vehicle unnecessarily.

Gas cylinders must be re-tested and stamped every 10 years. To check for gas leakage, spray soapy water on any suspect connection or hose and watch for bubbles. Never use a naked flame to detect a leak.

FIGURE 1.29 Intermediate bulk container

Source: iStockphoto/Joe Gough

Safety data sheets (SDS)

As described by SafeWork SA in its 'Preparation of Safety Data Sheets for Hazardous Chemicals Code of Practice' (http://www.safework.sa.gov.au), a safety data sheet (SDS), previously called a material safety data sheet (MSDS), is a document that provides information on the properties of hazardous chemicals and how they affect health and safety in the workplace. Safety data sheets provide information on hazardous substances and dangerous goods additional to those on a label.

Methods to control risks, including exposure risks and relevant exposure standards, can also be found in the SDS. A code of practice has been developed by Safe Work Australia to guide in the preparation of an SDS, which must include:

- identification of the material and supplier
- hazards identification
- composition/information on ingredients
- first aid, accidental release and fire-fighting measures
- handling and storage
- exposure controls/personal protection
- physical and chemical properties including stability and reactivity
- toxicological and ecological information
- disposal considerations
- transport and regulatory information
- other relevant information as required.

SUMMING UP...

- Hazardous substances are classified using the Globally Harmonised System based on medium- to long-term effects. A substance can be identified as hazardous by using the Hazardous Substances Information System (HSIS) at the Safe Work Australia website.
- Dangerous goods are classified by the Australian Dangerous Goods code based on short-term effects and safe transport requirements of the substance. Details of dangerous goods can be found at the Australian Transport Commission website.
- Two systems of pictograms are used to identify hazardous substances and dangerous goods. The GHS is used to identify hazardous substances on labels and packaging. Transport pictograms are used to identify vehicles carrying dangerous goods.
- Hazardous substances and dangerous goods should be stored in small amounts to achieve a minor storage classification and to reduce risk.
- Incompatible substances should not be stored without segregation.
- Safety Data Sheets (SDS) provide detailed information on all substances, not just those considered hazardous.

1.5 Working at heights

A large amount of electrical work must be done at heights. Lighting, cable tray, data and fire services are generally located in the ceiling space of most buildings. Low voltage distribution and high voltage transmission conductors are installed at great height to place them 'out of reach'. When working at heights the 'National Code of Practice for the Prevention of Falls in General Construction' provides guidance to both PCBUs and workers on working at heights.

Dangers of working at heights

Slips, trips and falls in any work environment make up a significant percentage of injuries. Where height is involved the consequences of an event at even a small height can be significantly greater than at ground level. Injuries can range from mild muscle strains to ligament damage, broken bones, spinal damage and at worst death. Injuries involving ladders and scaffolds frequently cause permanent disability. The hazards associated with ladders and scaffolds include:

- falls from ladders or scaffolds
- being struck by falling ladders
- being struck by materials falling from ladders or scaffolds
- tripping over ladders (erect or lying on the floor)
- lifting heavy ladders
- striking persons or objects when carrying ladders
- contact with live electrical equipment.

FIGURE 1.30 Categories for working at heights

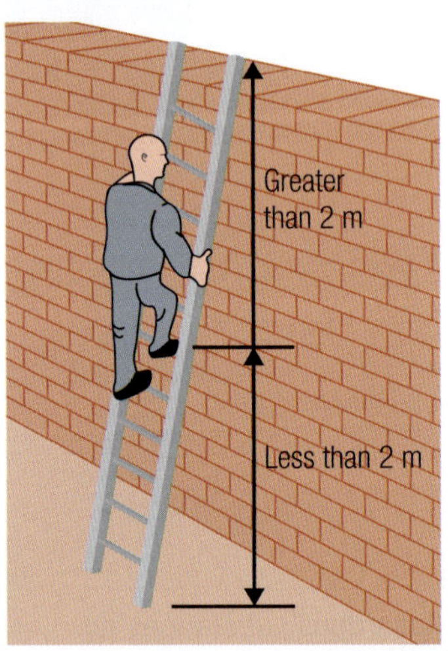

Source: 123rf/6kor3dos

Identifying and managing height risks

Under the code of practice, the risks associated with working at heights are divided into two categories: heights up to 2 m and heights greater than 2 m as shown in Figure 1.30.

Working at heights up to 2 m

PCBUs and workers are required to follow general risk management procedures that include:

- define risk criteria
- assess the risk
- monitor and review
- consultation.

A PCBU must also ensure that any information, instruction and training relating to working at heights of less than 2 m is provided.

Working at heights greater than 2 m

WorkCover NSW's *Safe Working at Heights Guide 2006* (http://www.workcover.nsw.gov.au) explains that a PCBU must ensure that physical fall prevention measures are provided so far as is reasonably practicable. If it is not practicable to provide physical fall protection, the PCBU will need to document why they have not been able to do so and document how they have reduced the risks to the lowest possible level by the use of other means in accordance with the hierarchy of control measures shown in Table 1.19.

If possible it is best to completely eliminate risks caused by working at height by using a level 1 control. If that is not reasonably practical, then control measures 2 to 5 may be used.

TABLE 1.19 Control measures for working at heights greater than 2 m

Level	Control measure
1	Work on the ground or from a solid construction.
2	Work using a passive fall prevention device. These include scaffolds, perimeter screens, step platforms, perimeter guard railing, elevated work platforms (EWPs), guard railing on edges of roofs and safety mesh.
3	Work using a work positioning system. These include travel restraint systems and rope access systems.
4	Work using a fall injury minimisation system. These include catch platforms and fall arrest systems.
5	Work from ladders, and/or implement an administrative control.

Source: 'Managing the Risks of Falls in the Workplace Code of Practice', Workplace Standards Department of Justice, Tasmanian Government, http://worksafe.tas.gov.au/__data/assets/pdf_file/0004/192064/managing_falls_code.pdf

Ladder selection and safety checks

Ladders are named by their construction. As shown in Figure 1.31, they are: step, platform, single and extension.

FIGURE 1.31 Ladder types

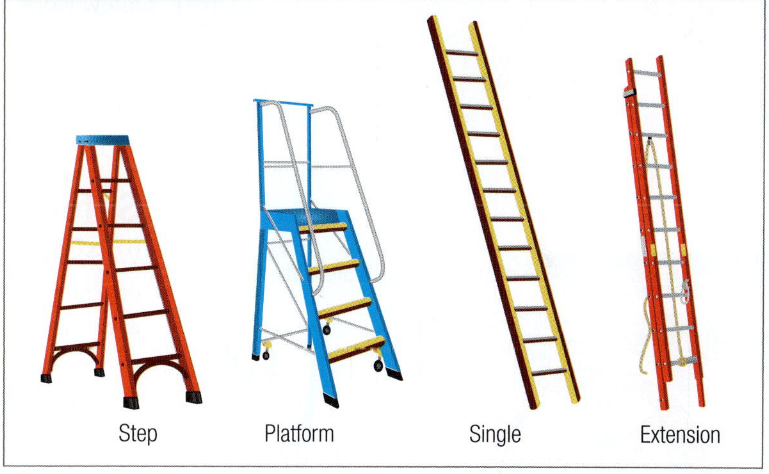

There are also a number of variations which are combinations of these. Typical materials used in the construction of ladders are metal (i.e. aluminium or galvanised steel), wood and fibreglass.

Aluminium ladders (as shown in Figure 1.32) are low-cost, lightweight, strong for their weight and require little maintenance. They are suitable for general construction work. Aluminium ladders are not safe to use near electricity as aluminium is an electrical conductor. If contact is made with a live electrical source the ladder will provide a path to earth. This may cause an electric shock. Electric shocks are potentially fatal, and as the worker is at a height the shock can also cause a worker to fall.

Timber ladders are less popular these days due to their high cost compared to aluminium. They have the advantage of being non-conductive so they are suitable for electrical work. They are much heavier to carry than aluminium and require much more maintenance. Fibreglass ladders are non-conductive, low maintenance, lightweight, very strong, but brittle. If left in an unsecured position it is possible that a strong wind gust can blow the ladder to the ground, causing the ladder to shatter.

Only use ladders which are marked to indicate they are constructed to an appropriate AS/NZ standard. Ladders are also marked to indicate the maximum safe working load they can support. If a ladder is rated to 140 kg and the typical mass of an adult is 80 kg to 100 kg this means that no more than one person should occupy the ladder at any one time. The weight of tools and materials should also be considered.

FIGURE 1.32 Ladder parts

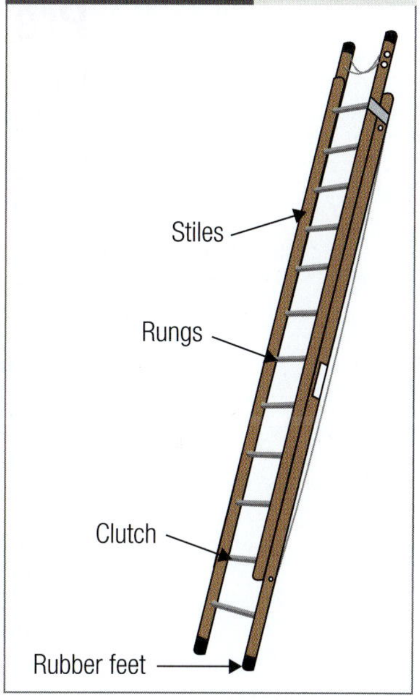

Working at heights up to 2 m

Where possible consider using alternatives to portable ladders, such as trestle and plank combinations (Figure 1.33). These may be more suitable after considering the length of time that workers will be accessing them.

If a ladder has to be used, select one that is long enough so the user can stand no higher than the second-last rung, as shown in Figure 1.34. A platform ladder is preferable to a step ladder (see Figure 1.34).

Keep step ladders a safe distance from floor openings or other fall hazards. Maintain a 'buffer zone' of the person's height + the work height + 1.5 m. So if a person is 1.6 m tall and

FIGURE 1.33 Alternatives to portable ladders

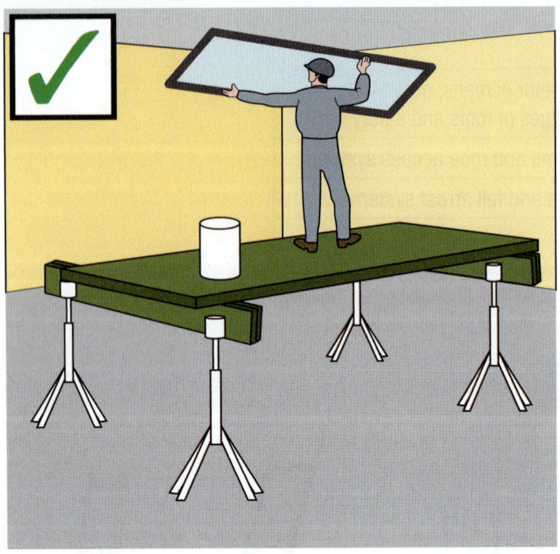

Source: Safe Work Australia

FIGURE 1.34 A platform ladder is preferable

the work height is 1.4 m, maintain a distance of 1.6 + 1.4 + 1.5 = 4.5 m from the fall hazard.

Working at heights greater than 2 m

At heights greater than 2 m ladders should be the last choice of a number of alternatives. Scaffolding either fixed or mobile, elevated work platforms such as scissor lifts or boom lifts (cherry pickers) are preferable over ladders.

When working higher than 2 m, the same rules concerning ladders apply. Only timber or fibreglass ladders are suitable for electrical work. Aluminium or steel reinforced timber ladders must not be used. Timber ladders must use a non-conductive fibreglass core as reinforcement to be safe. The length of the ladder should allow work to be done without standing less than three rungs from the top. If the ladder is being used to access a roof the ladder should be long enough to extend at least 1 m past the point of contact with the structure. This will give you something to hang on to and will help you maintain balance.

Safety checks

WorkSafe Tasmania's 'Preventing Falls in Housing Construction Code of Practice' (http://worksafe.tas.gov.au, p. 26) outlines that before every use of a ladder it should be visually checked for:

- cracks or splits in the rungs and stiles
- rungs, steps, treads or top plates which are missing or loose
- twisted, bent, kinked, crushed stiles or cracked welds or damaged feet in metal ladders
- worn, damaged or loose ropes, braces, or brackets which are missing
- broken or worn timber members.

Ladders that are found to be faulty should not be used and danger tagged to warn others.

Precautions for working on ladders

If it is not practicable to use the control measures shown in Table 1.19, a ladder may be used to carry out lightweight work for a short period. When using portable ladders, the following precautions should be considered:

- use only industrial rated ladders, do not use domestic ladders, use the correct ladder for the job
- ladders should only be erected on flat solid surfaces, beware of soft ground, ladders should be adequately supported at the base
- set the ladder at a slope of 4 in 1. That is, ladders must be angled 1 m out for every 4 m up (shown in Figure 1.35)
- ladders should extend at least 1 m above the access point
- ladders should be firmly secured or tied off, when tying off the ladder the ties should be attached to the stiles of the ladder and not the rungs
- when working near doorways or pedestrian traffic areas, lock the door shut or erect a barrier.

When working on a ladder:

- when ascending or descending the ladder always keep your body within the stiles and keep at least three points of contact with the ladder
- never stand on a rung closer than 900 mm to the top of a single or extension ladder

- never stand higher than the second tread below the top plate of any stepladder.
- using power tools (electric or battery, air or hydraulic) may require the operator to brace themselves against the high level of torque exerted by the tool
- never carry out work such as arc welding or oxy cutting from a ladder unless step platforms or other temporary work platforms are not feasible and the task is of short duration
- never work over other people, or allow anyone else to be on the ladder at the same time.

Except where additional and appropriate fall protection equipment is used in conjunction with the ladder, any person using a ladder should not face away from the ladder when going up or down, or when working from it.

FIGURE 1.35 Ladder safety

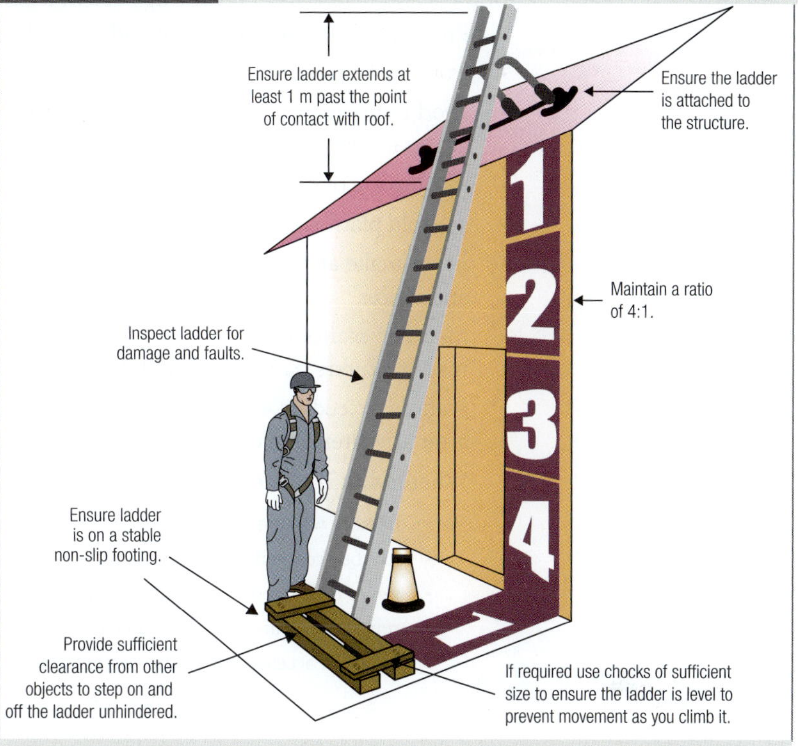

Source: Adapted from http://www.thesafetybloke.com

Precautions with scaffolds and elevated platforms

Scaffolds are a common means of providing a safe platform for working at height. There is a wide variety of scaffold systems available. Scaffolds are regulated by federal and state government laws. Fixed scaffolding must be erected and dismantled by a qualified scaffolder. Scaffolding is shown in Figure 1.36.

FIGURE 1.36 Fixed scaffolding

Source: Ed Hawkins

According to the 'Preventing Falls in Housing Construction Code of Practice' (WorkSafe Tasmania, p. 13), scaffolding must conform to the **AS/NZS 1576 Scaffolding** series of standards. Scaffolds are generally rated as light, medium or heavy duty:

- Light Duty – up to 225 kg per bay. Examples include painting, electrical work, many carpentry tasks and other light tasks. Platforms should be at least two planks wide (approximately 450 mm).
- Medium Duty – up to 450 kg per bay. This is suitable for general trades work. Platforms should be at least four planks wide (approximately 900 mm).
- Heavy Duty – up to 675 kg per bay. This is what is needed for bricklaying, concreting, demolition work and most other work tasks involving heavy loads or heavy impact forces. Platforms should be at least five planks wide (approximately 1000 mm).
- Special Duty – has a designated allowable load as designed.

Source: WorkSafe Tasmania, 'Preventing Falls in Housing Construction Code of Practice', p. 13.

These safe load limits include the weight of people, a nominal 80 kg, plus the weight of any materials, tools and debris on the working platform. A properly constructed mobile scaffold with a light duty platform can only safely support one worker and 145 kg of tools and material, or two workers and 65 kg of tools and materials ('Preventing Falls in Housing Construction Code of Practice', p. 13).

Precautions when working on and around a fixed scaffold should include:

- workers are provided with training to ensure they understand what loads the scaffold can safely take
- good housekeeping practices are observed; working platforms need to be kept clear of debris and access is kept free of obstacles
- incomplete or defective scaffolds must never be accessed
- scaffold platforms should be a minimum of 450 mm wide
- never make any unauthorised alterations to the scaffold such as removing guardrails, planks, ties and braces
- modular scaffolds should be of the same type, not mixed. Mixing components from different manufacturers have incompatibilities and can cause failures
- scaffold exceeding 4 m in height should be inspected and tagged by a competent person before use, after any alteration or repair, and at intervals of not greater than 30 days

Where work is carried out using mobile scaffolds, the scaffold must:

- be kept well clear of powerlines, open floor edges and penetrations
- not be accessed by climbing on the scaffold, workers must only use internal ladders
- never be moved while anyone is on it
- remain level and plumb at all times
- never be accessed until the castors are locked to prevent movement.

Source: WorkSafe Tasmania, 'Preventing Falls in Housing Construction Code of Practice', p. 13.

In most cases when using scaffolding, work will be performed at a height of greater than 2 m. Workers are required to wear additional fall protection such as travel restraint systems or fall arrest systems.

Elevated work platforms

Elevated work platforms (EWPs) allow easy access and an increased level of safety when working at heights. As long as access is possible the cost of the EWP is outweighed by increases in productivity and safety. Workers using EWPs must be trained in their use and many states require operators to hold licences for their operation.

EWPs fall into two categories – either scissor lifts or boom lifts. Scissor lifts are limited to providing access to work areas directly above the vehicle. They may be battery powered or internal combustion engine types. Some are designed for hard flat surfaces only, while others are designed for operation on rough terrain. Figure 1.37 shows a variety of EWPs.

FIGURE 1.37 Elevated work platforms (EWPs)

Sources: iStockphoto/Bluekite, iStockphoto/Gutzemberg, shutterstock.com/Cynthia Farmer

Boom lifts are more versatile allowing the boom to be extended to the side of the vehicle and in some cases even below the level of the vehicle. EWPs usually have two sets of controls. One set in the platform and another at ground level for emergency use.

Precautions when working on and around elevated work platforms should include (Worksafe Tasmania):

- only operating the EWP on a flat level surface unless the EWP is designed to operate on 'rough terrain'
- when using 'rough terrain' units on uneven surfaces operators must follow manufacturer's instructions
- operators must know the safe working load (SWL) and never exceed it.
- the work area should be checked to make sure that there are no penetrations or obstructions that could cause the EWP to roll over
- ensuring the floor area's SWL will support the combined weight of the EWP and load
- EWPs must not be used as cranes or lifting devices
- be aware of fumes from internal combustion engines in enclosed spaces
- when operating in EWPs workers must wear an anchored safety harness in accordance with AS 2550.10 or most recent equivalent ('Preventing Falls in Housing Construction Code of Practice', pp. 14–15).

Harnesses and personal fall arrest systems

When scaffolds and EWPs are used to access to work areas at heights of greater than 2 m additional fall protection is provided through personal fall arrest systems. The fundamental principles behind fall protection are shown in Figure 1.38.

For fall prevention to be effective a secure anchor point is required. Working on the assumption that the nominal weight of a person is 80 kg, a fall of just 1 m combined with gravity will result in a load of nearly 800 kg on the anchor point as shown in Figure 1.38. This load will also be applied to the body harness and connectors. Figure 1.39 shows a typical harness and connectors. If a personal fall arrest system is used, as part of the SWMS a rescue plan must be developed to recover the worker in the event of a fall.

FIGURE 1.38 Anchor points

FIGURE 1.39 Harness and connectors

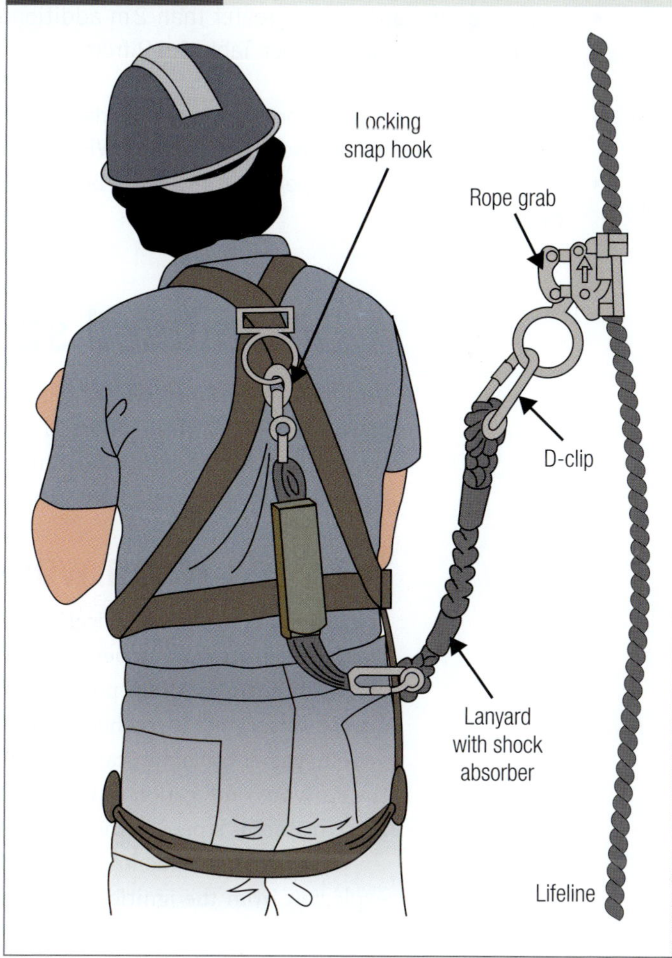

Case study

Fatal fall from height greater than 2 m

A painting contractor failed to secure the ladder he was using to a roof. The ladder was required to gain access to a wall above that roof section which required painting. A gust of wind blew the unsecured ladder to the ground, and as the painter was working alone he found himself trapped on the roof. While attempting to cross a 40-year-old section of corrugated asbestos cement roofing to an open window, the painter fell through the sheeting to his death. The building was constructed before codes required underpinning mesh. The coroner reported that a simple measure such as securing the ladder would have prevented his death. The coroner also found that scaffolding should have been used in preference to a ladder and that a catwalk should have been retrofitted to the building in lieu of the underpinning mesh.

SUMMING UP...

- Working at heights dramatically increases risks.
- General risk management practices and safe work method statements are used to control risk sources when working at heights of up to 2 m.
- When working at heights greater than 2 m additional precautions are taken as per Table 1.19 from levels 1 to 5.
- Never exceed the safe working load of ladders, scaffolding or elevated work platforms (EWPs).
- All ladders must be inspected for defects before every use.
- Never use aluminium or steel ladders for electrical work.
- Fixed scaffolding should only be erected or dismantled by a qualified 'scaffolder'.
- Scaffolds must only be accessed by the internal ladders, not by climbing the scaffold itself.
- EWPs should only be operated by trained and qualified workers.
- Harnesses must be worn when working from an EWP.
- If a personal fall arrest system is used, a rescue plan must also be developed to recover the fallen worker.

1.6 Confined spaces

Confined spaces in an electrical context are usually thought of as the roof spaces of buildings and the crawl space beneath the floors of houses. While these locations are uncomfortable, they do not meet the legislative criteria for a confined space. A common example of a confined space in an electrical context is a below-ground pump station.

Hazards when working in a confined space

The WorkSafe Victoria publication *Your Health and Safety Guide to Confined Spaces* (http://www.vwa.vic.gov.au) identifies a number of risks when working in confined spaces. Confined spaces have limited means of entry and exit, poor ventilation and the presence of toxic gases and vapours. Workers can also be suffocated or crushed by materials stored in the space. A contaminant is any dust, fume, mist, vapour, gas or other substance in liquid or solid form, the presence of which may be harmful. Some of the risks are: suffocation caused by a lack of oxygen, or suffocation or crushing after being engulfed by loose materials stored in the space such as sand, grain, fertiliser, coal or woodchips.

Other risks include loss of consciousness, injury or death due to contaminants in the air, fire or explosion from the ignition of flammable contaminants, enhanced combustibility and spontaneous combustion. Incidents involving confined spaces can often result in multiple fatalities. Other workers may not realise an additional hazard exists. They may enter a space to rescue a victim and be overcome by toxic vapours or gases.

Identifying confined spaces in workplaces

Safe Work Australia's 'Confined Spaces Code of Practice' (p. 9; http://www.safeworkaustralia.gov.au) recommends that to determine if the area you are working in is considered a confined space you should ask these four questions:

1 Is the space enclosed or partially enclosed?
2 Is the space not designed or intended to be occupied by a person?
3 Is the space designed or intended to be at normal atmospheric pressure while any person is in the space?
4 Does the space present a risk from:
   a harmful airborne or flammable contaminants
   b an unsafe oxygen level
   c engulfment? (Engulfment occurs when a worker is surrounded and overcome by a granular substance such as soil, sand, gravel, sawdust, seed, grain or flour or if submerged in a liquid such as water or a chemical.)

If the response to all of questions 1 to 3 is yes, and yes to any of the options in question 4, the work area is a confined space. Table 1.20 provides examples of the definition of a confined space for three work situations: a roof space in a domestic installation, the interior of a wheat silo and the drainage pit of an underground car park.

TABLE 1.20 Identifying enclosed spaces

IDENTIFYING ENCLOSED SPACES							
Space	Question			Question 4			Confined space Y/N
	1	2	3	A	B	C	
Roof space of house	Y	Y	Y	N	N	N	N
Wheat silo interior	Y	Y	Y	Y	N	Y	Y
Drainage pit	Y	Y	Y	Y	N	N	Y

Source: Safe Work Australia, 'Confined Spaces Code of Practice', http://www.safeworkaustralia.gov.au/sites/SWA/model-whs-laws/model-COP/Documents/Historical model Codes of Practice/Confined_SpacesV1.doc

Examples of confined spaces include vats, tanks and silos, pipes and ducts, ovens, chimneys and flues, shafts, trenches, tunnels and pits. Underground mines are not considered as confined spaces and are covered by separate legislation.

Control measures for working in a confined space

The Safe Work Australia 'Confined Spaces Code of Practice' (pp. 17–21) recommends the following control measures for working in a confined space.

Elimination

If it is practicable to do so, work should been done outside of the confined space by using a hook, long-handled clasp or magnet on a string to retrieve an object dropped into a confined space, or installing fixed or temporary cleaning devices.

Other methods include using remote cameras or a mirror attached to a probe for internal inspection of piping and vessels, using remote controlled devices to clear blockages.

Isolation

Where practicable, risk sources should be disconnected before anyone enters the confined space. These include piping, ducts, vents, drains, conveyors, service pipes and fire protection equipment that may introduce contaminates into the confined space.

Other risk sources that should be isolated are machinery and electrical equipment in the confined space that may release stored or potential energy that may create a crush hazard; and plant or services that could affect the space by heating or cooling from outside the confined space.

Pipe work supplying liquids, gases or vapours to the confined space should be isolated by closing valves or blanking off pipe work (Figure 1.40). Electrical isolation switches should be opened to disconnect electrical supply to the confined space.

FIGURE 1.40 Valve lockouts

Source: shutterstock.com/ArtboySHF

FIGURE 1.41 Electrical lockouts

Source: iStockphoto/Alacatr

Equipment or devices with stored energy, including hydraulic, pneumatic, electrical, chemical, mechanical, thermal or other types of energy, should be reduced to a zero energy condition. All isolating devices should be locked and tagged to prevent accidental operation until the work is completed. A typical multiple lockout for multiple workers is shown in Figure 1.41.

Purging makes the atmosphere safe by using exhaust fans to extract flammable, explosive, toxic or poisonous fumes and vapours. The confined space must then be tested with a gas detection device (see Figure 1.42) to ensure that no hazardous contaminates are present. Then a supply of fresh air can be blown into the confined space to ensure air quality.

Flammable gases have lower and upper explosive limits. The lower explosive limit (LEL) is the lowest ratio of fuel to oxygen mix that will allow combustion. The upper explosive limit (UEL) is the highest ratio of fuel to oxygen mix to allow combustion. Above the UEL there is too much fuel, the reaction is starved of oxygen, and therefore combustion cannot occur. If the concentration of any flammable gas, vapour or mist in the atmosphere of the confined space is more than 5% of its LEL the level must be monitored by a suitable gas detector. If the gas level rises above 10% of LEL all workers must immediately evacuate the confined space.

Ventilation, either natural or mechanically forced, may be used to keep the LEL of combustible substances below a hazardous level. If the flow of fresh air is restricted, carbon dioxide (CO_2) from the

FIGURE 1.42 Gas detector

Source: iStockphoto/Opreaistock

FIGURE 1.43 Ventilation

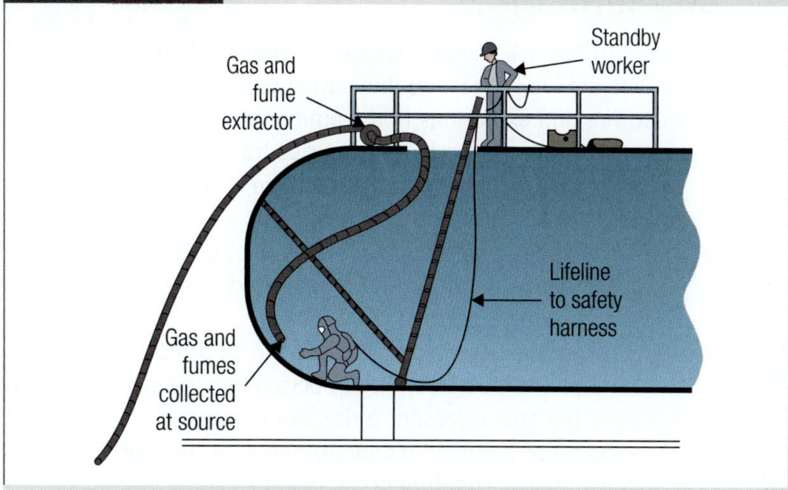

Source: Confined Space Program, http://www.ccohs.ca/oshanswers/hsprograms/confinedspace_program.html, Canadian Centre for Occupational Health and Safety (CCOHS). Reproduced with the permission of CCOHS, 2014

worker can build up within the confined space. While CO_2 is neither poisonous nor toxic, it cannot support life. Ventilation is achieved by blowing fresh air in, or extracting contaminated air out, of the confined space as shown in Figure 1.43.

A method is needed to enable communication between people inside and outside the confined space and to summon help in an emergency. Communication can be achieved by voice, radio, hand signals or other suitable methods (Safe Work Australia, 'Confined Spaces Code of Practice', p. 25).

Administrative

A standby person must continuously monitor the worker inside the confined space and initiate appropriate emergency rescue procedures if necessary. The standby person must understand the hazards inside the confined space and always remain outside the confined space even in an emergency. The Safe Work Method Statement must also include a procedure to rescue the worker if they lose consciousness without the standby person. This is normally achieved through the use of body harness and safety lines.

An entry or access permit system must be used to record the names of all workers in the confined space. Sign-in sign-out or tag systems are commonly used to ensure isolated hazards are not restored to the work area or machinery is not accidentally started until workers have left the confined space.

Signs 'must be erected to prevent entry of persons not involved in the work. Signs must warn against entry by people other than those who are listed on the confined space entry permit' (Safe Work Australia, 'Confined Spaces Code of Practice', p. 26). Suitable training must also be provided before workers enter a confined space.

Personal protective equipment

As well as normal PPE, confined spaces may require the use of respiratory protection. Dust masks are rated by the size of particles they can prevent from entering workers' lungs, but are not suitable in the case of either toxic or poisonous gases/vapours or low oxygen levels. In these latter situations air supplied and self-contained breathing apparatus must be used. Selection of appropriate respiratory protective equipment should be based on the level and type of contaminants and the work to be done (Safe Work Australia, 'Confined Spaces Code of Practice', p. 25).

SUMMING UP...

- Some of the risk sources associated with confined spaces are lack of oxygen, inhalation of fumes and dust, fire and explosions from flammable liquids, gases and vapours, and crush injuries from machinery.
- An area is considered a confined space if 'yes' is the response to all of these four questions:
  - Is the space enclosed or partially enclosed?
  - Is the space not designed or intended to be occupied by a person?
  - Is the space designed or intended to be at normal atmospheric pressure while any person is in the space?
  - Does the space present a risk from: harmful airborne or flammable contaminants; an unsafe oxygen level; engulfment?
- If possible, work should be done from outside the confined space.
- All risk sources such as electrical, hydraulic and mechanical must be isolated before entering the confined space.
- Hazardous atmospheres must be purged before entering the confined space.
- Ventilation and monitoring maintains a safe atmosphere within the confined space.
- The SWMS must include a rescue procedure for workers.
- Only workers listed on an access or entry permit may enter a confined space.

1.7 Physical and psychological hazards

Physical and psychological risk sources are rarely life threatening, but can severely decrease the quality of life of those exposed. Safe Work Australia provides a number of codes of practice to assist PCBUs and workers to control these risk sources.

FIGURE 1.44 Decibel chart

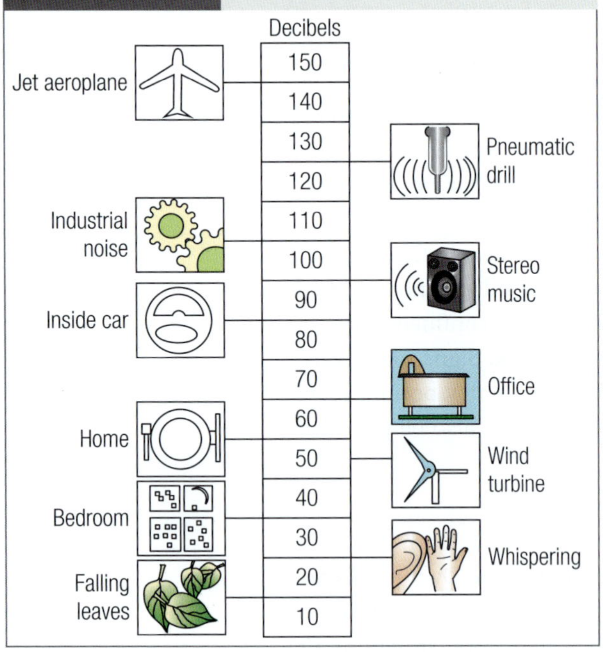

Excessive noise

Noise at work can cause hearing loss. This may be temporary, but long-term exposure to low-level noise or short-term exposure to very high noise can cause permanent damage. As well as hearing damage, exposure to high levels of noise can cause tinnitus (ringing, whistling, buzzing or humming in the ears). Working in a noisy environment makes communication difficult and means people may not hear warnings and alarms.

As well as the expected risk sources from machinery, electronic equipment can produce sudden, unexpected loud noises called *acoustic incidents*. These occur during telephone headset use and include crackles, hisses, whistles, shrieks or high-pitched noises. Acoustic shock is not caused by the loudness of a telephone, as all phone noise is electronically limited to a peak noise level of 123 decibels, but by a sudden rise in noise levels.

Sources of excessive noise are not only found at work. Recreation activities such as motor racing, listening to loud music on personal music players or amplified music at concerts can also contribute to hearing loss. Examples of various noise levels are shown in Figure 1.44.

The degree of hearing damage depends on the length of exposure to noise and the noise levels. Table 1.21 shows maximum lengths of time that workers can be exposed to noise without the use of hearing protection.

Noise at lower levels will not damage hearing, but can have other health effects. Noise reduces concentration and communication. Constant noise increases fatigue and the risk of cardiovascular disorders such as high blood pressure and heart disease.

WorkSafe Tasmania, in its 'Managing Noise and Preventing Hearing Loss at Work Code of Practice' (p. 10; http://worksafe.tas.gov.au), advises that to avoid adverse health effects noise should be kept below 50 dB where the work involves high concentration or conversation, and 70 dB where routine work that requires speed or attentiveness is performed or where it is necessary to carry on conversations.

TABLE 1.21 Noise exposures

Noise level dB	Exposure time	Noise level dB	Exposure time
80	16 hours	106	3.8 minutes
82	12 hours	109	1.9 minutes
85	8 hours	112	57 seconds
88	4 hours	115	28.8 seconds
91	2 hours	118	14.4 seconds
94	1 hour	121	7.2 seconds
97	30 minutes	124	3.6 seconds
100	15 minutes	127	1.8 seconds
103	7.5 minutes	130	0.9 seconds

Hierarchy of risk control

The WH&S regulations ('Work Health and Safety Codes of Practice 2011'; available at http://www.comlaw.gov.au/Details/F2011L02804/Html/Volume_2) require PCBUs to use a hierarchy of control to choose the control measure that most effectively eliminates or minimises the risk in the circumstances. Risk control may involve a single control measure or a combination of two or more different controls.

- *Elimination* is always the best control measure. Remove the source of noise completely by ceasing to use a noisy machine or changing the way work is carried out so hazardous noise is not produced.
- *Substitute* the hazard with plant or processes that are quieter.
- *Isolate* the source of noise from people by using distance, barriers, enclosures and sound-absorbing surfaces.
- *Engineering* controls. Purchase or hire equipment that can demonstrate a low-noise design, with noise control as a standard part of the machine. Modify plant and processes to reduce the noise.
- *Administrative* controls. Limit the time workers are exposed to noise as shown in Table 1.21.
- *Personal protective equipment* such as suitable earplugs or muffs must be provided and used.

Effects of vibration on the human body

Workers can be exposed to two types of vibration: hand-arm and whole-body. Hand-arm vibration can occur when using vibrating tools or equipment such as impact tools, including jackhammers, vibrating plates, chippers and pavement breakers, pneumatic and electrical rotary tools such as concrete breakers, grinders, sanders and drills.

Whole-body vibration occurs when the worker is seated or standing in plant or equipment such as tractors and heavy transport vehicles, cranes, forklifts and road-making plant. Operating this kind of plant may also expose workers to hand-arm vibration if the controls of the plant are vibrating as well.

Different parts of the body are affected by different types of vibration. The abdomen is affected by vibration of 4 to 8 times per second, while the eyes are affected by vibration at 20 to 90 times per second. The dissipation of vibration energy within the human body can cause various effects such as back pain, decreased grip strength and dexterity, plus increased pulse and respiratory rate.

The risk of harm depends on the characteristics of the vibration such as its magnitude, frequency, duration and direction. Control measures to reduce exposure to hand-arm vibration may involve finding alternative ways to do the work that eliminates the need to use vibrating equipment or to purchase tools that produce less vibration ('Work Health and Safety Codes of Practice 2011', http://www.comlaw.gov.au/Details/F2011L02804/Html/Volume_2). Control measures for full whole of body vibration include:

- *Elimination* – sending goods by train instead of employing several truck drivers.
- *Substitution* – a vibration source can be replaced by another machine with lower vibration magnitude, for example, machines with rotating instead of reciprocating parts, or belts instead of chains.
- *Isolation* – this could mean relocating operators or others to positions away from the vibrating source. Placing vibrating machines on separated slabs or spring mounts to isolate the source from the surroundings is another example of isolation.
- *Engineering* – vibration controls can be divided into two categories: controls at source and controlling the transmission.
- *Administrative* – reducing vibration exposure by reducing the duration of exposure. This can be done by rotating operators so each one would spend less time being exposed to vibration.
- *Personal protective equipment* – shoes with absorbing soles have been tried but only offer a very small reduction. Protective belts are another kind of PPE that have been tried by drivers of off-road vehicles and motorcycles.

Effects of thermal stress

Take care when working in hot conditions.

Thermal stress is defined as the physical and physiological reactions of the worker to temperatures that fall outside of the worker's normal comfort zone. A normal body temperature is around 36.5–37.5°C. If core body temperature rises or falls outside of this range, normal body functions and metabolism will be affected. Above 37°C heat stress may occur, and if core temperature falls below 35°C this can result in hypothermia.

Safe Work Australia's 'Managing Risks in Construction Work Draft Code of Practice' (http://www.safeworkaustralia.gov.au) warns that 'heat stress can arise from working in high air temperatures, exposure to high thermal radiation or high levels of humidity' (p. 30), including working in or on roof spaces, formwork deck and fit-out work in an enclosed non-air-conditioned structure. 'The symptoms of heat stress include dizziness, fatigue, headache, nausea, breathlessness, clammy skin or difficulty remaining alert' (p. 30).

Hierarchy of risk control

If elimination of exposure to heat and cold is not practicable, there are a number of other lower level control measures. These are:

- *Substituting* by performing the work in an indoor location instead of an outdoor location.
- *Isolation* of workers from sources of heat.
- *Engineering* controls including installation of air-conditioning or ventilation fans in indoor locations. Outdoor workers should be provided with sunshades and windbreaks, and protection in adverse weather conditions such as sheds, caravans or tents.
- *Administrative* controls including task rotation and rest breaks. Generally, workers involved in outdoor work are at greater risk from adverse climatic conditions. Activities should be organised so that workers carry out alternative tasks or work in shade during hot periods of the day. Workers must have access to water.

The Public Service Association (PSA) of NSW 'Policy on heat and work' (p. 5; http://psa.asn.au) recommends that the 'trigger temperature' for remedial action for outdoor work is 30°C. If temperatures reach and are sustained at this level for two hours or more the procedures shown in Table 1.22 are recommended to be followed.

TABLE 1.22 Working at temperatures of greater than 30°C

Temperature °C	Remedial action
30–32	10-minute break per hour from outside work
32–35	15-minute break per hour from outside work
35–36	30-minute break per hour from outside work
37+	Cease outside work until temperature decreases

Source: PSA Public Service Association of NSW, 'Policy on heat and work', p. 5, http://psa.asn.au/wp-content/uploads/2013/06/HEATPSApolicy.pdf

The guidelines further state:

> The above temperature/rest formula applies to light or very moderate physical activity. If the work involves any type of heavy work and/or rapid physical activity, the rest ratio in the formula must be increased by a minimum of 50% for each temperature range. . . . 37°C is a critical temperature... as workers who are carrying out physical work at this temperature, or above . . . will be at *significant* risk of increasing their body temperature to the point where they are in danger of heat stroke. (p. 5)

PPE

Protection against solar ultraviolet (UV) exposure is also important, for example, by providing personal protective clothing and equipment, such as a wide brim hat, long-sleeved and collared shirt, long pants, sunglasses and sunscreen, and hard hat attachments. As outlined in 'Managing the Work Environment and Facilities Code of Practice' (Safe Work Australia, p. 15), if it is not possible to eliminate exposure to extreme cold, hazard sources can be controlled by providing:

- protection from wind and rain, such as a hut or the cabin of a vehicle
- localised heating, for example, cab heaters for fork-lift trucks used in cold stores
- opportunities for workers who are not used to working in cold conditions to acclimatise, for example, job rotation and regular rest breaks
- protection through warm (and, if necessary, waterproof) clothing.

You should provide immediate assistance if any worker shows any of the following warning signs of hypothermia:

- numbness in hands or fingers
- uncontrolled shivering
- loss of fine motor skills (particularly in hands – workers may have trouble with buttons, laces, zips)
- slurred speech and difficulty thinking clearly, and irrational behaviour – sometimes a person will even begin to discard clothing.

Effects of ultraviolet (UV) radiation on the human body

Radiation is energy travelling as waves of electromagnetic radiation or subatomic particles. Ultraviolet (UV) radiation has three main sources: sunlight, electric arc welding and accidental exposure to arc flash.

Sunlight

The harmful effects from exposure to ultraviolet radiation can be classified as acute or chronic. The acute effects of UV-A and UV-B exposure are both short-lived and reversible. These effects include mainly tanning and sunburn. The chronic effects of UV exposure can be much more serious, even life-threatening, and include premature ageing of the skin, suppression of the immune system, damage to the eyes, and skin cancer.

UV rays can also damage the eyes as more than 99% of UV radiation is absorbed by the front of the eyes. Corneal damage, cataracts and macular degeneration are all possible chronic effects from UV exposure and can ultimately lead to blindness. Melanoma, a type of skin cancer, can also develop within the eye. In the longer term, repeated exposure to the sun can result in keratoses (sun spots), skin cancers, premature skin ageing and injuries to the eye.

The factors which affect the intensity of solar UV radiation include: proximity to the equator, season of the year, time of day, cloud cover, altitude (higher altitude results in increased exposure), extent of reflection and extent of shade.

For untanned skin, exposed to the summer sun, the effects are:

- 12 minutes of sun exposure will result in mild sunburn
- 30 minutes of sun exposure will result in appreciable discomfort
- 60 minutes of sun exposure will result in peeling and blistering
- 120 minutes of sun exposure will result in permanent damage.

Solar UV radiation is at its greatest intensity between 10 a.m. and 2 p.m. or 11 a.m. and 3 p.m. during daylight saving times. Cloud cover will reduce the intensity of solar UV radiation; however, sunburn can still occur on a cloudy day.

Photosensitisation of the skin and eyes can be caused by exposure to specific chemicals. For example, exposure to tar and sunlight can cause precancerous and cancerous skin lesions. Exposure to coal tar or creosote fumes can cause simultaneous inflammation of the cornea. Certain plants such as carrot, celery, chrysanthemum, fennel and fig can also increase the effects of solar UV radiation.

Sunburn can still occur in shaded areas, due to the scattering of solar UV radiation by clouds and reflection from brightly coloured or shiny surfaces such as metal roofing, concrete, sand, water, glass and snow. Control of solar UV radiation is shown in Table 1.23.

TABLE 1.23 Hierarchy of control measures of solar UV radiation

Level	Hierarchy	Control
1	Elimination	Perform work indoors
	Substitution	N/A
2	Isolation	N/A
	Engineering	Shade from trees, buildings, canopies, tents, screens and other portable structures
3	Administrative	Outdoor work not programmed between 10 a.m. to 2 p.m.
	PPE	• Long-sleeved shirts with collars and long trousers with a UPF rating of 50+ • Hats with brims of at least 80 mm • Hard hats and other protective hats, fitted with broad brims or attachable brims and neck flaps

Source: University of Minnesota, School of Public Health, 'Harmful effects of ultraviolet radiation', http://enhs.umn.edu/current/5103/uv/harmful.html

Electric arc welding

Electric arc and laser welding equipment emits ultraviolet, visible light and infra-red radiation. The potential effect of radiation on the body depends on the type, intensity, the distance you are from it and the duration of exposure. Eye disorders and skin burns may be caused by exposure to intense ultraviolet and infra-red radiation in welding. Exposure to the eyes causes 'arc eye' or 'welder's flash' which is a painful inflammation of the cornea. The cornea can repair itself in one to two days; however, if the cornea becomes infected it may lead to some loss of vision.

The effects of ultraviolet and infra-red radiation are not normally felt until some time after exposure. Radiation from laser welding is less obvious than from electric welding arcs, but both are serious hazards. Workers directly involved in the welding are at greatest risk; however, other workers in the workplace and passers-by could also be exposed to radiation. Table 1.24 shows examples of control measures used to reduce exposure to UV radiation produced by arc and laser welding.

FIGURE 1.45 Administrative control of arc welding risk sources

TABLE 1.24 Hierarchy of control measures of arc welding UV radiation

Level	Hierarchy	Control
1	Elimination	Don't weld
	Substitution	Bolt parts together
2	Isolation	Remove the welder and nearby workers from the hazard or isolate or screen the hazard from the welder
	Engineering	Install non-flammable screens and partitions
3	Administrative	• Use signs to warn that welding is occurring (see Figure 1.45) • Restrict entry into the work area where welding is being carried out
	PPE	• Filter shades for goggles and face shields to protect the eyes from radiation • Wear gloves and other protective clothing to cover exposed skin

Source: WorkSafe Tasmania, 'Welding Process Code of Practice', p. 9, http://worksafe.tas.gov.au/__data/assets/word_doc/0009/204777/welding_processes_code.doc

Dangers associated with laser-operated equipment

Lasers are commonly used for alignment, marking out, levelling and survey tasks for construction work. In electrical work they are a modern replacement for string lines and tape measures. Work Health and Safety (WH&S) regulations impose requirements in relation to the safe use of laser equipment in the workplace. Beam exposure can cause serious eye damage, blindness or burns to the skin. The extent of eye damage is increased if the iris is dilated, so extra care must be taken in dark locations or when using a laser at night.

Exposure to the laser beam is not limited to direct beam (intrabeam) exposure. Particularly for high-powered lasers, exposure to beam reflections may be just as damaging as exposure to the primary beam. The beam may be reflected off a surface into the eye.

The majority of control measures are administrative. The laser beam must never be deliberately aimed at personnel and must be prevented from being directed towards the eye of a person. Lasers are labelled

FIGURE 1.46 Laser class product label

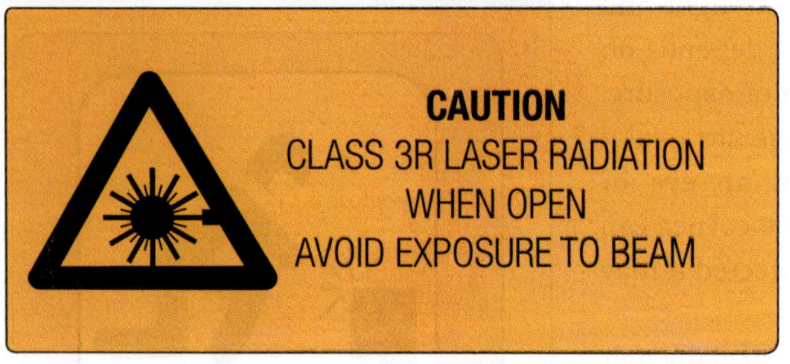

FIGURE 1.47 Laser warning sign

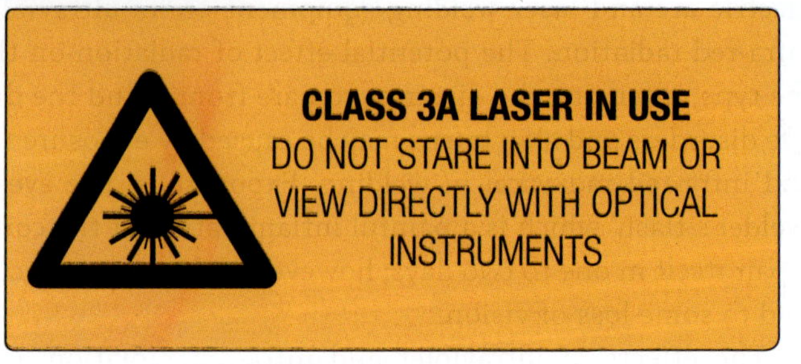

to show a classification from 1 to 4, as shown in Figure 1.46. A higher number indicates a greater risk. **AS 2397:1993 Safe use of lasers in the building and construction industry** provides guidance on the safe use of lasers.

Table 1.25 shows which classes of labels are suitable for use in building and construction sites. Where class 2 lasers and above are used, a certified Laser Safety Officer (LSO) has the authority and responsibility for ensuring that other people do not receive harmful exposure to laser radiation.

Warning signs must be displayed to indicate the use of lasers class 2 and above (see Figure 1.47).

TABLE 1.25 Suitability of laser classes for use in construction

Class	Description	Example	Use
1	Safe for use under all conditions of exposure.	• laser printers • DVD players	Yes
2	Low-powered lasers that present little hazard (eye protection is provided by normal blink and aversion responses).	• laser pointers • barcode scanners • distance measure	Yes
3A	These lasers emit higher levels of light. They are not used with optical instruments (a builder's level or theodolite) which would concentrate the beam so that it would all enter the eye.	• rotary levels	Yes
3B Restricted	Same power levels as class 3A but have higher levels of irradiance (power density). Should *not* be used in dimly lit building or construction applications (that is, less than approximately 100 lux).	• alignment lasers • earth moving equipment	Yes
3B	These lasers emit either invisible or visible radiation potentially hazardous to the eye and skin.	• physiotherapy treatments	No
4	High power output. Skin exposure to the direct beam is hazardous and capable of producing diffuse reflections hazardous to the eye.	• projection displays • surgery devices • metal cutting devices	No

Source: Safe Work Australia, 'Laser classifications and potential hazards', http://www.safeworkaustralia.gov.au/sites/swa/about/publications/Documents/685/Laser_classification_and_potential_hazards.pdf, p. 1

Classes 3A and 3B (lasers of the 'fixed beam' type) must be screened to prevent the beam extending beyond the work area. The 'beam stop' must be made from a suitably sturdy, opaque and rigid material, and it should not produce unwanted reflections. Earth walls, wood, plywood, brick, mortar, concrete and plaster board are all excellent beam stop materials. Table 1.26 provides a summary of laser risk source control measures.

TABLE 1.26 Laser risk source controls

Control	Class of laser permitted			
	1	2	3A	3B restricted
Classification marked on laser	✓	✓	✓	✓
LSO allocated for site	N/A	✓	✓	✓
Operator trained in use	N/A	✓	✓	✓
Warning signs on entry points	N/A	N/A outdoors ✓ indoors	N/A	✓
Area clear of mirror-like (shiny) surfaces	N/A	✓	✓	✓
Stationary lasers terminated by beam stop	N/A	✓	✓	✓
Class 3B (Restricted) laser marked to warn against use in low light conditions	N/A	N/A	N/A	✓
Beam blocked at source when laser unattended and power on	N/A	N/A	✓	✓
Laser beam located well above or well below eye level	N/A	N/A	✓	✓

Occupational overuse syndrome

Occupational overuse syndrome (OOS) is also known as repetitive strain injury (RSI). The term repetitive strain injury is not a specific medical diagnosis and is more of an umbrella term for a number of musculoskeletal diseases (MSD) that affect the soft tissues (muscles, tendons and nerves) of the neck, upper and lower back, chest, shoulders, arms and hands. It is caused by repeating the same motion over and over, such as repeatedly cutting cables. If the task involves repetitive or sustained postures, movements or forces, and long duration, this will increase the risk of MSD.

If a manual handling task involves a risk of MSD, environmental factors can increase that risk. Environmental factors include vibration, heat and cold. Some of the known risk factors associated with OOS are:

- awkward body postures
- poor design of equipment, machinery and tools including the effects of vibration and sudden impact forces
- poor design of tasks, employee position, forces required and placement of equipment
- required work output, duration and variation of tasks, number and duration of pauses
- poor job design, i.e. the requirement to perform the same repetitive movements
- new employees, or those returning to work after an extended absence, being required to perform repetitive movements without a period for adjustment.

The first stage in the three-stage approach to safe manual handling is to identify tasks which involve awkward postures, repetitive or forceful movement. Risk identification is carried out by:

- analysing the injury and incident records of the workplace
- consulting with the employees doing the task and employee representatives
- direct observation or inspection of the task or workplace.

If any of these three steps indicate the need for further action, a risk assessment is conducted. After consultation with workers a range of control measures, as shown in Table 1.27, can be applied.

Improper use of hand tools can be a major cause of OOS. Hand tools should be: appropriate to the task, appropriate to the work surface height, properly proportioned to the body dimensions of the worker, appropriate to the strength and endurance of the person, maintained in good working order, available in both left- and right-hand versions, and selected with the aim of minimising vibration and sudden impact forces.

TABLE 1.27 Hierarchy of control measures of occupational overuse syndrome

Level	Hierarchy	Control
1	Elimination	Change the work process to remove the need for continuous repetitive actions
	Substitution	Use power tools instead of hand tools
2	Isolation	Remove worker from source of vibration, heat or cold
	Engineering	• use mechanical aids • modification of workplace layout
3	Administrative	• monitor frequency and duration of task • task-specific training
	PPE	N/A

FIGURE 1.48 Stress

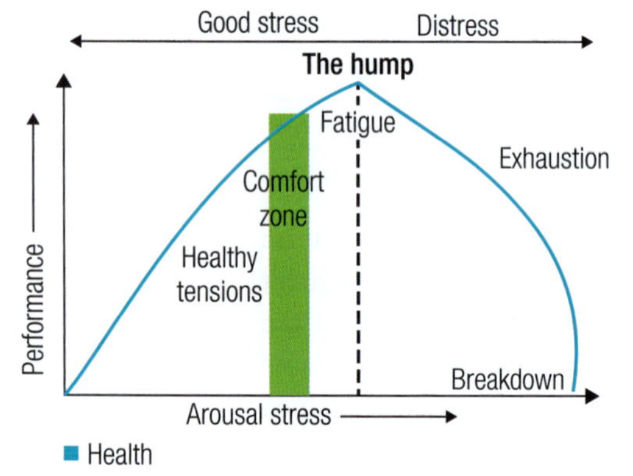

Source: The American Institute of Stress, http://www.stress.org/topic-definition-stress.htm

Stress in the workplace

Workplace stress on employees has been linked to a wide range of mental and physical health conditions. High levels of workplace stress can cause nervousness, tension, strain, anxiety and depression. Workplace stress can lead to accidents, high staff turnover and absenteeism. It creates a risk to personal and workplace health and safety. Figure 1.48 shows how different levels of stress affect people.

PCBUs must take steps to ensure that factors in the workplaces under their control don't cause or contribute to stress. Workers also have a responsibility to take reasonable care for their own health and safety. This means that if they recognise they are stressed, they must ask their employer/manager for workplace solutions and support, and take care of their own wellbeing.

Effects and dangers of drug and alcohol use

The effects of both alcohol and drug use during and outside of work hours can have a significant negative impact on workplace health, safety and productivity. Energy drinks containing high doses of caffeine can also cause health issues.

Alcohol is a depressant drug, slowing down activity in the central nervous system, including the brain. It affects concentration and coordination, and slows response times. Short-term risks include risk-taking behaviour, falls, confusion, drowsiness, blurred vision, poor muscle control, nausea and death from asphyxia (choking on vomit).

The physical and psychological effects of illicit drugs depend on the type of drug use. The most common illicit drugs consumed are cannabis, ecstasy, amphetamines and cocaine. Consumption of small quantities of these drugs can result in effects which include impaired coordination, affected thinking and memory, increased heart rate and low blood pressure.

Consumption of larger quantities of these drugs can result in effects such as distorted perception, confusion, restlessness, anxiety and panic, decreased reaction time and paranoia. Other drugs such as ecstasy can cause anxiety, increased heart rate, and higher blood pressure and temperature, with more dangerous side effects being convulsions, vomiting, irrational behaviour and hallucinations. Amphetamines also produce increased heart rate, breathing and blood pressure, restlessness, anxiety, irritability, hostility and aggression. Higher doses can cause headaches, dizziness, blurred vision, tremors and loss of coordination. Cocaine can produce symptoms of exhilaration, anxiety, poor concentration and judgement and intolerance to pain and fatigue ('All drugs and their effects', Blue Mirrow, http://www.bluemirrow.com). Repeated use can lead to extreme agitation, paranoia, hallucinations, tremors and loss of concentration and coordination.

As these drugs are illegal and therefore unregulated, there is no 'standard' that can be applied to the amount consumed; When illicit drugs are taken in combination with alcohol and/or prescription medications and/or energy drinks the detrimental effects will be exaggerated. Consideration should also be paid to the legal costs. Drink or drug driving convictions resulting in a loss of licence can affect employment possibilities, as can convictions for drug use and possession.

SUMMING UP...

- To avoid adverse health effects noise should be kept below:
  - 50 dB where work involves high concentration or conversation
  - 70 dB where routine work that requires speed or attentiveness is performed or where it is necessary to carry on conversations.
- For prolonged periods of exposure at 80 dB or above hearing protection must be used.
- There are two types of vibration: hand-arm and whole-body. Exposure to vibration can cause back pain, decreased grip strength and dexterity, increased pulse rate or respiratory rate and other effects.
- A normal body temperature is around 36.5–37.5°C. The symptoms of heat stress include dizziness, fatigue and headache.
- Ultraviolet radiation has three main sources: sunlight, electric arc welding and accidental exposure to arc flash.
- Just 12 minutes of sun exposure will result in mild sunburn to untanned skin.
- When arc welding, remove nearby workers from the hazard or isolate by screening the welder hazard source.
- Use signs to warn that welding is occurring.
- Restrict entry into the work area where welding is being carried out.
- Only class 1, 2, 3A and 3B (restricted) lasers are suitable for construction use. Class 3B and 4 are not.
- Never point a laser at a person's eyes.
- Warning signs must be displayed when class 2 lasers are used indoors and when class 3A or 3B (restricted) lasers are used indoors or outside.
- RSI or OOS is caused by repeating the same motion over and over.
- Improper use of hand tools can also be a major cause of OOS, and improved task design can control OOS.
- High levels of workplace stress can cause nervousness, tension, strain, anxiety and depression.
- Workplace stress can lead to accidents, high staff turnover and absenteeism.
- Consumption of alcohol, illicit drugs and some medications can result in some people not being fit for work.

1.8 Working safely with electricity

Electrical work presents a series of unique challenges different to the risk sources of any other type of work. Electrical energy is invisible, odourless and silent. Therefore, extreme care must be taken when performing electrical work. A phrase you should keep in mind is 'there are old electricians and there are bold electricians, but there very few old bold electricians'.

Effects of electric shock

There are four effects of electric current flow: thermal (heat), magnetism, chemical and physiological. When electric current flow finds a path through living tissue the physiological effects cause what is commonly known as electric shock. **AS/NZS 60479.1:2010 Effects of current on human beings and livestock** provides guidance on the effects of electric shock. An electric shock can cause muscle contraction, ventricular fibrillation (heart), cardiac arrest, asphyxia (lungs) and burns.

Ventricular fibrillation is when the heart takes on an irregular rapid beat that does not pump blood effectively through the body. Cardiac arrest caused by an even higher current is when the heart stops altogether. Asphyxia is a result of contraction of the lung muscles. The victim is unable to breathe in, and therefore is also unable to speak. Any of these conditions will cause loss of consciousness. Brain damage will occur at approximately two to three minutes followed by death.

The severity of an electric shock depends on three variables:

- the path the current flows through the body
- the amount of current that flows through the body
- the length of time the current flows.

Other factors that may affect the severity of the shock are the:

- presence of moisture in the environment
- phase of the heart cycle when the shock occurs
- general health of the person prior to the shock (Dong Zhao, Jason Lucas and Walid Thabet (2009), 'Using virtual environments to support electrical safety awareness in construction', Proceedings of the 2009 Winter Simulation Conference, p. 2679).

For an electric shock to occur a person must be in contact with at least two points of contact, one of which must be at a raised voltage. This is called a touch voltage. Most electric shocks occur between hand and foot, or hand to hand. The amount of current that will flow through a body will depend on the difference in touch voltage between the two points and the resistance of the current path through the body. This is called the touch current.

Currents larger than 200 mA and lasting more than a few seconds will cause deep burns to muscle and internal organs. Permanent damage to the nervous system and spine is also possible. Arcing will cause horrific burns to the skin and eyes. The heat generated by an electric arc has enough energy to melt metal switchboards.

FIGURE 1.49 Age of electrocution victims

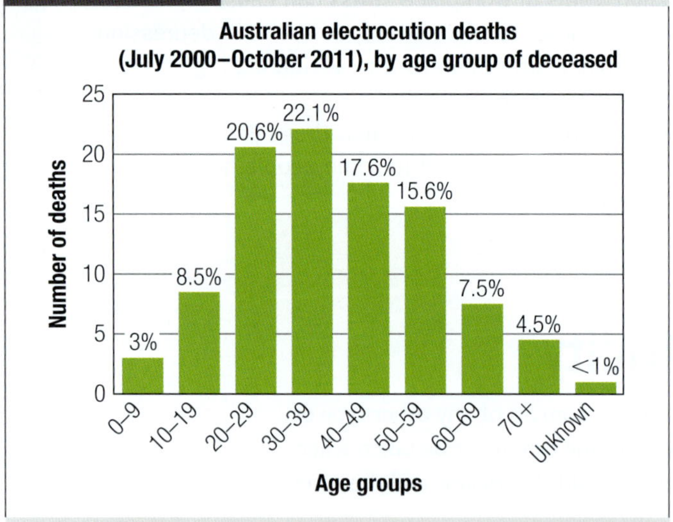

Source: National Coroners Information System (NCIS) Fact Sheet. January 2012 Electrocution Related Deaths, M D Rossetti, R R Hill, B Johansson, A Dunkin and R G Ingalls, eds, http://pandora.nla.gov.au/pan/126744/20120807-0704/Feb%202012%20FACT-SHEET%20-%20Electrocution%20related%20deaths.pdf

Common causes of electrical accidents

A statistical fact sheet released by the National Coronial Information System showed that between 1 July 2000 and 31 October 2011 there were 321 deaths in Australia caused by electrocution (National Coroners Information System (NCIS) Fact Sheet. January 2012 Electrocution Related Deaths, p. 2; see Figure 1.49).

The most common causes involved overhead power lines (22.1%), electrical appliances (16.6%) and direct contact with electrical wires (11.6%). Table 1.28 shows the locations where the electrical fatalities occurred. The highest number of incidents occurred not in workplaces but in the home. Unlicensed, unskilled workers or home handy persons taking short cuts by not following safe work methods are included in this number ('NCIS Fact Sheet. January 2012 Electrocution Related Deaths', p. 3).

TABLE 1.28 Australian unintentional electrocution deaths (July 2000–October 2011), by incident location

Location of fatal incident	Deaths
Home (including farmhouses)	96
Industrial or construction area	29
Commercial area	25
Farm or other place of primary production	19
Transport area – public roadway	11
Countryside	6
Other place of occurrence	13
Total	199

Source: National Coroners Information System (NCIS) Fact Sheet. January 2012 Electrocution Related Deaths, http://pandora.nla.gov.au/pan/126744/20120807-0704/Feb%202012%20FACT-SHEET%20-%20Electrocution%20related%20deaths.pdf, p. 4.

Direct contact

Direct contact is defined in **AS/NZS 3000:2007** clause 1.4.34. When direct contact is made with an active conductor (see Figure 1.50), a hand to both feet path is most commonly created for a current to flow through the victim to the ground (earth). Under **AS/NZS 3000:2007** the majority of electrical installations in Australia and New Zealand use the Multiple Earthed Neutral System (MEN) of earthing. Figure 1.51 shows by the red line how the MEN system provides a return path from the feet to the supply. Direct contact is the most common cause of electrical accidents.

FIGURE 1.50 Direct contact

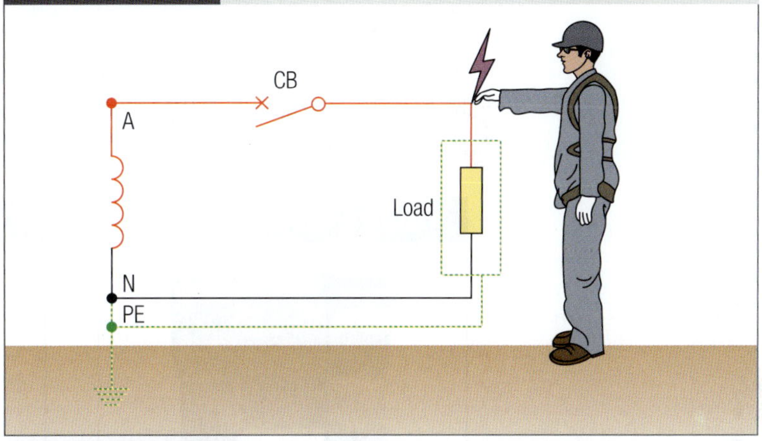

Indirect contact (fault protection)

Indirect contact is defined by **AS/NZS 3000:2007** clause 1.4.35. It can occur due to a cut in a cable's insulation or a live wire breaking off a terminal. If it touches a metal casing or conductive building material such as a metal wall stud, the metal is now also live. If a person makes simultaneous contact (touches both at the same time) with the stud and the ground a shock current will flow. Figure 1.51 shows by the red line, how current will flow from hand to both feet due to indirect contact.

FIGURE 1.51 Indirect contact caused by a fault condition

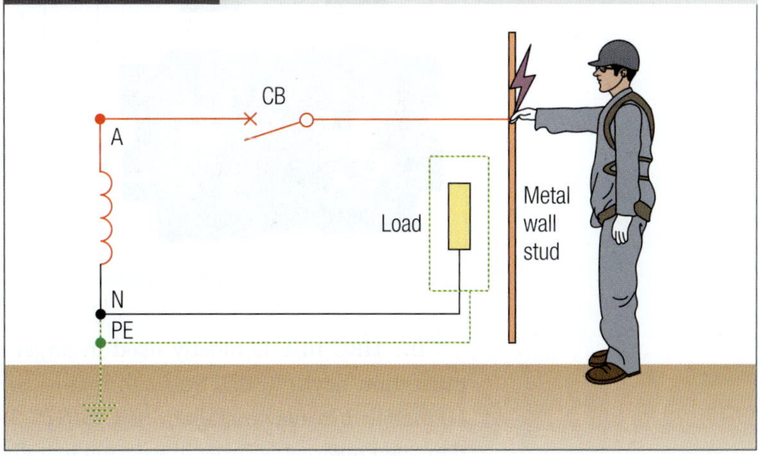

Minimising the chance of electric shock

Earthing

Earthing conductors are used to control the risk of electric shock as a result of indirect contact. Earthing conductors achieve this by electrically connecting exposed conductive parts of an installation together and the general mass of the earth. The aim is to prevent a difference in potential (voltage) between any exposed conductive parts and the ground, as shown in Figure 1.52.

FIGURE 1.52 The MEN earthing system

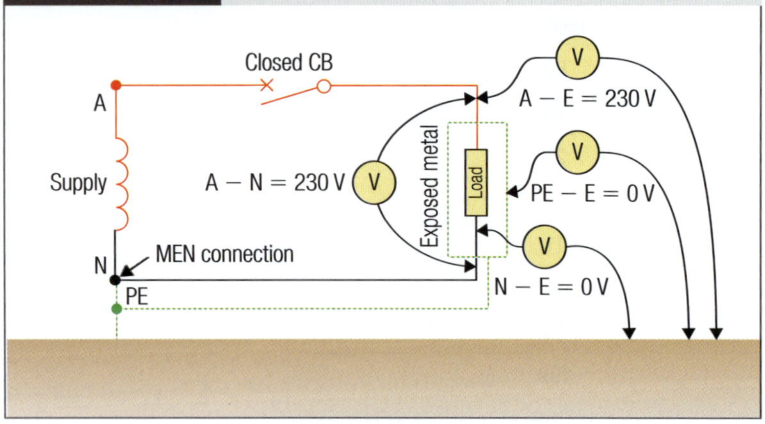

The earthing system also provides a path of very low resistance so that if an active to earth fault occurs, the current will be high enough to cause the circuit protection device (fuse, circuit breaker or RCD) to operate and disconnect the circuit very quickly. This is known as automatic disconnection of supply.

Extra low voltage

Many devices such as pool lights and down lights are supplied at extra low voltage (12V) to reduce the risk of electric shock. AS/NZS 3000:2007 clause 1.4.98 defines extra-low voltage (ELV). The idea of ELV is to reduce the voltage to a level where the resistance of the body is enough to limit the current flowing through a person to a value that does not cause undesirable physiological effects.

Fuses

AS/NZS 3000:2007 clause 1.4.57 defines a fuse. The two most common types of fuses used are the high rupturing capacity (HRC) type, as shown in Figure 1.53, and the semi-enclosed rewireable fuse as shown in Figure 1.54.

FIGURE 1.53 The HRC fuse

Source: Ed Hawkins

FIGURE 1.54 The semi-enclosed rewireable fuse

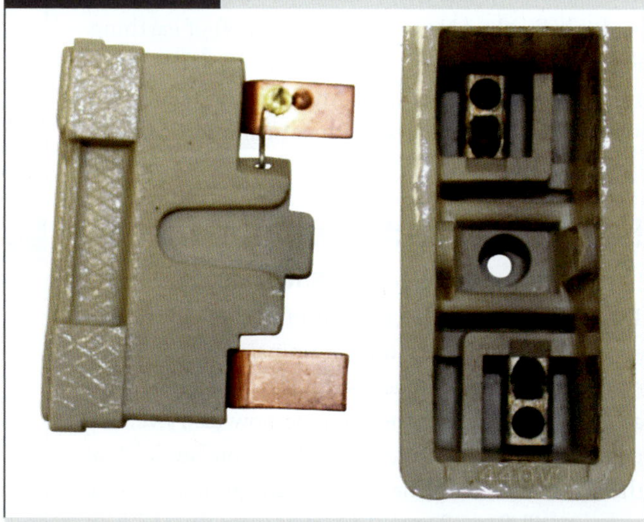

Source: Ed Hawkins

The HRC fuse is mostly used in larger protection applications such as consumer mains service protection device (SPD) and sub mains. Semi-enclosed rewireable fuses are no longer permitted to be installed in new installations (AS/NZS 3000:2007 clause 2.4.3); however, it is permitted to replace damaged rewireable fuses in existing installations.

A fuse is a circuit protection device that works on the 'weakest link in the chain' principle. When a current greater than nominal rating of the fuse flows through it, the fuse element melts as a result of the thermal effect of current flow. As the circuit's path through the fuse is now broken, current flow falls to zero, achieving automatic disconnection of supply. The rating of the fuse is selected so it is greater than or equal to the actual load current and less than or equal to the current rating of the cable it protects. The higher the current above the rating of the fuse, the shorter the time the fuse takes to operate (blow). This is known as an *inverse time current characteristic*.

Fuses can be used as protection against overloads and short circuits, and they provide fault protection (indirect contact) when used in conjunction with an effective earthing system. A fuse (when removed) can be used as an isolation device.

Circuit breakers

AS/NZS 3000:2007 clause 1.4.26 defines a circuit breaker. Miniature circuit breakers (MCBs), as shown in Figure 1.55, are the most common form of circuit protection due to their low cost, superior overload protection and ease to reset.

The basic operation of a circuit breaker is best described as a switch that provides automatic disconnection of supply when a fault condition is detected. Think of a circuit breaker as a cross between a switch and a mouse trap. Closing the circuit breaker is the equivalent of setting a trap's spring. That is why it is hard to close a circuit breaker, as the switch mechanism is closed against a spring being charged. A typical circuit breaker uses a bi-metallic strip to detect overload faults and a coil to detect short circuits. These devices trigger the circuit breaker's release mechanism in a similar way to a mouse trap, causing the circuit breaker to open the switch extremely quickly. Refer to Section 22 of *Electrical Principles* (3rd edition) by Peter Phillips for further information on circuit breakers.

Circuit breakers (CB) can protect against overloads, short circuit and indirect contact (fault protection) in a similar way to fuses. To indicate a CB has operated (tripped), the toggle moves to a position halfway between on and off. The big advantage of the circuit breaker over fuses is the ability to reset the breaker by turning the breaker to the off position and then back to the on position after the fault has been removed or repaired. A circuit breaker can also be used as an isolation device.

FIGURE 1.55 Miniature circuit breaker (MCB)

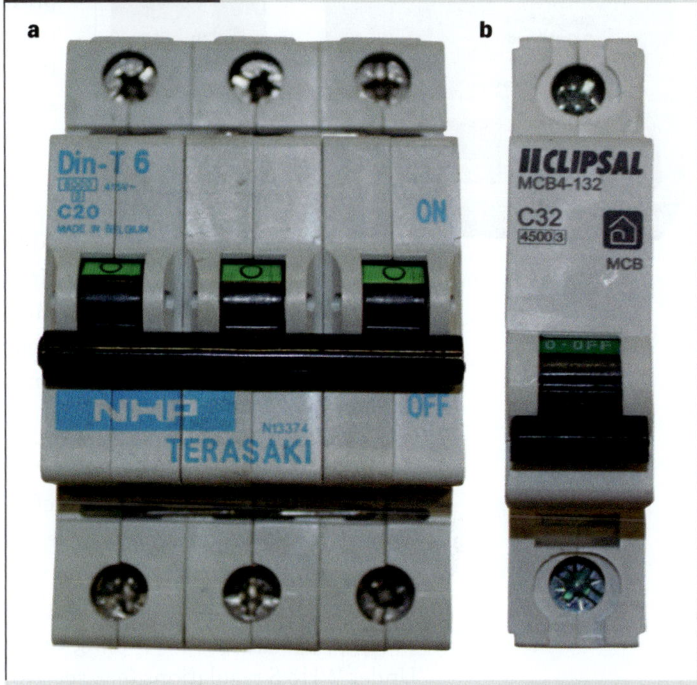

Source: Ed Hawkins

Residual current devices (RCDs)

AS/NZS 3000:2007 clause 1.4.80 defines a residual current device (RCD). Since 1991, when Australian/New Zealand standards made the protection of power circuits in new domestic installations compulsory, there has been a decline in the number of electrocutions. RCDs work on the principle of monitoring the current flowing in and out of a circuit. In a healthy single phase circuit, the current supplied to the load by the active wire should be the same as the current returning in the neutral wire. The current flow produces a magnetic field; if the current is equal and flowing in opposite directions, the resulting magnetic fields will also be equal and opposite and therefore the two opposing magnetic fields cancel each other out.

The majority of electrocutions occur as a result of a path through a person from active to earth. When this happens the active and neutral currents will no longer be equal. The difference between the active and neutral current is known as the *residual current*. Because of the residual current, the opposing magnetic fields will no longer cancel out. The residual magnetic field is detected by a 'toroidal' magnetic core which trips the RCD and automatically disconnects the supply. The preset sensitivity to protect persons and livestock is 30 mA. See Figure 1.56.

FIGURE 1.56 RCD operation supplementing protection against direct contact

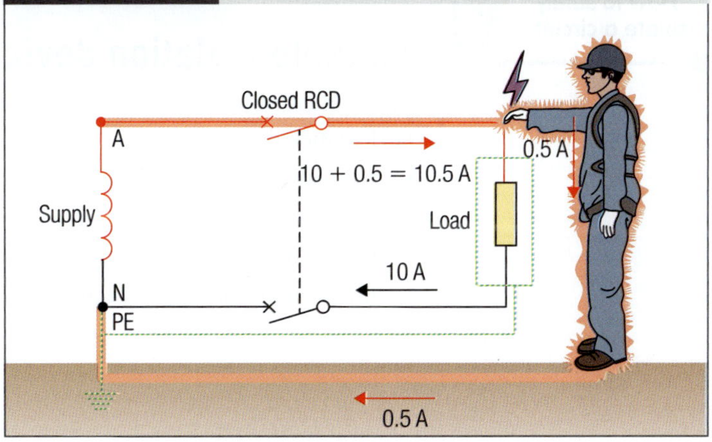

FIGURE 1.57 Standard RCD

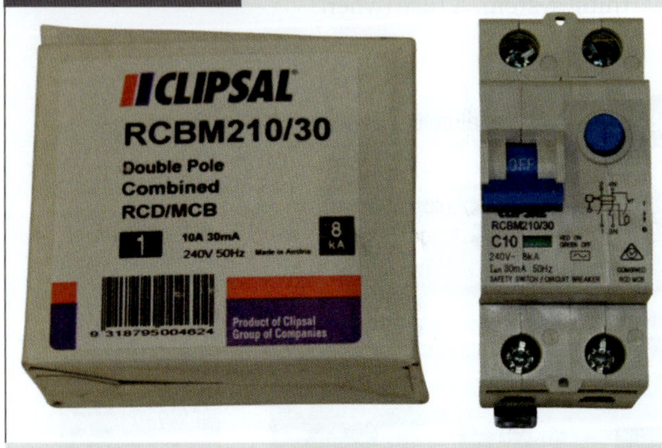

Source: Ed Hawkins

Protection offered by an RCD

An RCD must not be confused with a miniature circuit breaker, as standard RCDs only provide additional protection against direct and indirect contact (fault protection). They do not provide any protection for overload or short circuit. If an RCD is marked to indicate 63 A this means it is rated to carry and break 63 A, like an isolation switch. The RCD **will not** automatically disconnect the supply if the current flowing from active to neutral exceeds 63 A. Figure 1.57 shows a standard RCD. Note the test button, which is a good way to spot the difference between an RCD and a two pole miniature circuit breaker. A fuse or circuit breaker is normally added downstream from the RCD to provide overload and short circuit protection. A protection device called a combination RCD/MCB provides protection against overloads and short circuits, as well as additional protection against direct and indirect contact. An RCD can also be used as an isolation device.

Isolation of an electrical supply

AS/NZS 3000:2007 clause 1.4.62 defines isolation. Electrical installations are designed so that all or any part of the installation may be manually isolated from the supply. This ensures that parts of the installation that require maintenance, repair or testing can be temporarily disconnected without the inconvenience of cutting power to the entire installation.

All electrical work including testing for isolation must be carried out by a licensed electrician. Under WH&S regulations (e.g. see WorkSafe Tasmania's 'Managing Electrical Risks in the Workplace Code of Practice', p. 35; http://worksafe.tas.gov.au/home), energised electrical work (working live) is generally prohibited unless one or more of the following exceptions applies. A PCBU responsible for carrying out electrical work must ensure the work is not carried out on energised electrical equipment unless:

- other health and safety reasons outweigh the risk of working on energised equipment
- working energised is necessary for the work to be carried out properly
- it is necessary for the purposes of testing
- there is no reasonable alternative means of carrying out the work.

Convenience, cost or time considerations are not sufficient reasons to work live. These requirements do not apply to supply authorities.

There is no one-size-fits-all approach to isolation as every installation is slightly different and every piece of electrical equipment is also slightly different. However, the overall approach remains the same. Table 1.29 shows the appropriate steps to ensure safe isolation of equipment.

How to safely isolate a circuit

Suitable isolation devices

Fuses, circuit breakers and residual current devices can be used as isolation devices. Other suitable devices include:

- main and isolation switches
- plugs and socket outlets
- special terminals that do not require the removal of a conductor
- links – removal of links requires an insulated tool.

TABLE 1.29 Steps to ensure safe isolation

Consultation	• Consulting with workplace management
Isolation	• Identifying the circuit(s) requiring isolation • Disconnecting active conductors from the relevant source(s), noting there may be multiple sources and standby systems/generators/photovoltaic systems as well as auxiliary supplies from other boards • If a removable (or rack out) circuit breaker or combined fuse switch is used it should, if reasonably practicable, be racked out or removed then locked open and danger tagged
Securing the isolation	• Locking the isolating switch(es) where practicable or removing and tying back relevant conductors
Tagging	• Tagging the switching points where possible to inform people at the workplace
Testing	• Testing the testing device to prove correct operation • Testing to confirm the relevant circuits have been de-energised in the work area • Re-testing the testing device to prove it is still working correctly
Re-testing as necessary	• If the person carrying out the work leaves the area, checking to ensure that the electrical equipment being worked on is still isolated

Sources: Safe Work Australia, 'Managing Electrical Risks in the Workplace Code of Practice', http://www.safeworkaustralia.gov.au/sites/swa/about/publications/pages/managing-electrical-risks-in-the-workplace; WorkSafe Tasmania, 'Managing Electrical Risks in the Workplace Code of Practice', http://worksafe.tas.gov.au/__data/assets/word_doc/0019/203770/managing_electrical_risks_code.doc, http://www.comlaw.gov.au/Details/F2012L02522/Html/Volume_5

Lock out and tag out

Once the supply has been disconnected it must be locked out (secured) to prevent accidental re-energising. Lives depend on it. A typical lockout kit includes:

- lockout devices for use on cables, gate valves, ball valves and plugs
- clamp-on types of lockouts for circuit breakers
- universal lockout devices for multi pole circuit breakers
- group lockout hasps
- tags and cable ties
- padlocks.

A general lockout tag out procedure is:

Step 1 – Lock the isolation device so it can't be re-energised.

Step 2 – Tag the isolation device.

Step 3 – Try to reclose the isolation device. It should NOT be able to reclose.

Step 4 – If there has been a change in conditions repeat steps 1 to 3.

This will ensure the lock outs and tag outs are secure. Isolation switches have provision for padlocks, and devices such as CBs and RCDs can be locked out using devices similar to that shown in Figure 1.58.

Danger tags similar to those shown in Figure 1.59 should also be fitted to indicate the name of the electrical worker who isolated the circuit. The lockout and danger tag must not be removed until work is complete and the circuit is safe to re-energise.

It is critical that the isolation procedure is tested before commencing with any work. After a circuit has been isolated, locked out and tagged, it should be tested first to make sure it is de-energised.

FIGURE 1.58 CB lockout

Source: Ed Hawkins

FIGURE 1.59 Danger tags

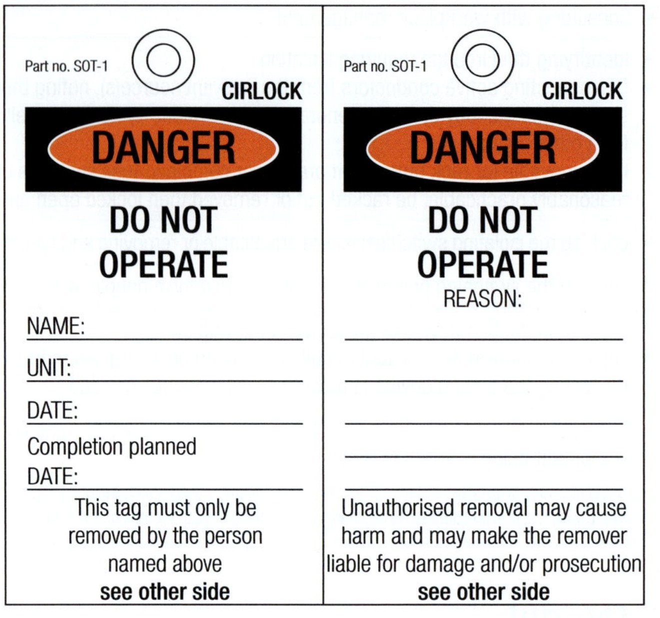

The Victorian WorkCover Authority's *Lock Out and Tagging of Plant* guidelines advise:

> It is not safe to assume an isolator has locked out an electricity source simply because it is in an open position. While normally this should open an air gap between contact points, it is possible for contact points to become welded together by the passage of electricity, and remain so even when the isolator appears to be open.

The operation of any instruments required to test the isolation procedure should be checked before and after isolation (Victorian WorkCover Authority, http://www.vwa.vic.gov.au).

Removing an electric shock victim from electrical contact

If proper isolation procedures are not followed a worker may come in contact with live conductors. Due to current flowing through them, the victim may not be able to let go of the live conductor to release themselves or call for assistance. Care must be taken so that the rescuer does not also become a victim. Before attempting any action, rescuers must be aware of the voltage involved. If it is **High Voltage** (greater than 1000 V_{AC} or 1500 V_{DC}) **supply must be isolated first** before any rescue attempt is made.

You should follow this rescue procedure:

1 Remain calm, and if possible involve others to call for an ambulance.
2 Isolate supply before attempting rescue.
3 If you are unable to isolate supply use an insulated rescue hook or object such as a timber or fibreglass shovel handle to drag the victim clear.
4 Attend to any burning clothing, and if possible smother flames by wrapping the victim in a fire blanket.
5 Apply first aid using DRS ABCD approach (explained in the following section) until assistance arrives.

A typical Low Voltage Rescue kit contains:

- 1000 V insulating gloves
- a low voltage rescue hook
- a torch
- a fire blanket
- an 'isolate here' sign
- burns dressing
- a first-aid hint booklet.

Case study

This case study illustrates the importance of correct isolation practice.

Coronial Information System – Fact Sheet. November 2012

The deceased made direct contact with electrical wiring while performing electrical work in the roof space of a commercial building.

The coroner recommended that when electrical work was to be performed in areas such as roof spaces electricians should either:

- work at a time when the power can be turned off, or
- that an electrician should never work alone and that the area be well lit if it was considered necessary to work while the power is on.

It was further recommended that a work plan be prepared to ensure that the work can be conducted safely and that written approval from the installation owner or operator be obtained for the live work.

The coroner also stated 'that in respect of the wiring being worked on, that there must be adequate isolation, testing and securing of the isolation. Danger tags should always be used, prominently displayed and applied so that the power is not turned on by a person unaware of the work being performed by the electrician.' (Source: 'NCIS Fact Sheet. January 2012 Electrocution Related Deaths', ISSN 1839-0528, p. 5.)

SUMMING UP...

- Possible effects of electric shock on a human body are:
  - muscle contraction
  - ventricular fibrillation (heart)
  - cardiac arrest
  - asphyxia (lungs)
  - burns.
- The electric shock will depend on the:
  - path the current flows through the body
  - amount of current that flows through the body
  - length of time the current flows.
- Direct contact is contact with live parts that are normally live.
- Indirect contact is contact with live parts that are *not* normally live such as earthed metal.
- Earthing conductors:
  - join exposed metal parts together and to the greater mass of earth
  - provide a low resistance path to ensure circuit protection devices operate quickly.
- Extra low voltage (ELV) is voltage not exceeding 50 V a.c. or 120 V ripple-free d.c.
- Fuses:
  - have a high rupturing capacity (HRC)
  - are rewireable – suitable for replacement of existing components only
  - operation – thermal energy melts fuse elements breaking path
  - provide protection against overload, short circuit and indirect contact.
- Miniature circuit breakers (MCB)
  - operation – thermal or magnetic trip mechanisms detect fault currents and open switch to break the path
  - provide protection against overload, short circuit and indirect contact.
- Residual current devices (RCD)
  - provide enhanced protection against indirect contact
  - *do not* provide protection against overload or short circuit.
- Combination RCD/MCD:
  - provides enhanced protection against indirect contact
  - provides protection against overload and short circuit.
- Electrical circuits must be isolated before work:
  - Consultation
  - Isolation
  - Securing the isolation
  - Tag
  - Test
  - Re-test as necessary.
- Lockout and tag out procedure
  - 1 – Lock
  - 2 – Tag
  - 3 – Try
  - 4 – Retry.

1.9 Life support – CPR in the workplace

First-aid equipment, personnel and facilities are important resources in assisting workplaces to manage workplace injuries and illnesses. Electrical workers receive first-aid instruction because of the added risk caused by the physiological effects of electric current. First-aid skills learnt in the workplace are useful also in everyday life. If you happen to be the first person at the scene, the skills you learn in this section may just save the life of one of your friends or family, as well as a work colleague.

First aid

First aid is the provision of initial care for an illness or injury. It is normally carried out by non-expert but trained personnel on an injured person until medical treatment arrives. Minor injuries may not require further care past the initial first-aid treatment. In the case of serious injuries the role of the first-aider is to preserve life until medical professionals such as ambulance officers arrive. First aid is a series of simple skills that an individual can be trained to carry out with minimal equipment. The aim of first aid is to preserve life, prevent further harm and promote recovery.

FIGURE 1.60 First-aid priorities

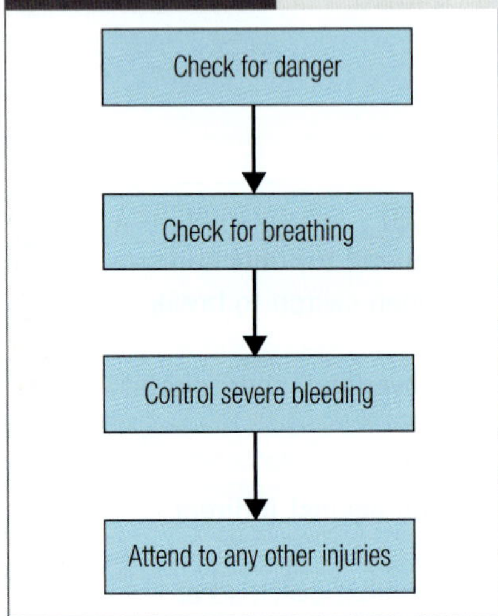

Responsibilities of the first-aider

St John Ambulance Australia (see 'Role of the first-aider', http://e-firstaid.stjohn.org.au) outlines a number of responsibilities the first-aider must be aware of:

- You must prevent further danger to yourself, to others or to the patient. Your own safety comes first. Then the safety of others, and then, only if it is safe to do so, you should assist the injured person. You will be no help to anyone if you also become a victim.
- You must not provide medical aid. Medical aid can only be provided by a doctor, registered nurse or ambulance officer.
- You must try to prevent the patient's condition becoming worse.
- You must prevent delay in the recovery of the patient. Recovery depends on how quickly you respond and how soon medical aid is obtained.
- You must prevent any harmful intervention. You must protect the patient from those who don't know what they are doing.

Priorities of first-aid management

First aid is a collection of procedures and priorities that seem to be overwhelming in terms of remembering what to do and when. It is important to prioritise your actions so tasks are performed in a logical sequence. There is no point splinting a fracture if the patient is not breathing. In an electrical context, someone receiving an electric shock must be removed from the supply before any first-aid treatment. Figure 1.60 shows the steps for a first-aider to follow.

When attending an accident scene

If an accident occurs, a rapid and appropriate response by the first-aider will drastically improve the prospects of survival and recovery for the victim. The first-aider should be prepared to:

- assess the situation quickly
- ensure safety
- identify the nature of the injury or illness
- manage the patient promptly and appropriately
- organise bystanders
- get help [dial AUS 000 or NZ 111]
- gather and provide necessary information and, where possible, keep a written record
- stay with the patient until help arrives.

Source: St John Ambulance Australia, 'Role of the first-aider', http://e-firstaid.stjohn.org.au/text_alt/topic_01/index.html.

Duty of care, legal and ethical issues

Under Australian WH&S regulations, workers should have access to trained first-aid personnel who undertake initial management of work-related injuries. PCBUs should provide a first-aid room for any workplace with more than 200 employees or where the workplace is considered high-risk and has more than 25 employees. Workers should have access to trained first-aid personnel who undertake initial management of work caused injuries or illnesses.

Those who provide first aid are not expected to be perfect under the law. It is not expected that every incident will turn out perfectly. When providing first aid you should be responsible and prudent and act in good faith for the best interests of the casualty and undertake first aid to the best of your ability. The four main legal considerations when providing first aid are:

- duty of care
- consent
- negligence
- recording (see Safe Work Australia, 'First Aid in the Workplace Code of Practice', p. 11, http://www.safeworkaustralia.gov.au/sites/swa/about/publications/pages/first-aid-in-the-workplace).

Duty of care

Where an employee has received recognised training as a first-aider in the workplace, and receives payment for that purpose, that employee has an obligation to render assistance as required. In other words the first-aid officer on a construction site has a legal obligation to provide treatment when required.

It should also be noted that once a person begins to render first aid, there is a legal duty of care for the physical wellbeing of the injured person. A first-aider's duty of care cannot be evaded by abandoning assistance halfway through.

In the case of the workplace, this duty of care takes precedence over any authority that an employer may have over the first-aid employee, or the injured person. If a medical professional or other person with better qualifications than you arrives at the scene, it is usual practice for them to take over and you to assist where appropriate.

Consent

Australian law is based on the premise that a person has control over their own body and a person can bring a charge of assault if touched without consent. An injured person has the legal right to refuse any assistance or reject any advice from a first-aider. The first-aider must obtain consent, where possible, before applying first aid. If the patient is unconscious, the law will imply consent.

In an emergency, the law implies consent of the injured person if they are unconscious and seriously injured. This consent will only apply to conditions that imperil the life or future health of that person. Where a patient is a child, the consent of the parent or legal guardian should be obtained, where possible (*The Complete Guide to Pre-Hospital Emergency Care*, First-Aid Safety Training, pp. 5–6, http://firstaidsafetytraining.com).

Negligence

The article 'Legal considerations for outdoor educators when administering first aid' by Danny Parkin (pp. 1–2, http://www.projectnatureed.com.au) describes some of the legal issues associated with first aid. It explains that a case of negligence against a person providing first aid would be established if:

- the first-aider owed a 'duty of care' to the injured person
- the standard of care required by that duty was breached
- further preventable injury was sustained
- the further injury was sustained because the first-aider has gone beyond their level of training.

Recording

It is good practice to record in writing the details of any first aid you provide. In the workplace, it is a requirement to record all incidents even where first aid was not applied. This will allow the PCBU to maintain accurate record keeping, review the safety procedures and implement new safety requirements as required. This information may become necessary as evidence in a workers' compensation case or if the incident is brought to court.

Examination of a casualty for injuries

Look, Feel, Observe and Listen:

- *Look* to see if the victim is in contact with an electrical supply. If so turn off the supply of electricity. Until the power is off, avoid direct skin contact with the victim or any conducting material touching the victim.
- *Look* for visible signs of injuries and vital signs. These consist of the pulse, respirations, skin colour and the level of consciousness.
- Does the casualty have a medical alert necklace or bracelet?
- Does the casualty have a tattoo on the inside of their wrist indicating a medical problem or allergy?
- Is the casualty pale, sweaty, blue?
- Is the casualty bleeding and are there any wounds, puncture marks or bruises?

If pain is felt in the back or the casualty complains of pins and needles or numbness in the legs or arms DO NOT MOVE THE CASUALTY (adapted from 'H&A Training First-Aid', http://www.hatraining.com).

Respiratory arrest and cardiopulmonary arrest

The lungs function in much the same way as a set of bellows, drawing in fresh oxygenated air and expelling carbon dioxide exchanged from the blood. Like the heart the lungs are muscle driven. Respiratory arrest is a failure of the lung muscles to contract properly. Respiratory arrest can occur as a result of cardiac arrest or from a lack of a suitable supply of air.

The heart is effectively a pump that supplies oxygenated blood from the lungs to organs and body parts. The flow of blood also removes toxins that build up in body parts and filters the toxins through organs such as the liver and kidneys. The heart works by the contraction and release of muscles through a one-way valve in much the same way as a squeeze siphoning pump. Cardiac arrest occurs when the normal circulation of blood stops due to failure of the heart muscles to contract properly.

Effect of cardiopulmonary arrest (CPA)

A decreased blood circulation and delivery of oxygen to the body results in a lack of oxygen to the brain which causes loss of consciousness. This then reduces the capacity to breathe. Brain injury is likely if cardiac arrest goes untreated for more than five minutes. The cells in the brain start to undergo change, brought about by the depletion of all their nutrients, and the production of various toxic materials which can no longer be cleared by the circulation since it is no longer functioning. These toxic substances cause significant cell damage. Due to a decrease in oxygen levels in the blood a blue-tinged appearance of the skin is noticeable on the lips and extremities such as fingers.

Managing airway obstruction

Airway obstruction may be partial or complete and may be present in a conscious or an unconscious victim. The Australian and New Zealand Resuscitation Councils' guidelines state that typical causes of airway obstruction may include:

- relaxation of the airway muscles due to unconsciousness
- inhaled foreign body
- injury to the airway
- allergic reaction.

> The symptoms and signs of obstruction will depend on the cause and severity of the condition. Airway obstruction may be gradual or sudden in onset and may lead to complete obstruction within a few seconds. Consequently the victim should be observed continually.

Source: Australian and New Zealand Resuscitation Councils, 'Guideline 4, Airway', p. 4.

The tongue is the most common cause of airway obstruction in an unconscious adult. A head tilt chin lift (shown in Figure 1.61) applies tension to the tongue through repositioning the head and creates a small pocket for oxygen to pass through the trachea (or windpipe).

FIGURE 1.61 Head tilt chin lift

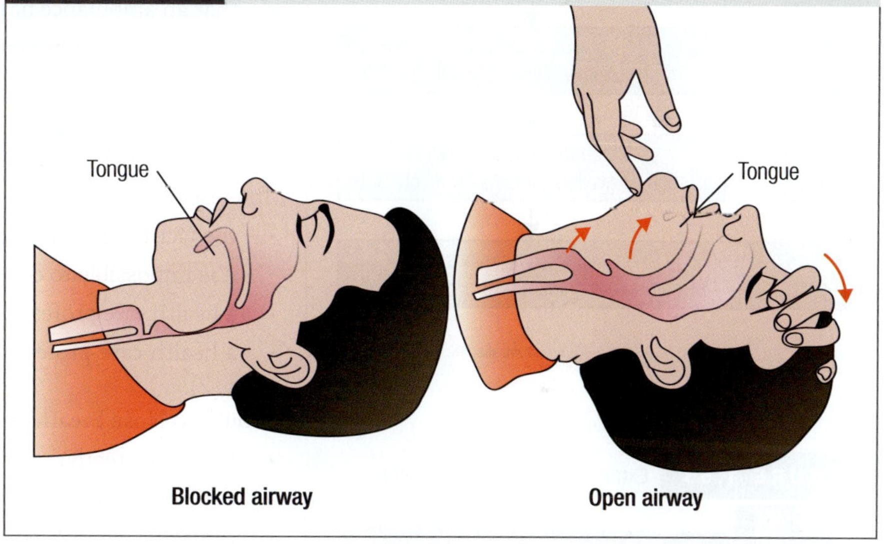

Head tilt chin lift

With this method, one hand is placed on the forehead or the top of the head, while the other hand is used to provide the chin lift. As noted by the Australian and New Zealand Resuscitation Councils, 'The head (NOT the neck) is tilted backwards. It is important to avoid excessive force, especially where neck injury is suspected' ('Guideline 4, Airway', p. 2).

Chin lift is when the chin is held up by the rescuer's thumb and fingers in order to open the mouth and pull the tongue and soft tissues away from the back of the throat ('Guideline 4, Airway', p. 3).

Cardiopulmonary resuscitation (CPR)

Over time the CPR process is reviewed and optimised for best chances of survival for patients. These procedures are adapted from the Australian and New Zealand Resuscitation Councils' 'Guideline 8, Pulmonary resuscitation', pp. 1–2. This guideline is applicable to adults, children and infants (see http://www.nzrc.org.nz).

> Cardiopulmonary resuscitation is the technique of chest compressions combined with rescue breathing. The purpose of cardiopulmonary resuscitation is to temporarily maintain a circulation sufficient to preserve brain function until specialised treatment is available. Rescuers must start CPR if the victim is unresponsive and not breathing normally. Even if the victim takes occasional gasps, rescuers should start CPR.

Source: Australian and New Zealand Resuscitation Councils, 'Guideline 8, Pulmonary resuscitation', p. 1.

Compression ventilation ratio

Thirty compressions followed by two ventilations are recommended for all ages regardless of the numbers of rescuers present. Compressions must be paused to allow for ventilations. No human evidence has identified an optimal compression-ventilation ratio for CPR in victims of any age.

FIGURE 1.62 Basic life support flow chart

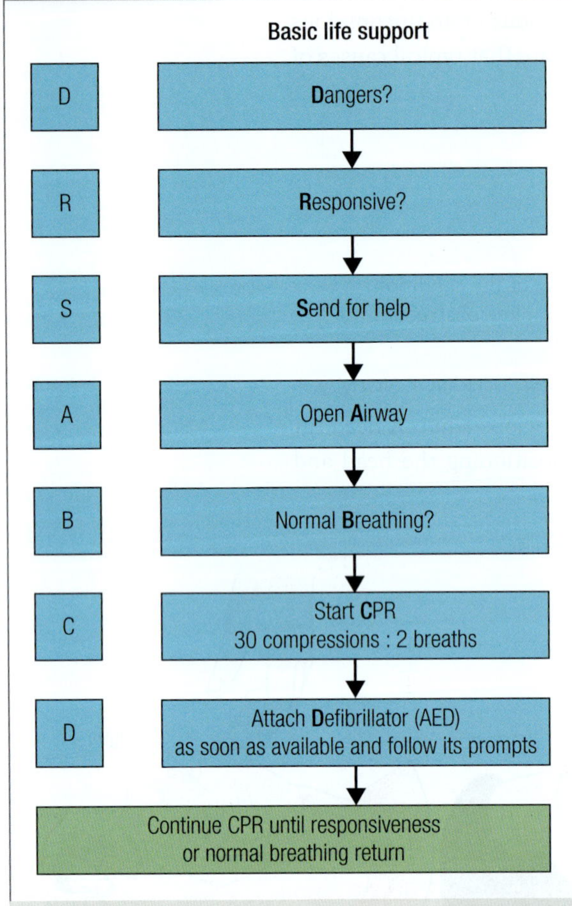

Source: Australian Resuscitation Council

FIGURE 1.63 External automatic defibrillator

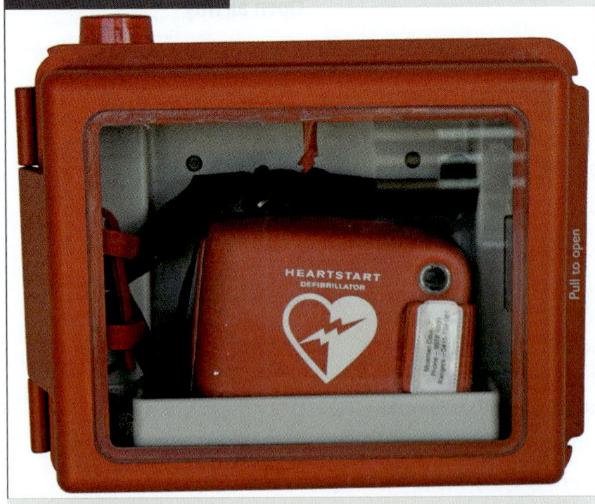

Source: Ed Hawkins

Steps of resuscitation

Initial steps of resuscitation are DRS ABCD as shown in Figure 1.62.

D – **Danger** – Check for danger (hazards/risks/safety)

R – **Response** – Check for a response (if unresponsive)

S – **Send** – Send for help call AUS 000 or NZ 111

A – **Airway** – Open the airway

B – **Breathing** – Check for breathing or abnormal breathing

C – **Circulation** – Give 30 chest compressions (almost two compressions/second) followed by two breaths

Basic Life Support D – **Defibrillation** – Attach an AED (automated external defibrillator as shown in Figure 1.63) if available and follow the prompts (see Australian and New Zealand Resuscitation Councils, 'Guideline 8, Pulmonary resuscitation', pp. 2, 4).

Multiple rescuers

When more than one rescuer is available ensure:

- that an ambulance has been called – dial AUS 000 or NZ 111
- all available equipment has been obtained (e.g. AED) ('Guideline 8', p. 2).

Duration of CPR

The rescuer should continue CPR until:

- the victim responds or begins breathing normally
- it is impossible to continue (e.g. exhaustion)
- a health care professional arrives and takes over CPR
- a health care professional directs that CPR be ceased ('Guideline 8', p. 2).

Mouth to Nose Breaths is used when:

- the jaw or teeth are broken
- the jaw is tightly clenched
- resuscitating in deep water
- resuscitating an infant or small child when your mouth can cover the nose and mouth together.

CPR compressions:

1. Kneel beside patient.
2. Locate the lower half of breastbone in the centre of the chest as shown in Figure 1.64.

FIGURE 1.64 Locating position for CPR

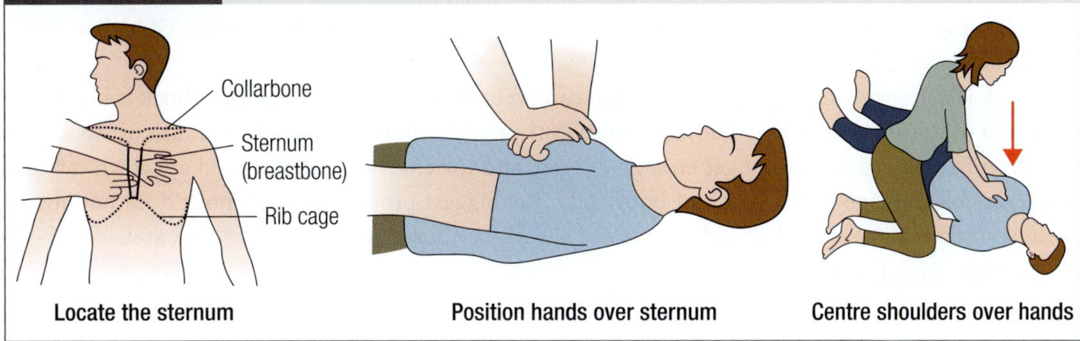

3 Place heel of hand on lower half of breastbone.
4 Place heel of other hand on top of first.
5 Interlock fingers and raise fingers.
6 Position yourself vertically above patient's chest as shown in Figure 1.64.
7 Press down on the sternum to depress it about 1/3 (or at least 4–5 cm on an adult) as shown in Figure 1.64.
8 Release the pressure.
9 Repeat 30 times at a rate of about 100/minute.
10 DO NOT do this on a conscious person.

Altered level of consciousness

The following information is an extract from the Australian and New Zealand Resuscitation Councils' 'Guideline 3, Recognition and first aid management of the unconscious victim' (pp. 1–2).

> Unconsciousness is a state of being unarousable. The causes of unconsciousness can be classified into four broad groups:
>
> - low brain oxygen levels
> - heart and circulation problems (e.g. fainting, abnormal heart rhythms)
> - metabolic problems (e.g. overdose, intoxication, low blood sugar)
> - brain problems (e.g. head injury, stroke, tumour, epilepsy).
>
> Combinations of different causes may be present in an unconscious victim, for example, a head injury victim under the influence of alcohol.... A person who fails to respond or shows only a minor response, such as groaning without eye opening, should be managed as if unconscious. ('Guideline 3', p. 1)

Managing a casualty with a low level of consciousness

Guideline 3 further advises that if the victim is unresponsive and not breathing normally, follow the Australian and New Zealand Resuscitation Councils' Basic Life Support Flow Chart (shown in Figure 1.62). 'With an unconscious breathing victim, care of the airway takes precedence over any injury, including the possibility of a spinal injury. All unconscious victims must be handled gently and every effort made to avoid any twisting or forward movement of the head and spine' ('Guideline 3', p. 2).

Signs and symptoms of shock

The following information is adapted from the Australian Resuscitation Council's 'Guideline 9.2.3, Shock', February 2009, pp. 1–2.

Shock is a life-threatening condition which can occur in any first-aid situation. Shock is basically a reduction of the amount of blood flowing through the body's vital organs. If the amount of blood flow drops too far, the organs cannot get enough oxygen, and toxic waste products build up which can lead to irreversible damage or the death of body cells. The biggest problem is that shock does not always present straight away, and when it does it is not always obvious. Some causes of shock are:

- fluid loss from burns
- intense pain
- damage to the heart, e.g. heart attack
- severe bleeding
- spinal injury
- severe vomiting and diarrhoea.

Immediately after an injury, there may be little evidence of shock. Initial symptoms and signs may include:

- collapse
- cool, sweaty skin that may appear pale
- rapid breathing
- confusion and deterioration of the level of consciousness
- rapid pulse which may become weak or slow
- vomiting.

Managing a casualty in shock

Management of shock:

1 Control bleeding with direct pressure.
2 Call an ambulance. Dial AUS 000 or NZ 111.
3 If the victim is unconscious manage as per the ARC Basic Life Support Flow Chart (see Figure 1.62).
4 Rest the victim in a position of comfort, ideally lying down.
5 Provide oxygen if available.
6 Maintain body temperature.
7 Provide reassurance.

SUMMING UP...

- The aim of first aid is to preserve life, prevent further harm and promote recovery.
- The responsibilities of the first-aider are:
  - to prevent further danger to yourself, to the patient and to others
  - you must not provide medical aid
  - you must try to prevent the patient's condition becoming worse
  - you must promote the recovery of the patient
  - to protect the patient from those who don't know what they are doing.
- Priorities of first-aid management:
  - check for danger
  - check for breathing
  - control severe bleeding
  - attend to any other injuries.
- First-aid legal and moral issues are:
  - duty of care
  - consent
  - negligence
  - recording.
- When examining a casualty for injuries: look, feel, observe and listen.
- Respiratory and/or cardiopulmonary arrest can result in:
  - decreased blood oxygen levels
  - increased blood toxin levels
  - damage to brain cells
  - death.
- The head tilt chin lift manoeuvre assists to clear a blocked airway.
- When performing CPR give 30 compressions to 2 breaths.
- Initial steps of resuscitation are DRS ABCD.
- The rescuer should continue CPR until:
  - the victim responds or begins breathing normally
  - it is impossible to continue (e.g. exhaustion)
  - a health care professional arrives and takes over CPR
  - a health care professional directs that CPR be ceased.
- The causes of unconsciousness can be classified into four broad groups:
  - low brain oxygen levels
  - heart and circulation problems
  - metabolic problems
  - brain problems.
- Unconsciousness is managed by DRS ABCD.
- Management of shock:
  - Control bleeding with direct pressure.
  - Call an ambulance. Dial AUS 000 or NZ 111.
  - If unconscious manage as per the DRS ABCD procedure.
  - Rest the victim in a position of comfort, ideally lying down.
  - Provide oxygen if available.
  - Maintain body temperature.
  - Provide reassurance.

REVIEW EXERCISES

Answer these questions and check your answers at the back of the book.

Review questions 1.1

1 What is the name of the state government agency that is responsible for work health and safety in your state?
2 Under work health and safety laws what is an apprentice classified as?
3 List three powers of a work health and safety inspector.

Review questions 1.2

1 Explain the meaning of the terms 'hazard' and 'risk'.
2 List three 'physical' hazards.
3 What class of fire is a fire fuelled by cardboard boxes and give three suitable types of extinguishers to fight the fire.

Review questions 1.3

1 List three types of injuries that can be caused by incorrect manual handling.
2 List the steps of a SMART lifting process.
3 What is the meaning of the symbol used in Figure 1.65?

FIGURE 1.65

Review questions 1.4

1 What is the difference between a hazardous substance and dangerous goods?
2 What is the maximum quantity of LPG that can be safely transported in a work van?
3 What is the purpose of packaging groups?
4 What is a Safety Data Sheet?

Review questions 1.5

1 List four things to consider when choosing and setting up a portable ladder in a workplace.
2 List three precautions that should be taken when working on a ladder.
3 What is the minimum class of scaffolding suitable for electrical work?

Review questions 1.6

1 List three risks of working in a confined space.
2 Give one example each of controlling risk sources associated with working in confined spaces by (a) elimination, (b) isolation, (c) engineering, (d) administration, (e) PPE.

Review questions 1.7

1 What is the approximate decibel output of a typical pneumatic drill and what is the maximum exposure time without wearing hearing protection?
2 What type of work can cause (a) hand-arm vibration, (b) whole-body vibration?
3 List three symptoms of heat stress.

Review questions 1.8

1 List three factors that determine the severity of an electric shock.
2 Under normal circumstances is it permitted for licensed electricians to work on live equipment?
3 List the six basic steps to ensure safe isolation of electrical equipment.
4 List the four basic steps of a general lockout procedure.

ONLINE RESOURCES

For more information about the contents of this chapter, visit the following websites:

Safe Work Australia
http://www.safeworkaustralia.gov.au

Comcare
http://www.comcare.gov.au

WorkSafe ACT
http://www.worksafe.act.gov.au

WorkCover NSW
http://www.workcover.nsw.gov.au

Northern Territory Work Safe
http://www.worksafe.nt.gov.au/

Work Place Health & Safety Queensland
http://www.deir.qld.gov.au/workplace/index.htm

Safe Work SA
http://www.safework.sa.gov.au

WorkSafe Victoria
http://www.worksafe.vic.gov.au

WorkSafe WA
http://www.safetyline.wa.gov.au

WorkSafe New Zealand
http://www.business.govt.nz/worksafe

Australian Resuscitation Council
http://www.resus.org.au/

St John Ambulance online
http://www.stjohn.org.au

2

ENVIRONMENTALLY SUSTAINABLE WORK PRACTICES

It is generally accepted that human activities are affecting the environment. The most significant effect is 'anthropogenic warming', otherwise known as 'climate change'. As a result, governments around the world are introducing laws to reduce this effect. The aim is to ensure a sustainable future, in which the present generation does not affect the ability of future generations to meet their needs. This chapter provides an introduction to sustainable work practices, environmental legislation and the principles of sustainable energy.

Chapter outline

2.1 Environmental risk management

Environmental risks can be described as elements of an organisation's activities, products or services that can interact adversely with the environment. In electrotechnology these risks include managing transformer oil storage and spills, generating and disposing of waste, greenhouse gas emissions, asbestos (handling and removal), air emissions (toxic and greenhouse), noise and vibration and dust generation.

FIGURE 2.1 The three pillars of sustainability

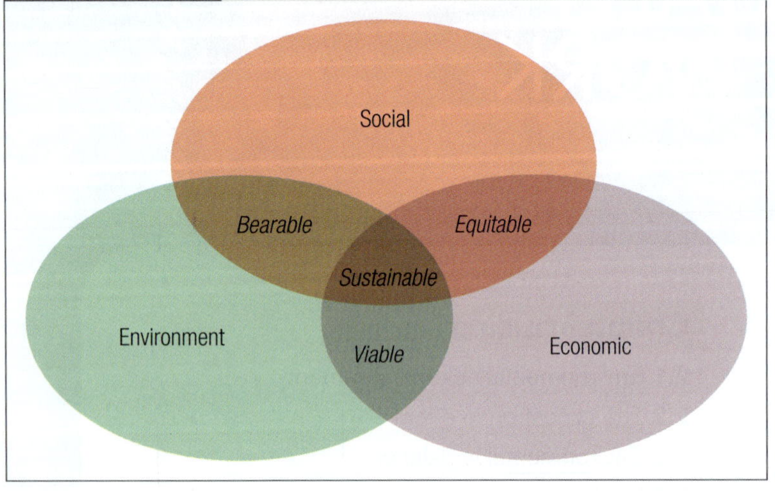

Environmental sustainability

Sustainability can be defined as 'improving the quality of human life while living within the means of the carrying capacity of supporting ecosystems'. The three components, or pillars, of sustainable development are economic development, social development and environmental protection.

Figure 2.1 shows what is known as the 'three pillars' of sustainability. Sustainable energy refers to the use of renewable energy sources and efficient use of electricity, as explained later in this chapter.

Environmental management

Governments around the world are legislating to promote sustainable work practices and energy efficiency. Financial penalties are a powerful motivator for the introduction of sustainable work practices.

Reducing a workplace's impact on the environment will save money in the long term. Employing sustainable practices such as choosing energy efficient office equipment and business vehicles can help the environment by preserving these resources. These practices can also reduce operating costs and lead to improved services, as well as improving a company's reputation for being environmentally friendly.

Environmental risks

The following ten environmental problems are significant issues faced by all nations:

- Climate change – global warming is generally considered to be the most influential cause of climate change in recent times.
- Energy – the generation of energy is a leading source of environmental damage, predominantly caused by the burning of fossil fuels.
- Water – as the world's driest inhabited continent, Australia is especially vulnerable to water pollution.
- Biodiversity and land use – Unsustainable land use has led to the degradation of many valuable ecosystems and the loss of irreplaceable biodiversity.
- Chemicals, toxins, and heavy metals – while chemicals and toxins exist in nature, the last 250 years have seen an explosion in man-made pollutants in the environment.
- Air pollution – when considering air pollution, the release of greenhouse gases is the most commonly quoted problem.
- Waste management – modern societies have considerably increased the amount of waste they produce as a result of manufacturing and packaging processes and the sheer increase in population size.
- Ozone layer depletion – depletion of our ozone layer has been mainly attributed to the release of chlorofluorocarbons or CFCs into the atmosphere.

- Oceans and fisheries – many of the world's oceans are overfished, with a number of previously valuable fish species experiencing catastrophic population declines.
- Deforestation – the clearing of forests has occurred at frightening speeds over the last 100 years.

Source: Adapted from CSglobe Expand Your Consciousness, '10 environmental problems', http://csglobe.com/10-environmental-problems.

Environmental management systems (EMS)

Organisations are required to have WH&S policies and procedures, and in many cases an environmental management system (EMS) to promote a healthy environment. The Commonwealth Government describes an EMS as

> a system or management tool to help an organisation reduce its negative impacts on the environment. Implementation of an EMS requires an organisation to:
>
> - devise a policy that explains the organisation's environmental commitments
> - appoint an environmental manager or management team responsible for the ongoing coordination of the EMS
> - identify the organisation's significant environmental aspects
> - identify legislative and regulatory requirements relevant to the organisation's environmental aspects
> - establish environmental objectives and targets
> - implement programs to achieve those objectives and targets
> - monitor and measure progress towards achieving those objectives and targets
> - take steps to continually improve the effectiveness and efficiency of environmental management within the organisation.
>
> Standards Australia has adopted the International Standards for Environmental Management Systems (ISO 14001 and ISO 14004) for use by Australian organisations. These are known within Australia (and New Zealand) as:
>
> - AS/NZS ISO 14001:2004 Environmental Management Systems – requirements with guidance for use
> - AS/NZS ISO 14004:2004 Environmental Management Systems – general guidelines on principles, systems and support techniques.
>
> These standards provide a helpful framework for organisations wishing to develop a high quality EMS, regardless of whether or not they intend to achieve certification to the EMS standard.

Source: Australian Government, Department of the Environment, 'Environmental management systems', http://www.environment.gov.au.

ISO 14001 standard

International Organization for Standardization **ISO 14001:2004 Environmental management systems – Requirements with guidance for use** specifies the requirements to enable an organisation to meet the criteria for an environmental management system. An EMS is built on the principles of a continual improvement process:

- Plan: establish the objectives and processes necessary to deliver results in accordance with the organization's environmental policy.
- Do: implement the processes.
- Check: monitor and measure processes against environmental policy, objectives, targets, legal and other requirements, and report the results.
- Act: take actions to continually improve performance of the environmental management system.

Source: Reproduced with permission from SAI Global Ltd under Licence 1501-c013

ISO 14001 was developed to help companies reduce their environmental impacts. Organisations can reap a number of economic benefits including conformance with legislative and regulatory requirements. As an internationally recognised standard, businesses can register as being ISO 14001 compliant.

Certification of an organisation's EMS means an independent auditing body has verified, through an audit, that the EMS conforms to the requirements. *Registration* means that the auditing body then records the certification in its client register. Once certified the organisation can display the ISO certified logo in its advertising and stationery. This will promote increased business from environmentally aware customers.

SUMMING UP...

- Sustainability is maintaining the quality of human life while living within the carrying capacity of supporting ecosystems.
- The three pillars of sustainability are environment, social and economic.
- Renewable energy is natural energy which does not have a limited supply.
- Reducing a workplace's impact on the environment has economic and social benefits.
- The top ten environmental problems are:
  - climate change
  - energy
  - water
  - biodiversity and land use
  - chemicals, toxins and heavy metals
  - air pollution
  - waste management
  - ozone layer depletion
  - oceans and fisheries
  - deforestation.
- An environmental management system (EMS) is a system or management tool to help an organisation reduce its impact on the environment.
- ISO 14001:2004 Environmental management systems – Requirements with guidance for use specifies the requirements to enable an organisation to meet the criteria for an EMS.
- An EMS is built on the principles of Plan Do Check Act and Continual Improvement.

FIGURE 2.2 The greenhouse effect

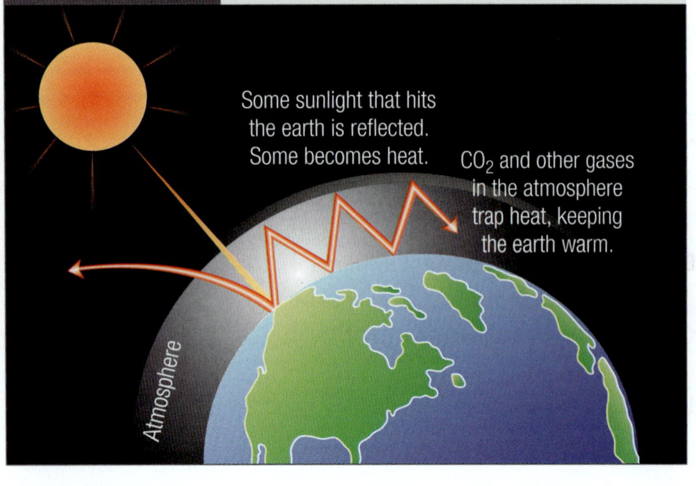

2.2 The greenhouse effect

The greenhouse effect refers to a natural phenomenon where heat energy from the sun is absorbed by gases in the atmosphere, with some of the heat being re-radiated back to earth (Figure 2.2). The absorbed energy warms the atmosphere and the surface of the earth to a temperature that allows life on earth to exist. Think of a car parked in the sun, where the temperature inside is much hotter than outside because the energy from the sunlight is trapped inside the car by the glass.

Causes of the greenhouse effect

A greenhouse gas absorbs and emits radiation within the atmosphere. The main greenhouse gases are water vapour, carbon dioxide, methane, nitrous oxide and ozone.

There are natural sources of the greenhouse gases in the form of carbon dioxide (CO_2) produced by vegetation, animals and volcanic or seismic events. At the same time, CO_2 is being removed by the oceans, green plants and in other ways. This is the carbon cycle. The balance in the cycle is critical in determining the earth's climate. A natural greenhouse effect is needed to maintain normal climatic temperatures.

Enhanced greenhouse effect

An *enhanced* greenhouse effect is caused by the additional release of CO_2 to the atmosphere by human activity. This CO_2 had been trapped underground for millions of years, and is now too much for the carbon cycle to process. More CO_2 is remaining in the atmosphere, trapping greater heat and causing global warming (see Answers™, 'Greenhouse effect', http://www.answers.com/mt/greenhouse-effect).

Sheep and cattle grazing are reported to make up to nearly 18% of greenhouse gas emissions. Land clearing for agriculture is another major contributor. Many of the forests that once absorbed and stored carbon have been cut down. This deforestation and land clearing is reducing the effectiveness of the carbon cycle.

Electricity consumed by households and industries in developed nations is another major contributor to greenhouse gas emission, as much of the electricity is generated using fossil fuels, such as coal.

Climate change

Global warming is causing climate change. Winters may become shorter, but harsher, and summers increasingly hotter. As described at http://www.answers.com, melting ice is already causing

> rising sea levels. Australia's climate has changed along with the global climate. In Australia, the average surface temperature has increased by about 0.7°C since 1960, with some areas having warmed faster and some showing relatively little warming [see Figure 2.3]. The warming has caused an Australia-wide average increase in the frequency of extremely hot days and a decrease in the frequency of cold days.
>
> Source: Answers™, 'Greenhouse effect', http://www.answers.com/mt/greenhouse-effect.

FIGURE 2.3 Rising Australian annual temperatures

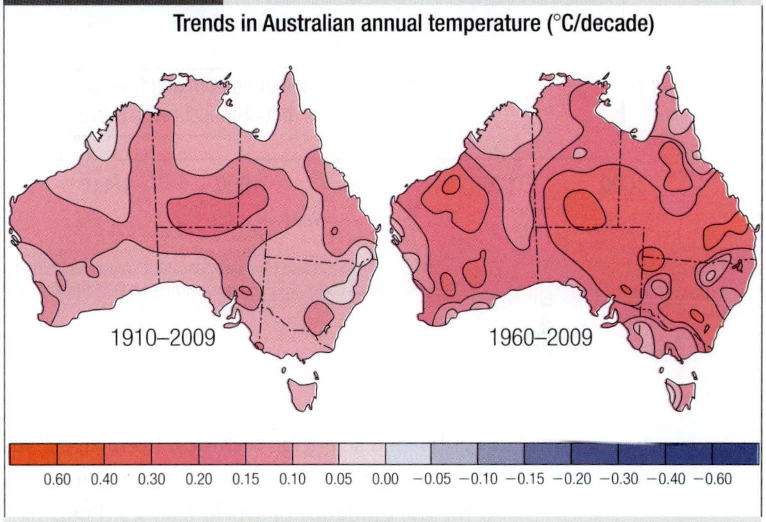

Source: Commonwealth of Australia 2014, Australian Bureau of Meteorology, http://www.bom.gov.au/climate/change/index.shtml#tabs=Tracker&tracker=trend-maps

While the longer term trends in rainfall are less obvious, there have been significant increases over north-western Australia, and decreases over south-western and south-eastern Australia since 1960 (see Figure 2.4). The warming and decreased rainfall over south-east Australia increase the risk of bushfires (*Report of the ACEN: 'Province of Australia' Presentation for Lima*, Peru Conference, August 2011, p. 26, http://acen.anglicancommunion.org).

FIGURE 2.4 Australian annual rainfall

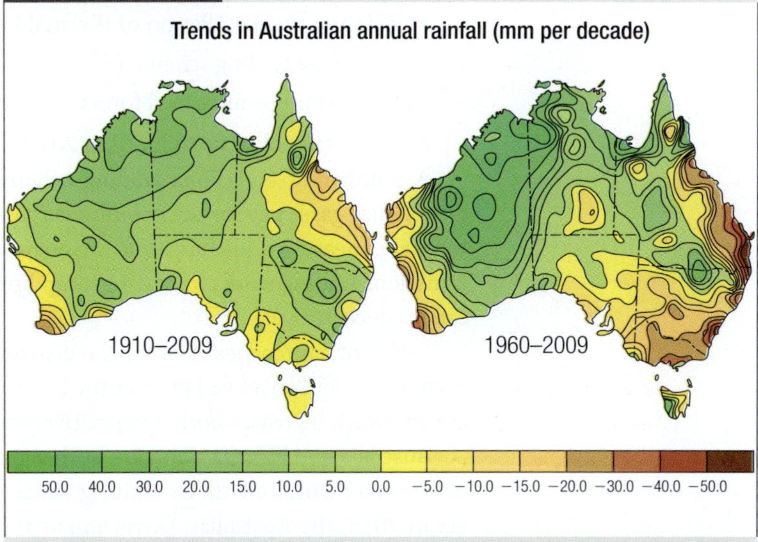

Source: Commonwealth of Australia 2014, Australian Bureau of Meteorology, http://www.bom.gov.au/climate/change/index.shtml#tabs=Tracker&tracker=trend-maps&tQ[map]=rain&tQ[area]=aus&tQ[season]=0112&tQ[period]=1970

International and national greenhouse imperatives

In 2011 the world's population was 7 billion people, having more than doubled from 3 billion in just 50 years. The challenge of sustainability is to use resources in such a way that the social and economic needs today do not exhaust the environmental resources of future generations.

FIGURE 2.5 Hockey stick graph

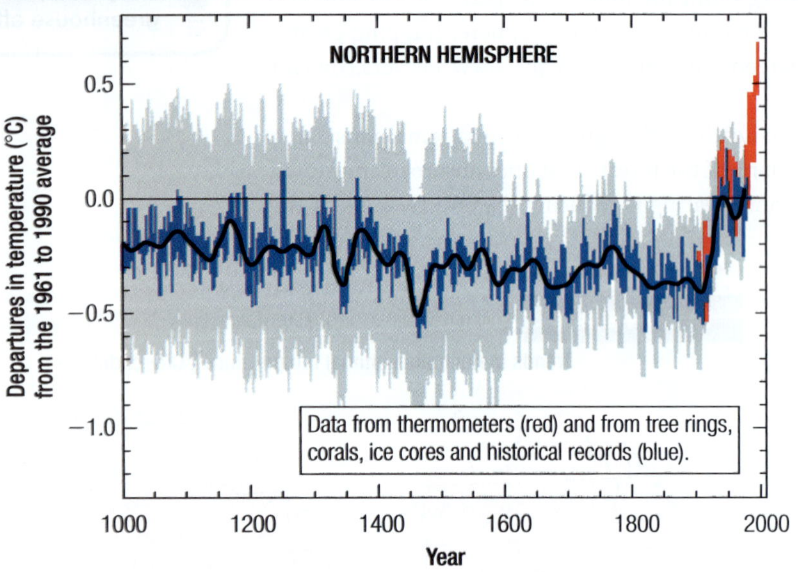

Source: IPCC, 'Climate Change 2001 – IPCC Third Assessment Report', http://www.ipcc.ch/ipccreports/tar/vol4/english/075.htmNational policies

International policies

In 1988 the Intergovernmental Panel on Climate Change (IPCC) was set up by the World Meteorological Organization (WMO) and the United Nations Environment Programme (UNEP) to provide world government policymakers with an objective source of information about:

- causes of climate change
- potential environmental and socio-economic impacts,
- possible response options.

The aims of the IPCC are to assess scientific information relevant to:

- human-induced climate change
- the impacts of human-induced climate change
- options for adaptation and mitigation (see IPCC, 'Working Group II', http://www.ipcc-wg2.gov).

As noted by the WMO, 'The IPCC does not carry out research or monitor climate related data. The role of the IPCC is to assess available information about climate change from published scientific technical literature' ('Summary of WMO co-sponsored programmes', http://www.wmo.int/pages/summary/cosponsored_summary_en.html). An assessment report is then prepared for the guidance of world leaders. Figure 2.5 shows a graph displaying the number of above average temperature days from the third IPCC Assessment Report, which was later used in Al Gore's award-winning documentary *An Inconvenient Truth*.

Australia and New Zealand have recently introduced a number of reforms relating to reducing climate change including:

- financial incentives to update items such as hot water systems to modern low energy consumption models and the installation of thermal insulation
- an emissions trading scheme (ETS)
- carbon farming – increased forestry activity to store carbon in trees.

A carbon tax policy introduced in 2012 in Australia has since been repealed.

A number of emissions trading schemes are being devised globally to make the emitters of greenhouse gases pay for the greenhouse gases they emit. To achieve this there needs to be a system to estimate, measure and charge for the environmental cost that an emitter places on the environment. Emissions trading works by setting a cap limit on the total amount of carbon pollution that a country can produce in a year.

Different approaches exist for the distribution of carbon permits in a carbon trading market. Some permits are sold on a fixed price at the beginning of each year while others are auctioned to the highest bidder, which increases both competitiveness and government revenue.

Australia and New Zealand are both signatories of the Kyoto Protocol. Australia and New Zealand have given commitments to limiting emissions to 108% of 1990 levels. Having repealed the carbon tax in 2014, the Australian Government is still to introduce alternative legislation to deliver on this commitment. The New Zealand ETS does not place a cap on emissions, but relies on New Zealand emission units (NZU) which can be traded.

SUMMING UP...

- Greenhouse gases in the earth's atmosphere are: water vapour, carbon dioxide, methane, nitrous oxide and ozone.
- Global warming is:
  - natural – caused by greenhouse gases emitted by plants, animals and volcanoes
  - enhanced – caused by greenhouse gases emitted by human activity.
- Consequences of global warming are:
  - rapid climate change
  - rising sea levels
  - decline of animal and plant species
  - reduction in the supply of fresh water.
- Climate change is caused mostly by the exchange of heat and water between deep oceans and upper ocean layers, which in turn impacts on the atmosphere.
- In Australia, the average minimum temperature has increased by about 0.9°C since 1960 (see CSIRO, 'State of the Climate 2012, Temperature', p. 2, at http://www.csiro.au/Outcomes/Climate/Understanding/State-of-the-Climate-2012/Temperature.aspx).
- The Intergovernmental Panel on Climate Change (IPCC) reviews published scientific information and reports on:
  - causes of climate change
  - potential environmental and socio-economic impacts
  - possible response options.
- Australian and New Zealand government global warming/climate change reforms include:
  - rebates on energy efficient electrical equipment
  - rebates on solar panel installations
  - the introduction of an emissions trading scheme
  - carbon farming and improved land management (restrictions on land clearing).

2.3 Environmental legislation

Environmental protection and reform is now embedded in law and all levels of government are responsible for its enforcement. The following outlines the responsibilities of Federal, state and local governments.

Australian Government legislation

At a federal level the key piece of legislation is the *Environment Protection and Biodiversity Conservation Act 1999* (EPBC Act) and its related regulations.

The EPBC Act ensures cooperation between the Commonwealth, state/territory and local governments in a national system to protect and manage important flora, fauna, ecological communities and heritage places that are defined in the EPBC Act as matters of national environmental significance.

As outlined by Environment Australia, the objectives of the EPBC Act include to:

- provide for the protection of the environment, especially those aspects of the environment that are matters of national environmental significance
- promote ecologically sustainable development through the conservation and ecologically sustainable use of natural resources
- promote the conservation of biodiversity
- assist in the cooperative implementation of Australia's international environmental responsibilities
- promote the use of Indigenous peoples' knowledge of biodiversity with the involvement of, and in cooperation with, the owners of the knowledge (Environment Australia, 'An overview of the Environment Protection and Biodiversity Conservation Act', http://auspost.com.au/media/documents/epbc-overview.pdf).

State/territory legislation

In the same way as workplace health and safety (WH&S) laws are being harmonised across Australia, the EPBC Act will eventually bring a consistent approach to environmental laws as well. All Australian state and territory environmental laws will be consistent with each other. The EPBC Act and associated regulations define what obligations and responsibilities state and territory governments have.

The Australian Constitution gives Australian states clear responsibility for resource management issues. The Environmental Protection Authority (EPA) is the government agency responsible for environmental legislation and enforcement. There is an independent EPA in each Australian state and territory as well as New Zealand. The activities of the different authorities vary, but an overview of the role of each EPA includes:

- audit and enforcement
- licensing and project approvals
- education
- cooperative partnerships
- economic mechanisms.

The main environmental issues of concern for an EPA include water, air and noise pollution, contaminated land, hazardous materials, waste and resource recovery.

Authorities are able to impose fines if individuals or businesses fail to meet their obligations under the relevant environmental legislation. If a pollution incident occurs, on-the-spot fines may be issued by either an EPA or local council officers, and fines may exceed one million dollars.

Environmental responsibilities

Responsibilities of corporations and individuals

Businesses have a responsibility to comply with all environmental obligations. Businesses need to be aware of restrictions on construction-related noise. If a site is close to residential areas the business will need to check with the local council for appropriate operating times. Contaminated building wastes such as asbestos and used fluorescent tubes must be disposed of so they don't end up in landfill.

There is a duty to notify relevant authorities of pollution incidents where material harm to the environment is caused or threatened. As outlined in legislation, material harm includes actual or potential harm to the health or safety of human beings or to ecosystems that results in actual or potential loss or property damage of a specified amount (for example, $10 000 in New South Wales). Failure to do so is an offence. The duty to notify applies to the person carrying out the activity, an employee carrying out the activity and the occupier of the premises where the incident occurs (see NSW EPA, 'About the POEO Act', http://www.epa.nsw.gov.au/legislation/aboutpoeo.htm).

Role of local councils

Local councils play a vital role in future sustainability. Any activities of, or development approved by, a council must comply with the EPBC Act.

The issues faced by a local council depend on the area, but often include:

- development control
- enforcement powers
- administrative responsibilities
- stormwater management and control
- risk control measures (pests, plant and animal)
- incentive programs
- management of local open space.

Source: Local Government & Shires Associations of NSW, 'Fact sheet 4', p. 1, http://www.lgnsw.org.au/files/imce-uploads/39/Wha_are_LGs_responsibilities_re_NRM.pdf.

Energy sector environmental policies and procedures

The energy sector has three main divisions: generation, transmission and distribution. Each division has its own challenge to meet government environmental legislation.

Energy sector companies have environmental policies and procedures that employees and contractors are required to follow. The policies are a key function of their environmental management system and **ISO 14001** compliance. A typical environmental policy would include efficient use of energy and resources to reduce consumption and emissions or discharges to the environment, waste management practices that include waste minimisation, recycling and approved waste handling and disposal, as well as regular environmental assessments. An important component is environmental incident response procedures for emergencies or other events which pose a risk to health, safety or the environment.

SUMMING UP...

- As outlined by Energy Skills Australia, environmental legislation means a company or individual must:
  - not cause water pollution
  - not cause air pollution (without a licence)
  - keep noise levels below certain limits
  - dump waste only at a licensed facility
  - not cause land pollution as a result of spills or illegal dumping
  - protect animals and plants in danger of extinction
  - manage dangerous chemicals responsibly
  - not start major construction works without appropriate approval.
- Local councils have an important role in ensuring compliance with environmental regulations. In some cases, local councils are empowered to act on behalf of the EPA (or other regulating authority).
- Energy sector companies often have environmental policies and procedures that employees are required to follow, especially regarding waste management and recycling.

2.4 Sustainable work practices

This topic identifies the various elements of environmentally sustainable work practices in the electrical sector. This topic will provide you with the skills and knowledge to be able to identify and adopt environmentally sustainable work practices in your own workplace to reduce harm on the environment and minimise wastage of resources.

Working in a sustainable way

Sustainable work practices involve:

- working safely by planning work beforehand
- taking advantage of alternative energies
- reducing the energy consumed by a workplace
- maintaining and repairing tools, equipment, plant and vehicles so they operate efficiently
- planning travel to reduce fuel consumption
- purchasing from suppliers that use minimum packaging to minimise environmental impact
- reducing waste products
- reusing and recycling materials
- responsibly disposing of waste products
- applying silt and erosion control measures.

Working sustainably is working smarter and more efficiently; by reducing waste, you also reduce costs. Figure 2.6 shows a method to improve sustainability in a workplace.

FIGURE 2.6 Improving workplace sustainability

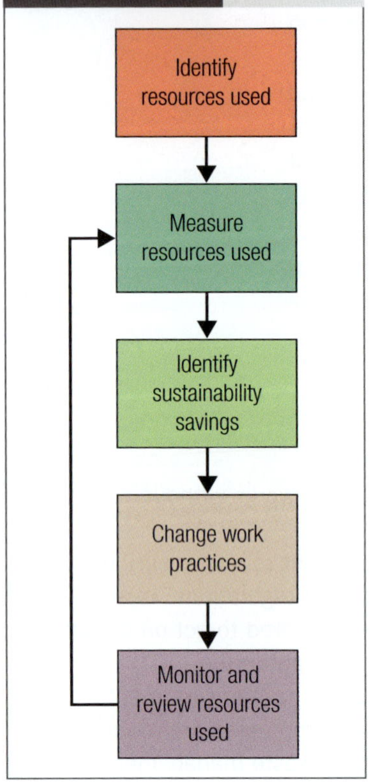

Environmental impact assessment

An environmental impact assessment (EIA) is an assessment of the possible positive or negative impact that a proposed project may have on the environment, taking into account the environmental, social and economic aspects. It is effectively the same as a WH&S risk assessment, but for the environment. It will identify possible risks to the environment and possible control measures to protect it. The EIA is a mandatory process in development applications for all large projects, and normally includes a consultation process with the local community and conservation groups.

The EIA will also provide information on how the project developers intend to comply with Commonwealth, state and local government legislation and any location-specific requirements. In regard to smaller projects, such as single residential housing, it is the responsibility of the on-site contractors to follow relevant environmental policies. Small projects do not require an EIA, but environmental issues are still considered in the project development application.

Sustainable outdoor work practices

The objectives of a sustainable outdoor work site are to reduce the resources used and to prevent pollution escaping from the site. Sustainable work practices for a typical outdoor work site, would include:

- ensuring that contractors and workers on the site are made aware of their environmental obligations
- separating different wastes before disposal, and reusing and recycling materials where possible
- reducing water wastage
- covering stockpiles and materials to reduce windblown dust
- general good housekeeping practices
- preventing paint and other chemicals from soaking into the ground or getting into stormwater drains
- installation and maintenance of sediment and erosion control devices on the site.

Substantial penalties apply to anyone causing contamination of stormwater drains and waterways.

Sustainable indoor work practices

The main things to consider when working indoors are to reduce the use of energy and other resources, and prevent contaminants leaving the work site. This may be achieved by:

- the installation of high efficiency lighting controlled by time clocks or motion sensors
- improvements in quality control to avoid waste of materials due to errors
- separating waste into groups for correct disposal and possible recycling
- responsible disposal of contaminated waste products.

Waste disposal and recycling

As outlined by the Commonwealth Department of the Environment, the aims of the National Waste Policy are to:

- avoid the generation of waste, reduce the amount of waste (including hazardous waste) for disposal
- ensure that waste treatment, disposal, recovery and re-use is undertaken in a safe, scientific and environmentally sound manner, and

- contribute to the reduction in greenhouse gas emissions, energy conservation and production, water efficiency and the productivity of the land.

Source: Australian Government, Department of the Environment, 'About the National Waste Policy', http://www.environment.gov.au/protection/national-waste-policy/about.

Waste classifications

Governments use waste classifications to aid in decision-making when issuing licences and permits for the treatment and disposal methods of waste. For example, Victoria's focuses primarily on Prescribed Industrial Waste (also known as 'hazardous waste') while New South Wales' classifications (see Figure 2.7) affect disposal options for specific wastes and use a risk-based approach. In Western Australia and South Australia, waste classifications are used primarily to direct wastes to disposal facilities best suited for those classes of waste. See the Australian Government's Waste Classifications at http://www.environment.gov.au > Topics > Environment Protection.

Waste can cause significant pollution unless dealt with properly.

FIGURE 2.7 Sample waste classifications, NSW

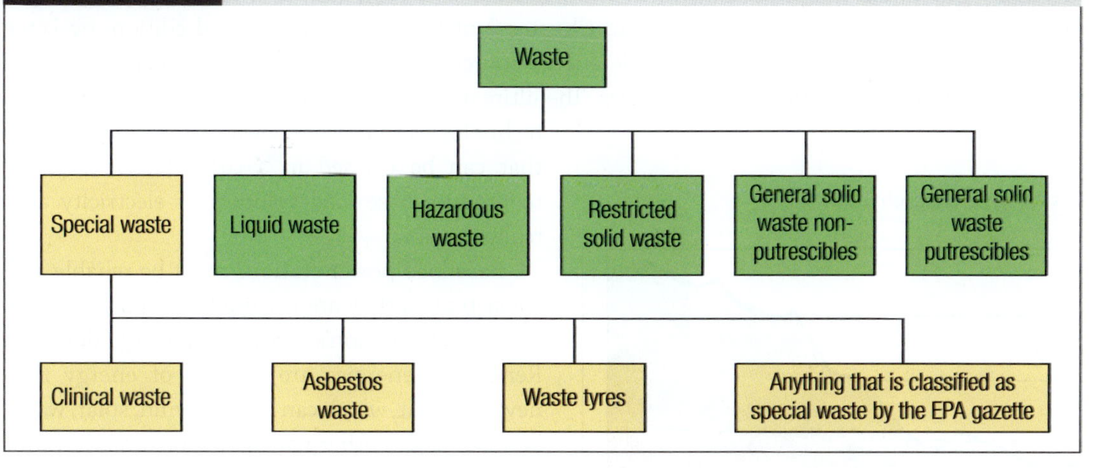

Recycling and recyclable materials

Waste recycling is now a common practice. In most energy sector organisations it is a method of processing used materials into new products. Examples of recyclable waste materials/products include:

- paper and cardboard
- transformer and motor oil
- metals (copper cables, aluminium, etc.)
- some types of light globes
- plastics.

Waste products that have economically salvageable or hazardous metals, and are therefore considered recyclable, include:

- electronic/computer circuit boards (rare metals, including gold)
- lead-acid batteries (lead)
- items such as some types of lamps, that contain hazardous metals (e.g. mercury) which can be removed for reuse.

SUMMING UP...

- Environmentally sustainable work practices reduce harm to the environment and minimise wastage of resources.
- Sustainable work practices can be achieved by energy efficiency, waste reduction, waste recycling and minimum use of resources.
- An environmental impact assessment (EIA) is an assessment of the possible positive or negative impacts that a proposed project may have on the environment, taking into account the environmental, social and economic aspects.
- The objectives of a sustainable workplace both indoor and outdoor are to reduce the resources used and to prevent emissions escaping from the site.
- Three ways to minimise waste products are avoidance, resource recovery and proper disposal.
- Waste classifications vary between states and territories.
- Recycling reduces waste going into landfill.

2.5 Sustainable energy

Sustainable energy is based on efficient use and using renewable energy sources.

Electricity generation is a process in which another form of energy, such as mechanical, chemical or radiant is converted into electricity (refer to Chapter 2 of *Electrical Principles*, 3rd edition, by Peter Phillips). As previously stated, a sustainable energy source is one which meets the energy needs of the present without depleting future reserves. The ultimate goal of a sustainable future is to develop technologies that use renewable sources of energy that can be utilised in a cost-effective manner to produce base load volumes of electricity at a realistic price.

Base load power stations must be reliable and consistent as they are required to supply some or all of a given region's continuous energy demand. Renewable energies are sources of energy that never run out, which can include wind, solar, wave, biomass and geothermal energy. Figure 2.8 shows a breakdown of Australian energy production.

FIGURE 2.8 Australia's energy production by fuel type

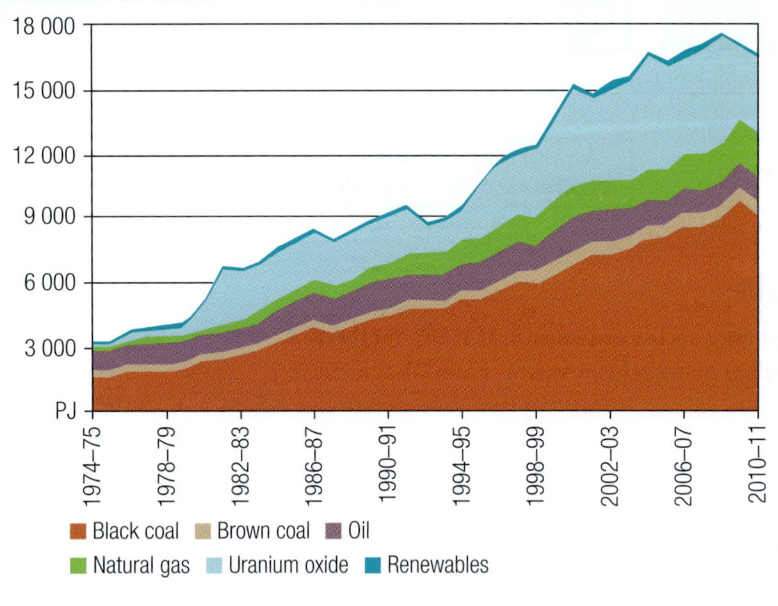

Source: BREE 2013, Energy in Australia 2013, Canberra, May. © Commonwealth of Australia 2013 http://bree.slicedlabs.com.au/sites/default/files/files/publications/energy-in-aust/bree-energyinaustralia-2013.docx

Minimum Energy Performance Standards (MEPS)

Minimum Energy Performance Standards (MEPS) set minimum efficiency percentages (refer to Chapter 4.2 of *Electrical Principles*, 3rd edition, by Peter Phillips) for a range of devices. MEPS programs are mandatory in Australia and New Zealand. Regulations specify the general efficiency requirements for MEPS for appliances, including offences and penalties for non-compliance.

It is mandatory for some products to meet the MEPS levels specified in the relevant Australian Standards, such as:

- refrigerators and freezers
- mains pressure electric storage water heaters
- three phase electric motors (under specified conditions)
- air-conditioners up to 65 kW cooling capacity

- compact fluorescent lamps
- distribution transformers (11 kV and 22 kV with a rating from 10 kA to 2.5 MVA)
- transformers and electronic step-down converters for ELV lamps.

The energy rating labelling scheme is a mandatory scheme for a range of appliances, which currently include refrigerators, freezers, clothes washers, clothes dryers, dishwashers, air-conditioners and televisions. Appliances must display a label that shows the star rating and other information about energy consumption (see 'Energy rating', http://www.energyrating.gov.au).

Figure 2.9 shows energy rating labels for two similar size and capacity refrigerators. The label on the right indicates a higher efficiency and therefore a lower annual power consumption than the label on the left.

Appliances such as air-conditioners require two ratings, one for heating and one for cooling, as shown in Figure 2.10. Annual energy usage is not shown for air-conditioners as the energy consumed will vary greatly with usage patterns and installation conditions.

FIGURE 2.9 Refrigerator energy rating labels (ERL)

Source: © 2014 energyrating.gov.au

Reducing energy usage

Electrical equipment that heats or cools consumes the greatest energy and should be the highest priority when establishing a strategy to conserve energy. A visual audit of a location and inspection of name or compliance plates tells the story. The larger the kilowatt rating and the longer the time the device is operated for, the more energy is used. Figure 2.11 shows the name plate and kilowatt rating of a heat pump style hot-water system, where the rated wattage is only 800 W compared to older element versions which were rated at 4800 W.

FIGURE 2.10 Air-conditioner energy rating labels (ERL)

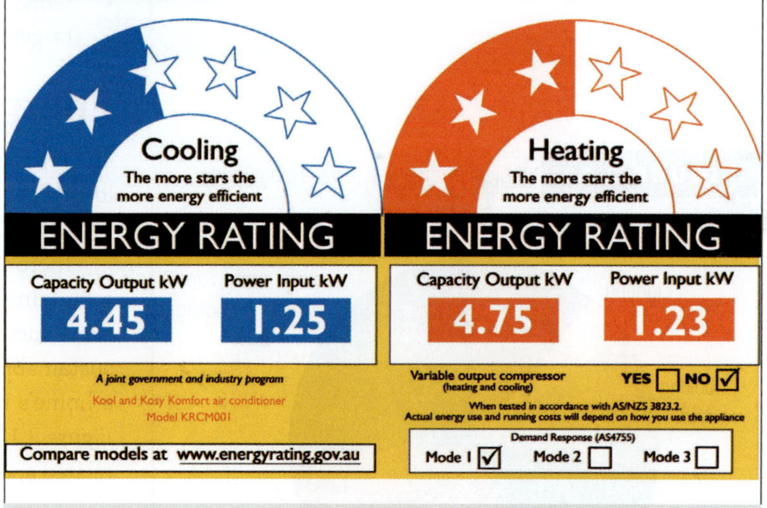

Source: © 2014 energyrating.gov.au

Significant energy cost savings are incentives for both homes and workplaces. Actions to consider include the following:

- regular servicing of heating and air-conditioning systems
- using energy-efficient appliances (machinery and motors should be MEPS compliant)
- making use of natural lighting, and installing local task lighting and high-efficiency compact fluorescent lighting
- establishing an energy reduction policy that requires the last person out of the workplace each day to turn out the lights, or using motion-sensor timers that switch lighting off when there has been no movement in a room for a certain period of time
- ensuring that all equipment is turned off at the end of the day rather than left on standby, i.e. computers are scheduled to automatically shut down at the end of the day
- adjusting air-conditioner thermostats so the temperature is set slightly higher than normal during summer and slightly lower in winter.

FIGURE 2.11 Appliance name plate and power rating

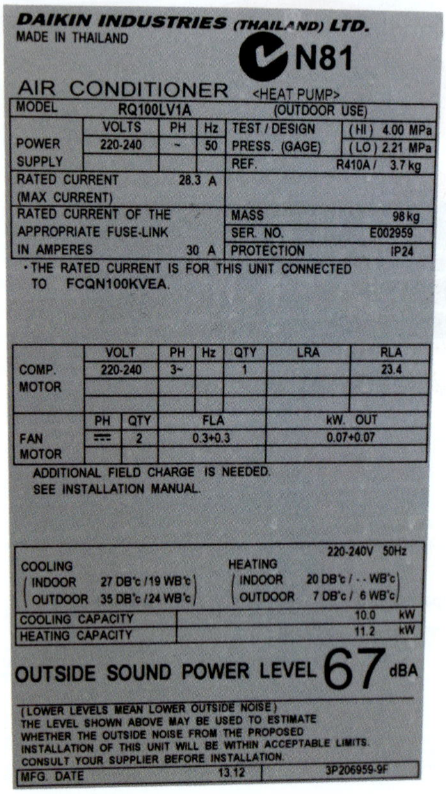

Source: Ed Hawkins

FIGURE 2.12 Typical energy use in WA households

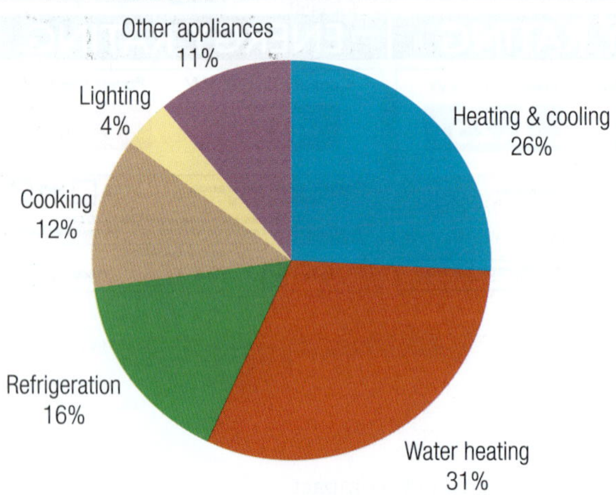

Source: Sustainable Energy Association of Australia, 'Your 6-Star guide to building an energy efficient home', http://www.seaaus.com.au/files/sea-6-star-booklet-240511.pdf

State/territory sustainable building regulations

The Building Code of Australia (BCA) sets minimum standards for the construction of all building types, including housing. The 2010 edition of the BCA increased the minimum energy efficiency standards for all new housing from 5 to 6 stars. Just like household appliances that have a star-rating system (energy rating labels), the energy efficiency of a home also has a nationally rated star system. The energy efficiency of houses is measured, with the worst-performing buildings achieving a zero rating through to a 10-star rating for a building that needs no artificial heating or cooling. The revised BCA 6-star minimum rating requires new minimum lighting efficiency standards plus increases in efficiency in heating and cooling (see http://www.commerce.wa.gov.au > Building Commission).

New Zealand and New South Wales require new residential single and multi-dwelling construction to comply with Building Sustainability Index (BASIX) standards. In New South Wales, BASIX replaces the BCA 6-star rating. BASIX is more comprehensive, it can also be applied to additions to existing buildings and swimming pools larger than 40 000 L.

BASIX sets sustainability targets for water and energy as well as minimum performance levels for the thermal comfort of the proposed development. BASIX targets typically include:

- reduction in potable water consumption
- reduction in greenhouse gas emissions (see BASIX, 'Basix targets', http://www.basix.nsw.gov.au).

Energy efficiency of new buildings

The energy consumed by a building will depend on the climate and location of where it is built. Figure 2.12 shows a percentage breakdown of pre-2010 West Australian homes built under BCA 5-star standards.

There are potentially significant reductions that can be made to a building's heating, cooling and hot water energy requirements. Construction designed for the climate zone will improve the sustainability rating. Here are just some suggestions that can increase a home's rating under both the BCA 6-star and BASIX standards. Figure 2.13 shows a summary of typical BASIX inclusions. Electrical related specifications include:

- use of natural lighting such as windows and skylights
- at least 80% of the light fittings in bedrooms/study, living/dining, kitchen, bathrooms/toilets, laundry and hallways lit by fluorescent or LED (light emitting diode) lamps, including use of sensors
- fittings in above areas are dedicated (only capable of accepting) fluorescent or LED lamps. Down lights score poorly and are discouraged
- use of gas or induction cook tops rather than electric
- utilising ceiling fans rather than air-conditioning
- use of high efficiency zoned air-conditioning
- solar, gas or heat pump style of hot water systems in preference to electric storage
- inclusion of photovoltaic (solar) panels.

FIGURE 2.13 Australian home built under BCA 5-star standards

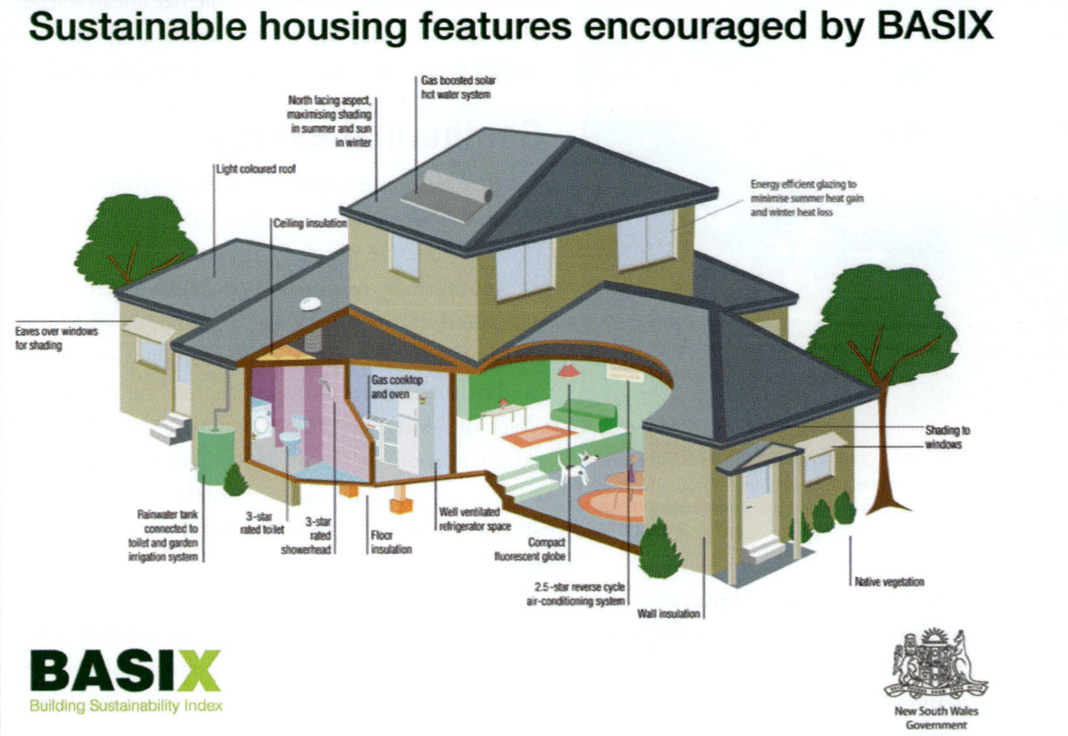

Source: http://www.basix.nsw.gov.au

Local councils are responsible for BCA and BASIX assessment in relation to development applications approvals.

Renewable energy sources (overview)

The majority of base load electricity is produced by the conversion of mechanical energy to electrical energy (refer to Chapter 2.1 of *Electrical Principles*, 3rd edition, by Peter Phillips). This is normally achieved by the rotation of copper conductors through a magnetic field. The energy to drive the prime mover (at present) comes from non-renewable sources such as coal and natural gas. Sustainable electricity generation in the future will need to substitute a renewable energy source for the prime mover or take a completely different approach to producing large quantities of base load electricity.

A typical coal-fired power station has multiple turbine generator sets capable of producing around 2000 MW of power per station. As of 2011, only 9.64% of generation in Australia used renewable energy (see Table 2.1).

TABLE 2.1 Australian renewable electricity generation

Fuel source	Estimated electricity generation per year (GWh)	Percentage contribution	Equivalent in households
Hydro	19685	67.2%	2772500
Wind	6432	21.9%	905800
Bioenergy	2500	8.5%	352100
Solar PV	680	2.3%	95800
Solar thermal	4.4	0.015%	620
Marine	0.75	0.003%	105
Geothermal	0.5	0.002%	70
Solar water heating	2115	N/A	297900

Source: Clean Energy Council Renewable Energy Database (ABARE), 2011

FIGURE 2.14 Hydrothermal–geothermal energy

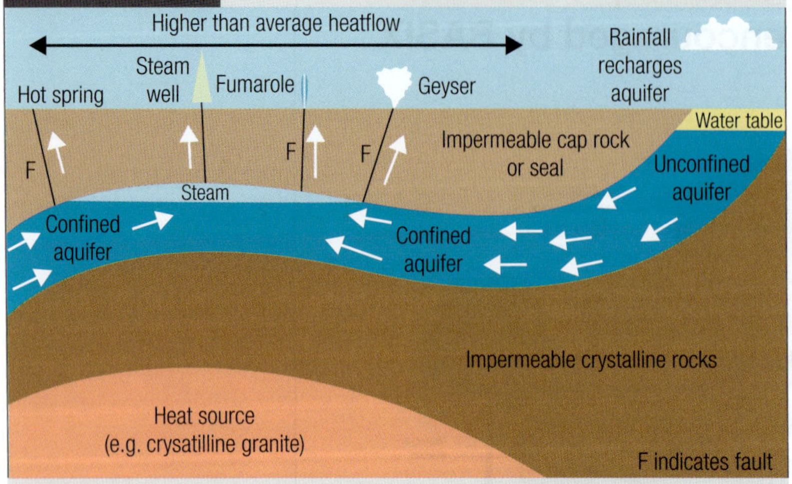

Source: Murdoch University online, http://www.see.murdoch.edu.au/resources/info/Res/geothermal/image003.jpg

New Zealand does significantly better because of natural seismic resources. The challenge ahead will be to develop technology that takes advantage of renewable energy sources that can supply base load demand.

Geothermal energy

The geothermal energy of the earth's crust originates from the original formation of the planet and from radioactive decay of minerals. The four types of geothermal resources are hydrothermal, geo-pressured, hot dry rock and magma, of which only hydrothermal is currently commercially exploited.

Hydrothermal plants situated in the Taupo Volcanic Zone, in New Zealand's North Island, produce about 13% of New Zealand's electricity supply. A typical modern hydrothermal power station such as the Te Huka station can produce up to 750 MW of electricity.

As described by Murdoch University online ('Geothermal energy', http://www.see.murdoch.edu.au/resources/info/Res/geothermal), hydrothermal resources require three basic components: a heat source (e.g. crystallised magma), an aquifer containing accessible water, and an impermeable cap rock to seal the aquifer (see Figure 2.14). The geothermal energy is usually tapped by drilling into the aquifer, and extracting the hot water or steam.

Conservative estimates of Australian geothermal resources suggest the energy available for electricity generation is 23 million petajoules, or 7500 years of Australian energy consumption at the current level. Over 80% of this resource is located in the Great Artesian Basin.

Figure 2.15 shows the hot dry rock (HDR) system. Here, water is pumped into fractures in rock deep below the surface and then recovered via a heat exchange unit to produce the steam required to power a turbine driving an alternator.

FIGURE 2.15 Hot dry rock

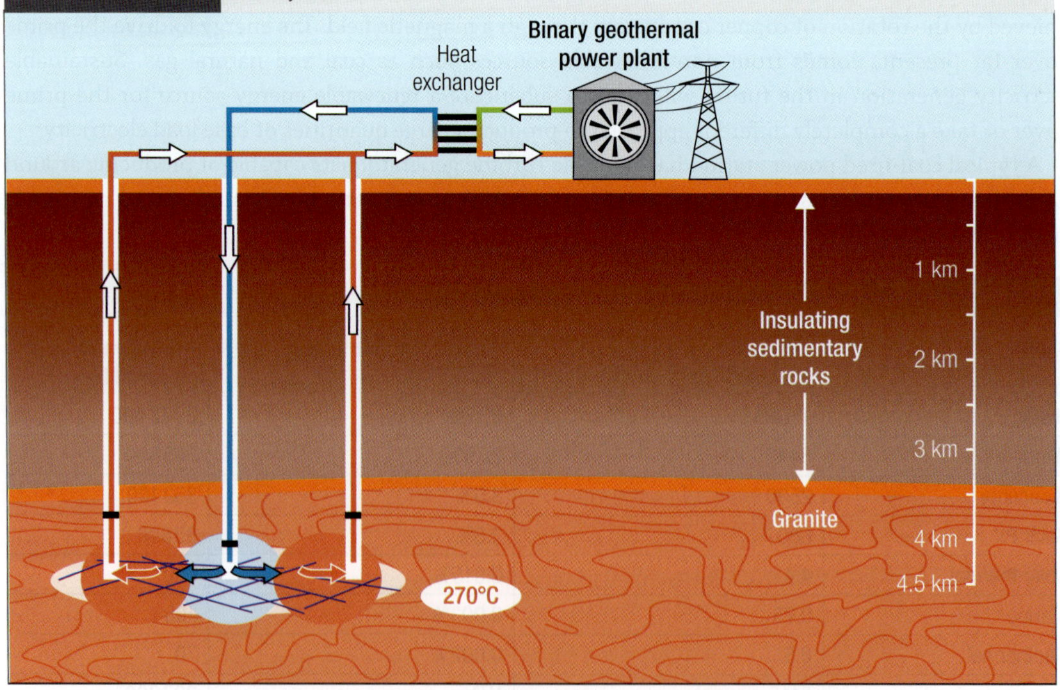

Concentrated solar power (CSP) systems generate electricity by using parabolic mirrors to concentrate solar radiation (sunlight) and capture it as extremely high-temperature heat. Energy can be captured in hot salt water storage or channelled into a turbine to create electricity, as shown in Figure 2.16.

Storing the thermal energy (the heated fluid), allows it to be converted into electricity when required. CSP can discharge from storage to serve high-price peak loads that occur outside the daylight hours, including the late afternoon and early evening.

There are four kinds of CSP systems – parabolic troughs, linear Fresnel lenses, towers and dishes. The USA's Solar Energy Generating Systems (SEGS) in California's Mojave Desert is currently the world's largest CSP project. It is a parabolic trough system comprising 936 384 mirrors and covering more than 647.5 hectares and can produce up to 354 MW of electricity.

FIGURE 2.16 Concentrated solar power

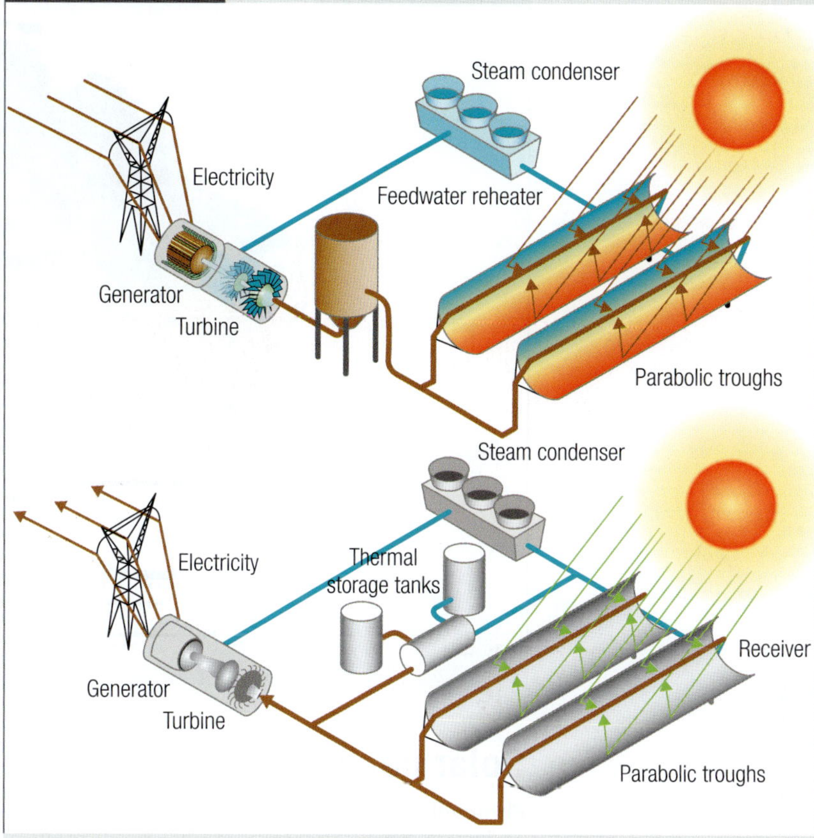

Source: Illustration by Ray David, National Renewable Energy Laboratory (NREL)

Hydroelectric

To create hydroelectric power, water is collected in a dam or reservoir and allowed to flow downhill through pipes (penstocks). The flow of water has significant kinetic energy such that as the water passes through a turbine, the kinetic energy causes the turbine to rotate. The turbine drives an alternator completing the energy conversion to electricity. Figure 2.17 shows the production of hydroelectricity.

Hydroelectricity is normally used to meet demand at peak times, as running the stations long term would deplete water reservoirs. Australia's largest hydroelectricity power station is Tumut 3 in the Snowy Mountains of New South Wales with a generating capacity of 1500 MW. Manapouri Power Station is the largest hydroelectric power station in New Zealand with a generating capacity of 800 MW.

Wind power

Wind is an endless resource created by high pressure air systems. Wind generators are located in situations where wind is frequent and speed is consistent (see Figure 2.18). Movement of air has kinetic energy, so as the wind moves past the pitch of the blades, the kinetic energy is transferred to the blade, causing it to turn. The turning force is then converted into electricity by an alternator mounted in the rear of the housing.

South Australia has the highest number of wind turbines in Australia and provides more than 50% of Australia's wind-generation capacity. Australia's wind-generating capacity has grown from almost zero in 2001 to more than 2000 MW, approximately equivalent to one coal-fired power station.

FIGURE 2.17 Production of hydroelectricity

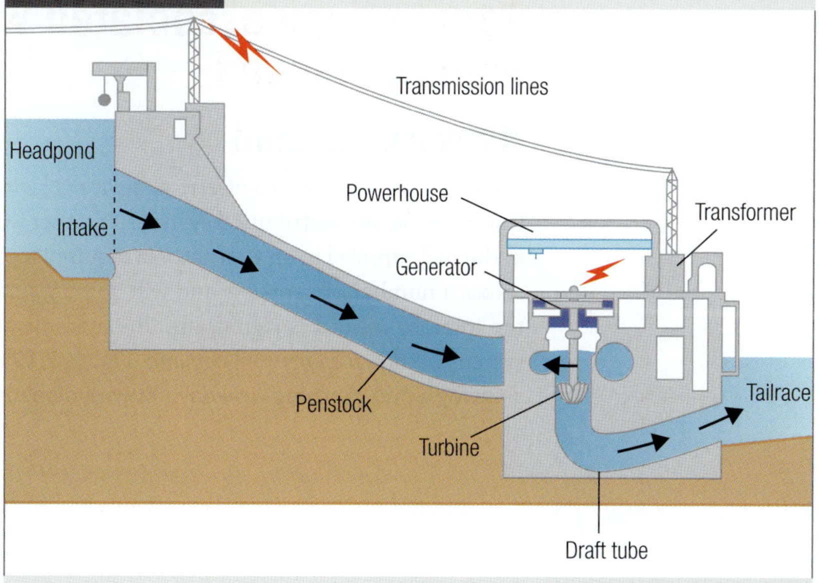

Source: © NB Power and Atlantica Centre for Energy, 2014

FIGURE 2.18 Locations with suitable wind speeds for electricity generation

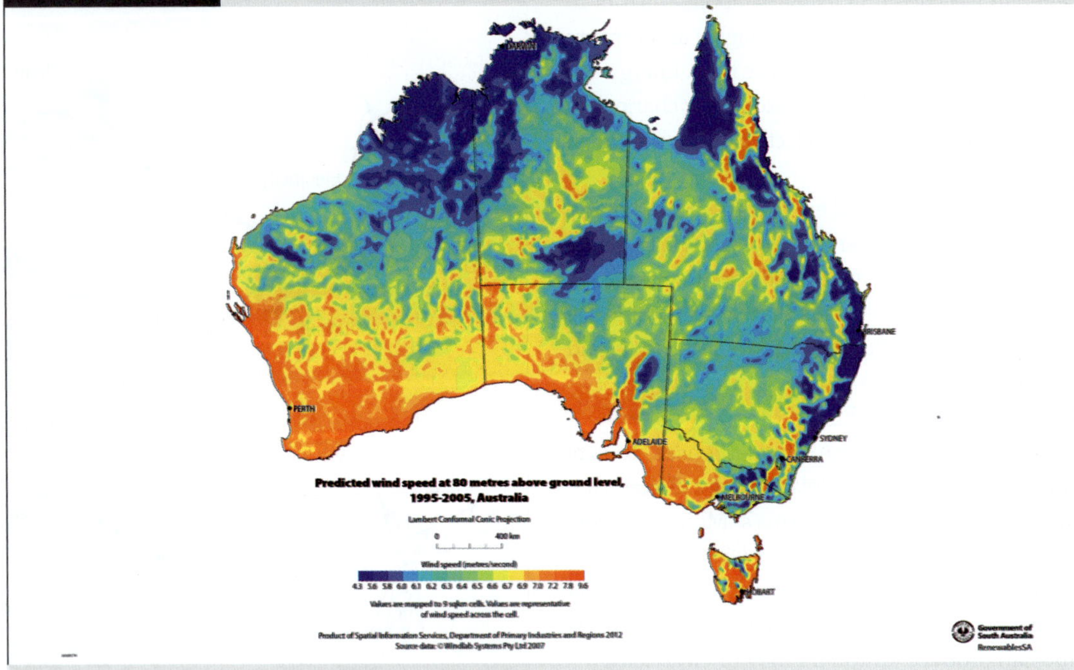

Source: RenewablesSA

Solar panels

Photovoltaic (PV) cells convert radiant energy (sunlight) to electrical energy (refer to Chapter 2.1 of *Electrical Principles*, 3rd edn, by Peter Phillips). Electronic inverters are then used to convert the d.c. produced by the PV cells to a.c. at 230 V and 50 Hz. Australia's biggest large-scale commercial photovoltaic system is at the University of Queensland with a capacity of up to 1.2 MW. Small-scale roof mounted systems between 1.5 kW to 5.0 kW are now common throughout Australia and New Zealand due to previous government 'back to grid' financial incentives. Locally situated back to grid connected solar panel systems feed excess electricity not used by the host into the neighbouring electricity distribution system. Local solar systems have less transmission losses than traditional coal-fired power stations.

Like wind power, Australia's PV solar capacity has risen dramatically in recent years. PV solar capacity has risen from just 0.4 MW in 2001 to over 1000 MW in 2011.

Terminology associated with electrical energy demand

Network demand

When any grid-connected electrical device is switched on, it places additional load on the generation, transmission and distribution systems. In other words, when a single light is turned on, a slight increase in the fuel supplied to an alternator is required to meet the additional load. The single increase is spread across a number of suppliers and will be very small; however, when millions of devices are turned on the increase in load is significant. The amount of electrical power that must be supplied across the system is known as *network demand*. Figure 2.19 is a screen capture of Queensland's Energex website showing a mid-morning weekday network demand of 2290 MW.

FIGURE 2.19 Mid-morning weekday network demand of 2290 MW

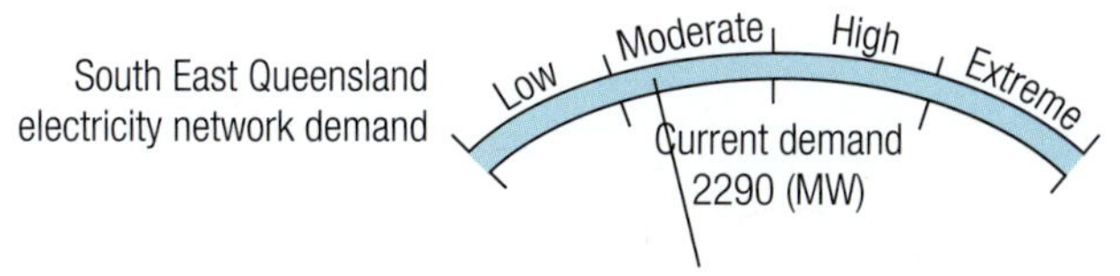

Source: Energex, *Queensland Electricity Connection and Metering Manual*, http://www.energex.com.au/contractors-and-service-providers/electrical-contractor/electricity-connection-and-metering-manual

The wholesale market price of electricity is market driven on a network demand basis. Australia has two distinct supply and demand electricity market systems: the National Energy Market (NEM). and the Wholesale Electricity Market of Western Australia (WEM), which supplies electricity to the South West Interconnected System (SWIS), the electricity system covering the south-west of Western Australia, including Perth. The National Electricity Market (NEM) is a wholesale market for the supply of electricity to Queensland, New South Wales, the Australian Capital Territory, Victoria, South Australia and Tasmania.

The NEM operates on the world's longest interconnected power system – from Port Douglas in Queensland to Port Lincoln in South Australia, a distance of around 5000 kilometres. More than $10 billion of electricity is traded annually in the NEM to meet the demand of more than eight million end-use consumers (NEMMCO, 'An introduction to Australia's electricity market', http://www.futureenergy.com.au/downloads/National Electricity Market Summary.pdf). The Australian Energy Market Operator (AEMO) was established to manage the NEM and gas markets from 1 July 2009.

The Energy Systems and Sustainable Living website describes this process:

> Wholesale trading in electricity is conducted as a spot market where supply and demand is instantaneously matched in real-time through a centrally coordinated dispatch process. Generators offer to supply the market with specific amounts of electricity at particular prices. Offers are submitted every five minutes of every day. From all offers submitted, the AEMO determines the generators required to produce electricity based on the principle of meeting prevailing demand in the most cost-efficient way.
>
> Source: 'Energy systems and sustainable living', http://ecoandsustainable.com/2013/04/01/australian-electricity-supply-ex-nt-and-wa.

Peak demand

Heat waves, cold snaps and other sporadic events can create major spikes in electricity usage, which are known as 'peak' or 'critical peak' demand. The magnitude of peak demand has risen over the last decade (see Figure 2.20), due to a growth in residential air-conditioning. Peak demand is a key driver of investment in generation and network capacity, and the costs are ultimately borne by all electricity users (Australian Government Productivity Commission, *Electricity Network Regulatory Frameworks Inquiry Report*, Chapter 9 'Peak demand and demand management', 2013).

Maximum demand

Not all the equipment in all electrical installations is turned on at the same time. This is referred to as *diversity*. An estimation of the maximum current expected to flow at any one time is made to determine the size of cables and distribution transformers required to supply parts or all of the installation. This

FIGURE 2.20 Peak demand of electrical usage

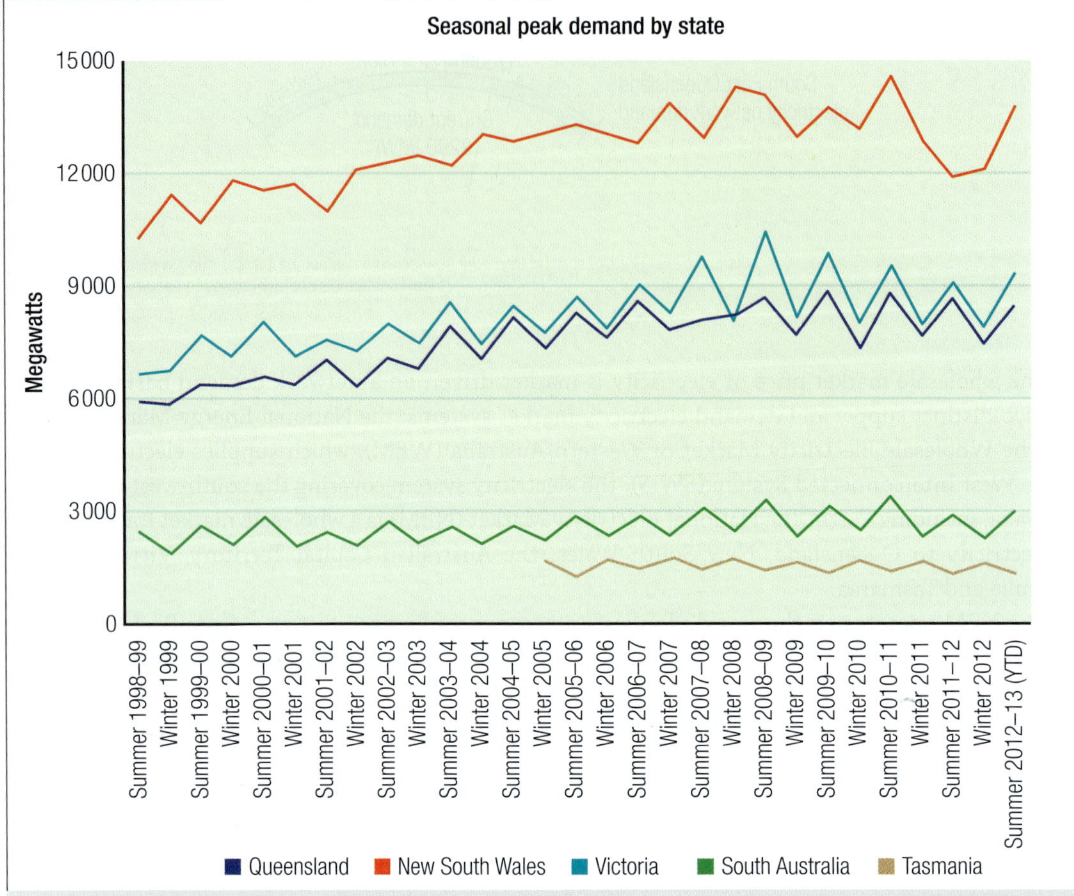

Source: Australian Energy Regulator, 'Seasonal peak demand (region)', http://www.aer.gov.au/node/9767

figure is far less than the total addition of all possible loads being turned on at the same time. Design based on full connected load is not practical as it will result in unnecessarily large current ratings of cables, switchboards and distribution transformers. Figure 2.21 shows a maximum demand indication (MDI) ammeter.

FIGURE 2.21 Maximum demand indicator (MDI)

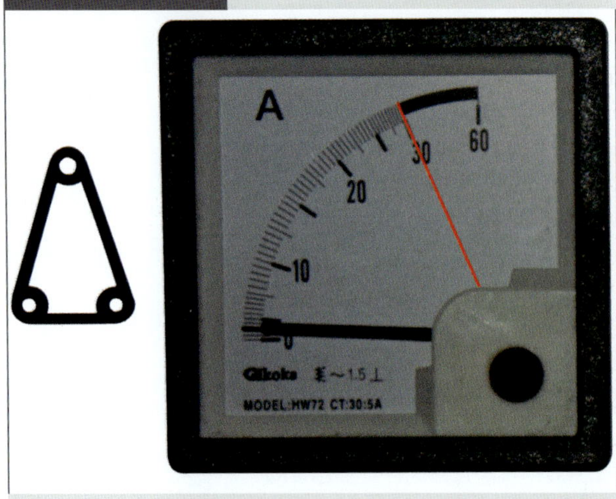

Source: Ed Hawkins

The black pointer is a real time indicator, rising and falling with load current. The red pointer only rises to indicate the maximum current that has occurred since the meter was manually reset by the operation of the black knob at the bottom right of the meter.

Tariff

Electrical energy usage is measured with a kilowatt hour meter, where one kilowatt hour (kWh) is the energy used over one hour by an appliance rated at one thousand watts. Individual electricity retailers have different tariffs. Tariff structures and pricing are set to help the retailer manage demand based on the type of customer and connected load. Retailers try to avoid large peaks in demand as the wholesale price they pay is much higher. Examples of tariffs include:

- domestic – residential general light and power
- commercial – shops, offices, factories

- off peak – electric hot water systems
- time of use – different prices are set for peak, shoulder and off peak times.

Electricity retailers are now fitting 'smart power meters' to measure the power taken by an installation, as this allows varying tariffs to be applied, depending on the time of day. Smart meters enable two-way communication between the meter and the retailer's central system (see Energy Skills Australia, http://www.ee-oz.com.au). Figure 2.22 shows a smart meter.

FIGURE 2.22 Electronic smart power meter

Source: Ed Hawkins

SUMMING UP...

- Renewable energies used in Australia and New Zealand to generate electricity include hydro, wind, bio, solar thermal, solar PV and geothermal energy.
- In 2011 less than 10% of electricity generation in Australia used renewable energy.
- The two types of geothermal energy are hydrothermal and hot dry rock.
- Thermal solar converts radiant energy to steam that drives a turbine to produce electricity.
- Photovoltaic solar cells convert radiant (light) energy directly to electrical energy.
- Hydroelectricity is produced by the use of a turbine to convert the kinetic energy of moving water to electricity. This is Australia and New Zealand's largest source of renewable energy.
- Wind power converts the kinetic energy of moving air to electricity by causing rotating blades to move and rotate an alternator.
- Network demand is the amount of electrical power consumed by electricity grid customers at any given point in time.
- Peak demand is the maximum electrical power consumed by electricity grid customers over a defined period of time, i.e. daily, seasonally or annually. The annual peak demand normally occurs on the hottest or coldest day of the year.
- Maximum demand is the highest current expected to flow for a given part or the whole of an electrical installation.
- Electrical energy is sold to customers in kilowatt hours (kWh). Different prices (tariffs) are charged for different types of customers and loads.

REVIEW EXERCISES

Answer these questions and check your answers at the back of the book.

Review questions 2.1

1 What are the three pillars of sustainability?

2 What is the purpose of the ISO 14001 Environmental Management Systems standard?

Review questions 2.2

1 What is meant by the term 'greenhouse effect'?

2 Describe two main human activities causing an increase in the concentration of carbon dioxide in the earth's atmosphere.

Review questions 2.3

1 Under what circumstances must an individual notify relevant authorities of a pollution incident?

2 What are the general responsibilities that environmental legislation imposes on an individual?

Review questions 2.4

1 What are four typical sustainable work practices associated with an outdoor work site? Assume the work includes excavation, plus the production of dust, waste products and noise.

2 Why is the disposal of batteries and electronic circuit boards into landfill illegal, and how are these best disposed of?

Review questions 2.5

1 What are the two main pathways to achieving sustainable energy?

2 Explain the difference between an electrical appliance with low efficiency compared to the same type of appliance with higher efficiency. Provide an example.

ONLINE RESOURCES

For more information about the contents of this chapter, visit the following websites:

State/territory government websites

ACT Planning and Environment Directorate
http://www.environment.act.gov.au/

NSW Environment & Heritage
http://www.environment.nsw.gov.au

Northern Territory Environment Protection Authority
http://www.ntepa.nt.gov.au/

Queensland Department of Environment and Heritage Protection
http://www.ehp.qld.gov.au/

EPA South Australia
http://www.epa.sa.gov.au/

EPA Victoria
http://www.epa.vic.gov.au

EPA Western Australia
http://www.epa.wa.gov.au

EPA Tasmania
http://epa.tas.gov.au/epa

Australian Federal Government

Commonwealth Department of the Environment
http://www.environment.gov.au/

Energy Rating
http://www.energyrating.gov.au/

Australian Building Codes Board
http://www.abcb.gov.au

Australian Government
http://australia.gov.au/ enter the search term 'environment'

3

FIXING AND SUPPORT DEVICES/TECHNIQUES

The electrotechnology industry requires a wide variety of items and equipment of varying load weights to be suitably and safely secured in place to an equally wide variety of surfaces and situations. Devices used for this purpose should be cost-effective and quick to install, while still being able to maintain consistent and reliable results. This chapter outlines the selection and application of a range of commonly used fixing devices.

Chapter outline

3.1 Devices for securing and mounting electrical equipment in hollow walls

Modern construction techniques present a number of situations where a cavity or void is created between two surfaces. Typical locations include spaces between plasterboard walls and areas above ceilings. Electricians take advantage of these locations where possible to locate cables, as the hollow space provides some degree of mechanical protection and hides the cables from view. This section outlines the selection and application of a range of commonly used hollow wall fixing devices.

Types and safe application of fixing devices

Screws and bolts are classified by their *drive method*, *head*, *thread* and *point types*.

Drive method

The drive on a fastener is the feature through which rotational torque is applied. All threaded fasteners will have a drive feature or will have a retention feature to prevent rotation while the mating part is rotated (James Glen, 'Driving methods', http://www.jamesglen.com.au/driving-methods). A sample of typical drive methods is shown in Table 3.1.

Head types

Head types are designed to suit the application and material to be secured and are shown in Table 3.2. The thread type determines what type of material the screw is suitable to secure. A coarse thread is normally used for timber and a finer thread is used for metals.

TABLE 3.1 Drive methods

Sketch	Drive type	Uses	Tools
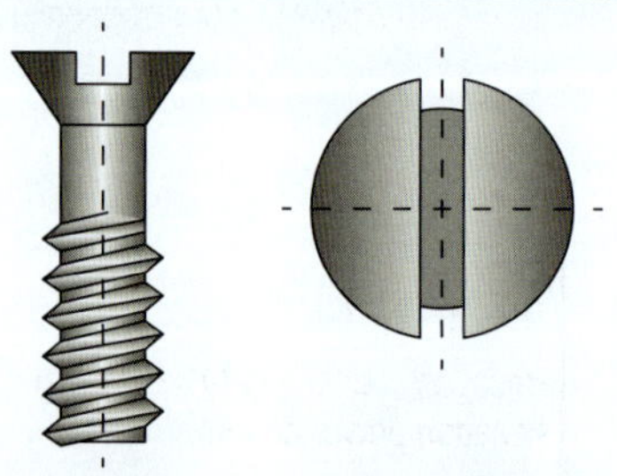	Slot	Commonly found on woodscrews and machine screws in domestic applications or where field retightening or removal may be required. Most suitable for hand operated tools. Oldest and simplest drive form.	Flat bladed common screwdriver.
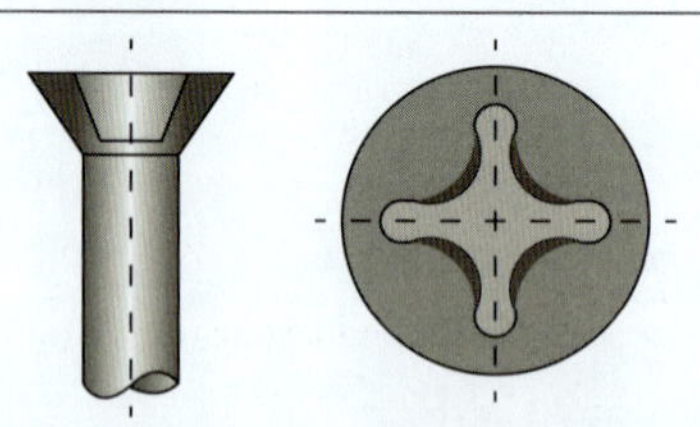	Phillips recess (Type 1) (X-recess) (Cross-recess)	Commonly found on self tapping screws and machine screws, particularly where they can be power assembled, e.g. on a production line for domestic appliances.	Ideal for power operated tools and hand tools.
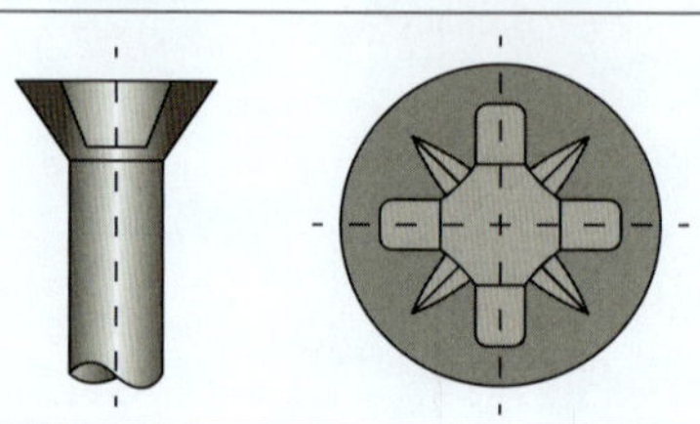	Pozidrive recess (Type II or IA) (X-recess)	As above, but is less prone to 'cam out' when drive tools are worn.	Power operated Pozidrive tools and Pozidrive hand tools.

(*Continued*)

Sketch	Drive type	Uses	Tools
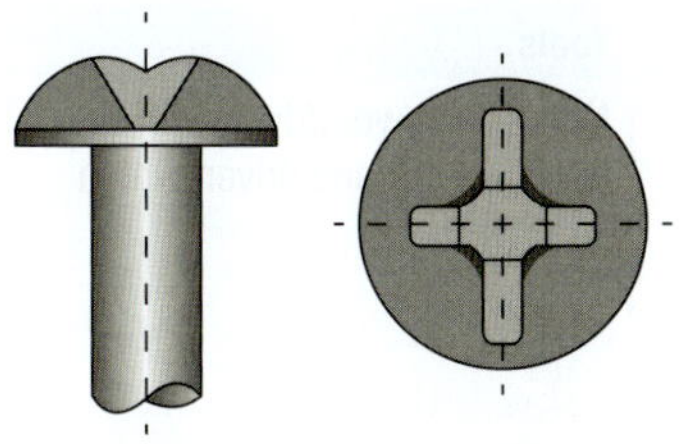	Combination (Phillips recess and slot) (X-recess or slot)	Most commonly found on mush head roofing screws and machine screws, allowing for power driven assembly and field removal or adjustment by common blade screwdrivers.	Power operated drive tools and hand tools, flat slot or cross recess.
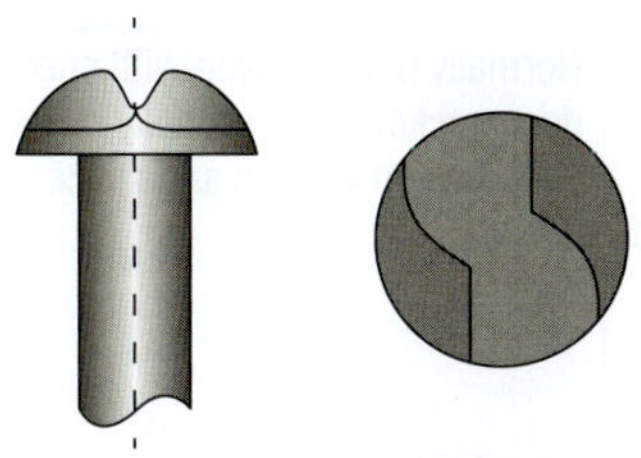	Clutch (One way) (Jail head)	Security applications usually on self-tappers, woodscrews or machine screws, e.g. domestic window locks, jail cell hinges, door lock exposed screws. Declining popularity.	Conventional flat blade screwdriver to tighten. Cannot be easily removed without drilling and using a special removal tool.
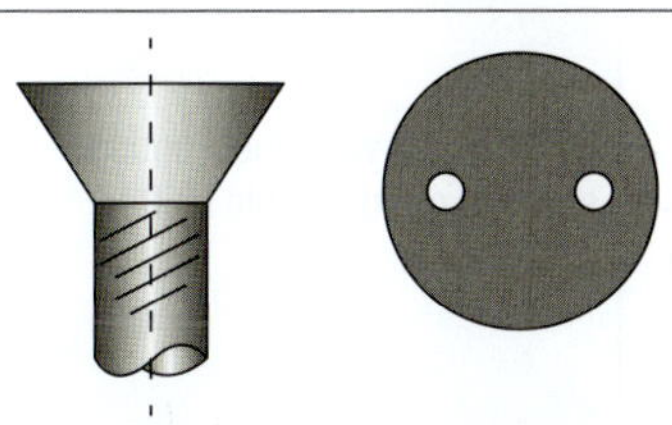	Eye drive (Snake Eyes)®	Security applications, usually on self-tapping screws or machine screws where high assembly torque is not required, e.g. into plastics.	Special hand screwdriver with 2-pin blade.
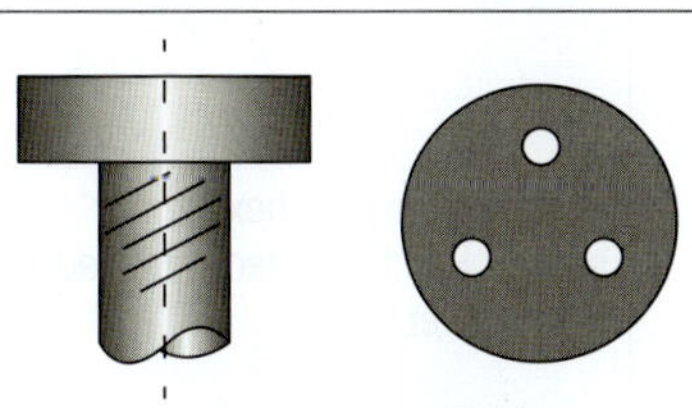	Tri-eye drive (3-pin drive)	As above. Also seen on golf shoe spikes or plastic inserts.	Special hand driver with 3-pin drive.
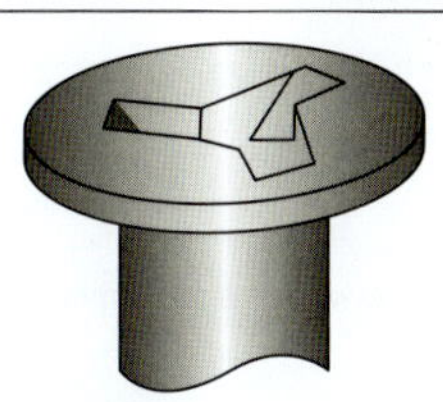	Tri-wing	Security applications usually confined to screw products, particularly in aircraft, public transport fittings and electrical appliances where fieldwork should only be carried out by authorised service personnel.	Special 3-bladed drivers both power and hand.
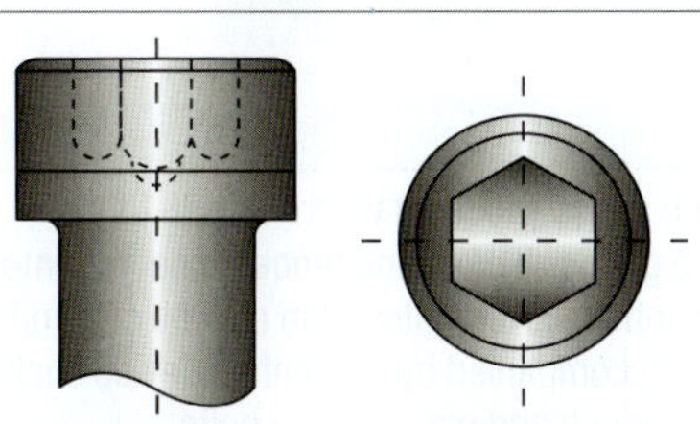	Hexagon recess	Principally used in high torque applications such as automotive, heavy equipment, tool die sets. Commonly associated with cap screws. (Also comes as square recess.)	Hexagon socket key (Allen key) and hexagon power drivers.
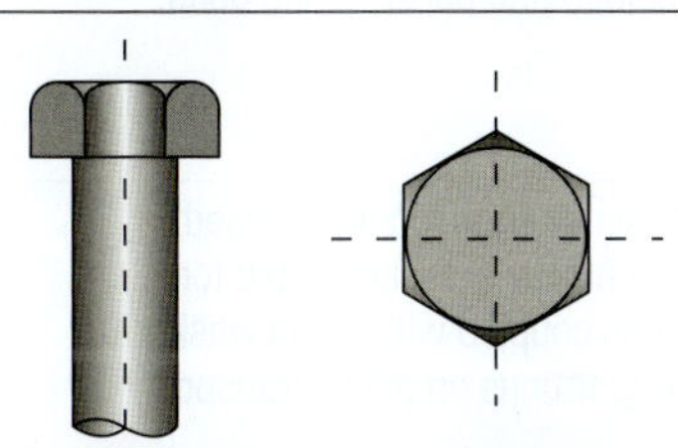	Hexagon	The most common drive on bolt products – very versatile in drive torque range; economical to produce.	Hand driven ring and open ended spanners, hand or power driven with socket drivers.
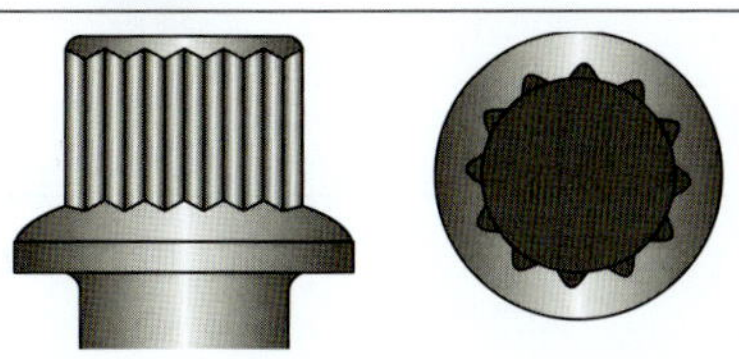	12-point (Double hexagon)	Usually associated with a flange head and in high torque applications where there is a restriction on head size or where space limits hexagon driving tools, e.g. inside a recess.	12-point socket driver. In some circumstances a ring spanner could be used.

(*Continued*)

TABLE 3.1 *(Continued)*

Sketch	Drive type	Uses	Tools
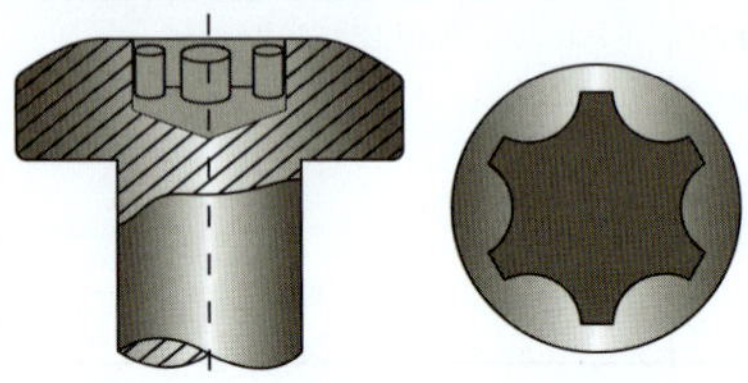	Internal Torx®	Gives very high driving torque capability with low risk of 'cam out'. Usually found in high production applications, particularly automotive and appliance industries. Usually restricted to screw products.	Normally power driven with special drive bits or hand driven with a Torx® key.
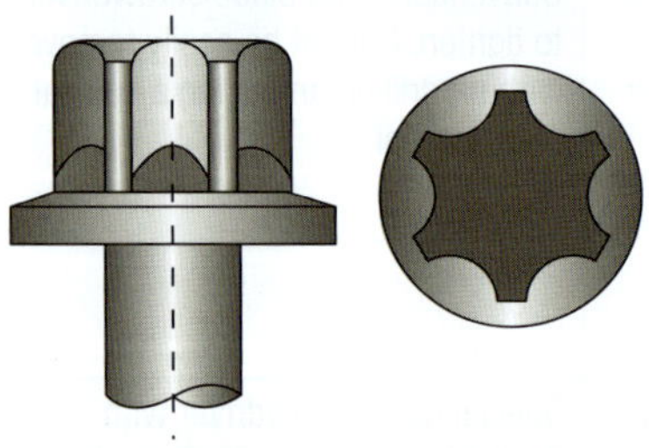	External Torx®	As with 12-point, usually associated with a flange head. Very high driving torque capability. Often found in automotive and aircraft engine applications. Usually restricted to bolt products.	Normally power driven with special drive sockets. Occasionally by hand with tension wrench and special sockets.
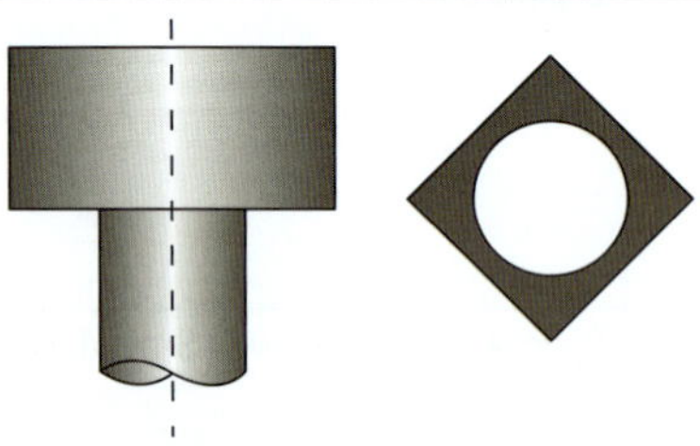	Square	Declining usage. Was often used in timber bridgework and where larger driving surface was required. Still used in a small version on square drive cap screws.	Hand driven with open-ended spanner or adjustable shifter.
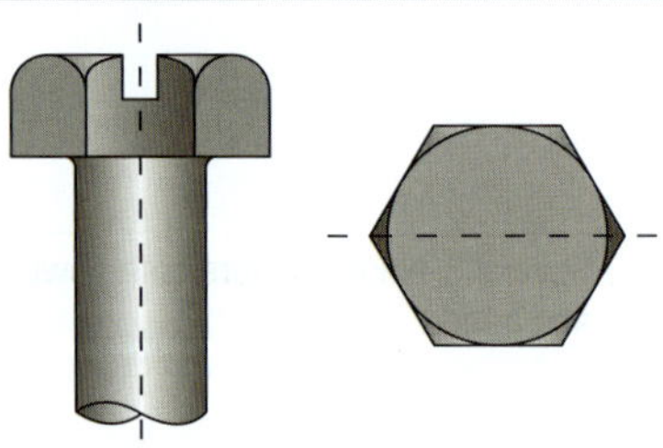	Hexagon slot combination	Usually associated with screw products. Is useful where it can be power driven on the assembly line and removed or adjusted in service with a blade screwdriver. Usually head is indent Hex and sometimes x-recess is also added.	Power driven with hexagon socket. Hand adjusted with blade screwdriver.

Source: James Glen, 'Driving methods', http://www.jamesglen.com/driving-methods

TABLE 3.2 Fastener head styles

Sketch	Title	Comments
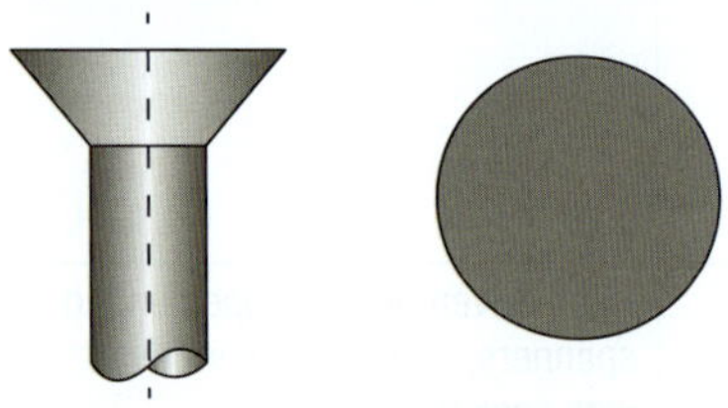	Countersunk head (Flat)	Most common usage is on screw products where a flush fit is required on the surface, e.g. door hinges and timber joinery or into steel applications, e.g. manhole cover plates. Can also be seen on bolt products, but usually accompanied by a retention device such as a square or a lug as in plough and earthmoving bolts.
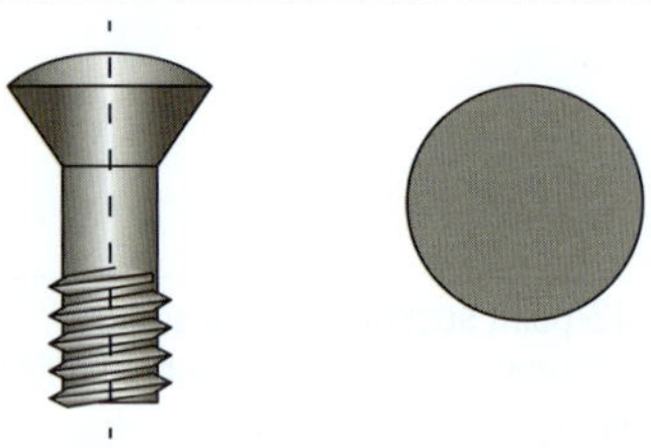	Raised countersunk (Oval and instrument)	Most often associated with screw products in exposed applications. Usually will be chrome or nickel plated for appearance and is sometimes coupled with a cup washer to enhance the appearance, e.g. fittings on public transport.

(Continued)

Sketch	Title	Comments
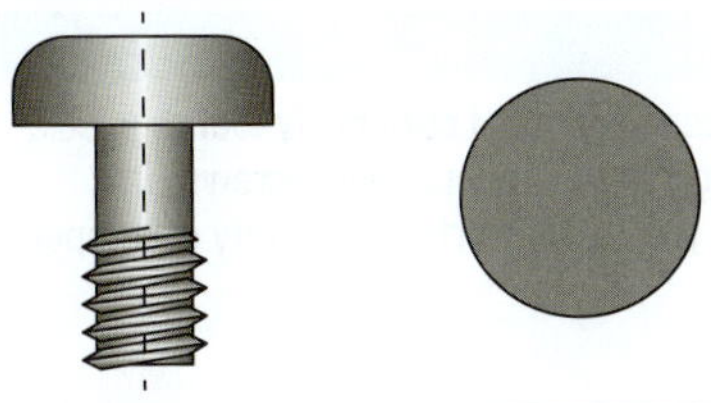	Pan head (American style)	Predominantly used on screw products where a flat bearing surface is required or conversely, where a countersunk is not required. More economically produced than countersunk.
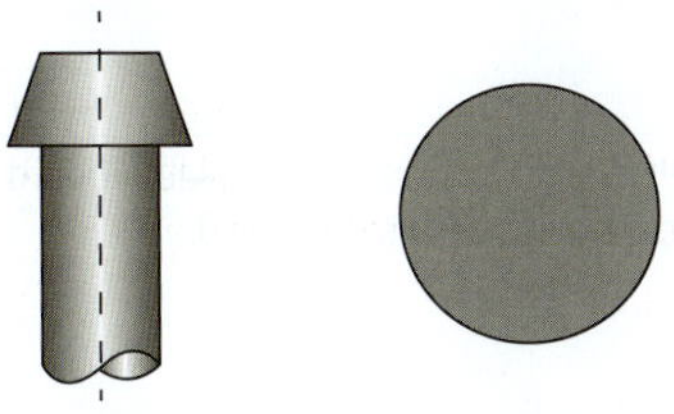	Pan head (English style)	Now normally only found on solid rivets and on some hot forged products.
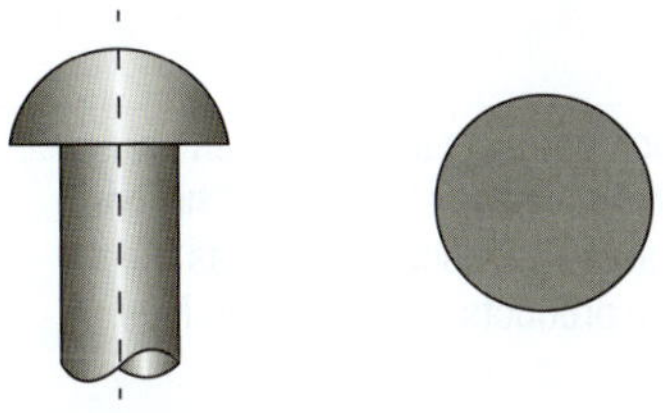	Round head (Cup, snap and button)	Declining popularity, but does provide deeper slot capability than pan and therefore, is more often used in woodscrews and machine screws. When recessed with Hex or x-recess becomes one form of button head.
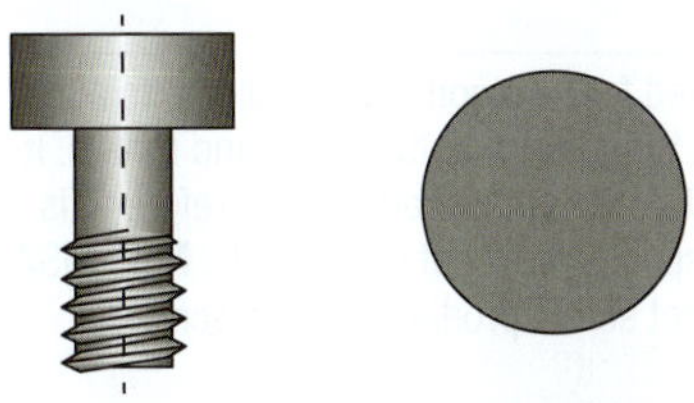	Cheese head	Less popular than pan, but again gives a deeper slot capability, so is most commonly seen as a slot product on machine screws.
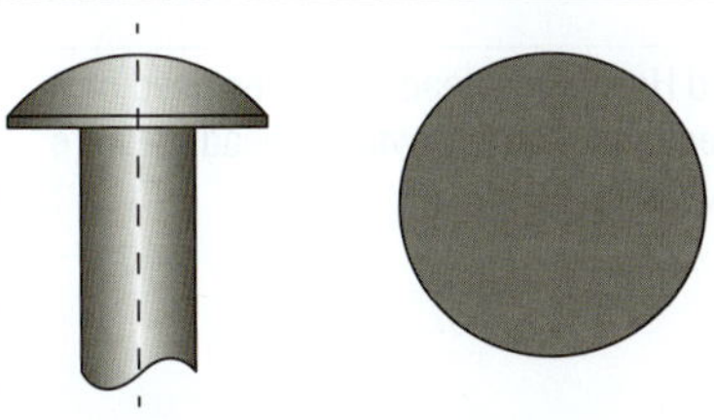	Truss head (Mushroom)	Larger diameter and lower profile head than round, pan or cup. Is most commonly seen with slot or combination slot x-recess on mush roofing bolt/nut product.
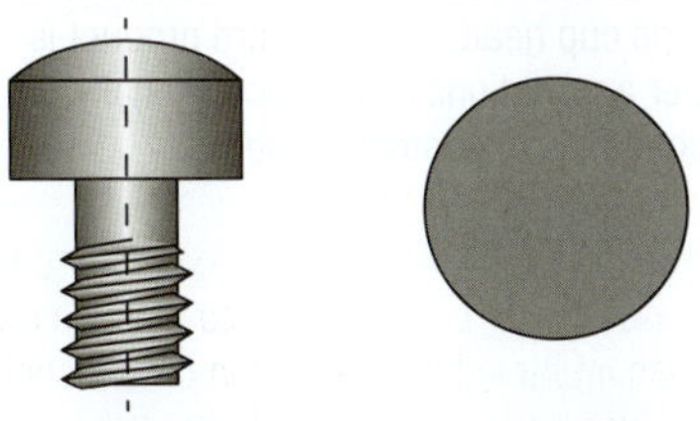	Fillister head (Raised cheese)	Rarely specified these days; was formerly seen predominantly on machine screws.
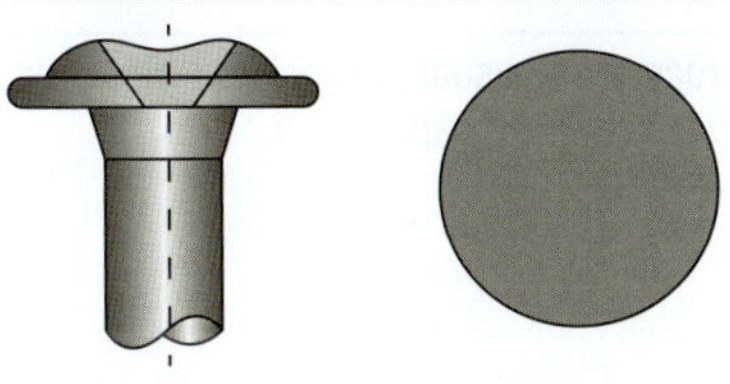	Wafer head	Usually incorporating a recess and most commonly found on self-drilling building fasteners.

(*Continued*)

TABLE 3.2 *(Continued)*

Sketch	Title	Comments
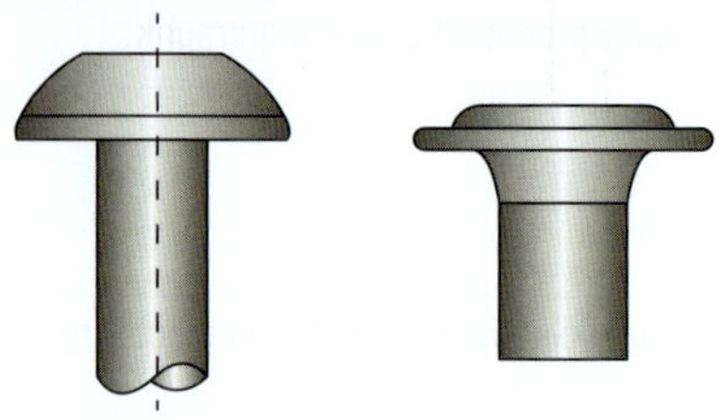	Button heads	Usually incorporating a recess and most commonly found on self-tapping, self-drilling screws and machine screws. Sometimes incorporates a post to convert to a security or tamper resistant recess.
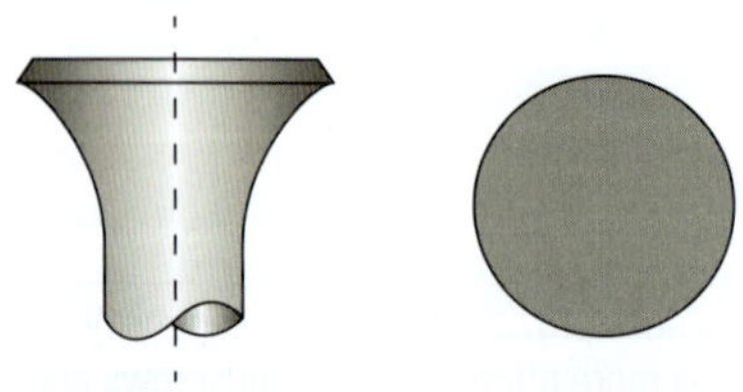	Bugle heads	Most commonly found on self-drilling screws used in plasterboard fixed to timber or steel frames. Designed to self-embed.
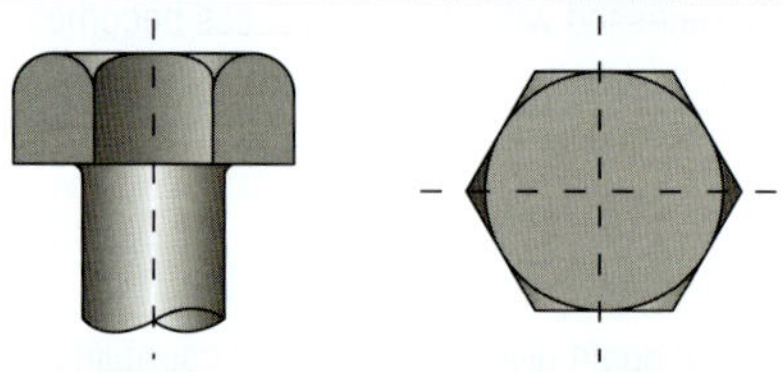	Hexagon heads	The most common head on bolt products and also seen on many screw products. Can come in several versions. This one is referred to as full bearing face. The manufacturer has the discretion to supply low tensile products as full bearing face.
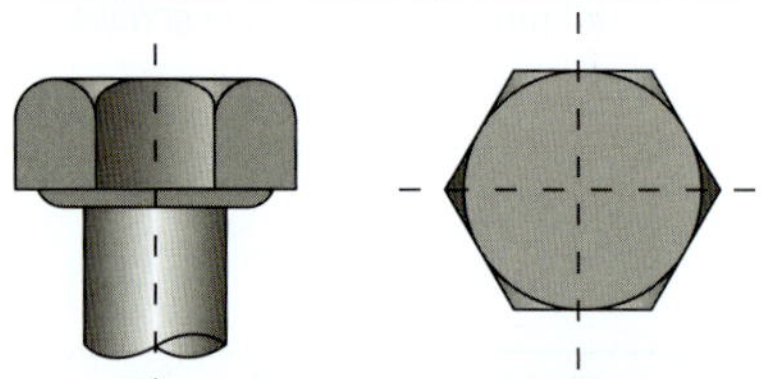	Hexagon washer faced head	Hex washer faced is specified for hexagon high tensile bolts and setscrews in the Australian Standard – both unified and metric. It is at the manufacturer's option for other products; therefore, it is not normally necessary to specify it except on specials. Note: USA market refers to bolts and set screw products as cap screws.
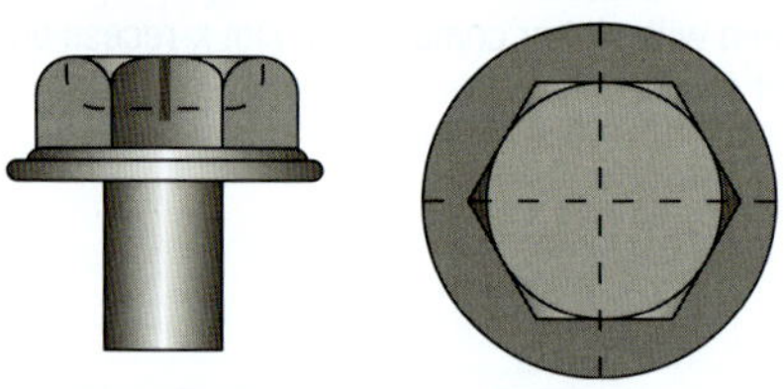	Hexagon flange head (Hex washer head)	On bolts, correctly termed Hex flange head and when used on self-drilling screws, correctly termed Hex washer head. Beware not to confuse with Hex washer faced.
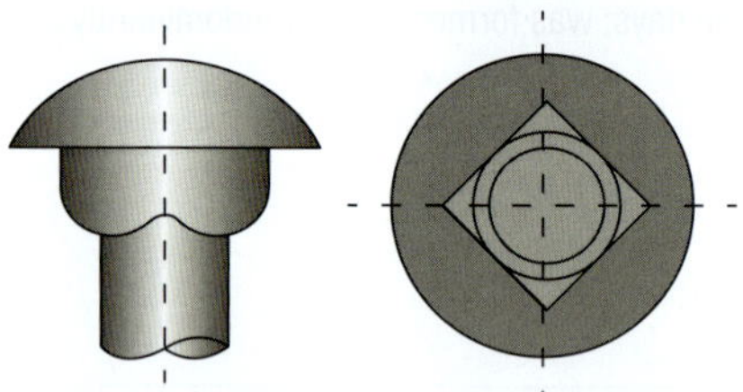	Cup square head (Coach head)	Normally referred to simply as cup head, this standard product is predominantly used in timber applications or very occasionally, in steel with a square punched hole, e.g. steel framed wheelbarrows.
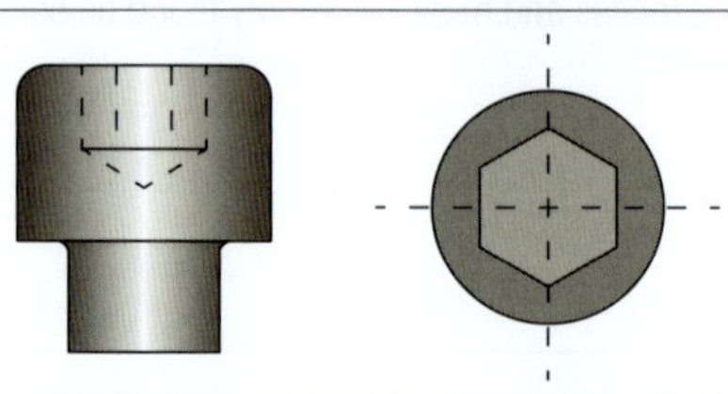	Cap head	Normally incorporating a recess and usually associated with very high tensile products, e.g. socket head cap screws.
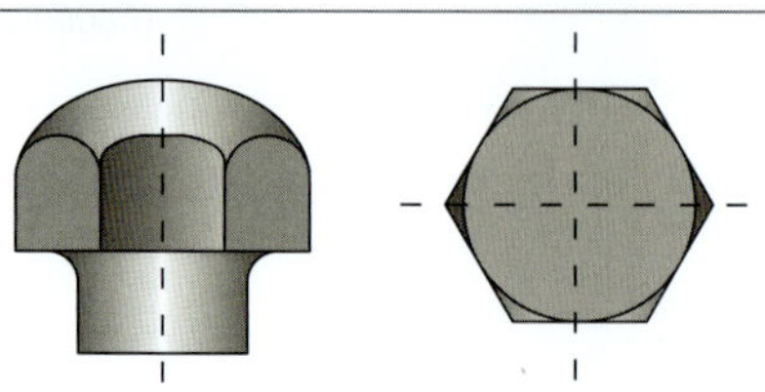	Domed hexagon head (Track head)	Sometimes found on special bolts for engine heads, but more often on earthmoving equipment as crawler track retaining bolts.

(Continued)

Sketch	Title	Comments
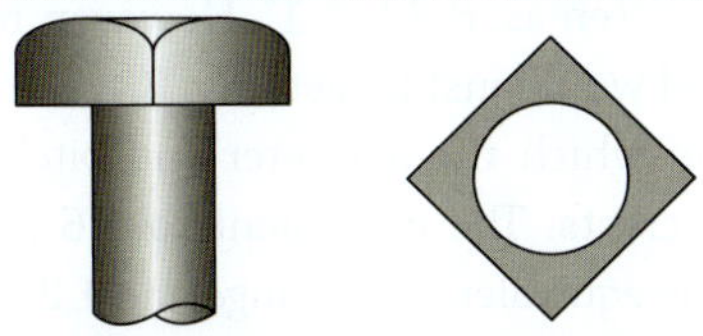	Square head	Rarely seen now except on some mining specials with very oversize square across flats. Were used on timber bridgework and where larger bearing surface was required. Modern cold forming methods producing flange heads have hastened its demise.
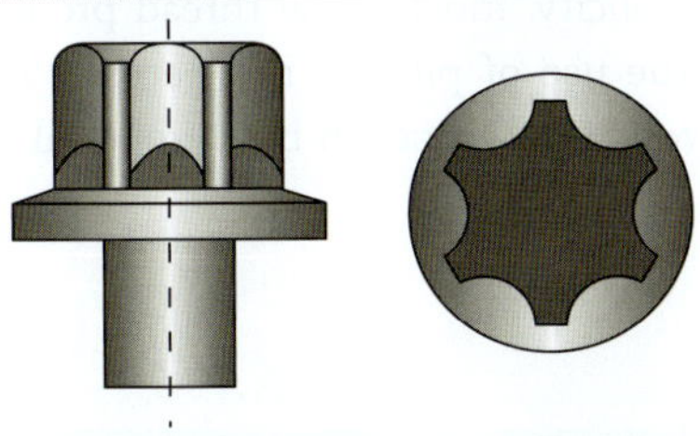	External Torx® flange	Normally confined to special high tensile products in high torque applications.
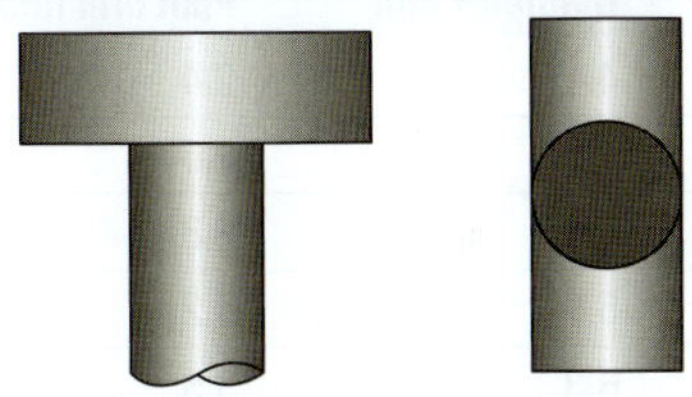	Tee head	Occasionally used as a bolt head which is retained in a channel or where the space for a head is restricted and narrow. The head is usually, therefore, retained stationary and tension is achieved by rotating the mating part, e.g. a nut.

Source: James Glen, 'Head styles', http://www.jamesglen.com/head-styles

All machine screw threaded products, bolts or screws have common technical terms as shown in Figure 3.1. When referring to the thread:

- lead or start of thread is the point at which the thread groove is visible
- threaded portion is the total section of the screw on which there is a thread
- thread run out is the point at which the thread and the plain shank meet.

FIGURE 3.1 Terms used with machine screw threaded bolts and screws

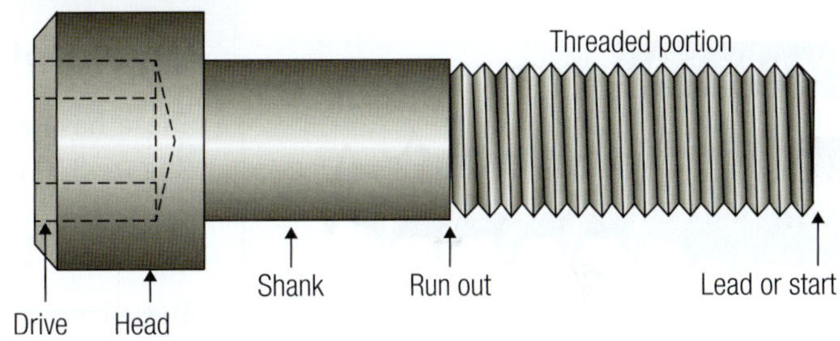

Threads

A thread is a sloping plane curled around a cylinder. External threads are on bolts or screws. Internal threads are on nuts. There are many forms of threads but two types are in common use on fasteners.

Spaced threads (timber)

Basic measurements are *major* (nominal) diameter, *minor* (root) diameter and *pitch* as shown in Figure 3.2. The major diameter can be measured with a caliper rule or slot gauge accurately enough to determine nominal diameter. The measurement is taken on the crests. The minor diameter and the pitch require special measuring equipment for technical accuracy. However, simple measurement at the crests will be accurate enough for most practical purposes in measuring pitch and determining thread designation.

The diameter of spaced threads is expressed as a gauge or a number (#). The pitch of imperial spaced threads is

FIGURE 3.2 Spaced threads

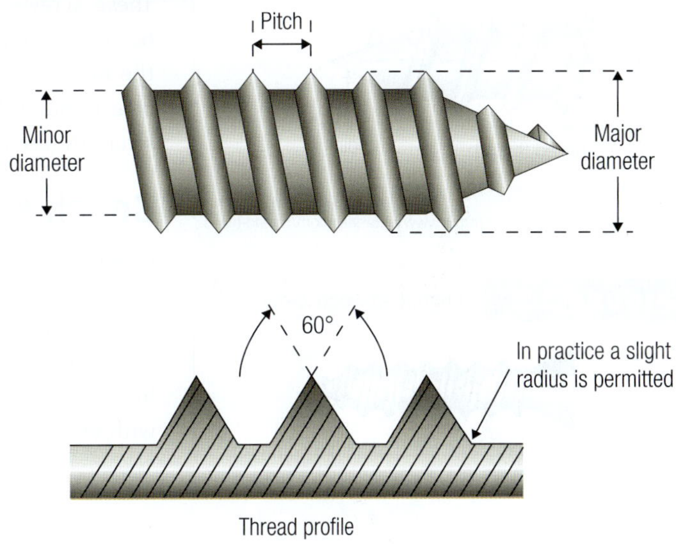

expressed as threads per inch (TPI). Screws are sized by gauge number and length. For example, an 8 gauge screw with 32 threads per inch and 1" in length would be written as: 8-32 × 1". However, most wood screws do not include the threads per inch measurement and would just be listed as 8 × 1".

Metric spaced threads are a soft conversion from imperial in which the diameter and pitch are expressed in millimetres, the pitch being the distance between crests. The equivalent to a 6 gauge 20 TPI imperial screw would be 3.5 × 1.27 mm where 3.5 mm is the equivalent of 6 gauge and 1.27 mm is the equivalent pitch for a thread of 20 TPI. To maintain simplicity, most screw thread products continue to be referred to in their imperial designations and the use of pitch is not necessary for standard woodscrews or self-tapping screws. Table 3.3 shows wood screw sizes in both imperial and metric equivalents.

TABLE 3.3 Wood screw sizes

Gauge number	Diameter mm	Pilot drill mm	Gauge number	Diameter mm	Pilot drill mm
0	1.6	–	8	4.17	2.4
1	1.78	–	9	4.52	2.6
2	2.08	–	10	4.88	2.8
3	2.39	1.5	12	5.59	3.2
4	2.74	1.6	14	6.3	3.5
5	3.1	1.8	16	7.01	3.8
6	3.57	2.1	18	7.72	4.2
7	3.81	2.3			

FIGURE 3.3 Wormed thread

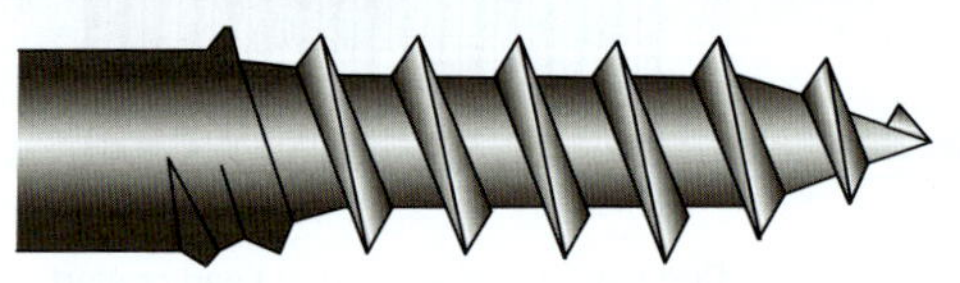

Source: James Glen, 'Thread types', http://www.jamesglen.com/thread-types

Wormed or cut thread woodscrews

These have sharply defined threads on a tapering shank to a gimlet point (Figure 3.3) and are used for general purpose timber fixing.

Rolled thread woodscrews

These have a spaced thread, a rolled taper point and are designed for quality cabinet making, furniture and joinery (see Figure 3.4).

FIGURE 3.4 Rolled thread

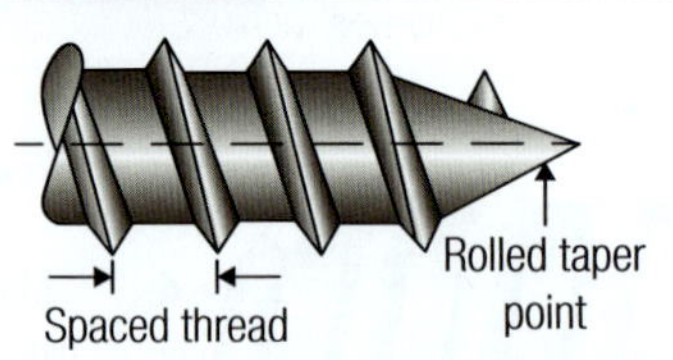

Source: James Glen, 'Thread types', http://www.jamesglen.com/thread-types

Longthread woodscrews

These screws have the same spaced thread as rolled thread woodscrews; however, the thread extends the full length of the screw to the underside of the head. These are designed for use in composite timbers such as particle board and craftwood, giving longer thread engagement and higher pull-out strength (see Figure 3.5).

FIGURE 3.5 Longthread woodscrew

Coach screws

Coach screws are designed to form their own thread in pre-drilled holes in timber (see Figure 3.6). They incorporate a woodscrew type rolled spaced thread which is dimensionally a soft conversion from imperial. However, designation of the size or nominal diameter is in millimetres, e.g. M6, M8, M10, and the hexagon dimensions are the same as for hexagon metric commercial bolts. As with woodscrews, there is no necessity to designate pitch in the description.

Coach screws are used for heavy duty fixing to timber. They have a hexagon head and therefore need a spanner to tighten them.

FIGURE 3.6 Coach screw

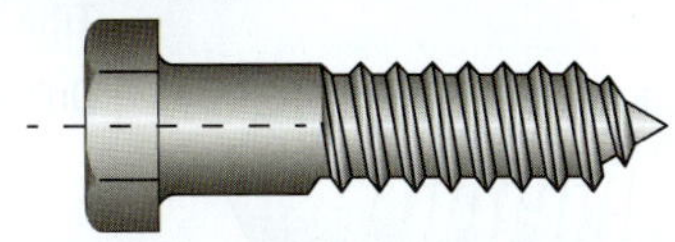

Self tapping screws

These are designed to cut a matching thread in the materials being joined, usually into pre-drilled or pre-punched holes in sheet metals. They are heat treated and hardened, are sometimes used with spring steel clips that act as nuts. They can also be used in aluminium castings, plywood, soft and high impact plastics and zinc die castings (see Figure 3.7).

FIGURE 3.7 Self tapper thread

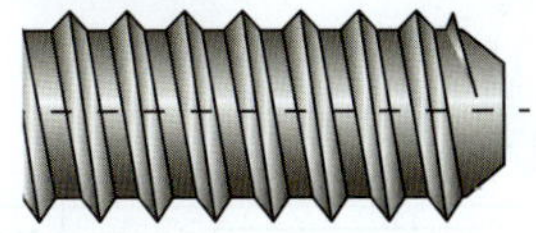

Self drilling

Tekscrews are designed for fixing sheet metal, fibrous cement sheet, aluminium sheet or timber panelling to timber supports (see Figure 3.8). Drilling of either the sheets or the supports is not necessary; the gimlet point will pierce and self drill the sheet and the milled slot will cut a pilot hole and thread while drilling.

FIGURE 3.8 Self drilling thread

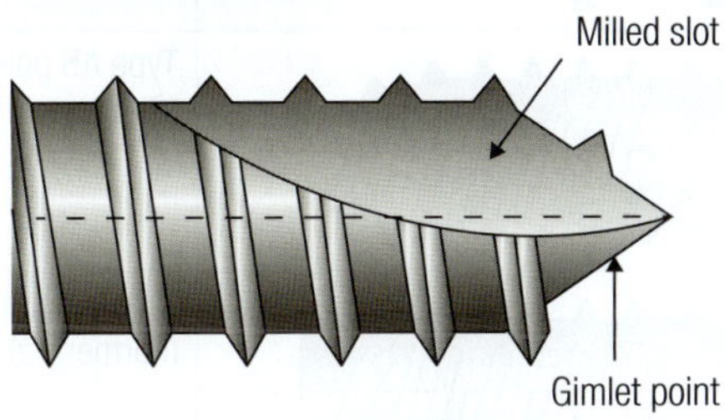

Points

The point of a screw or bolt can be used to locate the centre of the hole or nut it is being fixed to. Drill or form a hole for the fixing or in the case of a blind hole (a hole that does not go all the way through) fit flush against the bottom of the fixings hole. Table 3.4 shows point types.

TABLE 3.4 Point types

Sketch	Title	Comment
	Chamfer cut point (Boltmaker point)	The normal point found on most good-quality hexagon bolts/set screws and cap screws. The chamfer is applied in a pointing station on a boltmaker prior to ejection through the thread rolling plates.
	Rolled point	A point that is found mostly on machine screws and cup head bolts, where the cold header product is simply roll threaded with no special attention to the end of the blank shank.
	Radius point (Oval point) (Round point)	A radius point is kind to the surface of a mating part on which it bears. It may be found on a special screw which is being retightened regularly.
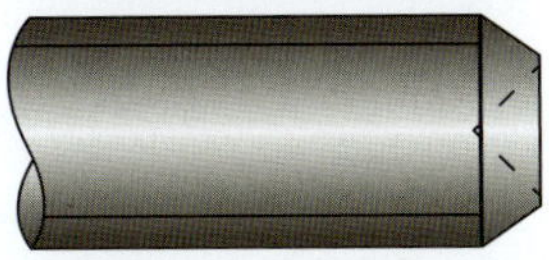	Cup point	Another point designed to bear on a mating part. Normally in a low tension assembly where the point contact assists retention. Sometimes has knurls on the tip circumference, known as knurled cup point.
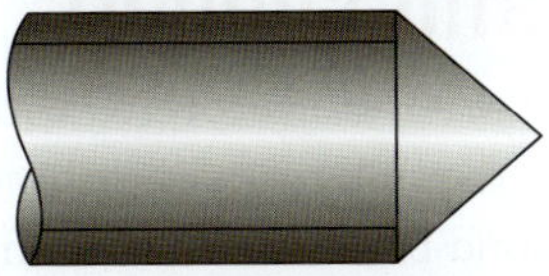	Cone point	Designed to positively locate a mating part by engaging in a drilled or prepared indentation. Only found on specials.
	Dog point	As with cone points, this point locates a mating part. A dog point would tend to be more accurate and withstand greater load. Again, specials only.

(*Continued*)

TABLE 3.4 *(Continued)*

Sketch	Title	Comment
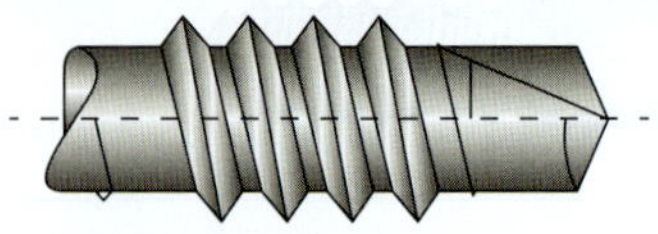	Drill point	A point very similar to a standard drill and designed to drill then tap or form a thread while driving. Some are milled points, some pinch pointed in special cold headers. There are a number of proprietary versions – Drill Kwik®, Teks®, Pias®, etc.
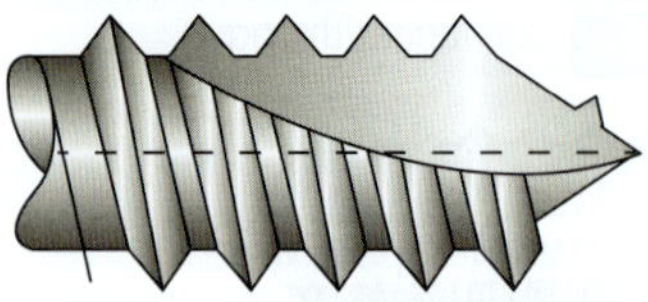	Type 17	Designed to pierce and self-drill sheet metal and drill timber supports, cutting threads while driving. The slot is applied by a milling shank slotter after thread rolling.
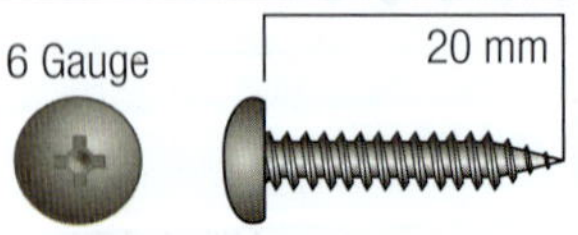	Needle point	A very sharp point made with special heading dies. Used to pierce light gauge sheet metals for self-tapper type applications without having to drill. Very often used in construction of steel-framed housing to secure plasterboard.
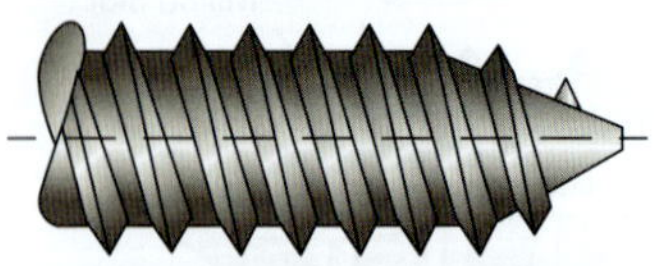	Type AB point (Gimlet point)	When on a self-tapper, it is called 'AB', on a woodscrew or coach screw is called 'gimlet'. A die produced point in the primary cold header, thread rolled leaving a thread start on the taper. This helps pull the screw into the hole and start the tapping groove.
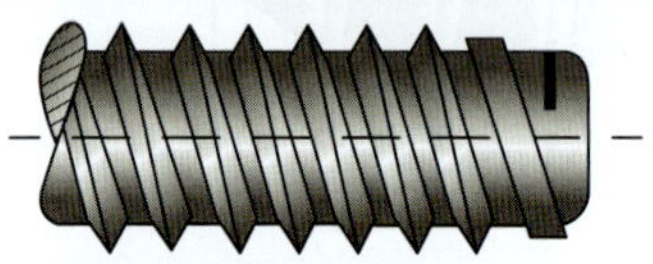	Type B point (Formerly z)	Generally only found on self tapping screws where it is either going into a blind hole where it will locate square more easily and give longer thread engagement, or where the point may protrude into an area where it could snag something or someone.
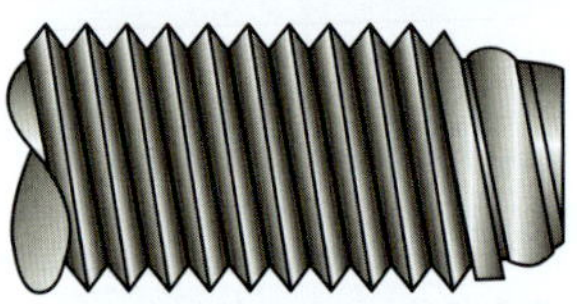	Taptite® (Type TT)	Tri-lobular point and body. Thread forming – refer to machine screw thread forms. Note other thread forming screws could include square-flow, and type F, each with similar point treatment.
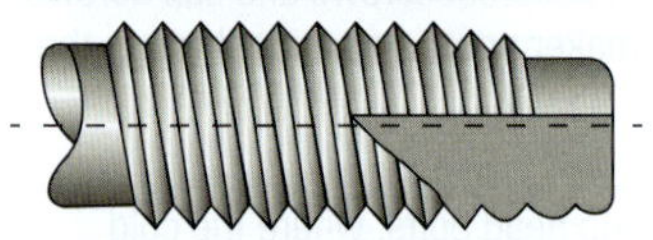	Type 23	Milled slot point, machine thread screw, thread cutting – refer to machine screw thread forms.
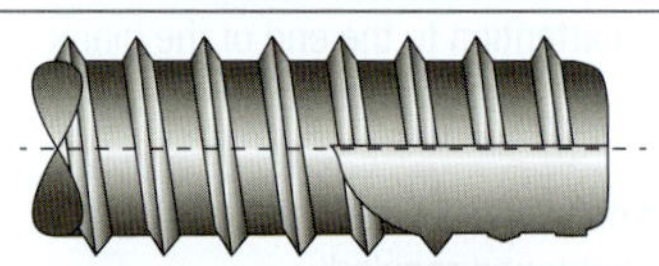	Type 25	Milled slot point, spaced thread screw, thread cutting – refer to spaced thread forms. Note: other spaced thread forming screws could include type BF and type Y.
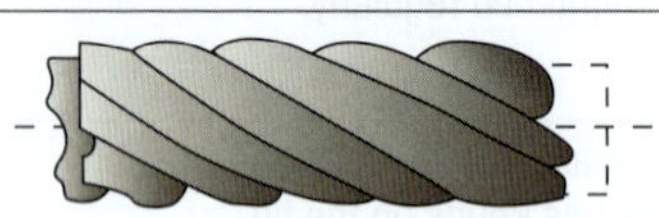	Type U (Hammer drive)	Very similar to a dog point to give a positive location in the prepared hole.

Source: James Glen, 'Point types', http://www.jamesglen.com/point-types

Types of devices for hollow wall fixing and support

Hollow walls are a cost-effective construction generally formed of timber or metal studs covered with plasterboard or other type of sheeting material. The plasterboard is secured to the studs with bugle head screws or clout nail. Adhesive is also used to increase the strength of the bond between the sheeting and the studs.

As described on the Buildex website ('Plasterboard screws', http://www.itwproline.com.au), the bugle shaped head countersinks neatly into plasterboard without crushing the core, or tearing the face paper. This device is also suitable for fixing accessories and general fixings such as saddles to timber and thin steel studs up to 0.6 mm thick, e.g. walls and ceilings, partitions, etc. For fixing plasterboard to steel from 0.8 mm to 2.5 mm thickness, self drilling plaster screws as shown in Figure 3.9 are used.

Hollow wall anchor devices are shown in Figures 3.10 to 3.14 and include the:

- hollow wall anchor
- WallMate™ plasterboard anchor
- toggle plasterboard anchor
- nylon toggle bolt
- metal toggle bolt.

The anchors reinforce the strength of the plasterboard and prevent the mount screws from pulling away from the plasterboard.

Always check that there are no cables or pipes hidden in the wall behind the area in which the anchor is to be inserted. The load to be supported by any of the anchors shown in Figures 3.10 to 3.14 should generally not be more than 2 kg as the plasterboard wall or ceiling may collapse.

FIGURE 3.9 Plasterboard is secured to wall studs using bugle head screws or clouts

Source: Ed Hawkins

FIGURE 3.10 Hollow wall anchor

Source: Ed Hawkins

FIGURE 3.11 WallMate™

Source: Ed Hawkins

FIGURE 3.12 Toggle plasterboard anchor

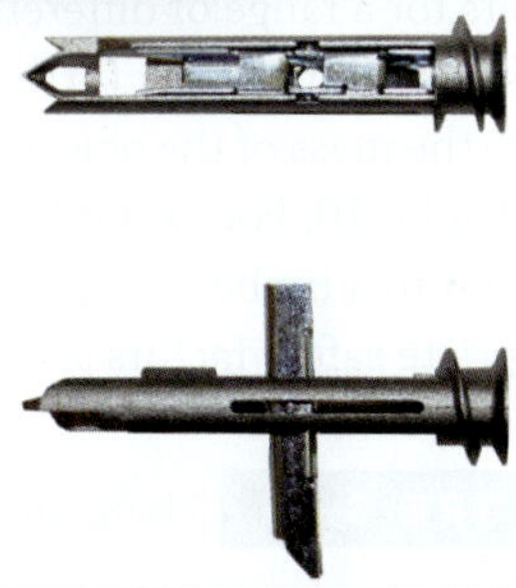

Source: Phill Chadwick

FIGURE 3.13 Nylon toggle bolt

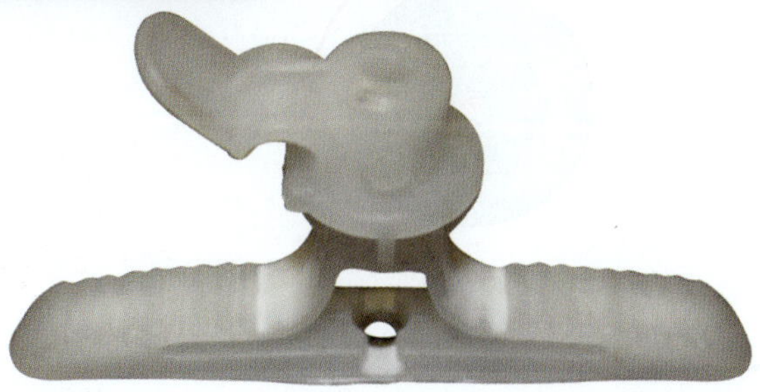

Source: Ed Hawkins

FIGURE 3.14 Metal toggle bolt

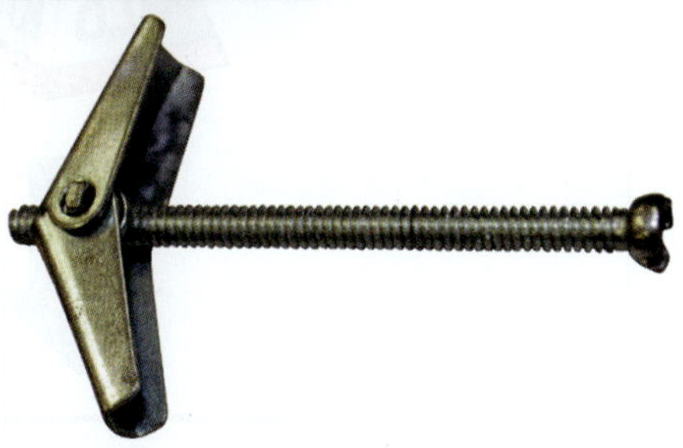

Source: Ed Hawkins

FIGURE 3.15 Common fixing ratings

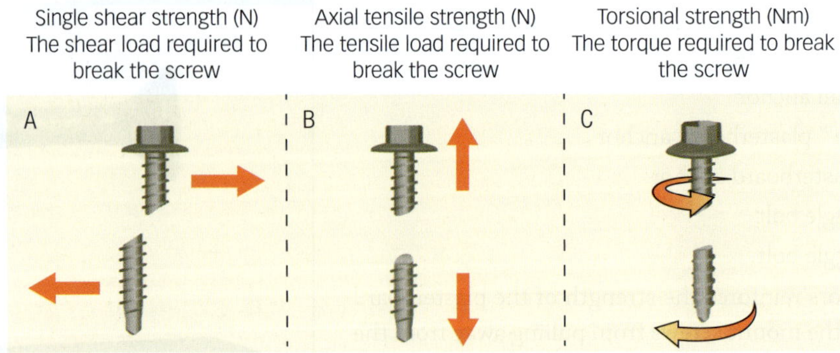

Heavier equipment must be secured directly to the timber or steel studs to prevent damage to the plasterboard. To locate a stud, use a stud finder, or gently tap the surface until you hear a muffled sound, which will indicate that there is solid timber behind. The distance between studs can vary between 450 mm to 600 mm. Screws are typically driven using a suitable tool through the plasterboard and into a wall stud, providing a very secure fixing.

Ratings of fixings

The load that a fixing can support is determined by the diameter of the fixing device; common ratings are shown in Figure 3.15. Ratings (in newtons N or newton metre Nm) include:

- single shear strength (N) – the shear load required to break the screw
- tensile strength (N) – the tensile load required to break the screw
- torsional strength (Nm) – the torque required to break the screw when turned
- pullout load (N) – the tensile load required to pull the fastener out of the material it is secured into.

The strength of the material the fixing is attached to must also be considered and in most cases will be the limiting factor. Manufacturers provide detailed technical specifications on the loading of fasteners for a range of different materials.

> The load bearing capacity of a fastener is around one tenth its pullout rating.

Figure 3.16 provides further explanation of fixing rating terms. Force in a downward direction is equal to the mass of the object being secured times 9.81 (gravity). To convert to mass the pullout rating is divided by 10. For example, if a 10 gauge 16 TPI screw has a pullout rating of 1860 newtons when screwed into a timber stud, it would pull out of the timber with a static load of approximately 168 kg. Appropriate safety factors should be applied for design purposes.

FIGURE 3.16 Fixing ratings

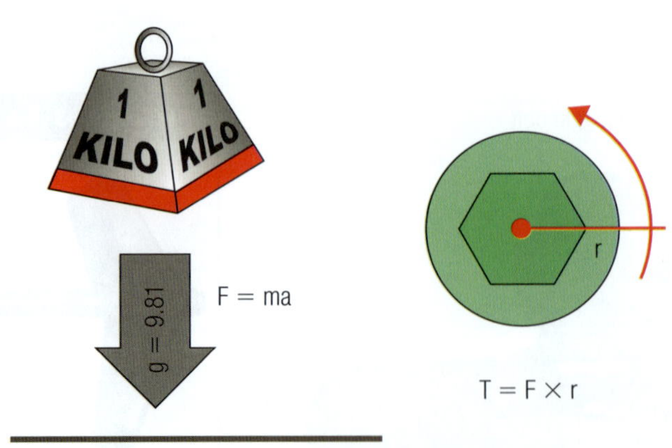

A falling object dramatically increases the load on the fastener. An object falling just 1 metre will increase the load on the fixing device by approximately 10 times.

Plaster and wall brackets

The plaster bracket as shown in Figure 3.17 is used where there is no wall stud or structural member available for support. The mounting screws sandwich the plasterboard between the wall plate and the bracket.

Wall brackets, such as that shown in Figure 3.18, must be fitted at the 'rough in' stage of construction, before the plasterboard is installed. They are nailed or screwed to the wall stud.

FIGURE 3.17 Plaster bracket or C-clip

Source: Ed Hawkins

FIGURE 3.18 Wall bracket, fitted before plasterboard is installed

Source: Ed Hawkins

Fixing techniques

Bolts and screws

Self drilling (tech) screws, as the name implies, can be drilled directly into sheet metal and timber. Thicker metal and larger bolts and screws will require pilot holes. The location of the hole is marked and centre punched before drilling.

The diameter of the pilot hole should be slightly larger than the minor diameter of the screw but smaller than the major diameter, as shown in Figure 3.19.

Hit nails (hollow wall)

Figure 3.20 shows the installation of hit nails. The Ramset 'Installation guide' (http://www.ramset.com.au) describes the following steps:

1 Drill a hole to the recommended diameter.

2 Tap the anchor into the hole until it is flush with the surface of the substrate.

FIGURE 3.19 Pilot hole size

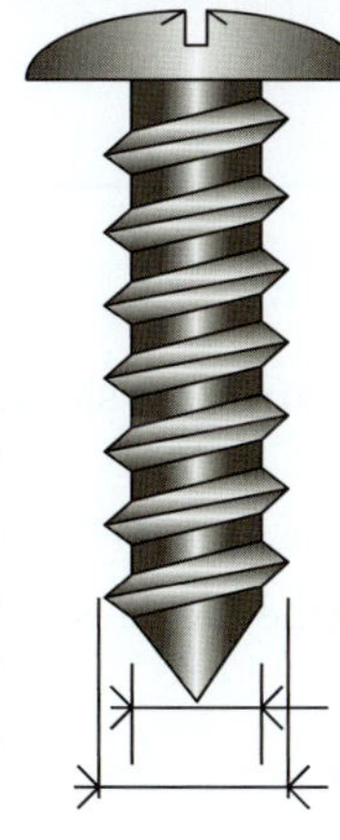

FIGURE 3.20 Installation of hit nails

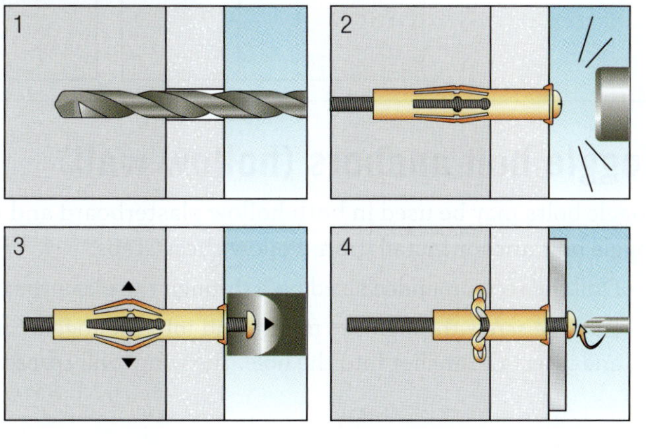

3 Attach the setting tool to the head of the screw. Squeeze the trigger of the setting tool to expand the anchor. (Small sizes may be set using a screwdriver.)

4 Remove the screw and place it through the fixture then back into the anchor. Tighten the screw until the fixture is fixed.

Wallmates (hollow wall)

Figure 3.21 shows the installation of devices known as Wallmates™. The Ramset 'Installation guide' (http://www.ramset.com.au) sets out the following steps:

1 Insert the point of a #2 Phillips head screwdriver into the recess of the WallMate™, place the drill point against the wall, lightly piercing the face paper, and screw the anchor into the plasterboard until the head is flush with the surface.

2 Place the fixture over the anchor, insert the appropriate gauge screw through the fixture, into the head of the WallMate™ and tighten the fixture in place.

FIGURE 3.21 Installation of Wallmates™

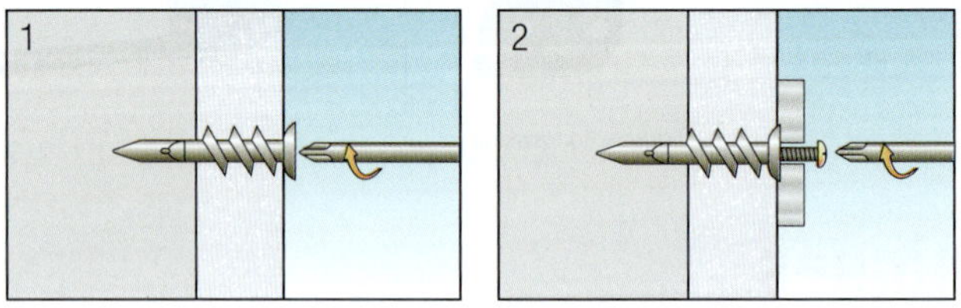

Toggle plasterboard anchors (hollow wall)

Figure 3.22 shows the installation of a toggle plasterboard anchor, using the following steps:

1 Insert the point of a #2 Phillips head screwdriver into the recess of the toggle plasterboard anchor, place the drill point against the wall, lightly piercing the face paper, and screw the anchor into the plasterboard until the head is flush with the surface.

2 Place the fixture over the anchor, insert a suitable size screw (typically 8 gauge) through the fixture, into the head of the toggle plasterboard anchor and turn to engage the clamp. Continue to turn the screw until the clamp is pulled tight against the plasterboard.

FIGURE 3.22 Installation of toggle plasterboard anchors

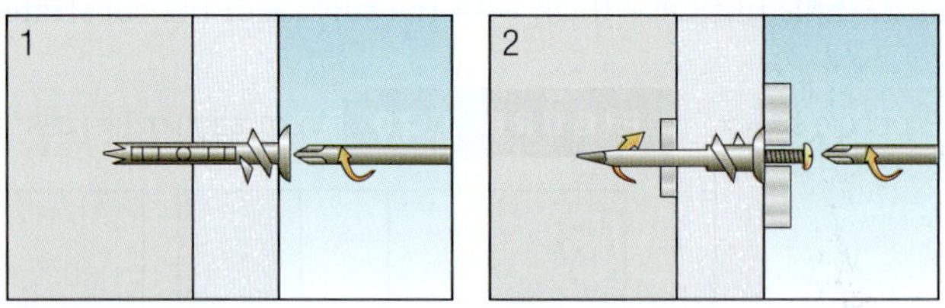

Toggle bolt anchors (hollow wall)

Toggle bolts may be used in both hollow plasterboard and masonry block walls. Figure 3.23 shows toggle bolt anchor installation. Follow these steps:

1 Drill the recommended sized hole through the plasterboard or masonry block.

2 Pass the screw through the fixture and into the toggle for a few turns. Compress the toggle wings and insert the anchor into the hole. The wings will spread behind the cavity wall.

3 Pull back on the fixture to hold the toggle in place against the reverse side of the cavity wall and tighten with a screwdriver, pulling the wings of the anchor firm against the cavity wall.

FIGURE 3.23 Installation of toggle bolt anchors

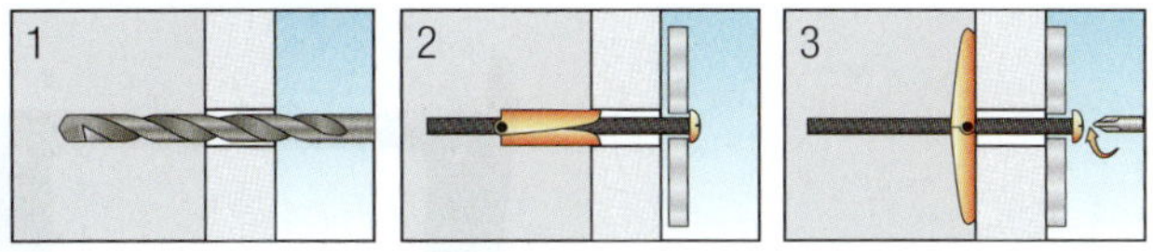

Plaster brackets (hollow wall)

Before marking the wall or cutting any holes, check that the location on the wall for the plaster bracket is free of wall studs, piping or other obstructions. Figure 3.24 shows the installation instructions.

FIGURE 3.24 Installation of plaster brackets

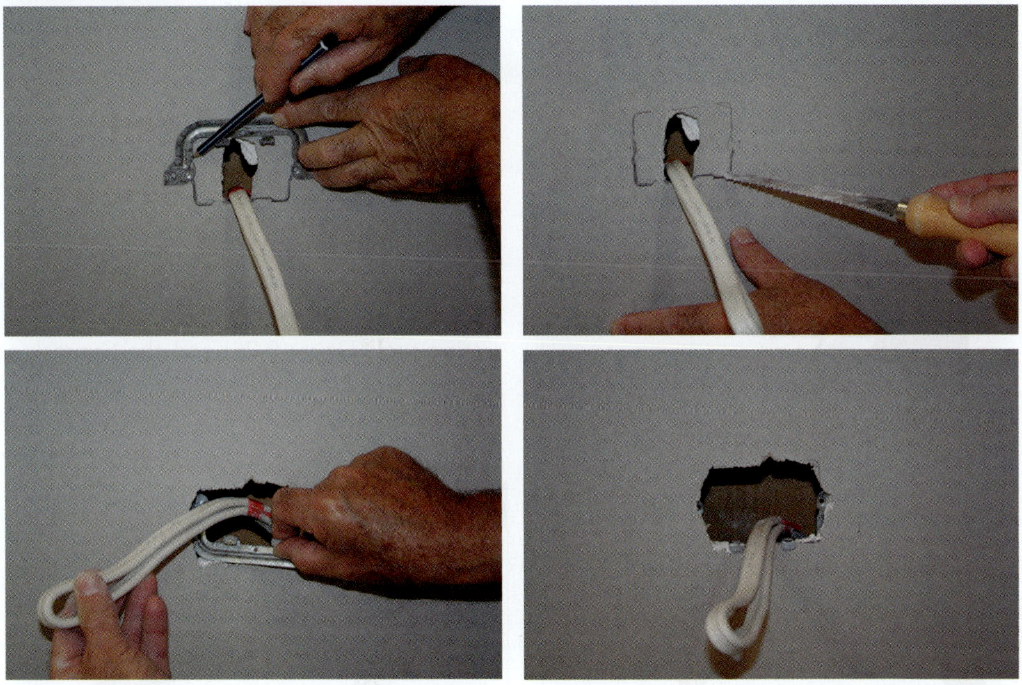

Source: Phill Chadwick

1 Mark on the wall with a pencil using the inside of the plaster bracket as a template. Mark the wall though the screw holes, then flip the bracket 180°, using the screw hole markings to centre the bracket. Mark the second half of the hole.
2 Starting in one corner of the markings, using the palm of your hand to push a Gyprock saw through the plasterboard. Complete cutting out the rectangular hole. Notches are then made in the plaster for the screw holes.
3 The main part of the bracket is inserted through the hole, so the screw holes in the bracket end up on the inside of the plaster and the two little metal tabs clip over the plaster through the hole and secure it at the bottom edge of the hole. (Tip: tie a bit of string onto the bracket so you can fish it out if you drop it in the process.)
4 Any standard wall plate accessory can then be screwed into place and checked with a spirit level.

Wall brackets (hollow wall)

Wall brackets must be fitted during the 'rough in stage', before the plasterboard is installed. Figure 3.25 shows the installation of a wall bracket for a power outlet.

1 From a floor plan or wiring schedule select and mark a wall stud to which to attach the wall bracket. Sockets are installed at approximately 350 mm from the floor to the centre of the wall bracket.

FIGURE 3.25 Installation of wall brackets

Source: Ed Hawkins

Brackets for light switches are typically installed at between 1200 to 1300 mm from the floor to the centre of the bracket.

2 Wall brackets are secured into place with nails or clouts. Cables are clipped or taped to a stud to hold them in place.
3 After the plasterboard is installed the wall bracket is located with a stud finder, strong magnet or from dimensions of a plan.
4 A Gyprock saw is then used to push through the plasterboard and locate an inside corner of the bracket. Expand the hole to a rectangular shape using the inside of the bracket as a guide.

FIGURE 3.26 Hollow wall anchor setting tool

Source: Ed Hawkins

Types of tools used for hollow wall fixing and supporting

Commonly used tools associated with plasterboard walls are the hollow wall anchor setting tool (Figure 3.26) and the Gyprock saw (Figure 3.27). A Gyprock saw has teeth that cut in both directions.

FIGURE 3.27 Gyprock saw

Source: Ed Hawkins

SUMMING UP...

- Screws and bolts are classified by their drive method, head type, thread and point.
- The most common drive types are slot (flat) and Phillips head.
- Screws for timber and plasterboard use spaced threads while metal screws use machine threads.
- Screw thread measurements:
  - major diameter, the outer diameter – clearance hole size
  - minor diameter, the inner diameter – pilot hole size
  - pitch – distance between crests.
- Common hollow wall fasteners include clouts, bugle headed screws, hollow wall anchors (hit nails), wall mates, toggle plasterboard anchors, metal and nylon toggle bolts.
- Fixings are rated by:
  - single shear strength in newtons (N)
  - axial tensile strength in newtons (N)
  - pullout load in newtons (N)
  - torsional strength (torque) in newton metres (Nm).
- Plaster and wall brackets provide a more secure support than hollow wall anchors.

3.2 Devices for securing and mounting electrical equipment in solid walls

Solid wall construction methods are found in commercial, industrial and residential buildings, and provide superior strength and speed of construction compared to hollow wall construction. Fixing devices are used to assemble components of the building itself as well as services such as plumbing and electrical. Figure 3.28 shows a typical solid wall construction.

The type of solid wall fixing that is selected will depend on the type of material it is being attached to and the load it is required to support. For example, a solid concrete wall will support more than a hollow brick or masonry block wall. Figure 3.15 earlier in the chapter provides a guide to a fixing's rating. Corrosion resistance is provided on metal parts of fixings by the use of zinc plating, galvanising or grade 316 stainless steel.

FIGURE 3.28 Solid wall construction

Source: Ed Hawkins

FIGURE 3.29 Nylon frame anchor

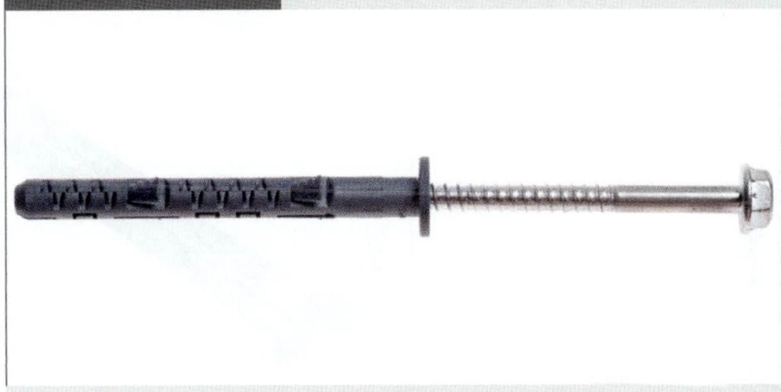

Source: Dreamstime/Nikolay Mikhalchenko

Light duty

Light duty solid wall fixing devices are designed for use in a variety of materials such as concrete, stone, solid brick, solid block, hollow brick, hollow block and hollow slab. These anchors take hold by either interference (a tight fit) or expansion once the insert is driven into place. Common solid wall anchors are shown in Figures 3.29 to 3.32 and include:

- nylon frame anchor
- nylon anchor
- metal drive anchor
- hammer-in anchor
- concrete nails
- plastic wall plugs (spaghetti when plugs are cut from one continual length).

The above solid wall anchors are suitable for a wide range of light duty electrical applications such as securing conduit saddles, plastic junction boxes and lightweight cable supports.

FIGURE 3.30 Metal drive anchor

Source: Ed Hawkins

FIGURE 3.31 Concrete nail

Source: Ed Hawkins

FIGURE 3.32 Wall (green) plugs

Source: Ed Hawkins

FIGURE 3.33 Screw in anchor

Source: Ed Hawkins

FIGURE 3.34 Sleeve anchor

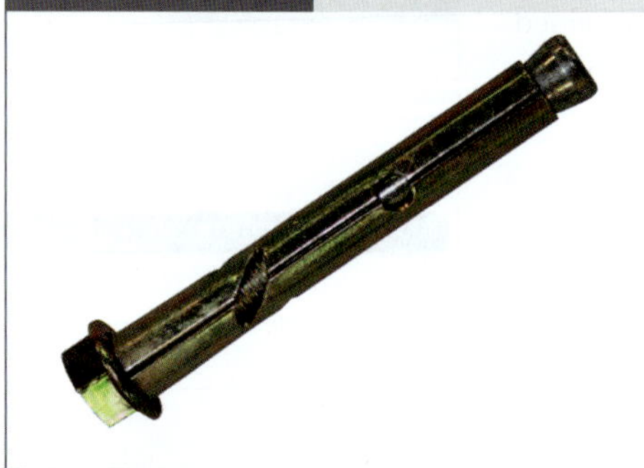

Source: Ed Hawkins

FIGURE 3.35 Shield anchor, medium duty, torque controlled, internally threaded, expansion anchor, also called Loxin™

Source: Ed Hawkins

Medium duty

Medium duty devices are available in thread sizes M4.5, M6, M8, M10, M12, M16 and lengths ranging from 26 mm to 157 mm. These devices are torque controlled, expansion type anchors, designed for medium duty anchoring of timber and steel fixtures to concrete, brick or block. The maximum ratings of fixings will rely on the class of concrete they are fixed to. The class of concrete is determined by its compressive strength (i.e. its ability to withstand pressure), measured in Mega Pascals (MPa).

Normal class concrete is rated between 20–50 MPa. Concrete used in house slabs, driveways, footings and footpaths is rated at 20 MPa or 25 MPa. Areas that will experience greater loads and traffic use 32 MPa, 40 MPa or 50 MPa rated concrete. Special class concrete with high strength ratings of 65 MPa, 80 MPa or 100 MPa is used for projects such as high-rise buildings. The ratings shown for common solid wall anchors in Figures 3.33 to 3.36 are when they are set in 32 MPa concrete:

- Screw in anchors (Figure 3.33) are suited to close to edge or close to anchor fixing, as the screw does not expand and therefore will not burst the surrounding material. Rated load maximum tensile strength ranges from 1.4 kN to 14.4 kN and maximum shear strength from 1.5 kN to 16.7 kN.
- Sleeve anchors (Dyna Bolts™) are shown in Figure 3.34. Rated load maximum tensile strength ranges from 2.7 kN to 16.6 kN, maximum shear strength from 2.5 kN to 15.6 kN.
- The shield anchor is shown in Figure 3.35. Rated load maximum tensile strength ranges are shown in Figure 3.15.
- Drop-in anchors (DynaSet™) have a rated load maximum tensile of 2.5–15.8 kN, with maximum shear of 2.2–13.1 kN.

Heavy duty anchors

Heavy duty anchors are available in thread sizes of M10, M12, M16 and M20 and lengths ranging from 75 mm to 170 mm. These devices are torque controlled, expansion type anchors, designed for high performance or high load applications (both static and dynamic) due to larger thread diameters. They are suitable only for concrete, as hollow brick and masonry block constructions do not provide the necessary support, i.e. the construction will not support the load on the anchor. The ratings for the common heavy duty solid wall anchors shown in Figure 3.36 are when set in 32 MPa concrete.

Drilled holes should be cleaned with a hand pump or vacuum so the dust does not cause pollution.

FIGURE 3.36 Stud anchor, rated load maximum tensile 3.4–32.5 kN, maximum shear 2.8–27.3 kN

Source: Ed Hawkins

FIGURE 3.37 Light duty anchor installation

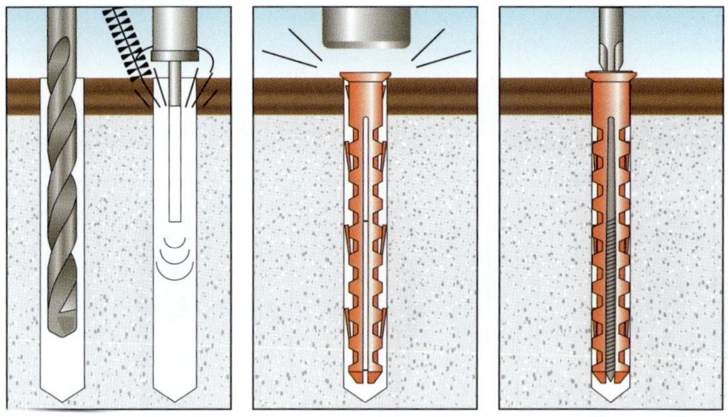

Fixing to masonry and concrete structures

Light duty anchor installation

Some light duty anchors are screw type and others use the driving of a nail to expand a plug. Figure 3.37 shows the common installation of plug type anchors. The Ramset 'Installation guide' (http://www.ramset.com.au) sets out the following steps:

1. Drill a hole using a masonry bit to the recommended diameter and depth using the fixture as a template. If the fixture thickness is less than the maximum, increase the hole depth accordingly. Clean the hole thoroughly with a hole cleaning brush. Remove the debris with a hand pump, compressed air or vacuum.
2. Insert the anchor through the fixture and tap with a hammer until the collar of the anchor contacts the fixture.
3. If the fixture is a screw type, tighten the screw until the head of the screw is flush with the collar of the anchor.

Figure 3.38 shows the typical installation method for wall plug type anchors:

1. Drill a hole to the appropriate diameter [as shown in Table 3.5] and depth. Remove the debris with a hand pump, compressed air or vacuum.

FIGURE 3.38 Wall plug anchor installation

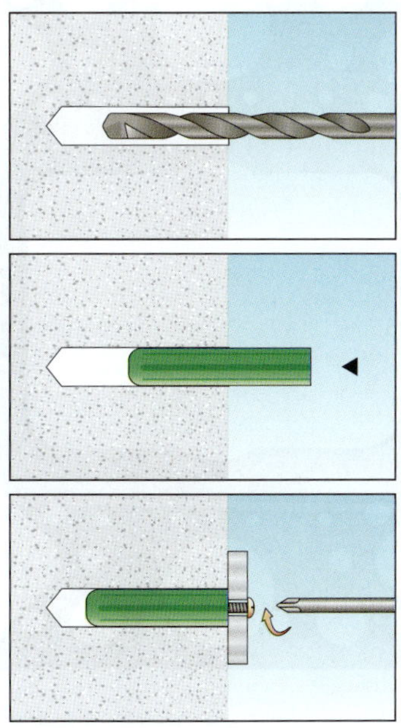

FIGURE 3.39 Heavy and medium duty anchor installation

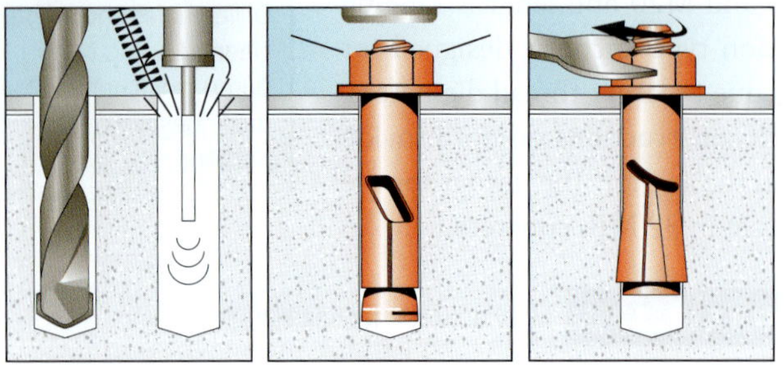

FIGURE 3.40 Drop in anchor installation

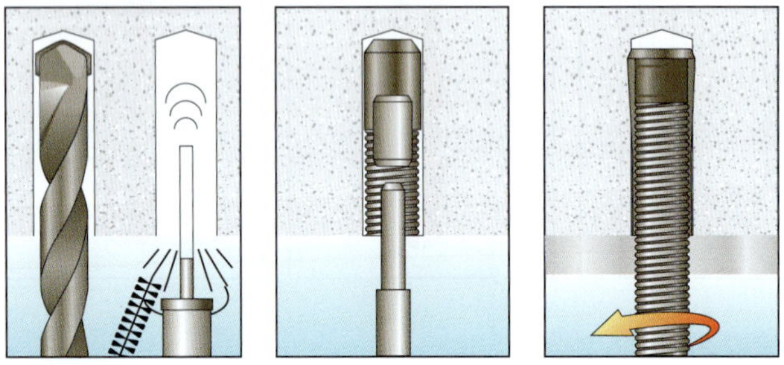

FIGURE 3.41 Masonry bracket

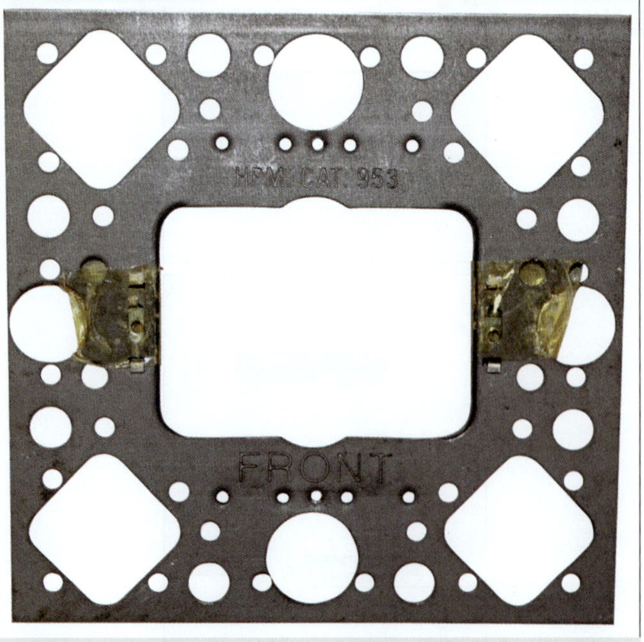

Source: Ed Hawkins

TABLE 3.5 Plastic wall plug anchor colour codes

Plug colour	Drill hole (mm)	Screw gauge (#)
White	5	4.5–6
Red	6	8–9
Green	6.5	10–12
Blue	8	14–16

Source: Adapted from Ramset, 'Plastic plugs & spaghetti', http://www.ramset.com.au/Product/Detail/75/Plastic-Plugs-Spaghetti

2 Insert the plug/spaghetti into the hole.
3 Screw the screw through the fixture into the plug/spaghetti until the fixture is firm against the substrate.

Heavy and medium duty anchor installation

A similar process is used to install heavy and medium duty anchors. Everything is simply upsized, and in the larger thread sizes a torque wrench is used to ensure the correct tension is applied. Figure 3.39 shows the installation of plug type anchors using the following steps:

1 Drill or core a hole to the recommended diameter and depth using the fixture as a template. Clean the hole thoroughly with a hole cleaning brush. Remove the debris with a hand pump, compressed air or vacuum.
2 Insert the fixing device through the fixture and drive with a hammer until the washer contacts the fixture.
3 Tighten the fixing device, allowing the sleeve to twist and pull down the fixture firmly onto the substrate. For optimum performance, a torque wrench should be used.

Figure 3.40 shows the installation of wall plug type anchors by way of the following steps:

1 Drill or core a hole to the recommended diameter and depth. Clean the hole thoroughly with a hole cleaning brush. Remove the debris with a hand pump, compressed air or vacuum.
2 Insert the anchor into the hole. Using a hammer and the appropriate setting tool, drive the expansion plug into the anchor until the shoulder of the tool rests against the surface of the anchor.
3 Position the fixture, insert the bolt or threaded rod and tighten with a spanner.

Fixing devices used with solid wall structures

A number of products have been designed specifically for the electrotechnology industry to fix cables or conduit directly to concrete, or aid in the construction of cable support systems such as cable trays and ladders. Masonry brackets, such as the example in Figure 3.41, are used to provide recessed mounting of accessories such as light switches and socket outlets to solid walls. Secured in place using concrete nails over the top of rectangular holes 'chased' into solid walls, cement render is then applied to conceal the bracket.

Cable tie systems, such as those in Figure 3.42, are attached using either any light duty anchor or clipped using a suitable gas actuated setting tool.

Suspension anchors are designed specifically for overhead threaded rod applications. Threaded rod is suitable for suspension of sprinkler pipe systems, metal and plastic pipes, steel ductwork, steel channel, cable trays and ladders. Suspension anchors can be used for overhead, horizontal and vertical applications and are suitable for close to edge or close to anchor fixing.

FIGURE 3.42 Cable tie system

Source: Ed Hawkins

Explosive powered tools

These tools are also known as actuated fixing tools, and are capable of rapid fixing to concrete and low carbon (mild) steel. They are not suitable for use with hard materials such as glazed bricks or tiles, marble, glass, slate, granite or high carbon (tool) steel, as these materials may shatter. These tools work in a similar way to a firearm where detonation of an explosive charge forces a projectile fixing into a work surface of concrete or steel. Extreme care must be taken when operating these tools as they are effectively a loaded gun.

Powder actuated tools

Powder actuated tools use a powder charge similar to the gun powder in a bullet. There are two classifications of powder actuated tools. In the older style *direct acting high velocity* tools the charge acts directly on the fixing. These are being replaced by *indirect acting low velocity* tools (see Figure 3.43), where the charge acts upon a captive piston within the device which transfers the energy to the fixing. These devices have fewer safety issues, with oversize charges pushing the fixing completely through the work surface, and ricochet.

FIGURE 3.43 Powder actuated tool: indirect acting low velocity type

Source: Ed Hawkins

Gas actuated tools

Gas actuated tools use a spark from a battery to detonate combustible aerosol gas within a chamber. The explosive force is transferred by a piston to the fixing. Gas actuated tools can attach hundreds of fixings from a single canister of gas which gives them a significant productivity advantage over powder actuated tools (powder actuated tools fitted with magazines can only hold around 10 charges). Both powder and gas actuated tools are activated by a trigger, so firing is only possible when the tool is cocked and pressed against a work surface.

Regulatory requirements for use of powder fixing tools

Persons conducting business or undertakings (PCBUs) are responsible to ensure that operators of powder actuated tools comply with Commonwealth, state and territory WH&S legislation, as outlined in Chapter 1 of this book. The Model Work Health and Safety Regulations (2011) is applicable to powder actuated tools and harmonises state and territory legislation. The relevant standard is the

Safety requirements with powder actuated tools

Do not blow dust from holes drilled in masonry with your mouth

series **AS/NZ 1873:2003 Powder actuated hand held fastening tools, fasteners, and explosive charges**. Regulatory requirements are summarised in **AS/NZS 1873.1:2003 Powder-actuated (PA) hand-held fastening tools – Selection, operation and maintenance**.

In New South Wales, effective from 1 February 2010, operators of explosive power tools are no longer required to hold a certificate of competency for the operation of powder actuated tools. This may not apply in all parts of Australia, or in New Zealand.

FIGURE 3.44 Rotary and demolition hammer drill

Source: Ed Hawkins

FIGURE 3.45 Masonry drill bits

Source: Ed Hawkins

FIGURE 3.46 Masonry chisel bits

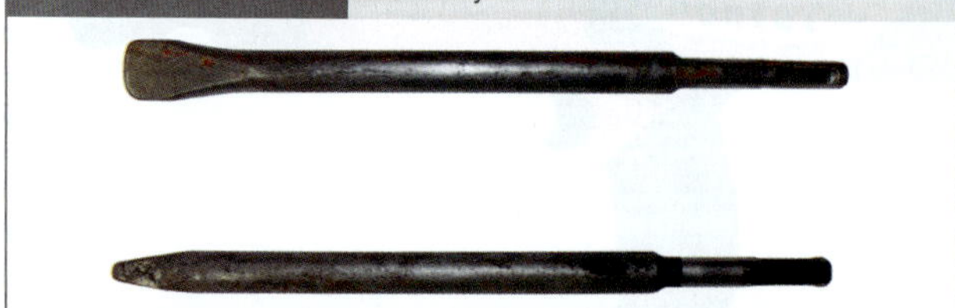

Source: Ed Hawkins

FIGURE 3.47 A 230 mm angle grinder

Source: iStockphoto/Difydave

Hand and power tools used in fixing and supporting

Manufacturers provide a number of specialised tools to assist in the installation of solid wall fixing devices. These are known as drop in anchor setting tools.

Reliable installation of fixing requires a close fit between the hole drilled in the work surface and the fixing device. Dust and other debris from drilling must be removed to allow sufficient clearance to allow installation of the fixing device. Dust and debris should not be removed by blowing directly into the hole as this presents a hazard to the eyes and lungs.

Conventional drills apply a turning force to the drill bit, while impact drills apply the force in short pulses. This mimics a sharp tug on a spanner to tighten or loosen a nut and bolt. Hammer drills apply a momentary forward driving force on the drill bit at the same time as rotating it. This mimics hitting the back of the drill bit with a hammer.

A rotary and demolition hammer drill is shown in Figure 3.44. These are suitable for drilling very hard work surfaces such as concrete. When drilling at speeds of up to 760 rpm, a hammer force is applied to the drill bit at up to 3600 hits a minute. In the demolition or chisel mode there is no rotation as only the hammer action is applied.

Masonry drill bits have a tough tungsten carbide chisel shaped head and spiral flutes to remove waste (see Figure 3.45).

High performance masonry drill bits and chisel bits are shown in Figure 3.46. These use an SDS chuck system that loosely translates from German to English to mean 'Insert – Twist – Secure'. It is a keyless quick change chuck system to fit a specially shaped tang of the bit. A small amount of special purpose grease is applied to the drill's SDS tang for lubrication to ensure effective operation and reliability of the chuck.

Angle grinders fitted with masonry blades are used to cut or 'chase' into solid walls. Solid wall construction does not provide a cavity to conceal wiring runs, so cables and accessories such as socket outlets and light switches must be chased into the brick masonry block or concrete surface.

Figure 3.47 shows a typical angle grinder suitable for concrete work fitted with a universal diamond cutting blade. Abrasive Type A is the choice for use into materials such as brick and masonry block. Larger grinders are fitted with electronic 'progressive starts' to reduce the kick when starting.

Specialty wall chasing tools used to 'chase' the slots for cable runs have two blades and are adjustable to cut a groove between 2.6 mm and 46.5 mm width to a maximum depth of 50 mm. Some tools use an attachable vacuum cleaner for dust suppression; other tools sometimes use a hose attachment and water.

Dust produced by a chasing tool should be collected rather than letting it pollute the air.

SUMMING UP...

- Typical light duty solid wall anchors include:
  - nylon frame anchors
  - nylon drive anchors
  - metal drive anchors
  - hammer in anchors
  - concrete nails
  - plastic wall plugs and spaghetti tubing.
- Typical medium duty solid wall anchors include:
  - screw in anchors – suited to close to edge or close to anchor fixing
  - sleeve anchors (Dynabolts™)
  - shield anchors (Loxin™)
  - drop in anchors (DynaSet™).
- Typical heavy duty solid wall anchors include:
  - safety anchors (SpaTec™ Plus)
  - coil expansion anchors
  - stud anchors.
- A simplified procedure for installing a typical solid wall anchor of any duty is to:
  - make or drill a hole to the recommended diameter and depth using the fixture as a template, and clean the hole thoroughly
  - insert the anchor
  - if the fixture is screw type, tighten the screw until the head of the screw is flush with the collar of the anchor.
- Typical fixing devices used in the electrotechnology industry for solid wall structures include:
  - masonry brackets – recessed mounting for accessories such as light switches and socket outlets
  - cable tie systems – securing cables within suspended ceilings
  - spring steel clips – securing cables and conduits to steelwork
  - suspension anchors – mounting steel work associated with cable tray and ladder supports.
- Actuated fixing tools (explosive power tools):
  - powder actuated – direct acting high velocity or indirect acting low velocity (the most modern and common type)
  - gas actuated – small explosion of air and gas forces a piston to drive fixing into the work surface.
- Typical hand and power tools used in the electrotechnology industry for solid wall structures include:
  - drop in anchor setting tool
  - hole cleaning brushes and pumps – do not blow out dust with your mouth
  - rotary and demolition hammer drills
  - masonry drill bits and chisels
  - angle grinders and wall chasers.

3.3 Devices for securing and mounting electrical equipment to metal

Typical fixing devices for metal include bolts, screws and rivets. Metal screws and bolts use a finer pitch than wood and plaster screws; this gives metal threads a greater mechanical advantage and the ability to achieve greater load ratings.

Threads

A thread is an inclined plane wrapped around a cylinder. When the threads of a nut and bolt are meshed and rotated in opposite directions the mechanical advantage produced creates a substantial clamping force that can be used to fasten a wide range of materials together. An advantage of nut and bolt fasteners over other joining methods such as welding, brazing, soldering and folding is the ease with which the process can be reversed.

FIGURE 3.48 Metal thread measurements

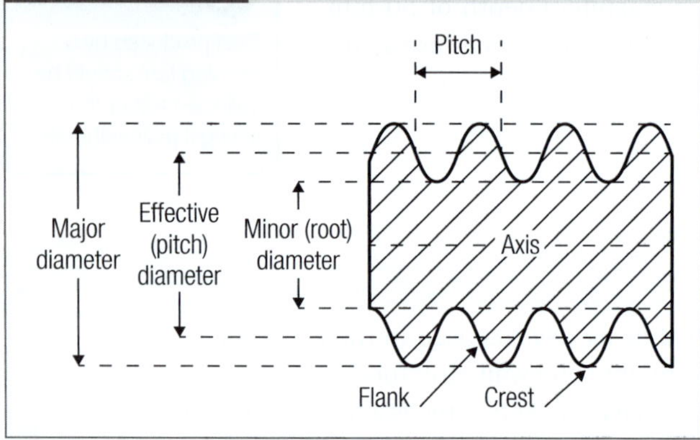

Machine screw threads (metal)

Metal screws are classified by their *drive method*, *head type*, *thread* and *point* as described for wood screws in Section 3.1 earlier in this chapter. The three commonly used threads are British Standard (England), unified (USA) and metric (Europe); all three have coarse and fine variations. Metric threads are now the most widely used due to international conventions.

Basic measurements are: Major (nominal) diameter, Effective (pitch) diameter, Minor (root) diameter, Pitch, Flank and Crest as shown in Figure 3.48.

The major diameter can be measured with a simple caliper rule or slot gauge to determine the nominal diameter. A bolt or screw is measured at the crests; a nut is measured at the thread roots. A thread's pitch can be measured between the thread crests to determine pitch and thread designation.

For metric and British Association (BA) threads (common in electrical terminals), the pitch is a single thread measured and expressed in millimetres, e.g. M10 × 1.5, being 1.5 mm from crest to crest. For imperial threads, pitch is expressed in numbers of threads per inch, e.g. ¼" – 20 means 20 threads per inch (TPI).

Metric, unified and Whitworth threads

In metric and unified threads, the crests and the roots theoretically should be flat; however, in practice, to aid manufacture and fit, they are rounded inside a maximum outline as shown in Figure 3.49.

The British Standard Whitworth (BSW) and British Standard Fine (BSF) threads' profile is more wave shaped, being a series of radius curves about the pitch line as shown in Figure 3.50.

Machine screw threads are symmetrical – the angle on both flanks being the same (see Figure 3.51). Flank angles for metric and unified threads (UNC and UNF) are 30°, and therefore have a total thread angle of 60°.

BSW and BSF are 27.5° and therefore have a total thread angle of 55° as shown in Figure 3.52.

FIGURE 3.49 Metric and unified threads

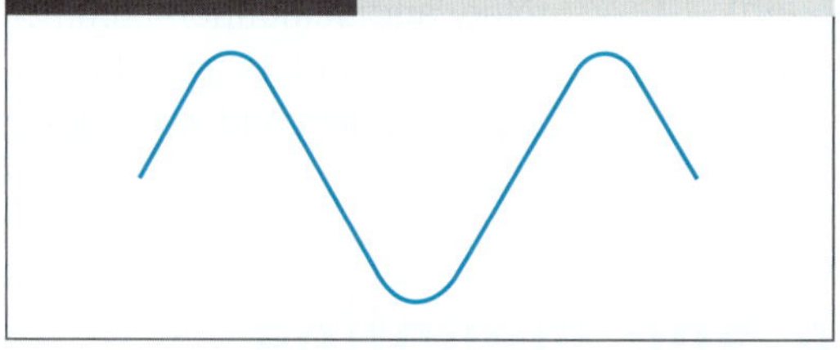

FIGURE 3.50 British Standard Whitworth thread

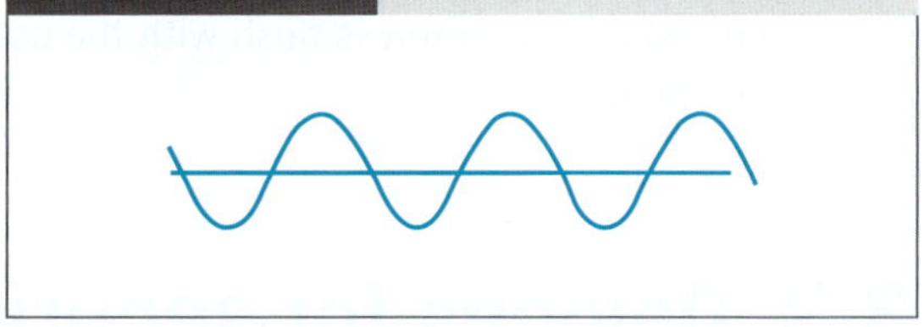

FIGURE 3.51 Metric thread flank angles

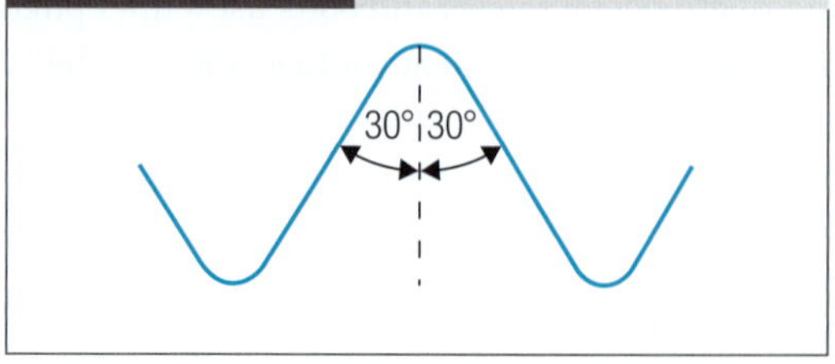

FIGURE 3.52 British Standard Whitworth thread angle

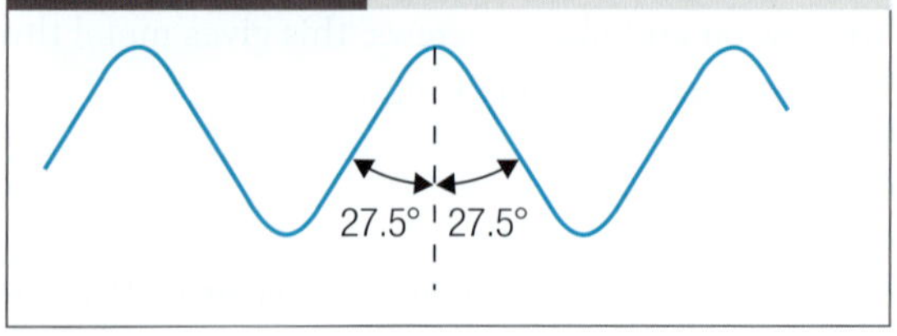

Sizing

Sizing of metric, Whitworth and unified threads is a measurement of the major diameter as shown in Figure 3.53 rather than a gauge as in wood screws. The length is measured from the point of the fastener to the head. An example of a BSW bolt sizing would be 3/8" × 24 TPI × 1" (major diameter, threads per inch and length). An example of a metric bolt sizing would be 10 × 1.25 × 33 mm (major diameter, pitch and length).

FIGURE 3.53 Metric bolt sizes

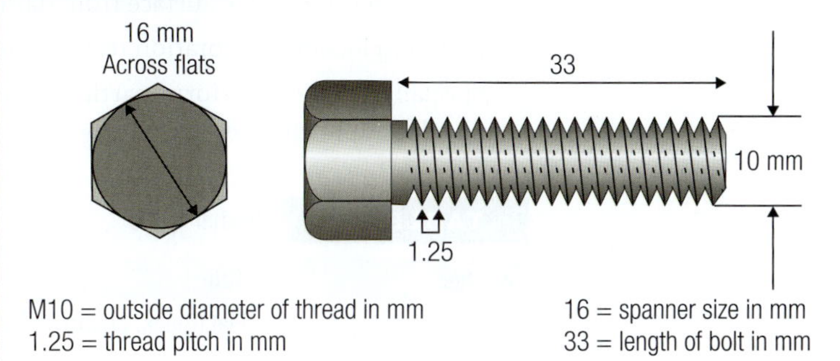

Load ratings

The major factor in determining the load a material can carry is its *tensile strength*, as shown in Figure 3.54. This is related to its hardness and is the maximum capacity of a particular material to stretch under tension load, prior to failure. It is normally expressed in kilo newtons (kN). *Ultimate tensile strength* (UTS) is the theoretical minimum point at which the material will fracture and is the common rating method.

The metals used in fastener manufacture are elastic materials which will stretch (elongate) under applied loads and return to their original shape when the load is removed. If excessive load is applied, the material will stretch beyond its yield point and enter a plastic zone, losing its elasticity and becoming permanently stretched. Further increase in load on the material will stretch its ultimate tensile strength and cause it to fracture.

FIGURE 3.54 Tensile strength

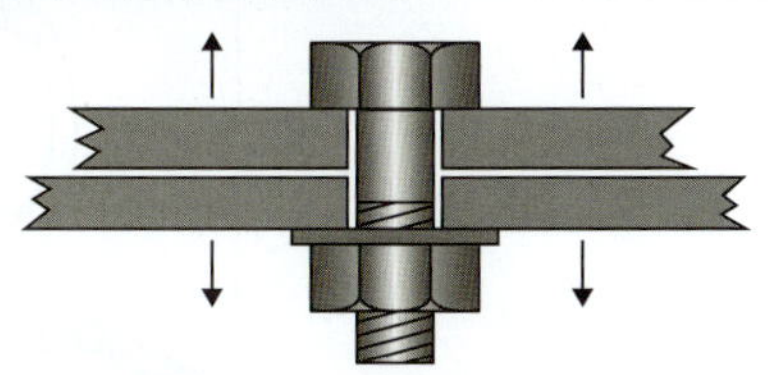

Load ratings must not be exceeded if failure is to be avoided.

The strength properties of an individual fastener are achieved by a combination of an appropriate base material selection and the manufacturing processes.

- Low tensile bolts/machine screws: Low carbon grades of steel are improved in hardness (strength) by cold working.
- High tensile bolts: Medium carbon grades of steel are improved in hardness (strength), after cold working, by controlled heat treatment and quenching.
- Self tapping/drilling screws: Medium carbon grades of steel are improved in case hardness, after cold working, by heat treatment and quenching.

Nuts

The primary function of the nut is to act as the device through which the tension is induced into the bolt or screw and to continue to retain that tension and the clamp load in the assembly. The vast majority of nuts have hexagon drive faces, and are made of carbon steel, stainless steel, brass, aluminium or nylon.

The corresponding nut must have the same thread pitch and diameter as the bolt. Because the pitch of some threads is common in the same diameters, it is possible to mate them, e.g. with BSW and UNC all diameters can be mated together except 1/2" (where UNC is 12 TPI, BSW is 13 TPI). However, because the thread angles and the profiles differ, the 'fit' will be loose and the mechanical requirements of the fastener will not be achievable. Therefore, mixture of thread forms *must* be avoided.

Washers

Washers come in a wide variety of designs, materials, dimensional variations and finishes.

Many washers can be used either under the head of the fastener, bolt or screw, or under the nut. Most washer sizes will be designated by the diameter (size) of the fastener with which they are to be used. Some of the most common are shown in Table 3.6. Washers are used to:

- spread the clamp load over a larger surface area
- cover an oversize or elongated hole

- reduce the friction of the rotated component
- protect the work piece surface from damage by the rotated component
- provide a locking or vibration resistant function.

Specialist washers perform particular functions in particular types of applications. These may be locking, load spreading, decorative, tension indicating, sealing or a combination of these functions.

TABLE 3.6 Washer types

Washer	Title	Comments
	Flat (Black, bright)	Common general purpose basic washer can come in various dimensional standards, quality levels, materials, hardness grades and finishes. Could also be available square. Often used in conjunction with a split spring washer.
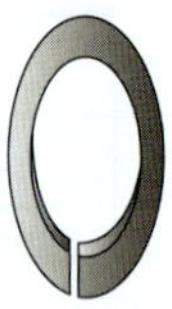	Split spring	Common locking washer, comes in a variety of thicknesses and sectional ratios. Available in various materials throughout a wide size range. Used to resist vibration loosening. Will damage surfaces it contacts.
	Belleville washer (Conical)	For use in high vibration applications where tension must be maintained if some loosening occurs. Can be used in stacks or series to increase the axial load or to increase the possible deflection length.
	Internal tooth lock washer (Shake proof)	Used with pan or cheese head machine screws to resist vibration. Minimal damage to surface.
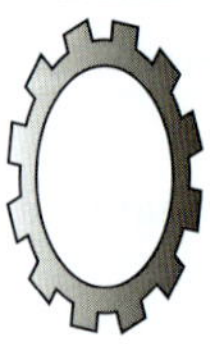	External tooth lock washer (Shake proof)	Same as above except slightly more damaging to surface. Also available in countersunk version.
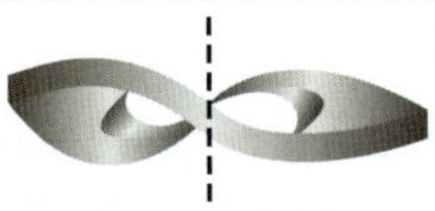	Wave	Can be available as full circle or split. Is used in place of spring washers where surface damage is to be avoided. Also used where some pressure is required on a free element of an assembly. Usually confined to small diameters. Similar products are crinkle washers.
	Curved	Generally confined to light applications similar to above, where only very light pressure is required.

Blind rivets (pop rivets)

Although bolting is the most common form of fastening, it is sometimes restricted in usage because of the need to screw a nut on the end. There are many applications where the back of the job cannot be accessed. In these cases, the most common solution is to use blind rivets, also known as 'pop rivets'. Blind rivets are so termed because there is no need for access to the rear or 'blind' side of the joint. Blind rivets are limited in their strength and should only be used when fastening relatively light gauge materials. Blind rivets are a two part fastener as shown in Figure 3.55.

TABLE 3.7 Blind rivet shell materials

Shell materials	Symbol	Characteristics
Aluminium	A	Lightweight yet strong. Good corrosion resistance.
Steel	S	Stronger than aluminium with a 'flash' coating of zinc.
Stainless steel	St	Good strength and excellent corrosion resistance.
Nickel copper	M	Excellent corrosion resistance and conductivity. The strongest blind rivet material.
Copper	C	Relatively soft. Excellent corrosion resistance and conductivity.

FIGURE 3.55 Blind rivet

Source: Ed Hawkins

Usually, the factors in material selection are strength, corrosion resistance and the material to be fastened. As a general rule, the shell material should be the same as the material under the head. Table 3.7 lists commonly used blind rivet shell materials.

As described in Ajax Fasteners' *Rivet Handbook* (p. 9; http://www.ajaxfast.com.au), the stem is usually steel, which is the most common, although stainless steel and aluminium stems are available in some rivets. Part of the stem remains in the shell after setting. In the case of steel, this part may rust, which is an undesirable attribute in many circumstances. In these cases, stainless and aluminium stems may be specified.

Break load

The load required to break off the stem is governed by the amount of shell deformation required and a groove cut to a specific diameter just back from the end. The break load is designed to prevent too much force being developed which may damage the material being fastened. Too little clamp force may fail to secure the joint.

Similar to a bolted joint, a blind rivet develops a clamping force to secure the joint members. It is crucial that the mandrel not break at a load lower than the clamping force. If this occurs the integrity of the joint may be compromised.

Grip range

Blind rivets are designed to clamp together specific thicknesses of material. The amount of deformation in the shell during setting depends upon the grip range. If the grip range is less than it should be (the rivet is too long), excessive material will be left on the blind side and more pulls will be required to break off the stem. If the grip range is too large (rivet is too short), insufficient material may be left on the blind side to secure the joint.

Galvanic corrosion

When dissimilar metals come into contact in the presence of an electrolyte, a galvanic action occurs which corrodes one metal at a faster rate and the other more slowly. The rate of corrosion depends upon the difference in electrical potential, the conductivity of the electrolyte and the relative sizes of the contacting areas. Table 3.8 illustrates which metals should and should not be used together.

i Corrosion and dissimilar metals

TABLE 3.8 Compatible metals

Rivet shell material	Metals joined				
	Aluminium	Steel zinc plated	Stainless steel	Copper	Brass
Aluminium	Yes	?	No	No	No
Steel zinc plated	No	Yes	No	No	No
Nickel copper	No	No	Yes	Yes	Yes
Stainless steel	?	?	Yes	Yes	Yes
Copper	No	No	Yes	Yes	Yes
? = Compatible in rural and mild environments. Some corrosion may occur in marine or industrial environments. Painting both metals could reduce the reaction.					

Solid rivets

Solid rivets are manufactured in mild steel, aluminium, copper, brass and stainless steel. A rivet is a permanent mechanical fastener; it can support tension loads; however, it is much better suited to supporting shear loads. Bolts and screws are better suited for tension applications.

Techniques for fixing to metal

Machine thread bolts

The use of nuts and bolts to secure metals such as mild steel requires the drilling of clearance holes in the material to be secured; typically an electric or pneumatic power tool is used. The clearance hole is slightly larger than the major diameter of the bolt to allow the bolt thread to pass through it without damage. Table 3.9 provides guidance on the selection of drill bit sizes.

TABLE 3.9 Metric thread drilling guide

METRIC – COARSE THREADS							
Thread size	Major dia. (mm)	Tap drill (mm)	Clearance hole (mm)	Thread size	Major dia. (mm)	Tap drill (mm)	Clearance hole (mm)
M1.6 × 0.35	1.6	1.25	1.8	M10 × 1.25	10.0	8.8	12.0
M2 × 0.4	2.0	1.60	2.4	M12 × 1.25	12.0	10.8	14.0
M2.5 × 0.45	2.5	2.05	2.9	M14 × 1.5	14.0	12.5	16.0
M3 × .05	3.0	2.50	3.4	M16 × 1.5	16.0	14.5	18.0
M3.5 × 0.6	3.5	2.90	3.9	M18 × 1.5	18.0	16.5	20.0
M4 × 0.7	4.0	3.30	4.5	M20 × 1.5	20.0	18.5	22.0
M5 × 0.8	5.0	4.20	5.5	M22 × 1.5	22.0	20.5	25.0
M6 × 1	6.0	5.0	6.6	M24 × 2	24.0	22.0	27.0
M8 × 1	8.0	7.0	9.0	M27 × 2	27.0	25.0	30.0

1 Determine the location from a plan or drawings and mark on the work surface with a pencil or scriber. Centre punch marking to prevent drifting.
2 Drill a pilot hole of a suitable size. The hole should be slightly larger than the web of the drill bit used for the clearance hole. Holes smaller than 6 mm will not need pilot holes.
3 Select a suitable size drill bit for a clearance hole. Enlarge the pilot hole to size and deburr the hole.
4 The nut, bolt and washer combination is then assembled. For consistency, if there is more than one bolt, all the heads should point in the same direction. Ensure all bolts are in place and finger tight before tightening by tool.

FIGURE 3.56 Bolt tightening sequence

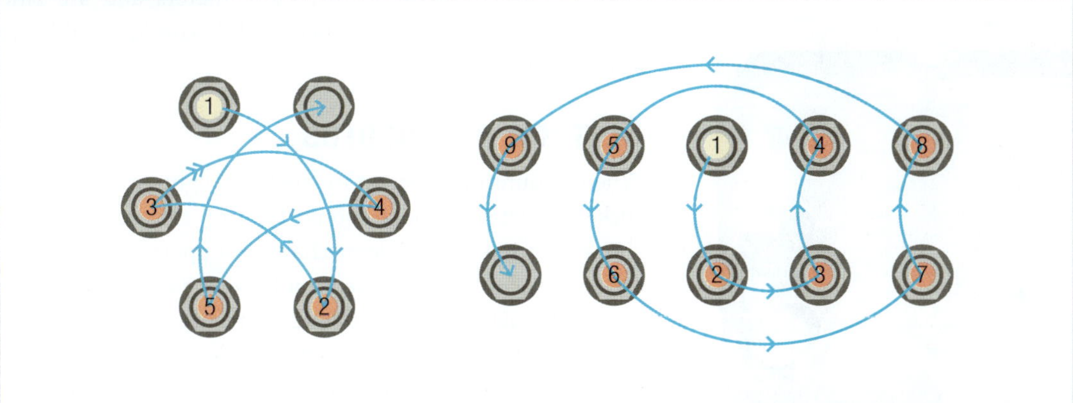

5 Whenever there are more than two fasteners securing a part, alternate the tightening order among them as shown in Figure 3.56.

6 Larger bolts may require the use of a torque wrench to tension the bolts to a specified value of torque.

FIGURE 3.57 Installation of a blind rivet

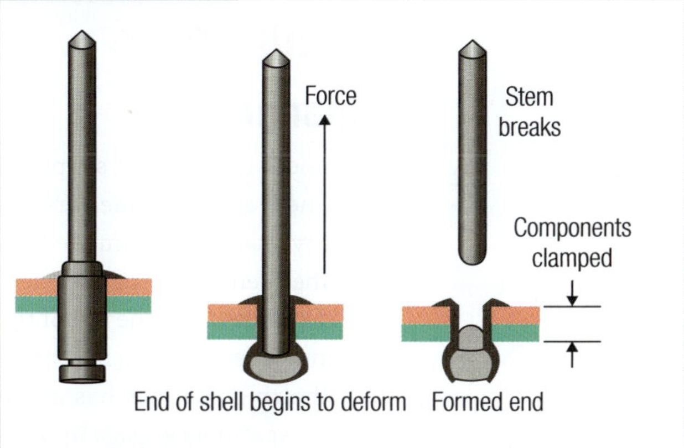

Blind rivets

Blind rivets are a two-part fastener consisting of a shell and a headed stem (mandrel) assembled so the shell can be placed into the work to be fastened. The installation is as follows:

1 A suitably sized hole is made in the materials to be joined and the rivet inserted. Clamping may be required.

2 The rivet is 'set' by drawing the stem through the shell with a special tool operated by hand, pneumatics or electricity. The shell deforms and clamps the material securely. After the desired clamping force is achieved, the stem breaks off and is discarded, as shown in Figure 3.57.

3 A small portion of the stem remains trapped in the bottom of the shell to ensure the clamping force is retained in the joint.

FIGURE 3.58 Installation of a solid rivet

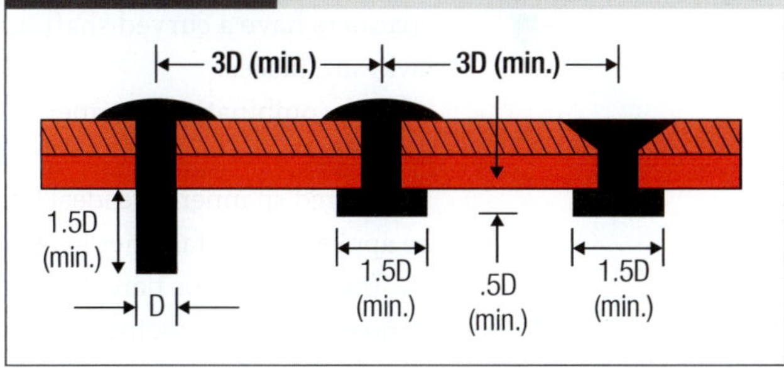

Solid rivets

The installation is as follows:

1 A suitable sized hole is made in the materials to be joined and the rivet inserted. Clamping may be required. In situations requiring a flush finish, the hole is counter sunk.

2 The head of the rivet is placed on a hard metal surface such as a riveting dolly.

3 The stalk of the rivet is then stuck with the ball of a ball-pein hammer to mushroom over the stalk as shown in Figure 3.58. This process is called upsetting.

4 Where a flush finish is required, the rivet's head is filed flush to the work surface.

Accessories that may be fixed to metal

Electrical accessories can be attached to metal construction components either through metal thread screws and bolts or pins inserted by explosive power tools. Spring steel clips allow cables and conduit to be quickly and easily secured to steel purlins and beams without the need of drilling.

FIGURE 3.59 Surface enclosures

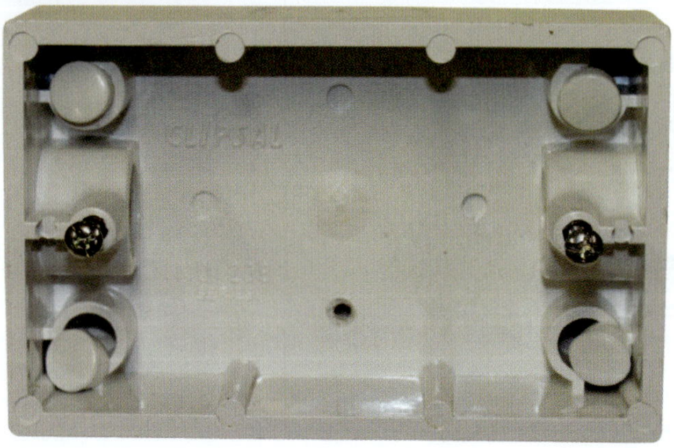

Source: Ed Hawkins

Half or single sided saddles are available in sizes to suit conduits from 12 mm to 50 mm diameters and are zinc plated for corrosion resistance. A full conduit saddle is also available.

Surface enclosures

When mounting accessories such as light switches and socket outlets on metal, some type of surface enclosure should be used to provide mechanical support and maintain double insulation requirements at the electrical terminations (see **AS/NZ 3000:2007** clause 1.5.5.4). Figure 3.59 shows how surface enclosures are provided with designated mounting holes. When securing the enclosure to conductive building materials (metal) plastic caps must be inserted over the screw heads to reinstate the supplementary insulation.

Fixing tools

Spanners

Spanners come in all shapes and sizes, many having been developed to deal with a specific job. Spanners and wrenches are hand tools used to grip and turn the heads of fixing devices. Modern quality spanners are made from chrome vanadium or chrome molybdenum steel.

The open ended spanner (see Figure 3.60) may have a single or double end. The head has its jaws offset by 15 degrees from the run of the shaft. This is so the spanner can be turned over to engage different flats of a nut when working in confined spaces.

The ring spanner has a completely enclosed head, and may have 6 or 12 flats. Better quality 12-flat spanners engage upon the corners of the nut and can engage both hexagon and square bolts. Ring spanners are stronger than the open ended type, but need access to fit over the nut and can take a little time to locate it correctly. An offset ring spanner has offset heads and half moon ring spanners have a curved shaft as shown in Figure 3.61, allowing the spanner to connect with nuts in awkward places.

The combination spanner shown in Figure 3.62 has both a ring and an open end spanner on the same tool.

Geared spanners are ideal for confined situations, as a reversible ratchet mechanism allows force to be applied to the fastener in as little as 5° increments (see Figure 3.63).

The split ring or flare nut spanner shown in Figure 3.64 is a hybrid, as it has a section of the ring removed so the six point jaws can be located like an open-ended spanner.

FIGURE 3.60 Open ended spanner

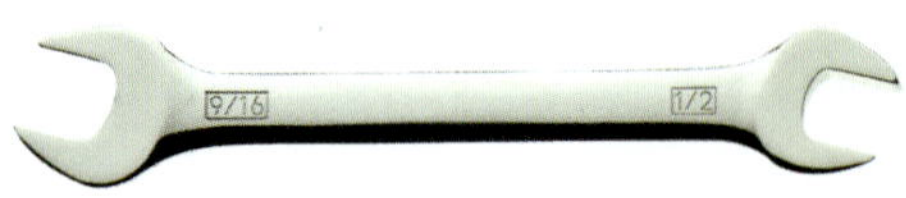

Source: iStockphoto/EHStock

FIGURE 3.61 Half moon ring spanner

Source: Ed Hawkins

FIGURE 3.62 Combination spanner

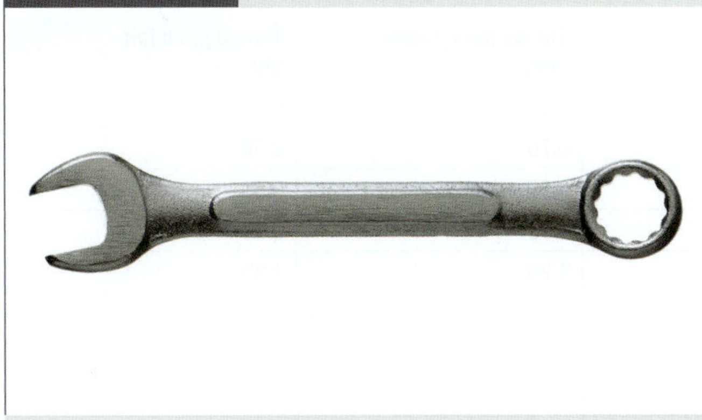

Source: iStockphoto/TokenPhoto

FIGURE 3.63 Geared spanner

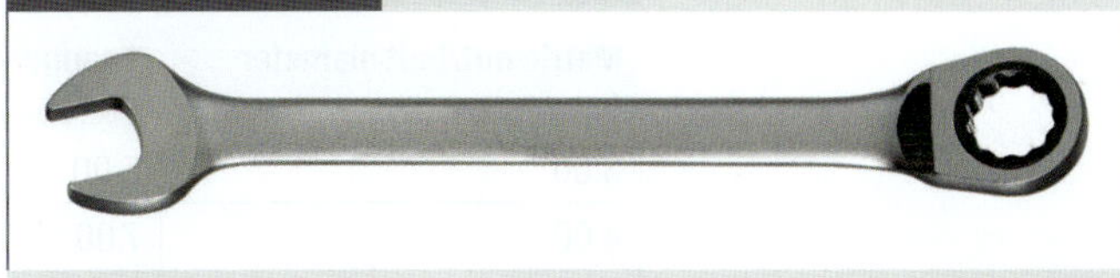

Source: iStockphoto/Prill Mediendesign & Fotografie

FIGURE 3.64 Flare nut spanner

Source: Ed Hawkins

The socket spanner in Figure 3.65 consists of a ring type head with a square hole at the rear designed to lock into various types of square drives. The square drive sizes range through 1/4", 3/8", 1/2" and 3/4" sizes. Sockets are generally needed when high torque is required to tighten nuts and bolts. They are available in sets, offering a range of heads, spacer drive bars and ratchet handles.

Adjustable or shifting spanners (see Figure 3.66) use a worm screw located close to the opening jaw, which has a rack engaged with the screw, making the spanner very easy to adjust with the finger and thumb of the holding hand. Adjustable spanners provide the least grip and should be considered only for low torque work. Otherwise, the spanner can damage the bolt head or nut by rounding its corners due to its imperfect grip. It can also slip off the bolt or nut unexpectedly.

FIGURE 3.65 Socket spanner

Source: Ed Hawkins

Spanner sizing

The most important consideration when using a spanner is to ensure that it fits the nut or bolt perfectly. Metric spanner and wrench sizes are measured in millimetres between the jaws (across flats), as shown in Figure 3.67, and have no relation to the major diameter of the thread. The spanner size of 15 mm shown is the distance between the flats of the spanner, not the major diameter of the nut or bolt.

FIGURE 3.66 Adjustable spanners

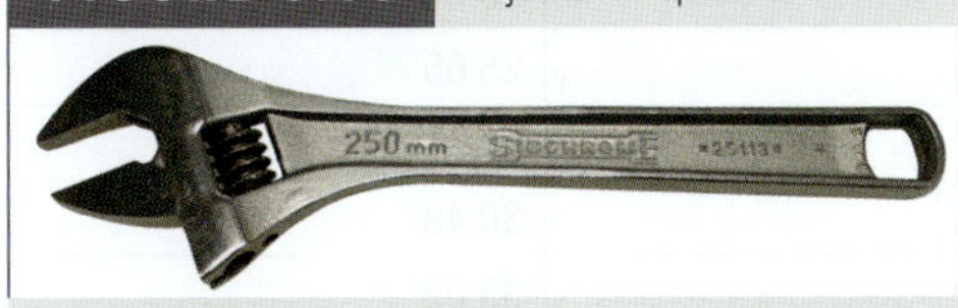

Source: Ed Hawkins

FIGURE 3.67 Across flats size measurement

Source: Ed Hawkins

Adjustable spanners can slip off the nut or bolt head unexpectedly if you apply too much torque.

TABLE 3.10 A guide to metric spanner selection

Metric nut/bolt diameter (mm)	Spanner size (mm)	Thread pitch coarse (mm)	Thread pitch fine (mm)
3.00	5.00	0.50	0.35
4.00	7.00	0.70	0.50
5.00	8.00	0.80	0.50
6.00	10.00	1.00	0.75
8.00	13.00	1.25	1.00
10.00	16.00	1.50	1.25
12.00	18.00	1.75	1.25
14.00	21.00	2.00	1.50
16.00	24.00	2.00	1.50
20.00	30.00	2.50	1.50
24.00	36.00	3.00	2.00

A guide to metric spanner selection is shown in Table 3.10. The spanner size is the same for both coarse and fine threads. A typical 14 piece metric spanner set would include all sizes from 6 mm to 19 mm spanner sizes. These are smaller increments than metric thread sizes, so not every spanner has a corresponding nut or bolt.

The American imperial UNF/UNC spanner size is also determined by the 'across flats' measurement. Be careful not to confuse the letters *AF* when used to mean 'across flats', with *AF* used to indicate 'American Fine' thread.

British Whitworth spanners are stamped with imperial sizes that bear no obvious relationship to the distance across flats. Guidance to Whitworth spanner selection is shown in Table 3.11.

TABLE 3.11 A guide to British Standard Whitworth spanner selection

Whitworth nut/bolt (inches)	BSW/BSF spanner (inches)	Across flats (inches)	Across flats (mm)
–	1/16" W	0.256	6.90
–	3/32" W	0.297	7.54
3/16	1/8" W	0.340	8.64
1/4	3/16" W	0.445	11.30
5/16	1/4" W	0.525	13.34
3/8	5/16" W	0.600	15.24
7/16	3/8" W	0.710	18.03
1/2	7/16" W	0.820	20.83
9/16	1/2" W	0.920	23.37
5/8	9/16" W	1.010	25.65
11/16	5/8" W	1.100	27.94
3/4	11/16" W	1.200	30.48
7/8	3/4" W	1.300	33.02
15/16	13/16" W	1.390	35.31
1.	7/8" W	1.480	37.59
1.1/8	1" W	1.670	42.42

Screwdrivers

Screwdrivers come in many types to suit specific tasks. Like spanners they are hand tools used to grip and turn the heads of fixing devices. Screwdrivers apply torque to a drive in the inside centre of the fixing's head rather than the outside of the head as is the case of a nut or bolt (see Figure 3.68).

Screwdrivers are classified by their head types. There are now literally hundreds of drive types available, and Figure 3.69 shows some of the common types.

The slotted (flat) drive head is common in woodworking applications and small electrical terminals. It is not suitable for power tool applications because the driver tends to slip out of the head, as it is not self-centring.

The Phillips head drive is designed to slip out of the screw head being driven if the torque required to turn the screw exceeds a certain limit. Phillips screwdrivers have an intentional shallow angle on the slots and rounded corners so they will fall out of the slot before a power tool will twist off the screw head. These are not suitable for high torque applications.

The Pozidriv drive head and screws are distinguishable from Phillips by the second set of indentations set 45 degrees from the cross recess. Phillips screwdrivers will fit in and turn Pozidriv screws, but will slip out if enough torque is applied. *The drive wings on a Pozidriv screwdriver will not fit a Phillips screw correctly*, and are likely to slip or tear out the screw head if use is attempted. They are commonly used on self tapping and drilling screws.

Torx drive heads and screws, also known as star bits, are commonly found on cars, motorcycles, bicycle disc brake systems, hard disc drives, computer systems and consumer electronics.

Security T drive heads and screws use a protruding obstacle to make the fastener more tamper resistant by inserting a pin in the fastener screw drive. The tool then requires a corresponding hole to drive the fastener.

Hexagonal or Allen drive heads and screws can be used with either a suitable screwdriver or Allen keys. Hexagonal drives can be rotated with spanners to gain extra leverage where space is restricted.

FIGURE 3.68 Screwdriver parts

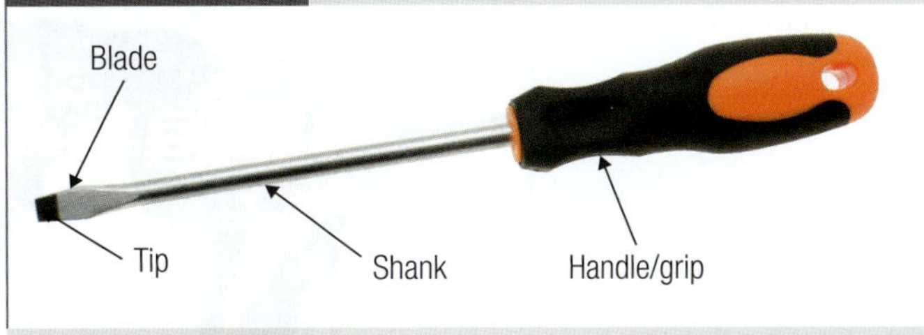

Source: iStockphoto/clubfoto

FIGURE 3.69 Screwdriver head types

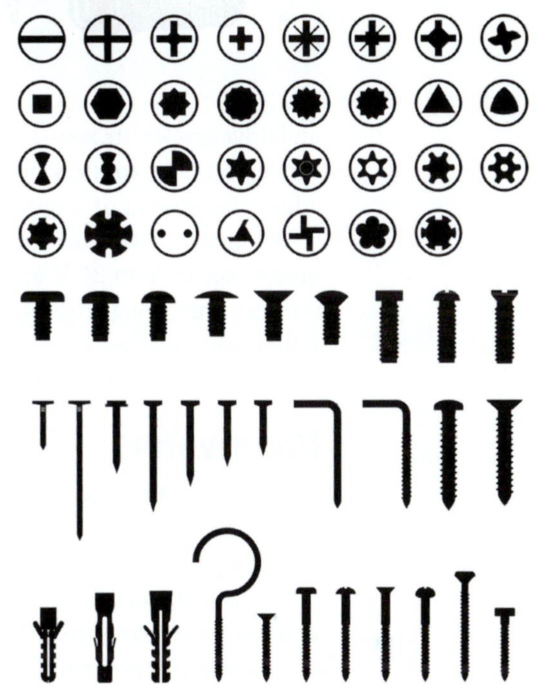

Source: Dreamstime/22bobe

Power screwdrivers

Power screwdrivers are rotary power tools used to grip and turn the heads of fixing devices. The power source can be pneumatic (air) or electric (mains supply or battery).

The battery model shown in Figure 3.70 has become the most popular due to its portability. They are smaller and typically lighter-weight versions of the cordless drill. They typically operate on a lower voltage than a drill, and have less torque. Power screwdrivers are for light duty assembling.

The drill driver is shown in Figure 3.71a. Drill drivers are medium duty, operating on usually 12 V to 18 V. A cordless drill/driver is a versatile tool that is designed to drill holes

FIGURE 3.70 Power screwdriver

Source: iStockphoto/Tim Abramowitz

FIGURE 3.71 (a) Drill (b) Impact driver

Source: Ed Hawkins

FIGURE 3.72 Hand blind rivet tool

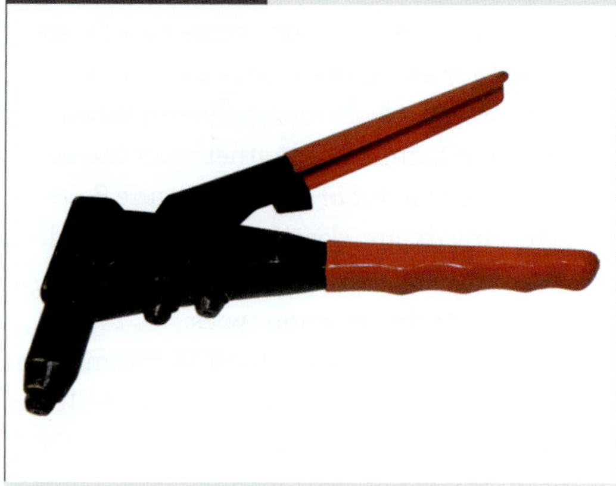

Source: Ed Hawkins

and drive screws. Its keyless chuck accepts a wide variety of round and Hex shank drill bits and screw driving bits. All cordless drill/drivers are also equipped with a slip clutch that allows the user to adjust the amount of torque for precise, consistent screw driving.

An impact driver is shown in Figure 3.71b. Larger than a drill driver in size, impact drivers generally operate on 18 V to 36 V, and provide considerably more torque and are useful for driving large lag screws or removing large nuts like the wheel nuts on a vehicle. They are designed to accept sockets rather than drill bits. The impact driver uses both bit rotation and rotational concussive blows to drive screws through thick, dense wood such as power poles.

Pop riveters

These are used to install blind rivets and are available in hand tool (Figure 3.72), electric (battery) tool or pneumatic (air) tool.

Basic riveters all come with a variety of nibs stored in the handle to accommodate different stem thicknesses. Riveters can also have longer handles, which provide extra leverage, making larger and stronger rivets easier to set. A 'lazy tong' has no levers; the setting force is created by the concertina body. The pushing action allows the operator to put their weight behind the tool.

Files

Files are a hand tool used to remove small amounts of material from a surface and to de-burr sharp edges. Files come in smooth, 2nd cut and bastard grades, as shown in Figure 3.73. The smooth file is the most suitable for removing sharp edges.

Figure 3.74 shows commonly used file shapes suitable for fixing work. The shape of the file selected will depend on the shape of the work surface. For example, the rat tail file would be used to remove sharp edges from a hole drilled in metal, where a flat or hand file would be used to de-burr a cable tray. There is a more concise coverage of the use of files in Chapter 4.

Always remove sharp edges caused by drilling or cutting metal.

Reamers

The most common type of reamer used in the electrotechnology industry is the pipe reamer, shown in Figure 3.75. It is a non-precision rotary hand tool used for enlargement of holes or for removing burrs.

FIGURE 3.73 File grades of smooth, 2nd cut and bastard

Source: Ed Hawkins

FIGURE 3.74 File shapes

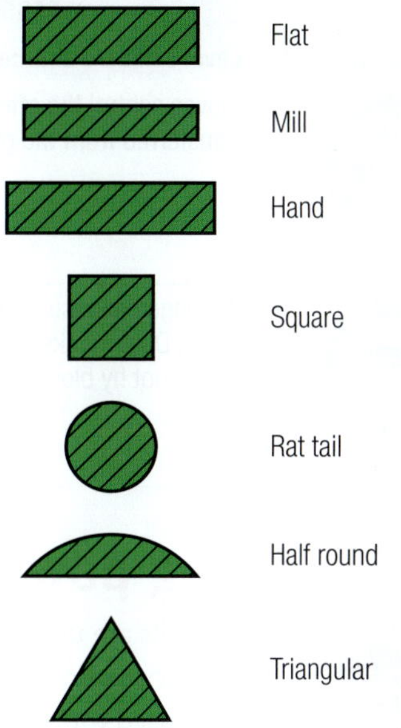

FIGURE 3.75 Hand pipe reamer

Source: Ed Hawkins

WH&S issues related to drilling and cutting

Risk sources (hazards) associated with drilling steel and masonry include noise, vibration, dust and sharp objects such as metal filings and swarf. There is also a very high possibility that drill bits can break, causing flying debris. If a larger sized drill bit jams, the torque from the power drill can cause a sudden, unexpected rotation of the power tool which can injure the operator. In locations close to walls, the operator's hands can be crushed between the wall and the drill handle.

Drilling and cutting metal can be dangerous.

In a construction situation, elimination (level 1 control) of drilling operations is rarely possible as the equipment has to be fixed in location. Pneumatic power tools are substituted (level 1 control) for electrical tools when working near flammable materials or on motor vehicles. Level 2 control by isolation and engineering control are achieved by ensuring other workers are at a safe distance from the drilling work and the appropriate type and size of drilling tool is used for the job, i.e. selecting a drill fitted with a support handle for drill holes larger than 12 mm. Administrative (level 3) controls restrict the time the operator uses the drill to shorter periods to control fatigue from noise and vibration. Personal protective equipment (PPE; level 3 control) specific to drilling operations are show in Table 3.12; other PPE items such as hard hats and high visibility clothing are assumed.

TABLE 3.12 Level 3 control measures

PERSONAL PROTECTIVE EQUIPMENT	
Risk source (hazard)	**Level 3 hazard source control measures**
 Noise	Hearing loss and increased fatigue from repetitive drilling operations (especially hammer drills) is controlled by hearing protection.
 Eye	Masonry chips, shattered drill bits or pieces of swarf (the metal waste) can be propelled and injure the eyes. Safety glasses are used as a control measure.

(Continued)

TABLE 3.12 (*Continued*)

PERSONAL PROTECTIVE EQUIPMENT	
Risk source (hazard)	**Level 3 hazard source control measures**
 Hand	Gloves control the effects of vibration and protect against sharp edges. Metal filings can still be transferred from the gloves to the eyes or nose if these are scratched or rubbed.
 Dust	Prolonged exposure to the dust created from drilling concrete and masonry can damage the lungs. Dust masks are used as a control measure. Holes should be cleared with a brush or air pump, not by blowing through the mouth.

FIGURE 3.76 Bench top drill press

Source: Ed Hawkins

FIGURE 3.77 Twist drill bit

Source: Ed Hawkins

Using power drills

Power drills are useful for driving screws, fastening bolts and drilling holes in wooden or metallic materials. Common power drills not already shown include the drill press, which is used for drilling precise holes at uniform depths and/or angles. Drill presses can be further classified as bench top drill presses (see Figure 3.76) or floor drill presses. The bench top type is of a compact size and can be clamped directly on to a work bench. Floor models are larger but are more versatile with more powerful motors and usually a number of attachments.

Right angle drills are useful when space is a restriction. They are ideal for drilling into studs and noggins.

Drill bits

Drill bits are made from high speed tool steel which is hard but brittle. The material for the drill bit needs to be harder than the work surface in order to cut it. There are different types of drill bits to suit a variety of applications and materials. Commonly used drill bits in the electrotechnology industry are explained below.

Drill bits have a shank by which they can be held by the drilling machine and spun at the appropriate speed and fed into the work. If the drill bit is below 12.8 mm in diameter, the shank is parallel, allowing the drill bit to be held in a three jaw Jacobs chuck.

The twist drill (Figure 3.77) is the most common type of drill. The flutes of the drill are spiral shaped and form the cutting edges of the drill as well as carrying the shavings away and providing a path for lubricant to the cutting edge. As with any other cutting tool, the drill has cutting angles, clearance angles and rake angles (these are further explained in Chapter 4). The auger bit is used for drilling into hard or soft wood. The bit shown in Figure 3.78 has four cutting faces and flutes to increase cutting speeds.

Spade bits or speed bores (Figure 3.79) are an alternative to auger bits for cutting into hard or soft wood.

Masonry drill bits (Figure 3.80) have tungsten carbide points to allow them to cut concrete without losing their shape. The body is heat treated to prevent compression to allow the energy of the power drill's hammer action to be transferred to the concrete.

FIGURE 3.78 Auger drill bit

Source: shutterstock.com/Albert Lozano

FIGURE 3.79 Spade bit

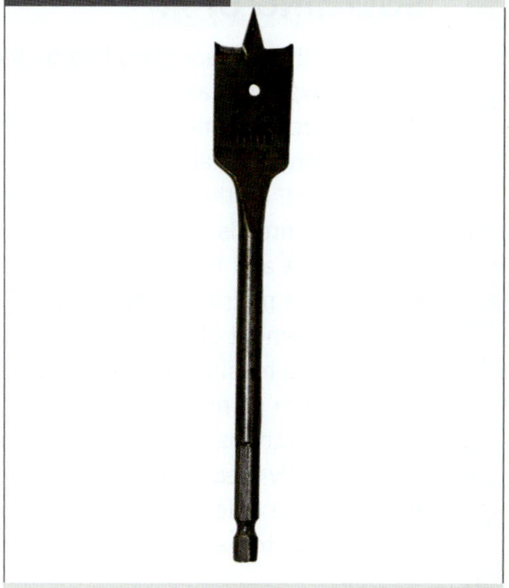

Source: Ed Hawkins

Drill speeds

Cutting speed is the speed at the outside edge of the tool as it is cutting. This is also known as surface speed. If two tools of different sizes are turning at the same revolutions per minute (RPM), the larger tool has a greater surface speed. Surface speed is measured in metres per minute (m/min).

Cutting speeds depend primarily on the kind of material being cut and the type of cutting tool being used. The hardness of the work material also has a great deal to do with the recommended cutting speed. If too high a cutting speed is used, the resulting heat will be excessive and damage the drill bit. As a general rule, the harder the work material, the slower the cutting speed. The softer the working material, the faster the cutting speed. Table 3.13 provides guidance on recommended drill speeds simplified to RPM rather than metres per minute. Lubricants used for drilling are discussed in Chapter 4.

FIGURE 3.80 Masonry bit

Source: Dreamstime/Noomcm

TABLE 3.13 Recommended twist drill speeds (RPM)

Size (mm)	Softwood	Hardwood	Plastic	Brass	Aluminium	Steel
1–5	3000	3000	2500	3000	3000	3000
6–10	3000	1500	2000	1200	2500	1000
11–16	1500	750	1500	750	1500	600
17–25	750	500	NA	400	1000	350

SUMMING UP...

- Metal threads are classified by their drive method, head type, thread and point.
- The three commonly used threads are British Standard (England), unified (USA) and metric (Europe); all three have coarse and fine variations. Metric threads are now the most common in Australia and New Zealand.
- Sizing of metric, Whitworth and unified threads is a measurement of the major diameter.
- Ultimate tensile strength (UTS) is the theoretical minimum point at which the bolt's material will fracture and is the common rating method.
- Commonly used rivets are the blind (pop) and solid rivets.
- Clearance holes are sized to allow the thread's major diameter to pass through it but not the head of the bolt.
- Tapping holes are sized to the thread's minor diameter to allow a tap to cut a thread.
- Pilot holes are a smaller hole drilled prior to larger holes and are sized slightly larger than the point of the larger drill.
- Spanners are identified by the style of head used to grip the nut or bolt.
- Metric spanner sizes are measured 'across the flats' of the spanner and not the major diameter of the nut or bolt's thread.
- Metric nuts or bolts are sized by the major diameter of the thread and the distance between the crests (pitch).
- Screwdrivers are classified by their head types, and there are now numerous drive types available.
- Power screwdrivers are rotary power tools used to grip and turn the heads of fixing devices. The power source can be pneumatic (air) or electric (mains supply or battery).
- Risk sources (hazards) associated with drilling steel and masonry include noise, vibration, dust and sharp objects such as metal filings, swarf and sudden, unexpected rotation of the power tool.
- Typical PPE required when drilling includes safety glasses, hearing protection, gloves and, if drilling masonry or concrete, a dust mask.
- Drill bits are made from high speed tool steel which is hard but brittle.
- The major parts of a drill include the shank, body, flutes and point.

3.4 Securing and mounting electrical equipment using fixing adhesives and tapes

In recent years significant developments in adhesives has seen the introduction of a wide range of products that allow quick and reliable fixing to a broad range of surface materials, and in some cases without the need to drill a hole. This section introduces the use of chemical adhesive fixings in the electrotechnology industry.

Adhesive and tape fixing devices

Adhesive tapes

Foam backed double sided tape adheres well to smooth to moderately textured surfaces, metals, most plastics and composites. Double sided tape is especially useful in rented premises where damage to walls and ceilings caused by the installation of screws is not permitted.

The 3M website describes industrial strength grades such as 3M VHB™ tapes as

> high strength bonding tapes and are a proven alternative to screws, rivets, welds and other mechanical fasteners. Unlike screws or rivets – which join materials at a single point – high strength bonding tape permanently adheres one substrate to another while spreading the stress load across the entire length of the joint. Once materials are joined, a virtually indestructible 'weld' is created.

Source: 3M, 'Industrial adhesives & tapes', http://solutions.3m.com/wps/portal/3M/en_US/Adhesives/Tapes/Brands/3M-VHB-Tape

Chemical anchors

FIGURE 3.81 Anchor stud

Source: Ed Hawkins

Chemical or resin anchors are a fixing system where steel studs, bolts and reinforcement bars (reo bar) are chemically bonded into a substrate, usually masonry and concrete, using a two-part resin-based adhesive system. Ideally suited for high load applications, typically the resulting bond is stronger than the base material itself and as the system is based on chemical adhesion, no load stress is imparted to the base material as with the mechanical expansion type anchors described in Section 3.2 earlier in this chapter. This makes chemical anchors ideal for situations such as close to edge fixing, close to fixing applications or for concrete of unknown quality or low compressive strength.

Chemical systems operate by the addition of a hardening agent to a base resin to begin a chemical curing process. The resin and hardening agent are mixed and then injected into a previously drilled, slightly oversized hole. Threaded metal studs (see Figure 3.81) or a reinforcing bar can be fixed into solid concrete, stone, hollow brick and hollow concrete block. Anchor studs are available in zinc plated or galvanised high tensile carbon steel and 316 stainless steel.

Load ratings of chemical anchors are measured by the tensile strength along the axis of the fixing and the shear strength at 90 degrees to the side of the axis in the same way as mechanical expansion anchors. Chemical anchoring adhesive load ratings are complex, as the rating varies widely depending on whether the anchor stud is high tensile or stainless steel and what grade of concrete the anchor is secured to. The examples provided are simple comparisons using high tensile steel studs. For a more detailed explanation refer to manufacturer's data.

Medium duty chemical anchor adhesive is suitable for anchoring of threaded studs and reinforcing bar into solid concrete, stone, hollow brick and hollow concrete block. The adhesive comprises two parts that are mixed together. The material sets to a grey colour and has a temperature range from −10°C to 80°C.

From Table 3.14 it can be seen that the load rating of the fixing is limited by the concrete compression rating and the length of the stud rather than the tensile strength of the steel anchor stud itself.

Hollow brick and block anchors and threaded inserts are secured using a medium duty chemical adhesive into hollow brick and block. A solid plug of adhesive in the brick or block cavity provides a secure anchoring point. Nylon sleeves are suitable for three hole brick and hollow concrete block. Fine metal mesh sleeves are ideal for ten hole bricks (see Figure 3.86). Load ratings of nylon sleeves, metal mesh sleeves and threaded inserts for hollow walls are much lower than those of anchor studs shown in Table 3.14. For a more detailed explanation refer to manufacturer's data.

TABLE 3.14 Medium duty chemical anchor ratings

WORKING TENSILE LOAD IN NEWTONS (kN)				
Anchor size	High tensile steel	Concrete compressive strength		
		20 MPa	32 MPa	40 MPa
M8	6.5	3.4	4.5	5.1
M10	10.3	4.5	6.0	6.8
M12	15.3	6.4	8.6	9.7
M16	29.4	10.1	13.5	15.3
M20 × 150 mm	44.3	16.6	22.3	25.2
M20 × 170 mm	44.3	18.8	25.2	28.6
M24 × 160 mm	64.2	24.4	32.7	37.0
M24 × 210 mm	64.2	32.1	42.9	48.6

Threaded inserts are steel threaded sockets that sit flush with the concrete surface and allow the use of various head style bolts. The bolts can be removed and reinstalled later.

Heavy duty chemical anchoring adhesive is suitable for anchoring of threaded studs and reinforcing bar into solid concrete, solid brick and stone. When the chemicals are mixed they take on a dark grey colour with a service temperature between −23°C and 60°C. Typical applications include fixing steel beams and columns, heavy plant, safety barriers, reinforcing bars and timber frame hold down.

Table 3.15 shows the increased load ratings achieved by heavy duty adhesive. The same anchor studs are used as in medium duty adhesives but the load ratings for all concrete grades are improved. Suitable only for solid concrete as hollow brick and masonry block constructions do not provide the necessary support, i.e. the construction will not support the load on the anchor.

For small one off fixings, glass capsule rapid cure acrylic resins are suitable for extra heavy duty anchoring of threaded studs into solid concrete and stone. They are ideal for overhead anchoring. The adhesive is the same as described previously but the chemicals are contained in a glass capsule which is absorbed into the chemical mix when broken. Each capsule is positioned in a hole and broken by the insertion of the stud, with no mixing, no mess and no waste. The adhesive cures in 20 minutes, whether applied to dry, wet or underwater.

TABLE 3.15 Heavy duty chemical anchor rating

WORKING TENSILE LOAD IN NEWTONS (kN)				
Anchor size	High tensile steel	Concrete compressive strength		
		20 MPa	32 MPa	40 MPa
M8	6.5	6.6	7.6	8.1
M10	10.3	8.6	9.9	10.3
M12	15.3	12.5	14.4	15.3
M16	29.4	16.9	19.6	20.8
M20 × 150 mm	44.3	24.3	28.1	29.9
M20 × 170 mm	44.3	29.3	33.9	36.1
M24 × 160 mm	64.2	28.8	33.3	35.5
M24 × 210 mm	64.2	43.3	50.1	53.4

FIGURE 3.82 Mini trunking

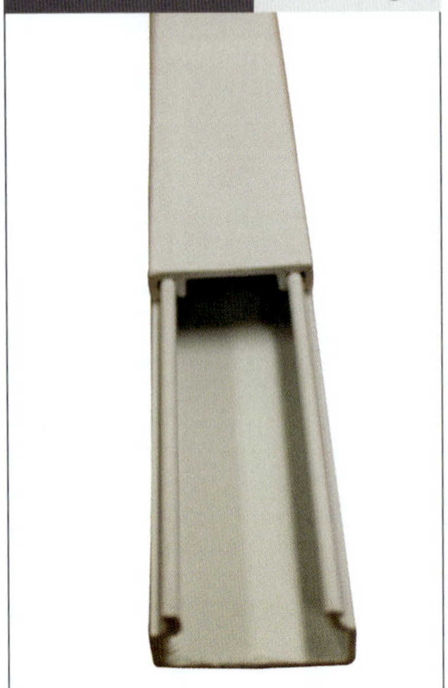

Source: Ed Hawkins

Accessories that may be fixed using adhesives and tapes

Adhesive tapes in regard to electrotechnology applications are usually restricted to light duty applications such as fixing small trunking (see Figure 3.82), labelling and signage to walls.

Chemical anchors have a major advantage, in that they can be fixed closer to edges and to other fixings than mechanical expansion anchors. Applications will be similar to those of mechanical anchors and may also include securing switchboards and enclosures to walls, cable tray supports, large motors, conveyer belt mountings and security fencing around transformers.

Techniques for the application of adhesives and tapes

A major advantage of adhesive bonding tape is its quick, simple and secure application. Figure 3.83 shows the fixing process for adhesive bonding tape, and the process is described in 3M's 'VHB™ tapes design guide':

1 Clean compatible surfaces with isopropyl alcohol and water mixture.
2 Apply tape.
3 Maintain pressure for bonding time.

The protective backing can now be removed and the second surface is attached by repeating steps 1–3.

At room temperature the tape will provide instant tack. Adhesive tape is best applied at temperatures between 10 to 26°C. The tape's bonding strength increases with time; 20 minutes gives 50% strength, 60 minutes gives 75%, 24 hours gives 90% and after three days full strength (100%) is reached.

FIGURE 3.83 Bonding tape application

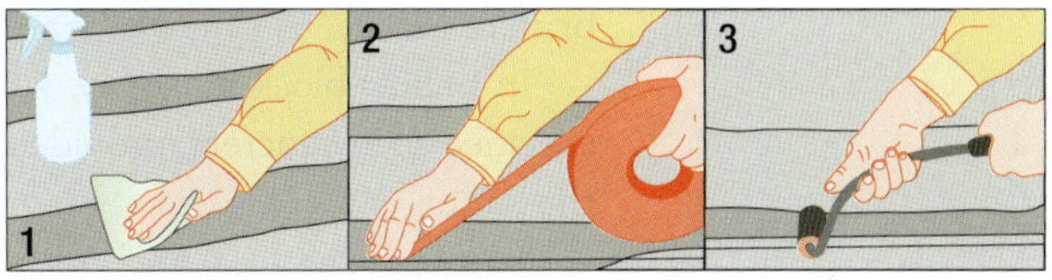

To achieve the medium duty load ratings shown in Table 3.14, chemical adhesive anchors have limits to how close to an edge or another fixing they can be installed. Figure 3.84 shows recommended minimum spacings in brick.

Figure 3.85 shows recommended minimum spacing in masonry block.

FIGURE 3.84 Chemical anchors in hollow brick walls – minimum distances

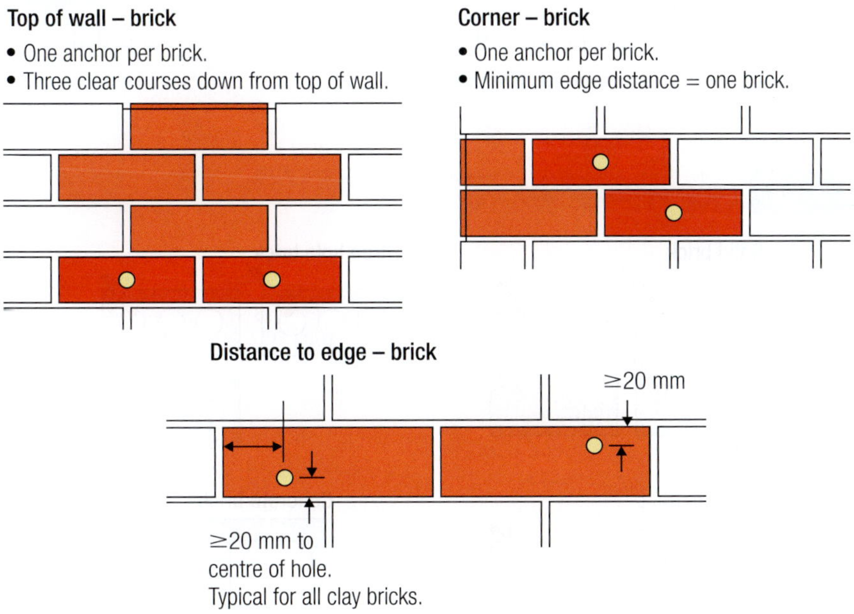

FIGURE 3.85 Chemical anchors in hollow block walls – minimum distances

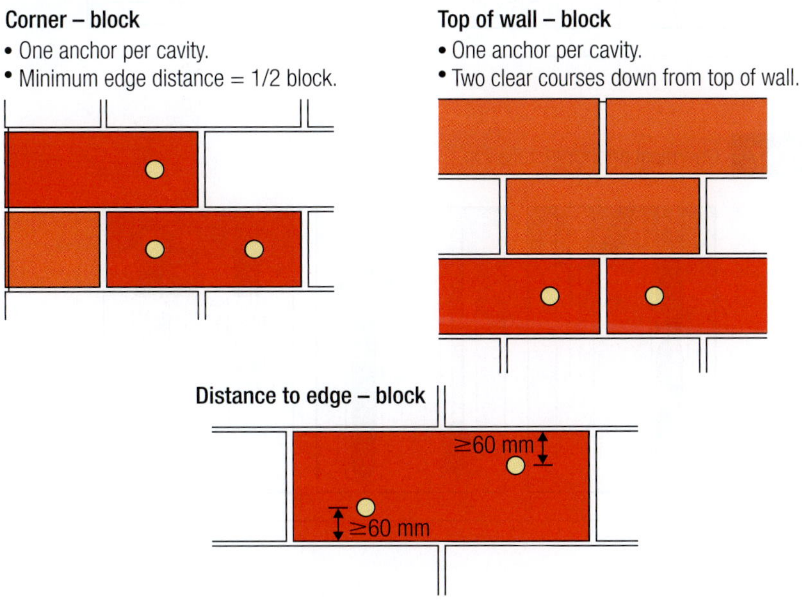

Too many fixings in the one brick or block will weaken the structural integrity of the brick or block. Figure 3.86 gives guidance to the recommended maximum number of medium duty anchor studs per brick or block.

In hollow block masonry, where the cores are filled with concrete, anchors may be designed and specified as in concrete, provided the designer assesses the effective strength of the masonry including the joints.

To install chemical anchors, as shown in Figure 3.87:

1 Drill or core hole to specified diameter and depth.
2 Remove dust and debris by brushing and blowing three times each (if hole is wet or flooded, remove water with wet/dry vacuum).
3 Screw mixing nozzle onto cartridge and dispense 2–3 trigger pulls of adhesive to waste until colour is dark grey with no streaks.
4 Insert tip of nozzle to bottom of hole and dispense adhesive.
5 Fill hole to about 2/3 full.
6 Insert threaded stud or reinforcing bar with rotating motion to release trapped air.
7 Wait until adhesive has fully cured before applying a load.
8 Clean up with acetone.

FIGURE 3.86 Fixings per brick or block

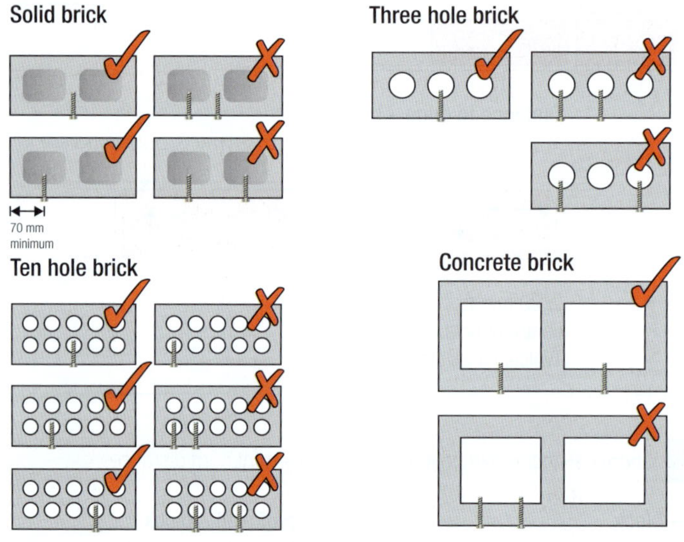

FIGURE 3.87 Chemical anchor installation

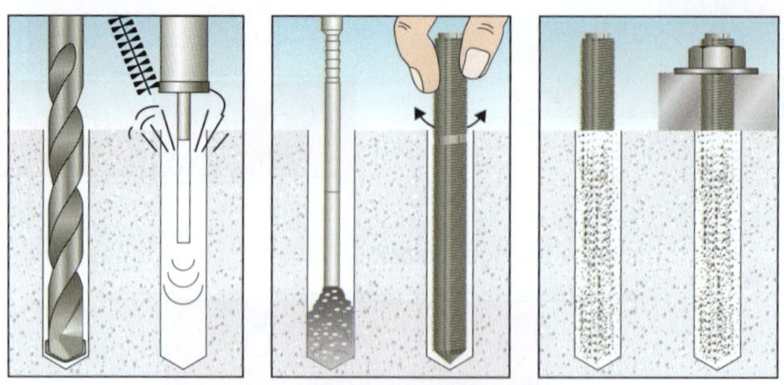

Tools used to apply and cut adhesives and tapes

Many of the same tools are used for both the installation of mechanical anchors and chemical anchors. They include rotary hammer drills and bits, cleaning brushes and air pumps. Figure 3.88 shows a caulking gun used to apply chemical anchor adhesive. Two part resins are mixed together as they are extruded from the packaging. Correct mixing of resins is indicated by a green/grey appearance with no streaks.

FIGURE 3.88 Caulking gun

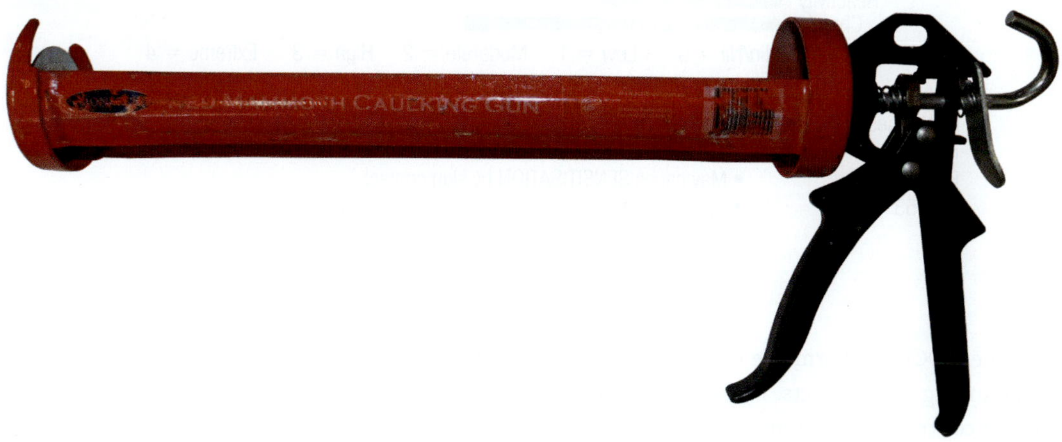

Source: Ed Hawkins

A pneumatic applicator is used when high volume application of chemical adhesive is required.

Mixing nozzles combine the base and hardener resins automatically. Always use recommended mixing nozzles for each adhesive. Using the incorrect nozzle will result in improper mixing and poor load resistance.

Hazards with adhesives and chemical fixing devices

As chemical anchors other than tapes require the drilling of a hole in the substrate the associated risk sources are the same as for mechanical anchors, i.e. noise, eye hazards, sharp edges, vibration and dust. Control measures are also the same as for mechanical anchors. Personal protective equipment should include hearing and eye protection, gloves and dust masks.

Chemical anchor adhesives present a number of risk sources as shown in the extract of the safety data sheet (SDS) in Figure 3.89.

As seen in the extract of the SDS, chemical anchor adhesives present risk sources which include flammability, toxicity, body contact, reactivity and chronic (long-term adverse effects on an aquatic environment). All of these risk sources are reduced (administrative control) by limiting the quantity of chemical adhesive and reduced sized packaging. If there is a fire or a spill, the resulting fire or contamination will be smaller and easier to control. The flammable vapours emitted are controlled through ventilation (engineering). The SDS also contains relevant information relating to treatment for accidental inhalation or ingestion.

Some adhesives release volatile organic compounds (VOCs) which are organic chemicals that have a high vapour pressure at room temperature. They have a low boiling point, which causes large numbers of molecules to evaporate and enter the surrounding air. Most of what we smell are VOCs.

Adhesive tapes and chemical fixings should only be used in work areas that have sufficient ventilation to provide fresh air and dissipate fumes. Care should be taken to prevent skin contact with adhesives, and solvents used in the clean up.

Chemical adhesives are an environmental risk; use carefully.

FIGURE 3.89 Chemical anchor safety data sheets

Section 2 – Hazards identification

Statement of hazardous nature
Hazardous substance. Dangerous goods. According to the criteria of NOHSC, and the ADG code.

Chemwatch hazard ratings

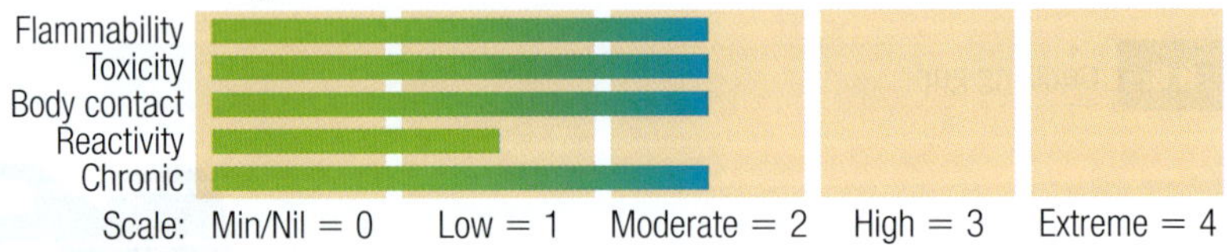

Risk

Risk codes	Risk phrases
R10	• Flammable.
R43	• May cause SENSITISATION by skin contact.
R52/53	• Harmful to aquatic organisms, may cause long-term adverse effects in the aquatic environment.

Source: *Ramset Chemset 801*, http://www.chemwatch.net

Some VOCs can harm the environment (ozone) and are dangerous to humans, causing long-term compounding health effects. The safety data sheet of the adhesive will indicate the level of VOC emissions. The best control measure (engineering) is to use a chemical adhesive that does not emit VOCs.

SUMMING UP...

- Adhesive tapes commonly used by electricians are:
  - foam backed double sided tape for light duty applications
  - bonding tape for heavy duty applications.
- Chemical adhesives use a two part resin to glue metal studs, fine wire mesh or plastic plugs to a substrate.
- Medium duty chemical adhesive fasteners are suitable for use in brick, hollow blocks and solid concrete constructions.
- Heavy duty chemical adhesive fasteners are only suitable for use in solid concrete constructions.
- Mini trunking (in some cases) and labels are commonly secured by the use of adhesive tape.
- To achieve the medium duty load ratings shown in Table 3.14 only one chemical anchor should be installed per brick or hollow block.
- When installing chemical anchors it is most important to ensure the hole is clean and dry before applying the adhesive.
- Commonly used tools for chemical anchor installation include rotary hammer drills, cleaning brushes or hand air pumps, caulking guns and mixing nozzles.
- General risk sources for chemical anchor installation include noise, eye hazards, sharp edges, vibration and dust. General PPE includes hearing and eye protection, gloves and dust masks.
- Installers should be familiar with the SDS (safety data sheet) for the particular chemical adhesive in use. Typical important information shown on the SDS will include chemical product and manufacturer identification, identification of risk sources, first-aid measures, fire-fighting measures and PPE.

REVIEW EXERCISES

Answer these questions and check your answers at the back of the book.

Review questions 3.1

1 What is the most suitable type of screw for heavy-duty fixings to timber?
2 What is the difference between self tapping and self drilling screws?
3 List four ratings of a fixing.

Review questions 3.2

1 List three types of light duty solid wall anchors.
2 What are the two classifications of powder actuated tools?
3 How is a hammer drill different to an impact drill?

Review questions 3.3

1 What are the three main types of metal thread screws?
2 List three uses for washers.
3 A square steel plate is to be secured using four M6 nut, bolt and washer sets by four holes drilled in each corner of the plate. Describe the process of assembly and tightening including the size of the spanner used.
4 Why should a 'shifting spanner' only be used on light work, and then only as a last resort?
5 List four risk sources (hazards) associated with drilling steel and masonry and four items of PPE that should be used when drilling.
6 What type of drill bit is commonly used for mild steel?
7 Name two types of drill bits suitable for timber.
8 What is the type of drill bit suitable for concrete, and what is special about this type of drill bit?

Review questions 3.4

1 What is an advantage of bonding tape over bolts, screws or rivets?
2 Why are heavy duty chemical adhesive anchors not suitable for hollow block walls?
3 What is the recommended tensile load rating of an M12 high tensile steel stud secured with a medium duty chemical anchor adhesive in 32 MPa concrete?
4 What preparation should be made to a surface before applying adhesive tape?
5 How long does it take for bonding tape to reach 100% of its full strength?
6 What is the minimum recommended distance to the edge of a solid brick for the centre of a hole to fit a chemical adhesive anchor?

ONLINE RESOURCES

For more information about the contents of this chapter, visit the following websites:

DeWalt tools and equipment
http://www.dewalt.com.au

EDL Fasteners New Zealand
http://www.edlfast.co.nz

James Glen Stainless Fasteners
http://www.jamesglen.com.au

Ramset tools and equipment
http://www.ramset.com.au

4

HAND AND POWER TOOLS AND THEIR APPLICATION

The electrical tradesperson is required to install and maintain a wide variety of equipment in an equally wide variety of situations. This chapter describes the competent use of commonly used hand and power tools and the techniques used in typical electrotechnology installation and maintenance tasks.

Chapter outline

4.1 Mechanical drawing interpretation and sketching

Drawings used in the electrotechnology industry take many forms. They are used to convey instructions and specifications of all types of work. They allow standardisation and consistent results across work sites. Think of the expression 'a picture is worth a thousand words'. To someone who knows how to interpret it correctly, a drawing can pass on a great deal of information and dramatically reduce the risk of errors.

Engineering drawing is the main method of communication between all persons concerned with the design and manufacture of components, a building, construction and erection of projects and the maintenance of equipment and installations. It shows shapes and dimensions of parts, how those parts fit together and how to connect electrical apparatus. A properly constructed drawing, using correct standards and conventions, can be understood by anyone who recognises the standards used, even though they may not speak the language of the person who did the drawing. In engineering the three main purposes of a drawing are for:

- communication between people involved in a project
- discussion and development of ideas and theories in design and production
- record keeping for future development, reordering and marketing.

Drawings may also be required for legal reasons. For example, house plans must be submitted to a council to meet their requirements or as part of a regulation or standard. Australian and New Zealand drawing standards are aligned with ISO (International Organization for Standardization) standards, so that they may be used in any country that uses these conventions. Common drawing standards used in the electrotechnology industry include:

- **AS 1100.101-1992 Technical drawing – General principles**
- **AS/NZS 1102 Graphical symbols for electrotechnology** (Tafe NSW, *Electrotechnology Practical LSEEE280A*, 2008, p. 3).

Line types

Drawings are pictorial representations of an object. For clarity, the standards require the use of different types and thicknesses of lines to represent various parts of a drawing. Table 4.1 and Figure 4.1 show a summary of commonly used line types in technical drawings.

TABLE 4.1 Line types

AS 1100 SERIES LINE TYPES			
Letter	**Line type**	**Purpose/use**	**Example**
A	Continuous – thick	Visible outlines, general details, existing buildings, landscaping in site plans, busbar and transmission paths	
B	Continuous – thin	Outlines, dimension lines, projection lines and leaders, hatching and revolved lines of sections, fold lines, short centre lines, general electrical conductors and symbols	
C	Continuous – thin free hand	Break lines	
D	Continuous – thin ruled	Break lines	
E	Dashed line – thick	Hidden outlines and hidden edges	
F	Dashed line – thin	Hidden outlines, jumper connections and magnetic or electric screening	
G	Chain line – thin	Centre lines	

FIGURE 4.1 Line types

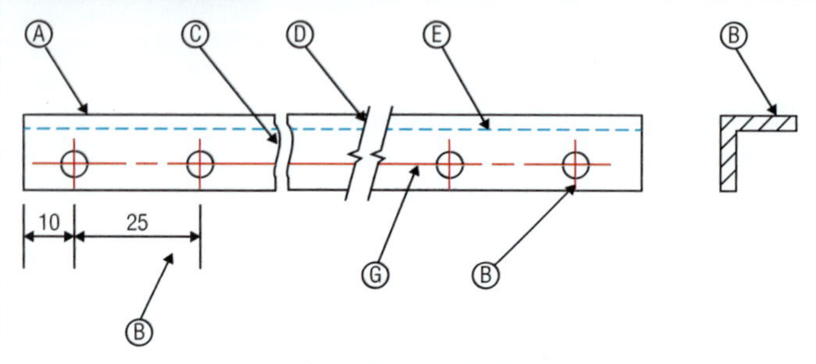

Medium thickness lines may also be included in some structural or electrical diagrams. However, for mechanical drawings it is desirable to limit the number of thicknesses of lines to just two, i.e. thick and thin.

Symbols and abbreviations

Symbols are used in technical drawings to reduce the amount of text in a drawing. This reduces clutter and makes the drawing easier to read. A sample of **AS 1100** symbols is shown in Table 4.2.

TABLE 4.2 Drawing symbols

AS 1100 SYMBOLS		
Symbol	**Meaning**	**Use**
∅	Diameter	Shows the dimensions of the diameter of a circle
R_	Radius	Shows the dimensions of the radius of a circle
□	Square	Shows that dimensions refer to the width across flats of a square section
⌲	Taper	Shows a taper and its direction
⌳	Slope	Indicates a slope and its direction
℄	Centre line	The centre line of a part, feature or group of features
S∅	Spherical diameter	Shows the diameter of spherical surface
SR_	Spherical radius	Shows the dimension of a radius of a spherical surface
⌵	Countersink	Indicates that countersink allows screw head to sit flush in surface
↧	Depth	Indicates depth of a feature such as a hole or groove
⌒	Arc	Shows that a dimension refers to the arc length
⌖	Position	Used to indicate positional tolerance
//	Parallel	Used to provide tolerance information on components which are parallel

(*Continued*)

TABLE 4.2 *(Continued)*

Symbol	Meaning	Use
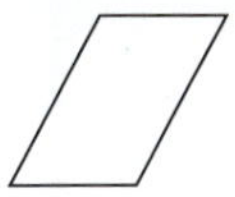	Flatness	Used to provide tolerance information on components which are flat
	Squareness	Used to provide tolerance information on components which are at 90° to each other
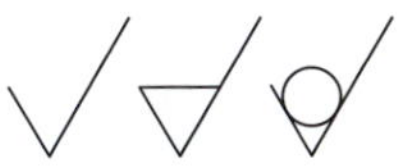	Surface finish	(Left) any convenient method (Centre) machined finish required (Right) removal of original material not permitted
	Dimension datum identification	A datum is used as a reference point to measure to. A datum can be a point, line, plane or cylinder depending on the object being drawn
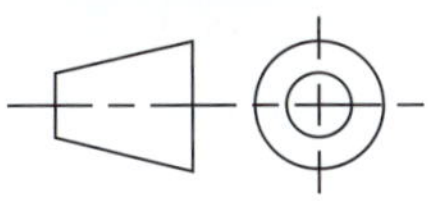	1st angle projection	When views are drawn in their relative positions, top view comes below front view, right side view comes to left of front view
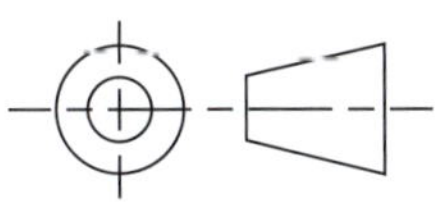	3rd angle projection	When views are drawn in their relative positions, top view comes above front view, right side view comes to right of front view

Abbreviations are also used to reduce the clutter in a drawing and simplify interpretation. For a complete list of acronyms and abbreviations refer to Table 1.1 of **AS 1100.101**.

Interpreting mechanical drawings

A drawing is effectively a picture converted to a series of single lines, like a cartoon is to a photograph. A drawing is less complicated and much easier to produce, while still providing the minimum required information. Industry requires the use of drawings prepared using standard conventions.

Orthogonal projections

The direction at which an object is observed is called a *view*. Think of how a dice has six sides and six possible views. Figure 4.2 shows how the six possible views may be laid out flat when viewing one side. Orthogonal drawings are a two-dimensional graphical representation of an object, showing as many views as required to provide the required detail to document the object. If an object contains circular shapes and is viewed at any angle other than 90° to the object, those circles will become eclipses and become difficult to reproduce with conventional drawing instruments such as a compass. Multiple views allow circular shapes to be drawn without distortion. In most cases only three views are required. These are called regular views, and normally only the front, right end and top views are shown. In some cases a cross-sectional view may be included to show any hidden detail. Figure 4.3 shows an example of an orthogonal drawing.

FIGURE 4.2 Orthogonal drawing views

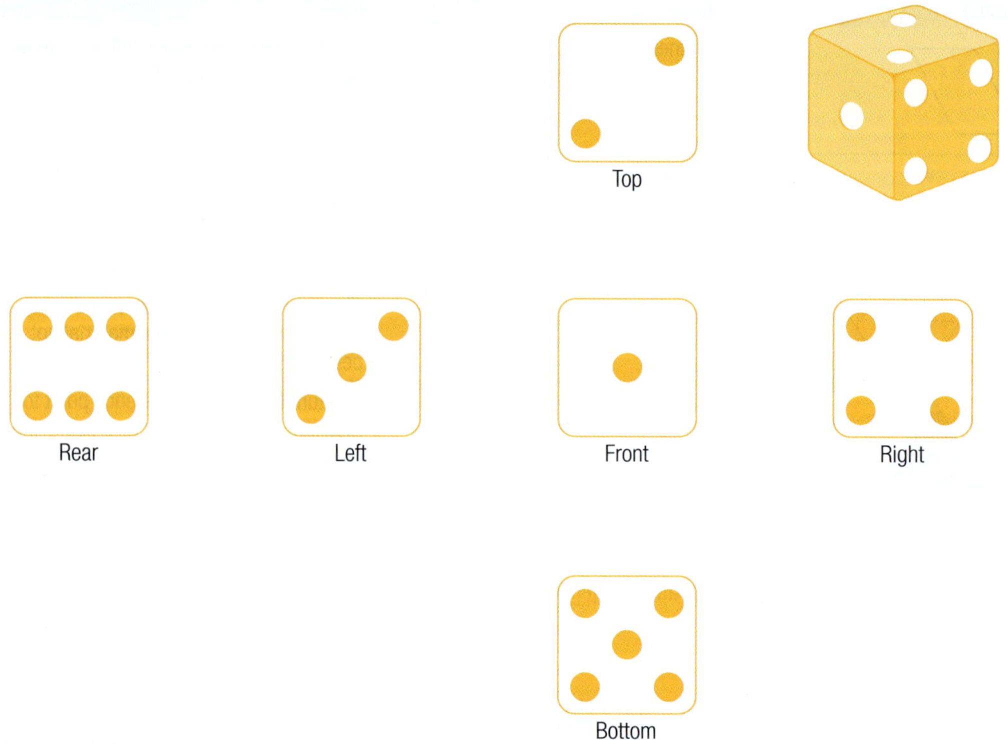

FIGURE 4.3 Orthogonal drawing 3rd angle of projection

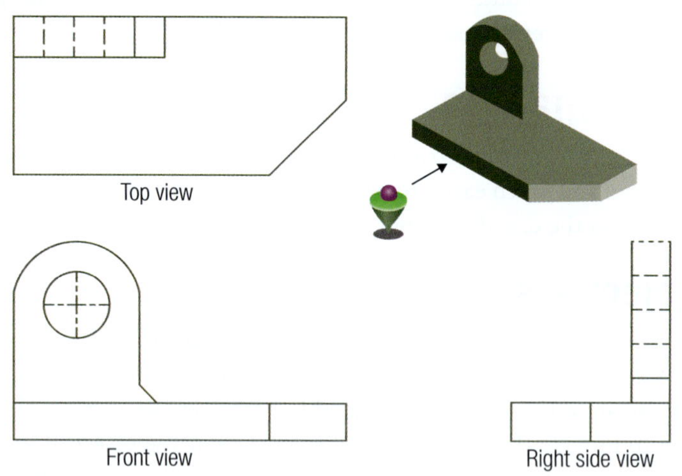

Angle of projection

In orthogonal drawings there can be two methods of angles of projection called 1st (Japan and Europe) and 3rd angles of projection (Australia and the US), as mentioned in Table 4.2. Drawings must identify which method is used in the title block. Symbols that identify the method used are shown at the bottom of the left column in Table 4.2.

An orthogonal drawing using 1st angle projection is shown in Figure 4.4. **AS 1100.101** recommends the use of 3rd angle of projection, as shown in Figure 4.3, due to its logical layout, that is, the side view is to the right-hand side of the front view.

FIGURE 4.4 Orthogonal drawing 1st angle of projection

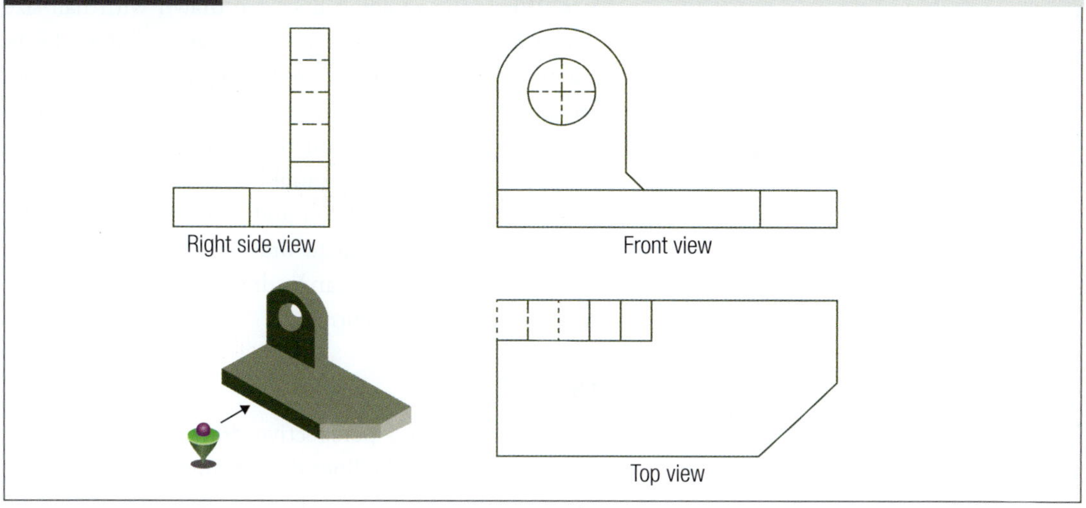

Pictorial views

When a three-dimensional representation of an object is required, pictorial views are drawn. This type of drawing allows designers to visualise the object before any prototypes are constructed.

Perspective

Imagine you are taking a photo of a building. If an orthogonal view was to be shown, the picture would be taken from directly in front of the building, and a number of pictures would need to be taken to show the whole building. In a perspective view the image is captured from one *corner*, usually at some height to show two sides of the building. From this view point the front side and top are distorted and appear to merge to a point somewhere in the distance. You will have seen this optical illusion, known as a vanishing point, on long straight roads and railway lines.

Perspective drawings are often used by architects (artist's impressions) to show the appearance of a complicated building. It is a type of drawing that realistically depicts three-dimensional objects in a two-dimensional plane, as illustrated in Figure 4.5.

Oblique drawings

These pictorial drawings use one face as a true shape drawn from directly in front of the object. This is useful if the surface contains circular objects. To create the illusion of 3D, lines receding into the

FIGURE 4.5 Perspective drawing's vanishing point

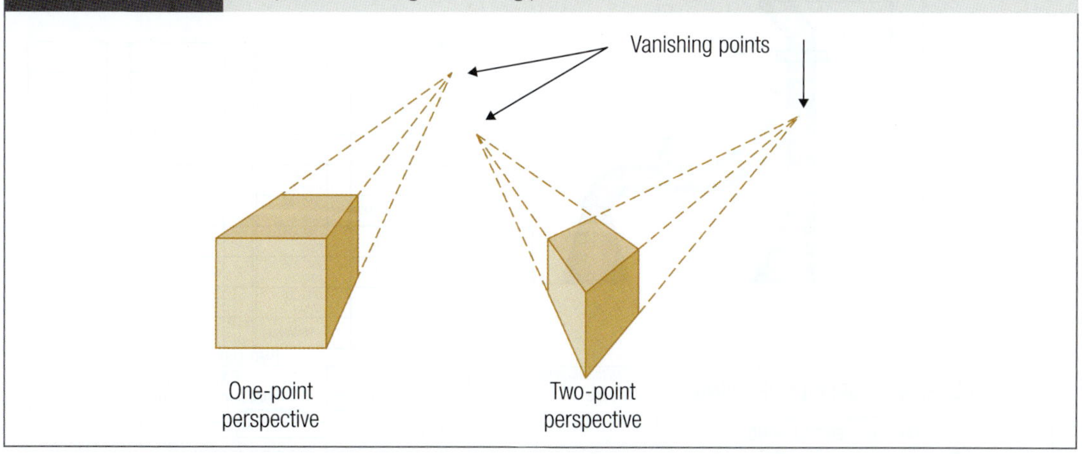

FIGURE 4.6 Oblique drawings

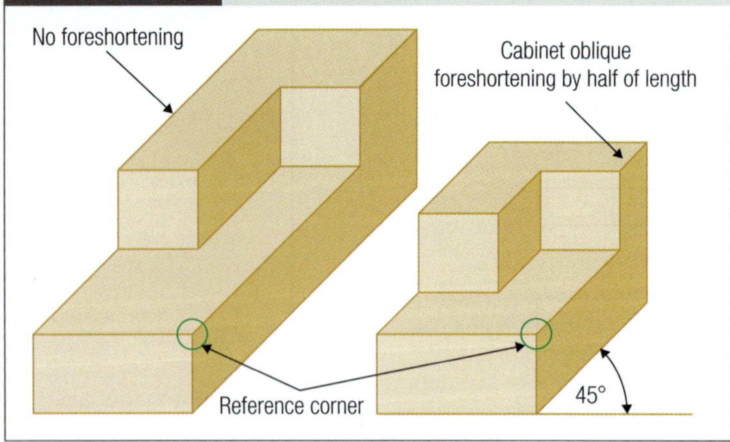

FIGURE 4.7 Isometric drawings

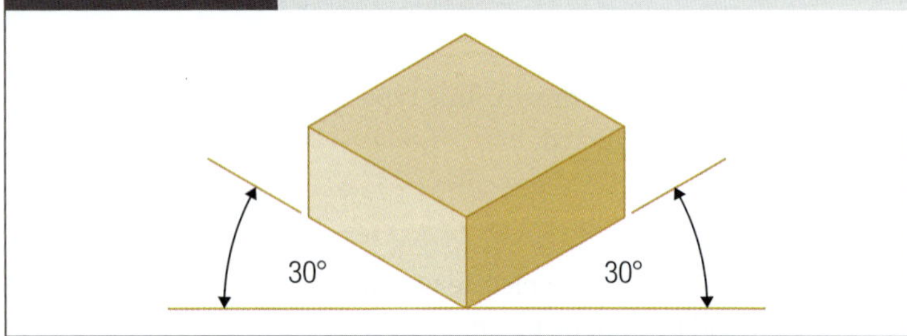

background are angled at 45° to the horizontal front true face. When the receding lines are drawn with half-size dimensions (foreshortening), as shown in Figure 4.6 to maintain proportionality, the drawing is called a 'cabinet oblique'. Notice how the image on the right-hand side of Figure 4.6 appears more realistic than the image to the left.

The reference corner should be chosen to show the maximum amount of detail and if possible any circular objects, to avoid having to draw distorted circles. The receding lines therefore can be drawn to either the left or right of the reference corner.

Isometric

This is similar to a perspective drawing, except in an isometric drawing the lines do not merge to a vanishing point, but are parallel to each other. This type of drawing is the most commonly used type of technical drawing. All lengths are drawn full size (no foreshortening) and all receding lines are drawn at 30° from a horizontal datum as shown in Figure 4.7.

Assembly drawings

Assembly drawings show a general overview of the completed, assembled work, with the arrangements and a list of parts. They show only overall dimensions or none at all, so they are not suitable for the manufacture of parts. They are also called general assembly (GA) or arrangement drawings when showing larger equipment or installations as in the example in Figure 4.8.

FIGURE 4.8 General assembly drawings

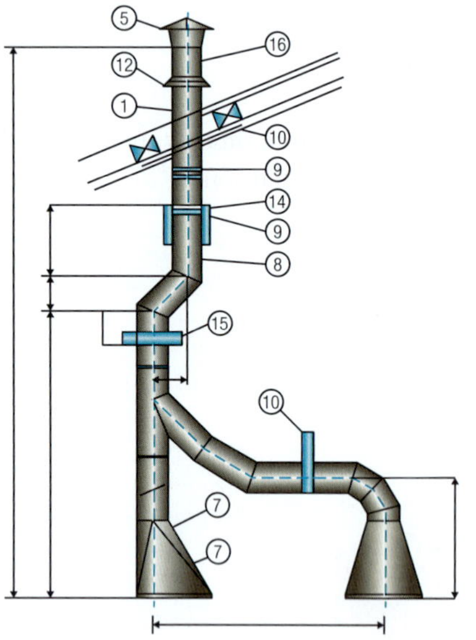

PART No.	DESCRIPTION	QTY	REMARKS	PART No.	DESCRIPTION	QTY	REMARKS
9	FLEXIBLE JOINT	1	ORG 3B2 ITEM 1				
8	ARIAL FAN	1	PURCHASE SEE NOTE				
7	RECTANGULAR ROUND HOOD	1	ORG 380 ITEM 1	16	DAMPER	1	ORG 230 ITEM 2
6	CONICAL HOOD	1	ORG 330 ITEM 3	15	DUCT SUPPORT	1	ORG 332 ITEM 3
5	COWL	1	ORG 381 ITEM 1	14	FAN SUPPORT SUB-ASSY	1	ORG 325
4	ANGULAR BRANCH	1	ORG 329 ITEM 1	13	DUCT SUPPORT	1	ORG 355
3	4-PIECE 80" ROUND ELBOW	1	ORG 326 ITEM 3	12	APRON	1	ORG 332 ITEM 4
2	ROUND OFFSET	1	ORG 326 ITEM 2	11	DUCT	1	ORG 329 ITEM 2
1	CYLINDRICAL FLASHING	1	ORG 326 ITEM 1	10	DUCT SUPPORT	1	ORG 332 ITEM 2

PART LIST

GENERAL ASSEMBLY DRAWING	ALL DIMENSIONS IN MILLIMETRES TOLERANCES	APPROVED ISSUD PRACTICE A.S. 1100	EXHAUST DUCT FAN SUPPORT - SUB ASSEMBLY
TAFESA - ENGINEERING	LINEAR ANGULAR		SCALE 1 : 1 — REQD. 1. off Brass DRN BY. NOH — UNIT No. 7.32A DATE 300906 — DWG No. 603.1

Source: TAFE SA Vocational Education & Training Resources, MEM30305 Certificate III in Engineering – Fabrication Trade – RPL Forms, http://creativecommons.org/licenses/by/3.0/au/deed.en

Sub-assembly drawings

These show the arrangements of a few parts of the general assembly drawing so it can be seen how a section of the total work is assembled (see Figure 4.9).

FIGURE 4.9 Sub-assembly drawings

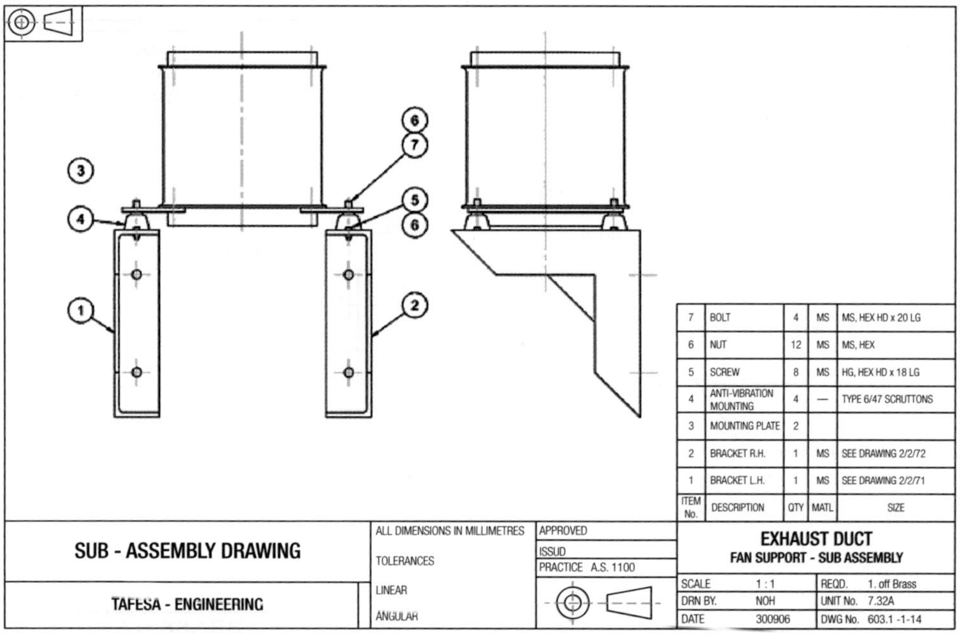

Source: TAFE SA Vocational Education & Training Resources, MEM30305 Certificate III in Engineering – Fabrication Trade – RPL Forms, http://creativecommons.org/licenses/by/3.0/au/deed.en

Detail assembly drawings

As illustrated in Figure 4.10, detail assembly drawings show how the work is assembled, together with the details needed to manufacture it.

FIGURE 4.10 Detail assembly drawings

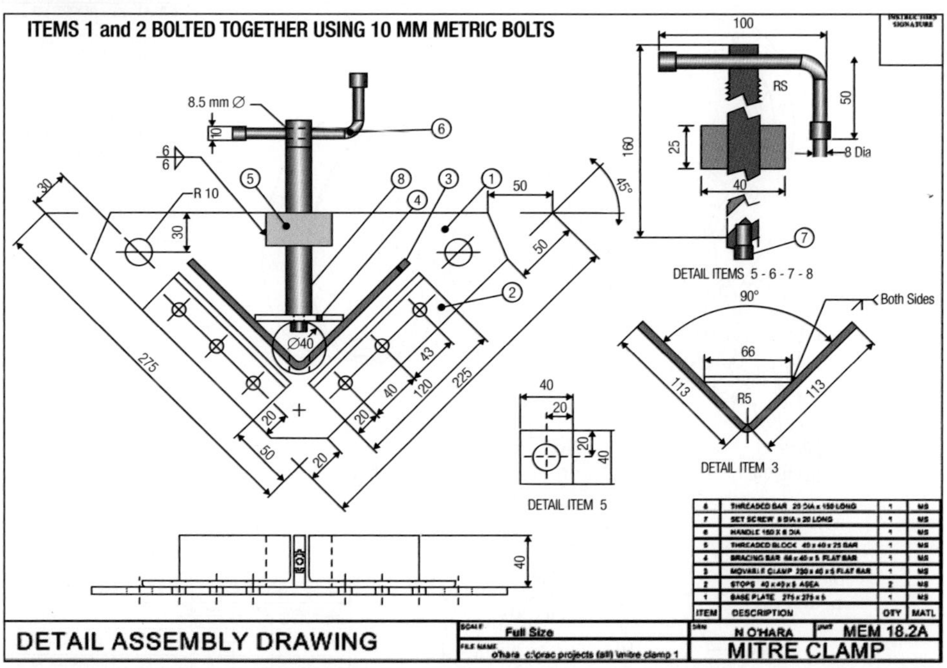

Source: TAFE SA Vocational Education & Training Resources, MEM30305 Certificate III in Engineering – Fabrication Trade – RPL Forms, http://creativecommons.org/licenses/by/3.0/au/deed.en

Detail drawings

These drawings contain all the information required to manufacture the item that is drawn. They may consist of a single part or a number of parts of the one assembly on the same sheet, as shown in Figure 4.11.

FIGURE 4.11 Detail drawings

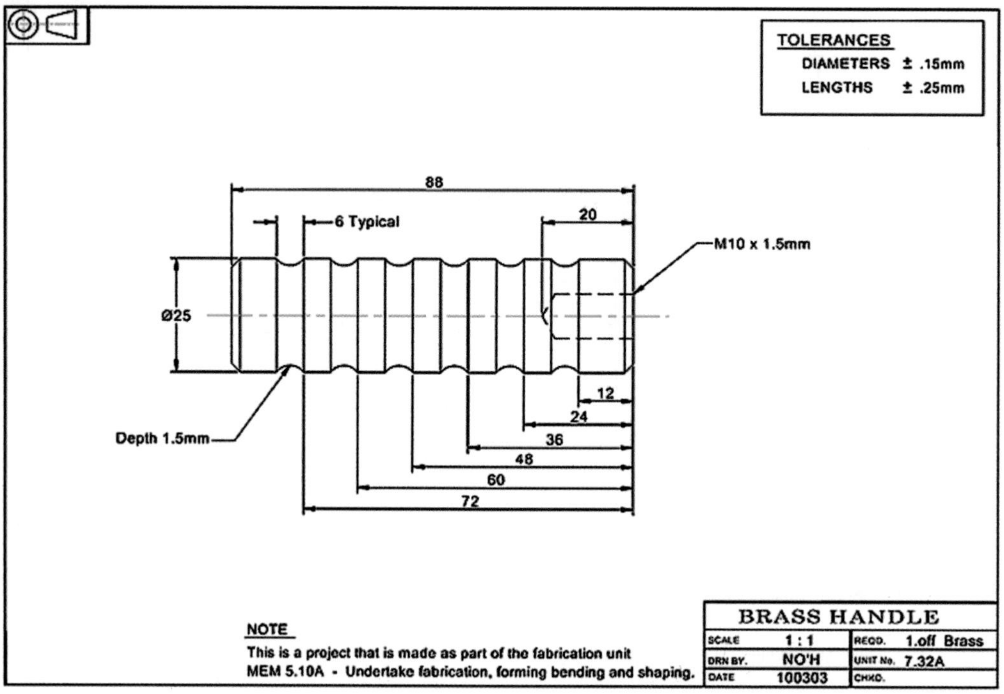

Source: TAFE SA Vocational Education & Training Resources, MEM30305 Certificate III in Engineering – Fabrication Trade – RPL Forms, http://creativecommons.org/licenses/by/3.0/au/deed.en

These diagrams are used to assist in assembling and disassembling an object. An example is shown in Figure 4.12. Parts are labelled and numbered to assist identification.

FIGURE 4.12 Detail drawings

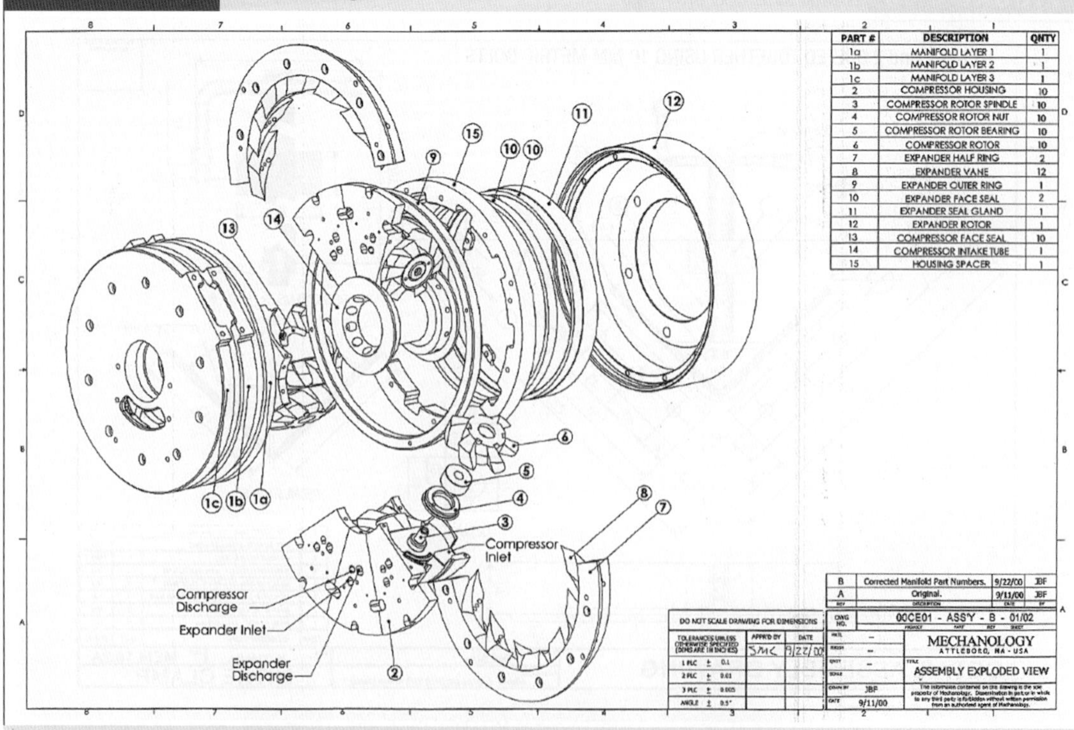

Source: Mechanology

Sectioning

If some detail is hidden inside the object, a pictorial drawing or a projected view is not enough to show a clear view of the construction of the object. Hidden outlines may be used to give a clearer picture if the drawing is quite simple. In more complex objects a cross-section is used to provide detail. If hidden detail outline can't be used, or would make the drawing harder to read, a sectional view or views may need to be drawn.

Figure 4.13 shows a cross-section of a fitting which has been imaginarily cut between the two points labelled 'A' on the end view drawing – the resulting section is therefore labelled 'A-A'.

FIGURE 4.13 Sectional view

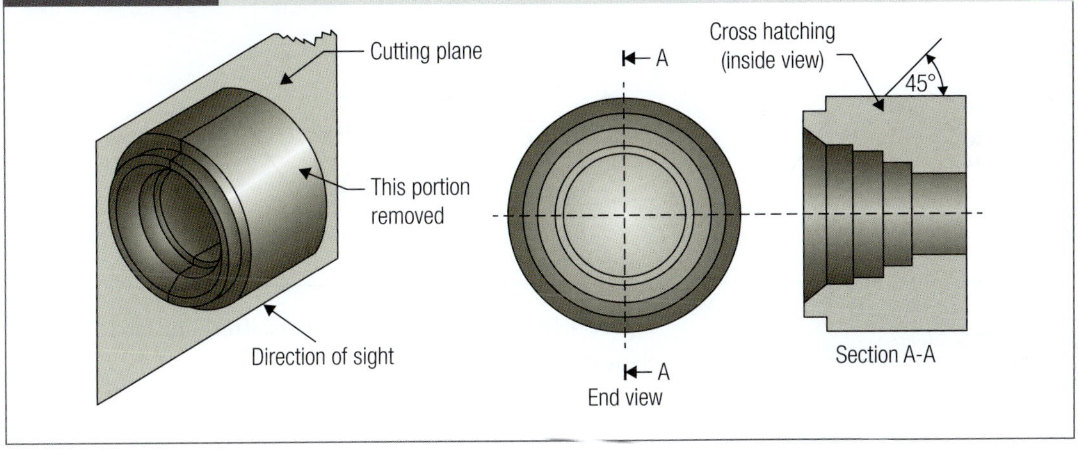

Freehand drawings

FIGURE 4.14 Simple freehand drawing

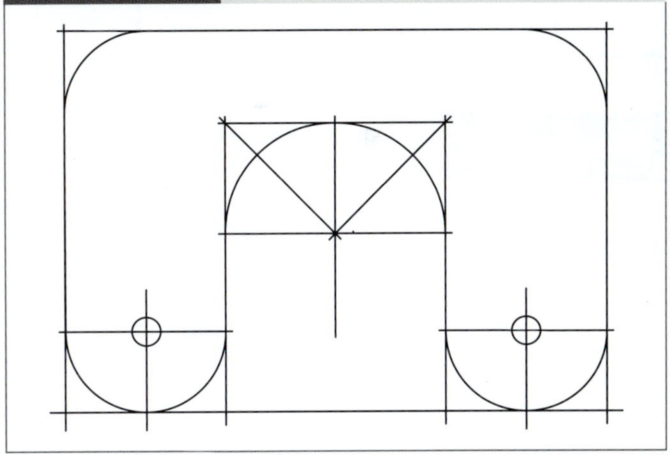

This refers to freehand drawings of mechanical components that show all the information needed for its manufacture/fabrication. Freehand drawings (see Figure 4.14) are drawn without the use of drawing aids such as rulers, set squares and compasses. They are intended for preliminary designs which may later lead to further drawings or one-off jobs that do not warrant the time required to produce technical drawings.

Freehand sketches should be drawn using standard conventions, that is, dimensioning and line types. NSW HSC Online describes such sketches as: 'In appearance, the completed freehand drawing is entirely different from the clean-cut precision of a mechanically ruled or computer generated working drawing. It may have a rough appearance, but it should be clear, concise and suitable for the intended purpose'.

It is standard practice to sketch the terminal connections on equipment to record electrical connections, polarity and phase sequence before dismantling, to enable correct reconnection. Electricians often need to sketch a part for identification when ordering or when replacing a broken part of a machine or installation.

Freehand sketching technique

When drawing straight lines, move your hand on the paper, and draw longer lines in parts, moving your hand each time. The paper may be rotated when drawing lines at an angle.

Figure 4.15 shows some of the techniques used in freehand sketching. These include:

1 When drawing horizontal lines, your hand slides along from left to right.
2 For vertical lines use a wrist movement from top to bottom.
3 For sloping lines from upper left to lower right, move your wrist, holding your hand above the line.
4 Draw lines from lower left to upper right by sliding your whole hand.
5 When drawing lines from upper right to lower left, use your wrist. Your hand will always be below the line.

Remember, hold your pencil lightly and hold it about 5 or 6 centimetres from the point.

FIGURE 4.15 Freehand sketching straight lines

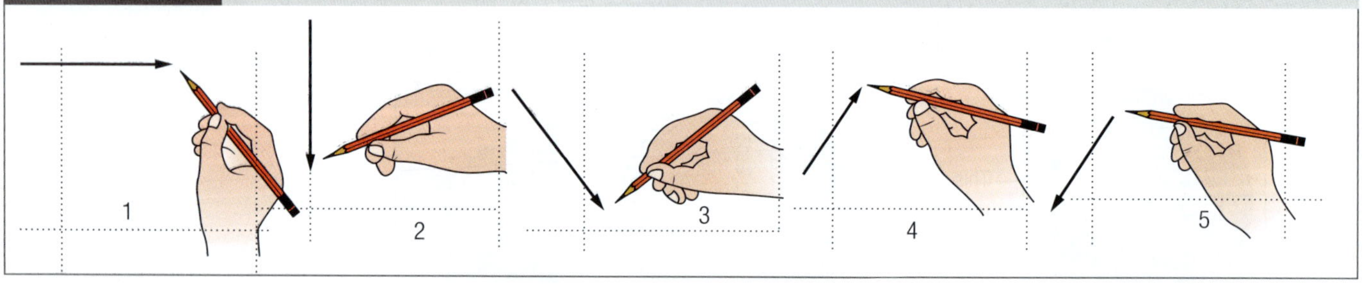

FIGURE 4.16 Freehand sketching circles

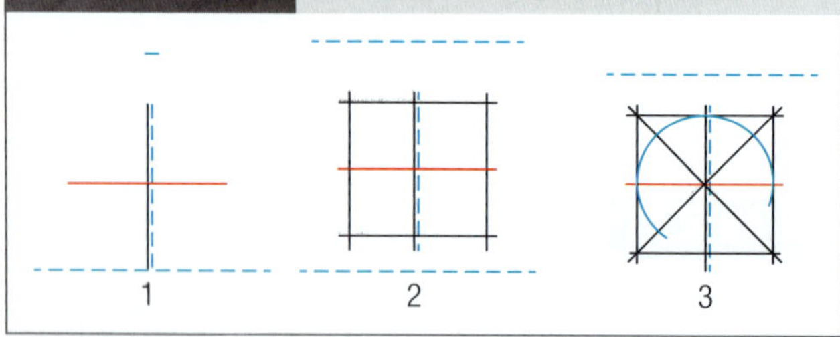

FIGURE 4.17 Simple freehand sketch techniques: large circles

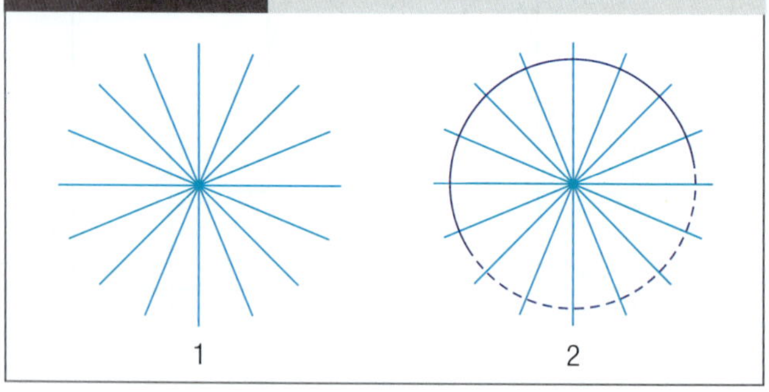

Figure 4.16 show the steps to sketch a freehand circle. That is:

1 draw light centre lines
2 mark the radius in the form of a box
3 join the points with a smooth even curve.

Larger circles can be draw as shown in Figure 4.17 by:

1 lightly sketching a number of radial lines marked with the radius of the circle, then
2 joining the points.

Drawing layout conventions

Drawings are produced on standard size sheets to aid in their storage and also as an aid to reproduction and copying. Typically A3 or A4 size sheets are used for drawings, although larger sheets are used for building construction plans and smaller sheets could also be used, depending on the type of drawing required. Figure 4.18 shows sheet sizes A0 to A4. Notice that each smaller sheet size is achieved by folding the larger sheet size in half, down the long side of the sheet.

A mechanical drawing does not only show the part being represented. There needs to be information enabling the drawing to be interpreted, identified and archived. This information is mainly located in the *title block*, which is found in the bottom right-hand corner of the drawing. The format and layout of the title will vary between organisations. The title block shown in Figure 4.19 is typical of industry standards.

FIGURE 4.18 Detail assembly drawings

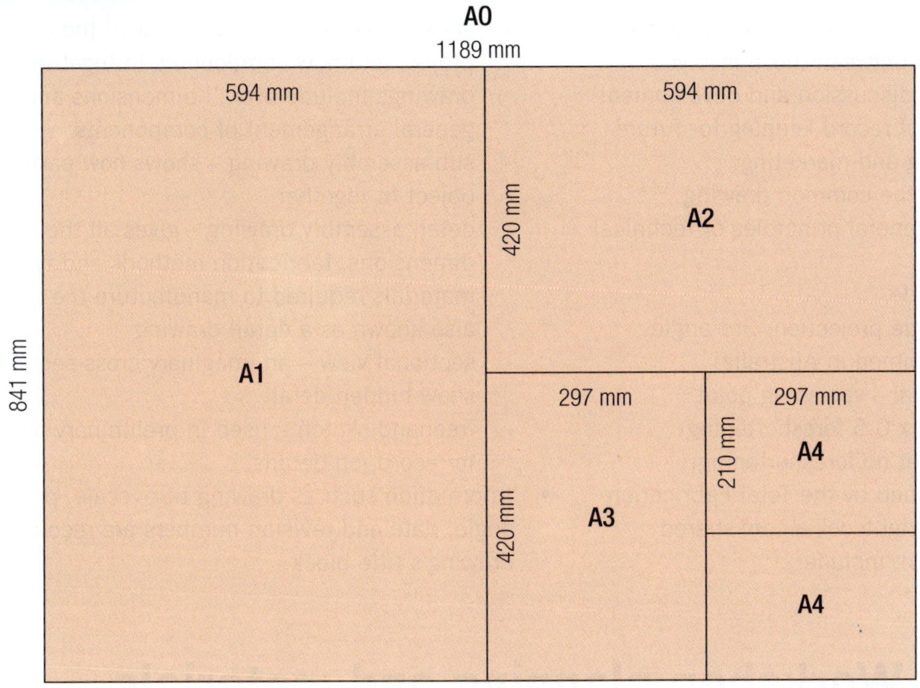

FIGURE 4.19 Example of a title block

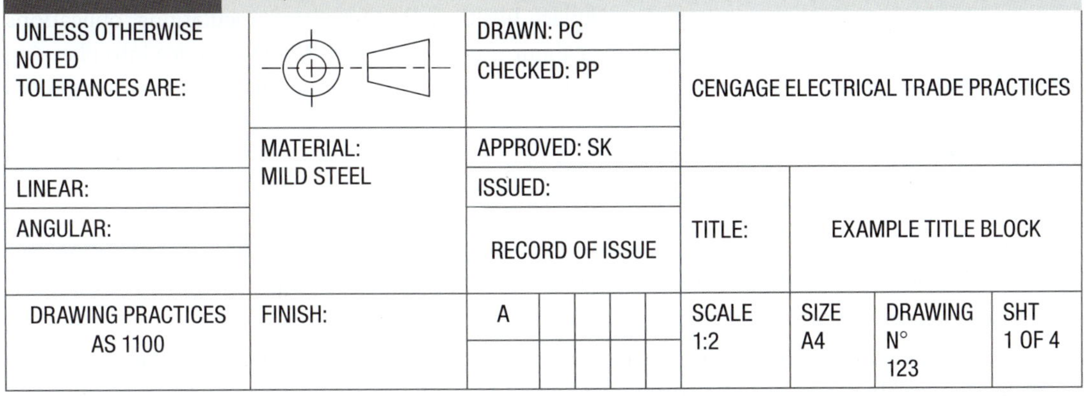

The information found in the title block is self-explanatory. Key items may include:

- part name, what is drawn on the sheet, item or part number(s)
- drawing number generally abbreviated to Drg N°
- the sheet number in a set of drawings
- scale of the drawing
- projection symbol indicating method of representation which is normally 3rd angle in Australia
- date drawn
- date issued and record of issue.

SUMMING UP...

- The main purposes of an engineering drawing are to provide communication between those involved in the project, to promote discussion and development of ideas and as means of record keeping for future development, reordering and marketing.
- AS 1100.101-1992 is the common drawing standard covering the general principles of technical drawing.
- Types of drawings include:
  - orthogonal – 1st angle projection; 3rd angle projection (most common in Australia)
  - perspective – pictorial – vanishing point
  - oblique – 45° offset x 0.5 foreshortening
  - isometric – 30° offset no foreshortening.
- Detail drawings, as defined by the Total Fabrication Glossary (http://metal.brightcookie.com/shared/glossary/glossary_a.htm), include:
  - general assembly – give a view of the completed project and how components fit together. These drawings include overall dimensions and the general arrangement of components
  - sub-assembly drawing – shows how parts of an object fit together
  - detail assembly drawing – gives all the dimensions, fabrication methods and types of materials required to manufacture the article; also known as a detail drawing
  - sectional view – an imaginary cross-section to show hidden detail
  - freehand sketch – used in preliminary design or to record job details.
- Information such as drawing title, scale, projection angle, date and revision numbers are recorded in a drawing's title block.

4.2 Workshop planning and materials

Before any work begins, some thought must go into planning the work about to be undertaken. The result of a properly planned job will see increases in both safety and productivity. Chapter 1 of this book provided a comprehensive view of workplace health and safety laws, regulations, obligations and procedures. This section provides guidance about integrating good WH&S practice into planning everyday work activities. This section also describes a variety of typical materials and their properties used in the electrotechnology industry.

The planning process

For work to be carried out in a safe and productive fashion it should follow the systematic process of Plan, Do, Complete and Verify. The planning stage has a number of key components which include:

- WH&S risk assessments, toolbox meetings and appropriate training of workers
- identifying possible environmental issues and establishing appropriate safeguards such as correct disposal of waste materials and control of sediment run-off
- development and interpretation of working drawings
- ordering and supply of required materials and tools
- determining the sequence in which the job is to be carried out.

Working safely

You will recall from Chapter 1 that under WH&S best practice, before a job begins (in the planning stage) persons conducting business or undertakings (PCBUs) should:

1. **define** (identify) risk criteria (see Figure 1.4)
2. **control** the risk using the hierarchy of WH&S control measures (see Figure 1.5)
3. **monitor** and review the process. The risk assessment is repeated until the risk is reduced to an acceptable level (see Figure 1.3).

PCBUs have a legal and moral obligation to ensure this process is carried out. Workers have the same obligations to participate and cooperate in the process. Safe working method statements developed in consultation with workers should be developed and used as procedure for all work activities.

Typical workplace non-electrical hazards

Hazards associated with using hand tools, such as hammers, chisels, screwdrivers, pliers, wrenches and cutting tools such as saws and knives include:

- laceration and puncture wounds
- eye injury
- noise
- manual handling
- being struck by a falling or flying object
- slips, trips and falls.

Similar risks (hazards sources) are present when using powered equipment with the addition of entanglement in rotating parts, injury if power tool jams, flying debris, dust, vibration and noise.

Risk assessment

Chapter 1 provides detailed guidance for undertaking a workplace risk assessment. See Figures 1.6 and 1.7 in Chapter 1, which show examples of a completed risk assessment form or 'risky'. The aim of a risky is to identify all risks likely to be associated with a given task. Once these are identified, the control measures to minimise each risk are determined.

Controlling identified hazards

Figure 4.20 shows instructions for the correct and safe use of a pedestal drill in the form of a standard operating procedure (SOP). The SOP provides a list of control measures for the safe use of the equipment. All equipment in a workshop environment should have an SOP specific to the equipment, and all users should be familiar with the SOP. Typically, an SOP is posted next to the associated equipment.

FIGURE 4.20 Safe operating procedures (SOP)

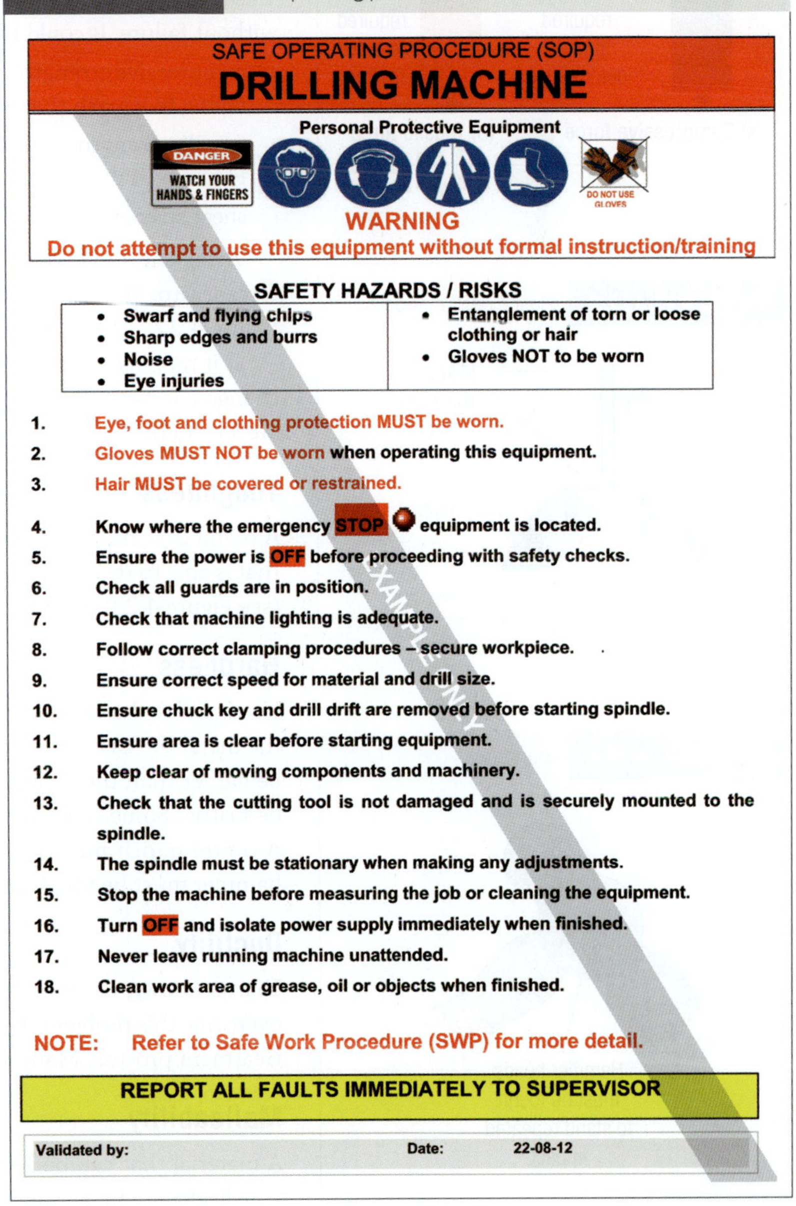

SAFE OPERATING PROCEDURE (SOP)

DRILLING MACHINE

Personal Protective Equipment

DANGER WATCH YOUR HANDS & FINGERS

DO NOT USE GLOVES

WARNING

Do not attempt to use this equipment without formal instruction/training

SAFETY HAZARDS / RISKS

• Swarf and flying chips • Sharp edges and burrs • Noise • Eye injuries	• Entanglement of torn or loose clothing or hair • Gloves NOT to be worn

1. Eye, foot and clothing protection MUST be worn.
2. Gloves MUST NOT be worn when operating this equipment.
3. Hair MUST be covered or restrained.
4. Know where the emergency STOP equipment is located.
5. Ensure the power is OFF before proceeding with safety checks.
6. Check all guards are in position.
7. Check that machine lighting is adequate.
8. Follow correct clamping procedures – secure workpiece.
9. Ensure correct speed for material and drill size.
10. Ensure chuck key and drill drift are removed before starting spindle.
11. Ensure area is clear before starting equipment.
12. Keep clear of moving components and machinery.
13. Check that the cutting tool is not damaged and is securely mounted to the spindle.
14. The spindle must be stationary when making any adjustments.
15. Stop the machine before measuring the job or cleaning the equipment.
16. Turn OFF and isolate power supply immediately when finished.
17. Never leave running machine unattended.
18. Clean work area of grease, oil or objects when finished.

NOTE: Refer to Safe Work Procedure (SWP) for more detail.

REPORT ALL FAULTS IMMEDIATELY TO SUPERVISOR

Validated by: Date: 22-08-12

Materials and their application

Metals

Pure metals commonly used in the electrical trades include iron, copper, gold and silver. A metal alloy is a combination of elements and other compounds, for example, mild steel, stainless steel, brass, bronze and aluminium. Metals that are used extensively in manufacturing include steel, aluminium, copper and zinc. Each metal has different properties and acts in a different way when subjected to the effects of a force, heat, chemical action, electricity or magnetism. For example, copper is a very good conductor of electricity, but does not have the mechanical strength or the magnetic properties of steel. It is, however, cheaper than materials with higher conductance such as silver.

Properties of a metal

The characteristics and behaviour of a metal determine its classification and its uses. How these properties change with external influences such as temperature is also important, for example, changing the properties of a metal to make it more useful. Temperature change, or heat treatment, is used to alter the properties of certain types of steel, making them suitable for use as cutting tools such as files or saw blades.

FIGURE 4.21 Compressive and tensile strength

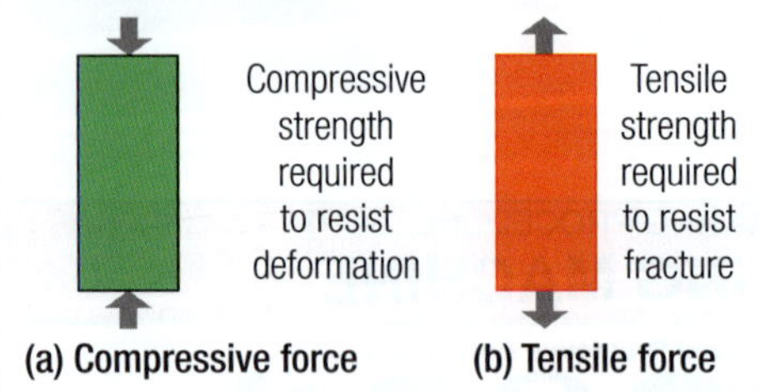

FIGURE 4.22 Plastic deformation

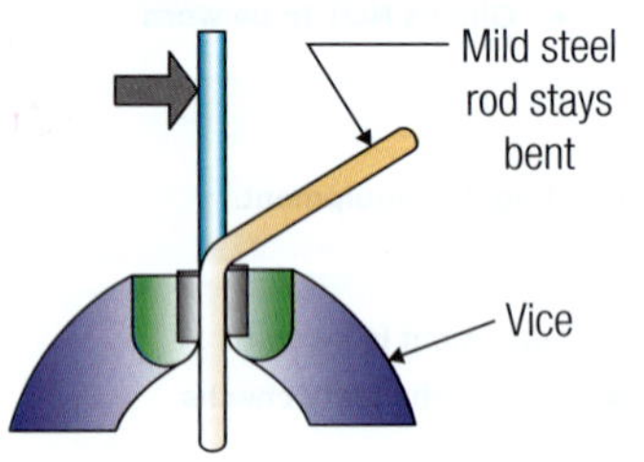

FIGURE 4.23 Toughness

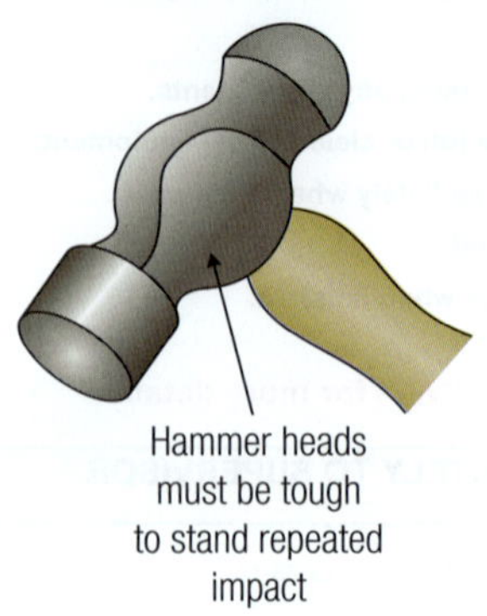

Mechanical properties of a metal

As explained in Chapter 3, whenever a force or load is applied to a piece of solid metal the force tries to change its shape and causes stresses to be present within the metal. The way the metal's shape changes depends on how the force is applied. This is said to be the strength of the metal, which is its ability to withstand stress without failure. It could be:

- compressive strength – the resistance to a squeezing or squashing force that is trying to squash the material, as shown in Figure 4.21
- tensile strength – the resistance to a pulling or stretching force as shown in Figure 4.21
- shear strength – the resistance to forces trying to cut through the material.

After the force is removed the material may remain in the shape it was forced into. If so, the material has suffered plastic deformation as shown in Figure 4.22.

If it returns to its original shape or springs back, that is, it is not permanently changed, it has suffered elastic deformation. The springs used in car suspensions are a form of elastic deformation.

Toughness

A metal that fractures under impact or shock loads is said to be brittle. A metal that is able to resist fracture under shock loads is said to be tough. A hammerhead (see Figure 4.23) must be tough to avoid shattering under impact.

Hardness

Hardness is the ability of a material to resist indentation and wear. The softer the metal the easier it is to form or cut to shape. The tools that cut or form must be harder than the material being cut. Generally, a material that is very hard will be brittle. Some tools are heat treated to become partly hard and partly tough. A centre punch must have a hardened point, but the head being struck by the hammer must be tough enough to resist damage.

Ductility

This is known as *work hardening*, as the copper will no longer be ductile. To overcome this problem the material is regularly annealed which is a softening heat treatment process. Many metals are affected by work hardening.

Malleability

A material that can be beaten into shape is said to be malleable. An example of a malleable metal is lead, which can easily be shaped with the used of mallets or hammers.

Types of metals used in electrotechnology

Copper

Common uses for copper include the manufacture of electrical wiring and busbars as well as pipes and tanks. Copper is a good conductor of electricity and heat, and is also corrosion resistant. It is easy to forge or beat into shape but is not easy to cast or machine. However, many of its alloys are readily shaped. These alloys resist corrosion even better than plain copper. Examples include:

- brass – an alloy of copper and zinc. Used for such things as electrical contacts and terminals, switch mechanisms and plumbing fittings
- bronze – an alloy of copper and tin. Used for bearings, castings and marine fittings.

Aluminium

Aluminium is relatively cheap and widely used in industry for such things as large cross-section electrical conductors, structural components and window frames. Aluminium is a good conductor of heat and electricity, it resists corrosion and has a density of about one third that of steel.

Alloying aluminium with other elements produces strong, lightweight parts. Magnesium, titanium and beryllium are common alloying elements. Aluminium alloys are used for castings and almost everything that steel can be used for.

Ferrous metals

These are metals that contain iron, and include steel, cast iron and stainless steel. They all vary in their composition, creating a wide, very versatile group of metals.

Steel

Steel is an alloy of mostly iron and carbon. The carbon content of steel determines its grade and uses. Table 4.3 provides a summary of typical steel grades and uses.

TABLE 4.3 Typical steel grades and uses

Grade	Carbon content	Comments	Uses
Low carbon (mild) steel	Up to 0.25%	Most common type of steel	Steel framework, cabinets, nuts, bolts, screws and nails
Medium carbon steel	Up to 0.5%	Stronger, harder and tougher than low carbon steel. Less ductile and more difficult to weld	High tensile bolts, gears and other machine parts, some hand tools
High carbon steel	Up to 0.8%	Even stronger, harder and more wear resistant than low and medium carbon steel	Springs, spanners, hand tools and with other alloying elements, railway lines
Carbon tool steel	up to 1.5%.	Easily heat treated to produce tools that are hard or tough or a combination of both	Tools such as files, chisels, scribers and punches are made from high carbon steel

Other specialist variants of steel include:

- high speed steel, which contains tungsten, cobalt and other additives as well as large amounts of carbon. Widely used for cutting tools such as drill bits, taps, dies, saw blades, etc.
- stainless steel, which also contains nickel and chromium. Various types are corrosion resistance with some being non-magnetic
- cast iron which contains between 2% to 5% carbon. Easily melted then poured into moulds to produce castings. Relatively cheap, strong in compression and resistant to wear. Brittle, but

this problem can be improved with suitable heat treatment. Used for such things as vices and machine bases.

Types of plastics used in electrotechnology

Most of the plastics in use today were developed in the 20th century, and are a by-product of petrochemical manufacture. Plastics are good electrical insulators. They are also excellent thermal insulators, being around four times better than asbestos and twice as good as wood.

Plastics are classified either as *thermosetting* or *thermoplastic*. All plastics soften when initially heated, to allow for moulding. Their classification depends on their behaviour upon cooling and subsequent reheating. Thermosetting plastics are those which *cannot* be remelted or softened by heat after they have been solidified. They undergo a chemical change when heated. Types of thermosetting plastics include:

- phenolic (Bakelite) – good heat resistance, good insulating properties. Uses are electrical equipment, distributor caps, grinding wheel bonds, etc.
- polyurethane – good strength, good adhesion and flexibility, suitable for uses such as cushioning materials, carpet underlay, safety padding in automobiles and safety helmets
- epoxy resins – tough, chemical resistant, good adhesive properties. Uses include metal to metal adhesives, ceramic gluing and surface coating (*Workshop Practice NE175/1*, p. 19).

Thermoplastics can be resoftened by heating and then remoulded by heat and pressure, providing that the heat does not damage or decompose the material's chemical structure.

Types of thermoplastics

Table 4.4 summarises the types and uses of typical thermoplastics.

TABLE 4.4 Types and uses of typical thermoplastics

Type	Comments	Applications
Polystyrene	Low heat resistance, good moulding qualities, low solvent resistance	Moulded food and drink containers, kitchenware, packaging
Cellulosics	High impact strength, poor heat conductor, resists oil and detergents	Screwdriver handles, photographic film, recording tape
Polyethylene	Tough, resists solvents, low moisture absorption, flexible	Packaging, bags, bottles and sheet wrap
Polyvinylchloride (PVC)	Non-toxic, high strength, abrasion resistance, rot proof	Cable insulation coating, electrical conduit
Fluorocarbons (PTFE)	Non-toxic, high thermal stability, low coefficient of friction, excellent electrical insulator, poor wear resistance and poor mechanical strength	Non-stick coatings, seals, gaskets, bearings, etc.
Polyamides (nylon)	High tensile and impact strength, low coefficient of friction, wear resistance	Brushes, gears, rope, washers, etc.
Acrylics (Perspex, Plexiglass)	Clear, transparent, good electrical resistance	Safety shields, windscreens, tail-light lenses
Acrylonitrile-butadiene-styrene (ABS)	High tensile and impact strength	Crash helmets, safety helmets, household appliances

Plastic identification code

You may have noticed a number surrounded by a chasing arrow symbol on consumer packaging like water bottles, jars, toys, etc. The symbol means the product can be recycled, and the number identifies which of the seven different types of plastic the item is made from. When recycled, different types of plastic must be sorted and separated into the number categories. Plastics of different types cannot be mixed in the recycling process as the batch will be contaminated and useless. Table 4.5 shows the resin identification codes for both thermosetting and thermoplastic classifications.

TABLE 4.5 The plastic identification code

Symbol	Type of plastic	Properties	Common uses	Recycled in
1 PET	PET Polyethylene terephthalate	Clear, tough, solvent resistant, barrier to gas and moisture, softens at 80°	Soft drink and water bottles, salad domes, biscuit trays, salad dressing bottles and containers	Pillow and sleeping bag filling, clothing, soft drink bottles, carpeting, building insulation
2 HDPE	HDPE High density polyethylene	Hard to semi-flexible, resistant to chemicals and moisture, waxy surface, opaque, softens at 75°, easily coloured, process and formed	Shopping bags, freezer bags, milk bottles, ice cream containers, juice bottles, shampoo, chemical and detergent bottles, buckets, rigid agricultural pipe, crates	Recycling bins, compost bins, buckets, detergent containers, posts, fencing, pipes, plastic timber
3 PVC	PVC Unplasticised polyvinyl chloride PVC-U Plasticised polyvinyl chloride PVC-P	Strong, tough, can be clear, can be solvent welded, softens at 80° Flexible, clear, elastic, can be solvent welded	Cosmetic containers, electric conduit, plumbing pipes and fittings, blister packs, wall cladding, roof sheeting, bottles Garden hose, shoes, soles, cable sheathing, blood bags and tubing	Flooring, film and sheets, cables, speed bumps, packing, binders, mud flaps and mats, new gumboots and shoes
4 LDPE	LDPE Low density polyethylene	Soft, flexible, waxy surface, translucent, softens at 70°, scratches easily	Cling wrap, garbage bags, squeeze bottles, irrigation tubing, mulch film, refuse bags	Bin liners, pallet sheets
5 PP	PP Polypropylene	Hard, but still flexible, waxy surface, softens at 140°, translucent, withstands solvents, versatile	Bottles and ice cream tubs, potato chip bags, straws, microwave dishes, kettles, garden furniture, lunch boxes, packaging tape	Pegs, bins, pipes, pallet sheets, oil funnels, car battery cases, trays

(*Continued*)

TABLE 4.5 *(Continued)*

Symbol	Type of plastic	Properties	Common uses	Recycled in
6 PS	PS Polystyrene	Clear, glassy, rigid, opaque, semi-tough, softens at 95°. Affected by fat, acid and solvents, but resistant to alkalis, salt solutions. Low water absorption, when not pigmented is clear, is odour and taste free	CD cases, plastic cutlery, imitation glassware, low cost brittle toys, video cases, foamed polystyrene cups, takeaway clamshells, foamed meat trays, protective packaging, and building and food insulation	Coat hangers, coasters, white ware components, stationery trays and accessories, picture frames, seed trays, building products
6 PS-E	PS-E Expanded polystyrene	Special types of PS are available for special applications		
7 OTHER	OTHER Letter below indicates ISO code for plastic type, e.g. SAN, ABS, PC, Nylon	Includes all resins and multi-materials (e.g. laminates). Properties dependent on plastic or combination of plastics	Automotive and appliance components, computers, electronics, cooler bottles, packaging	Automotive components, plastic timber

Source: QuickBooks DocStoc, http://www.docstoc.com/docs/document-preview.aspx?doc_id=72649119

SUMMING UP...

- Key components of planning a job include:
  - WH&S and environmental issues
  - drawings
  - materials and tools
  - determining the sequence.
- The risk assessment process is define, control and monitor.
- Typical non-electrical workplace hazards include:
  - laceration and puncture wounds
  - eye injury
  - noise
  - manual handling
  - being struck by a flying or falling object
  - slips, trips and falls.
- SOPs provide instruction for the safe use of plant and equipment.
- Properties of metals include:
  - compressive, tensiles and shear strengths
  - elastic deformation
  - toughness and hardness
  - ductility and malleability.
- Plastics are classified as thermosetting or thermoplastic.

4.3 Measuring and marking out

Since 1972 the SI metric system has been the standard measurement system in Australia, and is also used throughout Europe and Asia. The USA is the major exception and still uses the imperial system. While the imperial system is obsolete in Australia, because of the need to interpret the measurements of imported products, a basic understanding of it is still required.

Reasons for measuring and marking out

Measurement using a standardised system ensures accurate and consistent results across different locations and workplaces. Work instructions are mainly provided by working drawings, so the information recorded in the drawing needs to be transferred to the real world. This is done by interpreting the drawing, applying the prescribed scale to convert the drawing dimensions to real world measurements and then physically measuring and transferring those dimensions to the job.

Marking out transfers the information contained in a drawing to the work piece. Lines are drawn using tools such as scribers to indicate the size of the object and the location holes. Any errors made during the marking out process will reduce the accuracy of the finished work. Reasons for marking include to:

- inspect the material for faults
- check material is a sufficient size
- allow for variations in the material
- serve as an approximate guide for cutting and shaping
- assist in placement for machining.

Tools used for marking out

The measuring tool used will depend on the distance to be measured and the required accuracy. General electrical work related to the rough-in and fit-out stages of an installation would be measured with a tape measure. Work related to electric motor rewind and repair would require more precise measuring instruments, such as vernier calipers, which are shown later in this chapter.

Tape measures are used to measure long lengths to a reasonable degree of accuracy. They are available in a variety of lengths from 3 m upwards. Steel tape rules up to 10 m in length use spring steel for the tape and are semi-rigid. Some are marked in both metric and imperial scales. Plastic coated fabric tapes are available in lengths starting at around 30 m. The use of flexible materials for the tape means they have no rigidity and are non-conductive. Steel tape measures will rewind automatically but fabric tapes must be rewound manually. When retracting a tape measure, take care that it is not retracted too quickly, as this can damage the mechanism, and also cause the tape to move about rapidly, possibly hitting something or someone.

Steel tapes should never be used where they may come in contact with live electrical conductors.

To allow both outside and inside measurements to be made, tape measures usually have a moveable catch on the end of the tape. It moves an amount equal to its thickness which allows accurate measurements to be made.

Basic models of laser distance estimators have an effective range of 12 m with an accuracy of ±60 mm (0.5%). The top-end models are capable of measuring distances of up to 100 m with an accuracy of ±1 mm (0.001%). Laser devices are capable of calculating volume of an area as well as distance. Class 2 lasers are used in all measuring devices to comply with WH&S regulations.

Steel rules are the most common measuring tool used in industry. They are a quick, easily read tool that allows the measurement of lengths to an accuracy of approximately ±0.5 mm. They are available in standard lengths of 150 mm up to 3 m. The smallest graduation on a typical metric rule is 0.5 mm.

Scribers, as shown in Figure 4.24, are used to mark fine, accurate lines. They are normally made of heat treated high carbon steel.

FIGURE 4.24 Scriber

Source: Ed Hawkins

Scribed lines on some materials are often not very clear. For example, a scribed line on aluminium is as shiny as

FIGURE 4.25 Spring divider

Source: iStockphoto/travellinglight

FIGURE 4.26 Jenny calipers

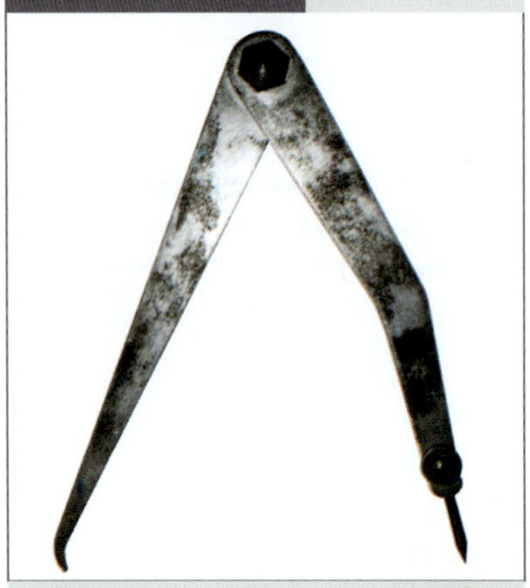

Source: Ed Hawkins

FIGURE 4.27 Surface gauge

Source: Ed Hawkins

the material. To make the line stand out, a marking medium is used. Commonly used marking mediums include:

- Marking blue, also known as 'Prussian blue', is a spirit based metal marking die which can be painted on to machined or finished surfaces. Once the job is complete it is easily removed with methylated spirits.
- Whitewash is a very basic paint. It consists of 'whiting' powder mixed with water or methylated spirits. It is most suitable for use on darker, rough surfaces.

Marking pens of almost any colour are suitable for marking shiny surfaces. They are easy to use, dry quickly and the writing is removable using basic solvents. They are only suitable when used on small surface areas as marking pen can become very time-consuming to apply over a large area.

Dividers serve the same purpose as a compass in drawing, that is, to draw circles or arcs, set out distances, divide circles, lines or angles, etc. but on hard surfaces rather than on paper. The types of divider are the spring divider, quadrant divider and trammel.

All types have two points (scribers) that are mounted in such a way that the distance between them can be adjusted. Spring dividers, as shown in Figure 4.25, use a *spring joint* to maintain their distance setting. They are the most common and are the easiest to adjust. For greatest accuracy the points of the divider must be kept sharp and the legs must be the same length.

The jenny caliper, shown in Figure 4.26, has one internal caliper leg and one divider leg normally pivoted with a *firm joint*. They are used for locating the centre of cylindrical stock, or for locating points or scribing lines parallel to a surface or shoulder. To find the centre of a shaft set the jenny to approximately the radius of the shaft and scribe arcs from four positions equally spaced around the circumference.

Scribing blocks and surface gauges are used for setting out, finding centres, marking heights or other marking out procedures. The scribing block has a solid base, machined flat on the bottom, with a fixed vertical column. A scriber is clamped to the column using a universal clamp that allows the scriber to be rotated to any position.

The surface gauge in Figure 4.27 is similar to the scribing block except the column is moveable to give a greater range of adjustment and provision is made for fine adjustments using a spring loaded rocker bar and adjusting screw. A V is machined in the base for locating against cylindrical surfaces and two locating pins can be made to project through the base for locating the gauge against an edge.

Marking-off tables are made from heavily ribbed cast-iron castings, stress relieved steel or granite. They provide flat and square top and side surfaces. These are used as the datum or reference surfaces when marking out. The reference surfaces must be kept in perfect condition to maintain accuracy. Do not use the marking-off table to support work where hammering is involved, such as stamping or centre punching.

A surface plate is basically a small, portable marking-off table and all the care taken with the marking-off table should be followed.

The engineer's try square is used for testing or marking out angles of 90°. It consists of a stock and a straight, flat blade with parallel edges. The blade may be fixed or adjustable for length (*Basic Training Manual 16-4 Mechanical Measuring*, p. 6).

The bevel gauge has a body and an adjustable sliding blade which can be set to any angle and clamped. It is used for transferring angles from a protractor to a job or from job to job.

The digital bevel gauge is capable of measuring, storing and transferring inside and outside angles between angles of 0° to 360° and is accurate to 0.05°.

Figure 4.28 shows the combination set, which consists of a blade to which one of three heads may be fitted. The heads of the combination set are:

- the square head, used to test or measure angles of 90° or 45° or as a depth or height gauge
- the protractor head, used to measure or test any angle
- the centre head, used to find the centre of cylindrical stock or for marking or testing 45° angles.

FIGURE 4.28 Combination set

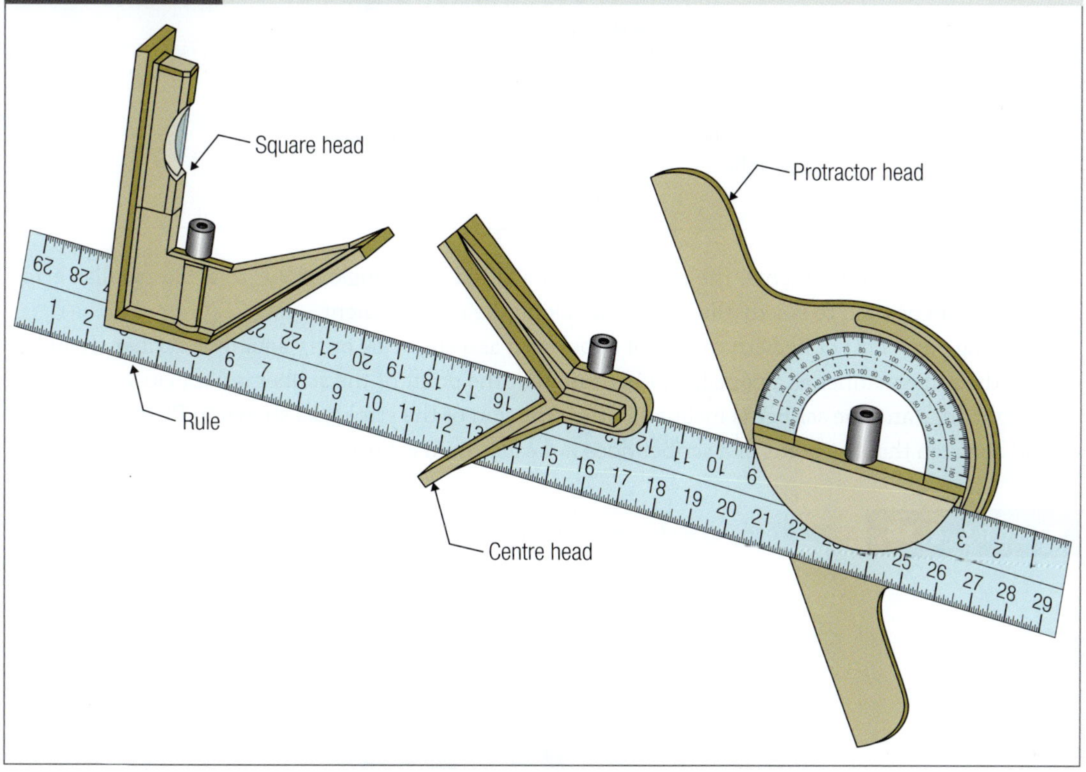

The centre punch and prick punch are similar except for the angle of the point. They are made from heat treated high carbon steel. The centre punch is generally used for marking centres ready for drilling. It is normally of heavier construction than the prick punch and has a point angle of 60° to 90° to make a large indentation.

The prick punch has a point angle of 30° to 60° and is used to make small indentations to locate a divider point or for witness marking. Witness marking is the process of highlighting with a series of light, permanent indentations.

Automatic centre punches have an internal spring and hammer mechanism. The user just has to apply a downward pressure with the palm of the hand to charge the spring; the stored energy is automatically discharged via the internal hammer into the work piece, forming an indentation.

Measuring and marking out

For measurements to be accurate, the measuring instrument must be accurate. To keep your measuring instruments accurate you must look after them and use them properly. Make sure you:

- store them where they can't be damaged
- clean them before and after use
- do not drop them, or place other tools on top of them on a work bench
- do not use precision instruments on rough surfaces.

When using a steel rule the end corners can become worn due to misuse. For this reason you should try to take measurements from the first main graduation, usually the 10 mm mark. Figure 4.29 shows the correct use of a steel rule. Starting the measurement at the 10 mm increment rather than the edge of the rule (0 mm), the value measured is 60 mm; after subtracting the initial 10 mm, the final reading is 50 mm.

FIGURE 4.29 Steel rule use

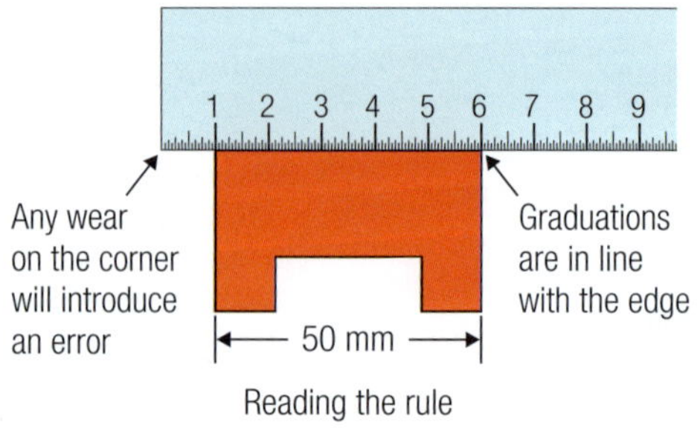

When taking measurements, the rule should be viewed at an angle of 90° to avoid parallax error. The thumb can be used to steady the rule while taking the measurement.

Marking out is best done with the aid of a tool such as a steel rule or try square. A scriber should be used in the same manner as a pencil, making sure that the point of the scriber is kept close to the rule or straight edge and the scriber is inclined in the direction of the stroke, as shown in Figure 4.30. Care should be taken that the point of a scriber does not cause injury during use or in storage.

FIGURE 4.30 Marking out with scriber and steel rule

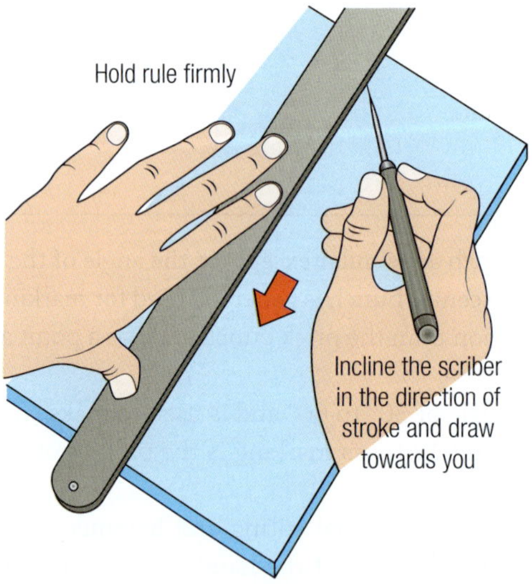

Marking blue applied to a work piece provides contrast to better show the marking out. The material should be checked for square with a try square. The two square edges are then marked. One side is selected as a datum and marked with an inverted vee. All measurements should be taken from the datum. Do not make a measurement from another marking, because if that mark is incorrect any further marking will also be incorrect; that is, the error will be exaggerated across the rest of the work.

When using a try square, make sure that the surfaces to be tested and the square itself are free from grit and burrs. Place the stock of the square against one surface with the blade just clear of the other surface. Check this surface for square by sliding the stock over to it and holding the job up to a light. If the surfaces are square no light should be visible between the job and the square.

When marking out with a try square and scriber, incline the scriber in the direction of movement, as shown in Figure 4.31. This prevents the scriber point digging into the surface being scribed.

FIGURE 4.31 Marking out with scriber and try square

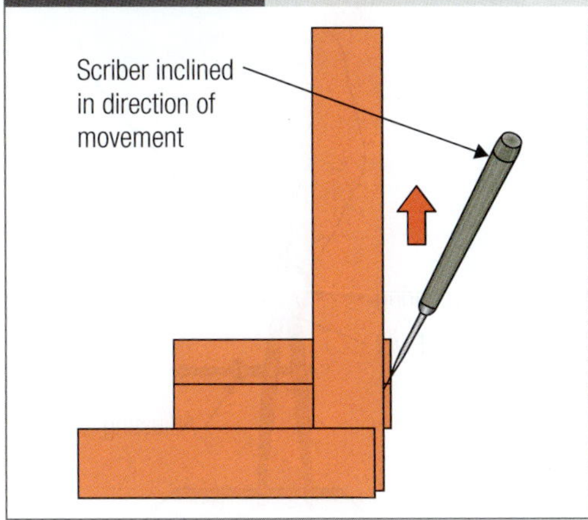

FIGURE 4.32 Outside diameter measurement with calipers

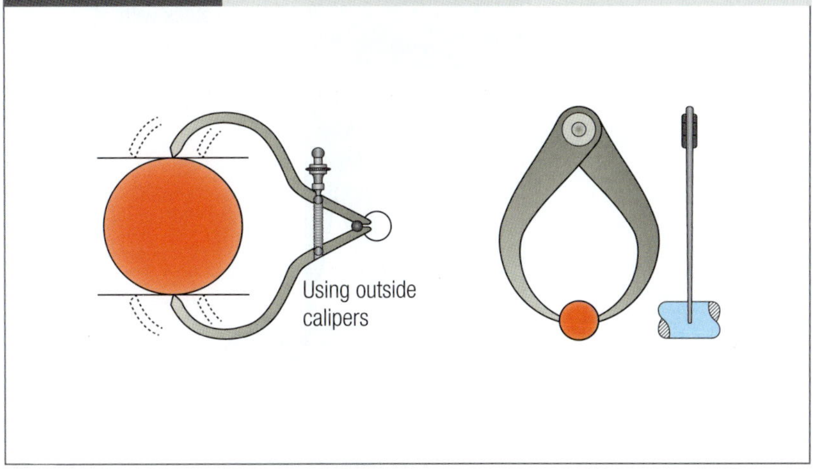

In general, calipers have no scale and are set by reference to a measuring tool such as a rule or micrometer. They are therefore a go or no go measurement device. It is possible when using calipers to detect differences in size of a few hundredths of a millimetre. Contact between the measuring surfaces of the caliper and the work surfaces must be very light; the desired 'feel' is best described as that point where the sense of contact with the work has just disappeared, as shown in Figure 4.32.

Most firm joint calipers are reversible. The method used to measure inside diameter is shown in Figure 4.33.

Jenny calipers are used to mark straight lines that are parallel to an edge. The distance is set into the jenny caliper with the use of a steel rule as shown in Figure 4.34.

Most rules have small indentations to emphasise the marking scale. The leg of the jenny caliper is held flush with the edge of the steel rule and the point leg is adjusted by hand until the point drops into the indentation of the required measurement. The leg of the jenny caliper is then held flush against the flat square edge of the work piece, while the upper part is tilted in the direction in which the marking out is to be scribed, as shown in Figure 4.35.

FIGURE 4.33 Inside diameter measurement with calipers

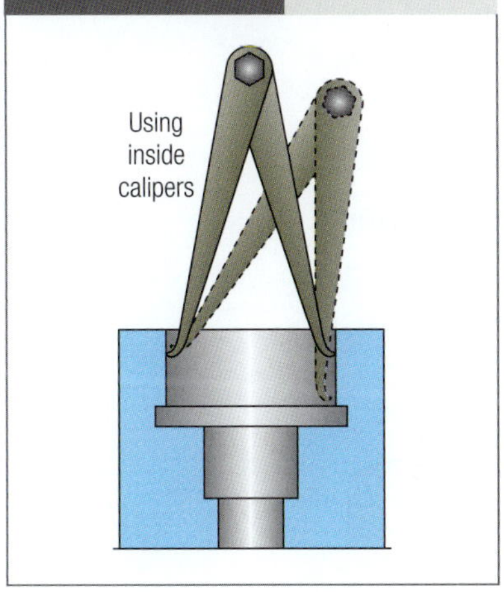

FIGURE 4.34 Setting a jenny caliper

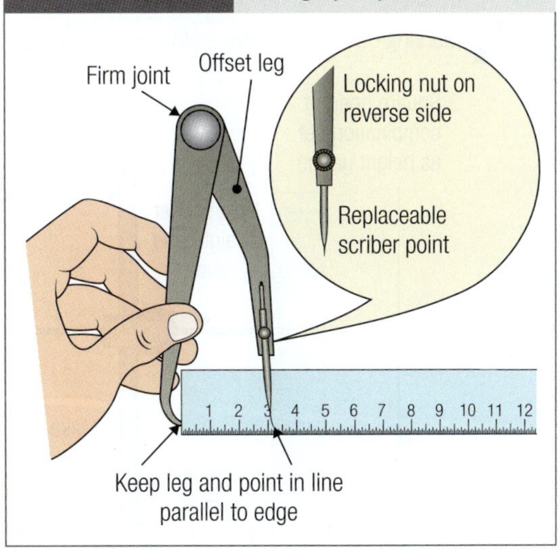

FIGURE 4.35 Marking out with a jenny caliper

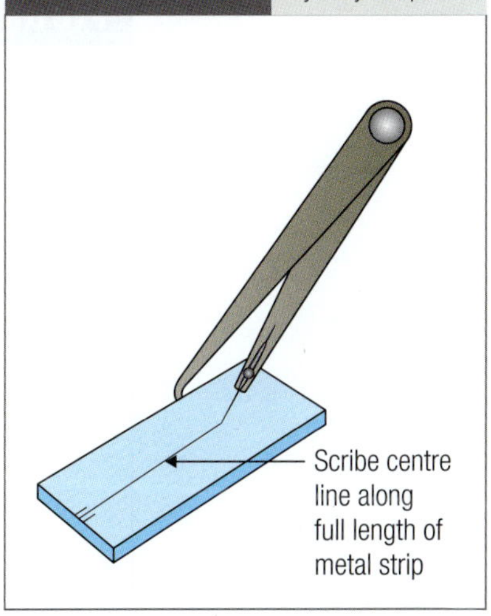

FIGURE 4.36 Locating the centre of round stock with a jenny caliper

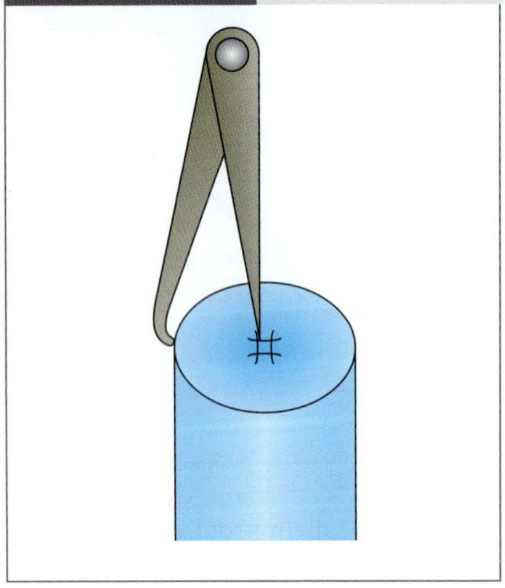

FIGURE 4.37 Marking out with spring calipers

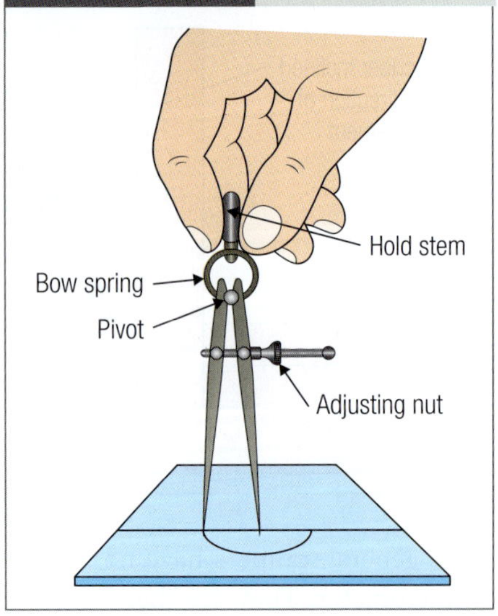

To find the centre of a shaft set the jenny to approximately the radius of the shaft and scribe arcs from four positions equally spaced around the circumference, as shown in Figure 4.36.

Circles are marked out with a divider as previously stated in a similar way to drawing a circle on paper with a compass. A guide point is made with a prick punch at the circle's centre to prevent the point of the divider slipping. Figure 4.37 shows marking out circles and arcs with the use of a spring divider (caliper).

For improved accuracy, the divider points should be set between indentations and not the end of the rule itself. The adjustment nut is rotated to allow the point of the divider to drop into the ruler's indented scale markings. When making large adjustments the legs of the divider should be compressed with the fingers to relieve the tension on the adjustment nut.

A surface gauge is set using a rule or square on a marking out table or surface plate as shown in Figure 4.38. Once set, the scriber is tilted in the direction of the line to be drawn and moved with a firm hand to produce the required marking out. The marking out table or surface plate acts as a datum or reference to a common point, so many lines can be drawn very quickly and accurately (*Workshop Practices NE175 Student Workbook*, p. 17).

FIGURE 4.38 Marking out with a surface block and marking table

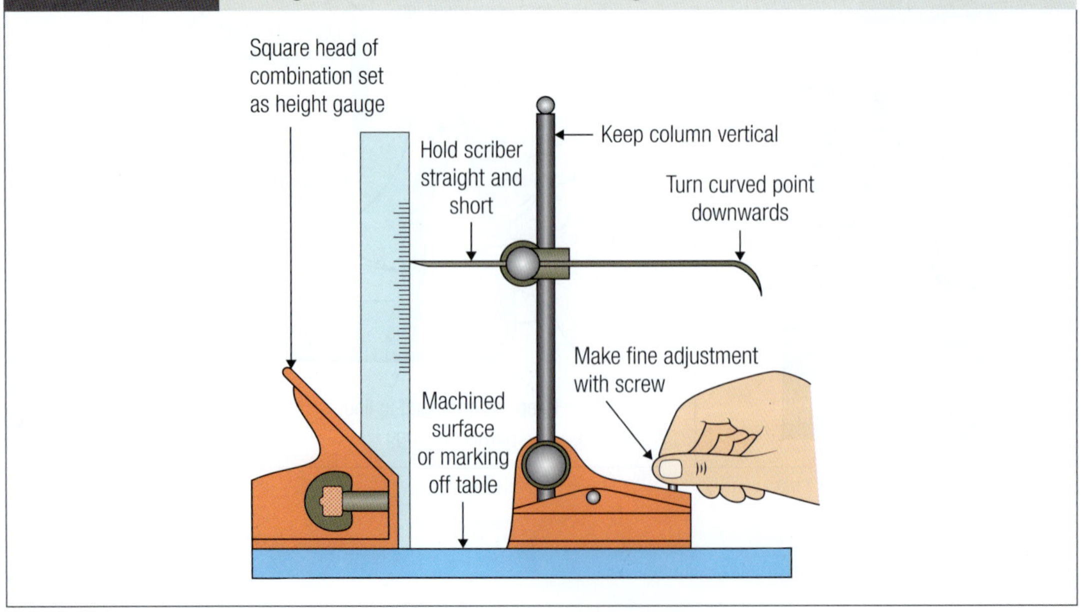

The blade of a combination square is designed to allow the different heads to slide along the blade and be clamped at any desired location. By removing all the heads, the blade may be used alone as a rule or a straight edge. The square head is designed with a 45° and 90° edge, which makes it possible to be used as a try square and mitre square. It may also be used as a height or depth gauge. The square head is also fitted with a level vial and sometimes a removable scriber. Figure 4.39 shows the uses of the square head.

The centre head can be used to locate and lay out the exact centre of round objects. A line is marked at the approximate centre of the object. The centre head is then rotated approximately 90° and a second line is marked on the object. The centre is where the two lines intersect, as shown in Figure 4.40.

The protractor head can be used to mark off or measure any angle through 180°. Angular graduations usually read from 0° to 180° in both directions to allow the angle to be read, as shown in Figure 4.41.

Once marking out is complete, locations that require drilling will need to be centre punched to ensure accurate positioning of the hole, as shown in Figure 4.42.

Witness marks, as shown in Figure 4.42, are made with a prick punch. These marks are a series of light, permanent indentations placed along the outlines produced when marking out. This may be necessary when there is a likelihood the marked lines may be worn away during transport, setting up, machining or any other work process.

The punching process is the same for both centre and prick punches. The work piece should be laid on a solid surface such as a steel work bench or block, as shown in Figure 4.43. The point of the punch is then drawn along the scribe mark until it drops into the intersection of the two marks. A lightweight hammer is used to tap the head of the punch to form the indent. The punch should only be struck once to avoid bouncing the point and accidentally forming a second indent in the wrong location.

Summary of marking out procedure

When marking out a job the following steps should be taken:

1 Study the drawing and familiarise yourself with it, making sure you understand what is required.

FIGURE 4.39 Various uses of the square head of a combination square

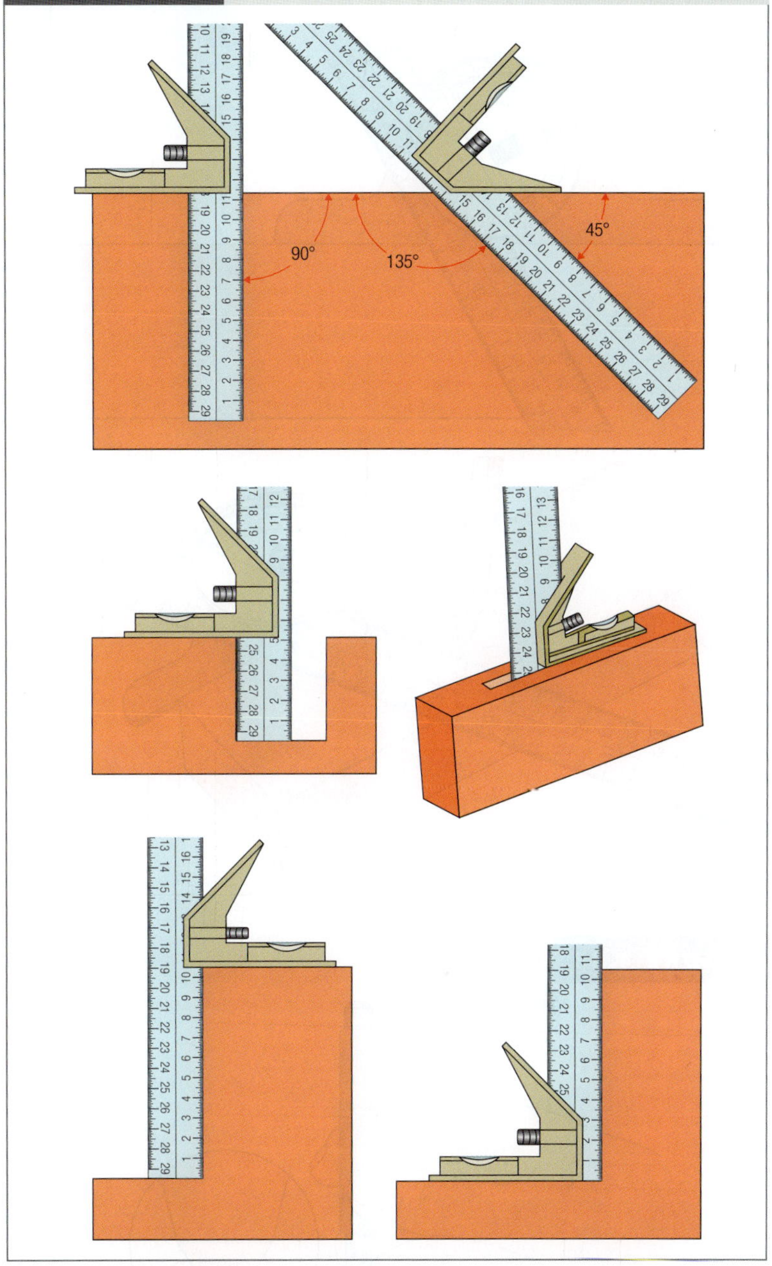

FIGURE 4.40 Locating a centre with a combination square

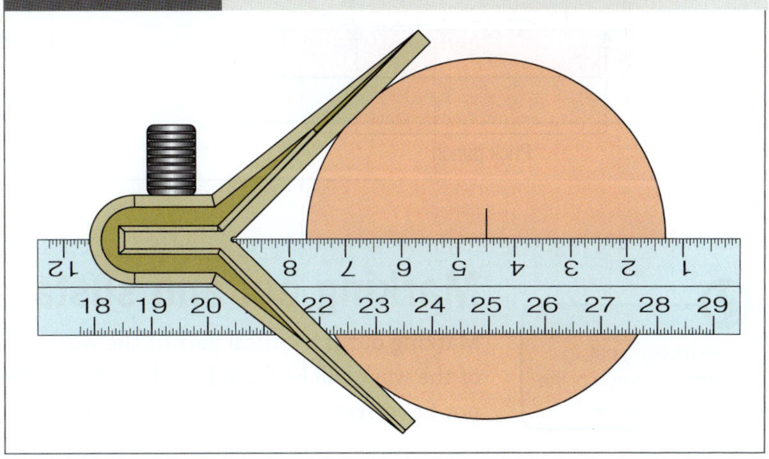

FIGURE 4.41 Marking an angle with a protractor head

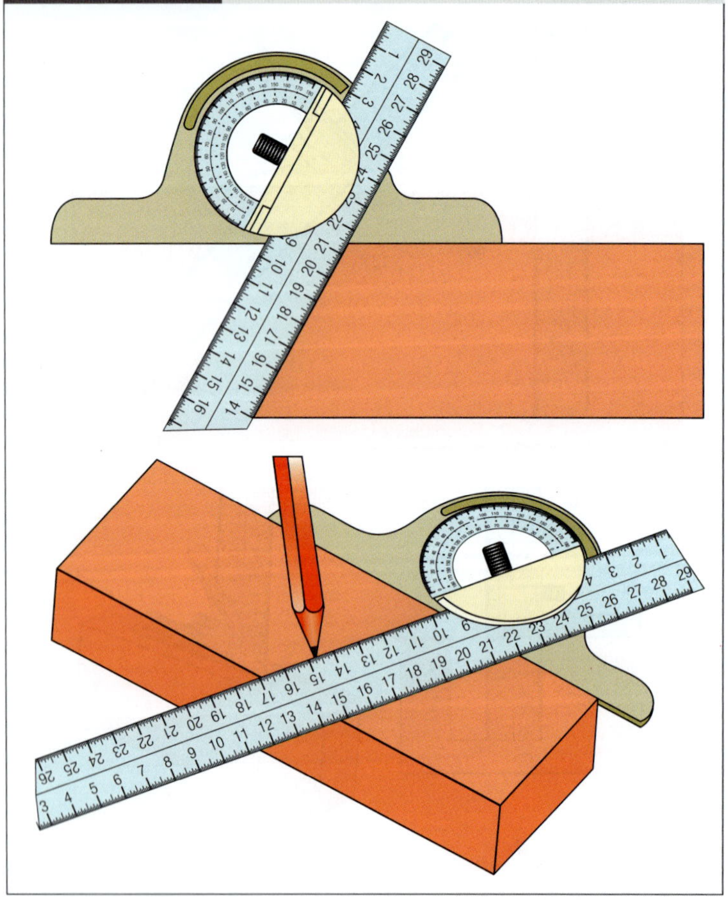

FIGURE 4.42 Centre punch use

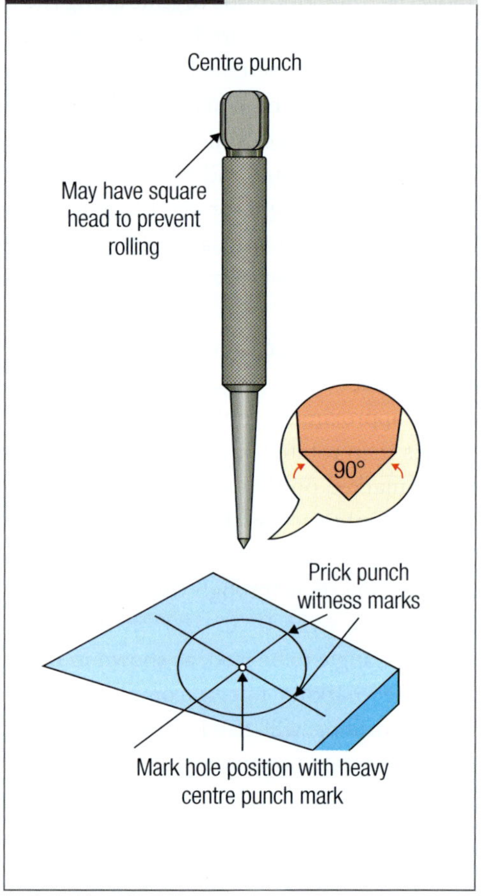

FIGURE 4.43 Prick punch use

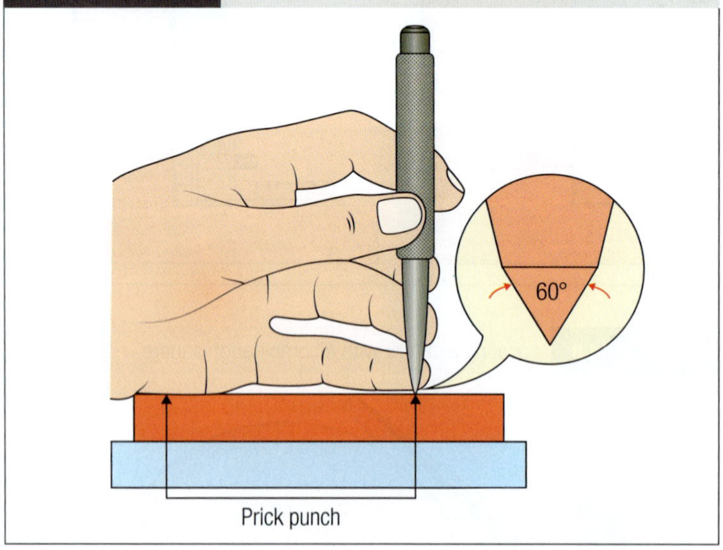

2 Select the tools required and lay them out on a well-lit area of the bench.
3 Check the raw stock for quality and size.
4 Lightly remove any burrs with a fine file. This improves the accuracy by giving a clean edge for measuring and laying out.
5 Coat one side with a marking dye to aid marking out.
6 Select one edge as a datum reference and mark it as such. This will be the edge from which all measurements are taken.
7 If necessary, square one end as a reference and mark it with a right angle symbol.
8 Accurately mark out the job, double checking all dimensions.
9 If the job requires more than very simple marking out and finishing, the important marks should be witnessed.
10 Centre punch holes, witnessing where required.

When marking out, try to minimise waste of the material being used.

Marking out and sustainable work practices

Marking out is a critical part of the work process – if a mistake is made during marking out the rest of the work will be flawed, so do not rush the process. Incorrectly marked parts when completed will not fit correctly with other components, resulting in waste and the need to start over. To avoid error all measurements should be checked or verified before proceeding to the next stage of work. In other words 'measure twice, cut once'. Error and waste is neither sustainable nor productive.

When purchasing materials, preference should be given to suppliers that are compliant with **ISO 14001:2004 Environmental management systems – Requirements for use** and certified as described in Chapter 2 (Section 2.1) and indicated by the use of the logo shown in Figure 2.3. Consideration should also be given to marking out parts to conserve material. Raw materials can be ordered pre-cut to a standard size that will allow efficient use of the material. The aim is to reduce off-cuts and waste, which is best achieved by working off the edges of the raw material, not the middle. Any off-cuts and waste will need to be sorted so they can be either recycled or disposed of in an appropriate manner.

SUMMING UP...

- Reasons for marking out include:
  - to inspect the material for faults such as insufficient material to complete the part
  - to allow for variations in the material
  - to assist in placement for machining
  - to serve as an approximate guide for cutting and shaping.
- Common tools used for marking out include:
  - tape measures (large work)
  - steel rules
  - scribers and marking pens
  - engineer's try square, bevel gauges and combination square sets
  - spring and jenny dividers
  - scribing blocks, surface gauges, marking out tables and surface plates
  - centre and prick punches.
- For improved accuracy always measure from a datum point or edge, not from other marks.
- Witness marks are made by a prick punch and are used to ensure a line or mark is permanent enough to prevent it being easily removed.
- Centre punches are used to mark a point in metal that locates a drill bit at the correct point.

4.4 Holding and cutting

Work pieces will often need to be temporarily held in a fixed position. This may be to hold the material in place while drilling a hole, securing a fixing device, cutting or bending the material. The object may be too hot or cold to hold with the hand or it needs to be held in place while welding is performed. This section describes the types of devices used to secure various materials, and the techniques used to safely and efficiently cut commonly used materials.

Common tools for holding

Vices

Vices vary in size and construction according to the purpose for which they are used. They are made of cast iron or fabricated steel, the size being stated as the width of the jaws. Cast-iron vices should not be used to hold jobs during heavy hammering, as cast iron is brittle and may fracture.

The bench vice, as shown in Figure 4.44, consists of a fixed jaw and base bolted to the bench, and a sliding jaw operated by a screw with either a square or acme thread. The jaws are normally case-hardened to reduce wear. A vice should be mounted over the bench's leg to absorb any shock loads on the work in the vice so that the top of the vice is at elbow height. The inside of the fixed jaw should be aligned just outside the edge of the bench. The offset vice has its jaws offset to allow long work to be held vertically within the full width of the jaws (*Workshop Practices NE175 Student Workbook*, p. 30).

FIGURE 4.44 Bench vice

Source: Ed Hawkins

FIGURE 4.45 Hinged pipe vice

Source: Ed Hawkins

FIGURE 4.46 Multi-grips

Source: Ed Hawkins

FIGURE 4.47 Vice grips

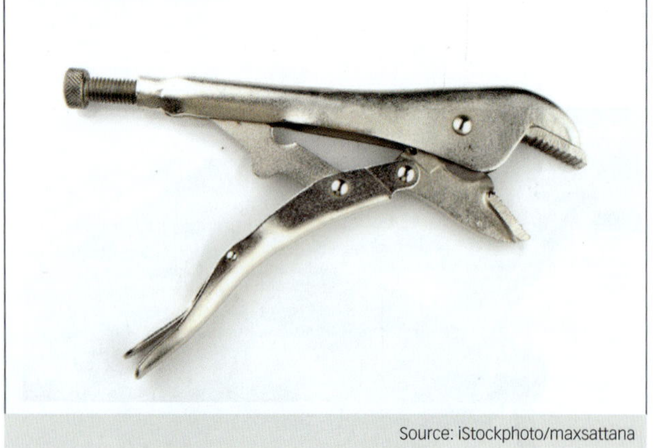

Source: iStockphoto/maxsattana

A pipe vice is used to hold round or tubular section materials, for example, electrical steel conduit, for threading applications. The two main types are the hinged pipe vice (Figure 4.45) and the chain pipe vice. When performing tasks such as bending or cutting pipe or conduit, a pipe vice is often attached to a portable stand.

The machine vice is able to be moved around on a machine table and then clamped in position. Machine vices are designed to hold jobs firmly during drilling, shaping and other machining procedures. Swivel base vices allow the vice to be turned for longitudinal filing or cutting operations (*Electrotechnology Practical LSEEE280A*, p. 55).

Hand tools

Multi-grips, as shown in Figure 4.46, are a hand tool which has a slip-joint adjustment to accommodate a wide variety of sizes. The multi-grip's serrated jaws are designed to grip curved or irregular surfaces and to hold small components. Multi-grips should not be used for tightening or loosening bolts or nuts, because they will damage the corners of the hexagon.

Vice grips, as shown in Figure 4.47, are similar to multi-grips as they are also adjustable in size and gripping force. The difference is they are lockable, making them suitable for use as a portable adjustable clamp, holding parts together during drilling or welding.

The two common types of wrench used to grip pipes, metal and round stock are the Stillson wrench and the footprint wrench. The Stillson wrench has a handle and spring loaded casing which carries a knurled adjusting nut. The nut engages a thread on the sliding jaw. A set of teeth fixed to the handle face outwards, and the teeth on the moving jaw face inwards. The wrench is adjusted to size and the handle pulled gently until the teeth grip. As more pressure is applied the grip on the pipe is increased.

The footprint wrench, so named because of its 'footprint' trademark, has a slip-joint adjustment between a folded handle with outward facing teeth and the solid handle with inward facing teeth. A series of holes along the fixed jaw allows the moveable jaw to be repositioned to accommodate various sizes of material. The footprint wrench is used in a similar way to the Stillson wrench.

Pipe wrenches should not be used on polished surfaces, nuts or bolt heads because of the damage caused by the serrated jaws of the wrench. Nor should they be used on hardened surfaces because of the damage the jaws would suffer. When work is complete, any burrs or indentations raised by the jaws should be filed off to prevent hand injury.

Common tools for cutting

The hacksaw

A hacksaw is a hand tool consisting of a frame and the blade, as shown in Figure 4.48. Most hacksaw frames are adjustable to take blades of either 250 mm or 300 mm long. The frame has a wing nut that is used to put the blade under tension. Hacksaw blades are suitable to cut metal and most plastics, but they are not suitable for cutting timber as the blade becomes clogged. Some hacksaw frames allow the blade to be set at a 45° angle so that cuts may be made in materials deeper than the height of the frame. When not in use the tension on the blade should be released by undoing the tensioning nut.

Hacksaw blades are made of alloy tool steel, high speed steel or a combination of both.

- For general work the flexible low tungsten steel blade is most popular. This blade is only hardened on the cutting edge with an unhardened body.
- The high speed steel blade, called an 'all hard blade' because the entire blade is hard, is brittle and easy to break if used when the work is not firmly clamped or supported. On the other hand, it will not wear out as quickly as other types of blade.
- A third type of blade is known as a 'bimetal blade' because it consists of two different materials. These blades are made by welding a high speed steel strip, on which the teeth are formed, to a low alloy body. They combine the flexibility of the low alloy blade and the cutting ability of the high speed blade.

Hand hacksaw blades are available in standard lengths of 250 mm and 300 mm and are available in standard pitches of 14, 18, 24 and 32 teeth per inch (or per 25 mm). See Figure 4.49 (*Basic Training Manual 16-3 Hand Tools Electrical Trades*, p. 71).

Soft materials normally require a coarse pitch (18 teeth per 25 mm) to allow sufficient chip clearance, while harder materials normally require a finer blade with a larger number of teeth in contact with the job. At no time should there be less than three teeth in contact with the job, otherwise teeth can break and the cut will be difficult to start.

The teeth on a hacksaw blade are 'set' to give clearance between the material being cut and the body of the blade. The groove cut while hack-sawing is called the 'kerf'. This clearance allows the direction of cut to be changed slightly and reduces friction between the blade and job. Three different methods are used to produce the set, as shown in Figure 4.50.

Worn teeth reduce the blade's clearance and increase friction. Worn blades should be replaced. Avoid starting a new blade in a cut made by a worn blade as this will damage the set on the new blade. Excessive wear on hacksaw blades can be caused by hard spots in the material being cut, or by excessive cutting speeds. Most manufacturers recommend a maximum cutting speed of from 50 to 60 strokes per minute.

Junior hacksaws are suitable for cutting smaller objects because the blades have a high TPI and allow greater access in tight locations.

FIGURE 4.48 Hacksaw

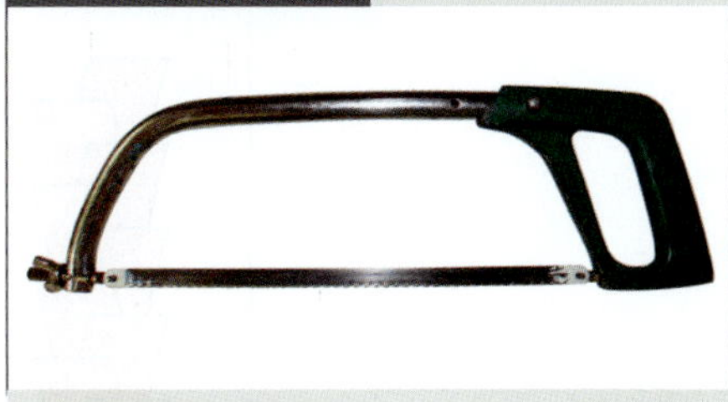

Source: Ed Hawkins

FIGURE 4.49 Hacksaw blade teeth

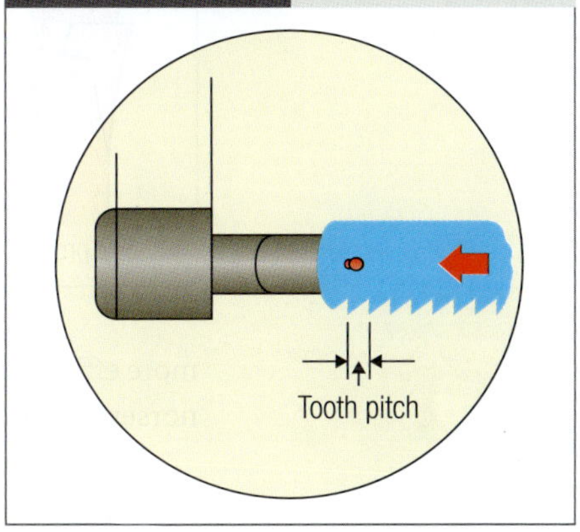

FIGURE 4.50 Hacksaw blade sets

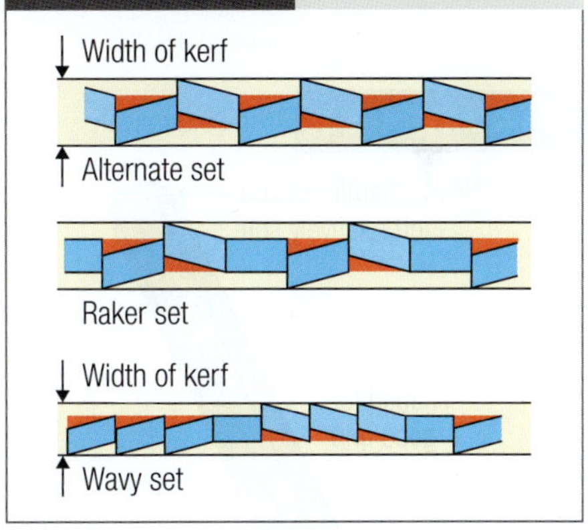

FIGURE 4.51 Timber saw types

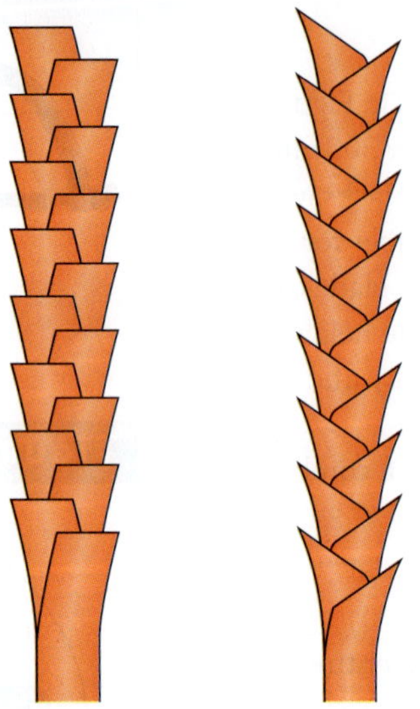

FIGURE 4.52 The firmer wood chisel

Source: Ed Hawkins

Timber saws

Timber saws have fewer cutting teeth per inch (25 mm) than metal saws. Therefore, they are not suitable for cutting metal. Saw lengths range between 10 (250 mm) and 26 inches (600 mm) with TPIs between 9 and 12.

A timber saw can have either rip or crosscut blades, as shown in Figure 4.51. A crosscut saw is for cutting wood across (or perpendicular to) the grain, and rip saws cut more efficiently with (or parallel to) the grain of the wood (Norsewood Smith, 'Handsaw basics', http://norsewoodsmith.com).

Like metal saws, rip and crosscut saws only cut on the forward stroke. A triple ground tooth (TGT) saw, optimises performance by cutting on both push and pull strokes, so less effort is required. This type of saw is suited for plasterboard, medium-density fibreboard (MDF), particle board, plywood and glued composite materials.

Chisels

FIGURE 4.53 The cold chisel

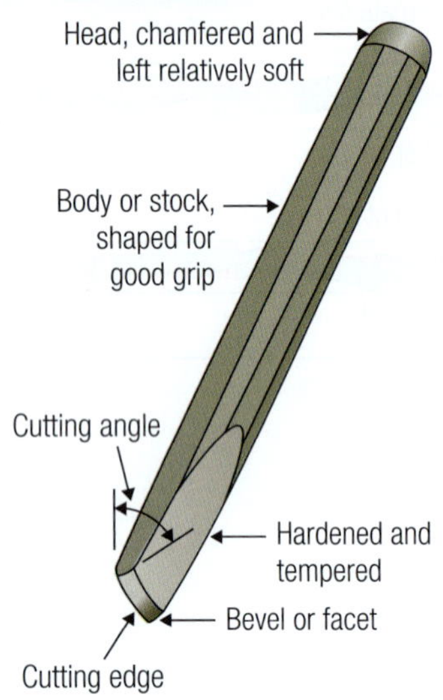

Chisels can be divided into two groups: wood chisels and cold chisels. As the name suggests, wood chisels are used for shaping wood. These tools must be kept sharp and treated with care. They are designed to be driven with hand pressure or by using a mallet. There are several types available including:

- The bevel edge chisel is used for cutting grooves or trenches in timber. The bevelled edges cause less friction in the groove.
- The firmer chisel, or square edged chisel, as shown in Figure 4.52, is stronger than the bevel type and used for heavier work.

Cold chisels are so named because they are meant to cut material that is not heated, and are used for the rough shaping of metal prior to finishing, for cutting out rivets or removing seized nuts or bolts. The cold chisel is made from an alloy of carbon steel (see Figure 4.53). It is heat treated at the tip for hardness but is left soft at the head to prevent chipping while in use. The head is also ground at a chamfer to reduce mushrooming due to the repeated striking with a hammer.

For general work, the flat cold chisel is used but other styles of chisel include side cutting, diamond point, round nose and crosscut. The cutting edge angle of a cold chisel depends on the metal to be cut. Softer metals need a sharper cutting angle than harder metals. The size of the cutting edge angle should be about 70° for mild steel, but it should be larger for harder materials and smaller for softer materials. The clearance angle labelled C in Figure 4.54 is formed by the user when chiselling. It determines the amount of material to be removed.

FIGURE 4.54 Cold chisel cutting angle

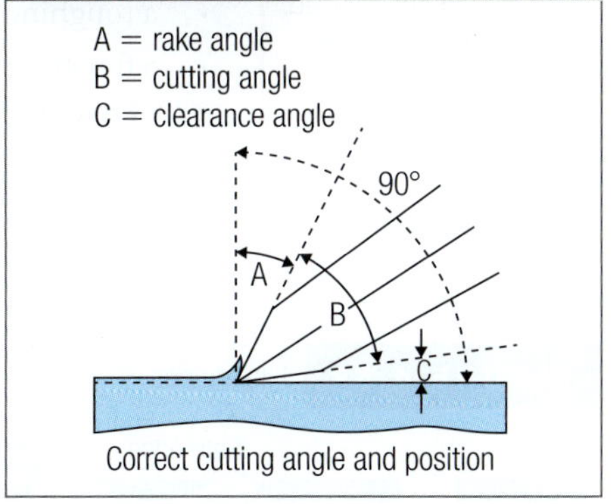

Pliers

Pliers are made from high carbon or alloy steel that is forged, machined and heat treated to a shape that provides efficient and comfortable hand grips and a jaw shape designed for cutting, gripping or shaping operations. Typical types of pliers used in the electrical industry include combination pliers, shown in Figure 4.55. These have a variety of uses such as gripping or shaping flat and round sections, cutting soft metal wire using the side cutting blades or cutting heavier wires using the joint cutters. They have flat serrated jaws, and round serrated opening and cutting edges.

Diagonal cutting pliers (side cutters) are shown in Figure 4.56 and are used mainly for cutting wires. Maximum cutting effort is gained by keeping the wire to be cut as close to the pivot as possible. Several types are available, some suited for cutting hard steel wire, others suitable only for cutting soft materials such as copper. If used for the wrong material the jaws will be damaged.

End cutting pliers (pincers) have similar applications to side cutters but the blades are 90° to the handles. They are also commonly used as an alternative to a claw hammer for removing nails from timber.

Long nose pliers, shown in Figure 4.57, are made with tapered jaws that have serrated gripping surfaces. They are useful for bending and folding wires or for reaching into confined spaces. They are not designed for the same type of work as combination pliers.

Round nosed pliers are used for forming loops in wires, etc. Flat nosed pliers are used to shape thin metal or harness wiring. Snipe nosed pliers, also known as mini long nosed pliers are used for gripping small components. Slip joint pliers work in a similar way to multi-grips by allowing for a greater range of sizes. These are used for holding jobs or cutting small wires.

FIGURE 4.55 Combination pliers

Source: iStockphoto/sinankocaslan

FIGURE 4.56 Diagonal cutting pliers (side cutters)

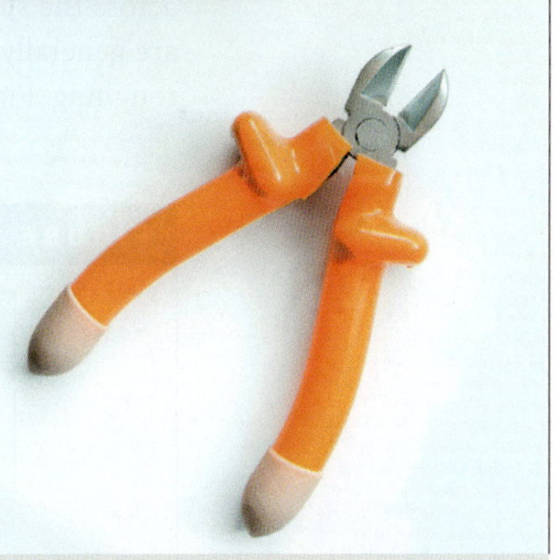

Source: shutterstock.com/beerlogoff

Files

Files are cutting tools with a large number of teeth and are used to remove material from a metal part. The teeth are formed on the body of the file, which is made from hardened tool steel. Being hard, the file is also very brittle. This means the file should never be used as a lever or struck with a hammer, because of the danger of pieces flying off and causing injury. The parts of a file are shown in Figure 4.58.

FIGURE 4.57 Long nose pliers

Source: iStockphoto/Erickson Photography

FIGURE 4.58 Parts of a file

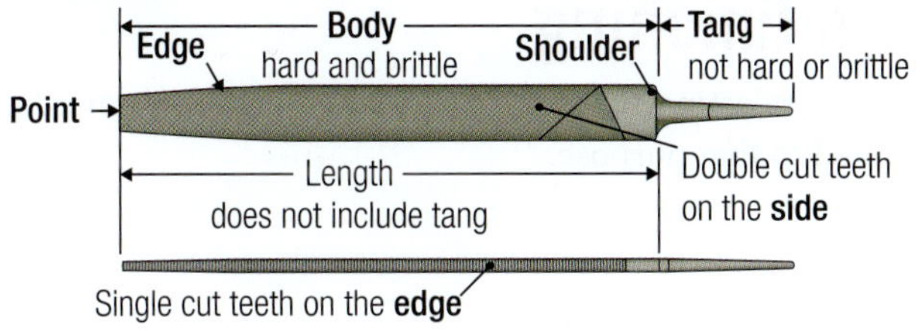

Files should not be used as a lever as they can break, causing sharp pieces to fly off.

The tang of the file is not hardened and is shaped to take a file handle. A handle must always be fitted to prevent the tang being driven into the palm of the hand. Filing can be:

- a roughing process to remove a large amount of material from the work piece
- a finishing process to give the work piece a smooth finish while removing only a very small amount of material.

In general, files are classified according to:

- length, as measured from the shoulder to the point, and may range from 100 mm to 500 mm
- cross-section or body shape.

FIGURE 4.59 File shapes

As mentioned in Chapter 3, there are many shapes of files (see Figure 4.59). For general use the most common is the flat file, which has parallel sides and edges for two thirds of its length and tapers slightly towards the point.

The hand file is similar to the flat file but is parallel throughout its length and has one safe edge, meaning there are no teeth formed on it. This is useful when filing up to a surface that is not to be damaged.

The mill saw file is similar, but is thinner and is single cut with either two square edges or one square and one radius edge. Square, round, half round and triangular files (slim taper) are tapered from the shoulder to the point, and have sections as suggested by their names.

The grade (coarseness of cut) or tooth size of files are given as:

- bastard cut (the coarsest), used for rough filing
- second cut, used when a finer surface finish is required
- smooth cut (the finest), used for finishing a surface to a high finish.

For example, a typical file specification could be 300 mm, hand, bastard cut file.

File teeth are produced as either single cut, which consists of long teeth extending at an angle across the side of the file, or double cut pointed, diamond shaped, individual teeth. Single cut files are generally better for finishing whereas double cut teeth bite into the surface and are better for roughing. Figure 4.60 shows double and single cut file types.

FIGURE 4.60 File cuts

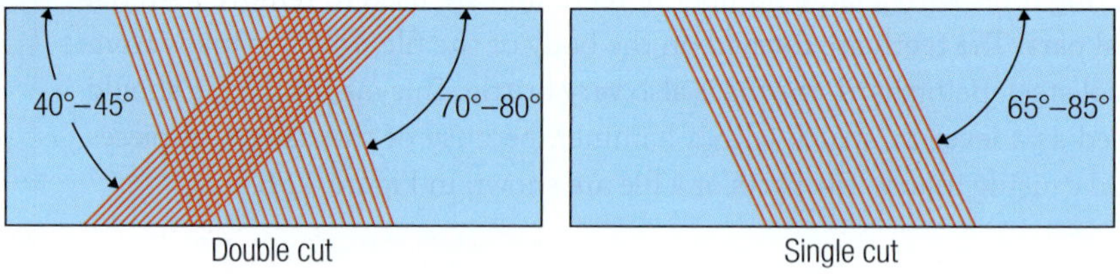

Cutting, shaping and finishing

The most important part of tool use is selecting the right tool for the job. Tools are designed for a purpose and their use should be limited to that purpose.

Hacksaws

The correct use of a hacksaw is shown in Figure 4.61. Always:

1 Check that the blade has the correct pitch for the job, is sharp and has sufficient set.
2 Check that the blade is placed in the frame correctly (i.e. the teeth are pointing away from the handle so that the blade cuts on the forward stroke).

3 Check the tension on the blade. Three full turns on the tension nut after the slack is taken up is a good rule of thumb. Over tensioning will warp the frame.

4 Start the cut using the thumbnail of the left hand (or right hand for left-handers) as a guide, or by filing a small notch in one corner of the job to prevent the blade from slipping.

5 Grip the hacksaw firmly in your right (or left) hand with the forefinger pointing along the frame in the direction of cut. Adopt a firm stance with the feet slightly apart and the left (or right) foot slightly in front of the other.

6 Start the cut slowly and concentrate on following the markings on the job. When the cut is well established, speed may be increased. Use long, steady strokes making use of the complete blade.

FIGURE 4.61 Correct hacksaw use

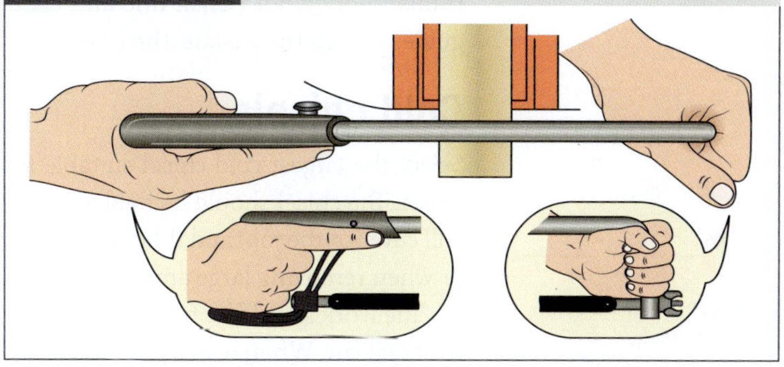

If a blade should break before a cut is finished do not start the new blade in the same cut; turn the job over and cut from the other side.

Timber saws

A timber saw usually has fewer teeth per inch (TPI), so being more coarse, extra attention is required to start the saw cut. The following steps have been adapted from 'Toolmanship: How to use a handsaw', at http://www.artofmanliness.com/2009/11/03/toolmanship-how-to-use-a-handsaw.

1 Measure where you want to cut and draw a line marking where you want the cut to be. The line will act as a guide to help you get a straight cut. Remember, measure twice, cut once.

2 When you make the first cut, use your thumb or the knuckle of your thumb on the hand holding the wood as a guide to ensure you cut along the cutline.

3 If you are using a crosscut saw, start your cut with the teeth nearest the handle. This will give you the best control. Make a few back cuts until you get a nice kerf (opening in the wood). If you are using a ripsaw, start the cut with the finer teeth furthest from the handle (near the point of the blade). Make a few short draw strokes to get a kerf going.

4 Don't start the cut right on the line you marked earlier with pencil. Instead, cut right next to the line on the waste side. The waste side is the part of the wood you're not going to use in the finished product. It's always better to have a piece of wood that's a bit long, than a bit short. You can always sand the wood down to the pencil line.

5 Angle the saw correctly. After you get your kerf going, you need to angle your saw correctly to get the best cut. For crosscut saws, the proper angle is 45° between the saw and wood. With ripsaws, it's 60°.

6 *Hold your elbows close to your body.* To counteract the natural tendency to angle the blade away from perpendicular, hold your elbows close to your body when sawing. This will also help prevent you from twisting and tilting the blade, thus ensuring a nice, clean cut.

7 To hold the saw, grip the handle so that your forefinger extends along the side of the handle. This helps you 'point' the saw along the line and ensures more accurate cuts. Hold on to the handle firmly, but not too tightly.

8 After you've started the cut, a few short forward strokes will deepen the cut so you can move your left hand away from the blade. Push the saw with an easy, free-running motion. Use long strokes so that each tooth does its fair share of the work. Short strokes dull the saw faster because only a few teeth are doing the work.

Resist the temptation to bear down on the saw. It won't do anything except tire you out. Let the saw do the work. If you feel that the saw isn't cutting properly, something might be wrong with the saw itself. For straight cuts and to ensure that you get a true and square cut, place an off-cut of timber along your pencil line and clamp it to the board you're cutting. The board will now act as a guide to keep the saw on the line.

Cold chisels

Select the largest cold chisel suitable for the job. The hammer used should be much heavier than the chisel. The chisel should be held at an angle that cuts the waste away cleanly. If held too high the chisel will cut into the job, and if held at too small an angle the chisel will not cut deep enough. For deep cuts or when removing large amounts of material, two or more cuts should be made.

Be sure to handle cold chisels safely.

The work should be firmly supported by a packing block in the vice and chipping should be towards the fixed jaw. When resharpening a chisel care must be taken not to raise the temperature of the cutting edge above the temper temperature. Ideally when grinding a chisel cutting edge no colour should show on the surface.

FIGURE 4.62 Cold chisel maintenance

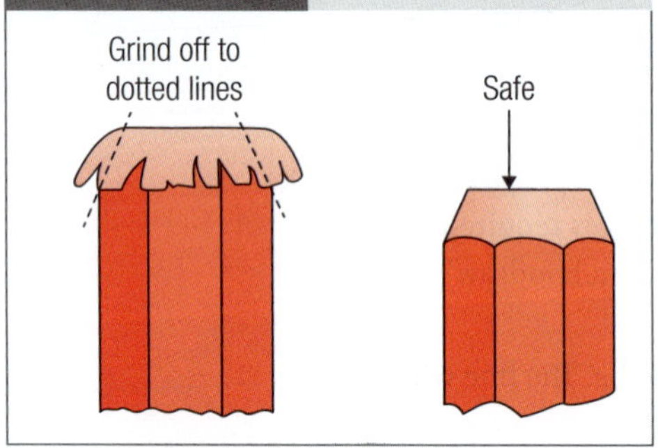

A major problem associated with chisels is the 'mushrooming' of the head due to the soft head and constant hammering, which spreads the material. If not removed the material may break away when struck, creating a hazard side. This deformation should be removed by grinding, as shown in Figure 4.62.

Pliers

Pliers intended for use in electrical work must be insulated. The plastic insulation on the handle does not mean it is safe to use the pliers on live electrical equipment or conductors; it is merely a safeguard in case of accidental contact. The type and size of pliers used depends on the material being dealt with. For heavy duty work, use combination pliers, while more delicate work calls for snipe nose pliers. Be sure to choose the right tool for the job.

All types of pliers are used in much the same way. Grip the pliers in one hand, using the little finger pressing on a handle to make the jaws open. Some types of pliers are spring loaded, causing them to open automatically. Grip the wire with the jaws of the pliers in whatever manner is needed to complete the task.

FIGURE 4.63 File selection

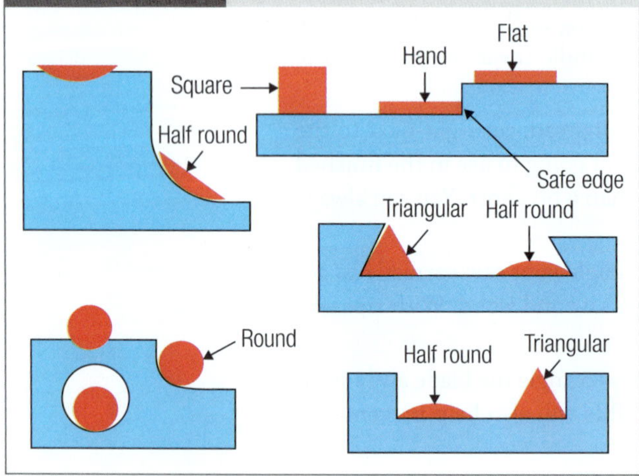

Files

Select a file shape and type that is suitable for the task, as shown in Figure 4.63. For roughing or cross filing, the file should be held with the handle in one hand and the point in the other. The work is best supported at elbow height. The feet should be spaced 0.3–0.5 m apart. During the cutting stroke, weight should be transferred from one foot to another. The file is pushed forward with a combination of arm and body movement. An average speed for filing is about 50 strokes per minute. It is essential that the flatness and squareness of the surface are checked regularly, using a square and a straight edge, and correcting faults as they occur.

By changing the direction of cut (crossing the cut) the point of contact can be checked, and corrected if necessary. A process known as draw filing may be used to remove file marks when close to size. The file is held by the body and moved along the length of the work. The file should be held as close to the work as possible and with the thumbs behind the body of the file (see Figure 4.64).

Pinning occurs when shavings of metals such as steel or aluminium clog the teeth of the file, causing scratches and score marks on the work surface. Pinning should be removed using a file card, as shown in Figure 4.65, or a stiff wire brush that is scrubbed across the face of the file in the direction of the tooth cut (*Workshop Practice NE175/1 Student Workbook*, pp. 40–1).

FIGURE 4.64 File techniques

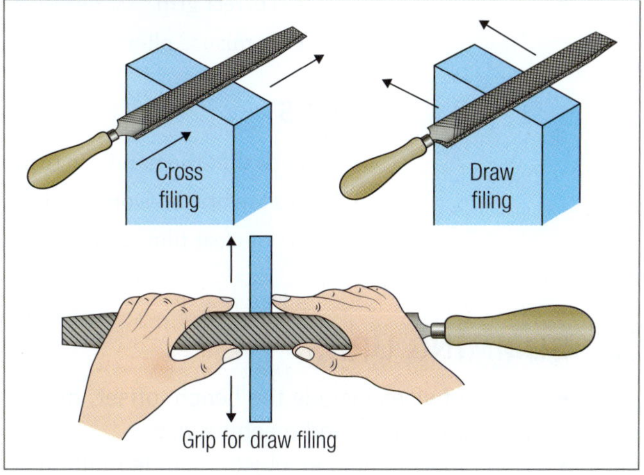

FIGURE 4.65 File cards

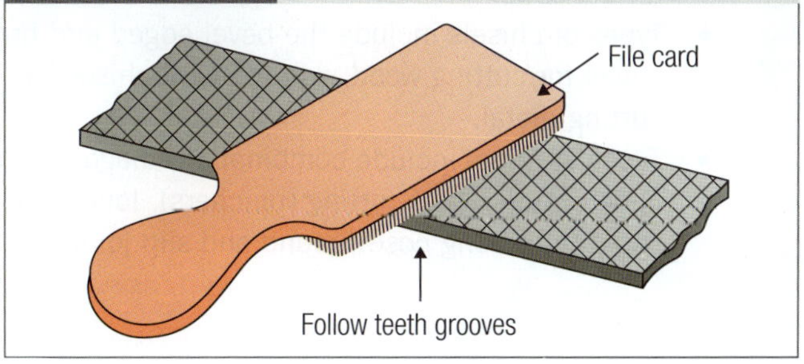

Safe use of cutting tools

Hacksaws

It is important to concentrate when using a hacksaw as the blade is easily broken and may cause injury. Use safety goggles if using high-speed blades, as they are very brittle.

- Always support the material being cut, and support both pieces on the last few strokes or the off-cut may fall on your foot or be damaged if it falls on the floor.
- Clamp thin, flat pieces requiring edge cutting.
- Keep saw blades clean and lightly oiled.
- Do not apply too much pressure on the blade as the blade may break.
- Do not twist when applying pressure.
- Do not use when the blade becomes loose in the frame.

Timber saws

Safety procedures are similar to those for the hacksaw, except additional care should be taken with a timber saw due to its sharp teeth.

- Hold stock being cut firmly in place.
- Wear appropriate PPE (gloves).
- Get assistance or a supporting bench or vice to support long stock if required.
- Keep teeth and blades properly set and sharpened.
- Protect teeth of the saw when not in use.
- Keep saw blades clean.

Cold chisels

Always wear appropriate PPE (safety glasses) when using a cold chisel.

- Do not use chisels with mushroomed heads as these may chip.
- Do not overheat the metal when grinding or sharpening a cold chisel.
- If working near others, use a chipping screen to stop chips from becoming a hazard to them, or keep other people clear of the area.
- Always ensure that the job is held securely (*Electrotechnology Practical LSEEE280A*, p. 59).

Pliers

- Select suitable pliers for the task.
- Inspect pliers for damage or wear.

- Ensure electrical supply is isolated correctly (test, tag, lockout and try).
- Use correct grip.
- Never misuse pliers, for example, they are not a hammer or a spanner.

Files

- Check that the handle is firmly fixed to the tang of the file. Never use without a handle.
- Files are brittle. Don't strike them or use them as levers.
- Do not clear filings with your hand as the chips and edges are sharp.

SUMMING UP...

- Types of vices include the bench, offset, hinged pipe, chain pipe and machine vice.
- Hand tools to hold an object include multi-grips, vice grips, the Stillson wrench and footprint wrench.
- Tools used for cutting include the hacksaw, junior hacksaw and crosscut, rip and triple ground timber saws.
- Types of chisels include the bevel edged and firmer chisel for cutting wood, and the cold chisel for cutting metal.
- Types of pliers include combination, diagonal cutting (side cutters), end cutting (pinchers), long nose and snipe (mini long nose), round and slip joint.
- Metalwork files are classified according to length (measured from the shoulder to the point) and cross-section or body shape.
- Grades of files are bastard cut (coarsest, used for rough filing), second cut and smooth cut.
- File shapes include flat, round, half round, triangular and square.
- Cutting, shaping and finishing tools include the hacksaw, timber saws, chisels, pliers or side cutters (to cut wire) and files.
- Safe use of cutting tools is described in the above section.

4.5 Drills and drilling

Cutting holes in a work piece is done by using a rotating cutting tool called a drill, or drill bit. Electricians need to be able to perform drilling operations in a variety of environments and situations. Drilling may be done as part of manufacturing in a workshop or on-site as part of the installation process.

Drills used in electrotechnology

Twist drill

The twist drill is usually made from high speed steel and is constructed for the efficient cutting of metals, wood and plastics. The drill bit needs to be harder than the material it cuts. For example, drills for cutting stainless steel are made of high speed steel alloyed with cobalt (HSSCo), with a titanium aluminium nitride (TiAlN) coating.

The parts of a twist drill are shown in Figure 4.66. The shank of the drill is generally straight for drills up to 13 mm diameter. Larger sizes often have a standard 'Morse' tape, like the drill bit in Figure 4.66. A straight shank allows the drill to be held in a standard drill chuck, while the Morse taper requires a mating tapered hole in the machine spindle. To remove the Morse taper drill from the spindle; a drift is pushed into a location hole in the spindle and forced against the tang of the drill. Shanks on large drills are sometimes reduced in diameter so they can fit into a standard 10 mm three jaw chuck.

FIGURE 4.66 Drill parts

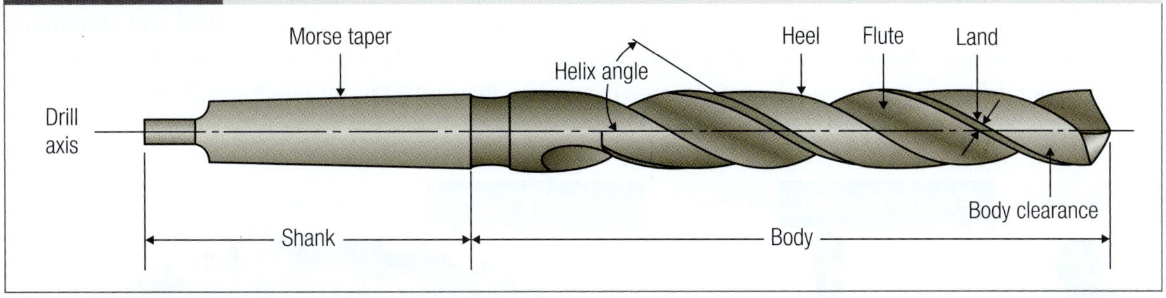

Sizing of drills can be metric (millimetres), imperial (fraction of an inch) or a gauge system:

- In the metric system (used in Australia and New Zealand), the diameter of the drill is given in millimetres. Starting at 1.0 mm, industrial use drill bits rise in increments of 0.1 mm up to 13.0 mm, and then in 0.5 mm increments to 51.0 mm.
- In the imperial system, fractional bit sizes are in increments of 1/64 inch up to 1 inch, 1/16 inch increments to 3 inches, 1/8 inch increments up to 3¼ inch, and one more ¼ inch increment to 3½ inches. An alternative is to convert the fraction to a decimal. For example, ½ inch is equal to 0.500 of an inch or 500 thousandth of an inch.
- Number and letter sizes start at size 1 (5.791 mm) through to 80 (0.343 mm) or a size A (5.944 mm) through to Z (10.49 mm). Wire sizes are another set of measurements starting at 1 and then increasing by whole numbers.

Drill bits are classified by their length:

- stub (shortest)
- jobber (most common)
- long
- extra long.

As well as the masonry drill bits discussed in Chapter 3 there is a wide variety of specialised drills available for specific uses.

FIGURE 4.67 Countersink drill

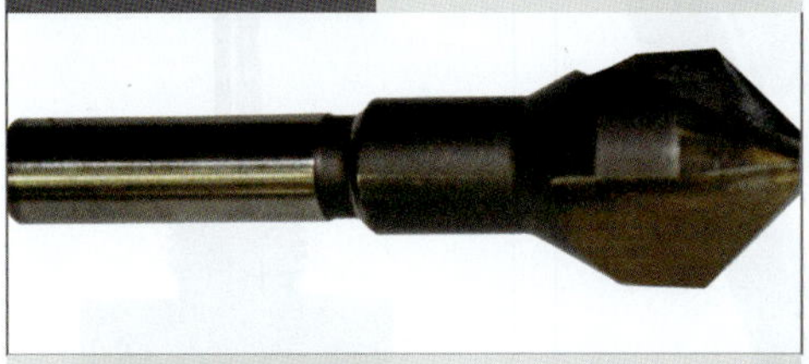

Source: Ed Hawkins

FIGURE 4.68 Countersunk screw head

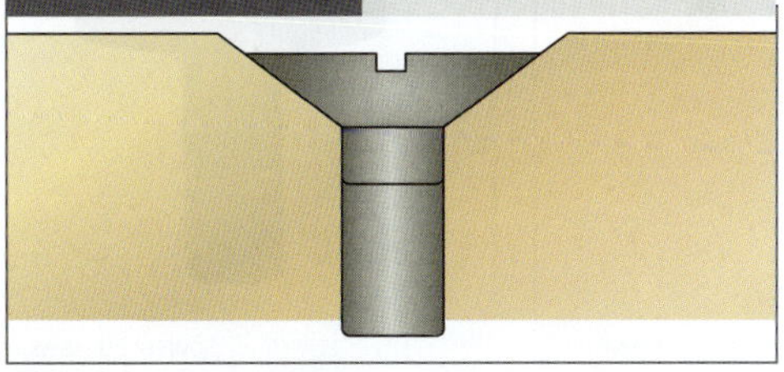

FIGURE 4.69 Cheese head bolt

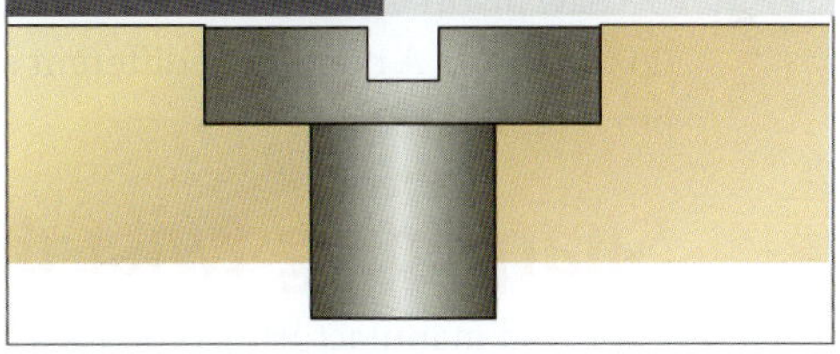

Countersink drills

A countersink drill is shown in Figure 4.67. They have an included angle of 90° and are used to countersink a hole for the head of a countersunk screw (see Figure 4.68).

Counter boring tools

Counter boring tools are used to counter-bore existing holes to recess cheese head screws, bolt heads or to provide a flat seat for a bolt head, as shown in Figure 4.69.

A similar process to countersinking is spot facing, which is done to flatten the surface of a material so a nut or bolt head can sit flush on the surface frame.

Centre drills

A centre drill (see Figure 4.70) is a special type of countersinking drill that uses a double ended stub twist drill that acts as a pilot drill. These drills are used to start drills when drilling in the lathe and to produce a location hole in work to be turned between centres in a lathe.

Hole saws

These are used to drill large diameter holes in sheet material such as Gyprock, steel and aluminium plate, because a large twist drill bit will bite into and damage the sheet material securely. Hole saws have a pilot drill mounted in an arbour to locate a centre (see Figure 4.71), and a high speed steel circular saw blade, shown in Figure 4.72.

FIGURE 4.70 Centre drill

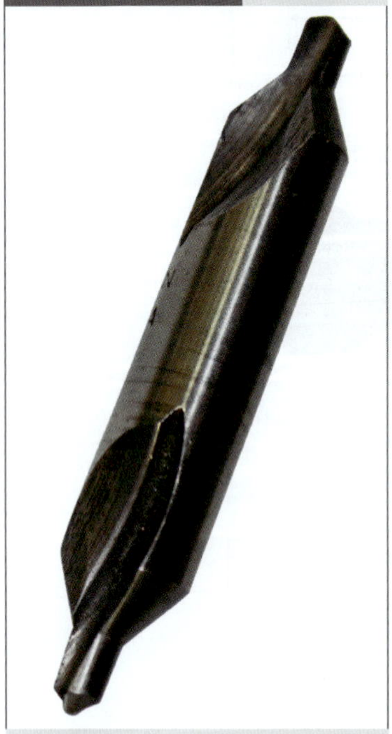

Source: Ed Hawkins

FIGURE 4.71 Arbour

Source: Ed Hawkins

FIGURE 4.72 Hole saw arbour

Source: Ed Hawkins

Saw blades are available in sizes ranging from 14 mm to 152 mm diameters and a cutting depth of up to 32 mm. A number of different sizes of blade may be used with an arbour, using a screw on-off and pin retaining system.

Sharpening twist drills

Improved manufacturing processes have reduced the cost of drills and as a result drill bits under 6 mm size have become almost throw-away items once they become blunt. However, the ability to sharpen a drill bit is still an important skill for an electrician, as larger drill bits will still require the occasional sharpening. The point of a drill bit, as shown in Figure 4.73, has a number of critical angles to allow the bit to cut effectively.

The standard point angle is 118° and is suitable for general purposes. When drilling specific materials an angle suited to the material being drilled should be used. A small error in the size of the point angle would not seriously affect the efficiency of the drill, but it is most important that the axis of the drill bisects the point angle.

The cutting edges are the working parts of the drill, and must be in contact with the job for it to cut. For this reason the flank is ground back to give a lip clearance. For standard twist drills this is usually 12°.

The chisel edge of the drill locates the drill and forms the pivot point for the cutting edges of the drill. The chisel edge does not cut and must be forced into the job. As drills become shorter, through regrinding, the length of the chisel edge increases, requiring more force to maintain the cut.

FIGURE 4.73 Twist drill

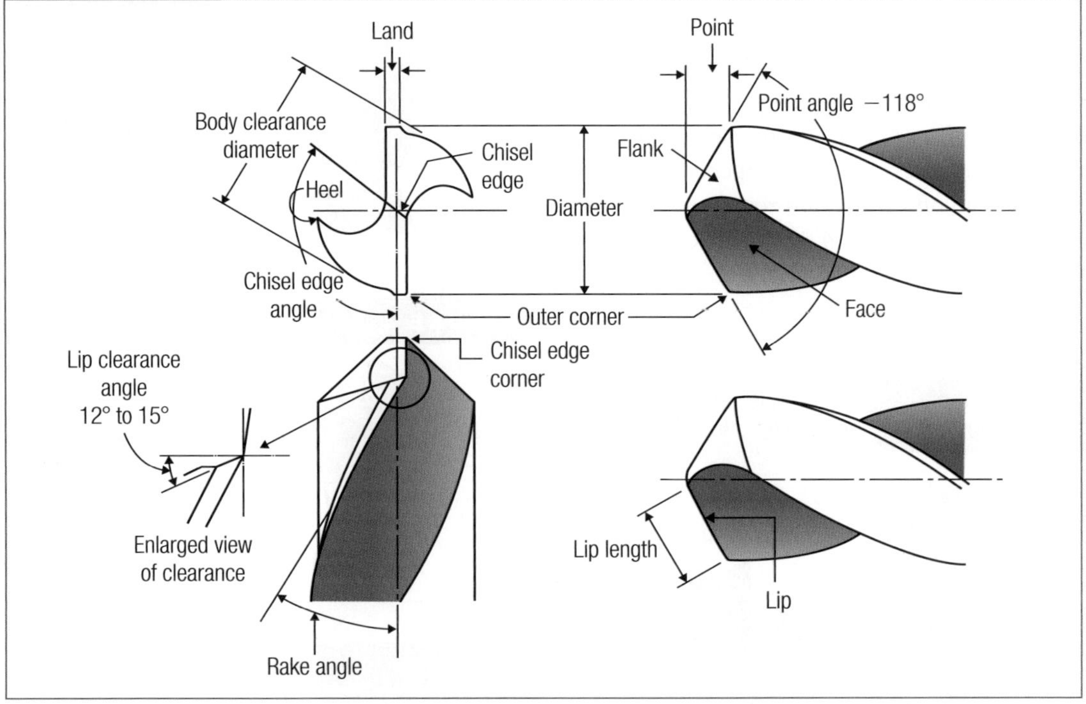

Flutes or helical grooves are included throughout the cutting length of the drill. They are used to carry away the cut material (swarf) from the drill point, and to allow the easy flow of lubricant or coolant to the drill point during drilling.

To reduce friction on the body of the drill, body clearance is provided by reducing the diameter of the body leaving only a small portion (the land or margin) in contact with the edge of the hole. Drill diameter is measured across the land. To prevent the drill from jamming, the diameter of the drill is reduced by about 0.01 mm per 100 mm of length (in J.F. Lancaster, *Metallurgy of Welding*, 6th edition, Abington Publishing, 1999).

Cutting clearances and angles

Most of the difficulties encountered in drilling are due to incorrect angles or clearances. The three important features to remember when sharpening drills are:

- the length and point angle of the cutting edges
- the lip clearance angle
- the correct location of the chisel point.

It is important that the half point angles are equal and the cutting edges are of equal length, as shown in Figure 4.74. For general work, the included angle of the point should be 118°. This angle needs to be ground within about ±5° or the cutting edge will become distorted, resulting in inefficient cutting. For general use, a lip clearance angle of from 12° to 15° is suitable, as shown in Figure 4.75.

Insufficient lip clearance is shown in Figure 4.76. This can cause overheating due to the flank rubbing on the work. Changing the lip clearance angle also changes the chisel point angle.

FIGURE 4.74 Drill point and lip clearance angles

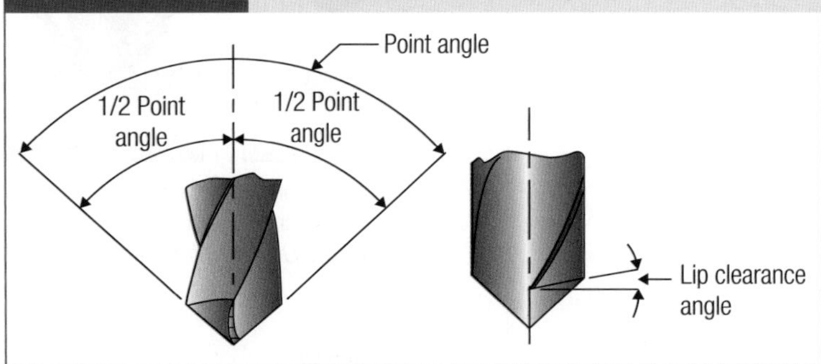

FIGURE 4.75 Correct lip clearance angle

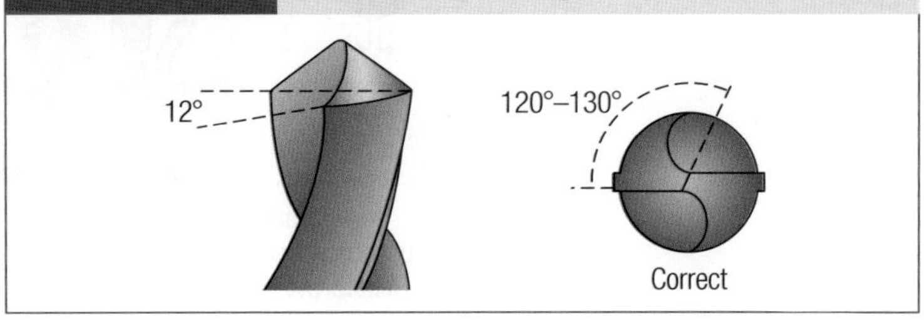

A large lip clearance angle, shown in Figure 4.77, will not support the cutting edge sufficiently, allowing it to chip and break.

A properly sharpened drill will cut evenly and with less effort. The chipping pattern should be even off both cutting edges of the bit. Uneven chipping indicates a poorly sharpened bit which will result in off-centre, non-round or oversized holes. Figures 4.78 to 4.80 show some of the effects of faulty drill sharpening.

FIGURE 4.76 Insufficient lip clearance angle

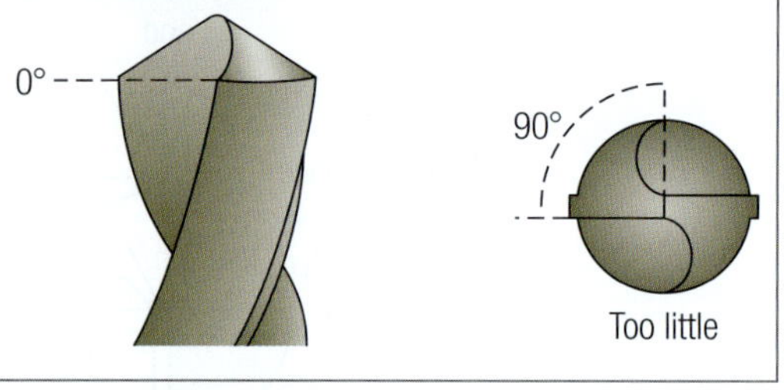

FIGURE 4.77 Excessive lip clearance angle

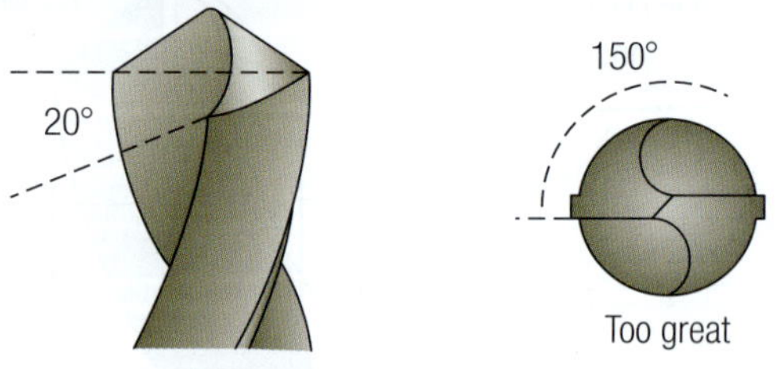

FIGURE 4.78 Results of incorrect drill sharpening: oversize hole

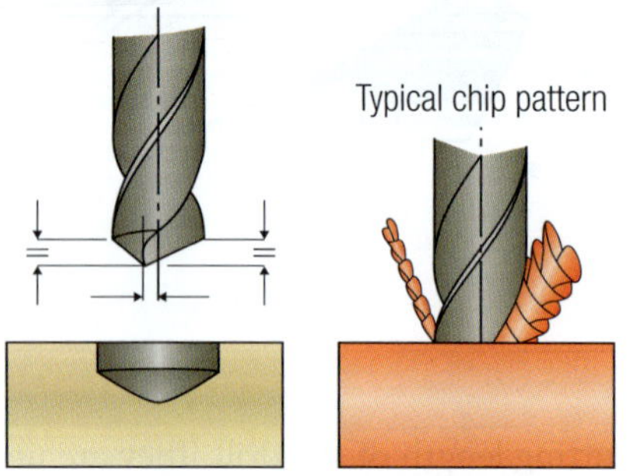

Unequal angles, chisel edge off centre, lip heights unequal:
Result: *Oversize hole.*

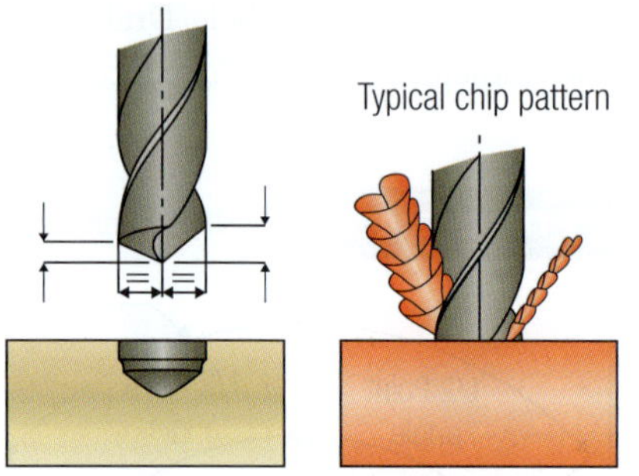

Drill point on centre, lip heights and lengths unequal:
Result: *Oversize hole, inefficient cutting.*

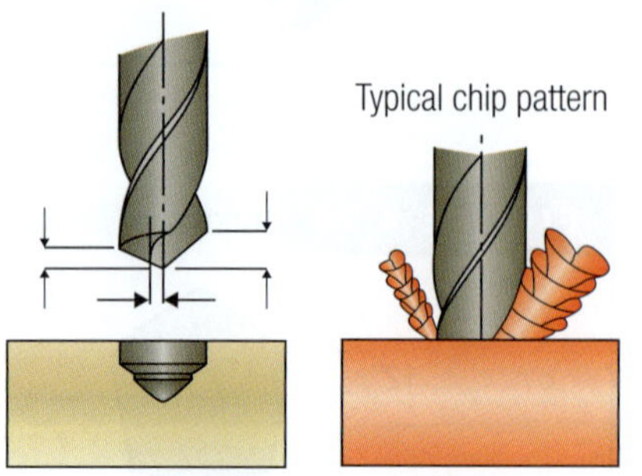

Unequal angles, lip heights unequal, chisel point off centre:
Result: *Oversize hole, stepped diameters.*

FIGURE 4.79 Results of incorrect drill sharpening: oversize hole, inefficient cutting

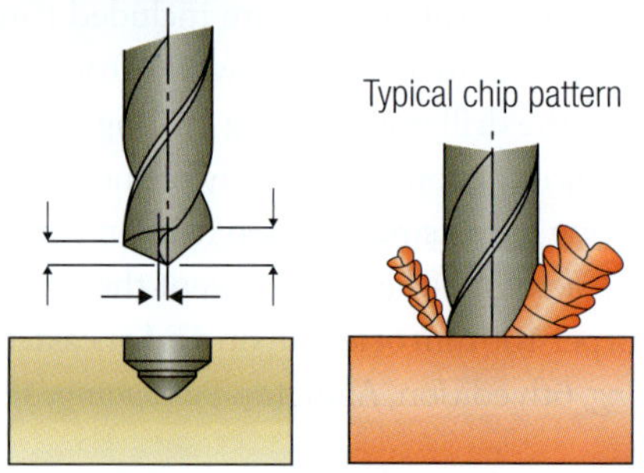

Unequal angles, lip heights unequal, chisel point off centre:
Result: *Oversize hole, stepped diameters.*

FIGURE 4.80 Results of incorrect drill sharpening: oversize hole, stepped diameters

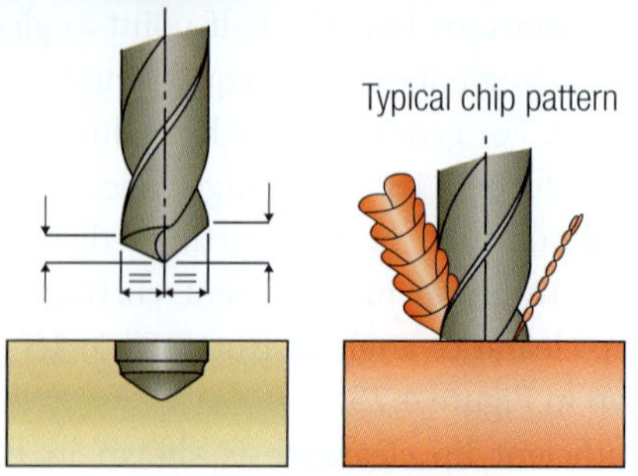

Drill point on centre, lip heights and lengths unequal:
Result: *Oversize hole, inefficient cutting.*

The angle of twist or helix angle of a drill is fixed by the manufacturer for general drilling of most materials. The helix forms the rake angle of the drill. The rake angle is important in the cutting process. Figure 4.81 shows examples of varying point angles for different materials. When drilling brass or gunmetal the rake faces of the cutting lips are ground parallel with the axis of the drill, giving a zero rake angle as shown in the bottom right of Figure 4.82. This is also useful when drilling thin metal sheet as the square edge prevents the drill from 'grabbing' as the drill breaks through.

An alternative method for drilling sheet metal is to increase the point angle. Figure 4.82 shows that with a standard 118° point angle, the point breaks through before the outer corners. By increasing the point angle to 140° the outer corners of the drill bit are engaged with the work piece, which prevents jamming.

FIGURE 4.81 Point angles for different materials

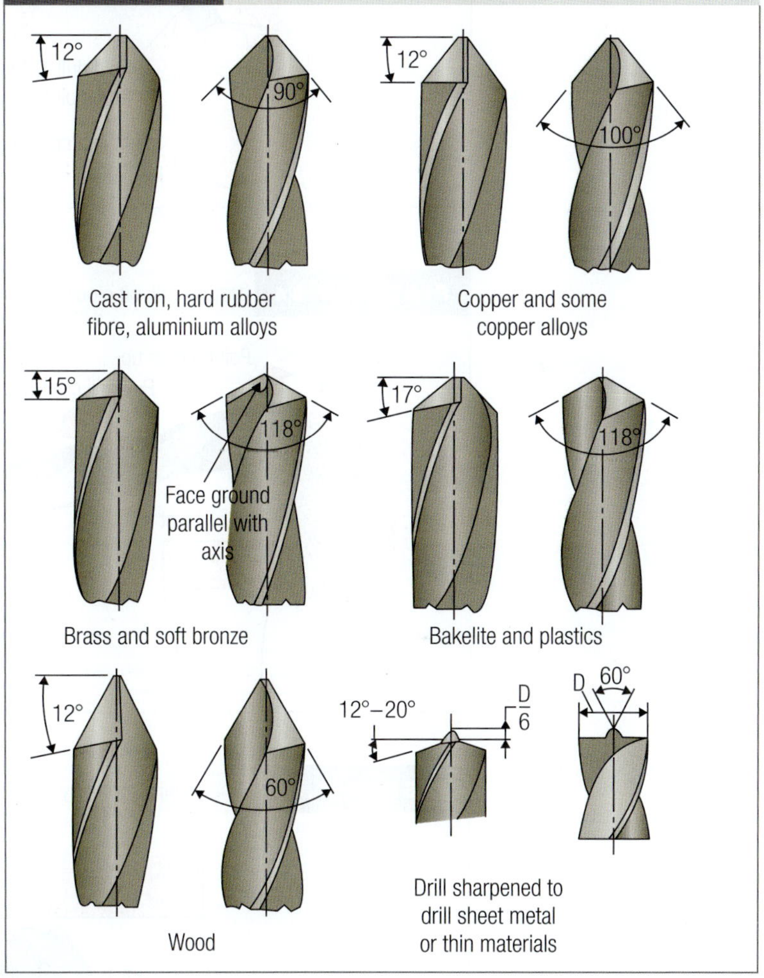

FIGURE 4.82 Changing the point angle to suit the material

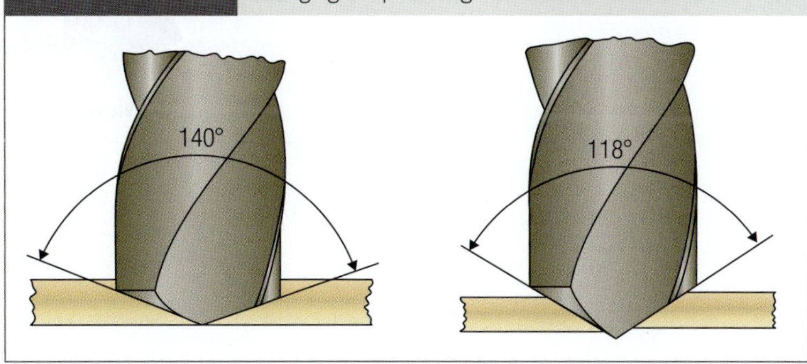

Sharpening drill bits

Drill bit sharpening is done through a process called off-hand grinding, which means the work piece is held by hand and ground using either a pedestal or bench grinder, and fitted with a wheel suitable for grinding tool steel.

When sharpening drill bits the following steps should be taken:

1 The twist drill is sharpened by holding it against the front face of the grinding wheel as shown in Figure 4.83 so that the cutting edge is horizontal and at such an angle that it will form half the point angle, approximately 59° (118° × 0.5) on one side of the drill. That is grind one lip.
2 As the grinding takes place the drill is twisted clockwise about its axis and the tang or shank is moved downwards to the left so as to form the lip clearance angle. Commence grinding with the drill facing slightly upwards and the cutting edge just engaging the wheel, as shown in Figure 4.84.

The twisting and downward movement needed to sharpen a drill can be performed by hand as shown in Figure 4.85, or by using a special drill sharpening attachment.

1 When sharpening by hand, it is more difficult to obtain the required angles and relative lip height. By setting the hands on the tool rest of the grinder as shown in Figure 4.85,and by carefully controlling the movement of the fingers, the action of the attachment can be copied. Whether the drill is sharpened by hand or by attachment, it is important to check that the height and length of each lip is the same and, particularly when hand grinding, that the axis of the drill bisects the point angle.
2 Do not grind heavily as this will cause the drill to overheat. Use several light grinding strokes, cooling the drill and checking the angles and clearances each time. When one edge is complete sharpen the other edge.

FIGURE 4.83 Drill bit sharpening

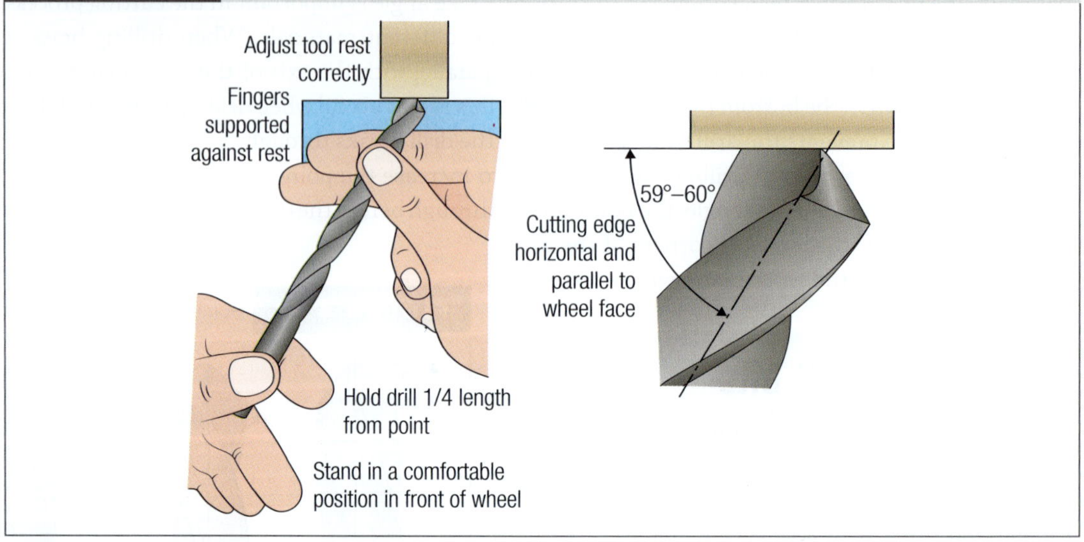

FIGURE 4.84 Rotate drill bit clockwise when sharpening

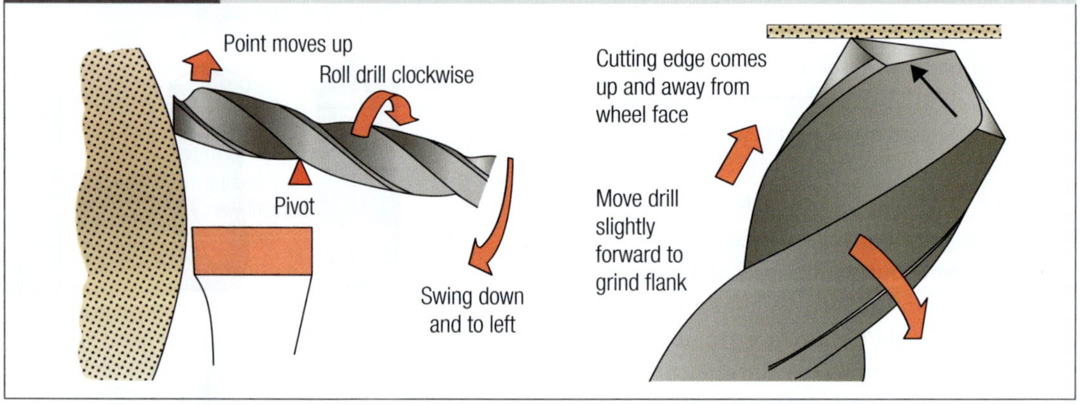

FIGURE 4.85 Sharpening a drill by hand

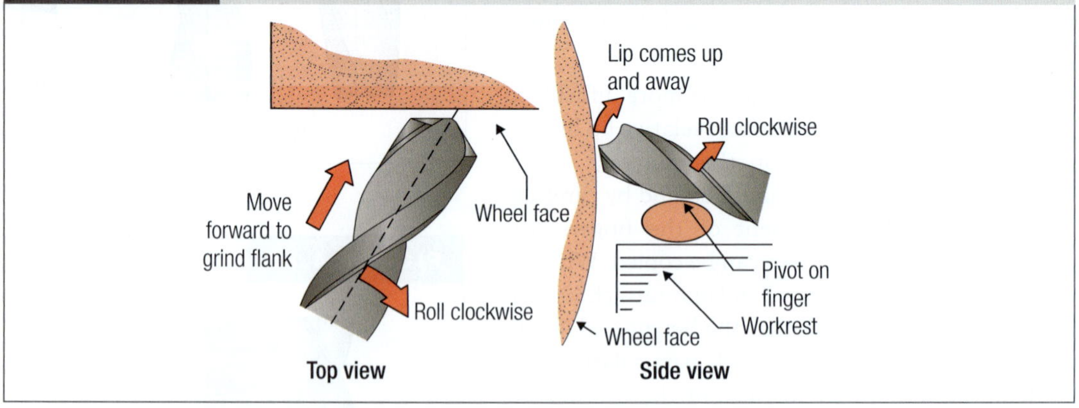

3 The cutting edge length and angle can be checked when sharpening with the aid of an attachment to a rule or a simple drill gauge as shown in Figure 4.86 (*NE175 Workshop Practice Student Workbook*, pp. 45–6).

FIGURE 4.86 Drill gauge

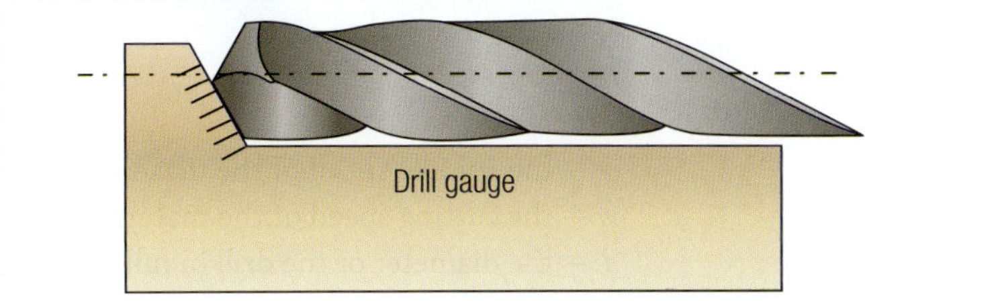

There is also a wide range of mechanical aids. These devices typically use a chuck to hold drill bits ranging in size from 4 mm to 20 mm and a series of cams to correctly position the drill bit for consistent sharpening.

When sharpening drills by hand in an off-hand grinder, first check the following:

1 Your eyes must be protected from flying particles. Always use safety glasses, goggles or a face shield, and with the eye shield fitted to the machine.
2 The grinding machine must not be used if the wheel is cracked, or if the wheel has received a severe shock such as being dropped in transport. The wheel must be removed and tested by a person competent and authorised to do so.
3 The wheel must be in good condition. A badly worn wheel grooved on the face or with worn sides must be dressed or replaced.
4 Do not remove the guards, work rests or face shields from the grinder.
5 The work rest should be set with no more than 1.5 mm to 2 mm clearance to the wheel face.
6 Do not wear gloves or hold the work with cotton waste or loose cloth. These materials may be caught in the wheel and drag your hand into the wheel (*NE175 Workshop Practices Student Workbook*, pp. 45–6).

Safety when sharpening drills

FIGURE 4.87 Sharpening marking out tools

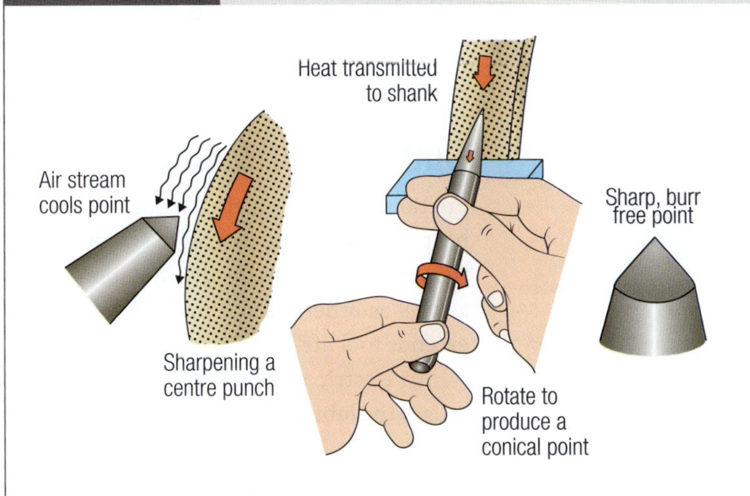

Sharpening scribers and punches

Scribers and punches are sharpened in a similar way to drills. The difference is that the point is rotated in a 360° motion while being held lightly against the wheel face, as shown in Figure 4.87. The point should be moved constantly backward and forward across the face to avoid wearing grooves into the wheel. Avoid overheating as this will soften the point and make it unsuitable for use.

Drilling metallic and non-metallic components

Drill speeds

For a drill to cut efficiently it must be rotating at the correct speed. There are three factors that govern this speed:

- Drill diameter – generally the smaller the drill the faster it should rotate.
- Drill material – this is normally high speed steel. Other cheaper alloy drills will need to rotate more slowly.
- Material being drilled – the common materials and their cutting speeds are shown in Table 3.12 in Chapter 3.

A simple equation can be used to determine the nearest correct speed.

$$n = \frac{300 \times v}{d}$$

Where:
n = the spindle speed of the machine in revolutions per minute (RPM)
v = the cutting speed of the material in metres per minute as shown in Table 4.6
d = the diameter of the drill in millimetres

TABLE 4.6 Recommended cutting speeds

RECOMMENDED CUTTING SPEEDS	
Material	**Cutting speed metres per minute**
Cast iron	20
Copper	20
Mild steel	30
Brass	50
Aluminium	100

For example, to drill a 6 mm hole in mild steel the correct speed is:

$$n = \frac{300 \times v}{d} = \frac{300 \times 30}{6} = 1500\,\text{rpm}$$

The machine should be set to the nearest matching speed (*Electrotechnology Practical LSEEE280A*, p. 71).

Cutting lubricants

Friction from the cutting action of drilling will heat both the drill bit and the work piece. Excessive heat will change the temper of the drill bit, possibly causing it to lose its hardness and becoming blunt, thus rendering it ineffective. Soluble oil is often used to cool and lubricate the cutting face of drill bits, as well as other machining tools. The cooling action prevents overheating and the lubrication action extends tool life.

An electric pump, a hose and a nozzle are used to spray lubricant onto the work piece. Gravity causes the lubricant to pool in a sump where it is collected by the pump.

Safe use of a bench drill

In Section 4.2 of this chapter, Figure 4.20 shows an example of typical safe operating procedures (SOP) for a pedestal drill.

When drilling using a pedestal drill you must:

- be aware of the correct and safe operation of the drill
- wear safety glasses to protect your eyes from chips
- not wear baggy or loose clothing, jewellery such as chains, and always make sure that long hair is covered by a cap or hair net
- ensure the drill bit is correctly sharpened to suit the job and the material
- check that the bit is fitted correctly into the chuck of the machine
- ensure the job is held securely by gripping in a vice or clamping
- check that the speed is suitable for the drill size and the material being drilled
- always remove the chuck key before starting the machine
- reduce the drilling pressure as the drill tip breaks through the opposite surface of the job

- use a lubricant such as soluble oil when required
- wait for the drill to stop of its own accord before taking hold of the chuck
- do not leave the drill running unattended
- do not adjust the job or work-table position while the machine is running (*Electrotechnology Practical LSEEE280A*, p. 75).

SUMMING UP...

- Twist drills are made from:
  - high speed steel for general use with materials such as mild steel and copper
  - cobalt grade of high speed steel (HSSCo) with a titanium aluminium nitride (TiAlN) coating for use with most grades of stainless steel.
- The main parts of a twist drill are the shank, body and flutes.
- The common systems used to size drills are metric, fraction of an inch and letter.
- The three most common lengths of drills from shortest to longest are stub, jobber and long.
- Specialty drills include counter sinking, counter bore and centre drills which are usually double ended. Hole saws are used to cut large holes in sheet metal and plasterboard.
- When sharpening a twist drill for mild steel the key angles for correct drill sharpening are the point angle (118°) and the lip clearance angle (12°).
- Drill sharpening can be performed 'off-hand' using a bench grinder and a suitable grinding wheel, sometimes with the use of a guide tool or with a purpose built tool.
- When drilling, the larger the drill diameter, the slower the drill speed. The harder the material the slower the drill speed.
- Lubricants such as soluble oil are used to cool drill bits and extend drill bit life.
- Always work safely when using any type of drilling machine.

4.6 Tapping and threading

Threads are commonly found on fixing devices such as nuts and bolts. A thread sometimes has to be cut into solid objects such as steel or copper bar, or onto the ends of round rod. This topic looks at the processes used to cut threads in a variety of situations.

Type and size of commonly used threads

Chapter 3 describes the common types of threads used in nuts and bolts. Tap and die sets are used to form threads by hand. Taps form a thread in pre-drilled holes and dies form a thread on round stock. The most commonly used threads are:

- metric
- imperial:
  - British Standard Whitworth (BSW)
  - British Standard Fine (BSF)
  - British Association (BA) commonly used on electrical terminals
  - American Unified (AF).

Taps and tap wrenches

Hand taps are made from high carbon or high speed steel which is hardened and tempered, similar to drill bits. This means taps are hard enough to cut softer materials such as mild steel, but just like drill bits, taps will shatter if flexed or jammed. While some taps are suitable for machine use, this section looks only at tapping by hand.

FIGURE 4.88 Parts of a hand tap

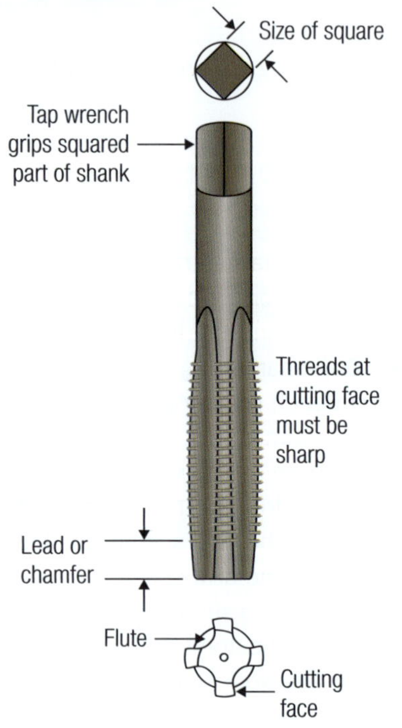

Figure 4.88 details the parts of a hand tap. The shank has a square drive ground on its end which allows the tap wrench to grip the tap. The flutes provide a path to remove waste. The leading threads on the point of the tap are tapered to assist centring of the tap.

The amount of taper at the point of the tap determines the tap type, as shown in Figure 4.89. The taper tap is used to start threads and may be used to cut threads that go through the work. The intermediate tap is used to complete deep through threads or in the second stage of tapping blind holes. The plug or bottoming tap is used to complete threads in blind holes. A blind hole does not pass all the way through the material. If a taper tap is used, the thread is not cut to the bottom of the hole.

Bright finish taps are suitable for general purpose use, TiN (gold colour) coated for tough materials and blue finish is suitable for ferrous materials.

Tap wrenches are used to grip taps and provide leverage on the tap to gain sufficient turning force. Smaller taps are usually operated with a chuck type wrench as shown in Figure 4.90. Larger taps are held using a bar type of tap wrench as shown in Figure 4.91.

FIGURE 4.89 Types of hand taps

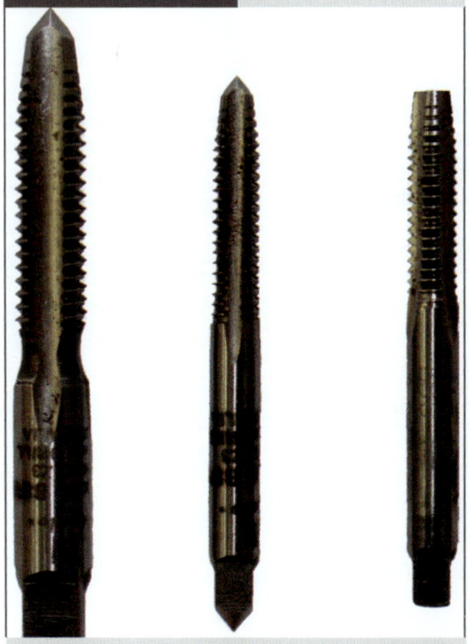

Source: Ed Hawkins

FIGURE 4.90 T bar tap wrench with chuck

Source: Ed Hawkins

FIGURE 4.91 Bar tap wrench

Source: Ed Hawkins

Tapping threads

Before tapping, a hole must be drilled the correct size for the type of thread that is required. A general rule is to tap not more than 75% of full depth of the thread. If tapping the full depth (100%) the force needed to turn the tap will be three times greater than at 75% depth, leading to tap breakage. At 75% the thread has only 5% less strength than 100%.

A general rule used to determine the size of the tapping hole for a metric thread is to subtract the pitch from the outside diameter. This will give approximately 75% thread depth.

Tapping hole size (mm) = nominal diameter − pitch

That is, for a tapping hole to give a 75% thread for an M14 × 2 thread, the hole diameter would be:

$$\begin{aligned}\text{Hole size} &= 14 - \text{pitch}\\ &= 14 - 2\\ &= 12\,\text{mm}\end{aligned}$$

Table 3.8 in Chapter 3 gives hole sizes for various sizes of metric taps. Charts are also available from tool manufacturers that give the tapping hole sizes for most standard threads. A cutting fluid or lubricant should be used when hand threading, to ease the cut and give a better thread finish. Table 4.7 shows examples of commonly used lubricants and cutting compounds. For hand threading only, apply a small amount of lubricant (*NE175 Workshop Practices Student Workbook*, pp. 57–60).

TABLE 4.7 Hand tapping cutting compounds and lubricants

Material	Lubricant
Aluminium	Kerosene or light mineral oil
Brass or bronze	Soluble oil or light mineral oil
Steel	Sulphur based oil
Copper	Soluble oil
Plastics	No lubricant needed

When tapping a thread:

1 The hole must be drilled to the tapping size.

2 The work must be held securely.

3 The tap and tap wrench should be chosen correctly. If the wrench is too large it is possible to break the tap because of the excessive force being applied.

4 Start the thread with a taper tap as shown in Figure 4.92. When starting to cut, the tap must be at 90° to the work in all planes. Use a small try square to test for squareness every quarter turn for the first full turn. If out of square, it is possible to straighten the tap if it has not entered the hole too far.

FIGURE 4.92 Tapping a thread

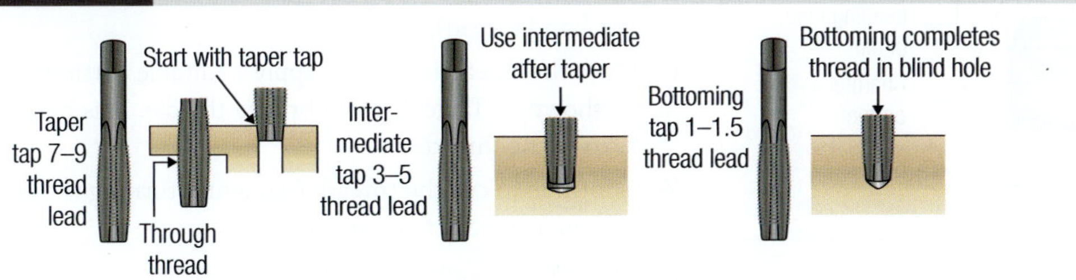

5 Apply the correct cutting fluid or lubricant as recommended in Table 4.7 and continue to cut the thread carefully, clearing chips from the tap regularly by turning backwards a quarter of a turn.

6 Complete through holes with the intermediate tap and blind holes with a bottoming (or plug) tap. When tapping blind holes, clean chips from the hole regularly (*NE175 Workshop Practices Student Workbook*, p. 57).

FIGURE 4.93 Parts of a die

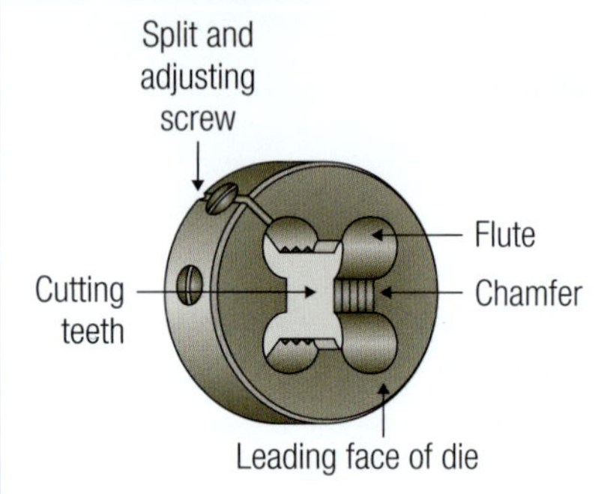

Stock and die tools

Dies cut a mirror image of the thread cut by a tap. Dies are made of hardened and tempered alloy steel similar to taps and drills. They are therefore hard, but brittle. Figure 4.93 shows the leading face of a die which is chamfered similarly to the taper on the point of a tap. Flutes provide a path to remove waste.

Types of dies include:

1. Button dies, elastic dies or divided disc type dies, generally used on threads up to 25 mm. These types are adjustable in size, although by a small amount.
2. Block dies are used on pipe or conduit threads up to 32 mm. These are not adjustable.
3. Chaser dies are used on larger diameter threads. Usually machine driven.
4. Die nuts are used to clean out damaged threads. Die nuts should not be used to form new threads.

An example of a stock used to hold a die is shown in Figure 4.94. These are made from mild steel or malleable cast iron. Stocks normally have a recess for holding a die, plus a means of adjusting the depth of cut and a guide to lead the die onto the job at the correct angle.

When electricians need to cut a thread on the ends of steel conduit, a ratchet handle stock is used.

FIGURE 4.94 Parts of a die stock

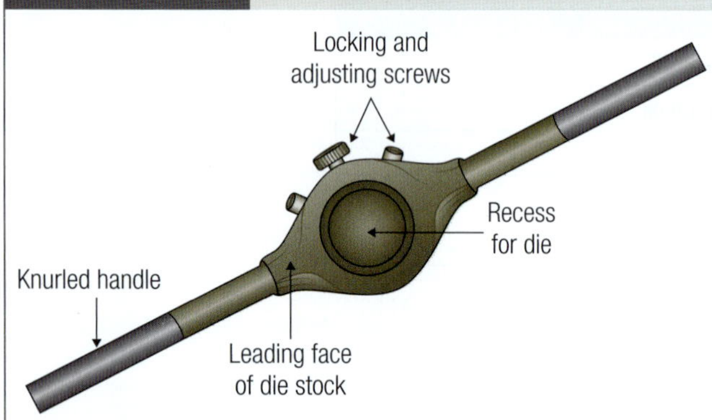

Threading with a stock and die

Dies are marked with the diameter and pitch of thread. Oversized rod should not be threaded as this will damage the die, but slightly undersized rod may be threaded providing the thread gives sufficient engagement with the nut.

When using stocks and dies:

1. Do not thread oversized rod or rod that is too hard to be cut easily with a file.
2. Hold the job firmly in a bench or pipe vice.
3. Chamfer the end of the rod or tube to an angle of about 20° as shown in Figure 4.95.
4. Check the die and guides for size and condition.
5. Slide the guide over the rod and apply pressure to the stock while rotating in a clockwise direction (anti-clockwise for a left-hand thread).
6. Once the thread is started, apply a suitable cutting compound as shown in Table 4.7 to lubricate the die. This helps prevent tearing the thread.
7. Continue to cut the thread, frequently turning the stock back one quarter turn to clear the chips.
8. Remove burrs and sharp edges with a file.

Never clean swarf from threads with your hand, as the chips and edges formed when threading are sharp and may cut or puncture the skin (*NE175 Workshop Practices Student Workbook*, p. 57).

FIGURE 4.95 Stock and die use

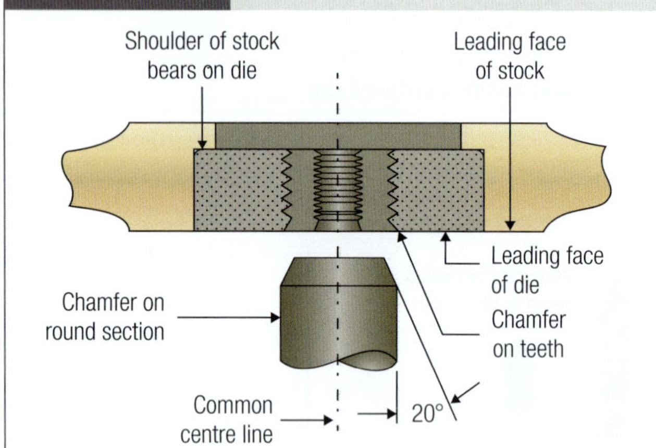

SUMMING UP...

- Taps and dies are made from high carbon or high speed steel which is hardened and tempered, similar to drill bits.
- The three types of taps are the taper, intermediate and plug.
- The two types of tap wrench are the chuck type and the T-bar type with a chuck.
- Tapping hole size (mm) = nominal diameter − pitch.
- In general do not tap more than 75% of full depth of thread.
- A lubricant or cutting compound should be used when cutting threads with a tap or die.
- The tapered side of a die should always face the work piece.
- Types of die include button or elastic dies, block dies, chaser dies and die nuts.
- Stocks fitted with ratchet handles are used with dies when threading conduit.

4.7 General hand tools

Hammers

The most common type of hammer is the engineer's or ball-pein hammer as shown in Figure 4.96. The ball-pein hammer has a flat face for general work and a ball-pein for special work such as riveting. These hammers are available with head weights from about 125 g to 1.4 kg to suit a wide range of applications. Light hammers are easy to control but are not suitable for applying heavy blows. Heavy hammers are difficult to use when striking light, accurate blows. For general work a 600 g hammer is ideal, while a 200 g hammer is useful for light work such as marking out. Traditionally, hammers have a timber handle; however, steel, fibreglass and carbon graphite composite materials that transfer less vibration during impact to the user's hand are now also commonly used.

Another common type of hammer, used in installation of wiring, is the claw hammer, shown in Figure 4.97. This type of hammer is used for driving and removing nails in timber. Care must be taken not to damage the timber surface when driving or removing nails. If nails are to be driven flush or recessed, use a nail punch.

A soft faced hammer or mallet as shown in Figure 4.98 is used to strike surfaces that might be damaged with a steel hammer. Soft faced hammers have an iron or steel body with replaceable soft metal, leather or plastic faces. A copper or brass face would be used to drive steel parts into position, a leather face could be used to seat coils in a motor winding and a plastic face may be used for

FIGURE 4.96 Ball-pein hammer 900 g × 380 mm handle

Source: shutterstock.com/dcwcreations

FIGURE 4.97 Claw hammer 0.45 kg × 330 mm handle

Source: iStockphoto/Osmany Torres Martín

FIGURE 4.98 Soft faced hammer 0.5 kg × 330 mm handle

Source: iStockphoto/Siraphol

FIGURE 4.99 Rubber mallet 0.5 kg × 340 mm handle

Source: iStockphoto/TokenPhoto

striking finished metal surfaces (*NE175 Workshop Practices Student Workbook*, p. 23).

A rubber mallet, as shown in Figure 4.99, can be used instead of a soft faced hammer.

A common type of hammer is the club, lump or mash hammer as shown in Figure 4.100. These are used in cabling installation. They are a heavy, short handled hammer for heavy duty chiselling or for breaking large objects. They are typically used with chisels or tools for chasing into brick or masonry, or for driving electrodes into the ground (*Electrotechnology Practical LSEEE280A*, p. 89).

Sledge hammers (see Figure 4.101) have a 4.5 kg head and an extended handle. They are generally used for demolition work and driving earth electrodes in hard earth.

When using any hammer, it should be held firmly at the end of the handle as shown in Figure 4.102. Do not hold the hammer close to the head. The face must strike the object squarely.

Light hammers may be used with a wrist action while heavy blows with a heavier hammer may require movement of the whole arm. When using a hammer for bending material, always apply force to the material so that it bends over the fixed jaw of the vice (*NE175 Workshop Practices Student Workshop*, p. 24).

Figure 4.103 shows the ball end of a ball-pein hammer being used to spread the shank of a rivet.

FIGURE 4.100 Lump hammer 0.9 kg × 250 mm handle

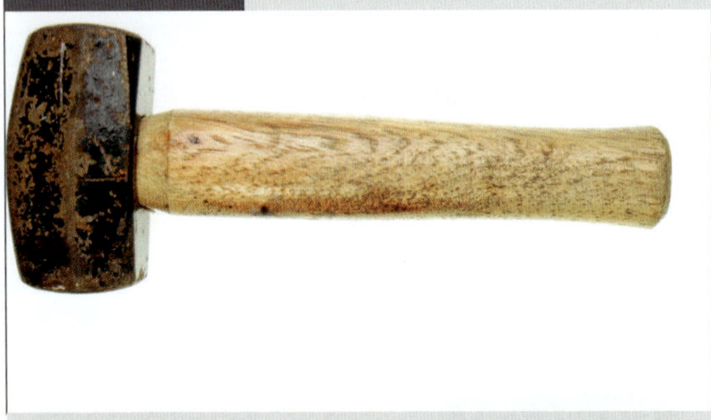

Source: iStockphoto/Richard Griffin

FIGURE 4.101 Sledge hammer 4.5 kg × 900 mm handle

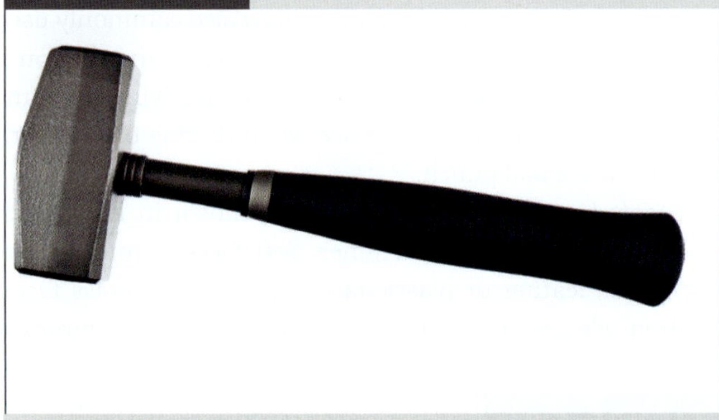

Source: shutterstock.com/Gemenacom

FIGURE 4.102 Correct hammer grip

FIGURE 4.103 Riveting with a ball-pein hammer

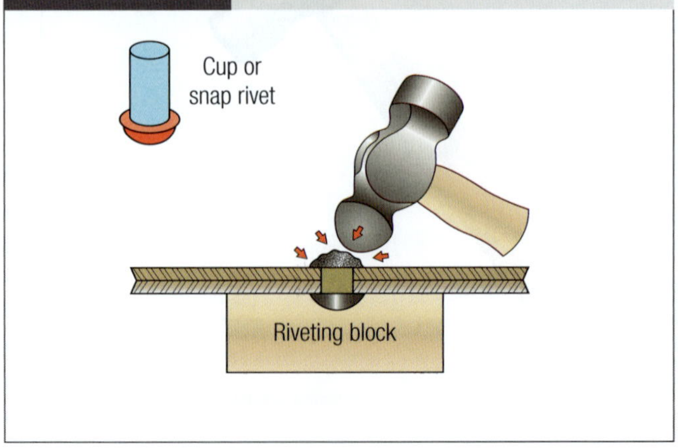

When removing nails, a suitable wood block should be used (Figure 4.104) between the hammer and the timber surface to increase the mechanical advantage of the hammer by moving the fulcrum point closer to the nail.

FIGURE 4.104 Wooden block used when pulling nails

Hammer safety

Figure 4.105 shows examples of common hammer faults and damage. Do not use a hammer which has a loose head, dirty, chipped or burred face, or splintered, split or greasy handle. Never strike hard metal (or other hammers) with a hammer, as this could chip the face of the hammer, causing chips to fly anywhere. Always wear safety glasses when chiselling or performing any work where flying particles are present (*NE175 Workshop Practices Student Workbook*, p. 25).

Component assembly using hand tools

Hammers

Hammers and nail guns are used extensively to assemble timber frames of buildings. Electrical components that are required to be attached to these timber frames may be nailed into place as described in Chapter 3. If using a hammer to tap a component into place, use a soft faced hammer, or place a piece of timber between the component and the hammer.

Safe use of hammers

Screwdrivers

It is important to select the size of the screwdriver so that the thickness of the blade makes a good fit in the screw slot. This prevents the screw slot from becoming burred and the blade tip from being damaged. Figure 4.106 illustrates both correct and incorrect screwdriver size selection.

FIGURE 4.105 Typical hammer damage

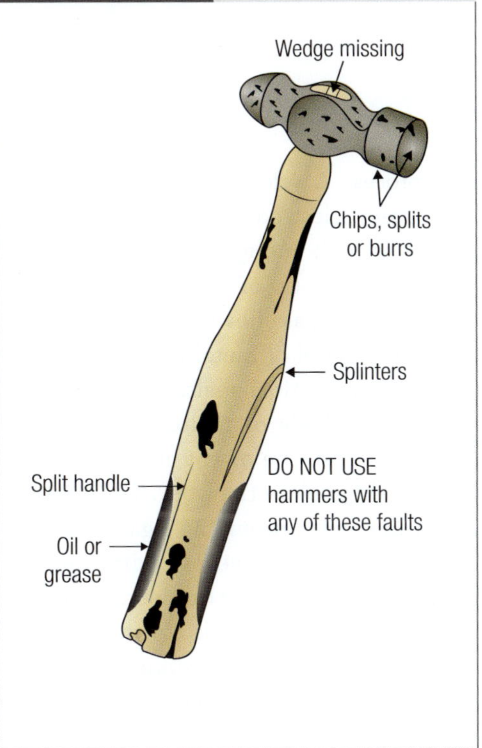

FIGURE 4.106 Screw driver selection

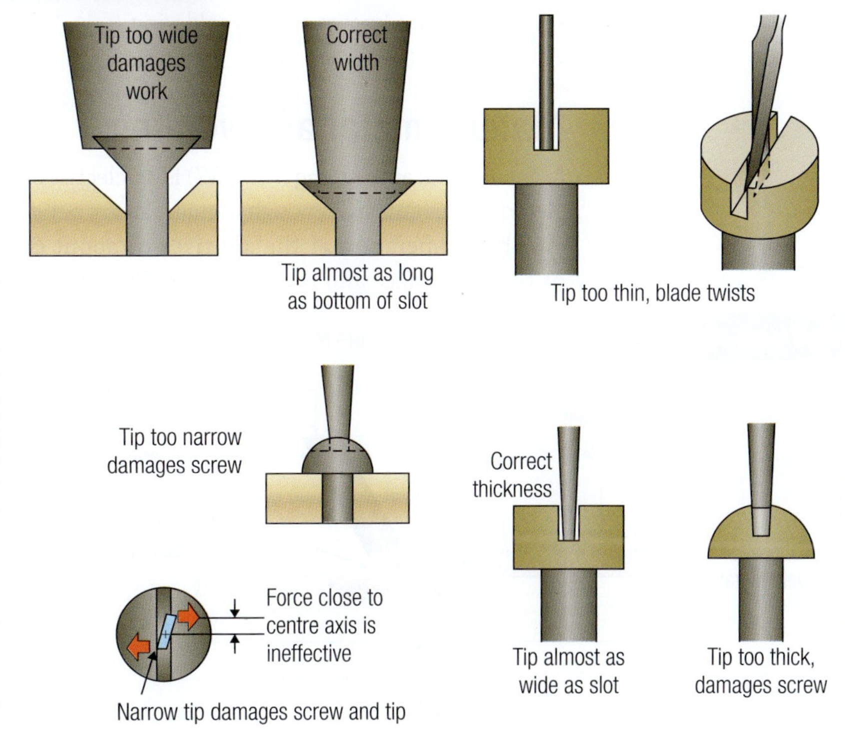

SUMMING UP...

- Common types of hammers used in the electrical industry include, ball-pein, claw, soft faced, mash (or lump or club) and sledge hammer, plus rubber mallet.
- When using any hammer it should be held firmly at the end of the handle.
- When removing nails with a claw hammer, a suitable wood block should be used to prevent damage to the work piece.
- Do not use any hammer which has a loose head, dirty, chipped or burred face, or splintered, split or greasy handle.
- Select a screwdriver so that the thickness of the blade makes a good fit in the screw slot.

4.8 Joining techniques

In Section 3.1, Chapter 3, we detailed a variety of fixing and supporting devices, the majority of which are easily removable. This section looks at joining techniques that are more permanent.

Forms of welding

Welding processes can be classified as fusion or non-fusion. With fusion methods the two metals to be joined are first melted, allowed to mix together to create an alloy and then cooled, forming a weld. Fusion methods of welding include:

- manual metal arc welding (MMAW)
- gas metal arc welding (GMAW or MIG)
- gas tungsten arc welding (GTAW or TIG)
- resistance or spot welding
- oxyacetylene welding (OAW).

Non-fusion methods of welding include:

- brazing, also known as hard soldering
- soft soldering.

Manual metal arc welding

Manual metal arc welding (MMAW) is an electric arc welding process. A welding machine provides the electrical current large enough to make an arc between a flux covered electrode and a metal work piece.

The temperature at the arc rises to about 6000°C. This is high enough to melt the surface of the work piece and the electrode. The heat is concentrated at the tip of the electrode and the point of welding, as shown in Figure 4.107. It melts a small pool of metal in the work piece and at the same time melts the end of the electrode into balls of molten metal, which are forced across the arc by magnetic forces into the weld pool on the work. The electrode flux coating melts, providing a gaseous shield around the arc to prevent oxidisation, and deposits a protective slag on the molten pool. The level of current can be controlled at the machine to suit different welding conditions, such as the materials being joined, material thickness, etc.

FIGURE 4.107 Manual metal arc welding (MMAW) process

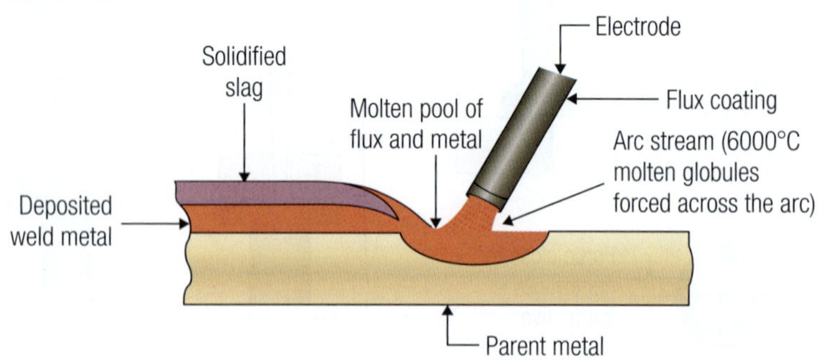

The MMAW process

All types of electric arc produce ultraviolet light that can damage the eyes and burn skin in a similar way to severe sunburn. Correct PPE is described later in this section.

Gas metal arc welding

Gas metal arc welding (GMAW or MIG) is a process where an electric arc is struck between the work and a continuous wire electrode fed through a gas shield, as shown in Figure 4.108. The gas protects the molten weld pool from contamination by nitrogen and oxygen from the atmosphere. Argon, helium, carbon dioxide and oxygen are the four most common shielding gases used in GMAW or MIG welding, with each providing unique benefits and drawbacks in any given application. A wide range of steels and non-ferrous metals are welded by this method, and in many cases GMAW or MIG has replaced MMAW.

Gas tungsten arc welding

Gas tungsten arc welding (GTAW or TIG) is also a fusion process and is shown in Figure 4.109. The heat is generated by an electric arc between a tungsten electrode and the work. The electrode is used to carry the current, which produces the arc and is not consumed to form part of the weld. A hand-held filler rod is used to supply additional weld metal. Argon is used as a shield gas to protect the weld from contamination. This type of welding is used mainly in light fabrication and general engineering. It welds to high standards and is used particularly for stainless steel, aluminium and other non-ferrous metals (*Miller TAFE College Welding and Thermal Cutting NBB09/2*, Tafe NSW, 1997, pp. 12–15).

Resistance spot welding

Spot welding is a form of electric resistance welding, which is a method of welding two or more metal sheets together by applying pressure and heat to the area to be welded. The process is most commonly used for joining sheet materials and uses shaped copper alloy electrodes to apply pressure and convey the electric current through the work pieces, as shown in Figure 4.110.

The heat is produced as a result of the current flow through the resistance of the work piece. The heat produced is proportional to the current flow squared ($P = I^2R$), meaning a small amount of current creates a lot of heat. The resistance of the work pieces is higher than the copper electrodes, so more heat is created in the work pieces than the copper electrodes for the same current flow.

The material between the electrodes, when heated, softens and is squeezed together. It then melts, mixing the metal of the two parts together. The current is switched off, but the material remains clamped until the molten material solidifies, forming a joint.

Spot welding has the advantages over other welding methods of speed, strength and minimal distortion of the work piece. It is commonly used to join sheet metal of less than

FIGURE 4.108 Gas metal arc welding (GMAW or MIG)

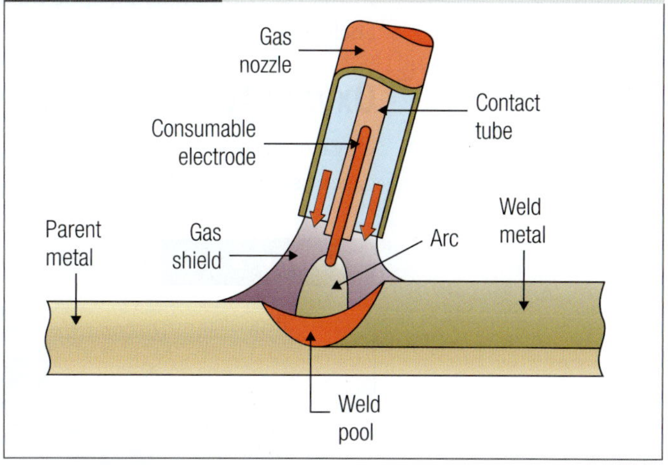

FIGURE 4.109 Gas tungsten arc welding (GTAW or TIG)

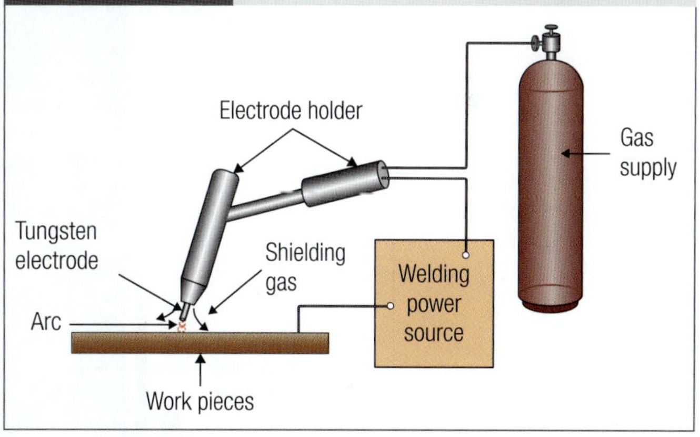

FIGURE 4.110 The spot welding process

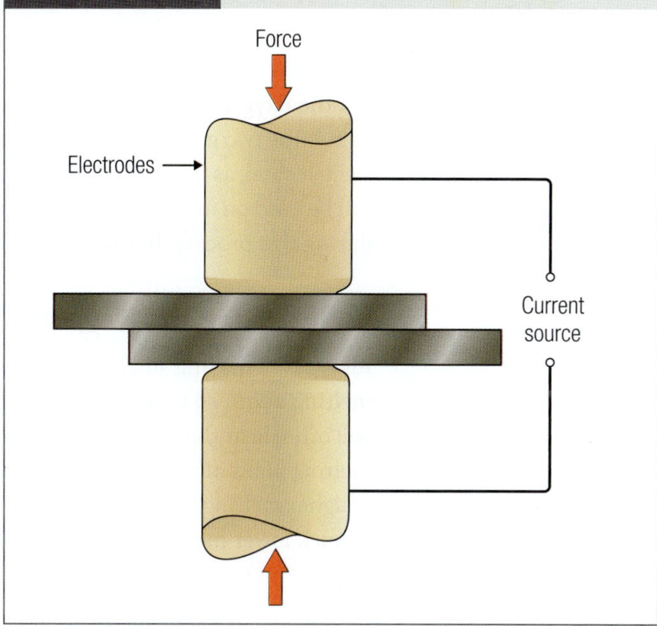

3 mm thickness, and is extensively used in the white goods and automotive manufacturing industries. In the electrical industry, spot welding is used in the construction of switchboards and other similar enclosures.

Oxyacetylene welding

Oxyacetylene welding (OAW) uses the heat from a gas flame to fuse or bond the parts together. The gases used for the flame are oxygen and acetylene, as shown in Figure 4.111.

FIGURE 4.111 Oxyacetylene welding (OAW)

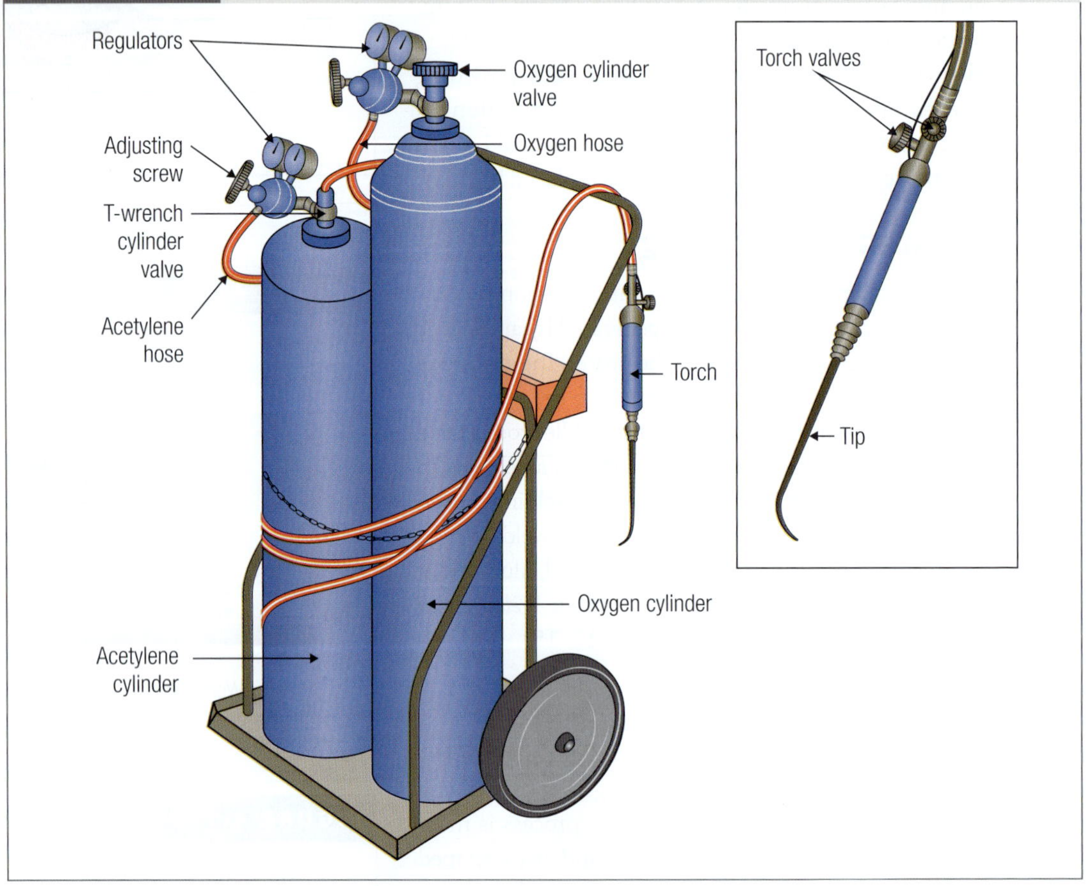

The oxygen accelerates the combustion of the acetylene, resulting in a much higher temperature flame. The two gases are combined in a torch and provide an intense flame of about 3100°C, hot enough to melt steel, as in Figure 4.112 on page 193. Oxy welding can be done with or without filler metals and can be used on materials of thin to medium thickness. Suitable goggles with tinted lenses must be worn during this process. It is used in light fabrication and general engineering.

Hard soldering and brazing

All types of brazing and soldering are non-fusion joints. The metal of the work pieces is not raised above melting point, so there is no mixture of the work piece materials. What actually melts is the filler. You will often hear the terms 'hard' and 'soft' associated with brazing and soldering. These terms refer to the melting point of the filler. Hard soldering, also known as silver soldering or brazing, means a filler with a higher melting point is used. Soft soldering typically uses a lead-based filler (called solder), which has a much lower melting point than fillers used in hard soldering.

Torch brazing (TB)

The melting point of the filler must be lower than the melting point of the work piece. Once the filler melts, capillary action draws the filler into small spaces between the work pieces to form an alloy of the metals upon solidification, as shown in Figure 4.113.

It is very important to make sure the work pieces are clean, and free of grease and oil, or the filler will not flow properly. In most cases a flux is applied to the work pieces to lift any contamination and help the filler flow. The filler metal is most commonly made from a silver, copper and zinc alloy with a commercially prepared flux, for cleaning the joint. To provide the high temperature required to melt the filler (usually above 450°C), oxy-fuel gas equipment is needed. Gas fuels are acetylene or liquid petroleum gas (LPG). Torch brazing is used for joining steel, copper, bronze and brass.

Braze welding uses an oxyacetylene flame, a flux to clean the surface and a filler rod made from a copper/zinc alloy. The filler rod melts at a much lower temperature than the parts to be joined, and the parent metal is not melted. Braze welding is a different process from brazing and uses different joint preparations, filler metal and welding technique heat. The parts are first heated, and then the flux is added to the end of the heated end of the filler rod and applied to the joint during welding. The bond strength comes from the filler metal 'sticking' to the surface of the joint preparation. Brazing can be used to join dissimilar metals, including non-ferrous and ferrous.

FIGURE 4.112 Oxyacetylene welding creates a molten weld pool

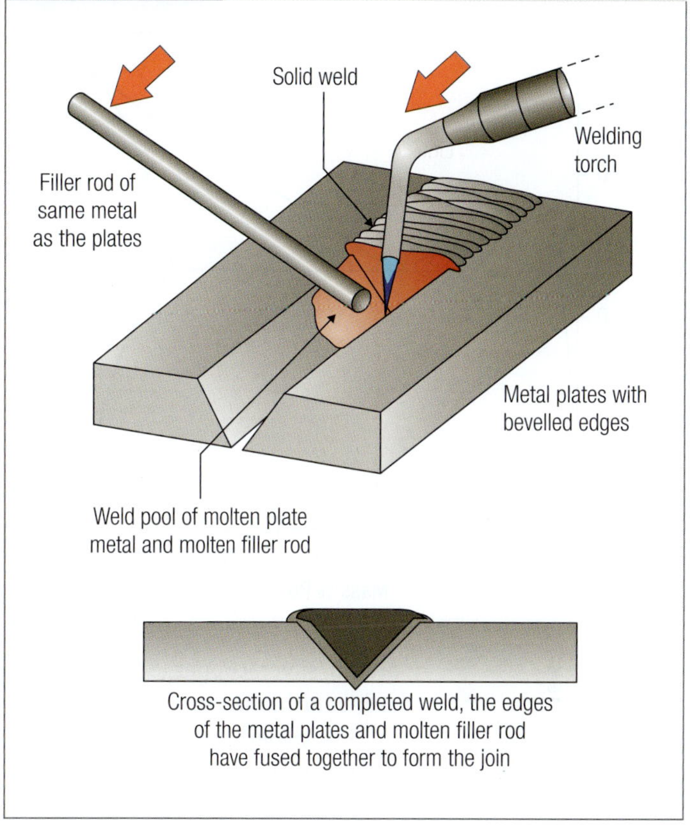

Soft soldering

Soft soldering is effectively the same process as brazing but uses a filler with a much lower melting point. Most soft solders, except lead-free solders, are an alloy of tin and lead, in various amounts. The tin content is always stated first.

The red line of Figure 4.114 shows how 60% tin and 40% lead solder changes from a solid to a liquid with just an increase of a few degrees. In between the liquid and solid states, the solder goes through a plastic range. It is this property of solder that makes it so useful.

Some types of solder include:

- 60/40 solder is known as fine solder. It solidifies quickly and because of its strength is used for good quality general purpose electrical work, and soldering copper tanks and heaters. It has a melting point of 185°C.
- 50/50 solder is used for general purpose and machine soldering and is not suitable for electrical connections due to its lower conductivity.
- 40/60 solder, known as coarse solder, is cheaper and often used on galvanised iron and is also not suitable for electrical connections.

FIGURE 4.113 Torch brazing

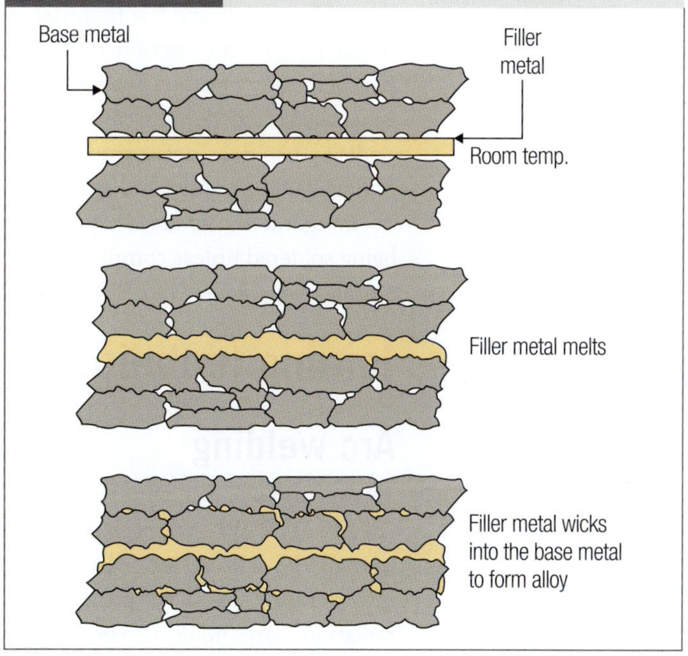

FIGURE 4.114 Tin lead melting point

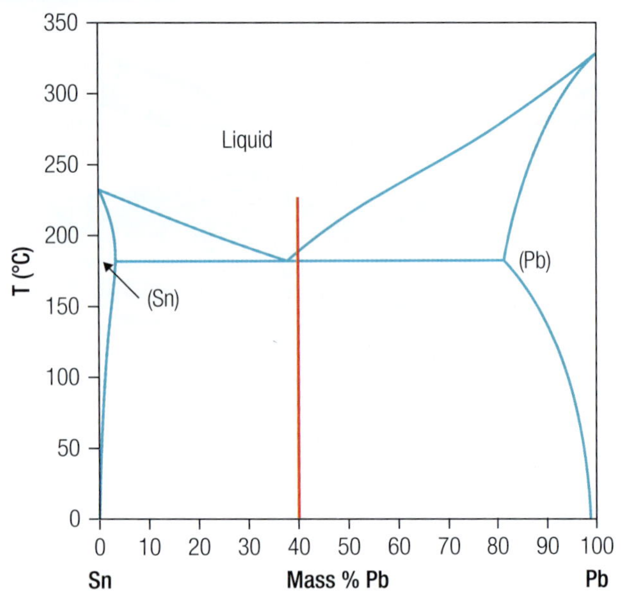

In hard (silver) soldering, the solder is an alloy of copper in varying proportions with zinc, silver, gold or cadmium and has a much higher melting point.

Lead free solder

Due to the health and environmental concerns associated with lead, the European Union Waste Electrical and Electronic Equipment Directive (WEEE) and Restriction of Hazardous Substances Directive (RoHS) restricts the addition of lead to most consumer electronics produced in the European Union (EU). Lead-free solder is now becoming more widely available. It reduces the danger of inhaling lead fumes during soldering, and reduces pollutants in the environment. It will become the normal solder for electrical work in the future.

Replacement solder alloys must be similar in properties to tin/lead alloy solders, such as having similar or improved mechanical, electromechanical and physical properties. The melting temperature should be similar, as should the cost of the solder. Two alloys that meet most of these criteria are tin/silver/copper (Sn/Ag/Cu) and tin/copper (Sn/Cu). Small quantities of other elements, such as bismuth, may be added to improve certain characteristics.

Soft soldering is used to join copper, zinc, steel, lead, galvanised iron, tin plate and alloys such as bronze and brass. A flux must be used to stop the surface from oxidising while heat is being applied and to help the solder to flow. Electrical joints are often made using soft soldering, although soldered joints are weaker than the base metals. For electrical applications, soft soldering is chosen for its electrically conductive properties rather than mechanical strength.

Flux

Fluxes fall into two categories: organic and inorganic fluxes. The flux melts at a lower temperature than the solder and cleans the material surfaces before the solder comes into contact, leaving the surface covered to prevent reaction with oxygen, and then helps the solder to flow into the joint. Organic fluxes are chemically inactive. They are applied to previously cleaned surfaces. For electrical work, resin flux is usually included in the core of the solder. This material is usually extracted from tree sap and is non-acidic and so does not corrode the parent metal. It leaves behind a non-corrosive residue that can be removed using methylated spirits. The residue can be left on the surface without any problems.

Inorganic fluxes are acidic (phosphoric acid or zinc chloride) and tend to corrode the materials being soldered unless completely removed after the joint has been made. They are commonly used by plumbers and other tradespeople for general purpose joints, but should be avoided for electrical work.

Joining techniques

Arc welding

This section will deal with the overall concepts of arc welding. Different welding processes – MMAW, GMAW or GTAW – will each require a slight variation of the technique to suit the process. In addition to normal PPE, a heavy long-sleeve welding jacket, hand protection (gloves) and a welding helmet or visor incorporating special eye protection must be worn to prevent permanent skin and eye damage from UV radiation. Care should also be taken to isolate the arc welding from other workers with suitable screens.

Some helmets are fixed, while the others provide automatic operation. The filter is transparent before welding, so the operator may observe the work surface clearly. When striking an arc, the filter immediately darkens. When the arc goes out, the filter will become transparent again. The switching time from dark to light may be set up within 0.1–0.9 seconds. The helmet is equipped with a continuous darkness adjusting unit, so the operator may select a dark shade number ranging from No. 9 to No. 13 (Ryan Equipment Solutions, 'Welding helmets', p. 1, http://www.ryanequipment.com.au). Table 4.8 provides guidance on lens shade selection.

TABLE 4.8 Lens shade number selection

Welding process	Welding rod size (mm)	Arc current (amperes)	Suggested shade number
Manual metal arc welding (MMAW)	< 2.5 2.5–4 4–6.4 > 6.4	< 60 60–160 160–250 250–550	7 10 12 14
Gas metal arc welding (GMAW)		< 60 60–160 160–250 250–550	7 11 12 14
Gas tungsten arc welding (GTAW)		< 50 50–150 150–500	10 12 14
Oxyacetylene welding (OAW) Light Medium Heavy	Plate thickness mm < 3.2 3.2–12.7 > 12.7		 4 or 5 5 or 6 6 or 7
Torch brazing (TB)			3 or 4
Torch soldering (TS)			2

When the electrode is moved along the work piece at the correct speed, the metal deposits in a uniform layer called a *bead*. The welding power source provides constant current (c.c.) and may be either alternating current (a.c.) or direct current (d.c.), depending on the type of electrode being used. The power in a welding circuit is measured in voltage and current. The voltage (volts) is governed by the arc length between the electrode and the work piece and is influenced by electrode diameter. Current is a more practical measure of the power in a weld circuit and is measured in amperes (amps). The amperage needed to weld depends on electrode diameter, the size and thickness of the pieces to be welded, and the position of the welding. Thin metals require less current than thick metals, and a small electrode requires less amperage than a large one.

It is preferable to weld on work in the flat or horizontal position. However, when forced to weld in vertical or overhead positions it is helpful to reduce the amperage from that used when welding horizontally. Best welding results are achieved by maintaining a short arc, moving the electrode at a uniform speed, and feeding the electrode downward at a constant speed as it melts.

Figure 4.115 provides guidance on welding techniques. That is:

- Make sure work piece is clean before welding.
- Place the work clamp as close to the weld as possible.
- Before striking an arc, insert a suitable electrode in the electrode holder. A small diameter electrode requires less

FIGURE 4.115 Arc welding technique

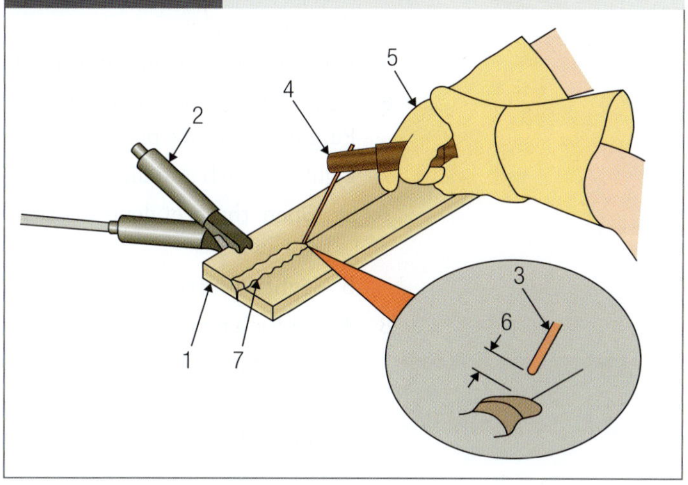

current than a large one. Follow recommendations of the electrode manufacturer when setting weld amperage.

- To strike an arc, simply tap or scratch the end of the electrode (rod) across the work, just like a match.
- As soon as the spark jumps, move the tip of the electrode slightly away from the work, then move the arc over to the weld and begin welding.
- A short arc with correct amperage will give a sharp, crackling sound. Correct arc length is related to electrode diameter. Examine the weld bead to determine if the arc length is correct. Arc length for 1/16" and 3/32" diameter electrodes should be about 1.6 mm; arc length for 1/8" and 5/32" electrodes should be about 3 mm.
- Use a chipping hammer and wire brush to remove slag. Remove slag and check the weld bead before making another weld.

It's good practice to lead the arc with the electrode just a little so you can see the puddle develop. Keep in mind that the electrode continues to melt away so you'll have to move it closer to the work as you weld. Move the electrode along the weld slowly, allowing the molten puddle to follow. If you move too rapidly the puddle will not fully develop and the weld will be weak.

Oxyacetylene welding (OAW)

FIGURE 4.116 OAW eye protection shade glass 4 to 7

Source: Ed Hawkins

The preparation required for oxyacetylene welding is the same as for arc welding. Protective clothing is the same, i.e. welding jacket, hand protection, etc. OAW does not produce dangerous UV radiation, so the eye protection as shown in Figure 4.116 uses a much lower shade glass number than for arc welding. See Table 4.8 regarding the selection of shade glass numbers.

To weld using oxyacetylene equipment:

1. Inspect all equipment for leaks and correct operation.
2. Check the torch handle to make sure it is closed by gently turning both knobs counter-clockwise to open and then clockwise to completely close.
3. Standing to the side of regulators, open both acetylene and oxygen shut-off valves slowly by turning counter-clockwise. Only open the acetylene shut-off valve half a turn. This is so the acetylene can be easily turned off in an emergency.
4. Open acetylene (red) and oxygen (green) regulator valves one quarter turn while directing welding tip towards a ventilated exhaust. The regulator pressure setting will depend on the tip size in use. For most general work a pressure of 5-8 PSI for acetylene and 15-30 PSI for oxygen is required. Acetylene becomes very unstable above 15 PSI or higher. NEVER set the regulator at this pressure.
5. Before the torch is lit, put on welding gloves and goggles.
6. Purge fuel by opening the torch knob and allow a 3–5 second flow. Then close the fuel gas valve. Purging must take place each time prior to igniting the torch flame.
7. Light the torch and direct the torch away from you, towards a ventilated exhaust.
   a. Open the acetylene control valve on the torch, approximately 1/8 turn rotation counter-clockwise.
   b. Light with striker. Never use lighters or other sources for ignition.
8. Adjust for a neutral flame.
   a. Adjust acetylene to form feather shaped flame.
   b. If there is too much acetylene, the flame breaks away from the tip and there is a noticeable amount of black smoke.

   c Slowly add oxygen. The flame will sharpen and elongate.
   d Continue to add oxygen until a neutral flame is established.

9 Welding
   a Focus the blue flame's tip (neutral flame) onto the area you are welding.
   b Maintain a consistent distance from the material, heating a narrow path between the two materials.
   c After a puddle is established, introduce the filler rod.
   d Melt the filler rod into the puddle and move along, establishing a weld path (bead).

Shutting down is the reverse process of lighting up. Always turn off the acetylene first and remember to purge the hoses of gas after closing the gas bottle's shut-off valves.

Torch brazing (TB)

The set up and preparation for brazing is very similar to OAW. In some cases acetylene can be substituted for liquid petroleum gas. The combination of oxygen and LPG still acts to increase the temperature of the flame but not to as high a degree, as only the filler rod is required to be melted. This can offer cost savings.

1 Select a suitable brazing alloy.
2 Select a suitable flux to suit the work piece material and the filler.
3 Adjust the torch for a neutral flame so it has a well-defined inner cone.
4 Heat the joint area. Always move the torch in short motions.
   a Work the flame alternately around the work pieces until reaching brazing temperature before applying the brazing filler metal.
   b When a flux is used, it will be a good temperature guide. Continue heating the work pieces until the flux passes the 'bubbling' temperature range and becomes quiet and completely fluid.
   c Sweep the flame back and forth along the work pieces to maintain uniform heat in all parts.
5 Apply the brazing alloy.
   a Feed the alloy into the joint between the work pieces.
   b The flame may be deflected momentarily to the tip of the filler metal to begin the melting process.
   c Always keep both the work pieces heated by playing the flame over them as the brazing alloy is drawn into the joint. The brazing alloy will diffuse into and completely fill all joint areas.
   d Do not continue feeding brazing alloy after the joint area is filled. Excess filler does not improve the quality or the dependability of the braze and is a waste of material.
6 When making horizontal joints heat the larger of the work pieces to be joined first. When adding filler alloy, make sure all work pieces are up to temperature.
7 Clean up after brazing. All flux residues must be removed before inspection and pressure testing.
8 Immediately after the brazing alloy has set, quench or apply a wet brush or swab to crack and remove the flux residues. Use emery cloth or a wire brush if necessary.

Torch soldering (TS)

As solder has a much lower melting point than other fillers, oxygen is not required to boost the temperature of the flame as with hard soldering or brazening. A simple LPG or burner is sufficient. LPG is also known as propane gas. The size of the torch will depend on the size and mass of the materials being soldered. For example, some torches are suitable for soldering large electrical cable joints such as earthing conductors in a main switchboard, while others are more suitable for tinning busbar.

When making electrical connections it is important to use 60/40 solder. The surface to be soldered must be cleaned and the flux applied before the surfaces oxidise. The solder should then run over the flux-coated surface when it reaches the correct temperature. If you are using resin-cored solder, this is done automatically as the solder is applied. There is usually no need for any additional flux. To make a soldered joint:

1 Clean the area to be soldered thoroughly with an abrasive such as fine emery paper or steel wool.
2 Hold or otherwise clamp the work pieces together.
3 Apply heat from an appropriately sized torch. Metals such as copper change colour slightly at the correct temperature to apply the solder.
4 Apply flux to the work area. If the flux splatters excessively this is an indication the work piece is too hot.
5 Apply solder, allow the solder to melt and flow between the work pieces.
6 Ensure the joint has cooled sufficiently before it is unclamped or moved.
7 Remove any excess flux with methylated spirits to prevent further corrosion of the joint.

The solder joint should have a slightly shiny, smooth consistent appearance. If grey or cracked in appearance, reheat and add additional flux and/or solder if the solder in the joint does not flow.

Tinning refers to coating a surface with a thin layer of solder, either prior to soldering or to protect the surface against corrosion. A soldering iron tip should be tinned before use. Sweating is the joining of surfaces that have previously been tinned. To do this, the surfaces are clamped or otherwise held together and heated beyond the melting point of the solder, so the solder runs together to form a bond when the metal cools.

Care must be taken not to damage insulating parts of conductors or adjacent equipment. Overheating causes the solder to run too freely and may leave insufficient solder for a good bond. Excess heat may also oxidise the solder, requiring re-cleaning and re-tinning the job. Be aware that the flame is a conductor of electricity and therefore should not be used in the vicinity of live exposed conductors.

FIGURE 4.117 Temperature controlled electric soldering iron

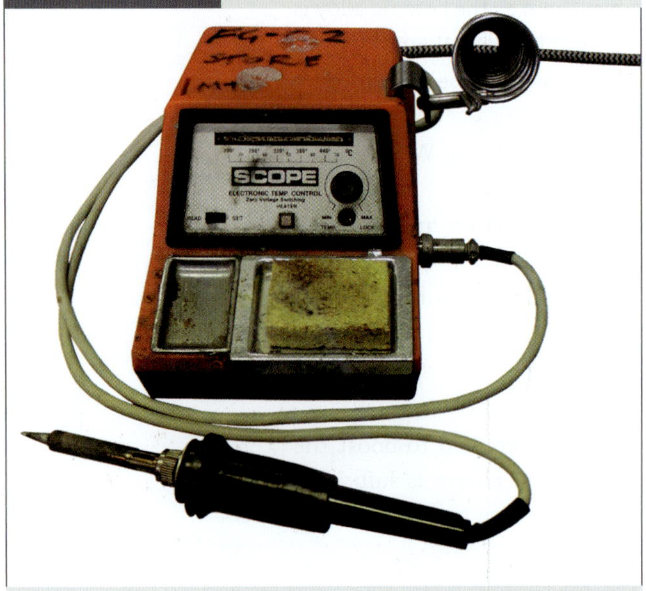

Source: Ed Hawkins

Soldering with a soldering iron

Electric soldering irons range from very small (10–20 watts), used for soldering circuit boards, up to very large irons (up to 1000 watts) used for general purpose soldering. Figure 4.117 shows a temperature controlled soldering iron suitable for soldering components onto a circuit board.

Two things must be considered when selecting a soldering iron:

- Tip temperature: If too low the solder may not bond with the metal being soldered and a dry joint may result. Such a joint in an electric circuit will offer a high resistance to an electric current, causing heat to be developed and further deterioration of the joint. If the temperature is too high, flux may burn away and the solder may undergo a permanent physical change.
- Heat capacity of the tip (related to the tip size): Small tips are suitable for circuit board work while a larger tip is needed to join large conductors.

Before using a soldering iron with a copper tip, the tip must be tinned. To do this:

1 Clean with an abrasive such as fine emery paper or steel wool. Copper soldering iron tips should be dressed by filing with a smooth file to remove all corrosion and pitting. Never file a plated tip.
2 Heat the surface, then coat with flux.
3 Apply solder to the entire surface.
4 Wipe excess solder from the surface with a rag leaving the surface completely covered with a thin, even coating of solder.

When soldering printed circuit boards, a soldering iron such as a 150 watt electric soldering iron is used. A typical PCB soldering process is:

1 Allow the iron to reach its correct temperature.
2 Check the tip is tinned, if necessary clean and re-tin.
3 Select the correct gauge of 60/40 resin-cored solder to suit the joint size.
4 Lightly wipe the tip on a damp sponge to remove any oxide, as shown in Figure 4.118.
5 'Wet' the tip to aid in heat transfer by applying a small amount of solder to the tip. Then apply the tip to the joint connection as shown in Figure 4.119, so the tip is in maximum contact. This allows the joint to heat quickly.
6 Apply a small amount of solder where the tip meets the joint so it forms a conducting heat bridge between the tip and all parts of the connection, as shown in Figure 4.120.

FIGURE 4.118 Cleaning soldering iron tips

FIGURE 4.119 Heating the joint

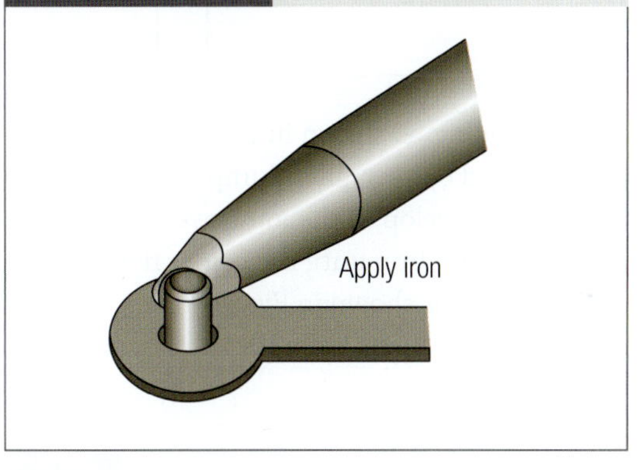

7 Maintain the heat bridge until the parts have reached a satisfactory soldering temperature.
8 Slowly feed the solder into the joint, starting close to the tip and moving towards the edge of the joint, as shown in Figure 4.120.
9 Continue applying solder to the joint until complete wetting of the joint has been achieved and the joint has concave fillets. Stop applying solder when this occurs.
10 Remove the iron and allow the joint to cool without moving it, as shown in Figure 4.121 (*Electrotechnology Practical LSEEE280A*, pp. 107–8).

FIGURE 4.120 Feeding solder into the joint

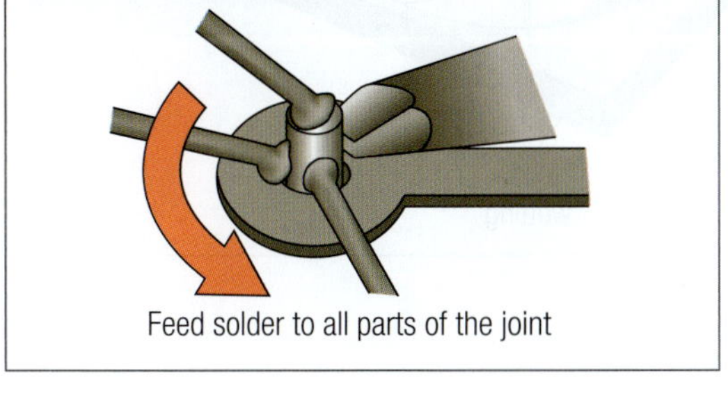

FIGURE 4.121 Allowing the joint to cool

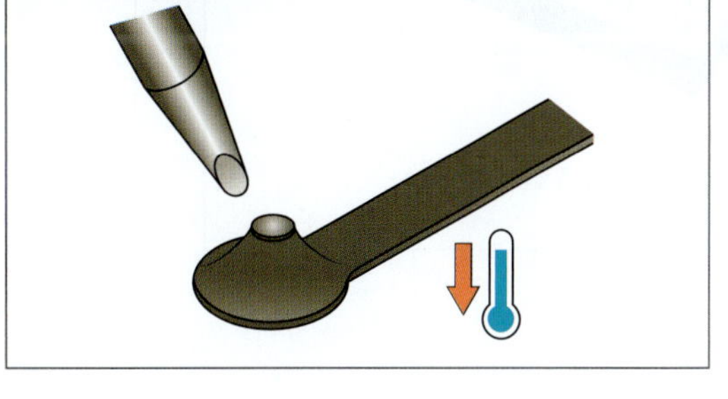

Once a joint is completed it should be visually inspected for defects. Soldering faults can be as simple as too much or too little solder, or too much or too little heat. Figure 4.122 shows an example of a well-formed joint.

If too much heat is applied to the joint, it will have a dull, grainy appearance to its surface as well as pits and voids in the solder itself, as shown in Figure 4.123. This joint will have poor electrical conductivity. The excess heat may also lift copper tracks of the PCB or damage electronic components. Remove the solder with a de-soldering tool and redo the joint.

FIGURE 4.122 What to look for in a good joint

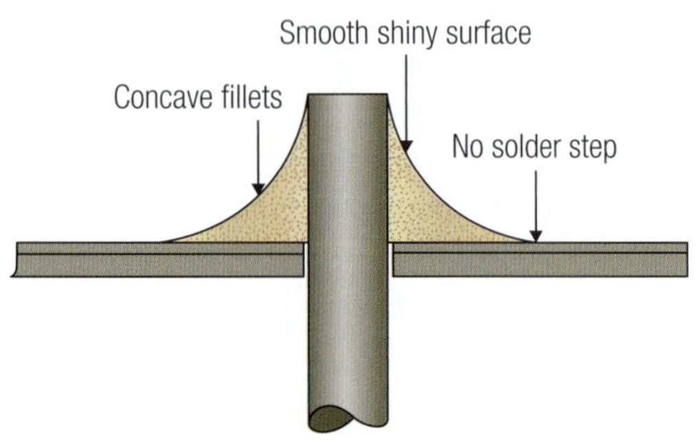

FIGURE 4.123 An overheated joint

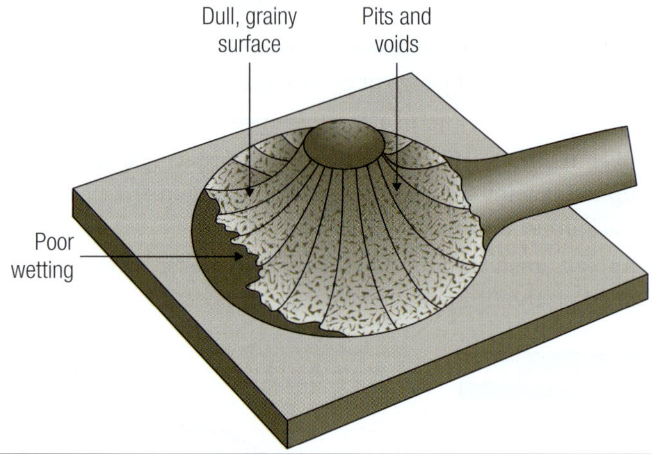

A cold joint, as shown in Figure 4.124, occurs when the solder did not melt completely. It is often characterised by a rough or lumpy surface. Cold joints are unreliable. The solder bond will be poor and cracks may develop in the joint over time. Cold joints can usually be repaired by simply reheating the joint with a hot iron until the solder flows.

A dry joint is shown in Figure 4.125. This is caused by movement of the component as the solder is cooling. It will cause a poor electrical connection and needs to be repaired by de-soldering as previously described.

FIGURE 4.124 A cold joint

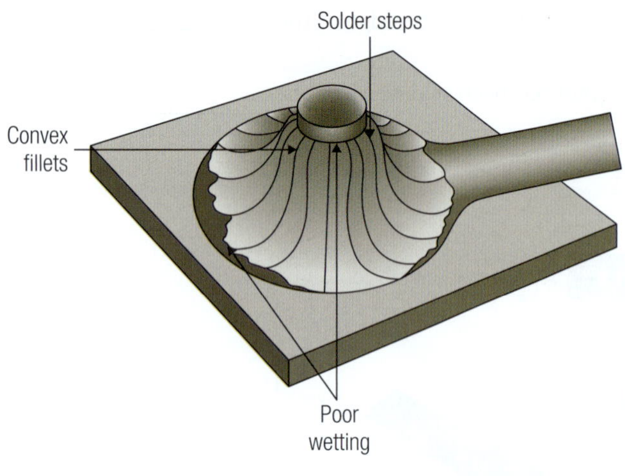

FIGURE 4.125 A dry joint

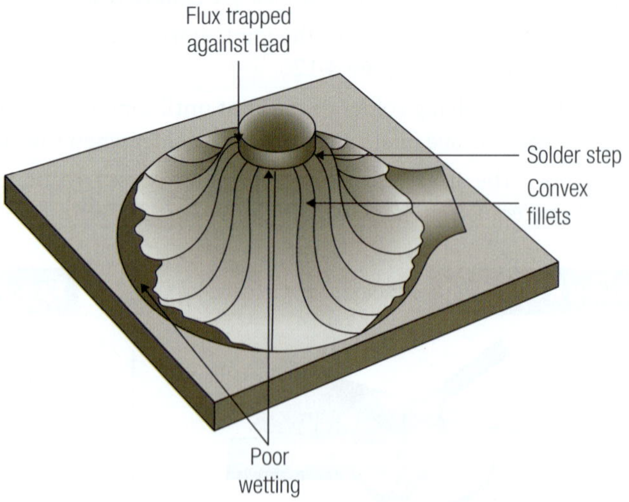

SUMMING UP...

- Welding processes can be classified as fusion or non-fusion.
- Fusion methods of welding include:
  - manual metal arc welding (MMAW)
  - gas metal arc welding (GMAW or MIG)
  - gas tungsten arc welding (GTAW or TIG)
  - resistance or spot welding
  - oxyacetylene welding (OAW).
- Non-fusion methods of welding include:
  - brazing or hard soldering
  - soft soldering.
- An electric arc produces a temperature of approximately 8000°C.
- An oxyacetylene torch produces a temperature of approximately 3100°C.
- When performing any type of arc welding, in addition to normal PPE, a heavy long-sleeve welding jacket, hand protection (gloves) and a welding helmet or visor incorporating special eye protection must be worn to prevent permanent skin and eye damage from UV radiation.
- 60/40 resin cored solder is the best type of solder to make electrical connections.
- Flux is used to remove oxides from the work piece.

4.9 Portable electric power tools

Electric power tools are either mains or battery powered and are used to make a manual task easier and more productive. In the hands of a skilled operator, a power tool can reduce the time and effort required to complete a task. However, any errors made by the operator will be significantly larger than if performed by hand tools.

FIGURE 4.126 A 115mm angle grinder

Source: iStockphoto/William Howell

Types of electric power tools

Grinders

An angle grinder is shown in Figure 4.126. These are used for cutting and finishing metallic and masonry surfaces. They typically use an abrasive disc available in diameters of 100mm, 115mm, 125mm and 230mm. However, some work sites have placed bans on the larger 230mm (9 inch) machines.

As the disc size increases, the RPM decreases to maintain a constant surface speed on the disc's cutting edge. Smaller grinders of 100mm in diameter run at around 12 000 RPM, and larger 230mm grinders therefore run at around 5700 RPM. A variation of the angle grinder is the straight grinder.

FIGURE 4.127 Die grinder

Source: Alamy/David J Green

Die grinders (Figure 4.127) are a smaller variation of the straight grinder, and operate at 19 000 RPM. The compact head allows access to tight locations and precision removal of material.

Discs suitable for grinding are always thicker than those used for cutting and are not interchangeable. Types of discs include metal cutting discs, metal grinding discs and masonry diamond blades. Wheels suitable for cutting and grinding a particular material, e.g. masonry, should never be used for cutting other materials. They will quickly wear out and may shatter.

When powered up, the high torque produced by a grinder may cause it to kick. Many larger grinders are now fitted with an electronic soft starter to reduce the torque produced and the electric current

Safety with hand grinders

taken during starting. Always wear hearing and eye protection when using portable grinders. Ensure people in the same area are protected from sparks and grit. Hot sparks can cause fires or injure others in the vicinity.

Sanders

Rotary sanders have a head that spins in a circular motion. They are capable of removing large amounts of material from a work piece, but they have a tendency to dig in and damage the work. Modern sanders use a vibrating action to remove material from a work piece. A fixed orbital sander is shown in Figure 4.128.

FIGURE 4.128 Fixed orbital sander

Source: shutterstock.com/Yotsatorn Laonalonglit

An orbital sander produces a similar sanding motion to that of hand sanding, as shown in Figure 4.129. The action is repetitive and tends to leave swirl marks in the work piece.

Another type of sander is the random orbital sander. The basic motion is similar to a Spirograph toy, as shown in Figure 4.130. It is the result of rotating one circle about another. To achieve a random motion, the sanding pad also has forced rotation as well as a swirling motion. The different directions of the two motions reduces swirl marks because the marks are more random than the circular pattern produced by orbital motion only.

FIGURE 4.129 Orbital sander motion

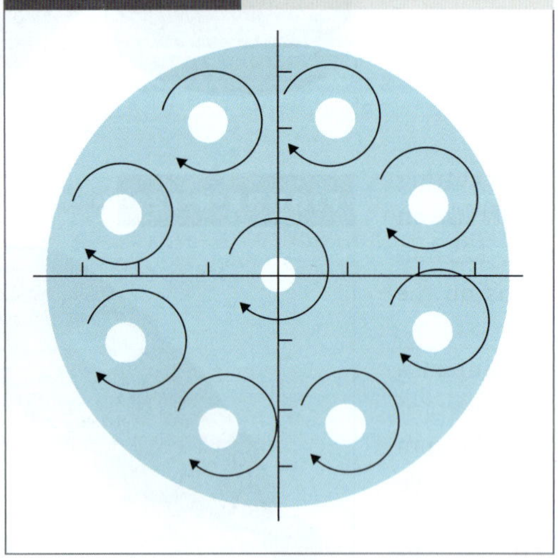

FIGURE 4.130 Random orbital sander motion

Sanding discs and paper are graded in numbers: the lower the number the coarser the grit. In summary:

- 50–30: Very coarse grit – rips off peeling paint and extraneous materials with ease. Also risks damaging the work piece.

- 80–60: Medium grit – general purpose grades of sandpaper.
- 180–100: Fine – used to finish work to a smooth surface.

A belt sander is used for quickly taking off large amounts of material from the face of a work piece. It is commonly used on timber and fibreglass.

Jigsaws

A jigsaw, as shown in Figure 4.131, uses a single blade to cut irregular shapes in materials up to 25 mm thick. The majority of jigsaws use a reciprocating (up and down) motion, but some high-end jigsaws use an orbital (forward and back as well as up and down) movement to improve cutting action and decrease blade wear. Just like handsaws and drills bits, the blade must suit the material being cut.

FIGURE 4.131 Hand-held jigsaw

Source: iStockphoto

Electric saws

A reciprocating or sabre saw, as shown in Figure 4.132, is a larger version of a jigsaw. The blade is stronger and more flexible than a jigsaw and has a typical stroke of around 29 mm. It is a versatile tool widely used for vertical or obscured places that no other types of saw can work in. The flexibility of the blade allows cuts to be made hard up against flat surfaces such as walls and floors, which is ideal for conduit work.

Most models have a quick change, keyless blade changing system. Like jigsaws, a number of different types of blades are available for a range of materials and uses. High-performance metal blades are capable of cutting steel up to 6 mm thick while demolition blades easily cut through timber embedded with metal nails.

FIGURE 4.132 Reciprocating saw

Source: shutterstock.com/Eugen Shatilo

A circular saw, as shown in Figure 4.133, offers portability on the job that can eliminate the need to manhandle large, heavy or awkward pieces of timber. Timber cut with a circular saw should be well supported clear of the ground, with support under the off-cut material.

The saw size is determined by the diameter of the saw blade that can be fitted to the tool. The power required increases with blade dimension. Circular saws are not limited to cutting timber. With suitable blades they are capable of cutting almost any material. Masonry blades are the most commonly used in the electrical trades, although because of the limited speed of this type of saw, this task is typically performed with an angle grinder.

FIGURE 4.133 Circular saw

Source: shutterstock.com/tale

Electric shears have the same cutting action as tin snips, that is, the material is cut as the two blades pass across each other. They tend to distort the material being cut.

An electric nibbler uses a reciprocating action to remove a small ‘c’ shaped section of material with each stroke. Nibblers have a major advantage over tools such as shears or tin snips as they do not distort the work piece while cutting (*Electrotechnology Practical LSEEE280A*, p. 118).

Portable power tools should be connected only to a power source that is protected with an RCD.

Using portable power tools

All types of electric power tools must always be supplied from a circuit that is RCD protected. If there is a failure of insulation in either the tool or extension lead the RCD will offer protection against both direct and indirect contact with the electrical supply. If you are not sure if RCD protection is available, use a portable power outlet device as shown in Figure 4.134. Ensure you are wearing sufficient PPE such as eye protection and, depending on the material to be cut, a dust mask. Do not blow waste dust away with the mouth; most power tools have a vent from the cooling fan to blow the dust from the work piece.

FIGURE 4.134 Portable RCD protected power outlets

Source: Ed Hawkins

Always select a suitable tool for the job to be performed. The most common reason for damage of power tools is using a tool which lacks sufficient power for the task. When using any power tool, always grip the tool so that the hands and fingers do not block the cooling vents as this will result in overheating the tool. Lower quality tools should not be used continuously as this will also result in overheating and tool failure. All power tools should be allowed to run up to speed before any load is applied. Overloading the tool will cause the speed to slow and current drawn by the power tool to increase, causing overheating and possible failure of the tool.

The following points are specific to a jigsaw but are consistent with the use of most power tools. When using a jigsaw:

- Check the marking out of the work piece is correct before you begin to cut. It can be costly to make a cut in the wrong place. Make a note of which side of the line you need to cut.
- Let the motor run up to speed before starting the cut. Never hold the blade against the work piece before squeezing the trigger, as the blade may jam or jump off position.
- Apply firm forward pressure to the work piece with the blade ensuring the jigsaw's guide shoe is flat against the work piece. This ensures the cut is made at 90°.
- The trigger action of most jigsaws has variable speed control. Greater movement on the trigger increases the speed of the blade.
- Never lift the jigsaw from the work piece while it is running, as the blade will fall out of the cut and strike the work piece, resulting in damage to both work piece and blade.
- When a cut needs to be made inside an object, one or more holes are drilled in the corners of the hole to provide a means of starting the cut.

When operating a belt sander:

> always use two hands (one on the trigger handle and the other on the forward handle for stability), ease the sander down with a slight forward motion. The rear roller should contact the wood first, and then set the rest of the sanding face down onto the wood.
>
> The belt will tend to grab the wood and drive the sander forward, so you'll need to hang on rather tightly. Work the sander with the grain of the wood as much as possible, and do not allow it to stay in one place. A belt sander, particularly one with a very coarse belt, can tear through a surprising amount of material very quickly, so keep the sander moving.
>
> Source: Macin Krawczynski & Krzysztof Kaczmarek, 'Belt sander', http://windturbinetechnology.wikispaces.com/Belt%20Sander2; http://creativecommons.org/licenses/by-sa/3.0/

SUMMING UP...

- Common types of grinders include angle, straight and die grinders.
- The type of disc used in a grinder will depend on the material being cut and include discs to cut metal, masonry, plastic and fibreglass.
- Common types of sanders include orbital, random orbital and belt sanders.
- Common types of power saws include the jigsaw, reciprocating saw and circular saw.
- Electric shears perform operations similar to tin snips.
- Nibblers are used to cut holes in sheet metal.
- Always use an RCD protected supply when operating electric power tools.

4.10 Sheet metal work

Sheet metals are used in the electrical industry for a variety of purposes, such as switchboard enclosures, cable trays and magnetic laminations. The selection of the most suitable metal will depend on the job for which the metal is to be used.

Types of sheet metals

When considering which particular metal best suits an application, some important properties to consider are conductivity, magnetic qualities, toughness, ductility, malleability, hardness, corrosion resistance and workability.

Mild steel is widely used for enclosing electrical equipment switchboard enclosures, control panels, cable tray, trunking, etc. It is inexpensive and easily worked, but must be coated to prevent corrosion where exposed to the weather. Some methods of corrosion proofing are painting, galvanising (hot dip and electro-plating), zinc annealing or coating with lead, tin or plastic. Zinc anneal sheet is galvanised iron heated in a furnace to 340°C and rolled. This causes the zinc coating and the surface metal to alloy. This process prevents the surface from flaking when bent. Mild steel sheets are easily worked, welded or soldered.

Brass is an alloy of copper and zinc and is used in sheet form in the manufacture of electrical equipment. It is stronger and harder than copper, is not magnetic, resists corrosion, is a fair conductor of electricity, is easily soldered and may be worked easily when annealed.

Aluminium is widely used for electronic housings, where heat must be conducted away from electronic parts quickly. Other popular applications include building panels, window frames and roofing. It is non-magnetic, soft, light, easily worked and a good conductor of heat and electricity.

Copper is soft, easily soldered, a good conductor of both heat and electricity and non-magnetic, but is more expensive than brass, which is often used in place of copper because of its cost and rigidity. Copper work hardens and requires annealing to maintain its workability.

Phosphor bronze is a tough copper-like metal with low electrical resistance. It is often used to provide a stiff backing to copper strap or in place of brass where high conductivity is required. Phosphor bronze has a grain, formed by rolling along its crystalline structure so that long crystals of metal are formed. Grainy metals are easier to bend along the sheet than across the sheet.

Muntz metal is an alloy of copper and zinc and is highly resistant to corrosion. It is sometimes used in place of sheet brass.

Silicon steel in sheet form is also called 'electrical sheet' and is used in the manufacture of laminated cores used in transformers and motors. There is a wide range of these steels to suit different requirements.

Sheet metals are manufactured in a standard preferred range of sizes as specified in **AS 1123-1972 Preferred metric sizes of wrought ferrous and non-ferrous sheet and strip**, which states that widths and lengths of sheets should be in steps of not less than 100 mm. Steel sheets are available in sizes of 1800 mm × 900 mm and 1200 mm, 2400 mm × 900 mm and 1200 mm, 3000 mm × 900 mm, 1200 mm and 1500 mm, and 3600 mm × 900 mm, 1200 mm and 1500 mm in the standard range of thicknesses from 0.3 to 3 mm. Other metals such as copper and brass are available in smaller sizes, due to the higher cost of these metals (*Electrotechnology Practical LSEEE280A*, pp. 120–1).

Tools used with sheet metals

The most common hand tools used for sheet metal work are tin snips. These are made of tool steel and are for cutting sheet metal by hand. They are available in a variety of shapes and sizes. Tin snips are not generally suitable for cutting metal thicker than 1 mm. Thicker metals need cutting by machine or by hand using a hacksaw or cold chisel.

FIGURE 4.135 Tin snips

Source: shutterstock.com/korolev-photo.com.ua

Tin snips are shears made in sizes from 150 mm to 400 mm with straight or curved blades for cutting thin sheet metal. When using tin snips, the line should be followed with the cutting face of the blade. Take care not to cut metal that is too thick. Care should also be taken with the sharp edges of the cut as well as not to catch fingers or hands between the handles of shears as the cut is completed. Figure 4.135 shows examples of universal, curved and straight tin snips.

- Universal tin snips – these have heavy section blades and narrow cutting edges.
- Curved snips are for cutting inside curves.
- Straight snips are for cutting straight lines or outside curves.

'Aviation' tin snips are used for cutting tight curves. Their double lever design increases mechanical advantage and allows them to cut materials up to 2 mm thick. They are available in right- or left-hand configurations as shown in Figure 4.136.

FIGURE 4.136 Aviation tin snips

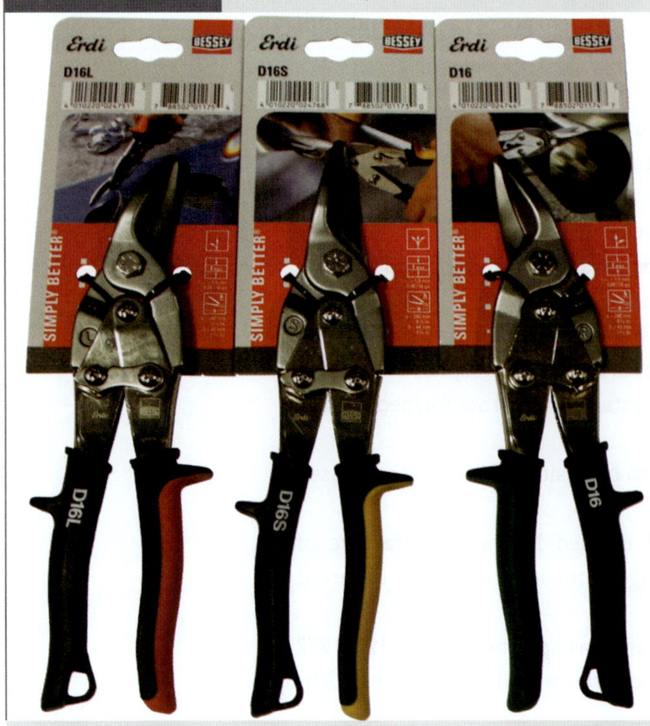

Source: Ed Hawkins

A *guillotine* is a machine for cutting sheet metal. It has a bed to place the sheet metal on, a clamp (hold) bar that usually acts as a safety guard to keep fingers clear of the cutting blade, and a cutting blade. The cutting blade operates with a shearing action. Smaller machines are operated manually, usually with a foot pedal. Larger machines are power driven. The machine may also have a backstop or gauge. This is useful when cutting a number of pieces of the same width.

For sheet metal it is usual to punch holes rather than drill them. It's a simpler process and the hole is cleaner. However, sizes are limited and punch and die sets can be expensive.

A *turret punch* is a more powerful form of metal punch, with a lever to push the selected punch through the metal sheet.

A *hand notcher* is a tool used to cut short lengths of stock, for cutting 90° notches and making trim cuts at various angles. The cutting blade has two faces at 90° to each other, located over a matching die. The material to be cut is placed under the blade and the cutting blade is pulled down, cutting a notch out of the metal. The notcher has minimal, if any, protection for fingers and hands, so it's very important to keep fingers clear when operating.

A *machine notching tool* is typically table mounted, and when operated by a suitable press it can cut multiple notches at preset distances.

Bending machines are used to produce three-dimensional sheet structures such as cabinets and boxes. They are also known as brake presses or bending brakes. Smaller models are hand-operated (*NE175/1 Workshop Practice Student Workbook*, pp. 112–14).

Fabricating sheet metal

When using tin snips:

- Grip the handles between thumb and first finger and allow the first and second fingers to curl around the lower handle. Place the third and small fingers between the handles but rest them on top of the lower handle.
- Check that the handles and blades open and close without problems.
- Place the sheet metal into the blades, with the junction of the blades aligned with the cutting line.
- To make a cut, squeeze the handles together but don't allow the blades to close completely or the metal immediately ahead of the snips will distort.

Sheet metal must be marked out before cutting or forming and the same tools as described previously are used for this process. Drilling is also done in the same fashion, using basically the same procedure as already described. Note that, because of the thickness, or lack of thickness, a drill bit may jam in the material. Precautions such as secure holding of the work piece must be used.

The general procedure to make a cut using a guillotine is to:

- Place the material in the machine with one edge located against the side gauge.
- Slide the material under the blade to either a marked line or to the back stop.
- Clamp the sheet metal in place.
- Make sure fingers, hands and loose tools or materials are well clear of the blade and clamp.
- Operate the machine.

Using a bending machine is similar to using a guillotine, that is:

- Set up the machine's fingers to required measurements.
- Place the material in the machine.
- Slide the material under the blade to either a marked line or to the back stop. Depending on the thickness of the material, some allowance may be required to allow for the radius of the bend.
- Clamp the metal in place.
- Make sure fingers, hands and loose tools or materials are well clear.
- Operate the machine to the required angle (*NE175/1 Workshop Practice Student Workbook*, pp. 112–14).

Joining and cutting mitred joints

When a flat section of sheet metal is to be folded into a three-dimensional shape, a portion of material in the corners of the fold is cut or 'notched' to make room for the fold. The left of Figure 4.137 shows a drawing of a mitre cut. Tin snips or a guillotine are typically used to make a cut at 45° across the corner. The right-hand side of Figure 4.137 shows a notch made at 90° to the corner, cut by tin snips or a notcher. The mitre cut is the easier to manufacture but the notched corner provides a closer fit that, when required, can be soldered to increase the strength of the joint.

FIGURE 4.137 Mitre and notched joints

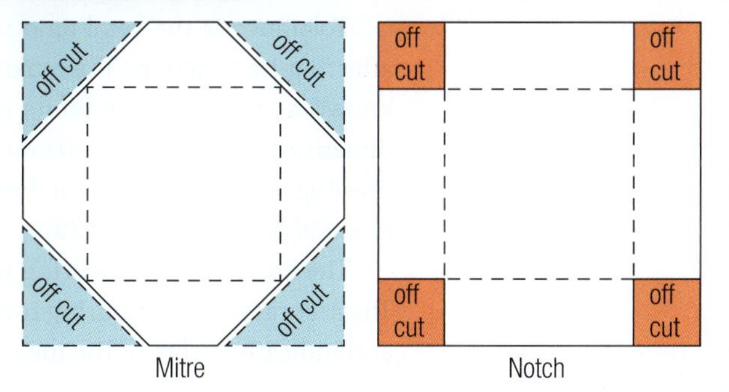

Sustainability can be achieved when working with sheet metal.

Sustainable work practices

When working with sheet metal, sustainable work practices are consistent with any other type of work performed. The goal is to avoid waste from errors during the marking out process and to use the least amount of raw materials and energy. Power tools should not be left running and lights should be turned off when the area is vacated. Adequate time must be dedicated to proper planning of the work process.

SUMMING UP...

- Common materials used to make sheet metal include:
  - mild steel
  - brass
  - aluminium
  - copper
  - phosphor bronze
  - muntz metal
  - silicon steel.
- The standard thicknesses of sheet metal range from 0.3 mm to 3 mm.
- Tools used to cut sheet metal include:
  - tin snips (universal, curved, straight and aviation)
  - guillotines (manual and powered).
- Punches are often used instead of drill bits to cut holes in sheet metal as they make a cleaner hole. Types of punches include the hand and turret punch.
- Notchers are used to remove material to prepare corners of work pieces for bending.
- Bending machines are used to fold sheet metal consistently and accurately. Finger or pan benders, as they are also known, have adjustable fingers which allow the folds of sheet metal to protrude into the bending machine.
- Care should be taken with all sheet metal tools as they can easily cut or crush the operator's fingers.
- Mitre joints are cut at 45° to the corner, notches are cut at 90° to the corner.

4.11 Low tolerance measurement

As described by the RMIT education website, 'Low tolerance measurement refers to measuring very accurately to within 0.01 mm. In the electrotechnology industry, low tolerance measurements may be required for measuring sizes of conductors, measuring bearings and shafts for an electric motor, or measuring circuit board drill bits'. The two common tools that are used for low tolerance measuring are the vernier caliper and micrometer.

Tolerance

Accuracy is how close a measured value is to the actual (true) value. Tolerance is the allowable variation either plus or minus to the nominal value or exact size. For example, a litre of milk is unlikely to contain exactly 1 litre of milk; the actual measurement is more likely to be a few millilitres more or less.

Tolerance is the total amount by which a size can vary around the ideal or design size, without affecting its function. The maximum and minimum sizes of each dimension are given and these are known as the tolerance limits of size. They are shown as a plus and minus amount. For example, a dimension might be 30.00 mm, with a tolerance of plus and minus 0.5 mm. This is shown on the drawing as 30.00 ± 0.5 mm. Precision is the ability to reproduce accuracy over time (*Electrotechnology Practical LSEEE280A*, p. 126).

The quality of a measuring tool can be determined by the accuracy and the resolution of the tool. The accuracy of a measuring tool is the difference between the actual measurement of the work piece and indicated value of the measuring tool expressed as a plus or minus percentage. Resolution is the number of increments or divisions for a given length. The higher the resolution the better.

In the manufacturing process a number of compromises are made to balance the cost of production. One of these is the degree of accuracy to which the part must be made. Making engineering parts usually involves two separate processes:

- Forming and machining the parts to shape.
- Assembling the parts to form the finished product.

In manufacturing, tolerances are included as part of standard dimensioning in engineering drawings. **ISO 286-1 Basis of tolerances, deviations and fits limit** provides guidance to the size parts must be if they are to fit together and function correctly. Figure 4.138 shows an example of tolerance dimensioning to indicate a tolerance of ±0.3 mm between the two surfaces labelled 'A'.

FIGURE 4.138 Dimension tolerance

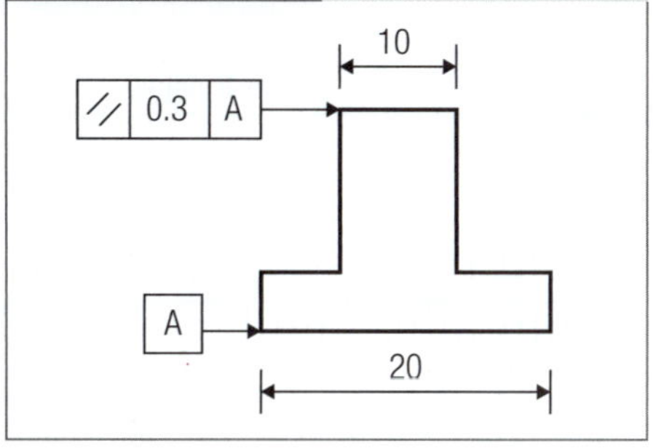

The degree of tightness between mating parts is called the *fit*. A clearance fit occurs when two tolerance mating parts will always leave a space or clearance when assembled. Clearance fits can be described as having a loose running fit, free running, close running or a sliding fit. An interference fit occurs when two mating parts fix or anchor one part into the other, as though the two parts were one.

To make something to an exact size is very expensive both in money and time. To reduce cost, most parts are allowed a tolerance on their designed size. For example, the outside dimension of a part that does not fit into anything else can be made to a wide tolerance, while parts which must be a running fit such as a bearing and shaft must be made to a fine tolerance (*Electrotechnology Practical LSEEE280A*, p. 126).

FIGURE 4.139 Vernier caliper parts

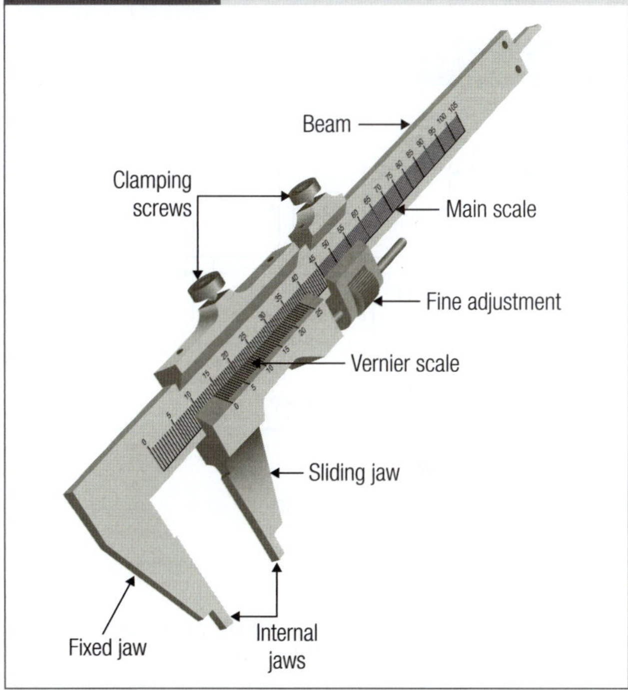

Vernier calipers

Vernier calipers are a commonly used tool that can make accurate measurements. They come in a variety of resolutions such as within 0.1 mm, 0.05 mm or 0.02 mm and an accuracy of ±0.02 mm. A typical vernier caliper consists of two scales: the main scale and the vernier scale. The main scale is graduated into standard divisions, in the case of a metric scale, in millimetres. The vernier scale is made such a length that when it is divided into a number of equal parts, each part represents a proportional length of the corresponding main scale division.

Figure 4.139 shows the construction of a typical vernier caliper. It is made up of a frame consisting of the fixed jaw and beam that is engraved with the main scale and a sliding jaw, engraved with the vernier scale. The sliding jaw can be moved along the beam. The part to be measured is placed between the jaws, the jaws are moved so they make light contact with the edges of the part, and the reading is taken.

Vernier calipers have clamping screws that can be used to hold the moving jaw in place while the vernier is read. Some models have fine adjustment by means of a knurled thumb screw. They often have provision for taking internal and depth measurements.

Vernier calipers are made in a variety of sizes. A common size allows a 200 mm maximum measurement, in which the main scale is graduated from 0 mm to 200 mm in 1 mm divisions, with every 10 divisions being numbered. The decimal part of the reading in tenths and hundredths of a millimetre is obtained from the vernier scale.

The most commonly used vernier scale is 49 mm long and divided into 50 equal parts, as shown in Figure 4.140. The 50 parts of the vernier scale are in total 1 mm shorter than the actual 50 mm on the main scale. Therefore each vernier scale division is one 50th of the total length of 49 mm, or 1/50 of 49 mm, which equals 0.98 mm.

FIGURE 4.140 Main and vernier scales

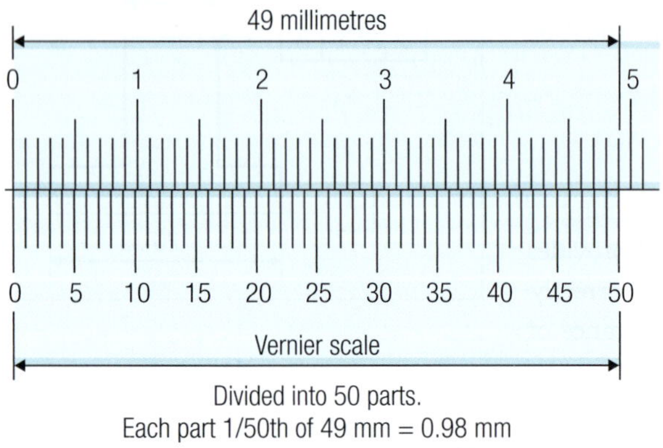

FIGURE 4.142 Main vernier scales

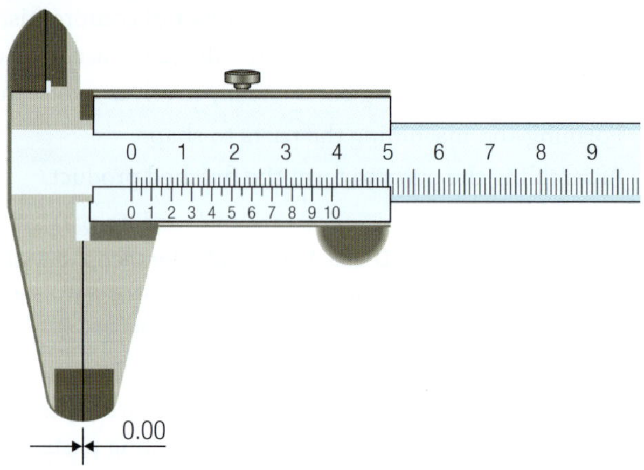

FIGURE 4.141 Main and vernier scales

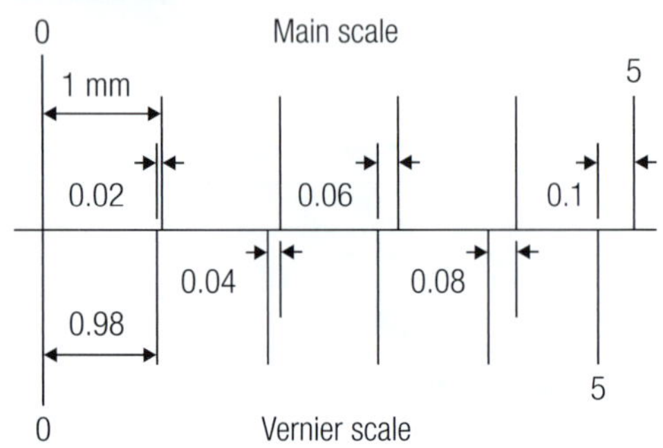

FIGURE 4.143 Digital vernier calipers

Source: iStockphoto/ajt

FIGURE 4.144 Reading vernier scales 1

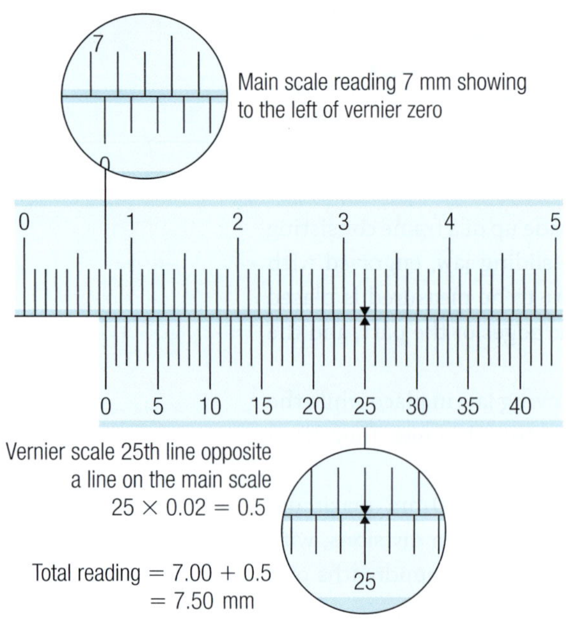

The main scale divisions are 1.00 mm long and the vernier scale divisions are 0.98 mm long. The difference between main and vernier scales is 1.00 mm minus 0.98 mm, so the vernier scale divisions are 0.02 mm shorter than the main scale divisions. In Figure 4.141, note that each vernier division is displaced by 0.02 mm as the reading moves along the main scale.

Figure 4.142 shows a vernier with a 0.05 mm resolution. The vernier scale has 20 graduations 1/20 of 1.00 mm which has a resolution of 0.05 mm.

A digital vernier caliper is shown in Figure 4.143. A typical digital vernier has a liquid crystal display that provides a resolution of 0.01 mm and an accuracy of ± 0.0025 mm.

Using vernier calipers

1 Move the vernier's jaws into light contact with the work to be checked ensuring it is square to the work.
2 Read the number of whole millimetres on the main scale, from the main scale zero up to the vernier scale zero mark. The upper magnified part of Figure 4.144 shows a reading of 7 mm on the main scale.
3 Look along the vernier scale to find a division that lines up with a division on the main scale. The lower magnified part of Figure 4.144 shows the vernier scale lining up on the 25th graduation. Multiply the number of the vernier scale line by 0.02 and add the result to the reading from the main.

For the reading of the vernier shown in Figure 4.145, there are 37 full divisions on the main scale, to the left of the vernier scale zero. This is equal to 37 mm. The 33rd line on the vernier scale is the only mark opposite a line on the main scale. Multiplying 33 by 0.02 gives 0.66 mm. Adding this to the main scale reading of 37 gives a total value of 37.66 mm.

The vernier calipers shown in Figure 4.146 have a scale 49 mm long, with each fifth line numbered from 1 to 10. As each division on the vernier scale represents 0.02, then the fifth line is equal to 0.1 mm, and is marked 1. The tenth line is marked 2 and so on to the end of the scale. The main scale reads 60 mm. The vernier scale aligns with the main scale at its fifth numbered division (which represents 0.5 mm) plus three extra divisions which represent 3 × 0.02 which equals 0.06 mm. The total reading is therefore 60.00 + 0.50 + 0.06 = 60.56 mm.

FIGURE 4.145 Reading vernier scales 2

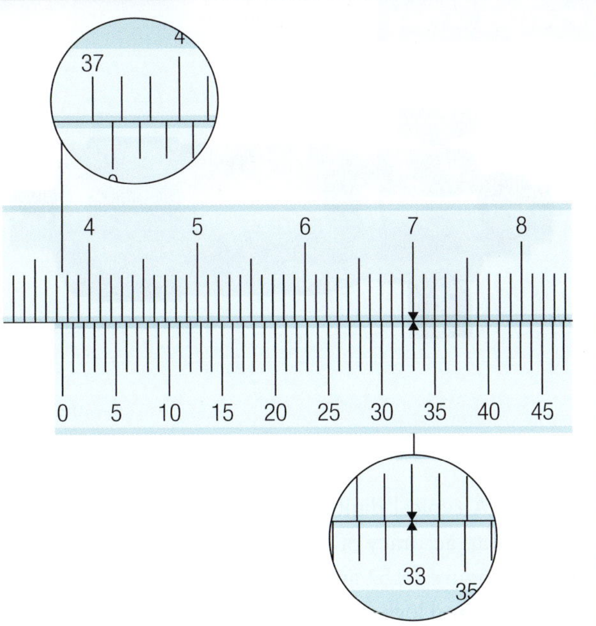

FIGURE 4.146 Reading vernier scales 3

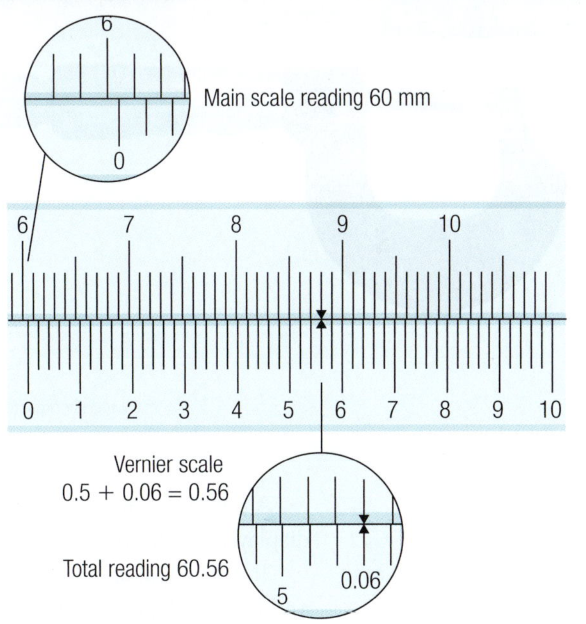

To measure an outside diameter, hold the calipers as shown in Figure 4.147. When measuring inside diameter the opposite inside jaws are used. A magnifying glass is of great assistance when reading a vernier. To achieve better accuracy if a magnifying glass is unavailable, hold the vernier so that you are looking at the scale at an angle and in line with the graduated line. Look along the line rather than at it, with the light coming from the back of the scale.

For the best safety and accuracy when using vernier caliper always:

- Check for zero and ensure that the jaws are clean and parallel.
- Set the calipers so that friction can just be felt between the caliper jaws and the job.
- Lock the jaws and remove from the job.
- Carefully read the calipers.
- Clean and lightly oil the calipers before replacing in the case.
- Never use the calipers on moving parts.

FIGURE 4.147 Holding vernier calipers

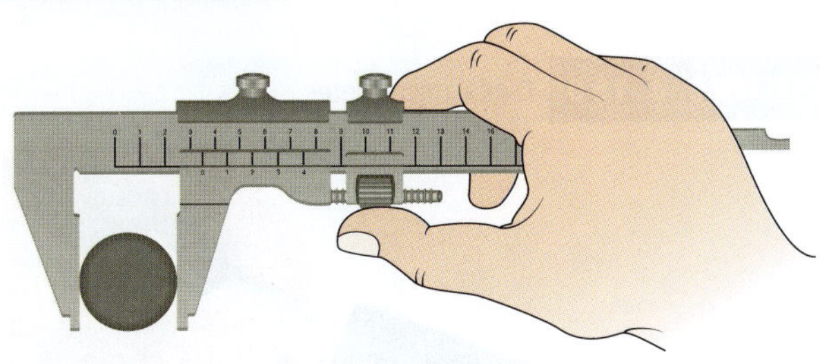

Measuring an outside diameter

Micrometers

A micrometer is a measuring tool that is capable of measuring to a higher accuracy than vernier calipers. There are three types of micrometer based on the type of measurement they are able to make. They are the outside, inside and depth micrometers.

An outside micrometer is shown in Figure 4.148. These are used to measure the outside diameter of rods, material thickness and part lengths. The size of an outside micrometer is determined by the frame size. All sizes, however, have a measuring range of 0 mm to 25 mm. So a 75 mm micrometer can measure accurately between 50 mm to 75 mm.

Figure 4.149 shows an inside micrometer. Uses include measuring pipe bores (inside diameter), holes and the like.

FIGURE 4.148 Outside micrometer

Source: iStockphoto/ajt

FIGURE 4.149 Inside micrometer

Source: Phill Chadwick

A digital micrometer is shown in Figure 4.150. A typical digital micrometer has a liquid crystal display, a resolution of 0.001 mm and in some cases, an accuracy of ±4μm.

The principal parts of a micrometer are shown in Figure 4.151 and include the frame, anvils, spindle and thread, sleeve or barrel, thimble, adjusting ratchet and lock.

Micrometers, as shown in Figure 4.152, have a thread with a pitch of 0.5 mm that moves the spindle a distance of 0.5 mm for each revolution. A main scale calibrated in half millimetres is engraved on the barrel of the micrometer.

FIGURE 4.150 Digital micrometer

Source: shutterstock.com/Florin Oprea

FIGURE 4.151 Parts of a micrometer

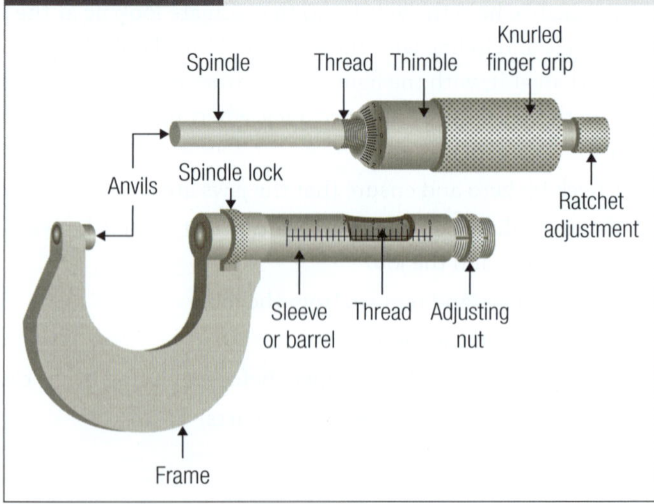

FIGURE 4.152 Micrometer pitch

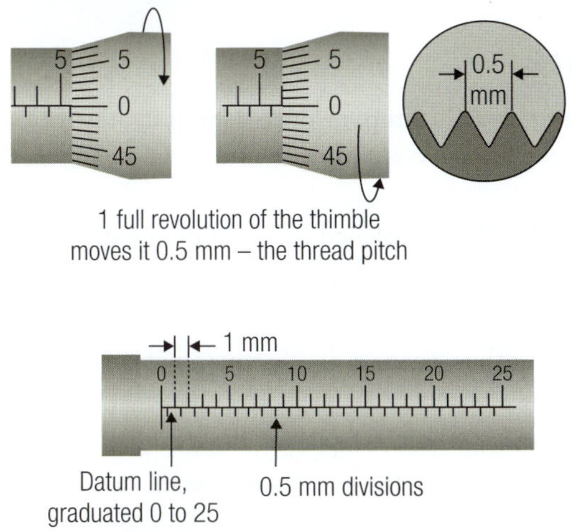

FIGURE 4.153 Micrometer scale

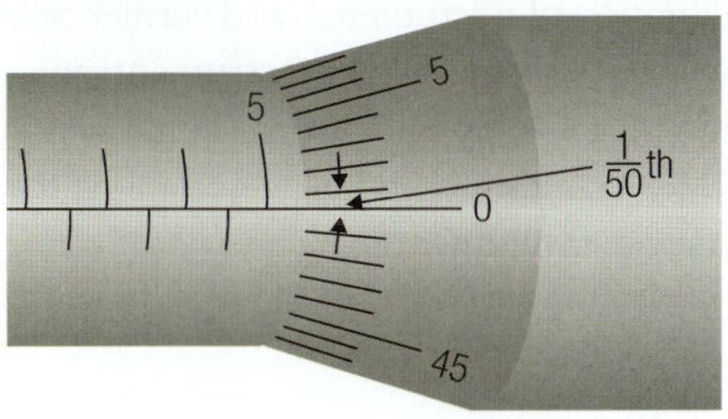

The thimble is divided into 50 parts

FIGURE 4.154 Reading a micrometer scale 1

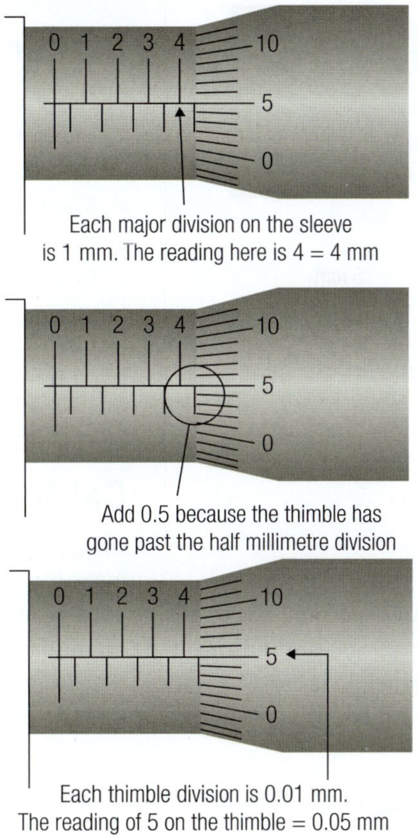

The engraved line along the barrel is the micrometer's datum line. Each millimetre is marked from 0 to 25. The circumference of the thimble is divided into 50 equal parts with each part equivalent to one fiftieth of 0.5 mm or 0.01 mm as shown in Figure 4.153. Every fifth division is marked 0 to 45 (50 in total).

Using a micrometer

First check the zero position by cleaning the anvil faces and closing the micrometer using its ratchet. When the anvils are just touching, the main and thimble scale zeros should be in line. If not, the error should be corrected by adjusting the sleeve position or noted and subtracted from the final reading.

1 Clean the surfaces to be measured and adjust the micrometer to the job using the ratchet. Bring the anvils lightly into contact with the work to be measured, ensuring they are square to the work.
2 Lock the spindle and remove the micrometer from the job.
3 Take the reading, that is, read on the barrel the number of millimetres completely visible. In the upper part of Figure 4.154, this is four major divisions or 4 mm.
4 Add any half millimetres that are visible. The example in the middle diagram of Figure 4.154 shows a reading 0.5 mm. Therefore the micrometer reading so far is 4.5 mm.
5 Note the number of the division on the thimble scale that is aligned with the datum line. The example in the lower diagram of Figure 4.154 shows a reading 5. As each thimble division is worth 0.01 mm, therefore the reading of the thimble is 0.01 mm × 5 which equals 0.05 mm.
6 The final measurement is the sum of all three readings, i.e. 4 + 0.5 + 0.05 = 4.55 mm.
7 Figure 4.155 shows another example of reading a micrometer. There are 5 mm visible between the zero and the thimble. There is also a graduation of 0.5 mm shown. The twelfth line of the thimble

FIGURE 4.155 Reading a micrometer scale 2

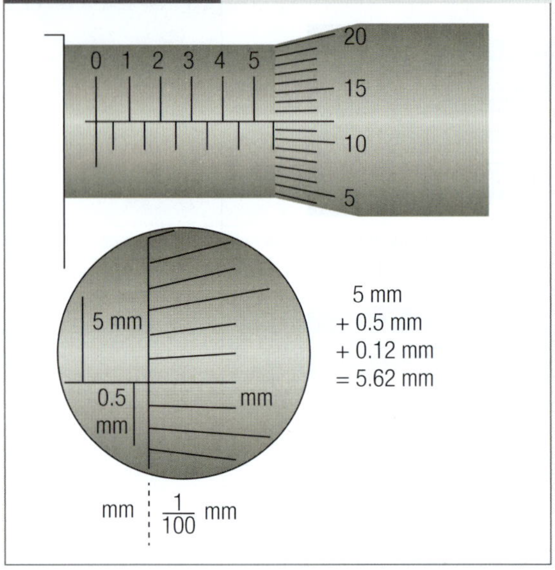

scale is level with the datum line. The total reading of the micrometer is 5 + 0.5 + 0.12 = 5.62 mm.

Micrometer care

- Never use a micrometer as a G-clamp, or over-tighten it. Always turn the barrel with the ratchet to avoid straining the frame or damaging the thread.
- Measure stationary work only, on finished or bright material.
- Don't drop or lay other parts on any precision measuring tool (micrometer or vernier).
- If using a micrometer other than a 0–25 micrometer, check its accuracy with test pieces.
- Unlock and clean a micrometer before replacing it in its box.
- Never adjust the micrometer while the spindle is locked.

Calibration

Gauge blocks and length bars are used to check the accuracy of precision measuring instruments. Calibration of micrometers and vernier calipers is an important part of quality control and ISO certification. Checking measuring instruments maintains precision and ensures parts manufactured in one plant fit correctly with other components from other factories. The National Measurement Institute (NMI) in Australia and New Zealand is the body responsible for biological, chemical, legal, physical and trade measurement.

SUMMING UP...

- The two common tools that are used for low tolerance measuring are the vernier caliper and micrometer.
- Accuracy is how close a measured value is to the actual (true) value.
- Tolerance is the allowable variation either plus or minus to nominal value or exact size.
- Precision is the ability to reproduce accuracy over time.
- A typical vernier caliper has a resolution of between 0.1 mm and 0.05 mm and an accuracy of ±0.02 mm.
- Check vernier calipers for zero and ensure that the jaws are clean and parallel.
- A typical micrometer has a resolution ranging between 0.01 mm to 0.001 mm and an accuracy of ±0.01 mm.
- Three types of micrometer are the outside, inside and depth micrometers.
- Check the micrometer for zero and ensure that the anvils are clean and parallel.
- Gauge blocks and length bars are used to check the accuracy of precision measuring instruments.

4.12 Dismantling and assembly techniques

To install, relocate, repair or test a machine it is sometimes necessary to dismantle some part of that machine. Like any other task it's important to follow a set procedure so work proceeds safely, productively and sustainably. Whenever possible use the manufacturer's data as a guide. Inspection of the machine before dismantling helps determine which parts are satisfactory, which need repair or need replacement. Inspection should be done during dismantling as well.

Tools used in dismantling and assembling

Bearing and wheel pullers

Components that have an interference fit will require some mechanical assistance to separate them. Bearing and wheel pullers come in two and three arm varieties, and they can be either external or internal types. Figure 4.156 shows an external puller that can be reversed to an internal puller by reversal of the threaded rod section. To remove a component, the claws of the arms of the puller grasp the bearing or wheel, and the threaded rod is then rotated with a spanner applying force to the shaft.

FIGURE 4.156 Bearing and wheel puller

Source: Ed Hawkins

Tension wrench

Bolts often need to be tightened to a specified torque, called tensioning. The classic style torque wrench has a micrometer style torque selection. The torque wrench indicates correct tension with a distinctive 'clicking' sound.

Figure 4.157 shows an example of an electronic scaled torque wrench. Correct torque is indicated by an audible tone and visually by a LED. Multiple torque settings may be preset to memory.

When using a torque wrench:

1 Check calibration and inspect wrench for defects before use.
2 Refer to component specifications to obtain torque setting.
3 Unlock the setting adjustment and set the torque wrench, approaching from the lower torque setting, lock the setting adjustment.
4 Check ratchet rotation is in the correct direction and fit appropriate extension and socket.
5 Apply wrench to nut or bolt, grip wrench in the middle of the handle and rotate until correct torque is indicated.

FIGURE 4.157 Digital torque wrench

Source: Ed Hawkins

Drifts and dollies

Drift punches, as shown in Figure 4.158, are used to apply force to objects such as a shaft without causing damage. Unlike a centre punch, which has a point, a drift punch has a flat face.

A dolly is a solid metal tool with a hole through its centre. These are used to remove components such as bolts and shafts from equipment. A dolly is selected with a hole larger than the bolt or shaft. The dolly supports the housing and a hammer and drift is used to tap out the bolt or shaft. A drift and dolly set can also be used to install components such as bearing races. The bearing is lightly lubricated and placed in position. A suitably sized dolly is bolted to the drift and a hammer is then used to tap the drift, forcing the bearing into the housing with an interference fit.

In trades such as panel beating a dolly can also refer to a small anvil shaped object that can be used to shape sheet metal with the aid of a hammer.

FIGURE 4.158 Drift punch set

Source: Ed Hawkins

Feeler gauges

Feeler gauges (Figure 4.159) are a 'go or no go' measuring tool. Typical metric sets consist of 25 blades ranging from 0.04 mm to 1.0 mm thick. When a distance between two surfaces is to be measured, the

FIGURE 4.159 Feeler gauges

Source: Ed Hawkins

smallest blade that will fit between the surfaces without force is considered to be the correct measurement.

Cleaning solvents

Electrical machines must be thoroughly cleaned before dismantling or reassembly to ensure that good electrical contact exists between the mating surfaces of moving and stationary parts. Electrical equipment needs to be free of contaminates and dirt which could possibly cause short circuits between the insulated components and earth.

Care should be taken when cleaning because some solvents will damage the insulation of the electrical conductors, present a fire hazard or affect moulded plastic components. The three surfaces of a typical electrical machine that require cleaning are external surfaces, internal surfaces and electrical mating surfaces. External surfaces with heavy dirt deposits can be cleaned by scraping or using a wire brush. Wash with a mild solvent such as kerosene and wipe with a lint free cloth. Clean electrical mating surfaces with compressed air and a solvent soaked lint free cloth. Always wear safety goggles and protective clothing such as rubber gloves when using cleaning solvents (*Electrotechnology Practical LSEEE280A*, pp. 139–40).

Safe treatment of dismantled components

Before commencing disassembly, establish the condition of the machine you are about to dismantle. Mark electrical connections as they are disconnected to identify where each one goes, as electrical polarity and phase sequence is often important. Phase sequence applies to three-phase supplies, in which swapping any two of the phases will cause a motor to run in reverse.

In the absence of manufacturer's details, you may be able to dismantle relatively simple machines by thoroughly inspecting the machine to determine the best way of dismantling it. Ensure the work area is clean and tidy to avoid contamination of the parts as they are removed. The area should also provide adequate space to carry out the work. As parts are removed to avoid loss they should be labelled and stored in a storage container.

Use witness marks on components such as end shields and frames to ensure correct orientation and alignment when reassembled. It is very important to document what you do at every step. A freehand sketch of an exploded view will greatly assist when it comes time to reassemble the machine. Remember it could be several months before you reassemble the machine.

Dismantling a piece of equipment

The following is an example of the typical procedure used to dismantle the three phase induction motor shown in Figure 4.160. The following example assumes that the motor has been correctly isolated and disconnected from its source of supply.

After cleaning and documenting:

1 Inspect the motor to determine readily removable items such as pulleys and gears. For d.c. machines and single phase squirrel cage motors it is important to remove the end shield at the drive end first to avoid damage to brushes or the centrifugal switch.

2 If the terminal block needs to be removed, identify and label the start and finishes of the three windings. Then remove the terminal nuts to disconnect wiring and remove the mounting screws to separate the terminal block from the frame.

3 Remove the cooling fan shroud. Next remove the fan from the rotor shaft then remove the four long bolts that hold the end shields on to the frame.

4 Next, if no witness marks are already present, centre punch the fan end shield with one dot on both the shield and the stator as shown in Figure 4.161. Do the same with the other shaft end shield and stator, this time using two centre punch dots.

FIGURE 4.160 Three phase induction motor exploded view

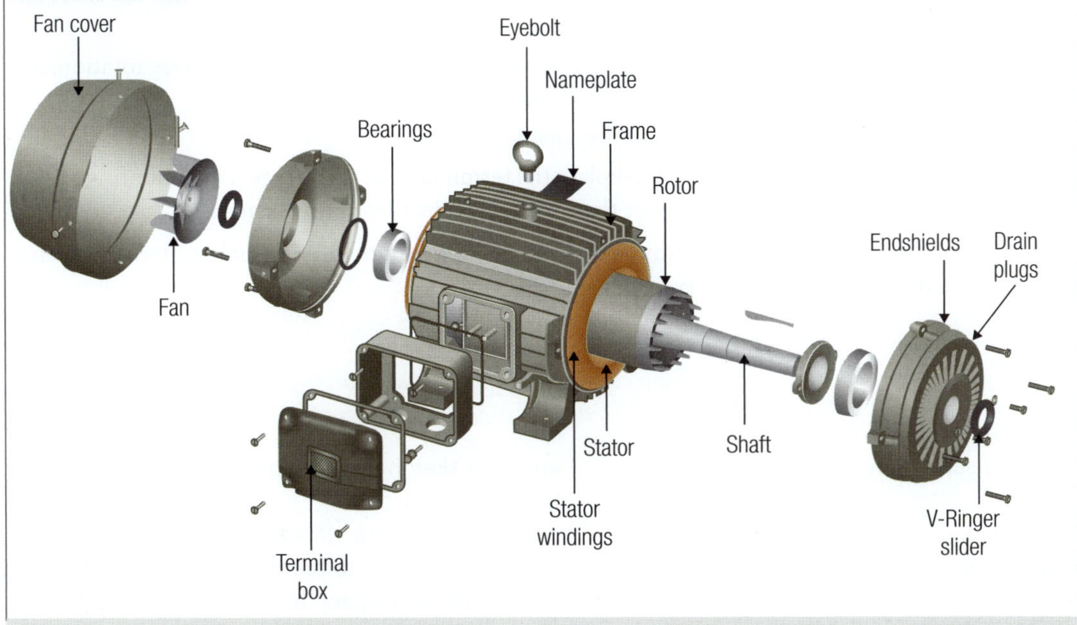

Source: Ed Hawkins

5 Separate the end shields from the stator on both ends. Now slide the rotor out of the stator from either end.
6 Bearings are carefully removed from the end shields with the aid of a bearing puller.
7 Recheck your assembly diagram and make any additional notes.

FIGURE 4.161 Witness marks

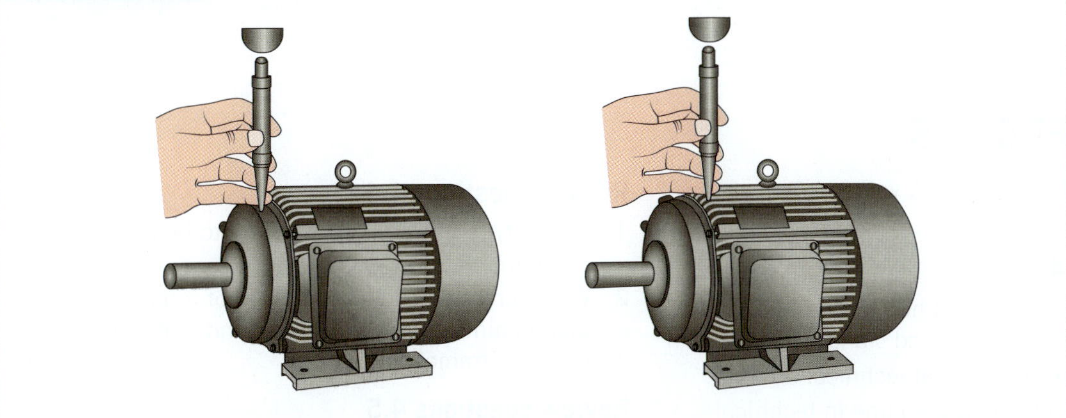

Assembling a piece of equipment

Reassembly usually follows the reverse procedure to disassembly. To ensure correct operation and avoid damage to components, care needs to be taken when assembling bearings, gaskets, seals, etc.

1 Recheck your assembly diagram and confirm all required parts are present and in a functional condition.
2 When installing ball or roller bearings it is important to note that only one race is a tight fit. Normally the revolving race is a press fit, so the bearing is pressed onto the shaft first. The stationary race is a light push fit, so it is pushed into place when the end shield is fitted.
3 The fan end shield is fitted into place ensuring the witness marks are correctly aligned.
4 The rotor is carefully inserted through the frame and aligned with the fan end shield. The outer race of the bearing can be gently pushed or tapped with a soft faced hammer part of the way home.

5 The shaft end shield is moved into position around the shaft and aligned with the witness marks. A gentle tap with a soft faced hammer may be required to seat the bearing race into the shaft end shield.

6 The four long bolts are replaced and partially tightened. The rotor is checked for free rotation, and the opposing bolts are gradually tightened forcing the bearing races home.

7 Fit the fan onto the rotor shaft. Replace the cooling fan shroud.

8 If the terminal block has been removed, replace the terminal bock and secure with mounting screws. Use the labelling as a guide to reconnect the winding conductors to the terminals.

If the manufacturer provides a recommended torque for the nuts in an assembly a tension wrench must be used for tightening to the correct tension.

SUMMING UP...

- Common tools used for disassembly and assembly include:
  - bearing/wheel pullers
  - torque wrenches (digital and analogue)
  - drifts and dollies
  - feeler gauges.
- Risk sources presented by the use of some cleaning solvents include:
  - electrical conductor insulation damage
  - fire
  - damage to moulded plastic components.
- Before commencing disassembly:
  - establish if the machine is functioning correctly or not
  - ensure a sketch or diagram of the machine has been recorded
  - ensure appropriate storage containers are available for parts to prevent loss or damage
  - ensure the work space is of sufficient area and is clean and tidy.
- During disassembly punch witness marks on parts to ensure correct alignment of parts.
- During reassembly, refer to your diagrams and witness marks to ensure parts are correctly assembled.

REVIEW EXERCISES

Answer these questions and check your answers at the back of the book.

Review questions 4.1

1 What is an ISO standard and how do these standards apply to Australian/New Zealand standards?

2 Name the standard for general technical drawing.

3 List two uses for a thick dashed line in technical drawings.

Review questions 4.2

1 Give two examples of pure metals.

2 Briefly define toughness.

Review questions 4.3

1 When marking out, how can scribed lines on metal be made to stand out more clearly?

2 Describe how you would locate the centre of a piece of round stock using jenny calipers.

Review questions 4.4

1 Which type of vice is best suited to hold steel conduit?

2 What direction should the teeth of a hacksaw blade point in relation to the handle of the hacksaw frame?

Review questions 4.5

1 What are the functions of the flutes in a twist drill?

2 How is a drill sharpened for drilling brass?

Review questions 4.6

1 State the three types of thread cutting taps, and where each type is used.

2 State the purpose of a die nut.

Review questions 4.7

1 What is the typical weight of an engineer's hammer suitable for general use?

2 List three defects of a hammer that would prevent you from using it.

Review questions 4.8

1 Give an example of the main two forms of welding and briefly explain the difference between them.

2 What is the purpose of the gas associated with gas metal arc welding (GMAW or MIG), and what types of gas are typically used?

Review questions 4.9

1 What is the advantage of a sander that has random orbital movement compared to just orbital movement?

2 What device is used to protect operators of electric power tools from direct and indirect contact with an electrical supply?

Review questions 4.10

1 Name three types of metals used to make sheet metal.

2 What type of machine is used to cut sheet metal?

Review questions 4.11

1 Which is the most accurate, a micrometer or vernier calipers?

2 Figure 4.162 shows a main scale and a vernier. What value is indicated?

FIGURE 4.162

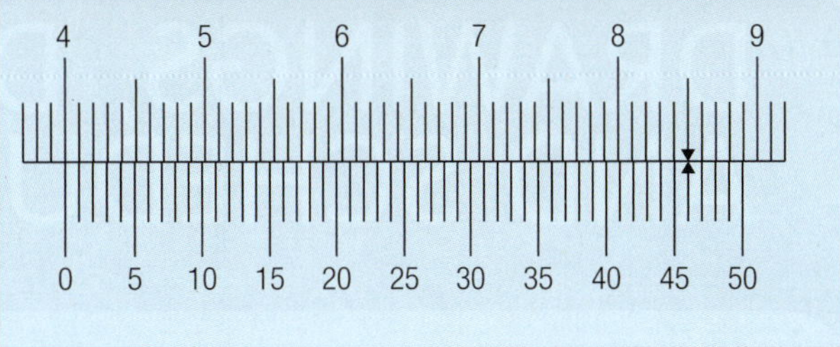

Review questions 4.12

1 How does a torque wrench indicate correct torque?

2 When disassembling an electric motor, how can you ensure that, on reassembly, the two end shields are correctly fitted?

ONLINE RESOURCES

Ramset
http://www.ramset.com.au/Document/Resources (power tools)

Stanley Tools
http://stanleytools.com

Stefanelli Engineering
http://www.stefanelli.eng.br (vernier caliper use)

5

DRAWINGS, DIAGRAMS AND SCHEDULES

Modern buildings require a variety of electrical services such as lighting, power, heating, air-conditioning, data and communications. Drawings must be created by designers to record the desired location of these services and provide guidance to the tradespeople involved in construction or maintenance. This chapter outlines the types of drawings used in the electrotechnology industry. As the person responsible for the installation and connection of electrical services, it is essential that you can correctly read, understand and interpret these types of drawings.

Chapter outline

5.1 Architectural drawings

As part of the construction process, architects and engineers must be able to communicate instructions to the tradespeople performing the work. Architectural drawings ensure the construction is consistent with the design. Quotations and costing of construction work are made from these drawings and they allow quality control by making a comparison of the finished product to the plan.

Types of architectural drawings

Standard drawings

Some project home builders use standard drawings extensively. They build the same basic structure on many different sites. Sometimes, because of the orientation of the site, the actual building is a mirror image of the standard drawing. The electrician must be able to work from standard drawings even though the structure is the reverse of that shown on the plan.

Working drawings

The working drawing is usually made up of a group of drawings, sometimes as a group of drawing sheets but often on the one large sheet. In order to understand the plan layout, it is necessary to know the parts of a complete working plan. As shown in Figure 5.1, these parts are:

- Site plan
- Floor plan
- Elevations (front, rear and side)
- Sectional
- Detail (adapted from Miller College Tafe, *Student Workbook*, 'Architectural drawings', Tafe South Western Sydney Institute, 2014, p. 2).

FIGURE 5.1 Working plans

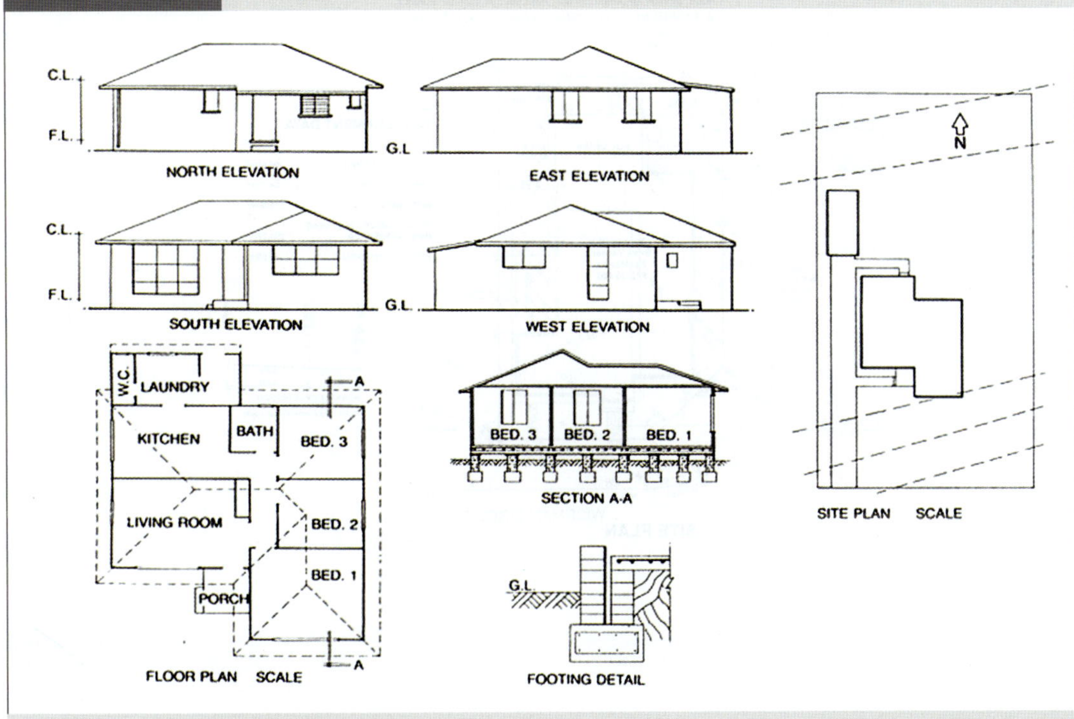

Source: Eden Brae Homes/A & N Design, Sydney

Site plan

The site plan in Figure 5.2 shows the outline of the building lot and the position (orientation) of the building on the lot. The external walls are shown in heavy outline with the eaves marked with a dashed line. A site plan typically shows details such as:

- scale of the plan
- surveyors datum location
- northerly direction (orientation)
- land contours
- street name
- lot number and the numbers of adjacent lots in order to clearly identify the correct site on which to construct the building
- deposited plan (DP) number
- exterior building dimensions and distances from the property lines
- site coverage of building(s)
- location and dimensions of any easements
- any existing buildings on the site.

A site plan will always indicate a north, which gives the orientation of the plan to the real world. Other details include particulars of the mains for services such as electricity, gas, water, sewerage and telephone as well as driveways, paths and existing trees (adapted from 'Architectural drawings', p. 3).

FIGURE 5.2 Site plan

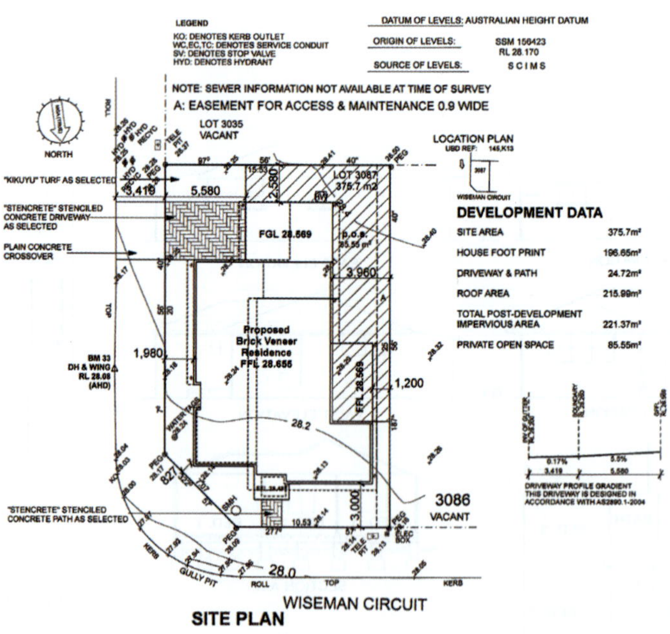

Source: Jandson Homes

Floor plans

This is a drawing viewed from above, with the roof, ceiling and part of the walls removed to provide a bird's-eye view of the house. It shows the outline and shape of the building and the position of all the internal walls. Floor plans, as shown in Figure 5.3, are the most useful type of drawing to an electrician and these generally show:

- all dimensions
- type of wall construction and thickness
- position of windows, doors and archways
- width of openings
- position of electrical accessories, appliances and other equipment

FIGURE 5.3 Floor plan

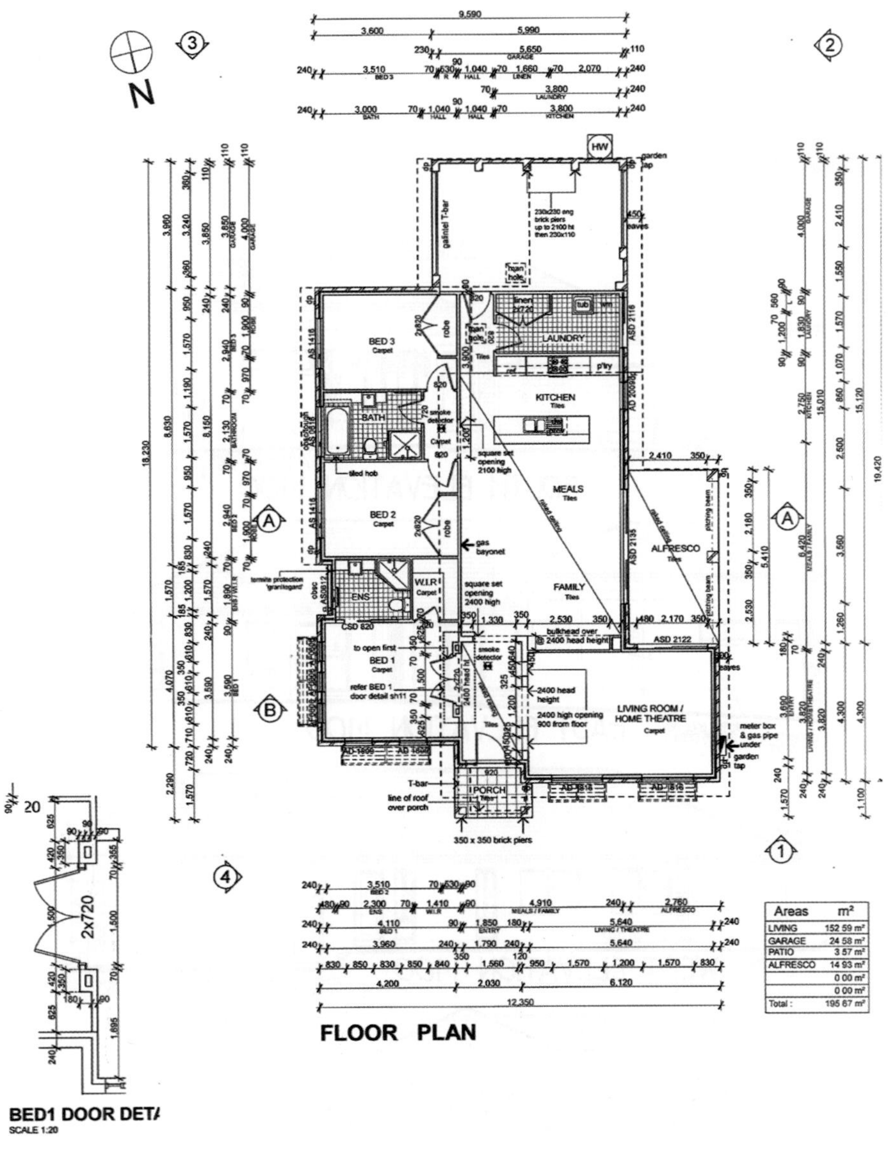

Source: Jandson Homes

- position of kitchen, laundry and bathroom fittings
- location of eaves, steps and downpipes.

To avoid congestion and possible confusion by superimposing electrical symbols over building fittings, a separate electrical floor plan is usually prepared. Other trades, for example plumbers, will often be supplied with a separate drawing that only shows plumbing services to avoid clutter in the drawing (adapted from 'Architectural drawings', p. 4).

Elevations

Drawings of the various sides of a building viewed at right angles (usually only four views) are referred to as elevations. The views are mostly named in relation to their geographic orientation. The top left of Figure 5.4 shows examples of the North, South, East and West elevations of a house. Elevations show:

- overall shape of the dwelling
- roof pitch and shape
- floor to ceiling heights
- position of doors and windows (adapted from 'Architectural drawings', p. 5).

FIGURE 5.4 Elevations

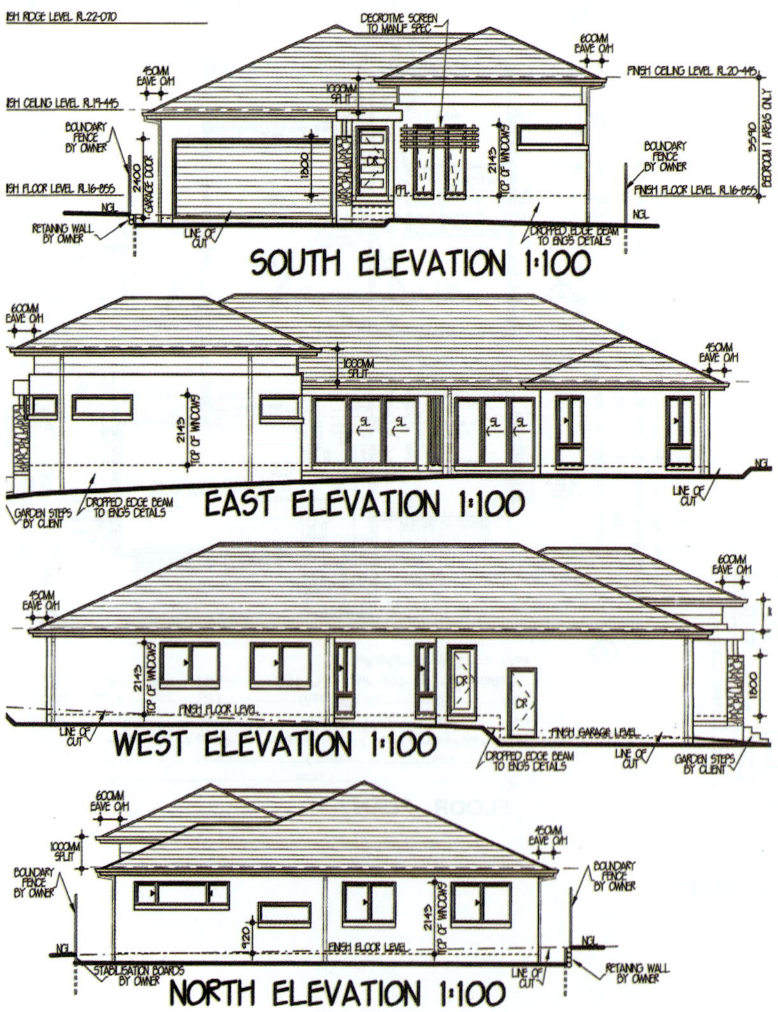

Source: Eden Brae Homes/A & N Design, Sydney

Sectional views

Sectional drawings show the internal features of the building as if it had been cut vertically and the structure closest to the viewer removed. These drawings are used to show constructional details which would normally be hidden in other views, such as the height of wall fittings, shape and height of arches, sub-floor and/or roof structures, etc. Figure 5.5 shows an example of the sectional view of a house (adapted from 'Architectural drawings', p. 5).

FIGURE 5.5 Sectional view

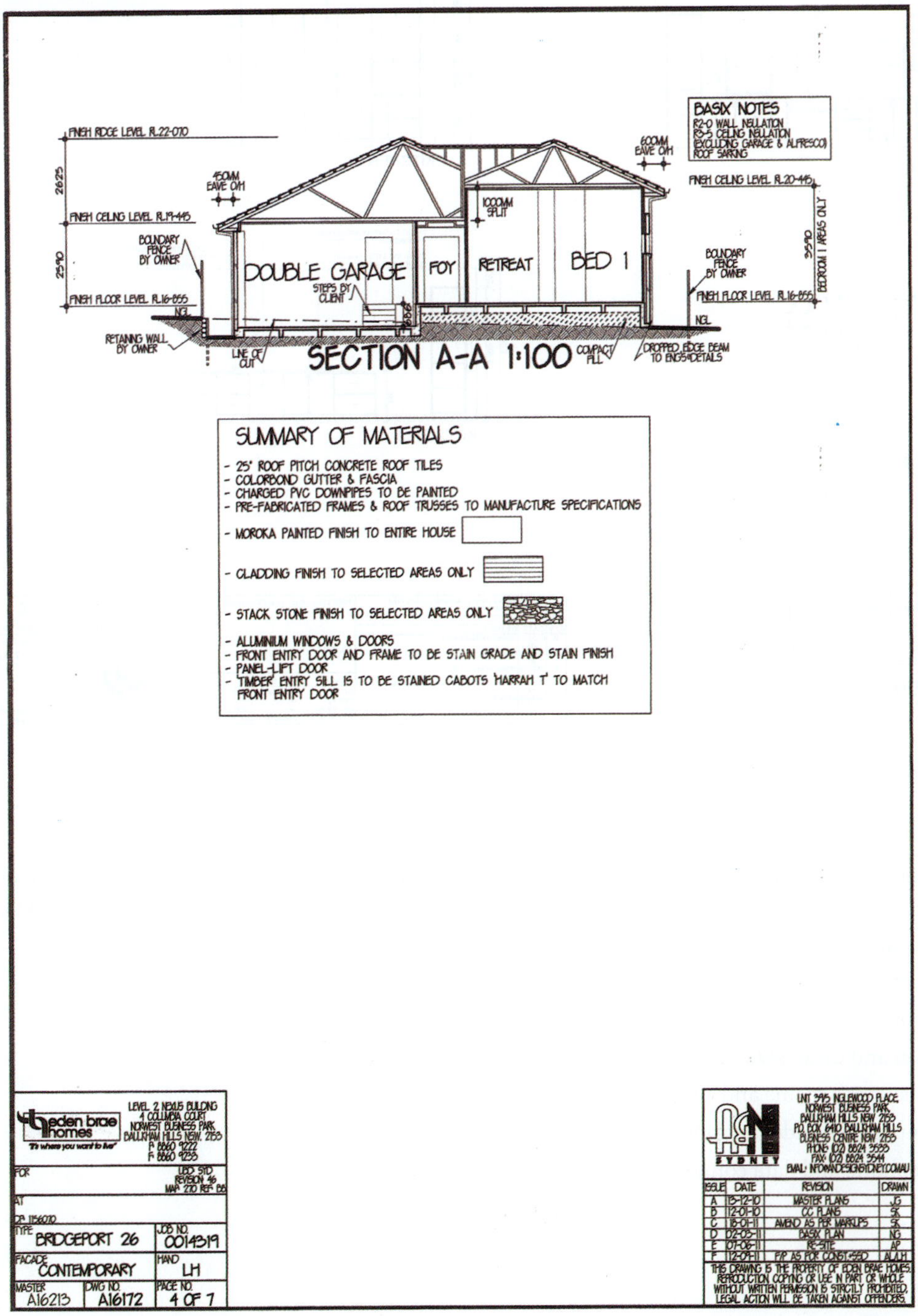

Source: Eden Brae Homes/A & N Design, Sydney

Detail drawings

These are larger scale drawings, used to clearly show details of specific areas of construction where greater detail needs to be shown. Figure 5.6 shows a detailed drawing of a kitchen (adapted from 'Architectural drawings', p. 5).

FIGURE 5.6 Detail drawing

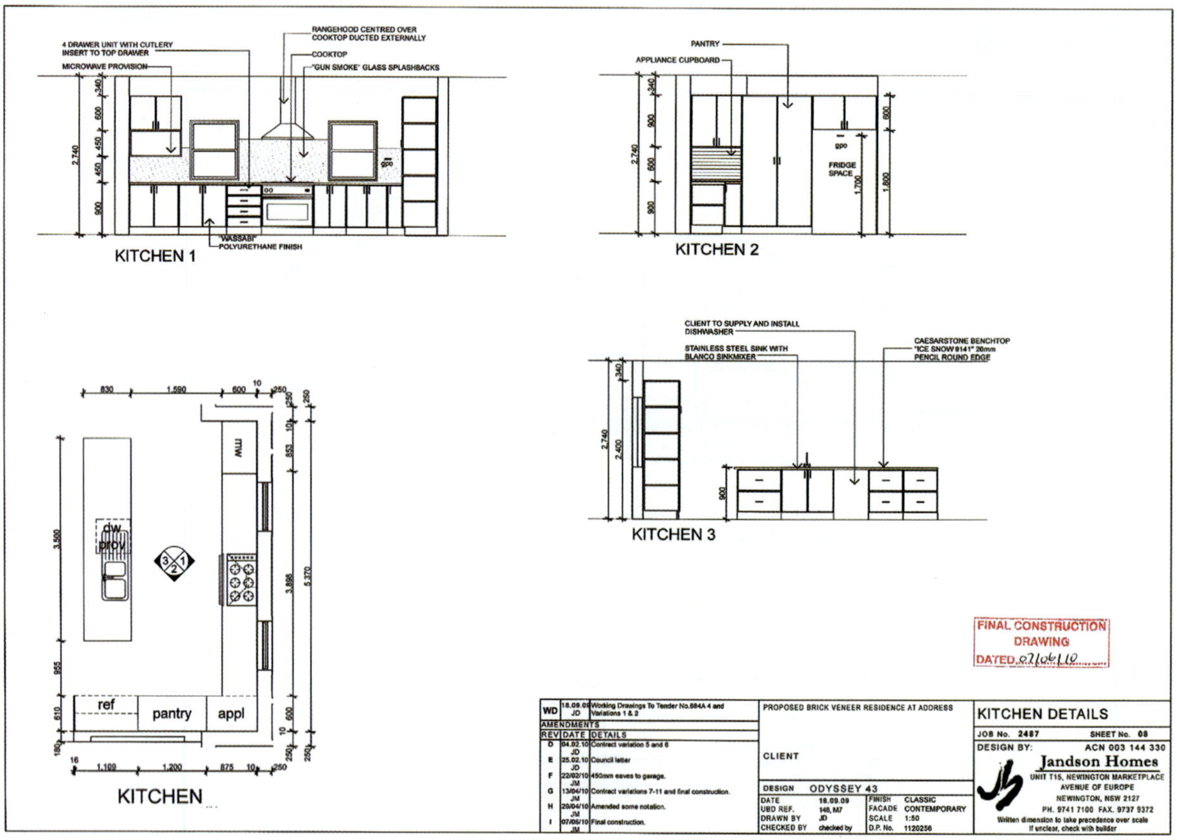

Source: Jandson Homes

Location of electrical services with a site plan

A site plan shows how a building sits on the block of land as well as the location and routes of services such as gas, sewerage, fresh and grey water supplies, as well as stormwater drainage pipes. Electrical services shown on a site plan can include the:

- point of supply
- point of attachment
- main and other switchboards
- consumer's mains and sub-mains
- temporary builder's supply if any
- telephone and data connections.

Accurate knowledge of the locations of services on a site can prevent costly accidental damage. Figure 5.7 gives an example of typical electrical services shown on a site plan.

FIGURE 5.7 Site plans electrical services

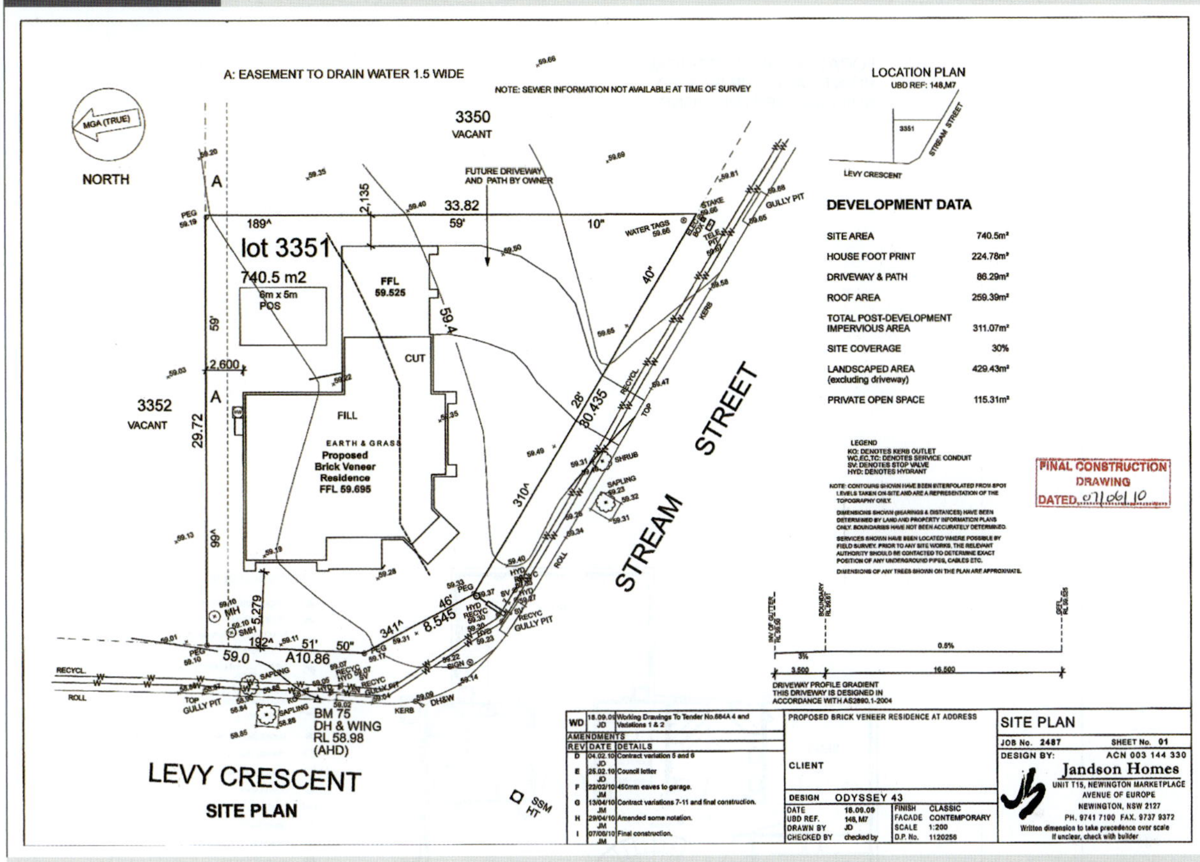

Source: Jandson Homes

Architectural floor plans

A graphical way to show the location of electrical and communication equipment is to mark an item's location on a floor plan as shown in Figure 5.8. It is much easier to read than the alternative, that is, an electrical schedule which is a list of the equipment and its location.

Typical electrical and communication equipment shown on an architectural floor plan includes:

- main and other switchboards
- power outlets, light points and switches
- appliances such as ranges, hot water systems, air conditioners, etc.
- telephone and television (aerial) outlets
- data and video outlets.

FIGURE 5.8 Architectural floor plan showing electrical equipment

Source: Prestige Electrical

AS/NZS 1102 series architectural location symbols

The location of all appliances and accessories to be installed are shown with the use of symbols on the floor plan. The symbols used are specified in the **AS/NZS 1102 Graphical symbols for electrotechnical documentation** series. Table 5.1 provides an example of commonly used symbols. These standard symbols are used to represent points in an electrical installation. A point may be a light, socket outlet or an appliance. Different symbols are used to represent lights and power points, and a variety of symbols are used for appliances, depending upon their type.

Electrical architectural symbols bear little, if any, resemblance to the items they actually represent, and are used to identify and show the location of accessories and other equipment. These symbols are called *location* symbols.

TABLE 5.1 Architectural electrical location symbols

EQUIPMENT DESCRIPTION	SYMBOL
Lighting	
Luminaire – general symbols or luminaire outlet	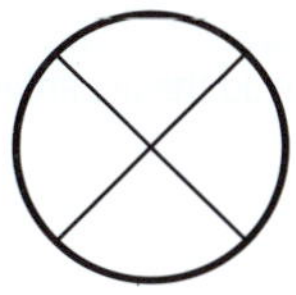
Luminaire fixed to a wall	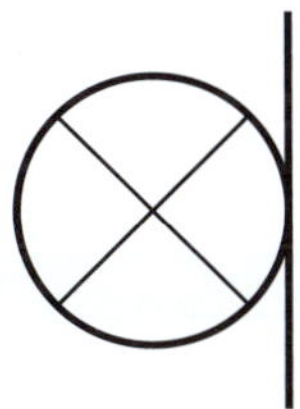
Number and power of lamps in a group may be specified	 3 × 15 W
Spotlight	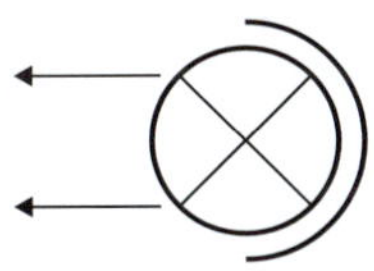
Floodlight	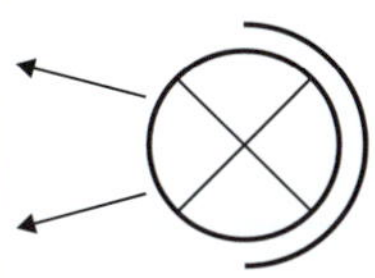
Tubular fluorescent luminaire – one lamp	
Tubular fluorescent luminaire – two lamp	
Switchboard	
General symbol: within the symbol the type of switchboard should be nominated. MSB = Main switchboard MB = Meter board	

(*Continued*)

TABLE 5.1 *(Continued)*

EQUIPMENT DESCRIPTION	SYMBOL
Appliances	
Fan	
General symbol: within the symbol is placed the accepted abbreviation for the appliance. HSW = Hot water system EF = Exhaust fan	
Air-conditioner (an example)	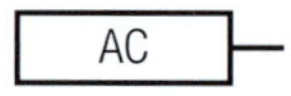
Electric heater	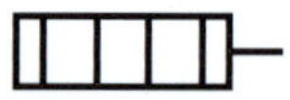
Switches	
Single pole (one-way) switch	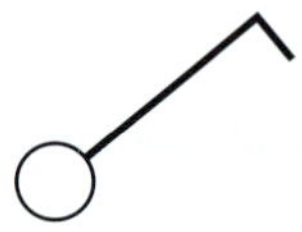
Single pole pull switch	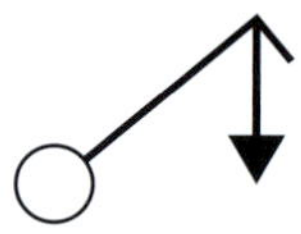
Two way switch	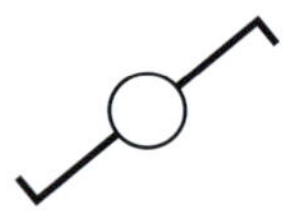
Intermediate switch	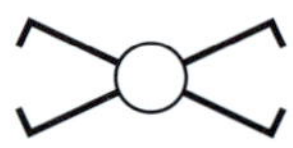
Dimmer control switch	
Time switch	
Push button	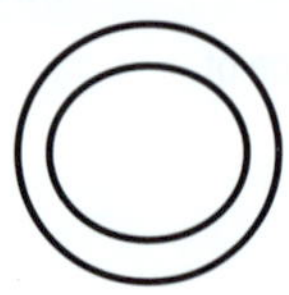

(Continued)

TABLE 5.1 *(Continued)*

EQUIPMENT DESCRIPTION	SYMBOL
Power	
Single 10 A socket outlet	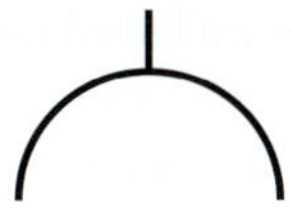
Multiple 10 A socket outlet (double)	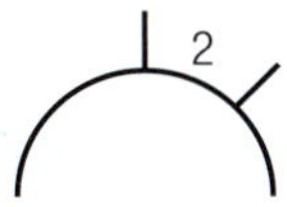
15 A plug socket	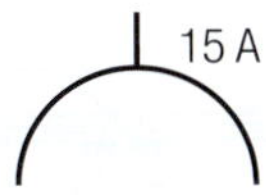
Special outlet (in this case – 32 V)	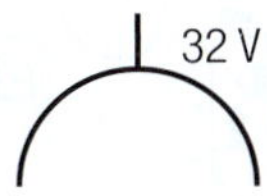
Multi-phase outlet (in this case – 3 phase)	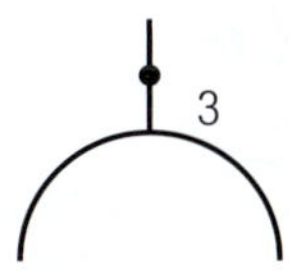
Communication	
Telecommunication outlet (use text adjacent to symbol) TV = Television	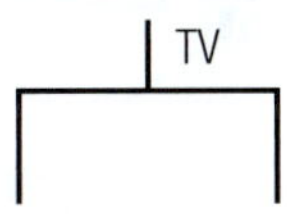
Wall telephone	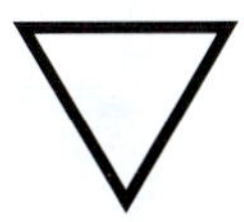
Floor telephone	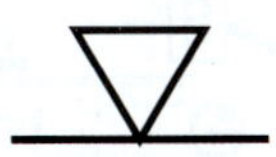
Electric bell	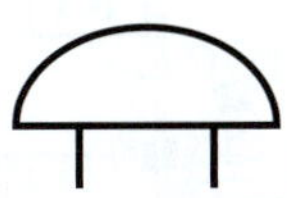
Electric clock	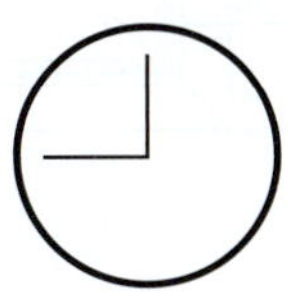

Reading floor plans to locate equipment

Location symbols are drawn so that the centre of the symbol is located at the approximate positioning centre of the accessory or appliance represented. The actual position of the equipment will depend on building and equipment construction. For example, in timber frame walls, light switches and socket outlets are usually positioned at the wall stud nearest to the symbol position. The final positioning of these accessories is left to the discretion of the installing electrician.

The dotted lines shown in Figure 5.9 indicate which light is controlled by each switch. It is important to understand this is not a wiring diagram or an indication of the routes of cables. This too is left up to the installing electrician to determine.

FIGURE 5.9 Architectural floor plan showing electrical equipment

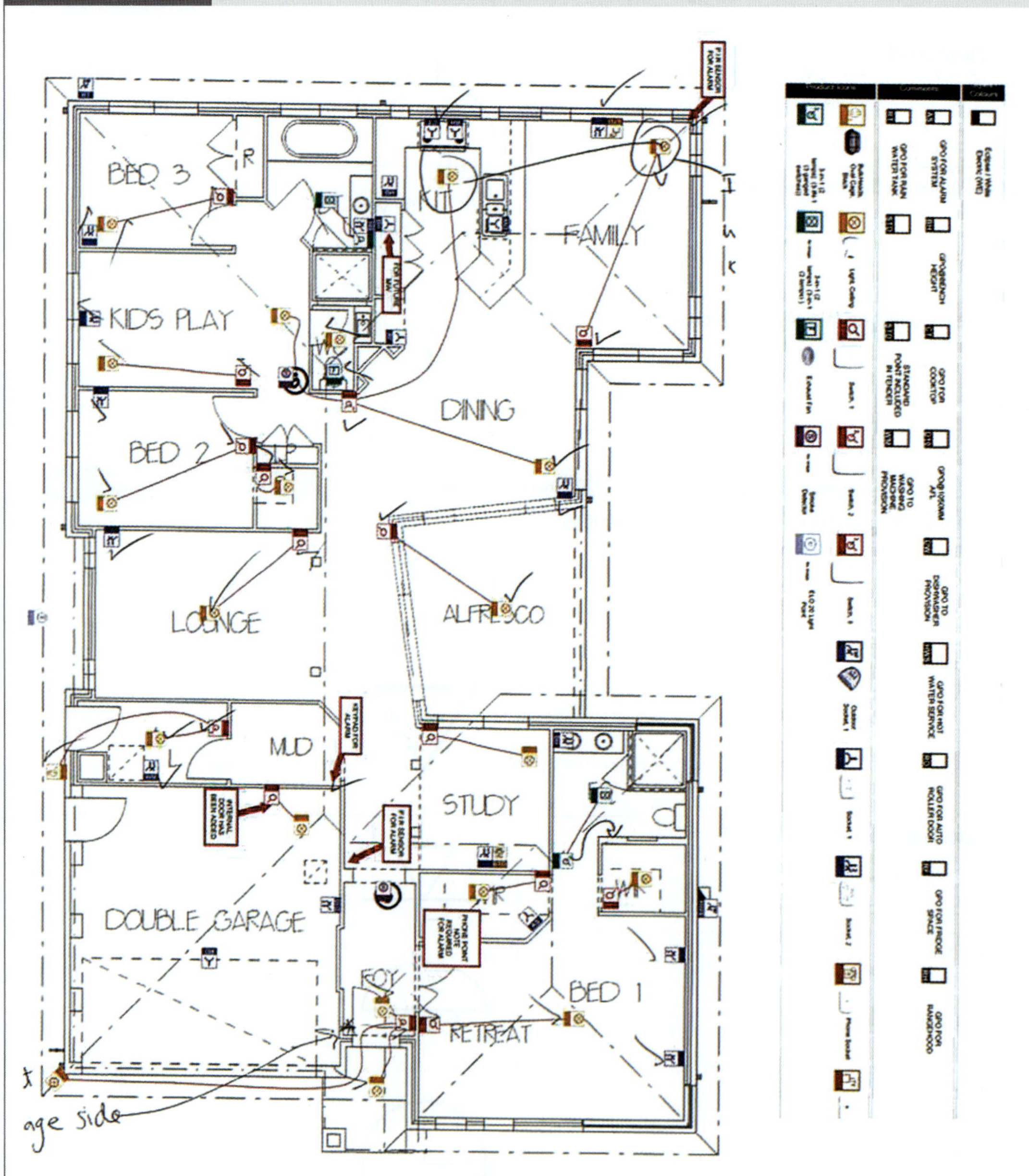

Source: Eden Brae Homes/A & N Design, Sydney

Many builders and architects use location symbols other than those indicated in the **AS/NZS 1102** series. Figure 5.9 shows an example of some non-standard symbols. You should always refer to the electrical legend on the working plan you are provided with and make yourself familiar with the location symbols on that particular plan.

Standard drawing scales

It is impossible to produce full size drawings of a building or its components. Building plans are drawn to a reduced size to a fixed scale. This means that on a single view or drawing every part is drawn to the same reduction ratio. The standard for linear measurement is millimetres, both on the plan and on the job. Working with measurements in units such as centimetres should be avoided.

The basic scale used for a drawing is expressed as a ratio of two numbers. The first number is always unity (1) and the second number is a multiple of 1. The measurement on the drawing is multiplied by the second number of the scale. For example, a scale of 1:100 means that 1 millimetre on the plan is equal to 1 x 100 millimetres on the job. A wall measured with a rule to be 34 mm in length on the plan will in actual construction be 3400 mm (3.4 m) long (adapted from 'Architectural drawings', p. 7).

To determine the required length on the drawing, the measurement of the job is divided by the second value of the scale. For example, if the scale is 1:500 a measurement of 6 m (6000 mm) on the job is divided by 500, which equals a distance of 12 mm on the drawing.

TABLE 5.2 Standard drawing scales

Plan type	Scale			
Site plans	1:1000	1:500	1:200	
Floor plans	1:200	1:100	1:50	1:20
Elevations	1:200	1:100	1:50	1:20
Sections	1:200	1:100	1:50	1:20
Details	1:10	1:5	1:2	1:1 (full size)

Working from photocopies can cause small errors. The plan may have been reduced or enlarged by photocopying or digital scanning. Plans, if not copied from an original, may become distorted when they are a copy of a copy of a copy, which could result in a false measurement when transferred to the actual job.

SUMMING UP...

- Working drawings typically contain:
  - a site plan
  - a floor plan
  - a number of elevations
  - a detail drawing and a sectional drawing.
- Electrical services shown on a site plan can include the:
  - point of supply
  - point of attachment
  - main and other switchboards
  - consumer's mains and sub-mains
  - temporary builder's supply if any
  - telephone and data connections.
- Architectural floor plans use symbols drawn to AS/NZS 1102 to show the location of electrical and communication equipment on a floor plan.
- Scaled drawings use millimetres as a unit of measurement.
- Scales are expressed as a ratio of the length on the plan to length in the real world, e.g. with a scale of 1:50:
  - a measurement of 20 mm on the plan converts to a job distance of 1000 mm
  - 1.5 m or 1500 mm on the job is drawn as a distance of 75 mm on the plan.

5.2 Electrical drawings

Different types of drawings are adapted to suit the information they are intended to convey. Electrical diagrams use a series of standardised symbols to represent electrical components and show how components are connected to each other. These diagrams allow instructions of the design, construction and operation of electrical equipment to be freely shared. Electrical tradespeople need to be able to quickly and accurately read and draw these diagrams. Six types of diagrams commonly in the electrotechnology industry are:

- block diagrams
- circuit diagrams
- wiring diagrams
- architectural diagrams
- ladder diagrams
- single line diagrams.

Types, purpose and application of electrical drawings

Block diagrams

A block diagram is a simple flow chart intended to explain the basic operation of a circuit or system. No detailed information is provided as block diagrams are normally used in the development of circuit diagrams. Figure 5.10 shows an example of a block diagram for a simple lighting circuit.

FIGURE 5.10 A block diagram

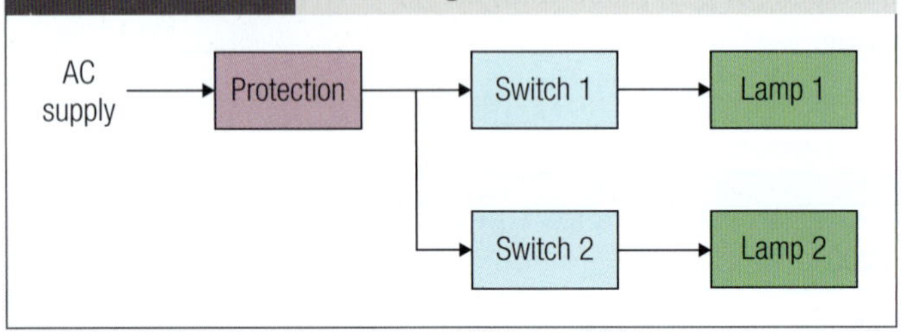

Circuit diagrams

The circuit diagram, also known as a schematic diagram, is a detailed diagram intended to describe the operation of a circuit. See the example in Figure 5.11. Circuit diagrams can be complex and in some cases may contain several circuits. All circuit diagrams contain symbols representing components, or items of equipment, interconnected by lines representing conductors. The layout of the diagram and position does not necessarily represent the actual physical layout of the equipment and wiring. Connections are shown by a dot. The purposes and applications of circuit diagrams are to:

- develop or design an original circuit from a block diagram
- determine the operation of a circuit
- assist in fault finding.

FIGURE 5.11 A circuit diagram

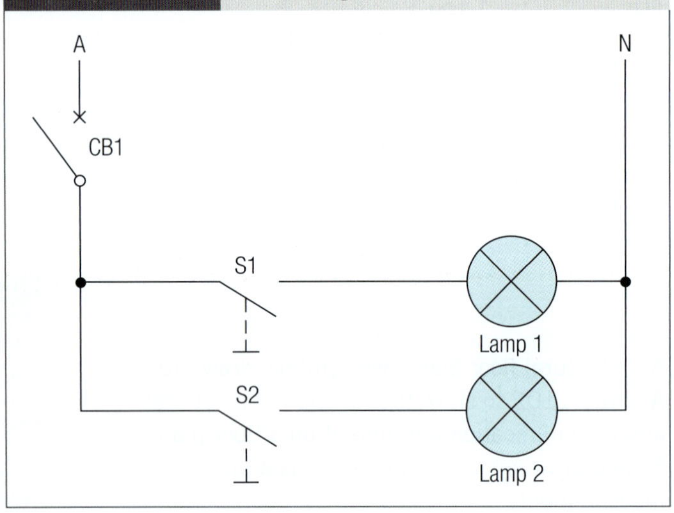

Wiring diagrams

A wiring diagram shows the exact location of all components and wiring. Components are typically drawn as an outline and are not necessarily drawn to scale, in order to reduce clutter. The location of the terminals at which every connection is made is clearly shown and must be strictly followed. Figure 5.12 shows an example of a wiring diagram for the circuit diagram of Figure 5.11.

Typical applications of wiring diagrams include:

- the installation of luminaires and appliances
- wiring of switchboards and control panels
- verification and fault finding in a circuit which has unauthorised or undocumented modifications made to it.

FIGURE 5.12 A wiring diagram

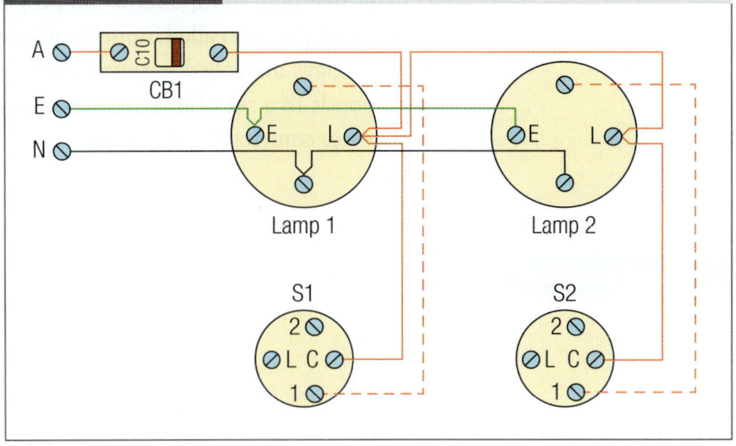

Ladder diagrams

Ladder diagrams are used as a graphical method of data entry for programmable logic controllers (PLCs). A PLC has a microprocessor that can be programmed to suit a particular application, such as in industrial automation. Ladder diagrams are virtual circuit diagrams, meaning there are no actual components to connect within the PLC. The symbols in a ladder diagram represent similar functions to that of a circuit diagram.

A ladder diagram is shown in Figure 5.13. Notice the similarity of the diagram to the rungs of a ladder, hence the name. Ladder diagrams were developed as a PLC programming method with a

FIGURE 5.13 Ladder diagram

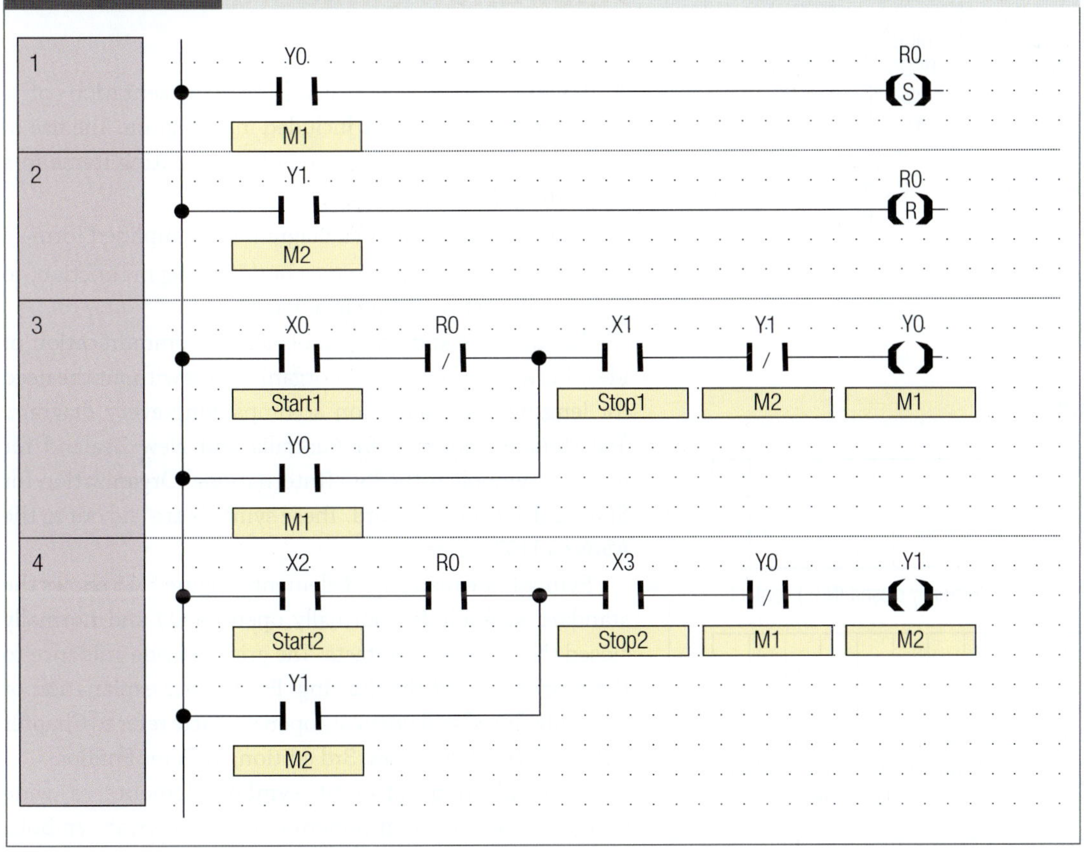

deliberate similarity to circuit diagrams so that electricians familiar with circuit diagrams could easily transition to programming a PLC. Two vertical lines represent the power rails (active and neutral) of a circuit diagram. Components are programmed between the 'rungs'. Symbols that represent devices such as switches are called inputs and kept to the left of the ladder. Symbols that represent load devices such as relays are called outputs and are kept to the far right of the ladder. Each rung of the ladder corresponds to a set of instructions that tells the PLC what to do in response to a change in status of each device connected to the rung.

FIGURE 5.14 Single line diagram

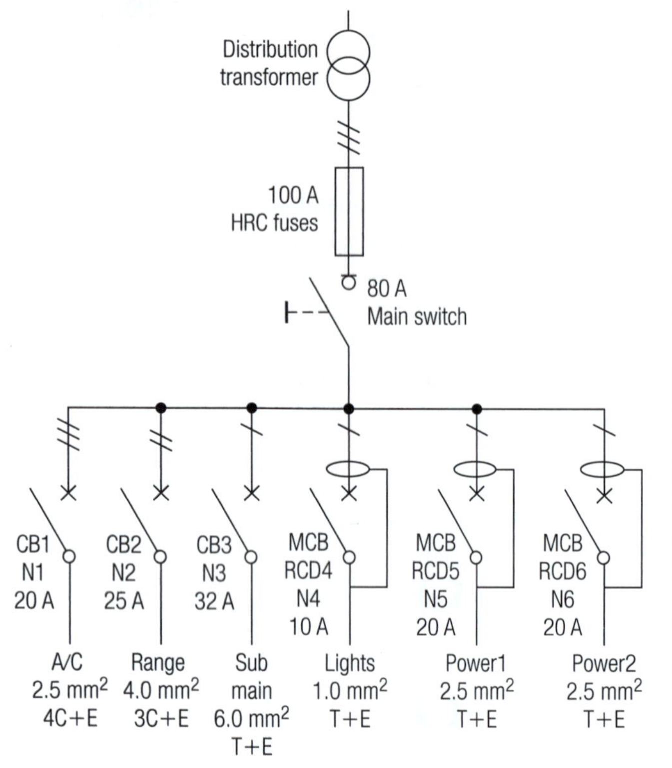

FIGURE 5.15 Contacts

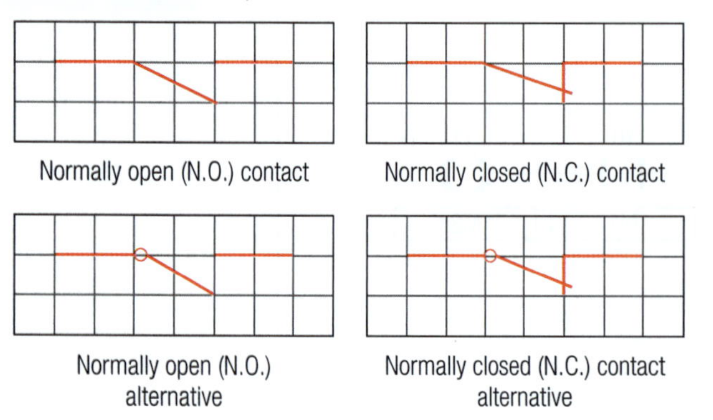

Single line diagrams

Single line diagrams show only the active conductors of a circuit; the neutral and active conductors are not shown. They are used when the position of and relationship between components is the important feature to be viewed. The reduction in the number of conductors makes interpretation of current flow through the circuit much easier. Figure 5.14 shows an example of a single line diagram. The number of lines that cross a conductor are an indication of the number of active conductors, which are known as *phases* of the circuit. One line indicates single phase, two lines indicate two phases and three lines indicate a three phase circuit.

These diagrams are commonly used to illustrate the connections of transformer sub-stations, isolation and protection devices in switchboards, and layouts for complicated installations.

Australian standard electrical symbols

A drawing symbol is a graphical representation of a component or item to be included in a diagram. The use of symbols is a convenient means of representing items in a diagram as this allows:

- items to be included in a diagram in a simplified form
- items, similar in appearance but differing in function, to be easily distinguished in a diagram.

The use of standard symbols allows communication of ideas between individuals or organisations without the need for lengthy documentation accompanying every diagram. The standard adopted in Australia and New Zealand for drawing symbols is the ISO (International Organization for Standardization) standard. These symbols are laid out in the **AS/NZS 1102** series.

Symbols are made up of elements. Figure 5.15 shows the standard symbols for normally open (N.O.) and normally closed (N.C.) switch contacts. The grids provide guidance to the proportions of the drawing. For further explanation of normally closed and normally open contacts refer to Chapter 1.9 of *Electrical Principles*, 3rd edition, by Peter Phillips.

By combining different symbol elements, a wide variety of electrical components can be shown as symbols.

Figure 5.16 shows a number of types of actuators for manual operation of a switch. The actuator general symbol when added to an N.O. contact forms the standard symbol for an N.O. manually operated switch, as shown in Figure 5.17.

If the same N.O. contact is combined with a push button actuator, the symbol for a momentary push button switch is formed, such as a door bell. If an N.C. contact is combined with a mushroom head actuator the symbol for an N.C. emergency stop button is formed.

The orientation shown is not mandatory, as symbols may be turned or mirror imaged as long as their meaning is not changed. Figure 5.18 shows the standard symbols drawn in correct proportions for a circuit breaker, a fuse and a lamp.

FIGURE 5.16 Actuators

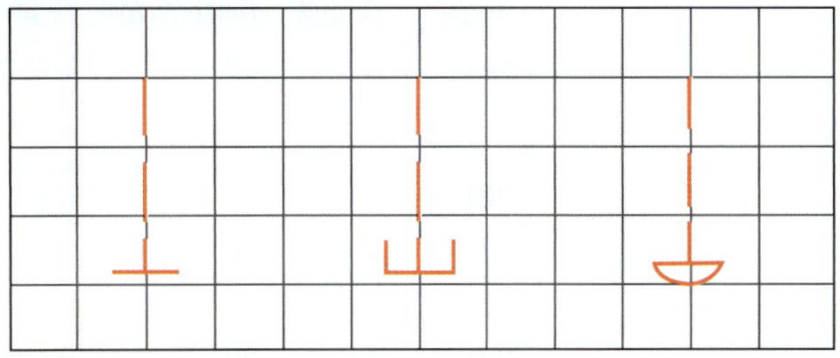

FIGURE 5.17 N.O. and N.C. switches

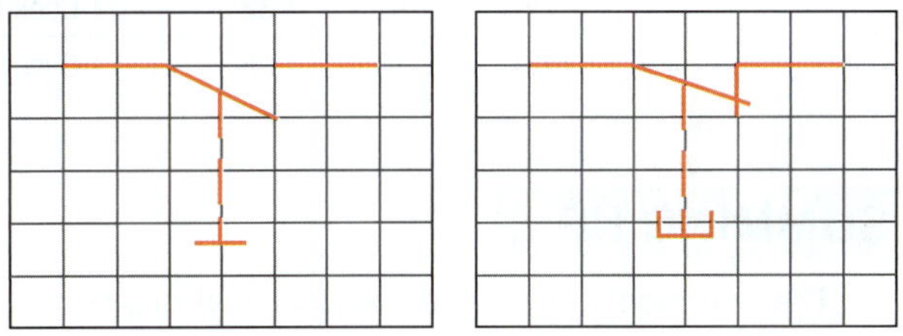

FIGURE 5.18 Standard symbols: circuit breaker, fuse and lamp

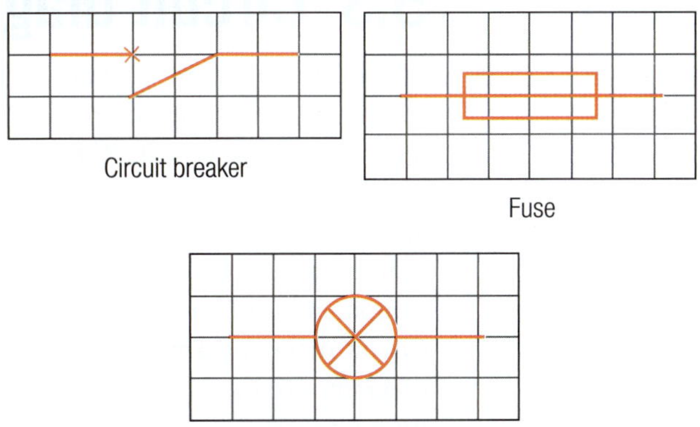

Cable schedules

Cable schedules are a written alternative to architectural electrical location diagrams. Cable routes and installation methods are the responsibility of the installing electrical contractor. Cable schedules are also used to prepare an estimate of the required materials and cost of the installation. Information provided includes the:

- switchboard at which the circuit originates
- circuit number
- circuit protection rating
- load description
- cable size and core types
- location of load
- approximate route length.

A cable schedule is developed by making a list of all electrical loads in an installation and then arranging those loads into circuits. Appendix C and Table C8 of **AS/NZS 3000:2007 Wiring Rules** provides guidance to the maximum number of points per circuit and the number of circuits required. Measurement of cable route length is estimated from architectural floor plans. Table 5.3 provides an example of a typical cable schedule.

TABLE 5.3 Cable schedule

Origin	Circuit	Current (A)	Load	Cable	Location	Route length (m)
MSB	1	20	Air-conditioner	2.5 mm^2 4C+E circ	South wall 5 m from east corner	20
MSB	2	20	HWS	2.5 mm^2 2C+E flat	South wall 7 m from east corner	25

(Continued)

TABLE 5.3 *(Continued)*

Origin	Circuit	Current (A)	Load	Cable	Location	Route length (m)
MSB	3	50	Sub-main	16 mm^2 4C+E circ	MSB to DB in kitchen southern wall cupboard next to wall oven	15
DB1	1	10	Light 1	1.0 mm^2 2C+E flat	Downstairs level	20
DB1	2	10	Light 2	1.0 mm^2 2C+E flat	Upstairs level	20
DB1	3	10	Light 3	1.0 mm^2 2C+E flat	External lights	30
DB1	4	20	Power 1	2.5 mm^2 2C+E flat	Upper level east	35
DB1	5	20	Power 2	2.5 mm^2 2C+E flat	Upper level west	20
DB1	6	20	Power 3	2.5 mm^2 2C+E flat	Lower level east + kitchen	20
DB1	7	20	Power 4	2.5 mm^2 2C+E flat	Lower level west + kitchen	20
DB1	8	20	Pool area	2.5 mm^2 2C+E flat	Pool area	35
DB1	9	32	Cook top	6.0 mm^2 2C+E circ	South wall kitchen	7
DB1	10	25	Oven	2.5 mm^2 2C+E flat	South wall kitchen	7

SUMMING UP...

- The commonly used types of electrical diagrams include block, circuit, wiring, architectural location, ladder and single line diagrams.
- Standard electrical drawing symbols are drawn to the AS/NZS 1102 series.
- Cable schedules are a written method of determining the location of electrical equipment in an electrical installation. Uses include determining cable routes, preparing cost estimates and material lists.

5.3 Circuit diagrams

Electrical equipment is made up of a number of components and conductors to connect those components together correctly. Block diagrams only provide basic information to the overall operation of the circuit. Wiring diagrams provide very detailed information for the construction or installation of the equipment, which for complicated items of equipment can be difficult to read and understand. A circuit diagram provides enough detail to describe the exact operation, but is written in an abbreviated form that reduces clutter, making it possible to quickly see how circuit components interact in their operation. This also makes them very useful when fault finding. Circuit diagrams are sometimes referred to as 'schematic diagrams'. For consistency, all circuit diagrams contain symbols representing components, or items of equipment, interconnected by lines representing conductors.

Conventions used in circuit diagrams

Electrical circuits function in a sequential order. In other words, if switch one is closed current will flow through any load controlled by that switch, resulting in lights turning on or motors starting, etc. Circuit diagrams are drawn to capture that sequence and flow. In a D.C. circuit, conventional current flow is assumed; that is, the current will flow from positive to negative.

There are two ways of laying out a circuit diagram to do this: horizontal and vertical orientations. Figure 5.19 shows a horizontally orientated circuit diagram, in which the current/energy flow is from left to right and the sequence of operation is from top to bottom.

Even though it is not mandatory, for best practice, general appearance and consistency, all like symbols where possible should be drawn in the same orientation and alignment. Switches and contacts are kept to the left and load devices such as lamps, coils and motors are aligned to the right.

FIGURE 5.19 Circuit diagram horizontal orientation

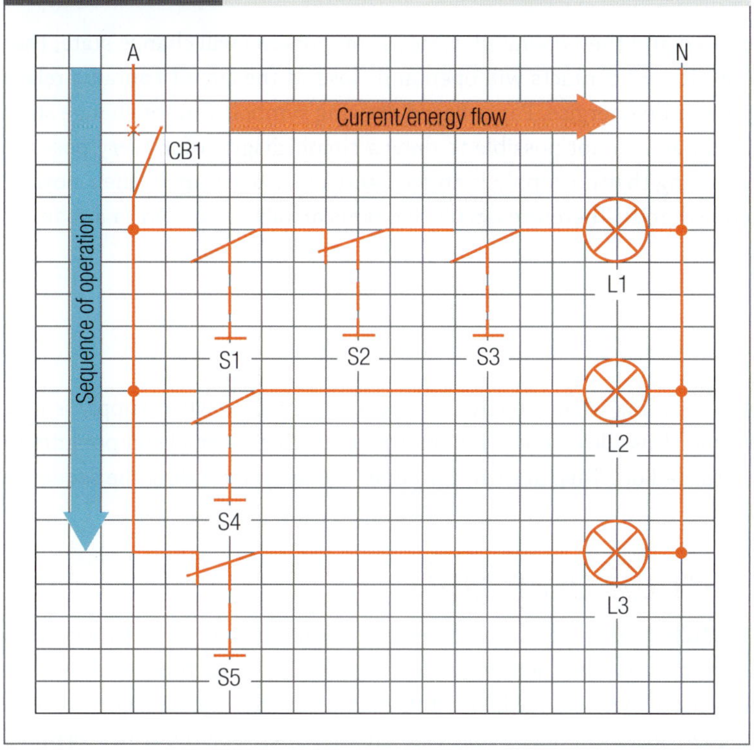

FIGURE 5.20 Circuit diagram vertical orientation

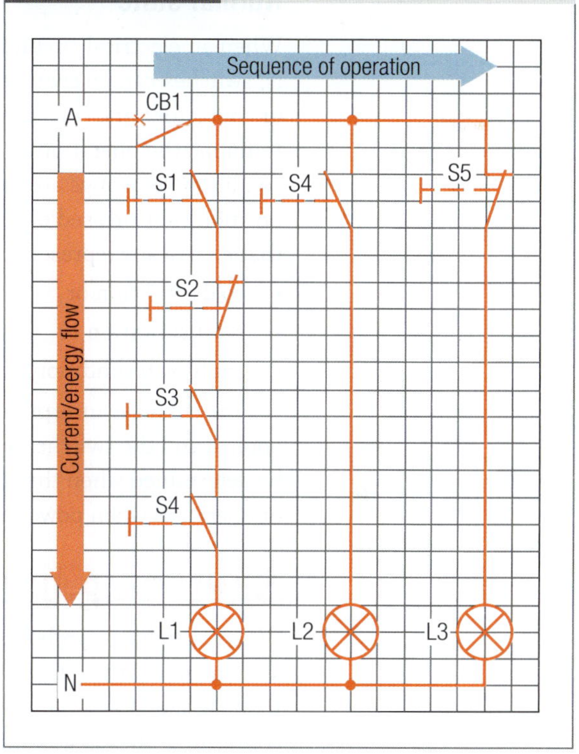

Figure 5.20 shows a horizontally orientated circuit diagram. Here the current/energy flow is from top to bottom and the sequence of operation is from left to right. Switches and contacts are aligned across the top of the drawing and load devices such as lamps, coils and motors are aligned at the bottom.

Conductors

The **AS/NZS 1102** series also prescribes methods of drawing conductors and terminations. Figure 5.21 shows examples of the correct method of drawing typical junctions and connections. If two conductors are shown intersected at a 'T', this is considered a connection. Another is to include the

FIGURE 5.21 Conductor conventions

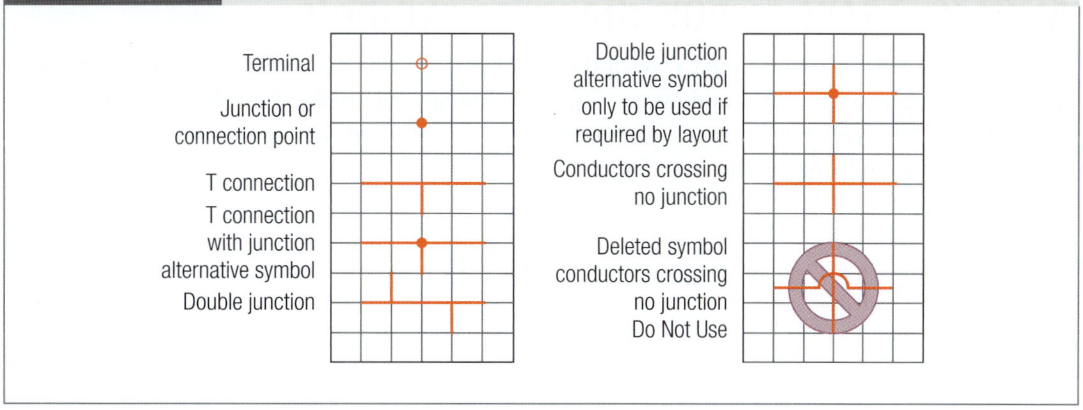

use of a dot to indicate the joint at the 'T' intersection. However, two lines drawn crossing with no dot indicate that no connection is made – the lines simply cross. Two conductors crossing with a semi-circular loop jumping one of the conductors is no longer used to show two conductors crossing but no connection.

Normal state

When an electrical circuit is energised and then operated, various components will change state; that is, switches can be switched on or off, and contacts will open and close. If the circuit contains relays or contactors, the coils can energise or de-energise. Paper based drawing symbols cannot change state to mimic the operation of the circuit as it is not possible to draw a circuit diagram for every possible condition. Drawings are made assuming that no supply is applied and all switches are in their normal position (not pressed or operated by hand). Therefore circuit diagrams are always drawn in a manner that electrically represents the:

- cold condition or
- reset condition or
- shutdown condition.

The circuit is interpreted by anticipating what the circuit components will do when supply is first applied and the switches are operated. This allows the state (on, off, open, closed) of all components to be determined when the circuit is turned off. This, in turn, simplifies the task of deciding the sequence of operation when power is applied.

Sketching basic circuit diagrams

When sketching symbols, the **AS/NZS 1102** series recommends that symbols are to be drawn to a size convenient for comprehension, and to give them suitable sizes relative to each other. A grid or module size M = 2.5 mm has been used. Symbols are laid out in such a way that the distance between connecting lines is a multiple of the chosen module to provide enough space for the usual terminal designations.

The sizes of symbols relative to one another may be changed to suit the circumstances of a given diagram or application. Also, if a symbol is used to qualify another, its size is often reduced. However, the relative proportions of the individual symbols should be retained when they are reduced or enlarged. In all cases, whether the size is enlarged or reduced, or dimensions modified, the thickness of the original line should be maintained without scaling.

Although there is no 'correct' size for a symbol, often because the diagram is to be drawn freehand, the correct shape or form of a symbol is important. It is also important to maintain the same size symbols for the same types of devices and equipment throughout a diagram. When freehand sketching a circuit diagram it is common to use 5.0 mm grid paper as an aid to developing accuracy; this means the symbols will be twice their normal size.

AS/NZS 1102 circuit diagram symbols

In the previous section we examined a few of the basic symbols, which included conductors, junctions, switches, circuit breakers, fuses and lamps. There are literally hundreds of symbols with thousands of variations. This section will expand your knowledge of the more commonly used symbols in the electrotechnology industry, as shown in Figure 5.22.

FIGURE 5.22 Circuit diagram symbols

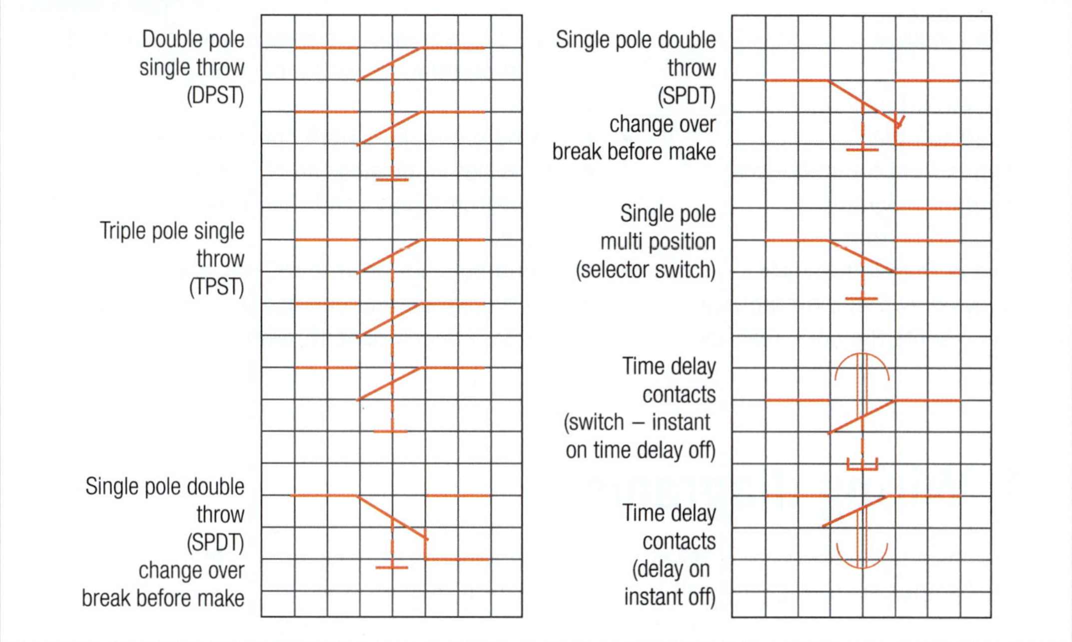

Break before make

The majority of contacts use what is known as 'break before make' operation. When the contacts change from their normal state, the normally closed contacts will open before the normally open contacts, so for a fraction of a second all the contacts are open. This prevents any components controlled by the change-over switch being energised at the same time.

Connecting equipment using circuit diagrams

Simple electrical circuits may be wired directly from a circuit diagram without converting the drawing to a wiring diagram. The key is to identify from the circuit diagram which terminal of a component a conductor starts, and which terminal of a component the conductor ends. A conductor may connect several terminals together. Wire numbers can be used to identify a location in a circuit and can assist in connecting components. From the example provided in Figure 5.23, it can be seen that wire 1 starts at the active terminal and ends at the top or line side of CB1. Wire number 2 starts at the bottom or load side of CB1 and then 'loops' to S1, S3 and S4. Conductors can be installed by using the numbers as a point-to-point guide. Conductor installation can be performed by either a number sequence, or line by line. To avoid errors, each conductor should be marked off on the circuit diagram to indicate it has been connected.

FIGURE 5.23 Wire numbers

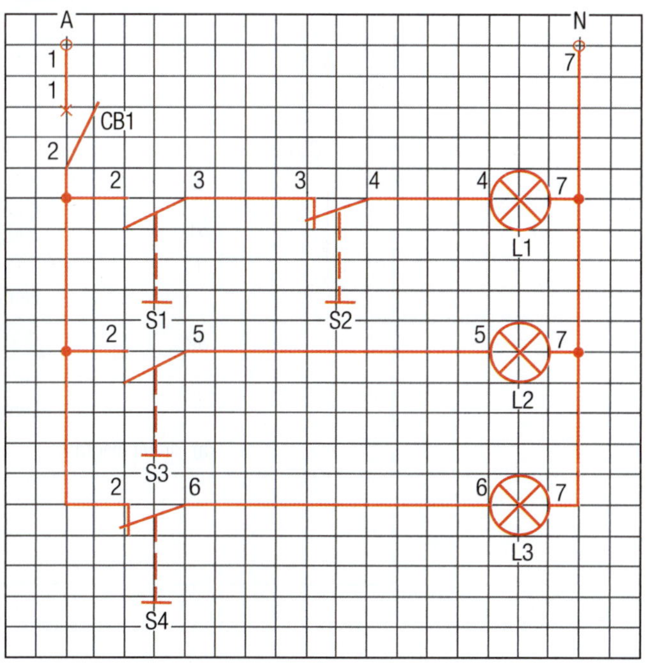

SUMMING UP...

- Conventional current flow (positive to negative) is assumed.
- In horizontally orientated circuit diagrams:
  - current flow is from left to right
  - sequence of operation is from top to bottom.
- In vertically orientated circuit diagrams:
  - current flow is from top to bottom
  - sequence of operation is from left to right.
- Circuit diagrams are always drawn in their 'normal state' that electrically represents the cold, reset or shutdown condition.
- Circuit diagram symbols should have consistent size and orientation, where possible, throughout the drawing.
- Break-before-make switch contacts open the N.C. contacts before closing the N.O. contacts, and are more commonly used than make-before-break contacts.
- Wire numbers are used to reference a location within a circuit diagram and can assist in the installation of conductors to component terminals.

5.4 Wiring diagrams

When wiring or installing complicated electrical equipment, wiring diagrams are preferred over circuit diagrams because of the higher detail in the information they provide. The increased level of detail in a wiring diagram shows the placement of all components, the routes taken by conductors and the location of the terminal where each and every connection is made. Figure 5.24 shows a simple circuit diagram for the control of two lights by two switches and two wiring diagrams, one being 'loop from the light', and the other 'loop from the switch' method of installation. From the circuit diagram alone there is no indication of which of the two methods is used.

FIGURE 5.24 Wiring diagram showing different looping methods

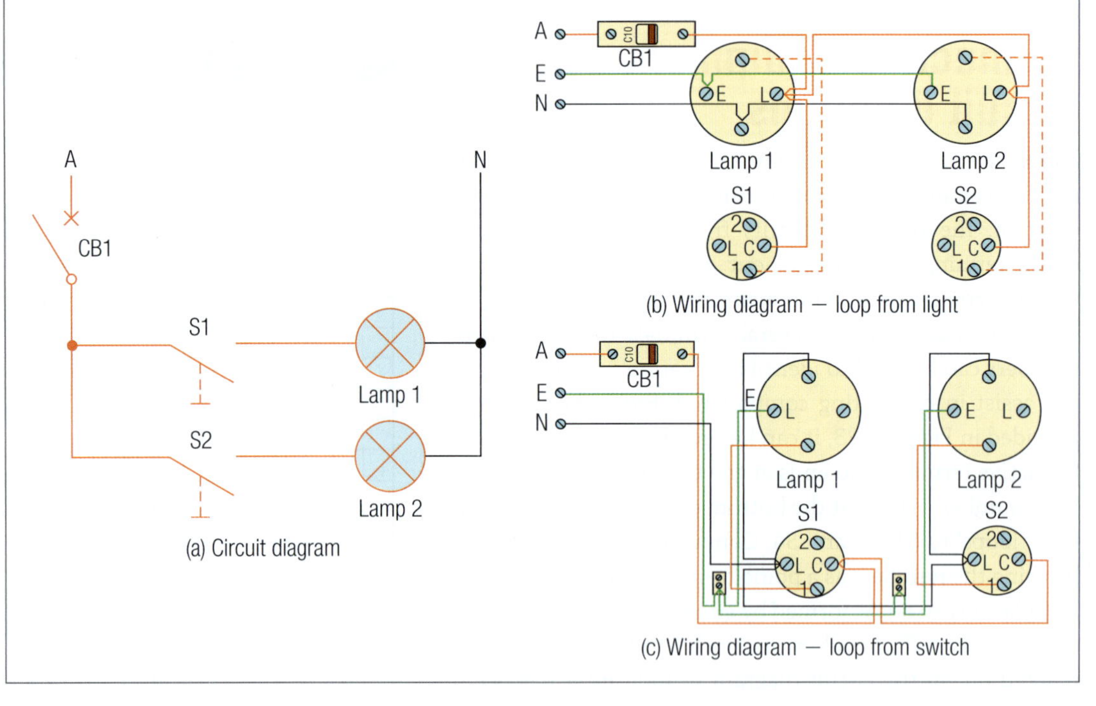

Wiring diagram conventions

Wiring diagrams are not necessarily drawn to scale, but components should be in proportion and positioned true to real life. There is no requirement to maintain any of the conventions for direction of current flow or order of operation. Unlike circuit diagrams, mid conductor connections are not used in wiring diagrams. Figure 5.25 shows the correct method to indicate a connection in a wiring diagram.

When laying out a wiring diagram, attention should be given to conserving cable. When conductors are looped from one terminal to the next, the route of the conductor should minimise the length of cable required, as shown in Figure 5.26.

FIGURE 5.25 Wiring diagram connections

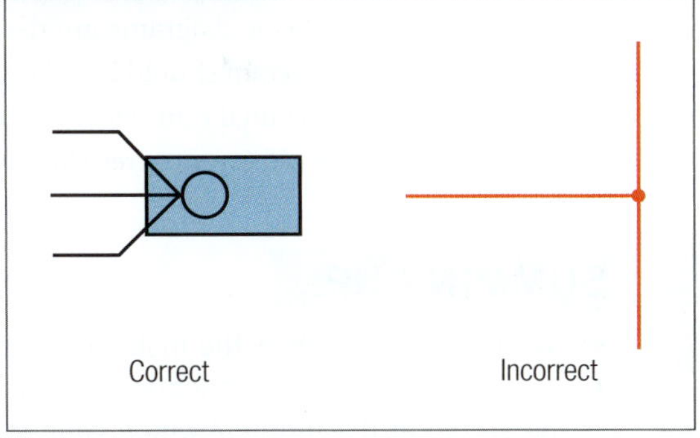

FIGURE 5.26 Wiring diagram location of terminations

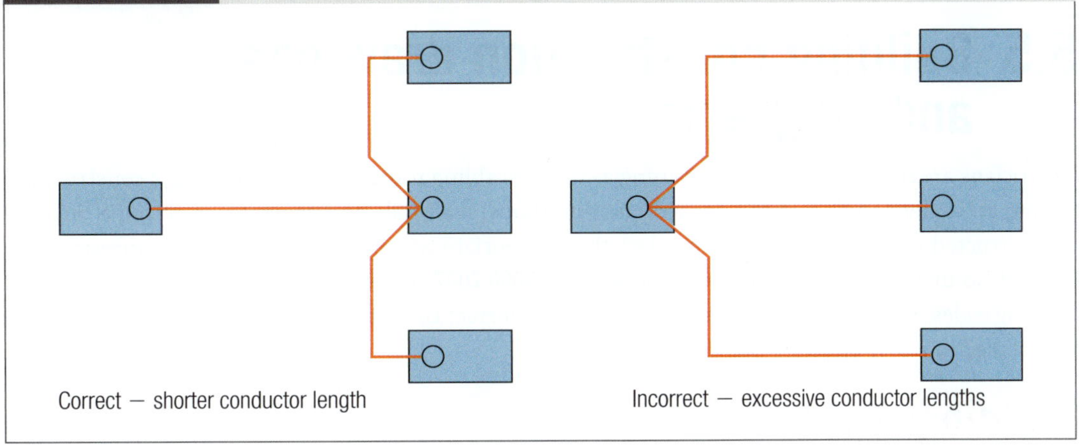

Common symbols used in wiring diagrams

Wiring diagram symbols are a combination of blocks representing the outline of components, and circuit diagram symbols to show what the block represents. Terminals are assigned numbers or letters to identify them, as shown in Figure 5.27. Wire numbers can also be included to assist in connection.

FIGURE 5.27 Wiring diagram terminal numbering

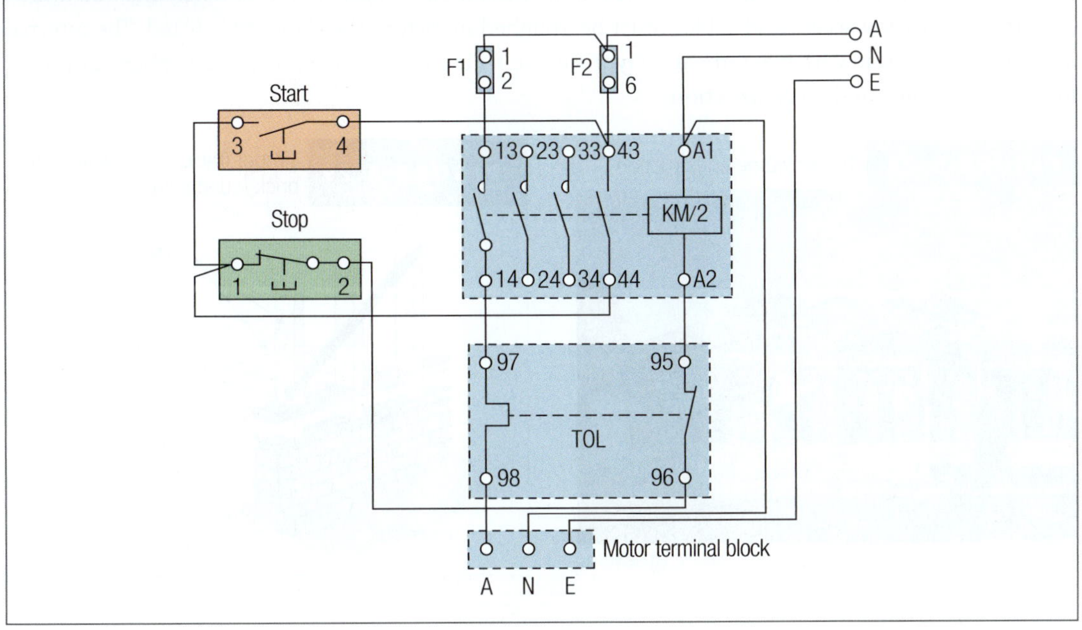

Connecting equipment using wiring diagrams

Wiring diagrams are designed to be a guide to assist in the connection of electrical equipment. The diagram should be followed exactly, paying particular attention to the route of the conductors and the terminal numbers to which conductors are connected. As when connecting from a circuit diagram, to avoid errors wires should be marked off to indicate their installation.

SUMMING UP...

- Wiring diagrams show the installation of conductors.
- In a wiring diagram, connections or joints must be shown at the terminal which conductors are connected to.
- The layout of a wiring diagram should minimise the length of conductors.
- Components should be drawn to show their true-to-life physical location.

5.5 Building construction drawings and diagrams

Electricians are responsible for determining a suitable cabling route through a building's construction. Having an understanding of the construction methods, materials and sequence in which a building is constructed enables electricians to select the appropriate wiring system, cables and accessories to be used, so installation complies with the **AS/NZS 3000:2007** series and Australian and New Zealand building codes. Electricians must also be aware of the correct timing in which the electrical services can be installed in a cost-effective manner.

Building types

The type of building construction must suit the environment it is to be located in and be suitable for its purpose. The factors that can influence house construction methods include climatic conditions, soil conditions, local, state and other building regulations, environmental and heritage regulations, cost, and services such as water, sewerage and electricity.

Timber frame

Timber frame construction employs a single timber frame, which forms the internal and external walls; that is, there is no cavity between the interior and exterior claddings. Therefore holes must be drilled in timbers for wiring access and cables must be 'roughed in' before the cladding is fitted. The internal walls are typically clad with plasterboard (Gyprock, etc.). Figure 5.28 shows a typical timber frame on a concrete slab home under construction.

FIGURE 5.28 Timber frame construction

Source: Ed Hawkins

FIGURE 5.29 Timber frame construction on brick foundations

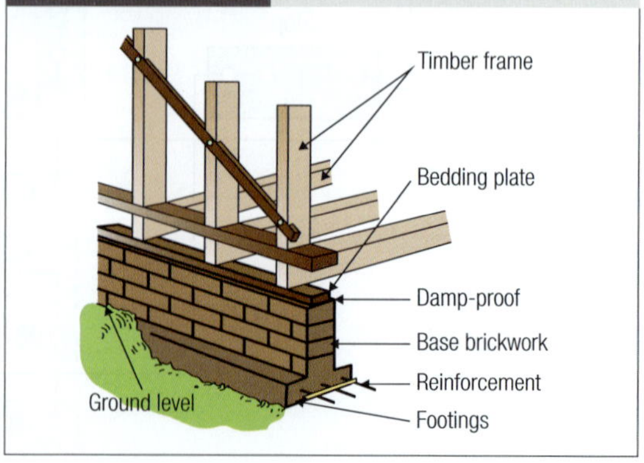

The outside of the frame is covered with some type of waterproof cladding such as weatherboards (timber), brick tile sheets, treated metal (aluminium or steel), fibre cement (sheets or planks) or PVC (sheeting or planks). Where the slope of the land does not permit the use of a concrete slab, the timber frame is supported on brick foundations, as shown in Figure 5.29.

FIGURE 5.30 Brick veneer construction

Source: Ed Hawkins

Brick veneer

Brick veneer, as shown in Figure 5.30, is now one of the most popular methods of construction. The construction has a non-load-bearing outer brick 'skin' and a load-bearing timber or steel frame, separated by a cavity.

Figure 5.31 shows a drawing of a brick veneer construction. The cavity between the outer brick skin and the frame is approximately 50 mm and serves two purposes:

- to prevent the transfer of moisture from the outer wall to the inner wall
- to allow for movement between the timber frame and the brickwork.

The outer brick wall is stabilised by tying it with wall ties embedded in the mortar to the inner frame, which in turn is anchored to the slab base. The cavity between the brickwork and inner lining can be used as a route for electrical wiring. The cavity provides sufficient mechanical protection for cables as they are considered 'not likely to be disturbed'. Thermoplastic sheathed (TPS) cables are commonly installed in this type of location without the need for further protection by enclosures such as conduit. Care should be taken during installation to prevent cables rubbing against the wall ties,

FIGURE 5.31 Timber frame brick veneer construction

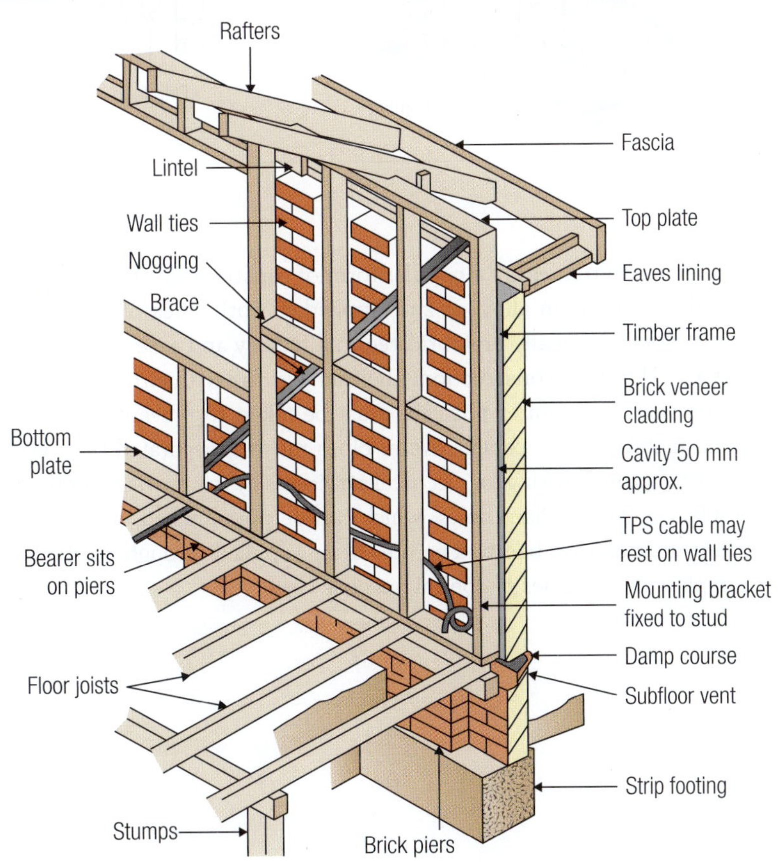

as this can tear the cable insulation and provide a path for moisture to flow between the brickwork and electrical accessories mounted on the inner wall. A drip loop, as shown in Figure 5.32, is used to prevent this.

Double brick

Double brick construction has two brick walls separated by a cavity of approximately 50 mm. The two walls are held together with wall ties of either steel strip or wire embedded in the mortar joints.

The inner skin of the brickwork is structural, i.e. the load-bearing part, while the outer skin provides protection from the weather as well as being designed for visual effect. The function of the wall cavity is to:

- provide a moisture barrier from the outside wall to the inside wall (most masonry is porous). Drip loops as shown in Figure 5.32 should be used to prevent moisture breaching this barrier
- provide air circulation to assist in drying out any moisture that has penetrated the external wall
- provide thermal and acoustic insulation.

The benefits of double brick construction are that the walls are almost finished once the bricks have been laid. The walls could be cement rendered on the interior and/or plasterboard can be fixed over the bricks. Double brick is a more expensive construction option.

FIGURE 5.32 Double brick veneer construction and drip loop

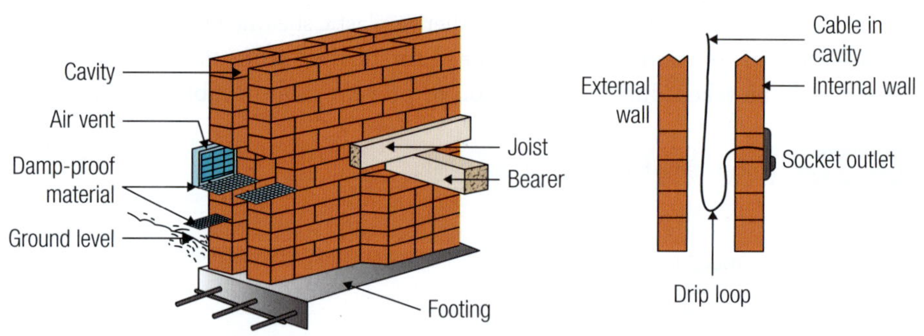

Metal frame

An alternative to timber, metal frames are a popular form of construction. The steel frame is protected from corrosion by a zinc aluminium alloy, or zinc applied to both sides of the steel by the hot-dip process. Frames and trusses are usually prefabricated in a factory and assembled on-site. Figure 5.33 shows an example of a steel framed construction. The frame is clad with either brick or steel cladding.

Steel frames are resistant to termites but care must be taken to prevent galvanic corrosion from contact with dissimilar metals such as copper. Water pipes must be separated from wall studs and top plates by grommets. Installation of wiring for electrical services is by the provision of pre-punched service holes which can be plain or flared. Plain holes, which can be pre-punched, hand punched or cut out with a hole saw, require grommets to be fitted before wiring is pulled through, as shown in Figure 5.34. Cables are secured in place by cable ties or special purpose clips.

FIGURE 5.33 Steel frame construction

Source: shutterstock.com/Lev Kropotov

FIGURE 5.34 Steel frame service holes and electrical grommets

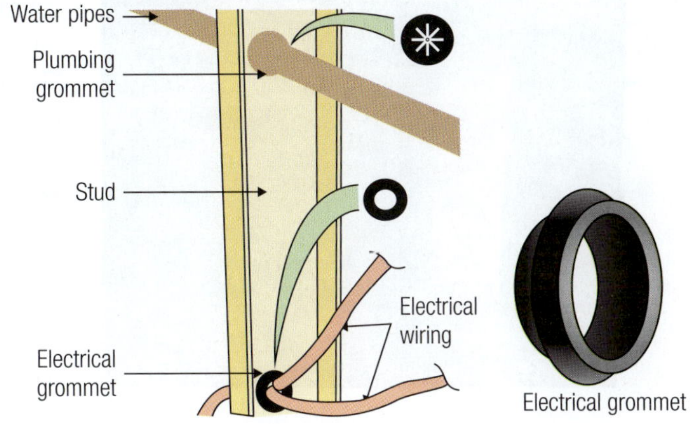

Brackets and service fittings are screwed to the frame with self-drilling screws, as shown in Figure 5.35. Holes which are cut into the steel frame do not require treatment to prevent corrosion, as a natural phenomenon known as sacrificial protection prevents corrosion of any steel exposed at cut edges and penetrations.

FIGURE 5.35 Steel frame construction wall brackets

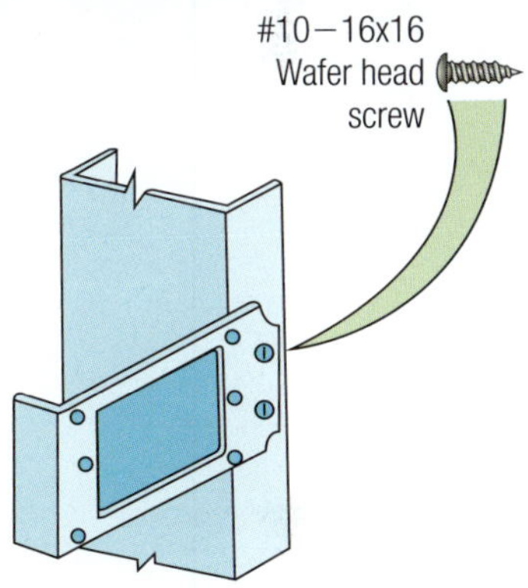

Parts of a construction

Footings

The type and size of footings used in a construction will depend on the:

- weight and type of construction
- class of the site, i.e. the type of soil, clay, sand or rock.

Double brick constructions will therefore require a more substantial footing than either a brick veneer timber frame or timber frame with cladding construction.

Stump footings

Stump footings are used to support timber framed houses, for which they are currently the most cost-effective. Materials commonly used for stumps are timber, concrete or steel. The stump sits on top of a concrete or treated timber pad to distribute the weight of the structure over a larger area.

Strip footings

Substantial footing supports of reinforced concrete are needed for buildings that have external or internal masonry walls – solid brick, cavity wall, brick veneer or concrete block. Reinforced concrete footings are commonly called 'strip footings' and are usually continuous around the entire perimeter of the building.

Pier and beam footings

When load-bearing soil cannot be located within 1.5 metres of the surface, or the site is on a steep slope, pier and beam construction is usually considered. The depth at which this load-bearing soil is located will determine the structural footing design and the cost of the footing construction. Holes are

FIGURE 5.36 Brick pier footings

Source: Ed Hawkins

drilled to the required depth and filled with reinforced concrete. A concrete slab or strip footing can then be poured on top of the pier.

Alternatively, an above-ground construction is achieved by building brick piers on top of the concrete as shown in Figure 5.36. The sub-floor area is then supported by steel or timber beams, depending on the span.

Slab footings

The stiffened raft slab is the simplest and most common slab construction (see Figure 5.37). Concrete beams support the slab floor. Typically, a stiffened raft slab consists of a concrete slab, edge beams, internal beams and steel reinforcement. The concrete is monolithic, that is, poured in one operation.

FIGURE 5.37 Raft slab footings

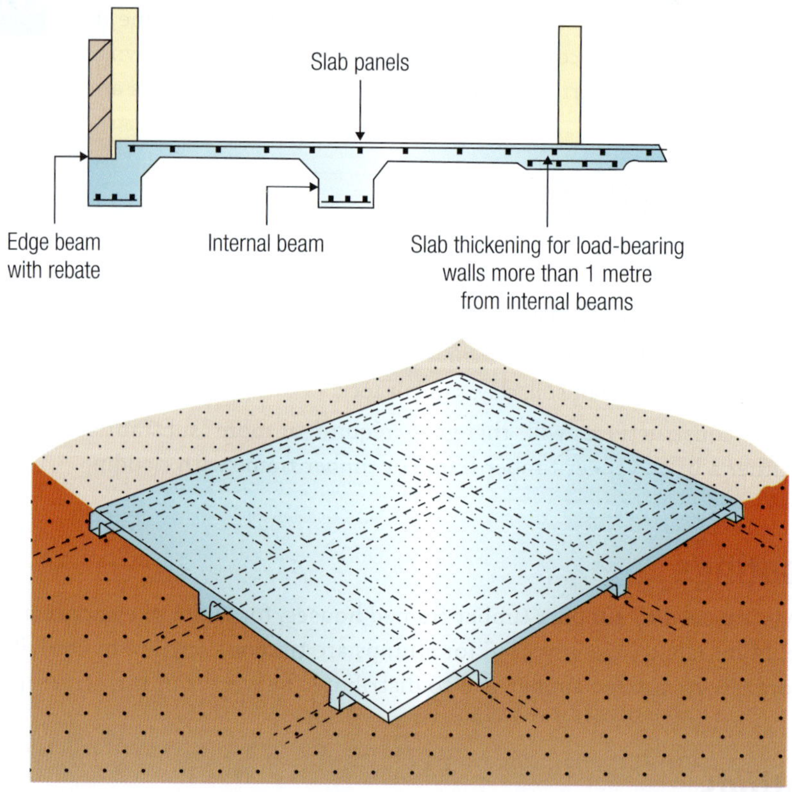

Waffle raft slab constructions, as shown in Figure 5.38, are formed on top of the ground using a grid of polystyrene void forms which are laid out on a levelled area. This construction method is best suited to very reactive clay sites.

FIGURE 5.38 Waffle slab footings

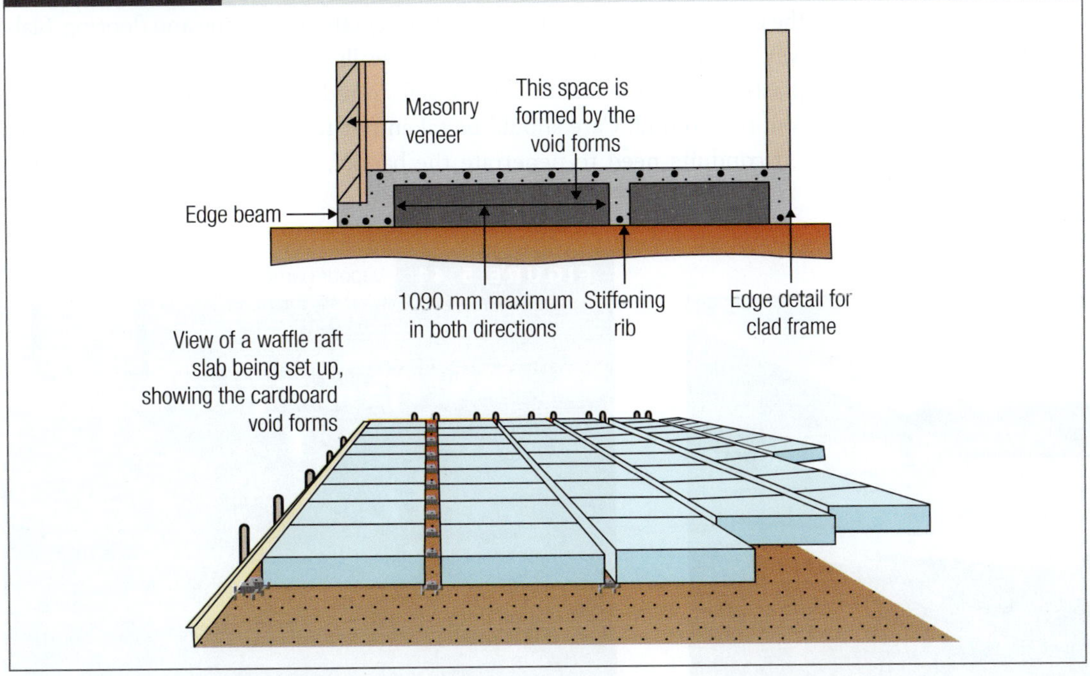

A damp course, shown in Figure 5.39, is a barrier of impervious material fitted into a wall or pier to prevent moisture rising from the ground to any part of the building. The damp course is built into base wall brickwork approximately two courses (two bricks) below the lowest timber member.

FIGURE 5.39 Damp course

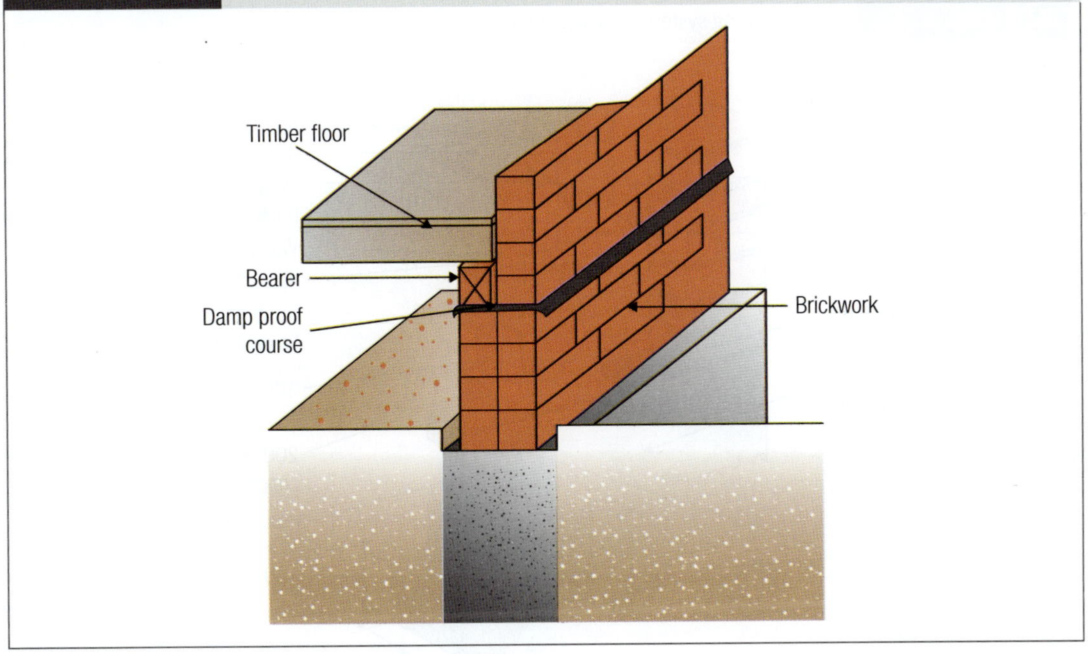

Floors

Concrete slabs poured on the ground, as shown in Figure 5.40, integrate the footing and flooring. Slabs can also be suspended above the ground on piers or load-bearing walls.

A vapour barrier is a plastic membrane laid under the slab to reduce moisture known as rising damp entering the slab from the ground. Care should be taken not to tear or puncture the vapour barrier (see Figure 5.41). If conduits need to penetrate the barrier, any tears or holes should be adequately taped.

FIGURE 5.40 Slab floors

Source: Ed Hawkins

FIGURE 5.41 Vapour barrier

Source: Ed Hawkins

Figure 5.42 shows an example of a typical raised flooring system. Construction will vary depending on the time available, the location and materials used.

FIGURE 5.42 Raised flooring system showing subfloors

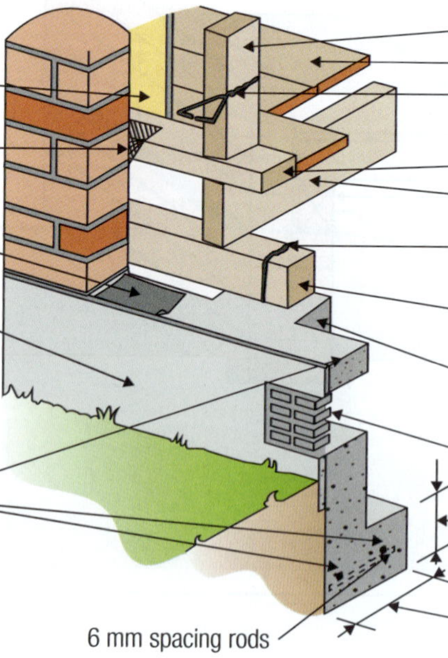

External walls

Figure 5.43 shows an example of the terms used to identify the typical components of an external wall frame. In modern open plan designs, external walls are load bearing, meaning they must support any upper floors and the roof of the building. Where sections need to be removed for windows, a lintel is installed to compensate for any loss of structural strength.

Brick veneer constructions provide a cavity between inner and outer wall linings which is suitable for cable installation. If holes for cables need to be drilled in the studs, top and bottom wall plates of timber frames, the hole must not significantly weaken the component. See **AS 1684.2-2010 Residential timber-framed construction – Non-cyclonic areas** for requirements on positioning and size of holes. In steel frames a typical hole size of 32 mm is pre-punched into frames during manufacture.

FIGURE 5.43 External walls

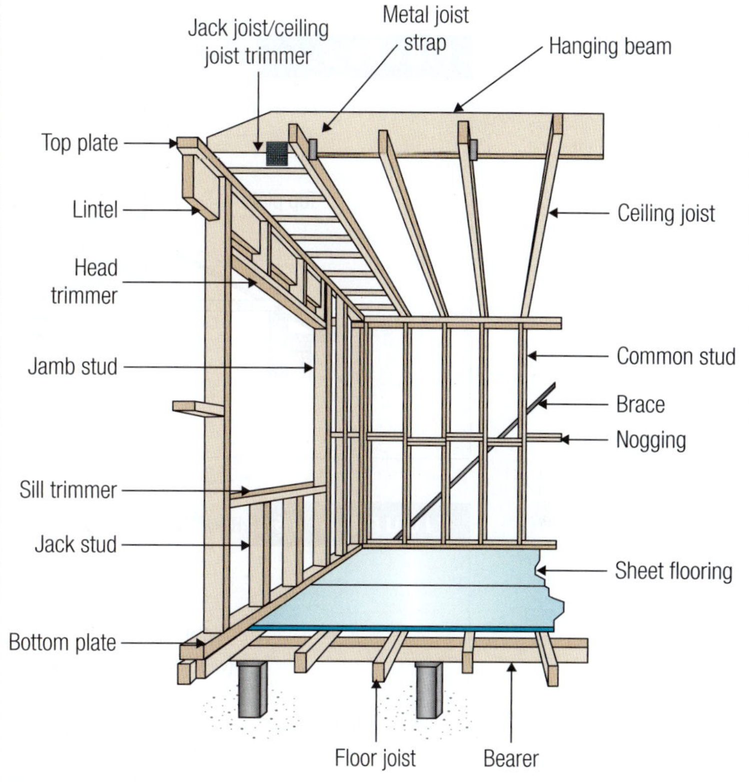

Interior walls

Interior walls can be either load bearing or non-load bearing. Constructions with larger roofing spans will require some of the internal walls to support the span. During renovations or remodelling these walls cannot be removed. Non-load bearing means they do not support any other building components; they are simply partitions between rooms. The terms used to describe internal wall components are the same as for external walls, as shown in Figure 5.43. As there is no cavity in an internal wall, holes must be drilled for cable access following the same rules as for external walls.

Roofs

Common roof styles are shown in Figure 5.44. They include the gable roof, hipped roof, hip and valley roof, hipped roof with Dutch gables and skillion roof. A roof may be a combination of any of these common styles.

A conventional roof is one that is constructed on-site. It uses ceiling joists, hanging beams, strutting beams, struts, underpurlins, collar ties, rafters and ridge boards. A conventional roof can be either coupled

FIGURE 5.44 Common roof types

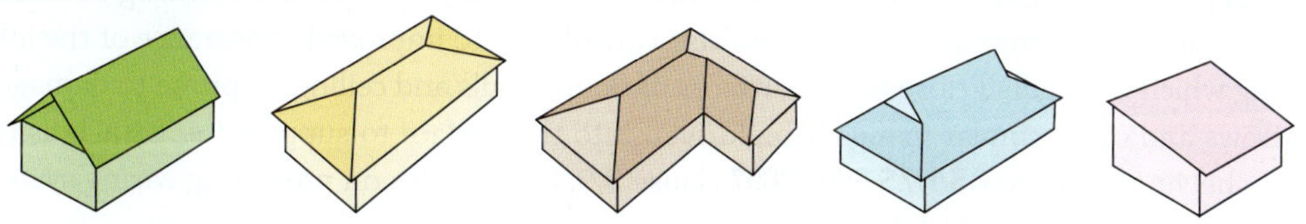

or non-coupled. A coupled roof uses a ceiling joist to couple the base of the two opposing rafters together, as shown in Figure 5.45. A non-coupled roof does not have a ceiling joist (or horizontal member).

Roof trusses are a prefabricated roof construction method used in buildings for all types of roofs. Figure 5.46 shows an example of the terms used to identify the typical components of a roof truss. They cost less than a conventional roof.

FIGURE 5.45 A conventional coupled roof

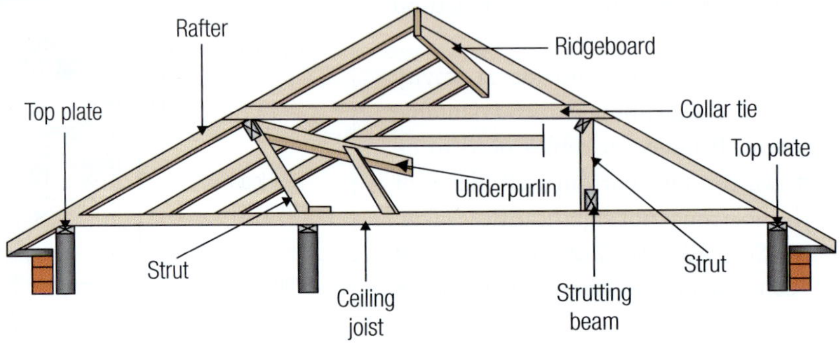

FIGURE 5.46 Roof truss components

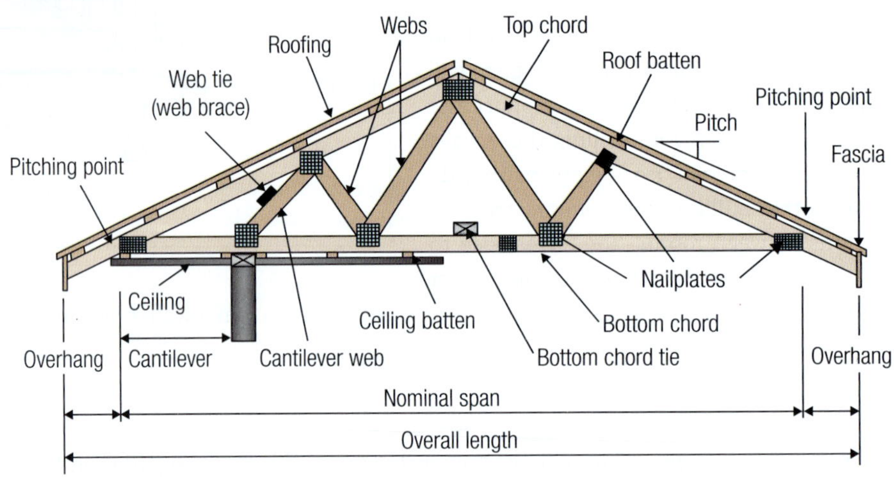

Typical cable routes

Cables must be properly protected.

The wiring system selected must be suitable for the route that a cable takes through a building. The route is normally chosen to allow ease of installation (cost-effective) and to take advantage of the mechanical protection offered by the construction of the building. Concealing the wiring within spaces such as wall cavities, ceilings and under floors also contributes to improved appearance of the job.

Wiring which is installed close to the surfaces of floors, walls and ceilings is prone to damage from nails, screws and other similar fixings being driven into the surface to support items such as picture hooks or shelving. Refer to **AS/NZS 3000:2007** clause 3.9.4.2 for rules on protecting wiring systems by methods outlined in clause 3.9.4.4.

Cables installed so that they are permitted to move freely to a point not less than 50 mm from the surface are exempt from these requirements. Compliance with **AS/NZS 3000:2007** clause 3.9.4.4 can be achieved by:

- choosing a route not identified by clause 3.9.4.2, or not securing the cable and allowing it to move freely

- providing circuits with RCD protection. In domestic installations this is mandatory for light and power circuits
- running cables fixed in these locations in earthed steel conduit or steel wire armoured cables.

Footings

AS/NZS 3000:2007 clause 3.9.7.2 requires TPI and TPS cables to be in suitable enclosure such as medium duty conduit when installed in concrete. The conduit of underground consumer's mains is required to enter the main switchboard of an installation; to do this the conduit must pass through the footing and slab of the building. Electricians must arrange for the conduit to be in place before the concrete of the footings and slab is poured. Where the conduit penetrates through the vapour barrier the membrane of the barrier must be appropriately reinstated. If termite protection mesh is installed a mesh collar is fitted to the conduit. This will also apply to any conduits leaving the building such as sub-mains to outbuildings.

Slabs

Conduits are installed in slabs to provide cable access between the two layers of steel reinforcing mesh. The first layer of mesh is laid; the conduit is then put in place and secured with tie wire. The conduit should be positioned at a depth greater than 50 mm from the finished surface of the slab to avoid the additional protection required by **AS/NZS 3000:2007** clause 3.9.4.4. The location of the ends of the conduit must be accurately positioned from the plan to prevent error. The open ends are capped to prevent cement entering. The second layer of mesh is then fitted.

Sub floors

Thermoplastic sheathed (TPS) cables supplying socket outlet circuits are typically installed in the crawl space between the underside of the floor and the ground, and are generally 'clipped' to either floor joists or bearers as shown in Figure 5.47. These cables are considered likely to be disturbed; see **AS/NZS 3000:2007** for standards applying to installation of these cables. Standards do not specify a required distance between supports, so clipping cables at a distance of approximately 300 mm to 400 mm. Figure 5.47 shows a clipped cable. Cables should not be located closer than 50 mm to the underside of the floor, to prevent damage from nails.

FIGURE 5.47 Cables clipped to floor joist

Source: Ed Hawkins

Walls

Brick veneer and double brick constructions provide a cavity in which cables may be easily installed. Internal walls or constructions where the cladding is fixed directly to the frame will require holes (not exceeding 25 mm) to be drilled in the frame to provide access. TPS cable installed in these locations is considered not likely to be disturbed and requires no further mechanical protection from an enclosure.

Cables such as flat TPS cables which are required to run on an exterior surface will require the additional protection of an enclosure such as a medium duty PVC conduit.

Due to the risk of damage from fixing screws, it is prohibited to install cables in the space between the exterior cladding such as weatherboard or Colorbond® sheeting and the supporting wall stud as shown in Figure 3.6 of **AS/NZS 3000:2007**.

Roofs

Cables in roof spaces of domestic installations are secured in place with cable clips in timber frames and cable ties on steel frames. Cables are typically run along ceiling joists, beams or the bottom chords of trusses. Within a ceiling space the area within 2 m of the 'manhole' entry to a roof space or anywhere that has a height of 0.6 m is considered likely to be disturbed. Cables that are likely to be disturbed are required to be fixed to prevent undue sagging. Within 2 m of a manhole, cables are typically clipped at intervals of 300–400 mm to prevent people in the roof from dislodging them.

Away from the area around the manhole, cables are secured at increased intervals sufficient to hold them to the timber or steel frame that supports them. In some cases additional supports must be installed to prevent undue sagging of the cables. In non-domestic installations, cables are typically supported on cable trays, racks, ladders or catenary wires, depending on the size and number of cables. The area known as the *eaves* in most types of roof construction is less than 600 mm in height and is therefore considered not likely to be disturbed. Cables laid flat on a continuous surface within an eave do not require fixing.

As with floors and walls, cables in a ceiling that are fixed in place or run through holes in structural members within 50 mm of the surface of the lining (Gyprock) will require additional protection as previously mentioned. Figures 3.4 and 3.7 of **AS/NZS 3000:2007** provide guidance as to locations within ceiling spaces that require additional protection.

Cables that are likely to be disturbed must be properly clipped in place.

Due to the risk of damage from fixing screws, it is prohibited to install cables in the space between the roof tiles or sheeting and the supporting rafter or top chord, as shown in Figure 3.6 of **AS/NZS 3000:2007**.

Construction sequence

There is no set standard for stages or sequence of construction. There are too many variables to consider, such as the type of construction, location and original condition of the site. Constructions need to be individually planned so that each trade works in cooperation with each other and components of the building are installed in a timely and efficient process. Stages are normally aligned with progress payments to the builder.

FIGURE 5.48 Construction stage 1 – Base

Source: Ed Hawkins

Stage 1 – Base

Base construction includes:

- Setting out the site:
  - Site is marked out to provide enough information to allow excavation works. The site is cleared of any rubbish, vegetation and topsoil. The site is then cut and filled to provide level 'platforms'. Any retaining walls that are required to support the footings are constructed.
  - Temporary fencing, a portable toilet, a temporary builder's power supply is erected. The builder's supply can be a standard meter box on a pole or positioned in its permanent location. Silt barriers of either hay bales or a geotextile fence are installed to prevent silt run-off.
  - Building set out – a detailed marking out and installation of site drainage which includes the installation of under slab ducts, pipes and conduits.
- Slab construction:
  - Base footings such as concrete piers are poured.
  - Sand/gravel slab base, termite mesh and waterproof membrane is installed.
  - Slab formwork is erected, waffle pods and reinforcement installed and slab is poured.

Figure 5.48 shows a typical cottage construction site up to the base stage. Early installation of the main switchboard removes the need to install a temporary builder's supply.

Stage 2 – Frame

Frame construction includes:

- standing of prefabricated wall frames
- erection of prefabricated roof trusses or construction of conventional type roofs on-site
- electrical first fixing, also called 'roughing in', which involves the installation of wall brackets, electrical cables and pipe work in wall cavities and ceilings
- wall and roof sarking applied or 'wrapped' to the frame and roof. This acts as a physical barrier to weather during the bricklaying, improves the environmental performance of the building by preventing drafts and allows moisture vapour to escape from the insulation after construction.

Figure 5.49 shows a typical cottage construction up to the frame stage.

FIGURE 5.49 Construction stage 2 – Frame

Source: Ed Hawkins

Stage 3 – Lockup

Work conducted to lockup includes:

- roofing is installed, either tiles or roof steel
- window frames are attached to the frame before the brickwork starts and the brickwork is then built against the frames and tied in
- exterior walls – brickwork is completed. Can also be cinder block for a rendered construction or cladding
- fascia boards, soffit boards (if construction has eaves) and guttering are installed. Downpipes are usually left until later
- installation of doors.

Figure 5.50 shows a typical cottage construction up to the lockup stage.

FIGURE 5.50 Construction stage 3 – Lockup

Source: Ed Hawkins

Stage 4 – Fixing

Work conducted during fixing includes:

- installing roof and wall insulation
- air-conditioning condenser unit and ducting is mounted in the roof space
- installation of Gyprock linings for ceilings, walls and cornices
- wet areas are waterproofed in preparation for wall and floor tiling
- carpentry – kitchen cupboards and bathroom vanities, doors, skirting boards and architraves and internal window sills are installed
- plumbing – baths, basins, sinks, shower trays and laundry troughs are installed and connected to the drainage system
- tiling – both wall and floor tiling are completed
- painting
- plumbing – taps, mixers and toilet bowls are usually left off until after tiling
- electrical second fixing – also known as 'fit out'. Lights, socket outlets, communications, security and electrical accessories fitted. Main switchboard is completed and installation is compliance verified and tested
- metering installed and supply connected
- carpet and other floor coverings laid.

Figure 5.51 shows a typical cottage construction up to the fixing stage.

FIGURE 5.51 Construction stage 4 – Fixing

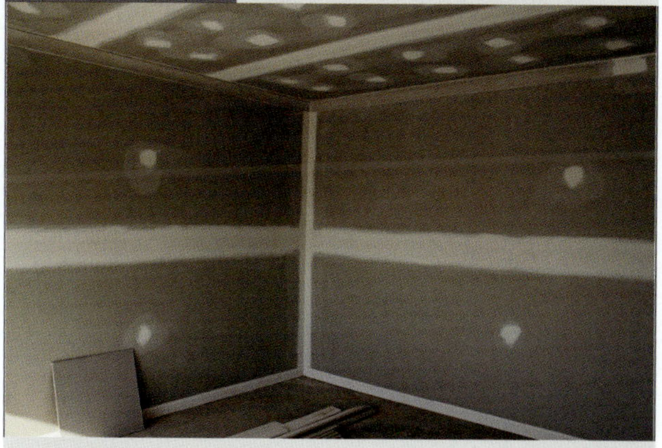

Source: Ed Hawkins

First and second fixings

An electrician must be familiar with the construction sequence and the structural details so that the installation of wiring can take the greatest advantage of the mechanical protection provided by the construction. Wiring can be concealed as much as possible and the time of the installation will be greatly reduced. It is vital to install any required electrical or communications cables, accessories and equipment at an appropriate point in the progress of construction of the building. A typical sequence of installation would include at the:

- base stage of construction:
  - setting out – a temporary builder's supply may be required to be installed in either a temporary or permanent position
  - footings – once formed up and reinforcement is in place, conduits for cables entering or exiting the footings such as consumer's mains must be secured in place before the concrete pour
  - slab – conduits for any cables to be run within the slab are best laid in between the first and second layers of reinforcement mesh
- frame stage of construction – first electrical fixing or 'rough in'
- fixing stage of construction:
  - second electrical fixing or 'fit out'
  - main switchboard
  - compliance verification and testing
  - metering installed and supply connected.

An appreciation of the role and timing of other tradespersons on the job can improve communication and cooperation. This is important to be able to determine when it is appropriate to install wiring, fit accessories and appliances. Figure 5.52 shows the sequence of construction for a typical cottage and the points in the sequence where an electrician may need to become involved. The diagram also shows which trades are involved in the various areas of construction.

FIGURE 5.52 Cooperation between building trades

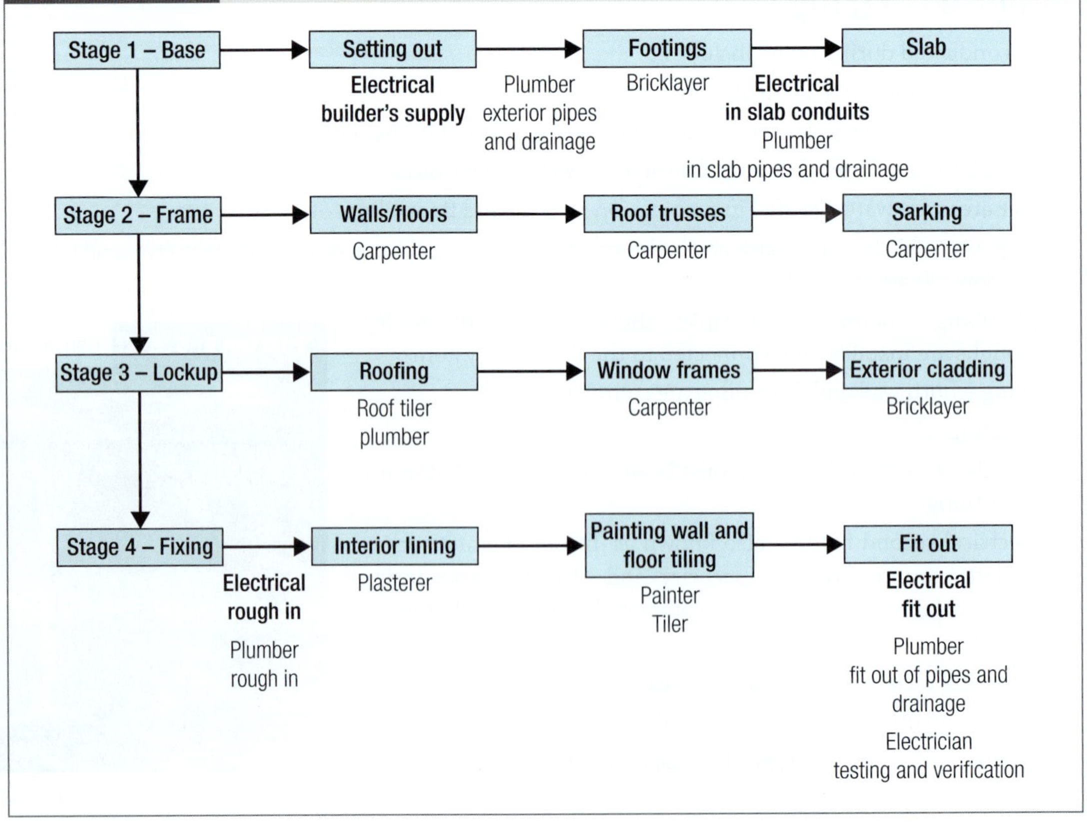

SUMMING UP...

- Common construction methods include:
  - timber or steel frame cladding veneer
  - timber or steel frame brick veneer
  - double brick.
- The major parts on a construction include:
  - footing
  - sub floor
  - floor – slab or raised
  - frame
  - walls – external and internal
  - roof.
- Typical stages of construction include:
  - base
  - frame
  - lockup
  - fixing.
- Critical stages of construction for installation of electrical services include:
  - base stage of construction – temporary builder's supply, conduits for cables entering or exiting the footings and conduits for any cables to be run within the slab
  - frame stage of construction – first electrical fixing or 'rough in'
  - fixing stage of construction – second electrical fixing or 'fit out'.

REVIEW EXERCISES

Answer these questions and check your answers at the back of the book.

Review questions 5.1

1 What type of drawing is used to determine the location of services such as gas and water, and why is it important?

2 Draw the architectural floor plan location symbols for a:
   a single light point
   b twin fluorescent
   c double socket outlet
   d hot water service (HWS).

3 If a drawing has a scale of 1:20 what is the length on the job of a 55 mm distance on the drawing?

Review questions 5.2

1 List five commonly used types of electrical diagrams.

2 Explain the difference between a circuit diagram and a wiring diagram.

3 Draw the AS/NZS standard symbols for:
   a a circuit breaker
   b a normally open momentary push button
   c a normally closed emergency stop button.

Review questions 5.3

1 Should load devices be drawn to the top or bottom of a vertically orientated circuit diagram?

2 What is meant by the abbreviation DPDT?

3 Draw the AS/NZS standard symbols for:
   a a triple pole single throw N.O. switch
   b two conductors crossing but not connecting
   c N.O. time delay contact, delay close instant open.

Review questions 5.4

1 When drawing a wiring diagram, how should components be positioned in the drawing?

2 Explain the difference between how terminations are drawn in a circuit diagram and a wiring diagram?

Review questions 5.5

1 What is the difference between brick veneer and double brick construction?

2 Can cables be clipped to timbers within 50 mm of flooring without any additional protection?

3 What are the requirements for cables installed in a position that are likely to be disturbed?

4 List the typical stages of construction in which electrical and communication services should be installed.

ONLINE RESOURCES

Steel frame construction
http://tradies.truecore.com.au > sparkies

6

DOCUMENT AND APPLY MEASURES TO CONTROL WH&S RISKS

Chapter 1 provided an extensive coverage of the legal, moral and social benefits of compliance with WH&S legislation. This chapter gives practical examples using a number of case studies to demonstrate the documents and control measures used to control risks when working in the electrotechnology industry.

Chapter outline

6.1 Assessment and management of risk sources

The principles of risk assessment and management, and the processes for conducting a risk assessment are detailed in Chapter 1, Section 1.2. As an example of the sort of task you may encounter, a medium size cable drum is to be lifted from the back of a vehicle to the ground. The shape of the cable drum pushes the centre of gravity away from the body of the person attempting to lift the drum. Among electricians, back injuries are common. These are caused by the improper manual handling of objects such as cable drums.

Recording hazards and assessing the risk

Table 6.1 shows the content of a typical hazard identification document.

TABLE 6.1 Sample hazard identification document

SAMPLE MANUAL HANDLING HAZARD (RISK SOURCE) IDENTIFICATION							
Manual task	**Identified by**	**Risk factor(s), e.g.**	**Location**	**Workers at risk**	**General comments**	**Order of priority**	**Date of risk assessment**
	• reported hazard • injury/illness • consultation • inspection • observation	• awkward postures • sustained postures • repetitive movement • forces and loads • vibration • working environment • systems of work • worker characteristics					
Transfer cable drum to new location	Inspection/ observation	• awkward postures • forces and loads • crush and pinch points • splinters	Vehicle	Electrical	Heavy, poor centre of gravity	1	17/12/2013

Table 6.2 provides an example of the content of a typical risk assessment document.

TABLE 6.2 Sample risk assessment

SAMPLE MANUAL HANDLING RISK ASSESSMENT FORM			
Location/group: Electrical	Manual task: Transferring cable drum		
Date of assessment: 18/12/2013	Assessors/people involved: Phillip Chadwick		
Risk factors to consider	**Level of risk**	**Sources of risk**	**Comments**
Holding or carrying a load away from the body requires more muscular effort and places more stress through the joints than when holding the same load very close to the body.	Estimate the level of risk each risk factor poses, i.e.: • N/A • low • medium • high	Possible source(s) of risk, i.e.: • work area/layout • nature of the load • nature of item, equipment or tool • working environment • systems of work, work organisation or work practice	

(*Continued*)

TABLE 6.2 *(Continued)*

Risk factors to consider	Level of risk	Sources of risk	Comments
Actions and postures: do the actions and postures involve:			
Holding loads or arms away from trunk	Medium	• work area/layout • nature of the load	Picking up a load further away from the body can mean the handling of the object is not controlled.
Reaching upwards or load handling above shoulder height	N/A	N/A	N/A
Bending the back or neck forward or handling below mid-thigh height	Medium	• work area/layout • nature of the load	Placing the load further away from the body will tire the muscles holding the load, due to the need for more careful control over its movement.
Twisting the back, neck or upper body	Medium	• work area/layout • nature of the load	
Forces and loads: are the forces and loads handled:			
Heavy	High	• nature of the load	approx. mass > 75 kg
Bulky, large or awkward	High	• nature of the load	Poor centre of gravity
Unstable, unbalanced or unpredictable	High	• nature of the load	Poor centre of gravity
Harmful or fragile	N/A	N/A	Only when dropped
Variations in levels or uneven ground	Low	• work area/layout	Area has been levelled

Identifying hazards by job safety analysis

Table 6.3 shows examples of the hazards associated with moving a cable drum. This type of analysis is known as job safety analysis (JSA), and is done by looking at the work task and thinking about the safest way to complete the task. A JSA should be done in the planning stage of a job. It is a way of becoming aware of the hazards involved in doing the job and determining the necessary actions to take to prevent an injury.

Job safety analysis worksheets

The Victorian WorkCover Authority outlines the procedures for job safety analysis (http://www.vwa.vic.gov.au > safety & prevention > job safety analysis (JSA)). Many tasks undertaken are done routinely and have probably been done the same way for years – sometimes safely, sometimes not. Before starting a task, it is essential to think of what is the safest and best way to do it. These JSA worksheets help in assessing the risks of doing the job. Each JSA must be site specific and include all workers in its development. The process is:

1. **Document the activity:** Assemble those involved in the activity and then, using the JSA worksheet, write down the tasks that make up the activity, step by step.
2. **Identify the hazards:** Next to each task, identify what part of the task may cause injury to those doing the work or to anyone else nearby.
3. **Document the control measures:** For each identified hazard, list the measures that need to be put in place to eliminate or minimise any likely risk of injury to those involved.
4. **Identify who is responsible:** Document the name of the person responsible for implementing the control measure.
5. **Monitor and review:** Make sure the activity is supervised to ensure the documented process is being followed. The JSA should be reviewed whenever a documented activity changes, when there is a change of personnel or after an appropriate length of time.

Source: Victorian WorkCover Authority, 'Job Safety Analysis',
http://www.vwa.vic.gov.au/safety-and-prevention/health-and-safety-topics/job-safety-analysis

TABLE 6.3 Example of a job safety analysis document (JSA)

SAMPLE JOB SAFETY ANALYSIS

Company name:	Cengage Electrical	Date:	1/12/2013	JSA No	1001
Site Name:	Sample Street	Permit to work requirement		Yes ☐	No ☐
Contractor:	Phill Chadwick	Approved by:	Peter Phillips		
Activity:	Transfer of cable drum from vehicle to road way				

Activity	Hazards	Risk control measures	Who is responsible?
List the tasks required to perform the activity in the sequence they are carried out.	Against each task list the hazards that could cause injury when the task is performed.	List the control measures required to eliminate or minimise the risk of injury arising from the identified hazard.	Write the name of the person responsible (supervisor or above) to implement the control measure identified.
Transfer of cable drum from vehicle	Awkward postures	Use mechanical aid (crane or hoist) 2 person lift use crow bar or similar through centre hole to lift	Phill Chadwick
	Forces and loads	Use mechanical aid (crane or hoist) 2 person lift use crow bar or similar through centre hole to lift	Phill Chadwick
	Crush and pinch points	Size up load and coordinate lift	Phill Chadwick
	Splinters	PPE leather gloves	Phill Chadwick

Source: Adapted from Victorian WorkCover Authority, 'Job Safety Analysis', http://www.vwa.vic.gov.au/__data/assets/pdf_file/0019/3628/jsa_jobsafetyanalysis.pdf

Safe work method statements (SWMS)

A high-risk job such as working at height or on or near high voltage equipment will require a more elaborate analysis. Table 6.4 shows a sample safe work method statement (SWMS). Preparing a SWMS is part of the planning of the work. The SWMS is designed to help persons conducting business or undertakings (PCBU) to think through the hazards and risks involved in the work, and to choose effective control measures. Employees, health and safety representatives (HSR), as well as contractors and their employees, must be consulted in the preparation of the SWMS so far as reasonably practicable.

TABLE 6.4 Example of a safe work method statement

SAMPLE SAFE WORK METHOD STATEMENT

This SWMS is a site-specific statement that must be prepared before any high-risk construction work is commenced.

Person responsible for ensuring compliance with this SWMS:		**Date:**	
High-risk job:		**Location:**	

(Continued)

TABLE 6.4 (*Continued*)

What are the tasks involved?	What are the hazards and risks?	How will hazards and risks be controlled? (describe the control measures and how they will be used)
Think about the work site and each stage of the project, including preparation and clean-up.		
Steps for filling out		
1. Discuss with relevant employees, contractors and HSRs what work will be high-risk, the tasks, and associated hazards, risks and controls.		
2. In the 'What are the tasks involved?' column, list the work tasks in sequence in which they will be carried out.		
3. In the 'What are the hazards and risks?' column, list the hazards and risks for each work task.		
4. In the 'How will the hazards and risks be controlled?' column, select the hazard or risk and then work through the control levels 1–4 from top to bottom. Choose a control measure (and how it is to be used) that is as close to level 1 as is reasonably practicable.		
Control levels		
1. **Eliminate** any risk to health or safety associated with construction work. 2. **Reduce** the risk to health or safety by any one or any combination of the following: • **substituting** a new activity, procedure, plant, process or substance • **isolating** persons from the hazard, such as barricading, fencing or guard railing, or • **using engineering controls**, such as mechanical or electrical devices. 3. **Use administrative controls**, such as changing the way the work is done. 4. Provide appropriate personal protective equipment.		
5. Brief each team member on this SWMS before commencing work. Ensure team knows that work is to stop immediately if the SWMS is not being followed.		
6. Observe work being carried out. If controls are not adequate, stop the work, review the SWMS, adjust as required and re-brief the team.		
7. Retain this SWMS for the duration of the high-risk construction work.		

Source: Victorian WorkCover Authority, 'Safe work method statement', http://www.vwa.vic.gov.au/__data/assets/pdf_file/0019/3628/jsa_jobsafetyanalysis.pdf

SUMMING UP...

- In the planning stage of a job, before work begins:
  - risk sources (hazards) must be identified and recorded
  - a risk assessment is done to prioritise the hazards and formulate control measures
  - for low risk tasks a job safety analysis (JSA) is developed by all workers on the site
  - for high risk tasks a SWMS is provided to workers.

6.2 Typical job hazards and control measures

This section gives examples of the typical documentation required for various work situations.

Working at heights

Tables 6.5 to 6.7 show examples of the typical WH&S documents required when working at heights.

TABLE 6.5 Example of WH&S document for working at heights: hazard identification

SAMPLE WORKING AT HEIGHT HAZARD (RISK SOURCE) IDENTIFICATION							
Task	**Identified by**	**Risk factor(s) e.g.**	**Location**	**Workers at risk**	**General comments**	**Order of priority**	**Date of risk assessment**
	• reported hazard • injury/illness • consultation • inspection/ observation	• Falls • Collapsing structures • Electrocution • Dropped equipment • Worker characteristics					
Ascending and descending ladders	inspection/ observation	• Falls • Electrocution • Collapsing structures	10 John Street	electrical		1 2	17/12/2013
Work on roof space	inspection/ observation	• Falls • Electrocution • Dropped equipment	10 John Street	electrical		2 1 3	17/12/2013

TABLE 6.6 Example of WH&S documents for working at heights: risk assessment

<table>
<tr><th colspan="4">SAMPLE WORKING AT HEIGHT RISK ASSESSMENT FORM</th></tr>
<tr><td>Location/group: Electrical</td><td colspan="3">Task: Solar panel roof installation</td></tr>
<tr><td>Date of assessment: 18/12/2013</td><td colspan="3">Assessors/people involved: Phillip Chadwick</td></tr>
<tr><th>Risk factors to consider</th><th>Level of risk</th><th>Sources of risk</th><th>Comments</th></tr>
<tr><td>Working at height greater than 2 m</td><td>Estimate the level of risk each risk factor poses, i.e.:
• N/A
• low
• medium
• high</td><td>Possible source(s) of risk, i.e.:
• Falls
• Collapsing structures
• Electrocution
• Dropped equipment
• Worker characteristics</td><td></td></tr>
<tr><td>Ascending and descending ladders</td><td>Medium</td><td>• Falls
• Collapsing structures
• Electrocution</td><td rowspan="2"></td></tr>
<tr><td>Work on roof space</td><td>Medium</td><td>• Falls
• Dropped equipment</td></tr>
</table>

TABLE 6.7 Example of WH&S documents for working at heights: SWMS

SAMPLE SAFE WORK METHOD STATEMENT			
This SWMS is a site-specific statement that must be prepared before any high-risk construction work is commenced.			
Person responsible for ensuring compliance with this SWMS:	Phillip Chadwick	**Date:**	1/12/2013
High-risk job:	Working at height greater than 2 m	**Location:**	10 John Street

What are the tasks involved?	What are the hazards and risks?	How will hazards and risks be controlled? (describe the control measures and how they will be used)
Think about the work site and each stage of the project, including preparation and clean-up.		
Ascending and descending ladders	• Falls • Collapsing structures • Electrocution	Stay within the rungs of the ladder Ensure ladder is properly secured and or roped off Only use non-conductive ladders Isolate supply before work
Work on roof space	• Falls • Electrocution • Dropped equipment	Fall arrest system or guard railing Isolate equipment before commencing work Use of tool bags and PEE hard hats

Working in confined spaces

Tables 6.8 to 6.10 provide examples of typical WH&S documents required when working in confined spaces.

TABLE 6.8 Example of WH&S document for working in confined spaces: hazard identification

SAMPLE WORKING IN CONFINED SPACE (RISK SOURCE) IDENTIFICATION

Task	Identified by	Risk factor(s) e.g.	Location	Workers at risk	General comments	Order of priority	Date of risk assessment
Replacement of float switch in a mixing vat	• Reported hazard • Injury/illness • Consultation • Inspection/ observation	• Restricted entry or exit • Harmful airborne contaminants • Unsafe oxygen level • Fire and explosion • Engulfment					
Entry and exit	inspection/ observation consultation	• Restricted entry or exit • Harmful airborne contaminants • Unsafe oxygen level	mixing vat 3	electrical		3 1 2	17/12/2013
Float switch replacement	inspection/ observation	• Electrocution	mixing vat 3	electrical		2 1 3	17/12/2013

TABLE 6.9 Example of WH&S document for working in confined spaces: risk assessment

SAMPLE WORKING IN CONFINED SPACES RISK ASSESSMENT FORM			
Location/group: Electrical	Task: Replacement of float switch in a mixing vat		
Date of assessment: 18/12/2013	Assessors/people involved: Phillip Chadwick		
Risk factors to consider	**Level of risk**	**Sources of risk**	**Comments**
Working in a confined space	Estimate the level of risk each risk factor poses, i.e.: • N/A • Low • Medium • High	Possible source(s) of risk, i.e.: • Restricted entry or exit • Harmful airborne contaminants • Unsafe oxygen level • Fire and explosion • Engulfment	
Entry and exit	• Low • Medium • High • High	• Restricted entry or exit • Harmful airborne contaminants • Unsafe oxygen level • Fire and explosion	
Float switch replacement	• Medium	• Electrocution	

TABLE 6.10 Example of WH&S document for working in confined spaces: SWMS

SAMPLE SAFE WORK METHOD STATEMENT			
This SWMS is a site-specific statement that must be prepared before any high-risk construction work is commenced.			
Person responsible for ensuring compliance with this SWMS:	Phillip Chadwick	**Date:**	1/12/2013
High-risk job:	Working in confined spaces	**Location:**	mixing vat 3

What are the tasks involved?	**What are the hazards and risks?**	**How will hazards and risks be controlled? (describe the control measures and how they will be used)**	
Think about the work site and each stage of the project, including preparation and clean-up.			
Working in a confined space	• Restricted entry or exit • Harmful airborne contaminants • Unsafe oxygen level	Use of entry permits, signage and communication with an observer Purge vat and use of positive ventilation Communication with an observer	
Float switch replacement	• Electrocution	Isolate equipment before commencing work	

Noise

Tables 6.11 to 6.13 show examples of typical WH&S documents required when working in areas exposed to noise.

TABLE 6.11 Example of WH&S document for working in areas with high noise levels: hazard identification

SAMPLE NOISE HAZARD (RISK SOURCE) IDENTIFICATION							
Task	**Identified by**	**Risk factor(s) e.g.**	**Location**	**Workers at risk**	**General comments**	**Order of priority**	**Date of risk assessment**
	• reported hazard • injury/illness • consultation • inspection/observation	• Hearing loss • Fatigue					
Operation of masonry hammer drill	• inspection/observation • consultation	• Hearing loss • Fatigue	10 John Street	electrical		1 2	17/12/2013

TABLE 6.12 Example of WH&S document for working in areas with high noise levels: risk assessment

SAMPLE JOB SAFETY ANALYSIS NOISE

Company name:	Cengage Electrical	Date:	1/12/2013	JSA No	1003
Site Name:	10 John Street	Permit to work requirement		Yes ☐	No ☐
Contractor:	Phill Chadwick	Approved by:	Peter Phillips		
Activity:	Operation of masonry hammer drill				

Activity	**Hazards**	**Risk control measures**	**Who is responsible?**
List the tasks required to perform the activity in the sequence they are carried out.	Against each task list the hazards that could cause injury when the task is performed.	List the control measures required to eliminate or minimise the risk of injury arising from the identified hazard.	Write the name of the person responsible (supervisor or above) to implement the control measure identified.
Operation of masonry hammer drill	Hearing loss	Use of PPE Hearing muffs	Phill Chadwick
	Fatigue	Rotation of workers and rest breaks	Phill Chadwick
	Hand injuries	PPE leather gloves	Phill Chadwick

TABLE 6.13 Example of WH&S document for working in areas with high noise levels: SWMS

SAMPLE NOISE RISK ASSESSMENT FORM			
Location/group: Electrical	Task: Operation of masonry hammer drill		
Date of assessment: 18/12/2013	Assessors/people involved: Phillip Chadwick		
Risk factors to consider	**Level of Risk**	**Sources of risk**	**Comments**
Noise and vibration	Estimate the level of risk each risk factor poses, i.e.: • N/A • Low • Medium • High	Possible source(s) of risk, i.e.: • Hearing loss • Fatigue • Loss of concentration	
Operation of masonry hammer drill	• Low	• Hearing loss • Fatigue	

Dusts

The term dust loosely applies to any particle which is small enough and lightweight enough to 'float' in the air. When materials are drilled or cut, a dust is often produced. Depending on the type of material the resulting dust may be hazardous when inhaled or swallowed.

Asbestos

Breathing in asbestos dust, even a small amount, is lethal. Asbestos fibres were used widely in building materials before the mid-1980s. Any construction that was built or renovated before the mid-1980s is likely to contain products made of asbestos cement. These products contain bonded asbestos cement, and include fibro sheeting (flat and corrugated as shown in Figure 6.1), water drainage and flue pipes, roofing shingles and guttering, even the backing of vinyl sheet floor coverings. If asbestos is disturbed it will release fine particles of dust containing asbestos fibres. Once inhaled, these fibres cannot be broken down by the body. Breathing in dust containing asbestos fibres can cause asbestosis, lung cancer and mesothelioma (a cancer which most often occurs in the lining of the lung). There is no cure.

A non-friable asbestos containing material (ACM) is any material that contains more than 1% asbestos, but cannot be pulverised under hand pressure. Non-friable ACM was commonly used as an electrical insulator on meter boards and panels in general, and as bases to busbars, spark arresters and flash guards. Timber meter boxes installed prior to 1990 were also commonly lined with asbestos-cement sheeting (fibro). Asbestos contaminated dust and debris from the various asbestos components can also be present within the cabinets (see Workplace Health and Safety Queensland, 'Asbestos within electrical installations' http://www.deir.qld.gov.au> workplace hazards> asbestos> electricity. A Zelemite (brand name for an asbestos product) switchboard backing is shown in Figure 6.2.

FIGURE 6.1 Corrugated roof sheeting containing asbestos

Source: Ed Hawkins

FIGURE 6.2 'Zelemite' switchboard backing

Source: Ed Hawkins

Before working on these types of meter boards, be very sure of the correct work procedures. Drilling into the board or cutting it can produce asbestos dust that must be contained, and not breathed in. It is your responsibility to make sure you follow the relevant legislation about working with asbestos. For example, in Australia if more than 10 square metres of bonded asbestos is to be removed, the legislation requires all work to be performed by a trained and licensed asbestos removalist.

Glass wool

Glass wool insulation is classified as non-hazardous because it is bio-soluble, meaning the product dissolves in bodily fluids and is quickly cleared from the lungs. Glass wool is composed of fine, stable and uniformly textured inorganic glass fibres bonded together by water soluble fire retardant thermosetting and heat resistant resin. It is manufactured by spinning strands of molten glass in a similar way to fairy floss. It is available in flexible blanket or semi rigid rectangular boards, slabs, or pre-formed pipe sections.

Glass wool dust and fibres can cause some temporary irritation to the skin, mouth, eyes and airways. Safety data sheets recommend moving to an area with fresh air and rinsing affected areas with fresh water.

Rock wool

A safe alternative to asbestos, rock wool insulation is similar to glass wool and is also bio-soluble. It is manufactured by spinning a molten mixture of natural rock and recycled products into fine wool-like fibres. It provides much better insulation qualities than glass wool. First-aid treatment is the same as for glass wool.

Fibreglass and carbon fibre

Fibreglass as a product is an extremely versatile material, being strong, lightweight and weatherproof. It is made by combining strands of glass fibre with a polymer resin. The dust is classified as non-hazardous, but is not bio-soluble. If the dust is inhaled into the upper respiratory tract, the fibres will not break down and must be expelled by mucus. Unprotected exposure can cause a rash, itching, conjunctivitis, coughing and sneezing. Carbon fibre has similar properties but is electrically conductive. Dusts must not be permitted to contaminate electrical equipment.

Tables 6.14 to 6.16 show examples of WH&S documents required when working in areas exposed to dust that does not contain asbestos.

TABLE 6.14 Example of WH&S document for working in dusty areas (non-asbestos): hazard identification

SAMPLE DUST RELATED HAZARD (RISK SOURCE) IDENTIFICATION							
Task	**Identified by**	**Risk factor(s) e.g.**	**Location**	**Workers at risk**	**General comments**	**Order of priority**	**Date of risk assessment**
	• reported hazard • injury/illness • consultation • inspection/ observation	• Asbestos • Explosive mixture • Toxic materials • Inhalation • Hearing loss • Electrocution • Foot injury • Fatigue					
Operation of masonry wall chasing tool	• inspection/ observation • consultation	• Inhalation • Foot/hand injury • Electrocution • Hearing loss • Fatigue	10 John Street	electrical		3 2 1 1 4 5	17/12/2013

TABLE 6.15 Example of WH&S document for working in dusty areas (non-asbestos): risk assessment

SAMPLE DUST RISK ASSESSMENT FORM			
Location/group: Electrical	Task: Operation of masonry wall chasing tool		
Date of assessment: 18/12/2013	Assessors/people involved: Phillip Chadwick		
Risk factors to consider	**Level of risk**	**Sources of risk**	**Comments**
Dust, noise and vibration	Estimate the level of risk each risk factor poses, i.e.: • N/A • low • medium • high	Possible source(s) of risk, i.e.: • Inhalation • Hearing loss • Fatigue • Electrocution • Hearing loss	
Masonry wall chasing tool operation	• high • medium • medium • low • low	• Electrocution • Foot/hand injury • Inhalation • Hearing loss • Fatigue	

TABLE 6.16 Example of WH&S document for working in dusty areas (non-asbestos): SWMS

SAMPLE JOB SAFETY ANALYSIS DUST					
Company name:	Cengage Electrical	Date:	1/12/2013	JSA No	1003
Site Name:	10 John Street	Permit to work requirement		Yes ☐	No ☐
Contractor:	Phill Chadwick	Approved by:	Peter Phillips		
Activity:	Masonry wall chasing tool operation				

Activity	Hazards	Risk control measures	Who is responsible?
List the tasks required to perform the activity in the sequence they are carried out.	Against each task list the hazards that could cause injury when the task is performed.	List the control measures required to eliminate or minimise the risk of injury arising from the identified hazard.	Write the name of the person responsible (supervisor or above) to implement the control measure identified.
Masonry wall chasing tool operation	• Electrocution	Chasing tool to be RCD protected	Phill Chadwick
	• Foot/hand injury	Use of PPE boots and gloves	Phill Chadwick
	• Inhalation	Use of water to minimise dust Use of PPE face mask	Phill Chadwick
	• Hearing loss	Use of PPE hearing protection	Phill Chadwick
	• Fatigue	Rotation of workers and rest breaks	Phill Chadwick

Gases

Tables 6.17 to 6.19 show typical WH&S documents required when working in areas where LPG gas is present.

TABLE 6.17 Sample WH&S document for sites where LPG is being used: gas related hazard identification

SAMPLE GAS RELATED HAZARD (RISK SOURCE) IDENTIFICATION							
Task	**Identified by**	**Risk factor(s) e.g.**	**Location**	**Workers at risk**	**General comments**	**Order of priority**	**Date of risk assessment**
	• reported hazard • injury/illness • consultation • inspection/ observation	• Explosive mixtures • Extremes of hot or cold pressure streams • Hot surfaces • Asphyxiation • Poisoning					
	• inspection/ observation • consultation	• Explosive mixtures • Hot surfaces	10 John Street	electrical	LPG gas	1	17/12/2013

TABLE 6.18 Sample WH&S document for sites where LPG is being used: gas risk assessment form

SAMPLE GAS RISK ASSESSMENT FORM			
Location/group: Electrical	Task: LPG burner		
Date of assessment: 18/12/2013	Assessors/people involved: Phillip Chadwick		
Risk factors to consider	**Level of risk**	**Sources of risk**	**Comments**
	Estimate the level of risk each risk factor poses, i.e.: • N/A • low • medium • high	Possible source(s) of risk, i.e.: • Explosive mixtures • Extremes of hot or cold pressure streams • Unsafe oxygen level – asphyxiation • Poisoning	
Explosive mixtures	high	LPG gas cylinder	
Hot surfaces	low	LPG torch	

TABLE 6.19 Sample WH&S document for sites where LPG is being used: job safety analysis

SAMPLE JOB SAFETY ANALYSIS GAS					
Company name:	Cengage Electrical	Date:	1/12/2013	JSA No	1003
Site Name:	10 John Street	Permit to work requirement		Yes ☐	No ☐
Contractor:	Phill Chadwick	Approved by:	Peter Phillips		
Activity:					

Activity	Hazards	Risk control measures	Who is responsible?
List the tasks required to perform the activity in the sequence they are carried out.	Against each task list the hazards that could cause injury when the task is performed.	List the control measures required to eliminate or minimise the risk of injury arising from the identified hazard.	Write the name of the person responsible (supervisor or above) to implement the control measure identified.
LPG gas burner operation	• Explosive mixtures	Adequate ventilation of work and storage areas	Phill Chadwick
	• Hot surfaces	Provide appropriate PPE and training	Phill Chadwick

Chemicals

Tables 6.20 to 6.22 show typical WH&S documents required when working in areas exposed to chemicals.

TABLE 6.20 Sample WH&S document when working with chemicals: chemical related hazard identification

SAMPLE CHEMICAL RELATED HAZARD (RISK SOURCE) IDENTIFICATION

Task	Identified by	Risk factor(s) e.g.	Location	Workers at risk	General comments	Order of priority	Date of risk assessment
	• reported hazard • injury/illness • consultation • Inspection/ observation	• Fire explosion • Inhalation of fumes • Burns from acid • Foot/hand injury • Manual handling					
Maintain lead acid batteries	• inspection/ observation • consultation	• Fire/explosion • Inhalation of fumes • Burns from acid • Foot/hand injury • Manual handling	10 John Street	electrical		3 2 1 1 4 5	17/12/2013

Source: Government of Western Australia, 'Code of Practice – Manual Tasks 2010', http://www.commerce.wa.gov.au; and Victorian WorkCover Authority, 'Job Safety Analysis', http://www.vwa.vic.gov.au/forms-and-publications/forms-and-publications/job-safety-analysis-worksheet-jsa

TABLE 6.21 Sample WH&S document when working with chemicals: chemical risk assessment form

SAMPLE CHEMICAL RISK ASSESSMENT FORM			
Location/group: Electrical	Task: Maintain lead acid batteries		
Date of assessment: 18/12/2013	Assessors/people involved: Phillip Chadwick		
Risk factors to consider	**Level of Risk**	**Sources of risk**	**Comments**
Explosive gases and corrosive fumes produced by sulphuric acid	Estimate the level of risk each risk factor poses, i.e.: • N/A • low • medium • high	Possible source(s) of risk, i.e.: • Fire/explosion • Inhalation of fumes • Burns from acid • Foot/hand injury • Manual handling	
	• high • medium • medium • low • low	• Fire/explosion • Inhalation of fumes • Burns from acid • Foot/hand injury • Manual handling	

TABLE 6.22 Sample WH&S document when working with chemicals: job safety analysis

SAMPLE JOB SAFETY ANALYSIS CHEMICAL					
Company name:	Cengage Electrical	Date:	1/12/2013	JSA No	1003
Site Name:	10 John Street	Permit to work requirement		Yes ☐	No ☐
Contractor:	Phill Chadwick	Approved by:	Peter Phillips		
Activity:	Maintain lead acid batteries				

Activity	**Hazards**	**Risk control measures**	**Who is responsible?**
List the tasks required to perform the activity in the sequence they are carried out.	Against each task list the hazards that could cause injury when the task is performed.	List the control measures required to eliminate or minimise the risk of injury arising from the identified hazard.	Write the name of the person responsible (supervisor or above) to implement the control measure identified.
Maintain lead acid batteries	• Fire/explosion	Adequate ventilation to prevent buildup of gas	Phill Chadwick
	• Inhalation of fumes	Use of PPE face mask Use of PPE	Phill Chadwick
	• Burns from acid	Use of PPE boots and gloves Provision of safety shower	Phill Chadwick
	• Foot/hand injury	Use of PPE boots and gloves	Phill Chadwick
	• Manual handling	Correct lifting techniques when moving batteries	Phill Chadwick

SUMMING UP...

- Hazards and risks when working on construction sites include:
  - manual and mechanical handling
  - working at heights
  - working in confined spaces
  - noise
  - dusts
  - gases
  - chemicals.
- Hazards should be:
  - identified
  - assessed
  - controlled
  - monitored.
- A job safety analysis (JSA) provides safety instructions for low risk activities.
- A safe work method statement (SWMS) provides safety instructions for high risk activities.
- High risk construction activities include:
  - risk of a person falling more than 2 metres
  - work on a telecommunication tower
  - demolition of an element of a structure that is load-bearing or otherwise related to the physical integrity of the structure involves, or is likely to involve, the disturbance of asbestos
  - involves structural alterations or repairs that require temporary support to prevent collapse
  - work in or near a confined space
  - work in or near a shaft, tunnel or trench with an excavated depth greater than 1.5 metres
  - the use of explosives
  - working near pressurised gas distribution mains or piping
  - working on or near chemical, fuel or refrigerant lines
  - working on or near energised electrical installations or services
  - working in an area that may have a contaminated or flammable atmosphere
  - working on, in or adjacent to a road, railway, shipping lane or other traffic corridor that is in use by traffic other than pedestrians
  - working in an area where there is movement of powered mobile plant
  - working in an area in which there are artificial extremes of temperature
  - working in or near water or other liquid that involves a risk of drowning, such as diving work.

6.3 Hazards with typical electrical installations

A range of voltages are used in electrotechnology. For example, a transmission line transfers huge amounts of electricity across vast distances at extremely high voltages, in some cases as high as 500 000 volts. The current is low in respect to the voltage level which reduces 'I^2R' losses in the cable. However it is neither safe nor economical to use voltages as high as this for house wiring, so the voltage is decreased by a transformer to a suitable level. Automotive systems run at even lower voltages to reduce the risk of electrocution in the event of a collision. This section explains how electricity is distributed to parts of an installation at suitable voltages.

FIGURE 6.3 Circuit arrangements

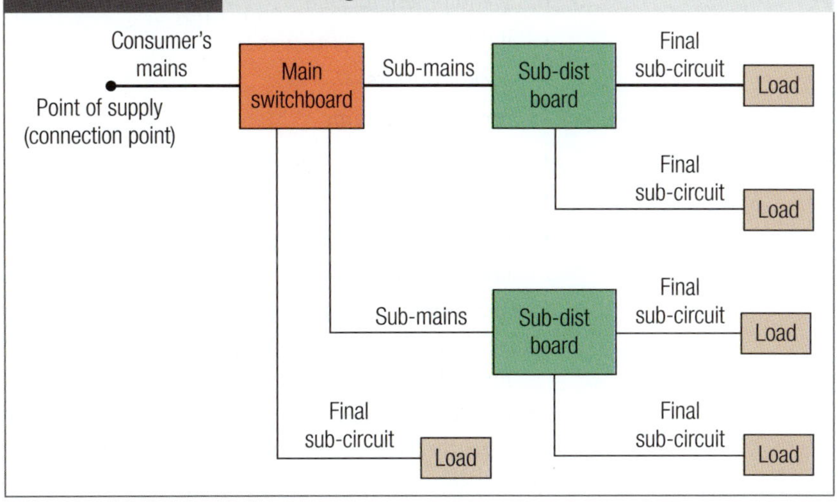

Power distribution and circuits in an electrical installation

Figure 6.3 shows an example of the circuit arrangement of a typical installation. Circuits can be classified as consumer's mains, sub-mains and final sub-circuits. Consumer's mains connect an installation's main switchboard to the supply authority distribution transformer or supply mains. In all but extremely large installations, only one set of consumer's mains is permitted, so quick isolation by emergency services personnel can be made. Sub-mains supply power to distribution boards in locations remote from the main switchboard. (A load is not directly connected to a sub-main.) Final sub-circuits originate at the circuit protection device in a main switchboard or distribution board, then run to the load itself and back to the neutral link at the board where it originated.

The load in a typical electrical installation is split into a number of final sub-circuits. Each sub-circuit usually has a similar type of load, such as lights and socket outlets. It is usually not practical to have a single circuit, as the cable size would need to be excessively large, which increases costs, and makes terminations very difficult. Imagine trying to connect 16 mm^2 cables at the rear of a socket outlet. In other words, loads must be split up and connected so that if a fault occurs or isolation is required for testing and or maintenance, there will be minimal disruption to the rest of the installation.

AS/NZS 3000:2007 clause C5.1 has specifications for loads which are greater than 20 A. **AS/NZS 3000:2007** clause 2.2.1.1 specifies the requirements for the arrangement of electrical circuits in an installation.

Systems or parts that operate at low-voltage and extra-low voltage

A typical electrical system is comprised of a source (supply authority distribution transformer), a path (conductors) and a load. In addition, fuses and/or circuit breakers are used to provide circuit protection and switches are used to provide control and isolation. Figure 6.4 shows a simplified single line diagram of an electrical distribution system. See **AS/NZS 3000:2007** clause 1.4.90 for the definition of low voltage.

FIGURE 6.4 Electrical system

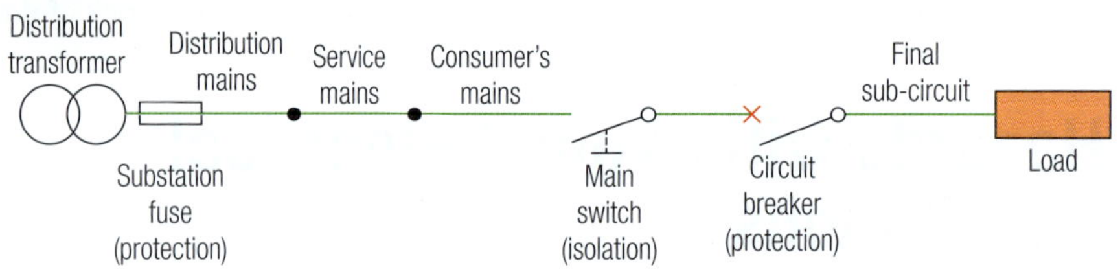

The nominal voltage for single phase low voltage equipment in Australia and New Zealand is 230 V a.c. The majority of electrical wiring systems and general equipment are rated at this voltage. However, there are other situations such as mines, aircraft and boats where different voltages are used. LV systems use less current to deliver the same power as extra low voltage, therefore smaller conductors can be used. Examples of LV equipment include:

- lights
- socket outlets
- hot water systems
- single and three phase motors
- air-conditioning
- ranges, ovens and hot plates.

Low voltage wiring systems and equipment must be selected and installed to provide basic protection (against direct contact) and fault protection (against indirect contact) under **AS/NZS 3000:2007** clauses 1.4.78 and 1.4.79 (http://biscosaudi.com/?p=471).

Extra low voltage wiring systems and equipment provide additional protection against direct contact (basic protection) and indirect contact (fault protection) as the reduced circuit voltage (≤50 V a.c.) greatly reduces any touch current that might flow under contact. **AS/NZS 3000:2007** clause 1.4.83 defines separated extra low voltage (SELV).

A transformer is used to reduce the LV supply to SELV. The primary and secondary windings of the transformer are electrically isolated, meaning there is no electrical path between the windings. As shown in Figure 6.5, without a connection to earth there is no possibility of a shock current flowing from the circuit to earth through a person.

Examples of SELV equipment include:

- down lights
- halogen and LED
- pool lights
- fish tank and fountain pumps
- doorbells
- roller door models
- alarm and security systems.

AS/NZS 3000:2007 clause 1.4.76 defines protected extra-low voltage (PELV). PELV systems have a protective earthing conductor connected to one side of the secondary winding of the transformer, as shown in Figure 6.6. The secondary winding is therefore not electrically separated from the primary side of the transformer, due to the earth wire.

Examples of PELV equipment include audio power amplifiers and high current d.c. power supplies.

FIGURE 6.5 SELV system

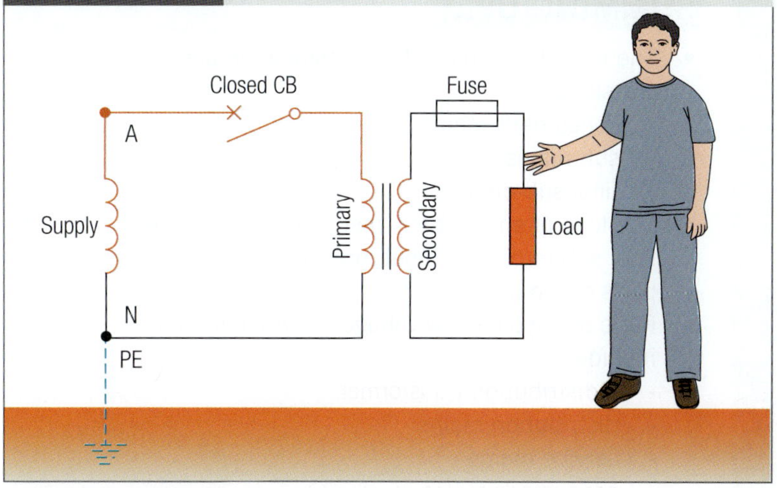

FIGURE 6.6 PELV system

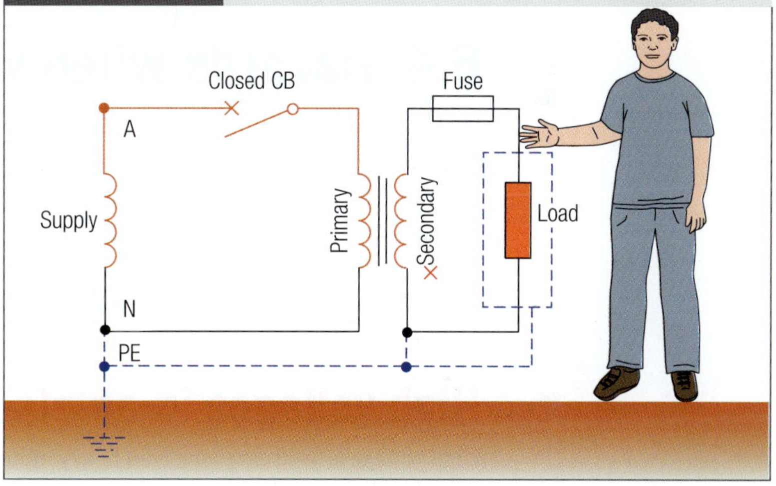

High currents in an electrical system

As there is usually only one set of consumer's mains and one main switchboard per installation, all of the installation's current must pass through these parts. The maximum current that is expected to flow under **normal** conditions through the consumer's mains, and the main switchboard is known as the *maximum demand* of the installation. If a fault that creates an over-current occurs anywhere else, that over-current will also have to pass through the consumer's mains and main switchboard. AS/NZS 3000:2007 clause 1.5.9 requires that all parts of an installation are protected against over currents; this is normally provided by automatic disconnection caused by fuses or circuit breakers.

Large installations taking a current of 800 A or more are considered to be heavy current under AS/NZS 3000:2007 clause 2.5.5. These switchboards are required to be provided with protection against switchboard internal arcing fault currents. The large conductors that supply these installations will have a very low resistance. Accidental contact between conductors and earth can cause an internal arcing at the switchboard. Any person working on or near the switchboard would be at severe risk of fatal injury.

Electric motors and equipment that contains electric motors can produce up to eight times their normal full load current when starting. The starting current and length of time the starting current flows will depend on the nature of the mechanical load connected to the motor.

When transformers are used to step voltages down, the voltage on the secondary winding is decreased, but it is possible to have a higher current flow in the secondary winding than the primary. This effect is used in electrical equipment such as arc and resistance (spot) welding machines.

SUMMING UP...

- The circuits of an electrical installation are classified as:
  - consumer's mains
  - sub-mains
  - final sub-circuits.
- The load of an electrical installation is arranged into final sub-circuits without interrupting supply to other parts of the installation.
- Parts of a typical low voltage electrical system include:
  - a distribution transformer
  - distribution mains
  - service mains
  - consumer's mains
  - sub-mains
  - final sub-circuits
  - main/isolation switch
  - circuit protection devices (fuses and or circuit breakers)
  - loads.
- Two commonly used extra low voltage systems are:
  - separated extra low voltage (SELV)
  - protected extra low voltage (SELV).
- Parts of an electrical system and equipment where high currents are likely include:
  - consumer's mains and main switchboards
  - electric motors and equipment that contains electric motors
  - secondary windings of a step down transformer.

6.4 Hazards when working with high voltage

High voltage is defined by **AS/NZS 3000:2007** clause 1.4.98(c). High voltage work is a very specialised field. The normal electrical licence does not qualify that person to work on high voltage equipment; additional certifications are normally required for workers to isolate, install, maintain and test high voltage equipment. High voltage does not require physical contact to be made for a current to flow. Close proximity is enough to cause air to ionise, that is, to change from behaving as an insulator and act as an electrical conductor resulting in arcing, extreme amounts of heat and ultraviolet light. This section describes the hazards associated with high voltage equipment.

High voltages in an electrical system

High voltage is used extensively in heavy industrial and mining situations. Equipment is often supplied at high voltage, as the losses in the conductors are reduced because less current is required for a given amount of power. This allows smaller conductors to be run over longer distances. Larger installations are supplied at high voltage, because the customer provides the infrastructure, such as distribution transformers. Installations like this will typically receive a discount on electricity charges.

Examples of HV equipment include:

- high power three phase motors.
- distribution transformers
- switch gear and protection devices
- neon signage. One metre of neon lighting requires approx 1 kV (supplied at LV).

Terms related to the hazards of high voltage

Touch voltage

Touch potential is the voltage between an energised object and the feet of a person in contact with the object. This occurs when your hands are contacting an earthed metal surface and your feet are contacting the ground, and there is a potential difference between them due to an earth fault current. Touch potential could be nearly the full voltage across a grounded object if that object is grounded at a remote point from where the person is in contact with it. A crane that is grounded to the system

FIGURE 6.7 Touch voltage and step voltage

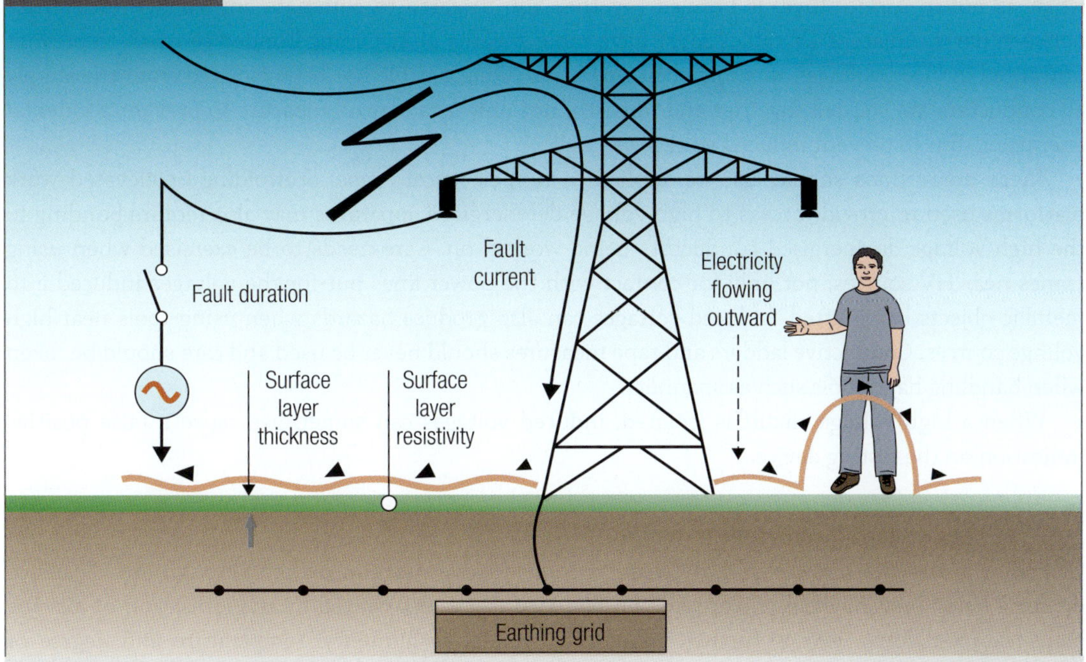

Source: E&S Grounding Solutions

neutral and contacts an energised line, for example, would expose any person in contact with the crane or its uninsulated load line to a touch potential nearly equal to the full line voltage. Figure 6.7 shows examples of touch and step voltages.

Step voltage

When an active to earth fault occurs on high voltage equipment, the fault current will flow from the equipment under fault to an earth grid below. The surface layer of the soil forms a parallel circuit with the grid. Most of the current will flow through the grid but some will also pass through the surface layer of soil; the amount will depend on the soil resistivity. Think of the ground as being a series of resistors with the fault current flowing through them. Each resistor will have a voltage drop across it.

As the distance from the point where the fault current enters the ground increases along the path, the greater the resistance between the two points and the greater the difference in voltage between the two points; that is, different points of the ground you are walking on are now at different voltages. The voltage between two points 1 metre apart, which is regarded as a typical step distance, is called the step voltage.

The voltage drop in the soil surrounding the grounding system can present hazards for personnel standing in the vicinity of the grounding system. It is recommended to only take small steps in areas such as HV switch yards.

Induced voltage

Whenever there is relative motion between a conductor and a magnetic field a voltage is induced. A live conductor carrying a current produces a magnetic field, and because the current is alternating, the resulting magnetic field constantly rises, falls and changes direction. Any conductors running next to this conductor will have a voltage induced into them. There is no need for any of the conductors to physically move, as the magnetic field is constantly moving.

HV systems can induce extremely large voltages into any conductive material. Any equipment which is not in service (live) is bonded together and to earth to eliminate any voltage differences between parts. Adjacent LV cables must have cable sheaths or screening bonded to earth to prevent unwanted induced voltages affecting their operation. Once a cable has been isolated from the supply all conductors are bonded together and to earth, not only as a control measure to prevent accidental energising, but to prevent induced voltages as well.

These precautions should also be applied to non-electrical items. Scaffolding or elevated work platforms used to provide access to high voltage de-energised apparatus may also require bonding to the high voltage de-energised conductors being worked on. Care needs to be exercised when using cranes near HV sources, not only for contact with the power lines but for the voltages induced into metallic objects being lifted. Induced voltages can also produce hazards when using tools near high voltage sources. Conductive ladders and tape measures should never be used and care should be taken when handling hand tools such as spanners.

When a high voltage circuit is isolated, induced voltages can sometimes cause a false positive indication on the testing device.

Creepage

Clearance is the shortest distance between two conductive parts measured through air. Clearance distance helps prevent dielectric breakdown between electrodes caused by the ionisation of air. The dielectric breakdown level is further influenced by relative humidity, temperature and degree of pollution in the environment (Source: Stack Exchange, 'Electrical engineering', http://electronics.stackexchange.com/questions/74244/why-route-air-gaps-for-voltage-isolation-on-pcbs).

Creepage is the shortest path between two conductive parts measured along the surface of the insulation. The tortuous path around the insulator discs effectively increases the path between conductors for the same given spacing. As explained by Stack Exchange, 'Electrical engineering', an

> adequate creepage distance protects against tracking, a process that produces a partially conducting path of localized deterioration on the surface of an insulating material as a result of the electric discharges on or close to an insulation surface. The degree of tracking required depends on two major factors: the comparative tracking index (CTI) of the insulating material and the degree of pollution in the environment.
>
> Source: Stack Exchange, http://electronics.stackexchange.com/questions/74244/why-route-air-gaps-for-voltage-isolation-on-pcbs.

Tracking can also create a major hazard when attempting to isolate HV equipment and cables. The isolation point is opened but it is possible for tracking across the insulators of the isolation point to maintain the connection to the HV supply. Any contact with the equipment is then potentially fatal.

Control measures for dealing with high voltages

Due to higher voltage levels, the risks involved when working with high voltage are also increased. Electrical equipment requires increased insulation integrity. Isolation procedures require a different type of test equipment, and an increased level of awareness of hazards is required of workers. The typical hazards associated with high voltage equipment can include:

- electrocution from contact with live conductors or accidental energising
- arcs, explosions and fire from equipment failure or improper operation
- voltage gradients (touch and step voltages) under fault conditions.

Proper design of the system will include engineering measures to control risks of equipment failure and voltage gradients. In the context of this section we examine a number of engineering and administrative controls to prevent contact with live conductors and accidental energising of equipment or conductors. Codes of practice have been developed by state and territory governments which specify the minimum requirements associated with high voltage electrical work.

Training

Workers must be provided with appropriate training for the type of work concerned, and then be able to demonstrate they can safely perform the work required to be undertaken. Staff required to perform switching and isolations must also receive instruction and demonstrate competency as high voltage switching operators.

Safe working distances

Table 6.23 shows the recommended safe working distances for authorised persons working near exposed HV equipment.

TABLE 6.23 High voltage safe working distances

HV SAFE WORKING DISTANCES	
Nominal voltage (U) volts	**Minimum safe working distance (mm)**
$U < 1000$	500
$1000 < U < 11\,000$	700
$11\,000 < U < 66\,000$	1000
$66\,000 < U < 132\,000$	1500
$132\,000 < U < 220\,000$	2500
$220\,000 < U < 330\,000$	3000
$U > 330\,000$	4000

Source: Government of NSW, Department of Water and Energy, 'Code of Practice – Electricity Transmission and Distribution Asset Management', Table 4, p. 25, http://www.resourcesandenergy.nsw.gov.au/energy-supply-industry/pipelines-electricity-gas-networks/electricity-networks/safety#_other-safety-codes-and-guidelines

Isolation and access permits/authorisations

All HV electricity works should be regarded as live until isolated and proved to be de-energised, earthed and short-circuited. An access permit, access authority or authority to work must also be issued. Access permits are normally issued by personnel that have received training as high voltage operators. Once the HV equipment has been made safe all members of the work group must receive instruction as to:

- what equipment they are permitted to work on
- where and how the equipment has been isolated and how it was proved not alive
- the locations of 'working earths'
- any special conditions or abnormalities.

All members of the work group must then sign the access permit to indicate they understand the conditions of the permit. To prevent accidental re-energising of the HV equipment, the working earths may not be removed or the supply restored until all members of the work group sign off and the permit is cancelled. Figures 6.8 and 6.9 show an example of a typical access permit.

FIGURE 6.8 High Voltage Access Permit

High Voltage Access Permit

High Voltage Access – PTW Number: ____________

Section 1 – General Details

Switching Sheet Ref No: .... **Access Permit No:** ....

Nominated Issue: Time: .... Day: .... Date: ....

Anticipated Surrender: Time: .... Day: .... Date: ....

Work Area Location: ....

....

Issued To: ....

Access to the following High Voltage Lines and Apparatus:

Work Details:

Description of Isolation Points with Do Not Operate Boards attached:

Location of Operator Earths with Do Not Operate Boards attached:

Other Precautions Taken:

- ☐ Barrier / Warning Tape
- ☐ Barrier Mesh / Hoarding
- ☐ Work Area Warning Sign
- ☐ Insulated Covers / Mats
- ☐ Other (specify): ....

Nearby Live HV / LV at the Work Area:

Section 2 – Issue of Access Permit

Issue Approval *(Switching Coordinator):*
I approve the issue of this Access Permit to enable safe access / work on isolated HV apparatus.

Name: .... Signature: .... Time: .... Date: ....

Issue Implementation *(Switching Operator):*
I issue this Access Permit to enable safe access / work on isolated HV apparatus.

Name: .... Signature: .... Time: .... Date: ....

Section 3 – Receipt of Access Permit

Recipient:
I acknowledge that I only have access to the lines and apparatus listed above while earthed, will have no difficulty in keeping clear of lines and apparatus not covered by this Access Permit and will not alter Isolation Points, Operator Earths or Other Precautions put in place.

Name: .... Signature: .... Time: .... Date: ....

Source: Port of Brisbane Pty Ltd, High Voltage Access Permit, p. 1, http://www.portbris.com.au/about-us/health-and-safety/inductions-and-permits

FIGURE 6.9 High Voltage Access Permit

High Voltage Access – PTW Number:	

Section 4 – Work Group Signatures

SIGN ON
I acknowledge that I only have access to the lines and apparatus listed on Page 1 while earthed, will have no difficulty in keeping clear of lines and apparatus not covered by this Access Permit and will not alter Isolation Points, Operator Earths or Other Precautions put in place.

SIGN OFF
I acknowledge that I no longer have access to the lines and apparatus listed on Page 1 and will regard the lines and apparatus as being live.

Name	Signature	Time	Date	Signature	Time	Date

Section 5 – Transfer of Access Permit

Outgoing Recipient:
I acknowledge that I have communicated with the Incoming Recipient and that I am no longer coordinating access/work relating to this Access Permit.

Name: .... Signature: .... Time: .... Date: ....

Incoming Recipient:
I acknowledge that I only have access to the lines and apparatus listed on Page 1 while earthed, will have no difficulty in keeping clear of lines and apparatus not covered by Access Permit and will not alter Isolation Points, Operator Earths or Other Precautions put in place.

Name: .... Signature: .... Time: .... Date: ....

Section 6 – Working Earth Schedule

Location of Each Set of Working Earths Placed:	On		Off	
	Time	Date	Time	Date

Section 7 – Abnormalities

Details:

Section 8 – Surrender of Access Permit

Recipient:
I acknowledge that all Working Earths placed have been removed except as specified in Section 7 Abnormalities, that the Work Group have fully signed off, that I no longer have access to the lines and apparatus listed on Page 1, and that I will regard the lines and apparatus as being live.

Name: .... Signature: .... Time: .... Date: ....

Source: Port of Brisbane Pty Ltd, High Voltage Access Permit, p. 2, http://www.portbris.com.au/about-us/health-and-safety/inductions-and-permits

SUMMING UP...

- Areas where high voltage is likely to be used include:
  - areas controlled by supply authorities
  - the mining industry
  - heavy industrial
  - large commercial buildings.
- Examples of HV equipment include:
  - high power three-phase motors
  - distribution transformers
  - power lines
  - switch gear and protection devices
  - neon signage.
- Touch potential is the voltage between an energised object and the feet of a person in contact with the object.
- A step voltage or voltage gradient occurs when a high voltage system experiences an earth fault. A voltage develops between different points on the ground adjacent to the fault. This can be hazardous to personnel as potentially fatal voltages can develop in the distance of a step.
- Clearance is the shortest distance between two conductive parts measured through air.
- Creepage is the shortest path between two conductive parts measured along the surface (tortuous path) of the insulation.
- The hazards of high voltage include:
  - electrocution from contact with live conductors or accidental energising
  - arcs, explosions and fire from equipment failure or improper operation
  - voltage gradients under fault conditions.
- Control measures used for dealing with the hazards of high-voltage include:
  - training of personnel
  - maintaining safe working distances
  - isolation, testing, working earths and access permits.

6.5 Hazards when working with low voltage equipment

Chapter 1 Section 1.8 'Working safely with electricity' gives detailed guidance as to the risks, control measures and procedures associated with working near or on low voltage electricity. This section looks at the documentation used to assist in managing and controlling these risks.

Risks and restrictions in working live

The Federal Model Work Health and Safety Regulations prohibit 'live' electrical work unless isolation creates a greater risk than working live. (Note that testing of live electrical work is exempted.) Excuses such as 'it is quicker' or 'cheaper' do not justify working on energised equipment or cables.

Safe Work Australia's 'Managing Electrical Risks in the Workplace Code of Practice' provides guidance to working on both de-energised and energised equipment and cables. Before work starts it must be established what cables and equipment are live and what are not live (p. 35). Refer to Chapter 1 Section 1.8 for the equipment required, locations, isolation and testing procedures to ensure equipment and cables are made safe before work begins.

The following example uses the scenario of a simple task involving the replacement of a faulty circuit breaker within a distribution board of a shopping centre. Table 6.24 provides an example of the appropriate Hazard Identification and Risk Assessment (HRA) documents.

TABLE 6.24 Sample Electrical Hazard Identification

SAMPLE ELECTRICAL HAZARD (RISK SOURCE) IDENTIFICATION							
Task	**Identified by**	**Risk factor(s), e.g.**	**Location**	**Workers at risk**	**General comments**	**Order of priority**	**Date of risk assessment**
	• reported hazard • injury/illness • consultation • Inspection observation	• electric shock causing injury or death • arcing, explosion or fire causing burns • toxic gases causing illness or death resulting from burning or arcing • electric shock from 'step-and-touch' potentials • fire resulting from an electrical fault					
Circuit breaker replacement	inspection/ observation	• electric shock causing injury or death • arcing, explosion or fire causing burns • toxic gases causing illness or death resulting from burning or arcing • electric shock from 'step-and-touch' potentials	shopping centre distribution board 2	electrical maintenance crew		1 2 3 4	17/12/2013

As previously demonstrated, once the hazards (risk sources) are identified they should be assessed, as shown in Table 6.25 – for simplicity only the electrical hazards have been included.

TABLE 6.25 Sample Electrical Risk Assessment

SAMPLE RISK ASSESSMENT FORM			
Location/group: Electrical	Task: Replace faulty circuit breaker distribution board 2		
Date of assessment: 18/12/2013	Assessors/people involved: Phillip Chadwick		
Risk factors to consider	**Level of risk**	**Sources of risk**	**Comments**
Live electrical conductors High fault current level	Estimate the level of risk each risk factor poses, i.e.: • N/A • low • medium • high	Possible source(s) of risk, i.e.: • electric shock causing injury or death • arcing, explosion or fire causing burns. • toxic gases causing illness or death resulting from burning or arcing • electric shock from 'step-and-touch' potentials • fire resulting from an electrical fault	
Working live			
• electric shock causing injury or death	high	• live sub-mains • live final sub-circuits • close proximity to earthed metal enclosures	
• arcing, explosion or fire causing burns	high	• live sub-mains • live final sub-circuits • possibility of tools making contact between live parts and frame	
• toxic gases causing illness or death resulting from burning or arcing	medium	• live sub-mains • live final sub-circuits • close proximity to earthed metal enclosures	

(*Continued*)

TABLE 6.25 (*Continued*)

• electric shock from 'step-and-touch' potentials	medium	• live sub-mains • live final sub-circuits • close proximity to earthed metal enclosures	
Working with supply isolated			
• Financial losses to shop	N/A	• Possible loss of trading time • Security issues from shop lifting	

Control measures for working live

The above scenario does not justify working live. In the rare case that it is found necessary to work live, Safe Work Australia's 'Managing Electrical Risks in the Workplace Code of Practice' recommends the following:

- Workers carrying out the electrical work must have or be provided with suitable, instruction and training in:
  - planning and preparation requirements for the carrying out of energised electrical work
  - safe work procedures, particularly those documented in safe work method statements
  - proper use of the relevant tools, testing equipment and PPE.
- First aid facilities must be provided at the workplace and they must be readily accessible.
- Emergency contact numbers should be made available at the workplace.
- Fire-fighting equipment that is suitable for electrical fires should be accessible.
- Management or those in control of the workplace must be consulted with before the electrical work is authorised.
- Energised conductors should be insulated where necessary to prevent inadvertent contact or flashovers.
- Unauthorised persons should be prevented from entering the work area, for example, through the use of barriers and signage.

Source: Safe Work Australia, 'Managing Electrical Risks in the Workplace Code of Practice' Section 7.3, p. 37.

As live work is considered a high risk activity the PCBU is required to develop a safe work method statement. The SWMS must:

- identify the electrical work
- specify the hazards associated with that electrical work and risks associated with those hazards
- describe the measures to be implemented to control the risks
- describe how the risk control measures are to be implemented, monitored and reviewed, and may include the risk assessment prepared for the relevant work.

Source: Safe Work Australia, 'Managing Electrical Risks in the Workplace Code of Practice', Section 7.3, p. 38.

An example of a safe work method statement for electrical isolation and testing is shown in Table 6.26.

TABLE 6.26 Sample electrical SWMS isolation procedure

Work activity:	**LOW VOLTAGE ELECTRICAL ISOLATION OF WIRING AND EQUIPMENT FOR MINOR WORK**	**SWMS Number**	**E-SWMS 001**
[Contractor Name, ABN, Office Address and Phone]		Client details	
Workplace location		Site Manager: Contact phone:	
Has the SWMS been developed based on a workplace-specific risk assessment?		☐YES ☐NO	
Have workers and their HSR(s) (if applicable) been consulted about the SWMS?	Workers☐ YES☐ NO☐ HSRs☐ YES☐ NO☐ N/A		
Person responsible for ensuring compliance with SWMS:	Phillip Chadwick	Date SWMS received:	
What measures are in place to ensure compliance with the SWMS?			
Person responsible for reviewing SWMS:		Date SWMS received by reviewer:	

(*Continued*)

TABLE 6.26 (*Continued*)

How will the SWMS be reviewed?			
SWMS review date:		Reviewer's signature:	

	TASK STEPS	HAZARDS and RISKS	RISK CONTROL MEASURES
1	Conduct/confirm workplace risk assessment HRA and secure area	Electric shock	Test before you touch Identify tasks and appoint competent personnel for work • Qualified licensed electrician • Instructed persons under supervision Inspect the work area and identify potential hazards Identify and review available SWMSs and SWPs Confirm suitability of work processes and record additional measures
2	Plan and ensure coordination of activities	Unscheduled disruptions to others Lack of coordination	Consult person/s with management control Consult with worker and other trades in the area Notify those potentially affected by power disruption
3	Is Safety Observer required?	N/A	Not required if not serious level of risk exposure
4	Safety Observer instruction (IF REQUIRED)	Unresponsive	Ensure Safety Observer is competent and trained in electrical rescue and cardio-pulmonary resuscitation (CPR) – current in last 12 months. Instruct Safety Observer regarding: • Planned work process • Emergency Procedures • Isolation point/s • Focus on the role of observing
5	Select tools, test equipment and PPE	Cuts, abrasions and shocks	Well maintained, suitably rated, checked and tested Rated for the application and calibrated (if applicable) Used properly
6	Identify ALL energy sources to be isolated	Electric shock	Test before you touch Check labelling, wiring schedules and drawings Confirm by switching where appropriate
7	Identify and assess isolation point/s	Electric shock Difficult access	Avoid touching exposed conductive parts Clearly mark or label Ensure unobstructed access of electrical workers Capable of being operated quickly
8	Keep work area clear of obstruction	Unnecessary distraction leading to electric shock	Clear area and use appropriate barricades and signage (where necessary) Maintain good housekeeping, tidy and clean
9	Confirm and record phase rotation if required for commissioning purposes prior to de-energising	Electric shock	Avoid touching exposed conductive parts
10	Switch power at isolation point on switchboard	Electric shock	Prevent inadvertent contact with exposed energised components
11	Secure the isolation point with lock and tag	N/A	Use lock and tag or personalised lock Provide contact details
12	Move to work area and confirm indication of power off		Confirm power off with minimal lifting of electrical covers, e.g. • Appliance not working • Lights off • Power point dead
13	Lift electrical cover/s as required	Electric shock from exposed electrical conductors	Test before you touch Use test meter, lamp or bell to detect energised Make positive contact with ALL the near exposed conductor Don't use a proximity sensor to prove de-energised
14	Check that testing meter/device is operational	Electric shock	Avoid touching exposed conductive parts
15	Undertake designated de-energised work		Be alert to possible changes to workplace conditions, including: • Possible new safety hazards or risks • Stop work if unsafe and immediately rectify or notify your supervisor TEST EVERY TIME BEFORE YOU TOUCH

(*Continued*)

TABLE 6.26 *(Continued)*

	TASK STEPS	HAZARDS and RISKS	RISK CONTROL MEASURES
16	Before leaving, make safe	Electric shock	Ensure that no isolated conductors can be energised by switching Terminate and test all conductors before energising Follow re-energisation procedure
17	On return, confirm integrity of isolation	Electric shock	Check switch/s and confirm isolation by re-testing, as above
18	Retain record for Supervisor/ Electrical Office	N/A	Retain if this SWMS is revised or a Risk Assessment was prepared Return to Supervisor/Office for ongoing review and improvement

Electrical worker's acknowledgement of instruction and training:

Name of worker(s)	Worker signature(s)

Risks when working on an electrical installation

Safety with using a multimeter on a live circuit

As outlined in Safe Work Australia's 'Managing Electrical Risks in the Workplace Code of Practice' (pp. 28–9), all electrical equipment and wiring should always be treated as live until proven de-energised by testing. Circuits can have interconnections with other circuits, or isolation points such as circuit breakers or switches can sometimes be incorrectly labelled. Cutting cables presents particular risks. Both ends of the cable should be checked for isolation prior to cutting. Depending on the situation, precautions may include:

- using a cable spiking or stabbing device to short the conductor to earth to prove the cable is de-energised
- a combination of proving the cable is de-energised and physically tracing out the cable.

Fault finding on circuits in some cases must be done live to prove or disprove operation of the circuit. Care must be taken to ensure any meters used are set to the correct functions and suitable ranges are selected before contact with the meter probes is made with live components. For example, a multimeter set to amperes and used to measure voltage will create a short circuit. Not only will the meter be damaged but arcing will also occur at the probe tips. In an enclosed space, such as a switchboard or terminal box, it is possible the air may be ionised (arcing), resulting in a much larger fault current.

When maintaining equipment, contact with electricity is not the only source of risk. Some types of equipment can store energy in springs, hydraulic reservoirs or through gravity. To prevent injuries such as crushing or entrapment, this energy must be discharged before work commences.

Control measures for electrical work

The primary control measure for electrical work is isolation, test and lockout. Other control measures include:

- training
- safe work method statements
- job safety analysis (JSA)
- PPE.

Table 6.27 shows an example of a JSA for the previous scenario involving replacing a faulty circuit breaker within a distribution board of a shopping centre. Once disconnected, tested and locked out, the work is considered low risk.

TABLE 6.27 Sample electrical job safety analysis

JOB SAFETY ANALYSIS SAMPLE

Company name:	Cengage Electrical	Date:	1/12/2013	JSA No	1006
Site Name:	Shop 2	Permit to work requirement		Yes ☐	No ☐
Contractor:	Phill Chadwick	Approved by:	Peter Phillips		
Activity:	Replacement of faulty circuit breaker shop 2 distribution board				

Activity	Hazards	Risk control measures	Who is responsible?
List the tasks required to perform the activity in the sequence they are carried out.	Against each task list the hazards that could cause injury when the task is performed.	List the control measures required to eliminate or minimise the risk of injury arising from the identified hazard.	Write the name of the person responsible (supervisor or above) to implement the control measure identified.
Notification of interruption of supply to shop staff	N/A	N/A	Electrician
Isolation test and lock out	Electric shock causing injury or death Arcing, explosion or fire causing burns Electric shock from 'step-and-touch' potentials	Isolate DB at MSB Use of suitable test equipment Use of suitable lockout device Appropriate PPE	Electrician
Replacement of faulty circuit breaker	Accidental re-energisation	Use of suitable lockout device	Electrician
Restoration of supply to shop and testing	Electric shock causing injury or death Arcing, explosion or fire causing burns Electric shock from 'step-and-touch' potentials	Removal of lockout device Close isolation device at MSB Use of suitable test equipment Appropriate PPE	Electrician

Source: Victorian WorkCover Authority, 'Job Safety Analysis', http://www.vwa.vic.gov.au/forms-and-publications/forms-and-publications/job-safety-analysis-worksheet-jsa

SUMMING UP...

- The model Work Health and Safety Regulations prohibit 'live' electrical work unless isolation creates a greater risk than working live.
- Safe Work Australia's 'Managing Electrical Risks in the Workplace Code of Practice' recommends control measures when working live including:
  - workers carrying out the electrical work must have, or be provided with, suitable instruction and training in planning and preparation, safe work procedures, and proper use of the relevant tools, testing equipment and PPE
  - first-aid facilities must be provided at the workplace and they must be readily accessible
  - emergency contact numbers should be made available at the workplace
  - fire-fighting equipment that is suitable for electrical fires should be accessible
  - management or those in control of the workplace must be consulted with before the electrical work is authorised
  - unauthorised persons should be prevented from entering the work area, for example, through the use of barriers and signage.
- As live work is considered a high risk activity the PCBU is required to develop a safe work method statement (SWMS).
- Risks when working on electrical installations include:
  - cutting live cables
  - contact with live conductors
  - arc faults caused by test equipment which is faulty or set to incorrect function or range
  - stored energy in equipment.
- Risk control measures for electrical work include:
  - isolation test and lockout
  - training
  - SWMS
  - job safety analysis
  - PPE.

REVIEW EXERCISES

Review questions 6.1

1 At what stage of work should risk sources be identified?

2 What is the name of a document that prioritises the hazards identified in a task?

Review questions 6.2

1 List three documents used to manage hazards (risk sources).

2 What is the difference between a job safety analysis (JSA) and a safe work method statement (SWMS) document?

Review questions 6.3

1 What is the voltage range of:
   a ELV
   b LV?

2 Name two types of ELV.

3 What is the purpose of a distribution transformer?

4 List three parts of an electrical installation that might operate at high current.

Review questions 6.4

1 What is the range of high voltage?

2 Explain the difference between 'clearance' and 'creepage' distances.

Review questions 6.5

1 Under what circumstances is it permitted to work live?

2 What type of document would be used to provide instruction to perform live electrical work safely?

3 Who is responsible for developing an SWMS and a JSA?

ONLINE RESOURCES

Sample safety forms

WorkSafe Victoria
http://www.worksafe.vic.gov.au > Forms and Publications > Safety and Prevention

Asbestos Awareness
http://www.asbestosawareness.com.au

Codes of practice: Managing electrical risks at the workplace, Confined spaces, etc.

WorkCover NSW
http://www.workcover.nsw.gov.au > Forms and Publications > Work, Health and Safety > Codes of practice

Documents related to HV work, including access permits

TransGrid
http://www.transgrid.com.au > Safety, health & environment > Health and safety > Power system safety rules

7

INTRODUCTION TO REGULATIONS, COMPLIANCE STANDARDS AND CODES

This chapter provides an introduction to the acts, regulations, compliance standards and codes which govern how you undertake or perform electrical work. While each state and territory has different acts and regulations they all have a common purpose. Regulations, acts, compliance standards and codes ensure safety for anyone associated with the generation, distribution, supply and consumption of electricity. The examples shown in this chapter relate to Queensland but all states and territories in Australia have parallel legislation. Rules that apply in tropical Queensland may not have relevance in cooler southern states.

Chapter outline

7.1 Regulations for undertaking electrical work

At present all the states and territories of Australia are working towards a national approach to occupational health and safety (OH&S; *Australian Work Health and Safety Strategy 2012–2022)* as well as a national approach to electrical safety and licensing through the Electrical Regulatory Authorities Council (ERAC). All of the acts, regulations, compliance standards and codes fall under the umbrella of OH&S (covered in more detail in Chapter 1).

Acts

Compliance with the *Electrical Safety Act*

An Act is a legal document voted on by Parliament and signed into law by the Governor. Regulation is the way or the roadmap on how to comply with the Act (see Figure 7.1).

A code of practice provides the nuts and bolts method for achieving compliance. It provides guidelines on how to meet the requirements.

In some cases you may not find any guidelines, so you will need to revert to hazard identification and risk assessment processes that may include a consultation process with manufacturers and experts in the field. This could be as simple as a phone call to a supplier or a full statement from an engineer with laboratory test results.

Compliance with acts and regulations is enforceable with penalties so you cannot choose to ignore them.

FIGURE 7.1 The legal framework, from legislation through to codes of practice

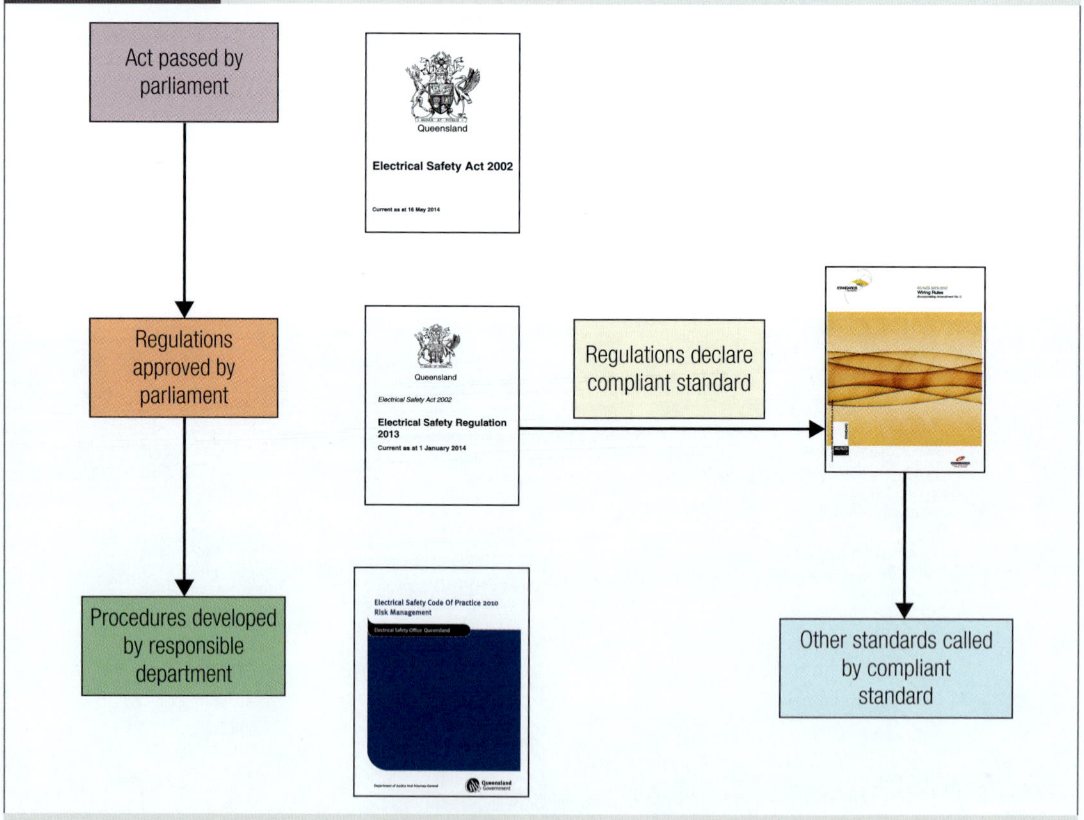

Sources: Government of Queensland, *Queensland Electrical Safety Act 2002*; Electrical Safety Office Queensland, 'Electrical Safety Code of Practice 2010 Risk Management'; Standards Australia, AS/NZS 3000:2007 Wiring Rules; Queensland Government, *Queensland Electrical Safety Regulation 2013* *© the State of Queensland (Office of the Queensland Parliamentary Counsel)*

The purpose of the Electrical Safety Act is:

- for installations and equipment to be inspected and tested to ensure they are safe
- to protect people and their livestock from being hurt or killed by electric shock
- to prevent damage to property from the effects of electricity.

To do this, the Act imposes obligations on electrical workers, supply industry workers, manufacturers and suppliers of equipment who may affect the electrical safety of others. These obligations include:

- establishing vicarious liability so that if an employee commits an offence, the employer is also guilty of that offence and liable to prosecution
- ensuring the standard of honesty and integrity demonstrated by the person in business activities (fit and proper person)
- setting standards for anyone, industry and/or the public through regulations and codes of practice for working on, near, adjacent to or around electricity (see Figure 7.2, which shows the Wiring Rules that are part of the regulations under the Electrical Safety Act)
- setting benchmarks for management systems for electricity entities, including power authorities, for the generation and distribution of electricity
- providing for the reliability and security of electricity supply
- providing a system for the oversight and issuing of licences for electrical workers and contractors
- providing penalties for breaches of the Act
- providing consumer protection against electrical work not being properly performed or completed
- establishing a consultative structure for industry, workers and the public to participate in improving electrical safety
- highlighting and implementing the efficiency of electrical equipment.

Regulations

The purpose of the regulations is to identify specific ways for meeting electrical safety obligations under the Act, or how to comply with the Act.

Briefly the regulations set out to define:

- safety and technical requirements for electrical work, electrical equipment and the works of electrical entities
- safety and technical requirements for work in contact with or near exposed parts
- requirements for safety management plans
- notification and reporting requirements for serious electrical incidents and dangerous electrical events
- all matters about electrical licences
- enforcing the electrical licensing provisions
- approvals of electrical equipment
- administrative matters pertaining to electrical safety regulation.

EXAMPLE 7.1

What do the regulations say about working on live equipment? Open a copy of the *Queensland Electrical Safety Regulation 2013*:

1 On the contents page you will find that this is covered in 'Part 3 Electrical work'.
2 In Part 3 you will find 'Division 1 Electrical work on energised electrical equipment'.
3 Follow down and you will find working live is mostly forbidden until you see '18 Electrical work on energised electrical equipment permitted in particular circumstances'. This essentially says that every effort must be made to avoid working on energised equipment.

You will notice there is not much usable detail, mostly cautions and warnings.

Each state and territory has different wording but the intent of the regulations is the same.

FIGURE 7.2 Electrical safety regulations reference the Wiring Rules

Part 6 Electrical installations

Division 1 Performing electrical work

70 Licensed electrical worker to comply with wiring rules

A licensed electrical worker who performs electrical work on an electrical installation must ensure the electrical installation, to the extent it is affected by the electrical work, complies with the wiring rules.

Maximum penalty—40 penalty units.

71 Person conducting a business or undertaking to ensure electrical installation complies

(1) A person conducting a business or undertaking that performs electrical work on an electrical installation must ensure the electrical installation, to the extent it is affected by the electrical work, complies with—

(a) if the electrical installation is to be used for construction work—the wiring rules and AS/NZS 3012 (Electrical installations—Construction and demolition sites); or

(b) if the electrical installation is not to be used for construction work—the wiring rules.

Maximum penalty—40 penalty units.

Note—

See section 22(3) of the Act for when a person conducting a business or undertaking is also a worker.

...

(b) a training program approved by the regulator that requires the person to perform electrical work.

training plan see the *Vocational Education, Training and Employment Act 2000*, section 13.

transfer date, for residential land, for part 6, division 4, see section 81.

transferee, of residential land, for part 6, division 4, see section 81.

transferor, of residential land, for part 6, division 4, see section 81.

type 1 safety switch, for part 6, division 6, see section 97.

type 2 safety switch, for part 6, division 6, see section 97.

unit contribution amount, for a financial year, for part 15, division 1, see section 273.

unsafe distance, for a person, operating plant or vehicle of an overhead electric line, see section 69.

untrained person, for an electric line, for schedule 2, see schedule 2, section 1.

vehicle, for part 5 and schedule 2, does not include—

(a) an aircraft; or

(b) a vehicle that is operating plant.

Examples for paragraph (b)—

1 A tip truck tipping a load would not be a vehicle for part 5 or schedule 2. However, a tip truck travelling between sites would be a vehicle for part 5 and schedule 2.

2 A vehicle that includes an elevated work platform being used for clearing vegetation would not be a vehicle for part 5 or schedule 2. However, when the platform is not being used for clearing vegetation, the vehicle would be a vehicle for part 5 and schedule 2.

WHS Regulation means the *Work Health and Safety Regulation 2011*.

wiring rules means AS/NZS 3000 (Electrical installations) (known as the Australian/New Zealand Wiring Rules).

Source: Queensland Government, *Queensland Electrical Safety Regulation 2013*, pp. 60–1, 243, https://www.legislation.qld.gov.au/LEGISLTN/SLS/2013/13SL213.pdf, © the State of Queensland (Office of the Queensland Parliamentary Counsel)

FIGURE 7.3 Electrical safety regulations require installation testing

Division 3 Testing of work

25 Testing of electrical equipment after electrical work

(1) This section applies if electrical work is performed on electrical equipment.

(2) A person who performs part or all of the electrical work, and is responsible for making the electrical equipment ready to connect to a source of electricity for use for its intended purpose, must ensure the electrical equipment is tested as required under subsections (6) and (7).
Maximum penalty—40 penalty units.

(3) A licensed electrical contractor or electricity entity who performs the electrical work of connecting the electrical equipment to a source of electricity for use for its intended purpose must ensure the electrical equipment is tested as required under subsections (6) and (7).
Maximum penalty—40 penalty units.

(4) If a training person performs the work mentioned in subsection (3) under the supervision of a licensed electrical contractor, subsection (3) applies to the licensed electrical worker and not the training person as if the licensed electrical contractor were performing the work.

(5) Subsection (4) does not require a distribution entity to ensure the testing of the electrical equipment if—

(a) the electrical work is, or is part of, electrical work performed on an electrical installation; and

(b) a licensed electrical contractor has given the distribution entity a notice under section 226 about the testing of the electrical work performed on the electrical installation.

(6) The testing of the electrical equipment must be directed at ensuring that the electrical equipment, to the extent it is affected by the electrical work, is electrically safe.

(7) If the electrical equipment is energised for testing, the person who performs the test must ensure persons not necessary for the testing are electrically safe.

Source: Queensland Government, *Queensland Electrical Safety Regulation 2013*, 'Division 3. Testing of work', p. 30, https://www.legislation.qld.gov.au/LEGISLTN/SLS/2013/13SL213.pdf, © the State of Queensland (Office of the Queensland Parliamentary Counsel)

Codes of practice

A code of practice does not specify everything that you should or should not do to meet your electrical safety obligations. However, you would fail to meet your obligations if you:

- did something that was against or inconsistent with a code of practice
- did not try to adopt a method as safe as, or safer than, that laid out in the code.

There are five codes of practice supporting the *Electrical Safety Act 2002* that give practical advice on how to meet your electrical safety obligations.

A code of practice expires 10 years from its commencement date. The making of a new code of practice must be gazetted.

It is important to note that the codes of practice do not cover everything you should do to meet your electrical safety obligations, but do provide you with a guide to getting started and practically achieving the desired outcomes.

EXAMPLE 7.2

You are about to test for voltage in a live switchboard with the front cover removed:

You will need to consult the 'Managing Electrical Risks in the Workplace Code of Practice' (Safe Work Australia).

1 Turn to the first page titled Scope And Application.
2 You will find that the information about live work is in Part B (Chapters 4–10).
3 Turn to the section titled Part B: Electrical Work.
4 Locate Chapter 7 Risk Controls – Energised Electrical Work.
5 The first item involves a caution – 7.1 Prohibition on energised electrical work.
6 You will see a reference to one of the electrical safety regulations (s18). Further reading reveals the procedures to be followed in great detail.

This shows how the codes of practice support the regulations but provide detailed additional information.

Codes of practice

The five relevant codes of practice are:

1 The 'Electrical Safety Code of Practice 2010 – Working Near Exposed Live Parts' applies to people such as:
   - crane operators
   - plant operators
   - painters
   - carpenters
   - people erecting or working on scaffolds
   - sign makers
   - agricultural workers
   - carriers of high loads
   - electrical repairers
   - builders working near exposed live parts
   - anyone who clears vegetation near overhead lines
   - electrical workers.

   If you do not fit into one of these categories, but are working around exposed live parts, this code of practice still applies to you.
2 The 'Electrical Safety Code of Practice 2013 – Managing Electrical Risks in the Workplace' applies to you if you are:
   - a licensed electrical worker
   - an employer or self-employed licensed electrical worker
   - a licensed electrical contractor.

   As a licensed electrician you will most likely be an electrical fitter or electrical mechanic. Other licensed electrical workers include:
   - restricted electrical licence holders
   - electrical work training permit holders
   - electrical lines people
   - cable jointers.

   Figure 7.4 shows how the code of practice lists risks associated with electrical work.
3 The 'Electrical Safety Code of Practice 2010 – Works' sets out practical advice for you or your employer on ways you can ensure electricity networks remain safe if you are involved in electricity generation, transmission or distribution.
4 The 'Electrical Safety Code of Practice 2010 – Electrical Equipment Rural Industry' applies if you or your employer are undertaking rural industry work in which case you will need to comply with special extra requirements.

5 The 'Electrical Safety Code of Practice 2010 – Risk Management' is for your special protection as an electrical worker. In it you will find how you should approach live work as well as codes of practice with practical advice on ways to manage your obligations under the Act and to help you comply with the regulations.

FIGURE 7.4 The 'Electrical Safety Code of Practice 2013 – Managing Electrical Risks in the Workplace' lists risks associated with electrical work

APPENDIX C – RISKS ASSOCIATED WITH ELECTRICAL WORK

Activity	Risks
Isolation and access	• Correctly isolating supply but not discharging residual energy e.g. a capacitive charge may be present in power supplies, single-phase motors or high power factor fluorescent fittings. • Insulation and equipment failing or partially breaking down. • Earth connection failing to stop an electric shock in earthed conductive parts when step and touch potentials exist. • Carrying out the task causes a person, something a person may be handling or something a person is in contact with to intrude into minimum safe approach distances. • A power system conducting fault current or being subject to high inrush currents. • Instructions or markings on the parts being inadequate, incorrect or both. • Using equipment not designed for, or capable of, an operation e.g. opening a 'no load – bus tie' under load conditions or relying on an open circuit breaker as an isolation point. • Another person energising circuits while a worker is working on them, or a vehicle hitting a pole. • Natural elements (i.e. lightning or wind) causing static charges, overhead mains to clash or a high-voltage circuit to fall onto a low-voltage circuit. • The inter-core capacitive effects of long multi-phase cables. • Changes to wiring not being reflected in drawings i.e. the drawings are not 'as built' e.g. a live control or supervision circuit being present though the drawing indicates otherwise. • If there has been an error in wiring, opening the isolator may not de-energise the switchboard e.g. if incorrect connection (incorrect polarity) occurred in the service to an installation, opening the main switch will open circuit the neutral rather than the active. • Intentionally disabling an interlock to perform a task e.g. opening the shutter of a 'rackable' circuit breaker test to prove de-energised in the orifice. • Inadvertently disabling an interlock while performing a task e.g. in a switchboard with an integrated circuit breaker, isolator and earth switch, the operator accidentally moving the isolator into the earthed position. • Poor direction and insufficient knowledge e.g. a worker is instructed to apply a set of earths and short circuits at a Ring Main Unit (RMU). The worker correctly observes the isolator is open, however they assume the earth switch can be closed because the isolator is open. As most RMUs are configured so the earth switch earths the cable, not the busbar, it is possible the worker would be earthing and short-circuiting a live circuit. • When applying a set of portable earths and short-circuits, accidental or inadvertent contact is made with live parts. If this occurs, the worker is using a device that is conducting fault current. • The threshold value (lowest level of indication or reading) of a test device causing a misleading interpretation of a test to prove de-energised. Depending on the device used, an indication that parts are not energised in a high-voltage situation does not mean that low-voltage and direct current voltages are absent. • Application of earthing and short-circuiting devices that depend on a conductive path through a fuse or circuit breaker that is not fit for purpose. • Ineffective connection to the general mass of the earth e.g. the electrode, grid or temporary electrode that the earth and short circuits relies upon in a situation where a single phase becomes energised. • Application of the short circuit portion of portable earthing devices prior to the earth tail being connected to the earth. • Arcing and splattering associated with the application of earths and short circuits, causing a risk. The arcing or splattering may result from using the device in situations that range from energised conductors to residual energy such as capacitance. If the parts are energised, the worker can draw the arc from one phase to the other, causing a phase-to-phase fault. • A potential electric shock path existing once the earth tail is connected to earth. A worker may touch another live part and the earthed connector at the same time, for example in a Common Multiple Earthed Neutral (CMEN) area, even when working on high-voltage, contact between the earthed connector and a low-voltage phase can cause an electric shock.

(*Continued*)

FIGURE 7.4 *(Continued)*

Activity	Risks

Working near sources of arcing, explosion or fires	Arcs, explosions and electrical faults can cause burns. Workers should be protected from the effects of burns. Examples include:
	• materials providing a conductive path between sources of potential, for example uninsulated tools falling across busbars
	• abnormal conditions on circuits such as:
	  • lightning striking mains
	  • circuits of different voltages touching each other e.g. high-voltage contacting low-voltage circuits
	  • high voltage in the secondary circuit of a current transformer if an open circuit occurs when current is flowing in the primary circuit
	• abnormally high voltages when synchronising different supplies. For example, if the waveforms are 180° out of phase, twice the peak-to-peak voltage may be imposed
	• voltage multiplication effects, including:
	  • ferro-resonance where the capacitive and inductive components of underground cables and transformers can significantly increase voltages when single-phasing occurs
	  • re-strike can occur if capacitors are energised, de-energised and re-energised in rapid succession
	• leakage or electrical discharge causing insulation to be compromised, for example a combination of a build-up of contaminants on insulators, wet weather or tracking through air voids in pitch filled insulating chambers
	• failure of insulating mediums.

Source: Electrical Safety Office Queensland, 'Electrical Safety Code of Practice 2013 – Managing Electrical Risks in the Workplace', pp. 50–1

SUMMING UP...

- Laws passed by parliament are enforced by penalties.
- The Electrical Safety Act aims to prevent fire and shock from electricity use.
- Regulations are passed by Parliament to implement the Act.
- Electrical regulations cover licensing and approved equipment.
- Codes of practice are documents made under the Act.
- Codes of practice give practical advice on ways to comply with regulations.
- Codes of practice are a minimum requirement for safe practice.

7.2 Standards philosophy and format

You will need to make sure you understand some of the methods or terms that are used to describe how standards and in particular technical standards apply. Standards have the following levels of importance:

- prescriptive – giving direction or telling you the way something must be done
- performance – particular way of doing things or aiming for a particular outcome
- descriptive – explaining what should happen or what the outcome should be
- normative – setting a norm or standard by prescriptive means
- informative – setting a norm or standard by being descriptive
- deemed to comply – by using or adopting a particular standard, especially for testing. Your work is deemed to comply if carried out to that standard, or the installation meets that requirement.

Electrical standards

The Wiring Rules

In the electrical industry you will use many standards but there are a couple of technical standards that you should be very familiar with. One in particular is **AS/NZS 3000:2007 Wiring Rules**, the technical standard for electrical installations.

This standard is also the one that has been mandated in every state and territory of Australia by legislation, in particular Electrical Safety Acts and regulations. This standard is a compliance standard and enforceable by regulation, in particular electrical safety regulations. Many other compliance standards are called up by **AS/NZS 3000:2007** and include:

- **AS/NZS 3004 Series Electrical Installations – Marinas & Recreational Craft** (*Normative*)
- **AS/NZS 3008.1.1:2009 inc A1. Electrical Installations – Selection of Cables** (Australia) (*Normative*)
- **AS/NZS 3012 Electrical Installations – Construction & Demolition Sites** (*Informative*)
- **AS/NZS 3017 Electrical Installations – Testing & Inspection Guidelines** (*Informative*)
- **AS/NZS 3760 In Service Safety Inspection & Testing of Electrical Equipment** (*Informative*)
- **AS/NZS 4836 Safe Working on Low Voltage Electrical Installations** (*Informative*).

Meaning of references

The meanings of the references in the Wiring Rules are:

- AS – applicable to Australia only.
- NZS – applicable to New Zealand only.
- IEC – International Electrotechnical Commission sets standards that apply specifically to electrical industries internationally and has representatives from electrical industries around the globe.
- ISO – International Organization for Standardization represents several countries and recommends best methods that apply internationally, especially where equipment is manufactured overseas and then imported or exported around the world. It can even apply where no applicable Australian Standard exists. These standards can also be used for accreditation purposes for dealing with other like-minded companies, for example, ISO 9001 for company accreditation or ISO 14001 for environmental accreditation.
- BS – British Standards are applicable to the countries making up the United Kingdom, but can be used where no applicable Australian Standard exists. BS88 sets standards for fuses as these are usually manufactured in Britain.

History

AS/NZS 3000:2007 Wiring Rules has been put together in agreement among stakeholders from various sectors including the:

- Australian Building Code Board
- Association of Consulting Engineers Unions
- Australian Industry Group
- electrical contractors associations (state and national)
- energy utilities
- communications unions and associations in both Australia and New Zealand.

Before 1931 various wiring and installation standards were developed to address the problems that insurance companies identified from the many building fires caused by faults in electrical installations. No fully uniform national wiring rules existed until 1931, when the Wiring Rules were first published as part of AS CC1-1931. After nine revisions, Australia and New Zealand agreed to the joint publication of **AS/NZS 3000:2000**. The most recent edition is **AS/NZS 3000:2007** and has had two amendments since publication.

Layout

The Wiring Rules are divided into two parts:

- Part 1: Scope, Applications & Fundamental Principles
- Part 2: Installation Practices & Appendices.

Part 1 has nine sections:

1 Scope
2 Application

Navigating the Wiring Rules

3 Referenced documents
4 Definitions
5 Fundamental principles
6 Design of an electrical installation
7 Selection and installation of electrical equipment
8 Verification (inspection and testing)
9 Means of compliance.

Part 2 has very specific sections which also contain subclauses under each of them. The nine sections in Part 2 are:

1 General arrangement, control and protection
2 Selection and installation of wiring systems
3 Selection and installation of appliances and accessories
4 Earthing arrangements and earthing conductors
5 Damp situations
6 Special electrical installations
7 Verification
8 Appendices A-M
9 Index.

> In the AS/NZS 3000:2007 index, a wiring enclosure can also be a called a conduit or a duct; a wiring system can be a cable or TPS or building wire in conduit; a bend can also be known as a change of direction.

When using the Wiring Rules you need to:

1 find the specific information
2 relate that information to the situation.

To assist you to find the information you need, you should use a combination of the contents page and the index. The contents page can be very useful to get you into the specific area or section you are after. Be aware when using the index that sometimes different names are used and if you understand this it will make your search easier.

EXAMPLE 7.3

Using AS/NZS 3000:2007

You need to know the size of protective earth (main earth) wire required for 16 mm² mains cable:

1 From the contents page you find that you have to go to Section 5 Earthing arrangements and earthing conductors.
2 Clause 5.3 has all the earthing system parts.
3 Sub-clause 5.3.3 has earthing conductor size, so you will go there.
4 There you will see sub-clause 5.3.3.1 is specific to protective earthing conductors.
5 Look down to 5.3.3.1.2 Selection.
6 Reading down a little further to sub-clause a, it refers you to Table 5.1.
7 Read down the column to find for 16 mm² mains you need a 6 mm² protective earth wire.

Service and installation rules

AS/NZS 3000:2007 is mandated as the minimum requirement for electrical installations across Australia and New Zealand. In addition to this minimum standard are added some more localised rules from supply authorities in each state and territory. These rules are the service and installation rules (SIR), also known as the 'Metering Manual' due to the inclusion of a lot of information around the installing of meters. Figures 7.5 to 7.8 show samples of the rules of different distributors within Australia. For web links to these documents go to the end of this chapter.

Building Code of Australia (BCA)

As communities become more aware of environmental issues they are becoming increasingly more concerned about ways to improve or lessen their impact on the environment. As an electrician you are responsible for ensuring that your customer's installation meets the requirements of the Building Code of Australia (BCA) that require, where possible, energy efficient appliances and fittings to be installed and used. These range from star ratings on appliances for water usage and energy efficiency to movement sensors for lighting and air-conditioning control.

You should always advise customers to install the most energy efficient systems possible.

FIGURE 7.5 Example of service and installation rules document for Aurora Energy

DISTRIBUTION BUSINESS

DOCUMENT TITLE:

DISTRIBUTION STANDARD:

SERVICE AND INSTALLATION RULES

DOCUMENT I.D. NUMBER:

DS-I-CA-01

REV NO.	DATE	REVISION DESCRIPTION	APPROVAL
0	Aug 01	Original Issue Prepared by P Findlater	Group Manager Asset Management
1		Total Review by C Nye and R Cooper	GMSAM
2	Jan 2012	Revision	
3	Mar 2013	Revise and Update	Metering Assets Manager

Source: Aurora Energy, *Distribution Standard: Service and Installation Rules*, March 2013

The building code lays out some very straightforward guidelines, and for your customer to obtain approval and/or occupancy these guidelines must be met. As with other standards, this code lays out minimum material and performance requirements. You will find the BCA in two volumes:

- Volume One – building classes 2–9 in nine sections
- Volume Two – building classes 1 and 10 in three sections.

Building codes

Volume One

The following details on Volumes One and Two have been adapted from the Wood Solutions Design and Build website.

FIGURE 7.6 Example of service and installation rules documents for SA Power Networks

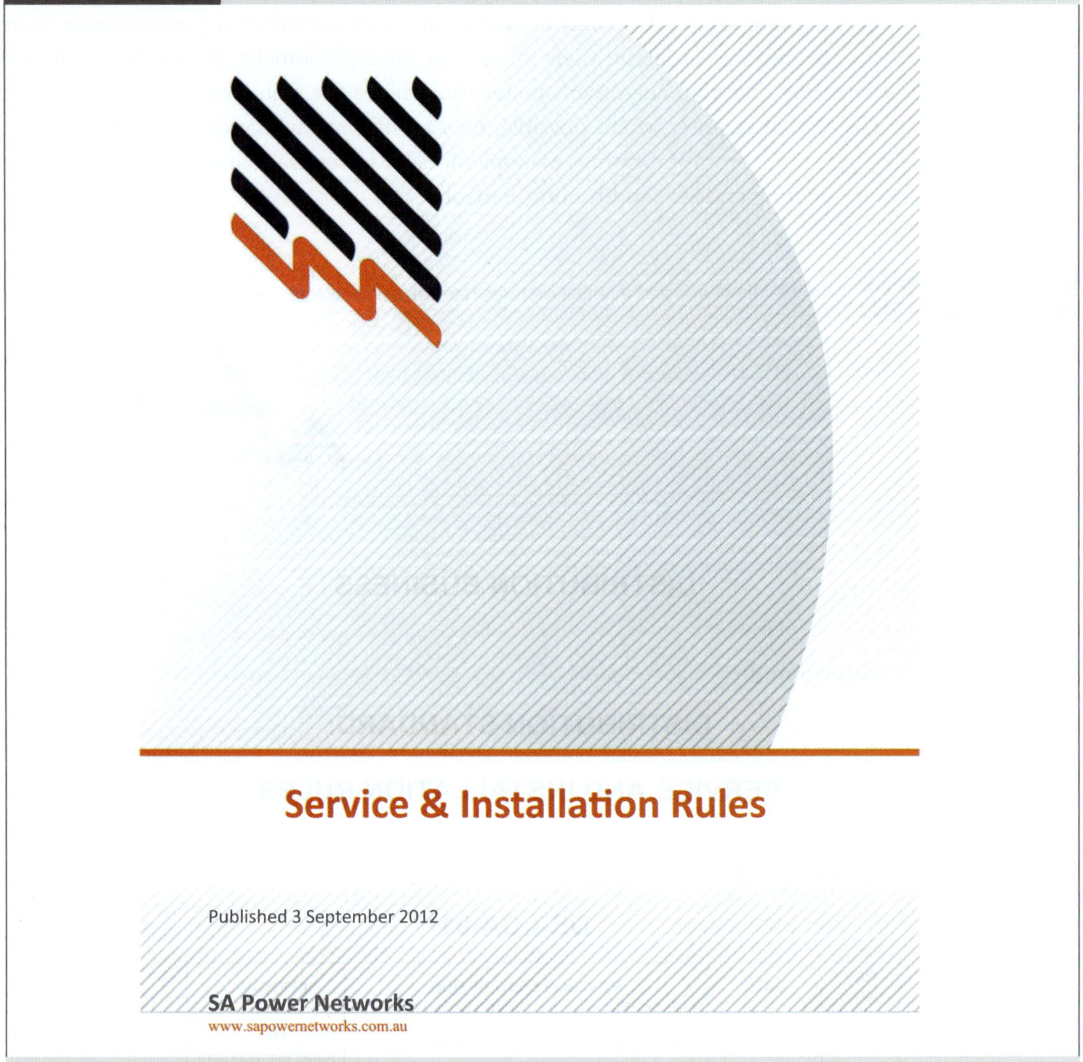
Service & Installation Rules

Published 3 September 2012

SA Power Networks
www.sapowernetworks.com.au

Source: SA Power Networks, *Service & Installation Rules*, 3 September 2012

The first section of Volume One is devoted to administrative provisions, while the remaining eight sections contain technical provisions. Each technical section has the following structure:

- Objective – a statement reflecting community expectations.
- Building Solutions – Practices used in the construction industry 'deemed-to satisfy' due to proven reliability. Innovative design is allowed but requires certification by an expert in the particular field.
- Functional Statements – how a building solution will comply with the objective.
- Performance Requirements – the minimum requirements a building solution must meet.

FIGURE 7.7 Example of service and installation rules document for Victorian Electricity Distributors

Source: Victorian Electricity Distributors, *Service and Installations Rules 2014*, 2014

Section A – General Provisions

Contains definitions, means of classification and the list of referenced documents.

Section B – Structure

Contains structural stability requirements such as resistance to:

- dampness
- insect invasion
- water penetration.

FIGURE 7.8 Example of service and installation rules document for ActewAGL

ActewAGL for you

SERVICE & INSTALLATION RULES FOR CONNECTION TO THE ELECTRICITY DISTRIBUTION NETWORK

INCLUDING

APPENDICES

Authorised when accessed at www.actewagl.com.au

(WHEN PRINTED THIS DOCUMENT IS UNCONTROLLED)

Note: User must be connected to the Internet to view linked drawings

Source: ActewAGL, *Service & Installation Rules For Connection to the Electricity Distribution Network*, September 2013

Section C – Fire Resistance

There have been many spectacular fires in history, two of them being the Great Fire of London (1666) and the Great Chicago Fire (1871). This section holds the accumulated wisdom from investigations into these historical events. There are details about preventing the start of a fire, the prevention of the spread of fire within a building, and the containment of fire within the affected building. Buildings are required to maintain their structural stability during a fire to allow evacuation of occupants and to provide relative safety for firefighters attempting to bring the fire under control. Spacing between buildings and construction of fire walls contributes to this effort, as does dividing the building into fire isolated zones including fire isolated stairways for exit and shelter.

This section contains terminology that you need to understand when working in the construction industry as an electrician to ensure you don't inadvertently leave fire barriers compromised when finishing at a work site.

Construction type refers to the level of fire resistance that particular elements of the building must achieve. There are three types of construction determined by the building's class and rise in storeys. These are:

- buildings that have higher risk such as high rise, high occupancy requiring most fire resistance
- buildings such as two storey units or boarding houses, and three storey buildings such as office blocks or shops
- one or two storey buildings that have lower risk and require least fire resistance.

Fire source feature is the part of the building that is the most likely entry point for a building fire spreading from another property. Such a point could be:

- the external wall of a unit or outbuilding on the same lot
- a side or rear boundary
- the far boundary of a road adjoining the lot.

Boundaries are regarded as fire source features because the owner of one property has no control over what happens on adjoining properties. Potentially a building might be built up to the boundary of the neighbouring property.

You must not compromise the fire safety of a building by leaving cableways unsealed.

Fire resistance level (FRL) is the resistance to the passage of fire achievable by a structural object in a building. Measured in minutes it comprises:

- structural adequacy
- structural integrity
- structural insulation.

Sole occupancy unit is a room or part of a building designed to be occupied by only one person on a continuous basis.

EXAMPLE 7.4

A structural element is described as having a 90/30/60 FRL. This means that it can achieve a fire level of:

- structural adequacy for 90 minutes
- structural integrity for 30 minutes
- structural insulation for 60 minutes.

Section D – Access & Egress

Emergency exits are a vital survival provision and the BCA is very specific about their:

- number
- size
- type
- separation of fire exits
- distance from fire exits.

The provisions are dictated by the building's class, size and occupation density. Important provisions relate to the accessibility of a building for people with disabilities.

Yellow fire extinguishers are illegal because they contain chemicals that destroy the ozone layer. If you find one you must advise the building owner.

Section E – Services & Equipment

You will notice the fire safety equipment installed in a building which should include:

- fire-fighting equipment (e.g. fire hydrants, hose reels, portable fire extinguishers, etc.)
- smoke hazard management (e.g. smoke detectors and alarms), emergency lighting and signs
- requirements for lift installations when one is installed in a building.

As a general rule all buildings require portable fire extinguishers to be installed, but provision of any other equipment depends on the building size.

Section F – Health & Amenity

Health and amenity are an important consideration for most buildings and include:

- damp and weather proofing
- sanitary facilities – determined by the occupancy level. Consideration must be given to sanitary facilities for people with disabilities
- room sizes
- light and ventilation – natural as well as artificial. Some buildings do not require natural ventilation or lighting
- sound insulation.

Section G – Ancillary Provisions

Ancillary structures are not necessarily part of the main building but add to the amenity and comfort. Such structures include:

- minor structures (swimming pools, vaults, cool rooms)
- heating appliances (fireplaces, flues and chimneys)
- atrium construction.

Buildings constructed in alpine areas and bushfire prone areas require special provisions to help them withstand such hostile environments.

Section H – Special Use Buildings

Some buildings are built for one purpose. Examples of these include:

- public halls
- cinemas
- bowling alleys.

Section I – Maintenance

As with all things in the built environment buildings must be maintained if they are to remain safe to use.

Volume Two

As previously mentioned, Volume Two is split into three sections. The first section contains administrative provisions while the second section sets out performance provisions and the third sets out acceptable construction provisions.

Section 1 – General Provisions

Here you will find information such as definitions, how buildings are to be classified and the list of guidance documents such as Australian Standards that have been referenced by the BCA as deemed-to-satisfy solutions.

Section 2 – Performance Provisions

This section is split into five subsections entitled:

- Structure
- Damp & Weatherproofing
- Fire Safety
- Health & Amenity
- Safe Movement & Access.

Within each subsection the Objectives, Functional Statements and Performance Requirements are stated.

Section 3 – Acceptable Construction

You will most likely hear this section referred to as the 'acceptable construction manual'. In it you will find the deemed-to-satisfy building solutions for Class 1 and 10 buildings. The section is split into 11 subsections:

- Site Preparation
- Footings & Slabs
- Masonry, Framing
- Roof & Wall Cladding
- Glazing
- Fire Safety
- Health & Amenity

- Safe Movement & Access
- Additional Construction Requirements and Structural Design Manuals.

SUMMING UP...

- A standard that is called up in an Act or regulation is mandatory.
- Standards are put together via consultation.
- Service installation rules are localised.
- Standards set minimum requirements for safety and operation.
- The Building Code of Australia (BCA) lays out minimum requirements for buildings.
- The BCA is divided into two volumes:
  - Volume One – nine sections
  - Volume Two – three sections.

7.3 Purpose, format and content of typical job specifications

Electricians and contractors are called upon to do varying types of electrical work. This can be from quoting for a job or undertaking the job after the tender has been won or the work assigned.

Job specifications

The detail of the job is often in the form of a contract or job specification.

Contracts

For smaller customers often a simple contract is all that is required. The contract needs to be signed by all parties before work commences. It usually includes:

- the names of all parties
- relevant addresses for job, contacts, etc.
- a brief title
- when work is expected to commence and when work is expected to be completed
- what work is to be done
- where the work is to be undertaken
- who is to be involved
- what standards, codes, drawings or compliance requirements apply
- what specifications are required
- how payments are to occur
- when payments are due
- any insurances required
- any warranty/maintenance periods that will apply and what is or is not covered in the warranty.

There may be other items added to the contract but the essential items will be the same. Above all don't forget to obtain written commitments setting out costs and payment methods. Include details such as halfway and completion payments including who bears the cost of any extra work that may develop as the project progresses.

> All contracts should contain details relating to the disposal of dangerous wastes such as asbestos and garden chemicals found on the premises during the project.

Electrical work specifications

The contract or clauses in the contract may call upon specifications for electrical work. The purpose of the electrical specification is to inform the contractor/electrician the details of the work to be covered and may apply to specific types of work such as:

- a new installation
- a renovation or extension to an existing building or job

- maintenance and overhaul
- shut down/strip out work.

The specification forms a legally binding document or contract and clearly sets out:

- whether it is a quote or tender for work
- who the parties are that tendered for the job and who the job has been given to
- who the job owner(s) is/are
- any drawings used or to be used
- how the job is to be run, e.g. planning the work, ordering materials
- who is responsible for materials arriving and stored on-site
- what compliance requirements are needed
- any issues with reference to inductions, amenities, etc.
- any fees applicable and who they go to and when
- particular details that clarify the work.

NATSPEC

Often electrical work specifications containing the detailed work sections are generated by computer software. Software is one method to streamline the processes while including all the relevant information. One such specification software is NATSPEC.

Electrical work sections in NATSPEC cover:

- electrical general requirements
- design and installation
- cable support and duct systems
- low voltage power systems and requirements
- stand-alone generating plants
- proprietary switchboards
- as-built or custom-built switchboards
- components of the switchboards
- any power factor correction requirements
- any UPS requirements
- lighting requirements
- specialised lighting
- communication, network and data cabling
- emergency lighting requirements
- fire/security alarms and detection systems
- emergency intercommunication and warning systems
- lightning or surge protection.

When using a computer you should always pay attention to the location and placement of keyboard, screen and seating.

The work sections are broken down into four specific sections similar to the sections or clauses in **AS/NZS 3000:2007** with subclauses or subsections where required. These sections are:

1 General:
    - Aims
    - Cross references
    - Standards
    - Interpretations
    - Inspection
    - Submissions

2 Products
3 Execution
4 Selections.

Before undertaking any electrical work, read the specifications carefully and understand clearly what is being said or required of you. Never make assumptions: ask the architect, engineer or builder.

SUMMING UP...

- Electrical specifications detail the work to be carried out.
- Watch out for details of specific work.
- Agree on payments before you start.
- Read the specifications and get clarification in writing.
- Once you take on the job you are responsible for your part of the work.

REVIEW EXERCISES

Answer these questions and check your answers at the back of the book.

Review questions 7.1

1 What is the purpose of Electrical Safety Acts?
2 What obligations are imposed on electrical workers?
3 What is the purpose of electrical safety regulations?
4 What is the purpose of a code of practice?

Review questions 7.2

1 State the year the standard AS/NZS 3000:2007 was jointly published.
2 State the volume and section of the building code that covers 'fire resistance'.
3 Describe the organisation that would publish service and installation rules.
4 What level of importance does the AS/NZS 3000:2007 have?

Review questions 7.3

1 List three types of work covered by an electrical work specification.
2 In what way is a contract legally binding?
3 List three important items that must be on a contract.
4 When does a tender become a binding agreement?

ONLINE RESOURCES

Building Code of Australia (BCA): Australian Building Codes Board
http://www.abcb.gov.au > About the National Construction Code > The Building Code of Australia

Codes of Practice: Queensland Government Department of Justice and Attorney-General
http://www.justice.qld.gov.au

Electrical safety forum: ICEweb Cool Engineering
http://www.iceweb.com.au > ICE web > electrical web > electrical safety > electrical safety for engineers and supervisors

NATSPEC: Construction Information
http://www.natspec.com.au

Standards International Australia: SAI
http://www.saiglobal.com

Service installation rules

Queensland: Ergon Energy
http://www.ergon.com.au

South Australia: SA Power Networks
http://www.sapowernetworks.com.au

Victoria: Victorian Electricity Distributors
http://www.victoriansir.org.au

New South Wales: NSW Trade & Investment
http://www.energy.nsw.gov.au

NATSPEC: Construction Information
http://www.natspec.com.au

Australian Capital Territory: ActewAGL
http://www.actewagl.com.au

Tasmania: Aurora Energy
http://www.auroraenergy.com.au

Western Australia: Government of WA Department of Commerce
http://www.commerce.wa.gov.au/publications/wa-electrical-requirements-waer

8

WIRING SYSTEMS TYPES, APPLICATION AND TERMINATIONS

This chapter introduces the concepts of wiring systems, cable types, enclosures and limits on the use of each type. When selecting cables you must ensure that the most economical system is installed without compromising the safety of the consumer or their property. The rules referenced in this chapter were correct at the time of writing but the standards are living documents and subject to continual revision. Reading this chapter should not be considered a substitute for reading and understanding the relevant Australian standards as amended.

While AS/NZS 3000:2007 Wiring Rules Section 3 Selection and installation of wiring systems is the main reference, the chapter refers to many other sections and standards such as Section 1 for definitions and AS/NZS 3008.1.1:2009 Electrical installations – Selection of cables for cable selection.

Chapter outline

8.1 Cable types and terminations

Preparing to select and install low voltage electrical wiring systems

Cables and their installation and protection will form an important part of your work as an electrical practitioner. The correct selection of cable and wiring systems is a vital skill and fundamental to the prevention of fire and electric shock.

Cable core

The basic component of the electrical cable is the cable core. A cable core comprises the conductor for carrying current, usually copper or aluminium, and the insulation to prevent unwanted circuit paths developing.

AS/NZS 3000:2007 Wiring Rules clause 1.4.19 defines a cable core, and **AS/NZS 3000:2007** clause 1.4.31 defines a conductor. For the purposes of the Wiring Rules a conductor only refers to the parts of the circuit between the supply and the load. This excludes the windings in transformers, motors and generators as well as heating elements.

Cables

A cable comprises one or more cable cores. The cable may include an outer protective sheath of plastic or metal for additional mechanical protection depending upon where and how the cable is to be installed. **AS/NZS 3000:2007** clause 1.4.17 defines a cable.

Typical single insulated cables (see Figure 8.1) encountered in the electrical industry include:

- PVC insulated copper cable known by industry practitioners as 'building wire'
- insulated aerial conductors
- PVC single flexible cables
- PVC twin flexible cables usually known as 'figure eight'
- glass fibre insulated cables
- telephone cable known as 'hook-up wire'
- bell wire
- instrument or data cable
- ribbon cable.

FIGURE 8.1 Single insulated cable

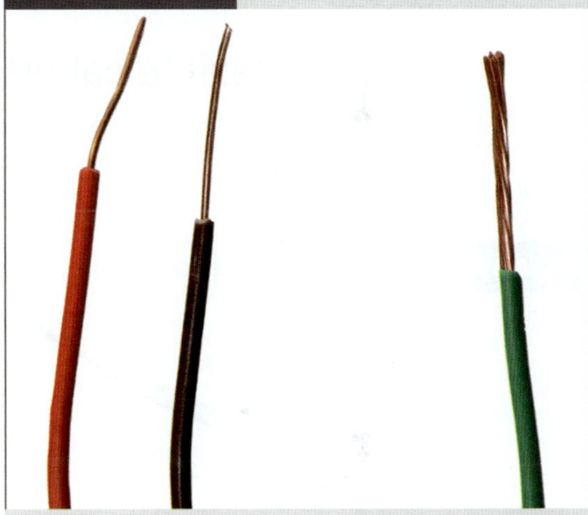

Source: Ed Hawkins

Most of these cables are not suitable for use as fixed wiring in low voltage installations. Building wire is acceptable as it is designed for use with low voltages and is made from reasonably coarse strands of conductor compared to flexible cable. This cable is commonly used in light and power installations. It is available as stranded or solid conductor.

Manufacturers specify the number and diameter of the strands as well as the cross-sectional area of the conductor; for example, 1 mm^2 (1/1.13) has one single strand 1.13 mm in diameter while 2.5 mm^2 (7/0.67) has seven strands 0.67 mm in diameter. This is equivalent to 2.5 mm^2 (1/1.78); the stranded conductor is physically larger but is rated at the same current capacity.

Cables are further classified by the type of protective covering provided with them. For example, **AS/NZS 3000:2007** clause 1.4.23 defines a sheathed cable (see Figure 8.2).

Other types of cable defined in **AS/NZS 3000:2007** are:

- clause 1.4.18 Armoured cable (see Section 8.8 in this chapter)
- clause 1.4.21 Mineral insulated metal sheathed (MIMS) cable (see Section 8.7 in this chapter)
- clause 1.4.22 Neutral screened cable where one of the conductors is in the form of a serving of conductor strands wrapped around the conductor cores. The screen is suitable to be used as an earth or neutral for the circuits supplied by the active conductors.

FIGURE 8.2 Single core sheathed cable

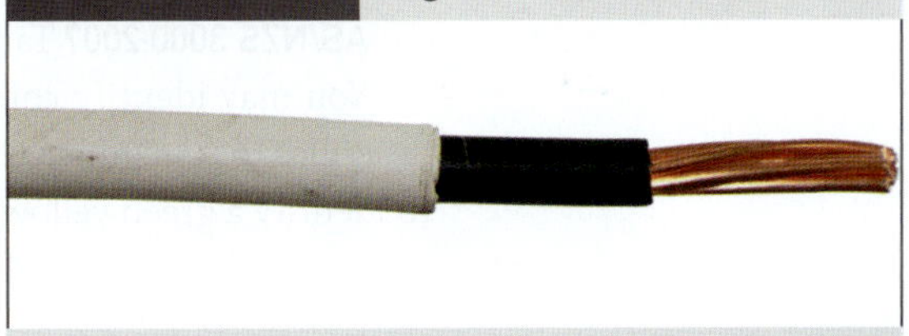

Source: Ed Hawkins

Cable selection

When selecting a cable for a particular application several features must be taken into account:

- The insulation must have temperature and voltage ratings sufficiently high to prevent breakdown between the poles of the circuit. **AS/NZS 3000:2007** Table 3.2 shows a range of low voltage cable insulation types and their temperature limits.
- The cross-sectional area of the conductor must be sufficiently large that the maximum expected current will not cause heating and compromise the insulation or cause fire to develop in adjacent materials. **AS/NZS 3000:2007** Tables C5 and C6 provide a simplified method of matching cable sizes and over-current protection devices.
- The voltage drop in any length of the conductor must not be large enough to cause equipment to operate in a dangerous manner or prevent protective and safety services from functioning. **AS/NZS 3000:2007** Appendix C4 provides a simplified method of determining voltage drop.
- The mechanical protection provided by thermoplastic, steel wire armour or other sheathing should be suitable for the location. Armoured cables and some metal sheathed cables are provided with a sheathing of PVC to prevent corrosive action especially when buried in concrete, in plaster or underground.

It is important to select cables that keep the voltage drop as low as possible since a drop of 1 volt in a cable carrying 10 amps will waste 1 kilowatt hour of electrical energy every 100 hours of running.

AS/NZS 3000:2007 Table 3.1 provides examples of commonly used cable types for low voltage fixed wiring and possible installation methods.

Flexible cables

In order to add flexibility to the cable the conductor can be made up of smaller conductors stranded together. Refer to **AS/NZS 3000:2007** clause 1.4.20 for the definition of a flexible cable.

Flexible cables are used where the amount of movement and flexing would cause metal fatigue in normal stranded cables. Typical uses include pendant control units for cranes and trailing cables to supply mobile equipment (see Section 8.8 in this chapter). If the strands are small enough the cable can be designated as a cord.

FIGURE 8.3 Single insulated twin flexible cord

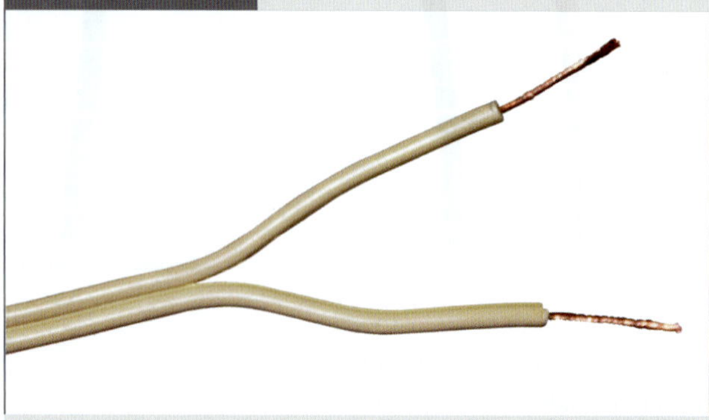

Source: Ed Hawkins

Flexible cords

A flexible cord is defined by **AS/NZS 3000:2007** clause 1.4.36. Cords have the most flexibility and are used primarily where you need to supply a piece of electrical equipment located some distance from the point of any fixed wiring (see Figure 8.3). **AS/NZS 3000:2007** clause 3.9.7.4 provides for the connection of flexible cords as installation wiring.

Colour codes

AS/NZS 3000:2007 clause 3.8.1 specifies that you must clearly identify conductors to indicate their intended function. If colours are to be used then you must use particular colours.

In fixed wiring you may use red, or any colour other than black, light blue, green, yellow or green/yellow combination for actives or switch wires, with black for neutral. You must only use green/yellow for earthing conductors.

You are allowed to use dark blue for active conductors in multiphase systems.

AS/NZS 3000:2007 Table 3.4 lists colours for installation wiring.

You may identify conductors with suitably coloured sleeving, and **AS/NZS 3000:2007** clause 3.8.2 describes the circumstances and restrictions that apply. For example you are not allowed to use sleeving to identify a green yellow core as an active conductor.

Three core flexible cords manufactured to old Australian standards used the same colour code as fixed wiring – red for active, black for neutral and green or green/yellow for earth. The colour coding of the cores of flexible cords is given in **AS/NZS 3000:2007** Figures 3.1 and 3.2. In summary, they are:

- 3 core – brown (active), light blue (neutral) and green/yellow (earth)
- 4 core – brown, red and blue (active) and green/yellow (earth)
- 5 core (AS/NZS) – brown, red and blue (active), black (neutral) and green/yellow (earth) (see Figure 8.4)
- 5 core (European) – brown, red, grey (active), light blue (neutral) and green/yellow (earth).

FIGURE 8.4 Colours in a five core flexible cable

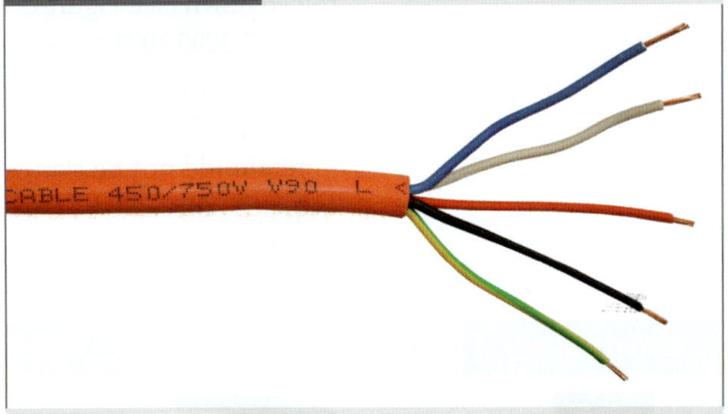

Source: Ed Hawkins

Joints and termination

All forms of cable need to be connected to an electrical supply at one end and to an electrical apparatus at the other end. Apart from unplanned downtime and repair costs, failure of the terminating points can produce excessive heat in the form of a high resistance point, or an electric arc as the circuit disconnects in an uncontrolled manner. To reduce this possibility, approved forms of termination have been devised for each type of cable or cord and the circumstances under which it is to be used.

AS/NZS 3000:2007 clause 3.7.1 specifies the general requirements for a low voltage join between conductors and between conductors and other electrical equipment.

The requirements are:

- Electrical continuity equivalent or better than the conductors being joined.
- Provision of insulation equivalent or better than existing insulation.
- Mechanical strength equivalent or greater than existing conductors.
- No undue pressure on the conductor or connection.

The most common terminations are:

- The tunnel type connector – the conductor end is restrained within a barrel and clamped in place by one or two screws depending on the current and size of conductor (Figure 8.5). **AS/NZS 3000:2007** clause 3.7.2.11 requires tunnel type connectors with two screws to be used for joining earthing conductors unless a single connector screw has an outside diameter of not less than 80% of the tunnel diameter. If the tunnel connector is one of the fixed terminals of an item of electrical equipment then a double screw connector is not required.
- The crimp lug – this is compressed onto the cable using a special tool (Figure 8.6); see **AS/NZS 3000:2007** clause 3.7.2.3.2. Some brands of wireman's pliers have a crimp attachment but as a general rule pliers are not a suitable tool for crimping lugs.

FIGURE 8.5 Tunnel type connectors

Source: Ed Hawkins

FIGURE 8.6 Crimp lugs in a control panel

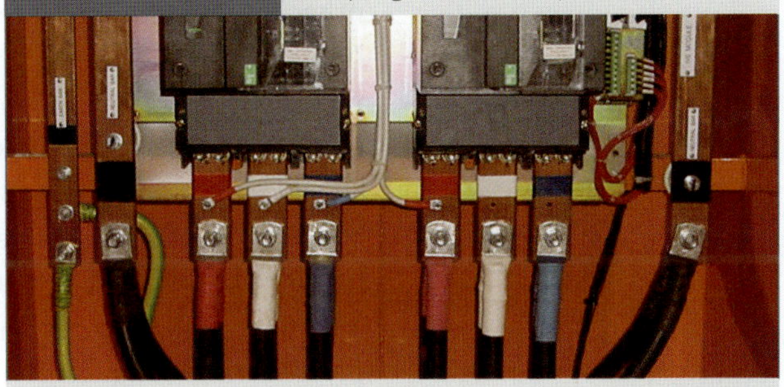

Source: Ed Hawkins

- Clamping – the cable is clamped under a screw and held in place by a washer that has the sides folded down (see Figure 8.7), to restrain the strands of the conductor from spreading as required by AS/NZS 3000:2007 clause 3.7.2.5. This group includes clamps and solder-less lugs.
- Soldered joint – the conductor is mechanically joined then soldered to exclude oxygen and moisture (Figure 8.8). The solder must not be relied upon for mechanical strength and the joint should not be clamped under a nut or screw as the solder may soften and loosen the joint (AS/NZS 3000:2007 clause 3.7.2.7). Clause 3.7.2.9 restricts the connections that can be made in aerial conductors.

FIGURE 8.7 Termination earth clamp

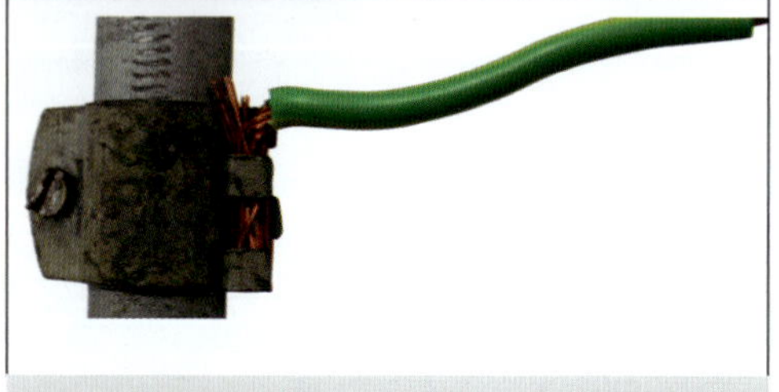

Source: Ed Hawkins

FIGURE 8.8 Soldered married joint in a large stranded cable

Source: Ed Hawkins

The majority of domestic fittings use tunnel type terminations but particular termination devices are required when flexible cords are connected to the fixed wiring (AS/NZS 3000:2007 clause 3.7.2.8b).

Aluminium conductors must be joined by suitable connectors to prevent oxidisation and electrolysis in the join. AS/NZS 3000:2007 clause 3.7.2.1.2 lists the requirements for joining aluminium cables. Also prohibited are any joints in cables that are under tension and between any pair of metals that would cause electrolysis to weaken or destroy the joint (clause 3.7.2.9.1).

Cable protection

AS/NZS 3000:2007 clause 3.3.1 provides for cables to operate safely and function properly under any conditions to which they might be exposed. These include:

- ambient temperature – wiring systems must be selected to withstand the highest and lowest ambient temperatures
- external heat sources – wiring systems must be protected from the effects of heat from external sources
- presence of water and high humidity – provision must be made to prevent ingress of water or provide drainage to remove it
- entry of solid or foreign objects – this includes building fastenings such as nails, screws or other metal building materials. Dust build-up around the cables may also affect the operating temperature
- presence of corrosive or polluting substances – this includes solvents, acids and alkalines such as encountered in soils by underground cables. Situations where dissimilar metals may cause corrosion should be kept separated by distance or suitable insulating materials
- impact – prevention against mechanical damage must be inherent to the wiring system itself or provided by additional enclosures
- vibration – the use of stranded or flexible cables and strain reduction loops in solid conductors to prevent metal fatigue

- other mechanical stresses – adequate support and fixings must be provided in areas where the cables are likely to sag or be disturbed. Special attention should be paid to the minimum bending radius of the cable and enclosure
- flora and fauna – special protection may be required if damage from plants and animals is possible. This is a problem that applies to equipment installed around farms and farm buildings
- solar radiation – exposure to sunlight can reduce the current carrying capacity of the cable by increasing the thermal load. Cables exposed to sunlight must be resistant to ultraviolet ray damage or installed in suitable enclosures
- hazardous areas – where cables are exposed to explosive dust, gas and vapours they must be installed in accordance with **AS/NZS 3000:2007** clause 7.7 and also **AS/NZS 2381.1 2005 Electrical equipment for explosive gas atmospheres** and **AS/NZS 61241.14 2005 Electrical apparatus for use in the presence of combustible dust**.

SUMMING UP...

- Cable cores comprise the conductor and insulation.
- Sheathed cables have added mechanical protection.
- Flexible cables have many fine strands to let them bend easily.
- Flexible cords are a subcategory of flexible cables.
- Conductors in cables are usually made from copper or brass.
- AS/NZS 3000:2007 states the definitions for types of cable.
- A cable comprises one or more cable cores.
- Building wire is a commonly used single insulated low voltage cable.
- Manufacturers specify the number and diameter of the strands as well as the cross-sectional area of the conductors.
- The temperature and voltage rating are selection criteria for insulation.
- The cross-sectional area is determined by the maximum circuit current.
- Armoured cable can be buried directly in concrete.
- Special fittings are required to connect flexible cords to the installation wiring.
- The cable cores are usually identified by colour.
- Green, yellow or green/yellow are only used for earth conductors.
- Black is only allowed to be used for neutral conductors.
- Cable joints must not have electrical qualities that are less than the cable being joined.
- Aluminium conductors must not be connected using brass or copper terminals.
- Conductors must not be soldered before fastening under a screw or clamp.
- Generally earth conductors must be connected under two screws.

8.2 Cords, cables and plugs

Your work as an electrical practitioner will involve the routine repair of flexible cords in many forms. You will need to be able to identify and remove dangerous cords from service as well as effect timely and cost-effective repairs.

Cord extension leads

Flexible cords are usually used to connect portable electrical appliances to the supply. Other cables which may be similar in appearance may be found in situations such as extra low voltage installations and instrumentation applications, but they may not be suitable for connecting to a 240 volt supply.

If you refer to **AS/NZS 3008.1.1:2009 Electrical installations – Selection of cables** Table 16 you can determine the maximum current carried by each core of a flexible cord.

AS/NZS 3000:2007 clause 3.7.2.8 specifically forbids you to tie a knot in a flexible cord to prevent undue stress on connections. This clause also tells you not to join flexible cord by any method other than using an approved connector. You must attach every flexible cord to an appliance, appliance connector, plug or cord extension socket or any other equipment in such a manner that no undue stress is imposed on its electrical connections. You achieve this by using a tortuous path, cord grip, clamp, pillar, post or other effective means.

> **i** Extra low voltage is defined by AS/NZS 3000:2007 clause 1.4.98 (a) as a voltage not exceeding 50 volts a.c. or 120 volts d.c. ripple free.

A single phase extension lead (or cord extension set) is a length of three core ordinary duty or heavy duty sheathed flexible cord with a three pin plug top on one and a three pin cord extension socket on the other. They are usually used to temporarily extend the length of the cord supplied with a portable appliance.

Single phase cord extension sets must have three cores, each core must have a cross-sectional area of at least 1 square mm, and the maximum permissible length is governed by the size of the cable.

If a cord extension set is stored on a drum or reel it should be unwound completely if it is to be used at its full current carrying capacity. If it is not unwound the current carrying capacity must be decreased (de-rated) in accordance with Table 16 of **AS/NZS 3008.1.1:2009 Electrical installations – Selection of cables**, to prevent the cable from being damaged due to overheating, as shown in the example in the following table:

Number of layers on the drum:	1	2	3	4
De-rating factor:	0.76	0.58	0.47	0.4

Selecting and installing low voltage electrical wiring systems

EXAMPLE 8.1

To calculate the de-rated current carrying capacity multiply the maximum current rating by the de-rating factor. If a 10 amp cable was used with four layers wound on a drum, the de-rated current carrying capacity of the cable would be 10 x 0.4 or 4 amps.

Preparation of cord ends

A 15 amp plug top can be distinguished by its wider earth pin that prevents it being accidentally inserted in a 10 amp socket.

When you are fitting a plug top or cord extension socket you must take care to observe the following points:

- Ensure that the flexible cord has the appropriate voltage, current and temperature rating.
- Remove the sheathing without marking or damaging the insulation around the cable cores in any way (see Figure 8.9). Do not remove any more of the sheathing than is absolutely necessary.
- Remove enough insulation to allow the conductor to be fastened in the terminal without any surplus conductor extending outside the terminal (Figure 8.10).
- Twist the strands of the cable cores and double them over, ready to insert them in the clamping terminal (Figure 8.11).
- Do not allow any insulation to be clamped between the metal surfaces of any terminal.

FIGURE 8.9 Outer sheath removed

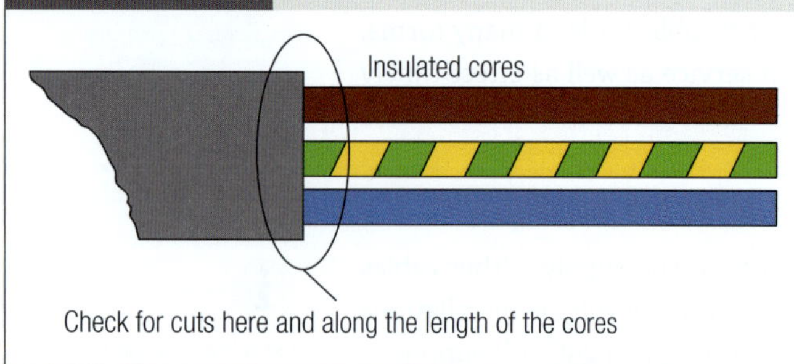

FIGURE 8.10 Conductors prepared

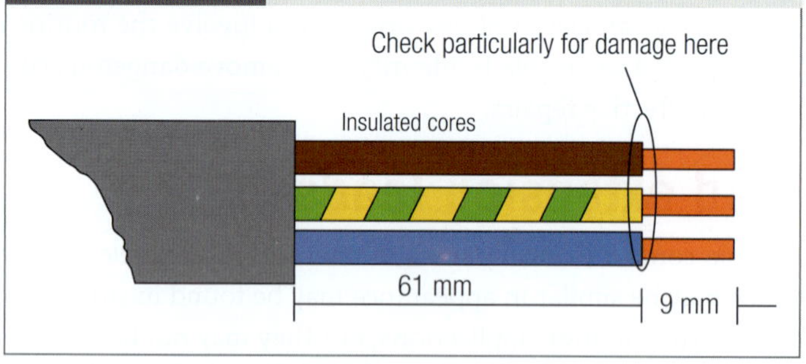

FIGURE 8.11 Doubled conductor

FIGURE 8.12 Socket outlet polarity

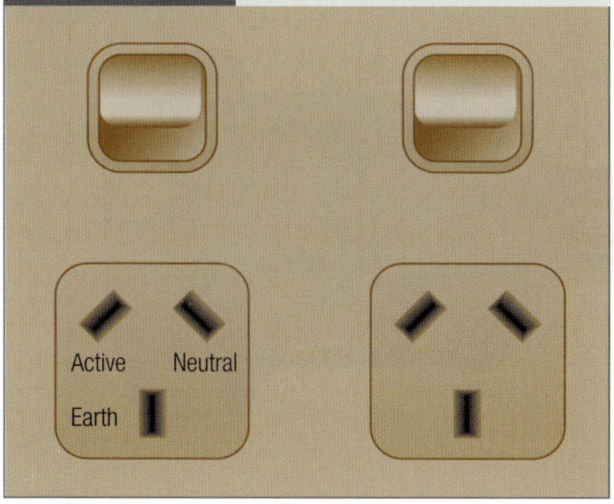

FIGURE 8.13 Plug top polarity

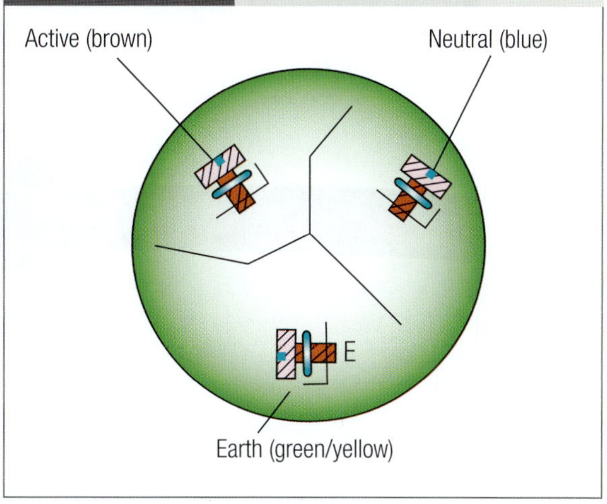

Fitting plug tops

When fitting plug tops to a flexible cord you must observe the following precautions:

- The three pin plug top must be polarised to match the standard single phase flat pin three pin outlet (Figure 8.12).
- If using internationally colour coded (IEC) three core flexible cord the earth is coloured green/ yellow, active is brown and the neutral is light blue (see Figure 8.13).
- Ensure that the insulation on each cable core is as close as possible to the clamping terminal.
- Ensure that the flexible cord passes through the tortuous path in the accessory. Figure 8.14 shows the plug top fitted.

> i
> AS/NZS 3000:2007 clause 4.4.5 specifies that when viewed from the front of the socket outlet the order of connection, starting from the slot in the radial line, must be EARTH, ACTIVE, NEUTRAL in a clockwise direction.

Fitting extension sockets

The same precautions should be observed when fitting an extension socket. The cord extension socket must be polarised as for standard single phase flat pin three pin outlets. The rear view of the extension socket is shown in Figure 8.15, the front view is shown in Figure 8.16, while the extension socket fitted is shown in Figure 8.17.

FIGURE 8.14 Plug top fitted

Source: Ed Hawkins

FIGURE 8.15 Rear view of the extension socket

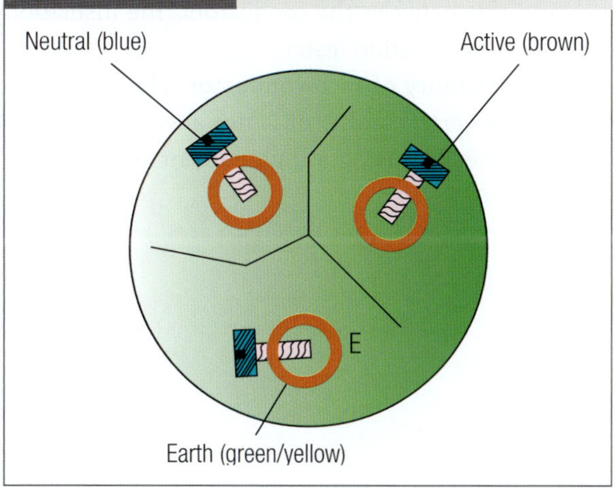

FIGURE 8.16 Front view of the extension socket

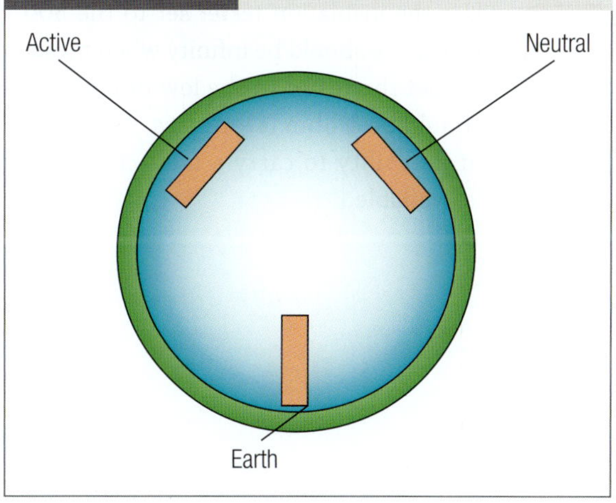

OH&S regulations require that if an extension cord is to be used on a construction site it must be of a type which cannot be rewired, or the plug and socket must be made of transparent material.

FIGURE 8.17 Extension socket fitted

Source: Ed Hawkins

Extension cords must be inspected by a qualified person at regular intervals and tagged to indicate that they are in a safe condition.

Final checks and tests

Visual and resistance checks need to be carried out to ensure the extension cord is safe to return to service. Faulty flexible cords are a serious safety hazard in a workplace.

Visual inspection

Carefully examine the finished connection and ensure:

- correct polarity of all connections
- correct colour coding of all conductors
- correct terminations at each clamping terminal
- no insulation was removed unnecessarily
- no cable cores are extending outside of the outer cover of the plug top or extension socket
- there are no nicks in the cable insulation or sheathing
- there is no insulation clamped under terminals
- the cable conductors are doubled over before connecting (where applicable)
- there is no mechanical damage to any of the accessories or the flexible cord
- the correct accessories and cable have been used.

Insulation and continuity tests

Ensure that the projecting plug pins are where nobody can come in contact with them unexpectedly.

Use the insulation tester set to the 500 volt range and test between the conductors. The insulation resistance should be infinity when measured with a 500 volt insulation tester.

Set the tester to the low ohms range and test the continuity of each conductor. The resistance should be 1 ohm (Ω) or less. The resistance check of the earth conductor is not always indicative of its ability to carry a fault current and it should be checked using a high current tester if one is available.

SUMMING UP...

- Flexible cords are usually used to connect electrical equipment to the installation wiring.
- Extra low voltage is lower than 50 volts a.c. or 120 volts ripple free d.c.
- The maximum current permitted in each core of a flexible cord is determined by AS/NZS 3008.1.1:2009 Table 16.
- Knots are prohibited for attaching flexible cords.
- Single phase extension leads must have three cores.
- If cords are used while rolled on a drum they must be de-rated according to the number of layers on the roll.
- A flexible cord must be fitted with a plug of the correct current rating.
- Transparent plug tops must be used on building sites.
- Flexible cords must be tested and tagged regularly by a qualified person.
- The polarity of plug tops and extension sockets must match the standard for socket outlets.
- All cords must be tested for polarity, continuity and insulation.

8.3 Flat TPS wiring systems

As an electrical practitioner you will encounter flat TPS wiring systems more often than any other as they are widely used due to the reduced cost of installation. You will be required to install TPS cables in many and varied types of premises and leave everything in a safe and functioning condition.

Ensure that PVC and copper are disposed of in a safe manner as some materials such as copper and plastics can be recycled.

TPS types and uses

A TPS cable comprises one or more single insulated cables surrounded by an additional layer of insulation called a sheath (see Figure 8.18). The sheath provides additional mechanical protection against damage to the internal cable insulation as well as providing the double insulation required in some situations such as swimming pools.

The colour of the sheath varies and usually indicates the type of installation the cable is suited to. The colours are:

- white – most common
- orange – usually circular
- grey
- black – usually multicore control cable
- red – fire systems
- blue – instrumentation.

Specially coloured sheaths such as red should not be used for any other purpose. You should not tap into a red coloured cable or you will set off the fire alarm or compromise the fire safety system. Sheaths cannot be coloured green or yellow as these are reserved for earth conductors.

The sizes of flat TPS cable are restricted by the cross-sectional area of the conductors and the number of cores due to the physical problems associated with installing multicore flat cable.

The predominant configurations are:

- single core – SDI, large size cables are only used in this format
- two core twin – either active and neutral or twin active (see Figure 8.19)
- two core and earth – twin and earth
- three core and earth or possibly four core and earth – available but not commonly used.

FIGURE 8.18 Flat TPS twin and earth cable

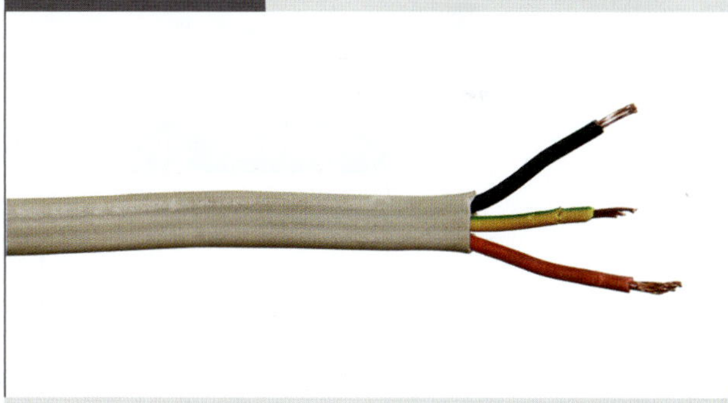

Source: Ed Hawkins

FIGURE 8.19 Flat TPS twin active cable

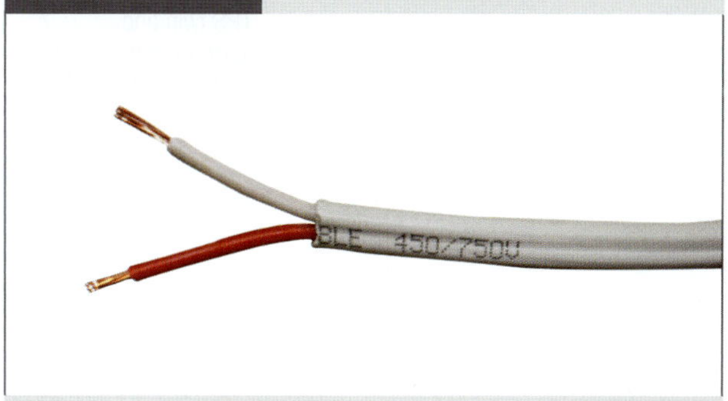

Source: Ed Hawkins

Fixing and support

There are various methods of fixing and support for TPS cables, the most common being clipping with pin clips or PVC clips.

AS/NZS 3000:2007 clause 3.9.3.1 requires the use of suitable clips, saddles or clamps or other means that will not damage the wiring system; where passing through structural parts of the building suitable openings or bushings need to be used. Special mention is made of suspended ceilings in clause 3.9.3.2 where cables may be prevented from being laid directly on the ceiling by weight restrictions.

Additional protection in the form of conduit or trunking should be provided if there is a possibility that the sheath may not provide full protection in a particular situation. Support for TPS cable is often required between where it is attached to the structure and where it enters the wired electrical equipment as in the case of a hot water cylinder or cooking range. The conduit provides the support for the cable but must itself be securely saddled at both ends. TPS flat cable is suitable for supplying socket outlets mounted on trunking to supply power outlets in offices and other work areas, as seen in Figure 8.20.

FIGURE 8.20 Trunking with socket outlets mounted

Source: Ed Hawkins

TPS flat cable in lighting circuits

When TPS flat cable is used to install lights the configuration of the wiring means that certain configurations must be used to make the most efficient use of materials and installation time.

The two most effective configurations for efficient use of materials and installation time are:

- Loop at the light – in this configuration the power feed is run to all light fittings in the circuit with special extra terminals provided for the un-switched active and the earthing conductor. The neutral conductors are connected into one of the lamp terminals and a twin active is run to the switch with the red conductor carrying the active voltage and the white returning to the second lamp terminal from the other side of the switch (see Figure 8.22). Loop at the light installation is shown in Figure 8.21.

FIGURE 8.21 Loop at the light installation

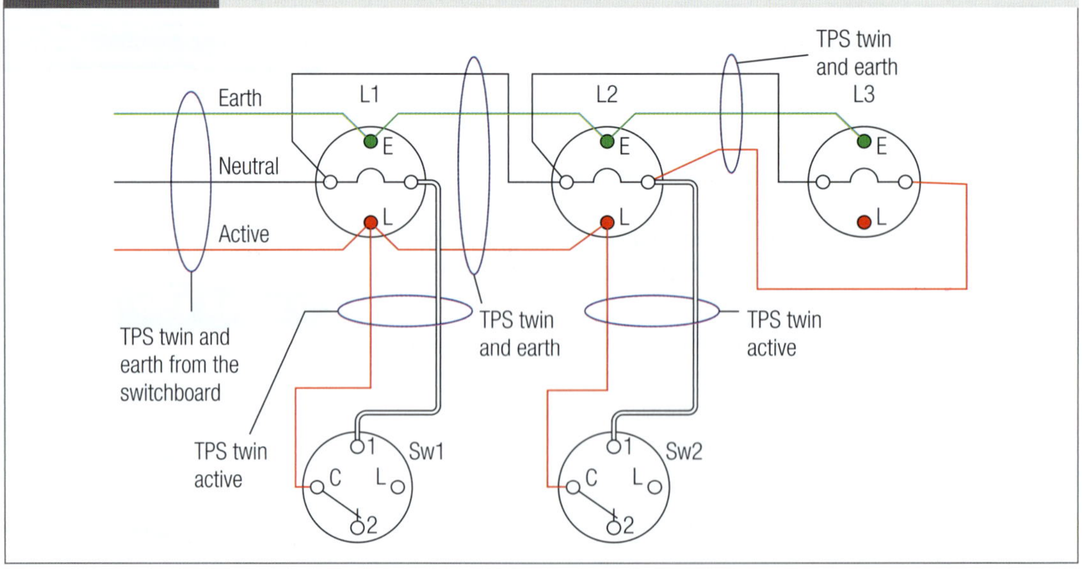

- Loop at the switch – this system is used to avoid using two different cables and is installed under the floor with the power feed joining in each switch. A twin and earth cable is run from the switch to the light. The active feed is connected to one side of the switch and the switched active to the lamp connects to the other. The neutrals are looped in an idle terminal on the switch and the earth is connected using an approved connector. Careful planning is required to avoid exceeding the number of conductors that can be joined behind the switches. Loop at the switch installation is shown in Figure 8.23.

FIGURE 8.22 Batten holder terminals for loop connections

Source: Ed Hawkins

Final checks and tests

Before powering up your TPS circuit, you will need to carry out some rigorous tests to ensure the integrity of the installation.

FIGURE 8.23 Flat TPS loop at the switch installation

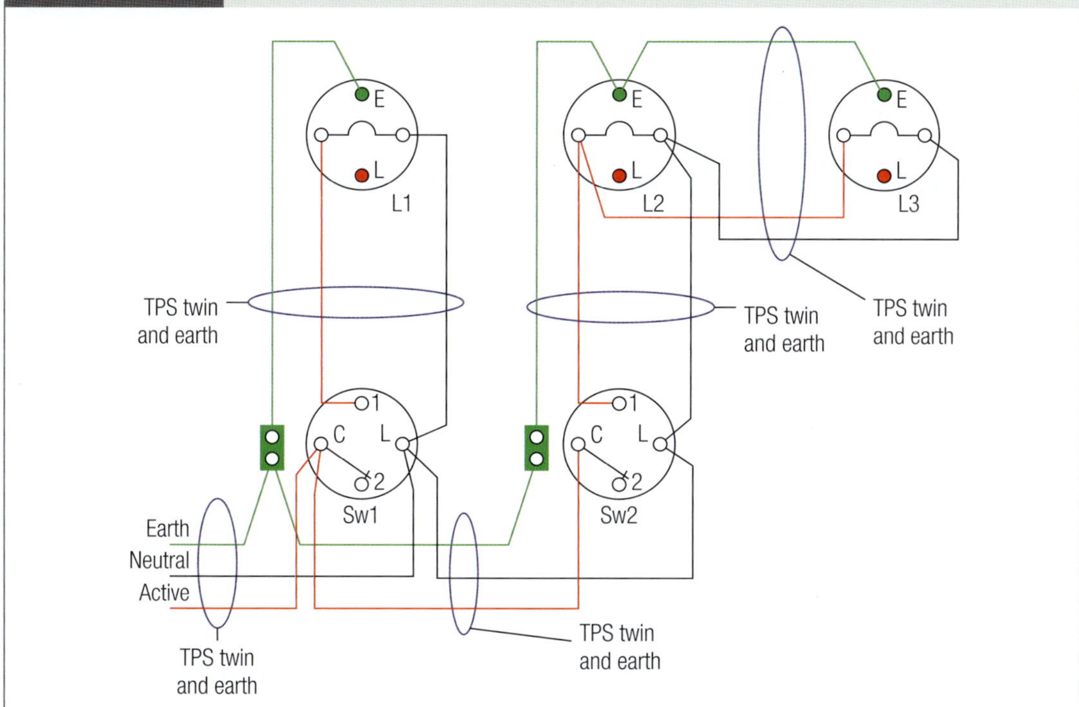

AS/NZS 3000:2007 clause 5.4.3 requires that an earth terminal be provided at all lighting points including transformers for supplying ELV lighting.

Visual inspection

Carefully examine the finished connections and ensure:

- correct polarity of all connections
- correct colour coding of all conductors
- no insulation was removed unnecessarily
- no cable cores are extending outside of the outer cover of the sheathing or enclosure
- there are no nicks in the cable insulation or sheathing
- there is no insulation clamped under terminals
- the cable conductors are doubled over before connecting (where applicable)
- there is no mechanical damage to any of the accessories or cable
- the correct accessories and cable have been used.

Ensure that you warn others working on the circuit so they do not come in contact with the conductors under test.

Insulation and continuity tests

Use the insulation tester set to the 500 volt range and test between the conductors. The insulation resistance should be infinity when measured with a 500 volt insulation tester.

Set the tester to the low ohms range and test the continuity of each conductor. The resistance should comply with **AS/NZS 3000:2007** clause 8.3.9.3.

SUMMING UP...

- A flat TPS cable comprises one or more cable cores surrounded by a sheath laid up in flat formation.
- TPS cables can provide double insulation when required.
- Flat TPS is the most commonly used wiring system.
- Systems are colour coded. Green, yellow and green/yellow are prohibited.
- Sizes of multicore cables are limited due to installation difficulties with heavy cables.
- Cables must be supported by clips to prevent disturbance or strain.
- Additional protection is required in high mechanical risk situations.
- Two configurations used in lighting circuits are loop at the light and loop at the switch.
- An earth terminal must be provided at all lighting points.

8.4 Circular TPS wiring systems

If your work as an electrical practitioner takes you into an industrial environment you will encounter circular TPS wiring systems. Slightly different skills are required when working with this type of cable compared with flat TPS. It is more difficult to strip the sheathing and you will need to be able to select the correct fittings such as cable glands and clips.

Where multiple core cables are required in an installation, circular cables are used as they are easier to install (Figure 8.24).

Circular cables are available in a wide range of conductor sizes and core numbers.

Commonly used sizes are:

- two core and earth (1.5 mm^2 and up)
- three core and earth (1.5 mm^2 and up)
- four core and earth (1.5 mm^2 and up)
- three and a half core (half size neutral for large applications).

Circular cables used in control and multiple circuit applications are available with large numbers of cores (up to 30 cores in both 1.5 mm^2 and 2.5 mm^2). The TPS sheath is usually orange in circular cables (as seen in Figure 8.25), but is also available in other colours, and frequently black.

FIGURE 8.24 Orange circular TPS with four cores

Source: Ed Hawkins

FIGURE 8.25 Circular TPS cable gland

Source: Ed Hawkins

TPS cables should not be installed when there is the risk of mechanical damage, the presence of excessive heat or the likelihood of solvents affecting the cable insulation.

When used with perforated steel cable tray, circular TPS provides a flexible system of cabling where changes in customer requirements can be readily met without needing to run new conduits. The cable can be simply laid and fastened onto the tray beside existing cables.

Circular cable can be sealed against ingress of moisture, dust and vapours using a terminating gland. The gland clamps the cable sheath and fastens into an accessory mounting box using a standard conduit thread.

Final checks and tests

Since circular TPS cables are most likely to be installed in an industrial environment, special attention must be paid to mechanical support and protection.

Visual inspection

Carry out the same visual checks as described for flat TPS, ensuring that cable glands have been tightened where the cable enters a fitting or piece of equipment.

Warn others working on the circuit to ensure they do not come in contact with the conductors under test.

Insulation and continuity tests

Carry out the same insulation and continuity checks as described in Section 8.3 above for flat TPS wiring.

SUMMING UP...

- Circular TPS cables are used where multiple cores are needed as they are easier to install.
- Circular multicore cables are available in a wide range of sizes.
- Circular TPS is used for control and multiple circuit applications.
- Circular TPS is used when the cable entry needs to be sealed against gas, dust or vapour.
- Special glands are required to terminate the cable in fittings.
- Circular TPS provides a flexible system when used with perforated cable tray.

8.5 Thermoplastic insulated cables in non-metallic enclosures

When you are required to install non-metallic conduit or trunking it is very likely that it will be open to view, requiring strict attention to detail. Your customers will appreciate the appearance of carefully measured and fastened trunking or PVC conduit.

Under **AS/NZS 3000:2007** clause 3.10.1.1 insulated and unsheathed cables must be enclosed in a wiring enclosure throughout their entire length. The enclosures can be non-metallic or metallic.

Non-metallic enclosures can be either trunking or conduit. Each method constitutes a wiring system of its own defined by suitably matched fittings and support systems.

Since the use of non-metallic materials provides no method for ensuring earth continuity it is important to ensure that an earth conductor is connected to any single insulated metal parts of the installation.

FIGURE 8.26 Socket outlet supplied by non-metallic trunking system

Source: Ed Hawkins

Non-metallic trunking

When using non-metallic trunking to enclose cables (as shown in Figure 8.26), you must comply with the requirements specified by **AS/NZS 3000:2007** clause 3.10.3.9:

- Covers must be able to be opened where practicable and continuous where passing through walls and floors.
- Trunking must be accessible through its entire length.
- Cables installed in trunking must not rely on a readily removable cover for support. This means that the cables will need further fastening within the trunking if it crosses the underside of a ceiling.
- The trunking must be non-hygroscopic, meaning that PVC trunking complies but untreated wood or unglazed ceramics would be unsuitable.
- Live parts of accessories such as switches or switch socket outlets mounted on cable trunking must be arranged so that basic protection is provided and that no contact can be made with live parts.
- When cutting conduit or trunking, sufficient length must be provided to allow full entry to the fittings and accessories being used.
- At no time must single insulation be exposed.

When cutting trunking, the eventual position relationship between fittings, trunking and lid must be taken into account since the trunking may need to continue a short way into the fitting while the lid may not.

Trunking as a rule cannot be set in the same way that conduit can, but manufacturers provide a range of stock bends to allow for most requirements (see Figures 8.27 and 8.28).

Multiple circuits may be installed in conduit or trunking provided they allow sufficient air circulation for the cable size. This is determined by referring to **AS/NZS 3000:2007** Appendix C6 and Table C9.

FIGURE 8.27 Non-metallic trunking right angle bend

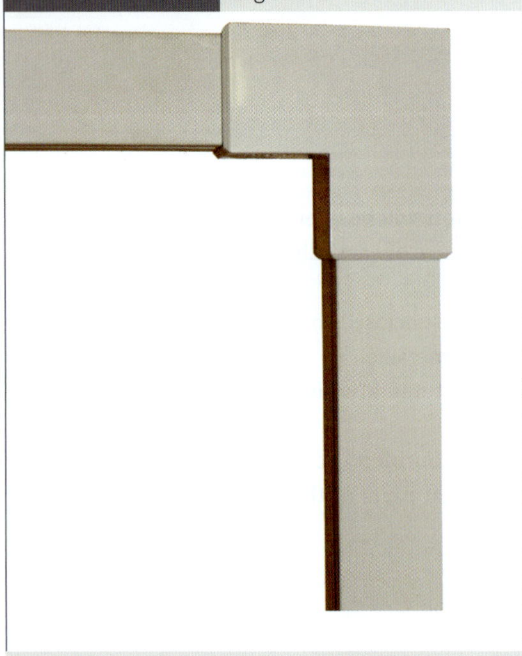

Source: Ed Hawkins

FIGURE 8.28 Non-metallic trunking tee fitting

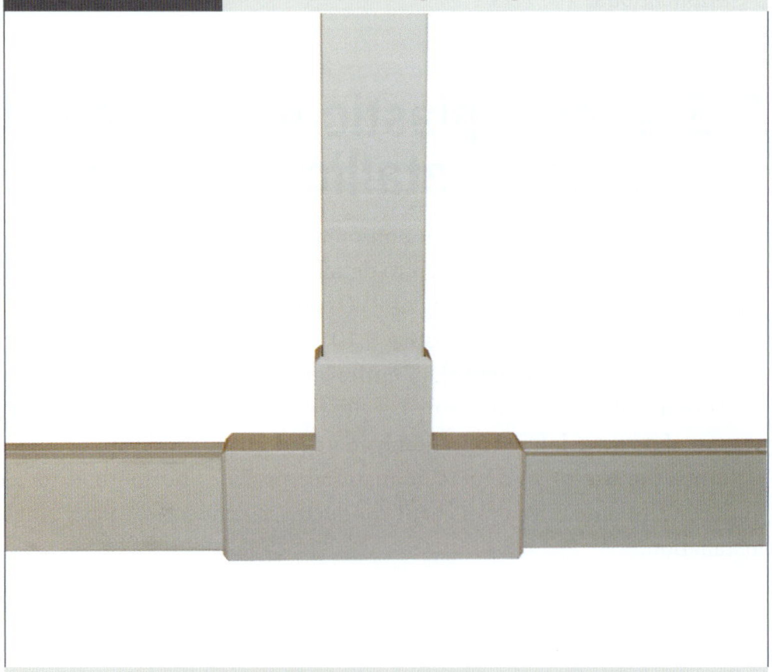

Source: Ed Hawkins

Non-metallic conduit

PVC rigid conduit is quick and easy to install. Joins can be made using approved glue and can be bent to any shape required (see Figure 8.31). To prevent the conduit from flattening while bending, a spring is inserted and removed once the bend is complete (Figure 8.30). PVC conduit is available in 4 and 6 metre lengths and diameters ranging from 16 mm, 20 mm and up; some examples are shown in Figure 8.29.

Medium duty conduit is used for most above-ground services and is coloured grey.

Heavy duty is available for above ground (grey) and below ground (orange).

PVC conduit is not suitable for exposure to UV rays and must be treated with a suitable acrylic paint.

Always use safety eyewear when cutting PVC conduit as small, sharp-edged particles can affect the eyes.

FIGURE 8.29 PVC conduit fittings

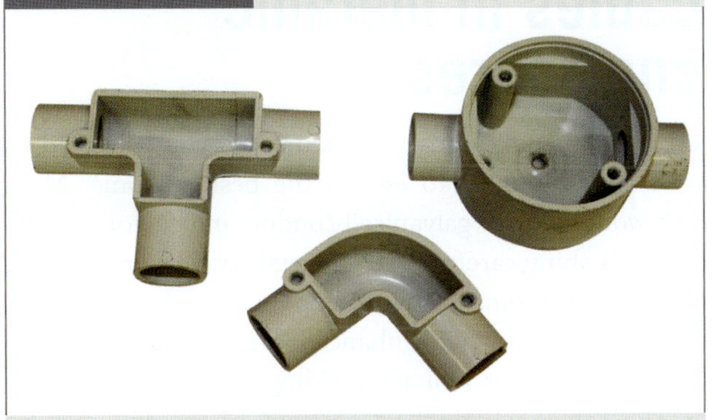

Source: Ed Hawkins

FIGURE 8.30 PVC rigid conduit bending spring

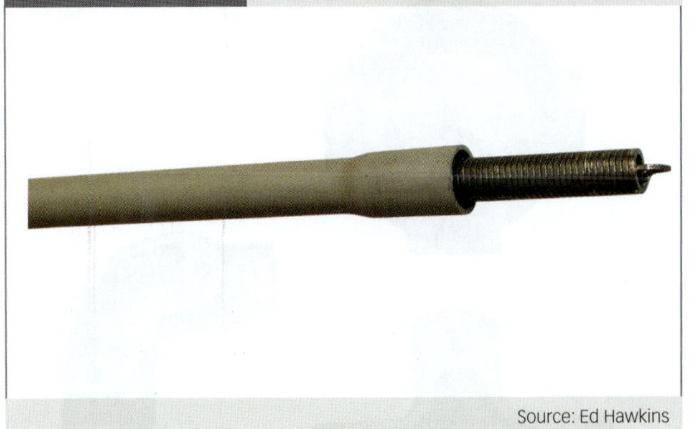

Source: Ed Hawkins

Final checks and tests

Non-metallic conduit is more vulnerable to mechanical damage so it is important to ensure it is adequately supported and protected.

Visual inspection

- Carry out the visual inspections on the cables as described in Section 8.3 earlier in this chapter.
- Ensure all fittings are glued.
- Check that no conduit has been flattened by undue strain or lack of adequate support.

FIGURE 8.31 Bends in PVC conduit

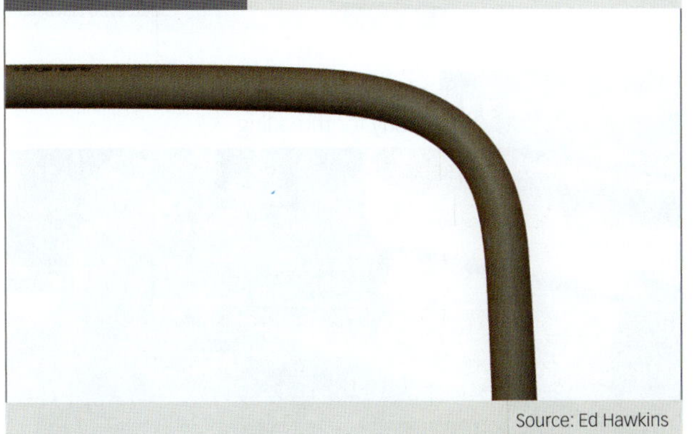

Source: Ed Hawkins

Insulation and continuity tests

Conduct insulation and continuity checks as described in Section 8.3 in this chapter.

AS/NZS 3000:2007 clause 3.10.2.2 provides that changes can be made between different enclosure types but must be done either at a switchboard or using a suitable adaptor made for the purpose.

SUMMING UP...

- Trunking must be installed in accordance with AS/NZS 3000:2007 clause 3.10.3.9.
- Single insulated cables can be protected by non-metallic trunking.
- Trunking must be non-hygroscopic.
- Protection must be maintained for switches mounted on the trunking.
- PVC trunking cannot usually be bent.
- Rigid PVC conduit has three categories.
- A bending spring is usually required when bending PVC conduit.
- Changes can be made between different enclosures using purpose made adaptors.

FIGURE 8.32 Galvanised steel conduit fittings

Source: Ed Hawkins

FIGURE 8.33 Steel conduit with cutting compound ready for threading

Source: Ed Hawkins

8.6 Thermoplastic insulated cables in metallic enclosures

To install cables in metallic conduit or trunking, you will need to develop metal working skills to ensure the best outcome. Most metal conduit work involves galvanised conduit and customers will be impressed with shiny carefully fitted metal conduit or trunking installed in their plant rooms and workshops.

Metallic enclosures provide enhanced mechanical protection to single insulated cables as well as providing extra protection and support for TPS cables where short sections of the run may be exposed to high levels of risk.

Rigid metal conduit is provided in 4 m lengths with a range of outside diameters of 16 mm, 20 mm, 25 mm and up. Figure 8.32 shows examples of galvanised steel conduit fittings. The conduit is galvanised inside and out and intended for use underground and in areas that are likely to be subject to water inundation and damage from moving vehicles, building fastenings and other objects.

The same requirements apply to metallic enclosures that apply to non-metallic with the additional requirement set out in **AS/NZS 3000:2007** clause 3.10.3.3 to maintain earth continuity throughout the system.

The same limits regarding interconnection of systems apply to metallic enclosures. If non-metallic enclosures are interconnected with metallic enclosures, using approved adaptors, strict attention must be paid to the continuity of earthing of the metallic enclosure.

Working with metal conduit

The ends of the metal conduit should be cut as square as possible using a hacksaw or tube cutter.

Gloves and safety eyewear should be used when working with metal products as sharp slivers can damage skin and eyes.

Threading metal conduit

A chain vice tightened down firmly should be used to hold the conduit to prevent damage while it is being worked, as shown on Figure 8.33.

The guide end of the thread die is placed over the end of the conduit and pushed on and held firmly. With the ratchet set for clockwise rotation, turn the die while ensuring that it stays square (see Figure 8.34). After two or three turns reverse the die and wind it backwards until you feel the chip break.

FIGURE 8.34 Cutting the thread on steel conduit

Source: Ed Hawkins

FIGURE 8.35 The thread length must be correct for the fitting

Source: Ed Hawkins

Continue to cut the thread, pausing to break the chip every two or three turns. Once the thread has reached the length you require, remove the die and clean the cutting compound off the thread. Make sure the thread goes all the way into the fitting but does not have any threads showing (see Figure 8.35).

i

Use generous amounts of cutting compound when threading metal.

Setting metal conduit

It is often necessary to bend or 'set' the conduit so that it will fit into switchboards or control panels without putting a strain on the frame. To make a double set to enter a switchboard or fitting you should:

1 First bend a piece of stiff wire to the shape that you want the bend to take.
2 Place the conduit in the bender with the hook just in front of the starting point for the bend (see Figure 8.36). Place your foot on the footrest and pull the handle up towards you until the desired angle is reached, and make regular comparisons with your wire pattern.
3 Turn the conduit over and place the hook where the next bend is to start (Figure 8.37) and once again pull the handle up towards you, making sure the conduit stays straight.

i

Screw a coupling onto the thread to make sure that it is not damaged during bending.

FIGURE 8.36 Setting steel conduit on the first bend

Source: Ed Hawkins

FIGURE 8.37 Setting steel conduit on the second bend

Source: Ed Hawkins

FIGURE 8.38 Checking the offset on the steel conduit double set

Source: Ed Hawkins

4 When the set matches your pattern remove the coupling from the end and carry out a trial fit (as shown in Figure 8.38). Make adjustments until the conduit fits firmly in place without having to strain any of the fittings or panels.

Installing circuits in metal conduit

When installing circuits in metallic enclosures you will need to:

1 Fully assemble the conduit system, placing draw wires in as you go.
2 Check each end of each piece of conduit and remove any burred edges with a round file.
3 Bundle the cable cores together, ensuring that there are no twists or kinks.
4 Join the cable cores onto the draw wire.
5 With one person pulling the draw wire and another feeding the cables into the conduit, draw the cables through, breaking out the cores that are required at each fitting along the run.
6 Leave sufficient core length at each fitting to make the connection to the electrical apparatus.
7 Fit off the circuit components and close all inspection covers.

Final checks and tests

Metallic conduit is usually installed in a location where the likelihood of mechanical damage is high. Particular attention should be paid to the length of threads and the mechanical support.

> Ensure that steel scrap is disposed of in a suitable manner as it can be dangerous for people and animals if left on the floor or the ground.

Visual inspection

Carry out the standard inspection as described in Section 8.3 earlier in the chapter and ensure the conduit is firmly fastened and no cable cores are exposed.

Insulation and continuity tests

An additional insulation test will be needed between active and neutral cores and the metallic enclosures.

Continuity tests must include continuity between each section of the conduit system (between screwed fittings) and the main earth at the switchboard. These are required by **AS/NZS 3000:2007** clause 5.5.4.2 to be electrically continuous and should comply with **AS/NZS 3000:2007** clause 5.7 and be no greater than the impedance of the circuit earth conductor.

SUMMING UP...

- Metallic enclosures provide cables with additional mechanical protection.
- Rigid metal conduits come in 4 m lengths.
- Galvanised conduit is galvanised on the inside as well as the outside.
- Metal fittings and saddles are provided to install metallic conduit in most situations.
- Metallic conduits need to be protected from corrosive materials.
- Precautions need be taken to preserve earth continuity when joining metallic and non-metallic enclosures.
- A chain vice and stock and die set are required when threading metallic conduit.

8.7 Fire protection cabling and systems

To fit fire protection (and, in fact, any safety systems into a premises) you will need to become familiar with the special types of cable and installation conditions that are aimed at preserving the electrical supply as long as possible.

The **AS/NZS 3000:2007 Wiring Rules** now include fire and smoke control equipment under clause 7.2.1.2. This is embraced by the broader Section 7.2 Safety services.

All the requirements for cable installation and connection apply to safety systems unless specifically altered by a clause in Section 7 Special electrical installations.

Fire protection cabling provides power and control to pumps for sprinklers, hose reels and hydrants as well as fire detection systems and air-handling systems.

The requirements that apply generally to safety services apply to fire protection cabling.

The first and most important is **AS/NZS 3000:2007** clause 7.2.2.1 Separation of supplies, which requires that each part of an electrical installation supplying fire protection equipment must have its own main switch separate from the rest of the installation. If a separate supply for the fire control equipment is used for other purposes it must not compromise the fire control system.

Another important requirement is discrimination required by **AS/NZS 3000:2007** clause 7.2.2.2 where failure of another safety system or another part of the installation must not affect the fire control system.

If an emergency generator is provided then a changeover switch must be installed. Ideally this could be installed close to the fire control equipment to save time during an emergency (clause 7.2.2.3).

The type of wiring system to be used is specified in **AS/NZS 3000:2007** clause 7.2.7.2 requiring that the wiring be of a type that is capable of maintaining supply to the equipment when exposed to either fire or mechanical damage. Other systems can be used provided they are protected from fire and mechanical damage by a suitable enclosure or isolated by location.

As a general rule the wiring system should be one that has a WS52 classification as specified in **AS/NZS 3000:2007** Appendix H2.3.

Mineral insulated metal sheathed (MIMS) cable is the ultimate in fire survival cables. It is able to maintain circuit integrity in a fire situation and will not propagate flame, produces no smoke and does not emit hazardous or acidic gas. MIMS cable has high impact resistance and high mechanical strength and is non-ageing.

The cable is made using copper conductors surrounded by a metal oxide powder insulation contained by a continuous copper sheath. The cable has special fittings to provide moisture exclusion and support (see Figure 8.39).

FIGURE 8.39 Termination of a MIMS three core cable

Source: Ed Hawkins

To terminate a MIMS cable:

1 Remove the outer sheath using a suitable tool.
2 Screw a brass pot onto the exposed end and separate the conductors carefully.
3 Press an insulating cold sealant into the pot without touching it with your fingers.
4 Press an insulating ferrule down onto the sealant.
5 At this stage use an insulation tester to ensure that no moisture has entered the cable before compressing the pot-seal in place with a screw clamp provided for the purpose.
6 Place insulating PVC sleeving onto the bare conductors and press each sleeve onto retaining nodules on the pot-seal.

FIGURE 8.40 Tools for preparing MIMS cable ends

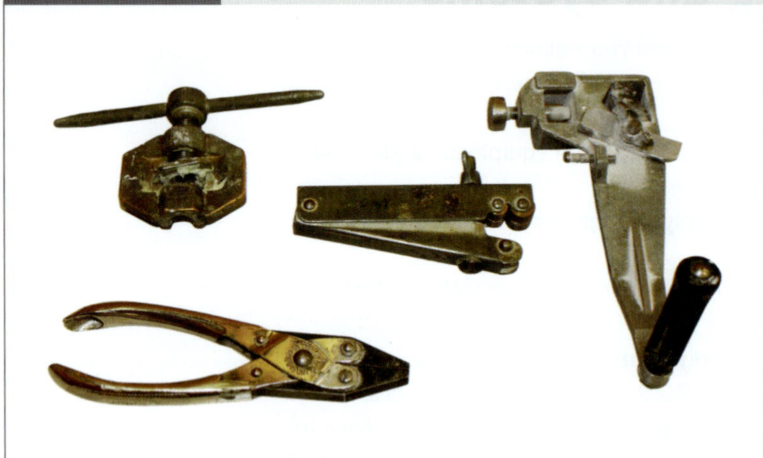

Source: Ed Hawkins

Tools for preparing MIMS cable ends are shown in Figure 8.40.

A range of polymetric cable is available and is designed to minimise the development of smoke, and acidic and dangerous gases, while maintaining circuit integrity in a fire situation. The main feature is a compound that causes the formation of a ceramic insulation around the conductors when it burns. As the outside sheath is destroyed by fire the insulation turns into a ceramic material that preserves the integrity of the circuit.

Be sure to comply with **AS/NZS 3000:2007** clause 3.9.9.3 Penetration of fire barriers – passing cables through fire isolation partitions. You must ensure the wiring is at least 50 mm from any other service and the hole must be close fitting and resealed with fire resistant sealing material. The conduits or ducts may be required to be internally sealed before penetrating the barrier.

Final checks and tests

MIMS cables may be used to supply safety systems and will need to be able to keep the safety systems working long enough for a building to be safely evacuated.

Visual inspection

Carry out your visual inspections, paying particular attention that MIMS glands are firmly fastened, and that no cable cores are exposed. Make sure that the radius of bends will not cause kinks.

Insulation and continuity tests

MIMS cable must have insulation tests carried out between active and neutral cores and the metallic sheath. Conduct similar continuity tests to the ones described in Section 8.5 in this chapter for metallic conduit.

SUMMING UP...

- All the requirements for other cabling systems apply to safety services.
- Fire pumps and sprinklers must be kept running as long as possible during a fire situation.
- Fire control systems must not be affected by failure of other parts of the installation.
- Changeover switches for alternative power supplies should be easily located and operated.
- MIMS cable is the most suitable system for fire control installations.
- MIMS needs to be worked with especially provided tools and fittings.
- MIMS cable can be provided with a PVC serving to prevent corrosion.
- A range of polymetric cables have been developed to maintain services even if they are burnt.

8.8 Steel wire armoured (SWA) cables

As an electrical practitioner you will find that SWA can be a difficult cable to work with if you do not employ the correct techniques right from the start. The techniques for removing the steel wire in preparation for connection are particularly important if you are not to waste cable.

Steel wire armoured cable has a steel wire armour usually covered with a PVC serving. The armour provides mechanical protection for the conductors.

There are two main types of SWA cable available:

- paper insulated lead sheathed
- PVC insulated, PVC sheathed.

SWA cable is primarily used in situations where there is risk of mechanical damage, for example, industrial and mining applications or hazardous areas involving explosive gases and dusts. Figure 8.41 shows an example of SWA cable and a cable gland.

FIGURE 8.41 Steel wire armoured cable and cable gland

Source: Ed Hawkins

FIGURE 8.42 Steps in preparing steel wire armoured cable for installation

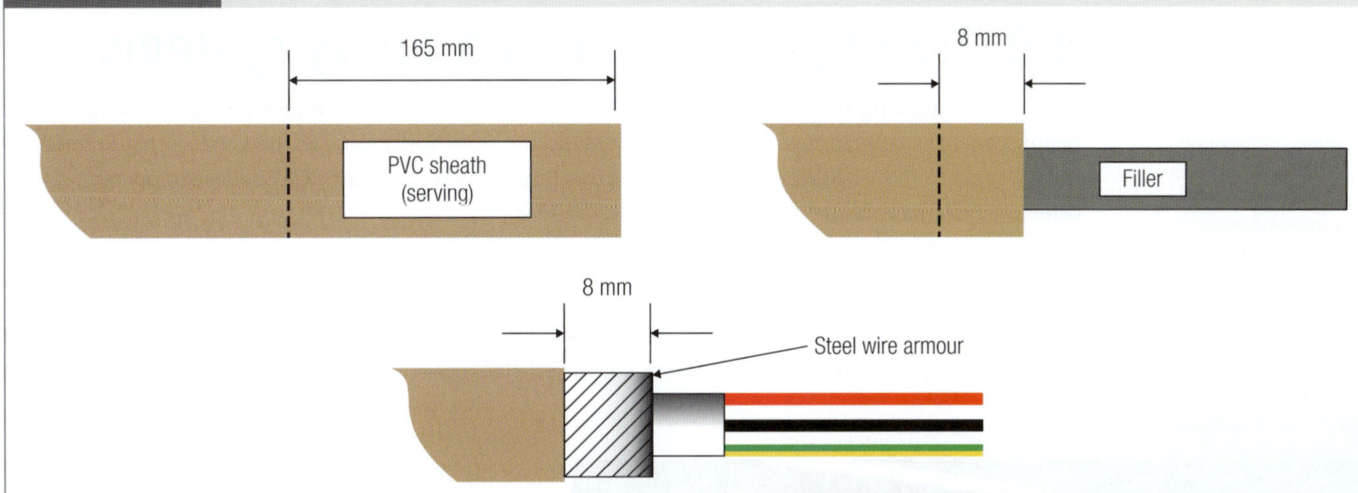

AS/NZS 3000:2007 clause 3.9.7.2 allows armoured cable to be installed in concrete, plaster or cement render without any further mechanical protection. Figure 8.42 outlines the steps in preparing SWA cable for installation, while Figure 8.43 shows the cable ready for installation.

As a general rule **AS/NZS 3000:2007** clause 5.4.1.1 tells you that you must earth metal sheaths on armoured cables since the metal glands that clamp the armour in place are connected to the rest of the fitting and through the threads to earthed situations within the connected equipment. The metal sheath need not be earthed if it complies with the exceptions listed in the clause.

FIGURE 8.43 SWA cable ready for installation

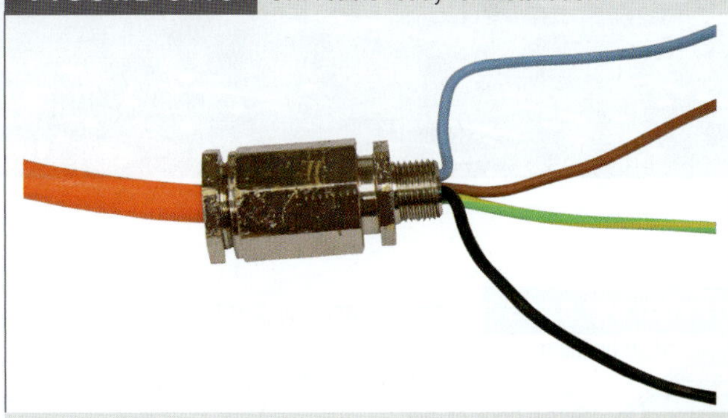

Source: Ed Hawkins

Final checks and tests

Steel wire armoured cable is used where other systems might be easily damaged. It is important to check the armour is clamped correctly and all metal fittings are earth bonded.

Visual inspection

Carry out the visual inspections as discussed under previous topics in this chapter and ensure the SWA glands are firmly fastened. Ensure that no cable cores or internal PVC sheathing are exposed.

> You will need to use safety eye-wear and leather gloves when working with SWA cables as the cut wire ends are sharp and jagged.

Insulation and continuity tests

Additional insulation tests between active and neutral cores and the armour must be conducted. You may have to do this with one end removed from the fitting unless the SWA is accessible from the outside of the gland.

In addition to the continuity tests carried out previously you will need to test the continuity between each section of the SWA system (between screwed fittings) and the main earth at the switchboard in a similar manner to metallic conduit.

SUMMING UP...

- SWA cable is highly rated against mechanical damage.
- SWA requires special fittings to clamp the wire and seal against dust and gases.
- Metal sheaths need to be earthed, usually at one or both ends.
- SWA cables can be buried direct in concrete or plaster.

8.9 Trailing cables and catenary systems

Before starting work on trailing cables you should ensure that all immobilisation procedures have been implemented to prevent unwanted movement of heavy machines.

You may be working on a straightforward overhead cable between a house and outbuilding or the reeling cable of a large mine reclaimer. Similar skills and rules apply, such as the need to ensure freedom for the cable to move without stress. Moving machinery such as cranes need cables supported on wire catenary systems to allow them to move with the machine.

Trailing cables

Trailing and reeling cables are electrical power cables used for mobile apparatus, such as travelling and slewing cranes, elevators, large excavators, draglines, stackers and reclaimers.

FIGURE 8.44 Trailing cable system on a travelling crane

Source: Ed Hawkins

The cable permits the mobile apparatus to move without disconnecting its electric power supply. A reeling cable is specifically designed to be frequently reeled on and off a cable drum or reeler (e.g. an iron ore rail-mounted reclaimer), whereas a trailing cable is specifically designed to be moved with the mobile apparatus (e.g. a crane cross travel, elevator car, coal mine dragline or mineral sands mobile processing plant); an example of a trailing cable system is shown in Figure 8.44.

Installation using trailing or reeling cables must comply with **AS/NZS 2802 Electric Cables – Reeling and Trailing – For Mining and General Use (Other Than Underground Coal Mining)**.

FIGURE 8.45 Catenary cable support between two buildings

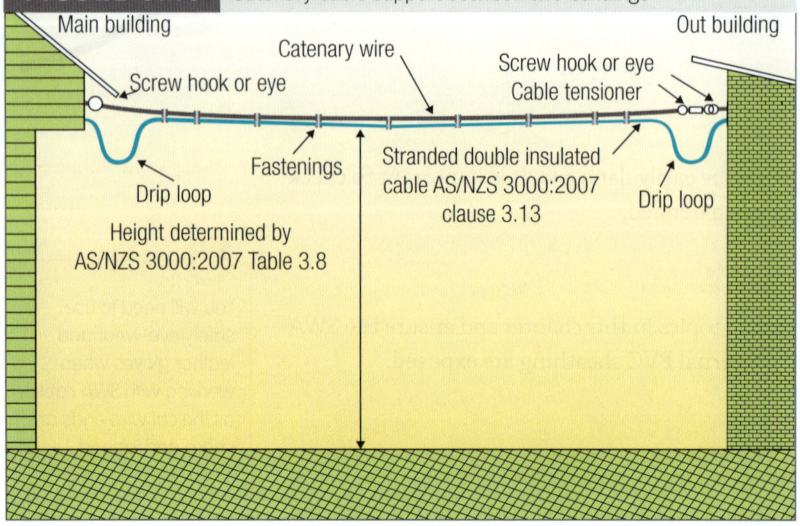

Catenary system

When a double insulated stranded cable is supported across a space on a stranded steel wire it is referred to as a catenary system (see Figure 8.45). The term catenary literally means the shape the cable takes under its own weight but the term now is used widely to identify the method of support.

The requirements such as clearances and height in **AS/NZS 3000:2007** clause 3.2, Aerial conductors, apply in a similar way to catenary systems. Additional requirements in clause 3.13, such as the need for stranded conductors and double insulation, apply to catenary cables.

You will need to form a drip loop to cause any water gathered on the cable to run onto the ground instead of entering the building. The drip loop has the additional benefit of preventing stress on the conductors during windy conditions. It is common practice to continue the cable and terminate it at the electrical equipment within the building; however, the cables can be terminated in suitable junction boxes on the outside of buildings. The tension on catenary supports can be controlled by including an adjusting screw.

Catenary supports can be used to suspend socket outlets in an area where underfloor wiring is not practical (see Figure 8.46). Sliding insulated mountings are available to allow the socket outlet to be moved in a limited direction along the supporting wire.

Final checks and tests

Cables supported by catenary and trailing cable systems are subject to considerable movement so require special attention to the suitability of fastenings and the freedom to move without stress on the cable cores (see **AS/NZS 3000:2007** clause 3.3.2.8).

Visual inspection

Carry out the visual inspections as discussed under previous topics in this chapter and ensure the catenary and cable supports are firmly fastened. Examine each connection point and ensure there is no possibility of terminals coming under tension from cable movement.

Insulation and continuity tests

When conducting insulation tests an additional test must be made between active and neutral cores and any structural metal that is likely to be in contact with the cable support system.

Carry out the standard continuity tests described in previous sections in this chapter. Test the continuity between each section of the support system that is required to be earthed and the main earth at the switchboard.

FIGURE 8.46 Suspended socket outlet

Source: Ed Hawkins

SUMMING UP...

- Trailing cables are used to allow mobility for machinery.
- Catenary cable systems are used to carry a cable across an open space.
- Cables supported by a catenary system must be double insulated and stranded.
- Specific height limits apply to aerial cables.
- Drip loops prevent water entering buildings and electrical equipment.
- Fastenings for cables must be weatherproof.

REVIEW EXERCISES

Answer these questions and check your answers at the back of the book.

Review questions 8.1

1 State the type of cable that is commonly known as 'building wire'.
2 List three requirements to be taken into account when selecting a cable.
3 State the maximum voltage that could be used on a 660/1kV PVC insulated cable.

Review questions 8.2

1 What colour is the earthing conductor in internationally colour coded three core flexible cord?
2 How can the earth pin in a three flat pin plug top be distinguished from the other two pins?

Review questions 8.3

1 'TPS cable' is the common term used to refer to which wiring system?
2 List the requirements for burying TPS cables underground.

Review questions 8.4

1 What is the main advantage of circular TPS cable?
2 What is the purpose of the cable gland?

Review questions 8.5

1 State the definition of trunking as provided in the AS/NZS 3000:2007 Wiring Rules.
2 Describe the extra support required if trunking with a removable lid is run across the underside of a ceiling.

Review questions 8.6

1 Describe the precaution you would need to take when joining PVC conduit to a metallic trunking system.
2 If the conduit has to make a 90 degree change in direction and it is not possible to bend the conduit, state the fitting that could be used.

Review questions 8.7

1 List two important systems that run on the fire control installation.
2 State the electrical tests to be conducted before commissioning an MIMS wiring system.

Review questions 8.8

1 List the protective clothing required when working with SWA cable.
2 When preparing the end of an SWA cable for termination, why should you cut halfway through the outer wire before removing the final piece of PVC serving?

Review questions 8.9

1 Describe the difference between a trailing and a reeling cable.
2 What are two functions of the drip loop?

ONLINE RESOURCES

Cables: Nexans
http://www.olex.com.au/eservice/Australia-en_AU/navigate_-34/Global_expert_in_cables_and_cabling_system.html

Cables: Olex
http://www.olex.com.au/eservice/Australia-en_AU/navigate_289423/Fire_performance.html

SWA cables and fittings: Eland Cables
http://www.elandcables.com > Electrical cable
http://www.elandcables.com > Armoured cable

Test and Tag Supplies
https://www.testandtagsupplies.com.au/zencart/

Trailing cables: Cavotec
http://www.cavotec.com

9

ELECTRICAL APPARATUS AND CIRCUITS

This chapter will help you understand electrical apparatus and circuits and how they are controlled and installed. As an electrical practitioner you will be expected to have a working knowledge of the most commonly used equipment and appliances and the installation rules that apply to them.

Chapter outline

9.1 Lighting circuits – looping at the light/switch

As an electrical practitioner you will need to be able to design and install lighting circuits as well as diagnose and repair faults.

Methods of wiring lighting circuits

Lighting control circuits

The method adopted for wiring lighting circuits will depend upon the wiring system employed. When using conduit or trunking the connection method will depend on the route taken by the conduit or trunking. When using thermoplastic sheathed (TPS) cable the choices are dictated by the range of stock cables supplied by the manufacturer. The most commonly used methods are 'loop at the light' and 'loop at the switch'.

The loop at the light method

Three main circuits are used to control lighting: one-way, two-way and three-way, or intermediate.

One-way switching

Figure 9.1 shows a TPS wiring diagram for a set of lights with switch S1 controlling Lamp L1 and switch S2 controlling Lamps L2 and L3. This is commonly referred to as 'one-way' switching as each light, or bank of lights, is controlled by only one switch.

FIGURE 9.1 One-way loop at the light circuit

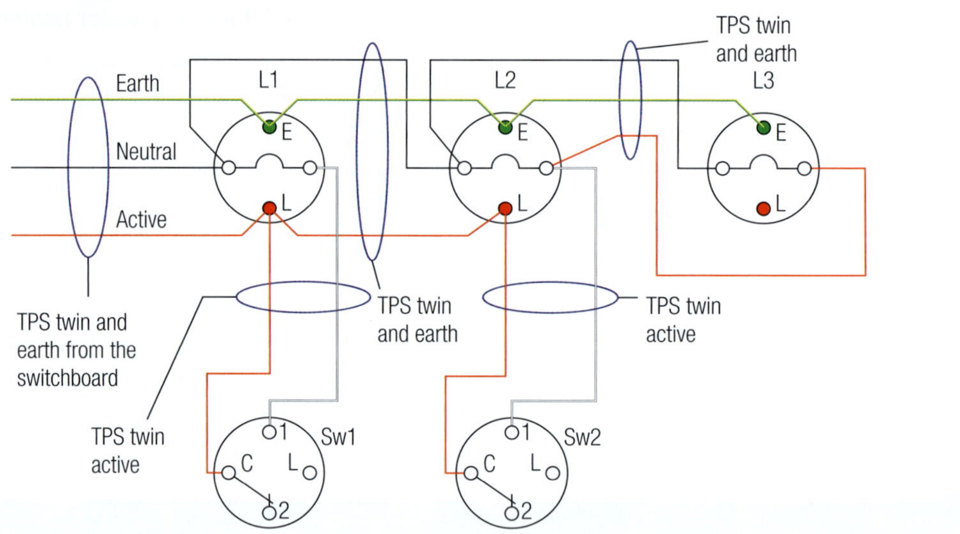

This method of using TPS cable provides a live supply feed to each light fitting and the twin active is run to the switch. All connections are made in the light fittings themselves, reducing the cost of junction boxes and installation time. Lamp L3 is on the end of the circuit and has the active conductor supplied from the switch wire at L2.

AS/NZS 3000:2007 Wiring Rules clauses 5.4.3 and 5.4.4 specify the earthing requirements for light fittings (luminaires). Take careful note that, with some exceptions, an earth conductor must be provided at the light point even if the fitting used does not require earthing as set out in clause 5.4.4.

There is a permanent active at the light fitting using this system. It is not safe to work on the light fittings unless the circuit is locked out at the source of supply.

Two-way switching

Two-way switching circuits allow the light to be turned on and off from each of two separate locations such as each end of a passage or stairway. The switches are standard switches that have the necessary extra contacts and terminals provided to allow them to be used for two-way switching if required. Additional cables are required in the form of single double-insulated (SDI) active conductors to act as the switch wire at one end and the active wire to the other, as shown in Figure 9.2. The first switch can select a 'strap wire' which it provides with an active potential. The second switch selects either an active or non-active strap wire, allowing the light to be energised or de-energised from either switch.

FIGURE 9.2 Loop at the light two-way with TPS cable

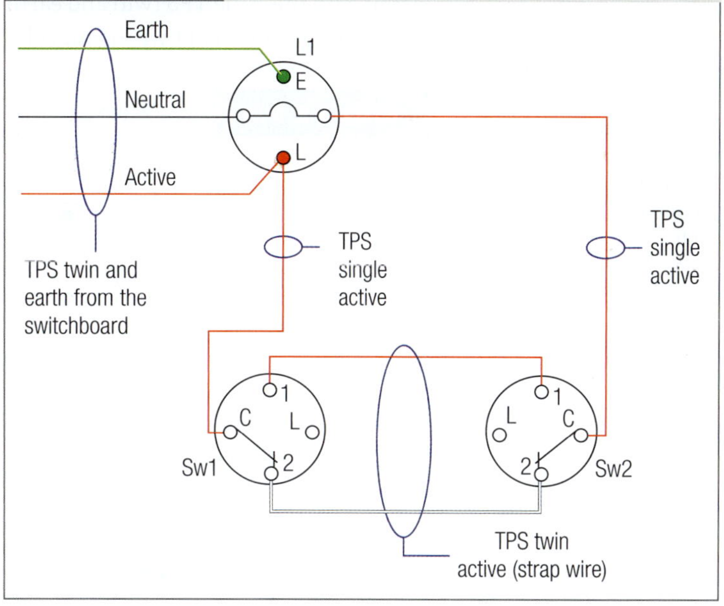

Three-way or intermediate switching

Connecting additional switch locations into the two-way circuit requires specially made 'intermediate' switch mechanisms as shown in Figure 9.3. Essentially a reversing switch, the intermediate switch reverses whatever selection (ON or OFF) is made by the two-way switches. Any number of switch locations can be added by inserting more intermediate switches into the circuit.

FIGURE 9.3 Loop at the light three-way with TPS cable

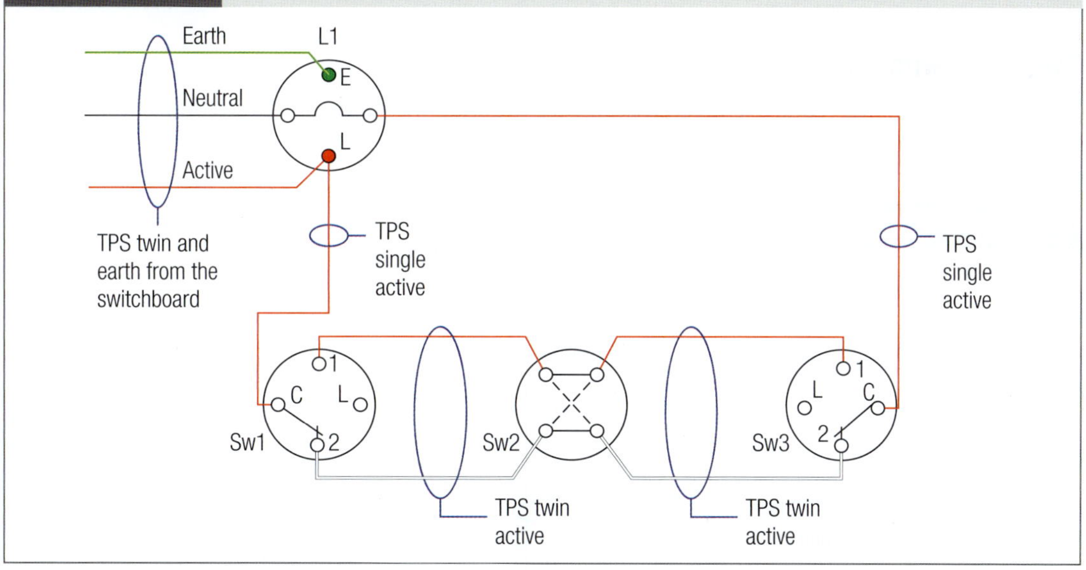

The loop at the switch method

The three lighting circuits can be used equally well with the loop at the switch method of connection.

One-way switching

The wiring diagram in Figure 9.4 shows the same three lamps from Figure 9.1 in a one-way circuit in TPS cable. This time the circuit is connected using the loop at the switch method. As with the loop at the light method this technique saves additional work and materials, but has the added advantage that

While there is no permanent active at the light fitting using this system, you should not rely on the light switch for isolation unless it is capable of being locked off.

only one type of cable is needed. The supply feed is run in TPS twin and earth cable to each switch in the circuit, and the same TPS twin and earth cable is run to the lamp. The loop at the switch method is used when the lighting circuit wiring must be installed under the flooring instead of through the roof space.

FIGURE 9.4 Loop at the switch one-way

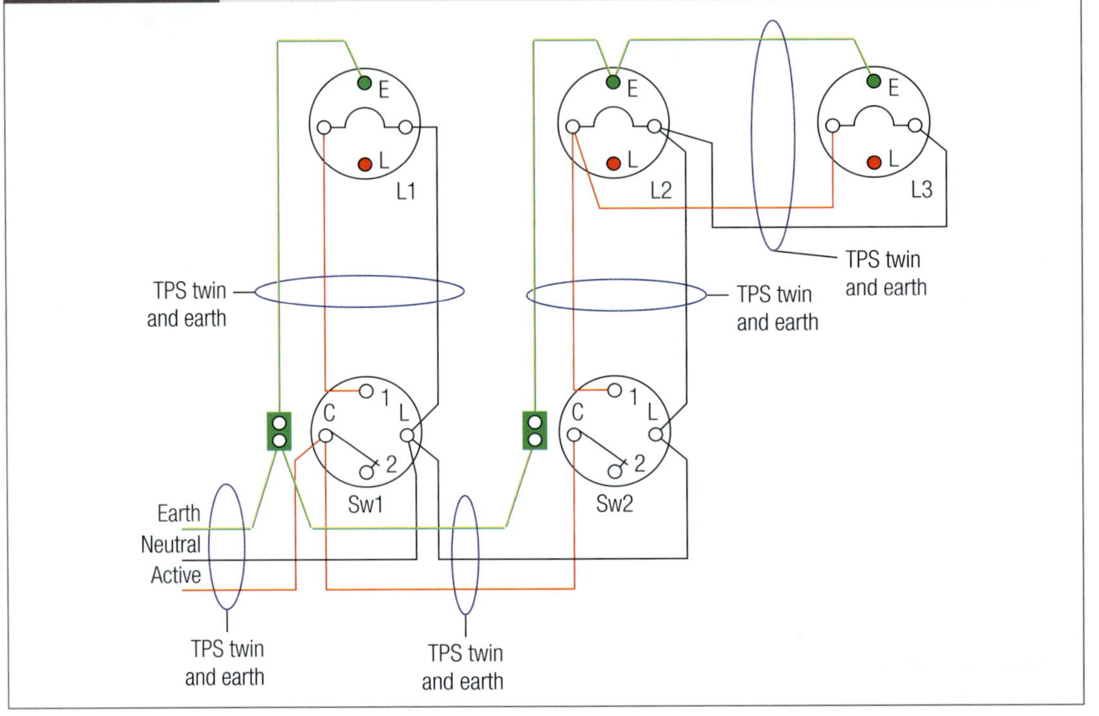

Two-way switching

The two-way switching requires twin active cable for the strap wire and one SDI to connect the active to the second switch as shown in Figure 9.5.

FIGURE 9.5 Loop at the switch two-way with TPS cable

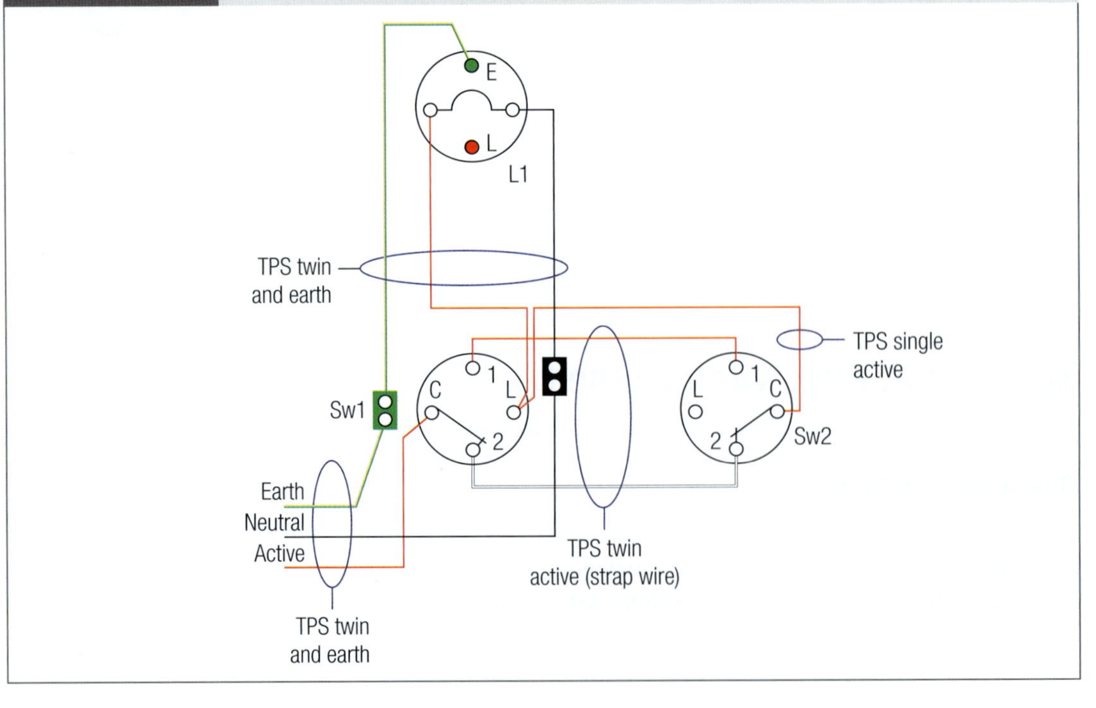

Three-way or intermediate switching of light points

As with the loop at the light method the intermediate switch is connected into the strap wire, as shown in Figure 9.6, without further complicating the circuit.

FIGURE 9.6 Loop at the switch three-way with TPS cable

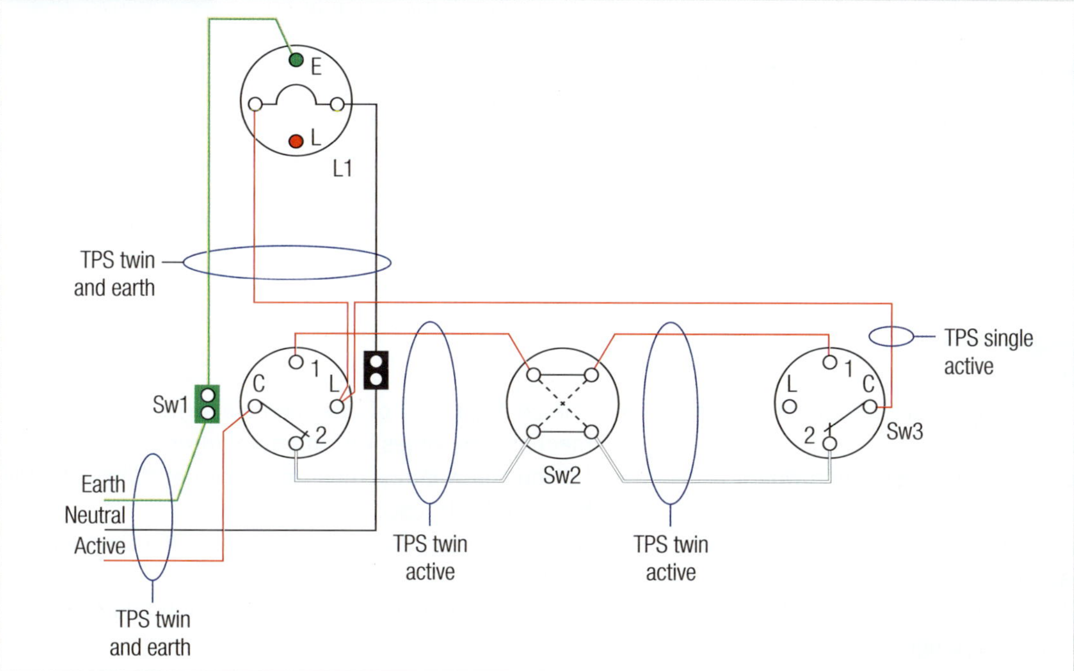

You will have noticed that no terminal numbers are provided for the intermediate switch. The major suppliers of switch mechanisms for the domestic market have terminal connections that are not the same. The diagram in Figure 9.7 shows the difference. You will need to carefully read the instructions when installing intermediate switches.

FIGURE 9.7 Differences among intermediate switches

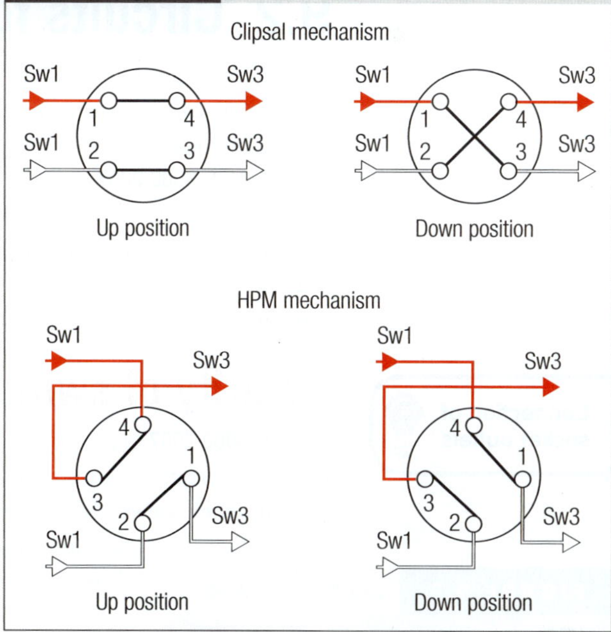

Checks and tests

Once your lighting circuit has been installed you will need to conduct a series of tests to ensure correct function and safety.

Visual inspection

Carefully examine the finished connections and ensure:

- correct polarity of all connections particularly Edison screw fittings where the active must be connected to the centre pin and not the thread
- correct colour coding of all conductors
- no insulation has been removed unnecessarily
- no cable cores are extending outside of the outer cover of the sheathing or enclosure
- there are no nicks in cable insulation or sheathing
- there is no insulation clamped under terminals
- the cable conductors are doubled over before connecting (where applicable)
- there is no mechanical damage to any of the accessories or cable

- the correct accessories and cable have been used
- lamps are installed to prevent undue temperature rise, ignition or deterioration of materials.

Insulation and continuity tests

Use an insulation tester set to the 500 volt range and test between the conductors. The insulation resistance should be infinity when measured with a 500 volt insulation tester.

Ensure that you warn others working on the circuit so they do not come in contact with the conductors under test.

Earth fault loop impedance

AS/NZS 3000:2007 clause 8.3.9.3 describes the method for measuring earth fault loop impedance with the supply not available. Table 8.2 in the Wiring Rules is provided to help you interpret your results.

Set the tester to the low ohms range and test the continuity of each conductor.

For a full description of the required tests and results, consult Chapter 13, Electrical installations – verification and testing.

SUMMING UP...

- Loop at the light circuit is most suitable for cables run in a roof space.
- Loop at the switch circuit is most suited where cables run through the under-floor space.
- Two-way switching allows operation of a light from two different locations.
- Intermediate switches allow any number of switch positions to control the one light.
- The configuration of intermediate switch mechanisms varies with manufacturer.
- Edison screw lamps must have the neutral connected to the outer thread.
- All light fittings must be provided with a functional earth.
- All circuits must be tested for function and safety.

9.2 Circuits for socket outlets

A socket outlet is a device for attaching a flexible cord having contacts where a removable connection can be made with the pins of a plug. The term 'socket outlet' also applies to a cord-extension socket attached to a flexible cord that is permanently connected to the installation wiring (see **AS/NZS 3000:2007** clause 4.4).

Socket outlets allow users the choice of location and type of appliance they wish to use. You will be required to assess the need and location of socket outlets and plan circuits to prevent overloading and nuisance tripping. Most of the references that follow use examples from single phase circuits but the same requirements apply to multiphase socket outlets.

Polarity of switched socket outlets

Connection of socket outlets

AS/NZS3000:2007 clause 4.4.5 specifies the sequence and polarisation of socket outlets. The most important requirement is that you must connect all sockets that accommodate three-pin/flat pin plugs so that, when viewed from the front of the socket outlet, the order of connection commencing from the slot on the radial line shall be earth, active, neutral in a clockwise direction as shown in Figure 9.8.

FIGURE 9.8 Socket outlet polarity

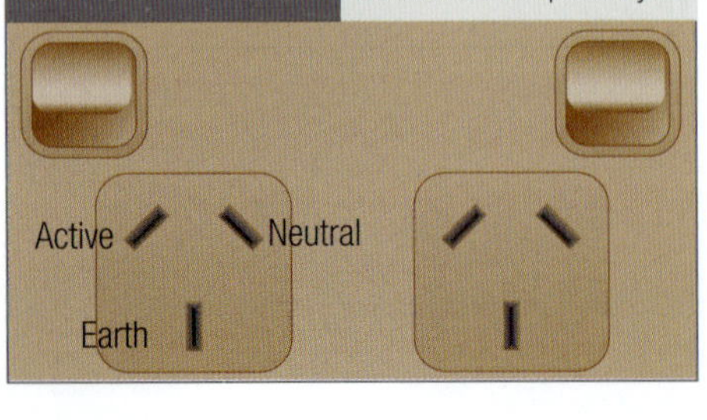

Other socket outlets in the same installation such as three phase and special outlets should all be connected in a uniform sequence.

Switching requirements

AS/NZS 3000:2007 clause 4.4.4 specifies the requirements for the provision of switches for socket outlets. You must make sure each socket outlet is individually controlled by a switch that operates in the active conductor and has a voltage and current rating to suit the rating of the socket outlet.

You are allowed use one switch to control two socket outlets only if the controlled sockets are immediately adjacent to each other and the switch current rating is not less than the combined current rating of the socket outlets or the circuit protective device (whichever is smaller). In every case, you must position the switch as close as possible to the socket outlet and mark it to indicate the socket outlet or equipment that it controls. You do not have to label the switch if the controlled socket is obvious by its location. A socket outlet rated at 10 amps or less and installed for a particular appliance or luminaire need not be controlled by a switch provided it is not readily accessible for any other use. You would usually install a special socket with a round earth pin, as shown in Figure 9.9, to prevent this type of point being used for other purposes.

FIGURE 9.9 Restricted access socket outlet

Source: Ed Hawkins

Earth connections

You must provide every socket outlet with an earthing contact (**AS/NZS 3000:2007** clause 4.4.3). You must connect an earth to the earthing contact except in separated circuits where you will connect it to the protective bonding system. This applies to all socket outlets except those supplied by extra low voltage circuits.

Circuits for socket outlets

In general the number of socket outlets that you can connect to a circuit depends upon:

- the number of points to be supplied in combination (lighting, socket outlets or appliances)
- the operating characteristics of each item of equipment
- the circuit current under expected operating conditions
- the effects of overload on other equipment on the circuit (security, emergency, medical and critical communications) – see **AS/NZS 3000:2007** Appendix C2.1 a, b, c and d.

To help simplify the process Table C8 is provided in **AS/NZS 3000:2007**.

To use the values in Table C8 there must be two power circuits in your installation. You must count multiple socket outlets as one socket outlet for each socket on the multiple combination. If you know that the socket outlet is to be used for a particular appliance such as a dishwasher you should use the load of the appliance. If the installation has permanent heating or cooling systems installed you can increase the number of socket outlets since it is less likely that any of the socket outlets will be for plugged-in space heaters.

The cable size to supply socket outlets must be rated to match the expected load, and must not be lower than the rating of the protective device. You are not permitted to connect multipurpose socket outlets using any cable smaller than $2.5\,mm^2$ or a circuit breaker smaller than 13 amps. A 15 amp or 20 amp socket outlet would usually be installed on a circuit of its own and supplied with a suitable size cable. You will find that it is not practical to run a large cable and connect a number of 15 amp socket outlets to it as the terminals in the fittings are not large enough.

You are only allowed to use a 10 amp socket outlet with a cable smaller than $2.5\,mm^2$ if it is to be used for a fan, lamp or appliance rated no more than 150 watts and is installed more than 2.3 m above the floor.

Installation of single phase socket outlets

Every type of wiring system has provision for installing single phase socket outlets. Thermoplastic sheathed (TPS) and conduit have mounting boxes, mounting blocks and fully enclosed socket outlets especially made for each system. Metal conduit fittings are usually suitable for use with steel wire armoured and mineral insulated metal sheathed (MIMS) wiring systems, while metal and PVC trunking usually have adaptors provided to allow conduit fittings to be used if required.

Thermoplastic sheath (TPS)

Two methods of using TPS wiring are:

1 Surface mounting. The cable and fittings are mounted on the surface of the structure and the socket outlets are fastened onto mounting boxes to provide protection from the live terminals and mechanical protection for the single insulated cables behind the fitting. If the building material is conductive the mounting box must include an insulating barrier between the live terminals and the conductive building materials. Alternatively, the conductive material should be earthed.

2 Concealed wiring. The cable is run in hollow spaces within the building or in metal or other ducts provided. The socket outlets are mounted on metal brackets supported by the building structure. You should pay careful attention to the degrees of protection required within walls and ceiling spaces as set out in **AS/NZS 3000:2007** clause 3.9.3. If the sockets are mounted on the inside of an external wall of brick veneer, cavity brick or masonry, it is classed as a damp situation. The live terminals of the socket outlet must be insulated from possible contact with a damp wall or conductive building insulation. If the structural material is conductive, such as steel studs, the building material must be earthed.

Conduit

When using steel or PVC conduit the equipment will usually be mounted on the surface of the structure. The fittings designed for use with each system should be used with the wiring system they are designed for. Mixing PVC and steel conduit fittings can compromise earth continuity and mechanical protection. It is common practice to use TPS mounting blocks with PVC conduit; this is quite satisfactory, but you should make sure that the mounting blocks provide adequate protection for exposed building materials.

Trunking

PVC or metal trunking can be used to supply socket outlets. The socket outlets can be surface mounted on metal or PVC boxes or mounted on the trunking itself, and in this case the trunking should be made for this purpose.

FIGURE 9.10 Phase rotation for a three phase socket outlet

Source: Ed Hawkins

Testing and compliance

All the following tests must be conducted with the supply locked and tagged out. Socket outlets must be tested for insulation integrity, polarity and phase sequence, and earth fault loop impedance as described in Section 9.1 earlier in this chapter.

Polarity and phase sequence

You must take care to ensure the correct polarity of each fitting and confirm that all three phase sockets have the same phase sequence. You will need a trailing lead that will reach to each socket outlet in the installation to test each connector and ensure the neutral, active and earth are not transposed anywhere and that each neutral conductor only serves the circuit it is intended for.

Three phase socket outlets must be tested to ensure they have the same phase sequence throughout the installation, as shown in Figure 9.10.

SUMMING UP...

- Socket outlets allow users to choose where and when they will use an appliance.
- Socket outlets must comply with uniform polarity and phase rotation standards.
- Switches for socket outlets must be clearly labelled or located in a way to indicate the socket outlet they control.
- The correct socket outlet mountings should be used to suit the installed wiring system.
- Socket outlets must be selected to suit the anticipated load current.
- Insulating barriers must be provided to prevent exposure of damp building materials to live metal.
- Adjacent conductive building materials must be earthed.
- The earth pin of a socket outlet must be connected to earth.
- Special socket outlets should be used to prevent incorrect use of low current and PELV socket outlets.
- The number of socket outlets connected to a circuit is determined by the rating of the over-current protective device and the anticipated load for each point.
- All socket outlets should be tested for polarity and phase rotation before commissioning.

9.3 Final sub-circuits and segregation

When planning an installation you must be able to organise sub-circuits supplying consuming devices and socket outlets and select suitable protection against overload and short circuit. Segregation of circuits refers to an electrical distribution that includes separated final sub-circuits not of equal size or capacity. It also applies to the electrical and physical separation of circuits forming safety systems.

Final sub-circuits

AS/NZS 3000:2007 clause 2.2 outlines the requirements for the most common distribution arrangement for a low voltage electrical installation, that is, radial branched distribution, as shown in Figure 9.11.

FIGURE 9.11 Distribution of sub-circuits

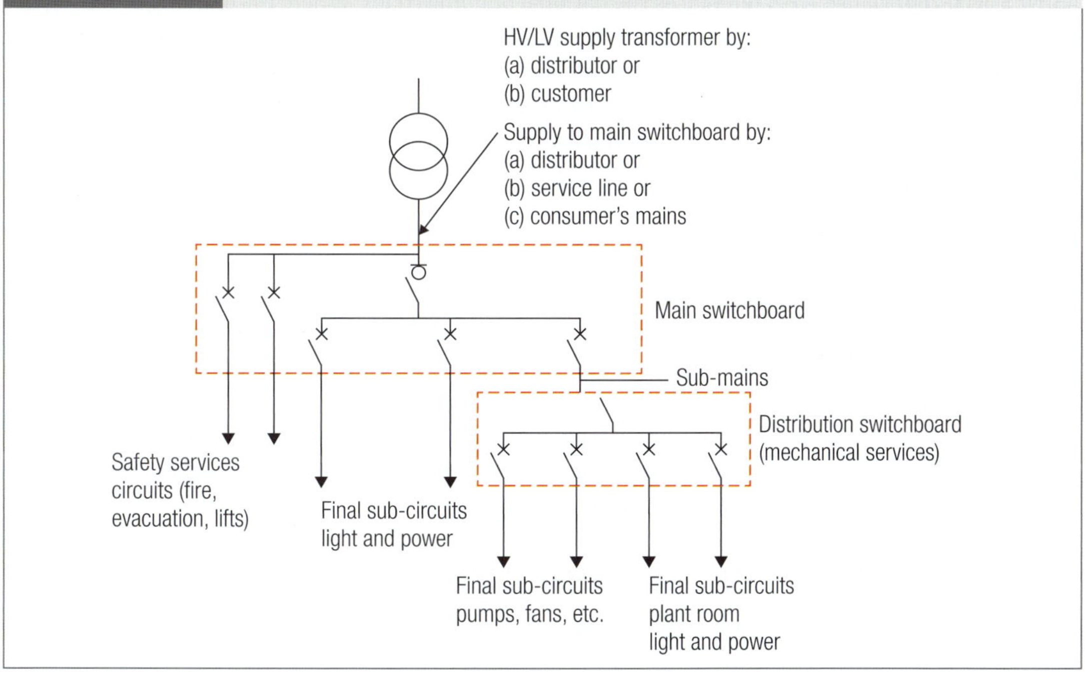

Division of circuits falls into several categories, each an individual circuit or group of circuits. Typical circuit groups are:

- lighting
- socket outlets
- heating and/or air-conditioning appliances
- motor driven plant
- auxiliary services, such as indication and control
- safety services.

AS/NZS 3000:2007 clause 2.3.1 specifies that every circuit must be capable of being isolated from the supply conductors:

- Active – requires a device for isolation, switch or circuit breaker as shown in Figure 9.12.
- Neutral – no switch to be installed in the neutral, except where the active is switched at the same time. The only exception is the control circuit of a fire pump (clause 7.2.9.4).
- Earthing conductors – must not be switched under any circumstances (clause 2.3.1.1(c)).

FIGURE 9.12 Circuit actives can be isolated at the switchboard

Source: Ed Hawkins

Advantages are that:

- one circuit only will be shut down in case of a fault
- location of a fault is simplified
- maintenance or extension to a circuit can be carried out leaving the remainder of the installation in service
- conductor sizes can be reduced at protective devices on switchboards to suit decreasing demand in sub-circuits.

Conversely, a fault on one of the circuits from the main switchboard will shut down supply to all circuits of related downstream distribution boards.

Segregation of circuits

AS/NZS 3000:2007 clause 7.2.8 references the national building codes. Where the building codes require emergency equipment to be installed, the switchboard must be constructed with partitions to prevent faults from spreading to the circuits that supply the emergency equipment.

Separate enclosures such as trunking and conduit must be provided for each safety circuit. If the safety circuits are run in multicore cables each safety circuit must have its own cable and not be included in the same cable with any other system.

Mixed circuits

Each item of equipment that has a current rating in excess of 20 amps per phase should be connected to a separate and distinct circuit. However, circuits of 20 amps and lower can be used to supply a combination of appliances, socket outlets, motors and lights providing the cable used is a suitable size for the protective device.

A common use for a mixed circuit is to supply light and power to a small outbuilding with a small number of lights and socket outlets. If the circuit is protected by a 16 amp circuit breaker then the cable must be rated at 16 amps throughout the entire circuit.

Circuit loading for a mixed circuit

When more than one item of equipment is to be connected to a mixed circuit, consideration needs to be given to:

- the number, distribution and type of equipment such as lighting points, socket outlets or appliances to be supplied in combination

- the operating characteristics of the different items of equipment, including seasonal or daily variations
- the circuit current under expected operating conditions and the coordination with cable and protective device ratings to minimise the risk of overload fault
- the effects of an overload fault on the circuit, including loss of supply to equipment that performs a special function, for example, security, emergency, medical or critical information and telecommunications services.

SUMMING UP...

- The most common distribution arrangement is the radial branched system.
- Sub-circuits are provided for each purpose and load type.
- Each circuit must be capable of being isolated from the supply conductors.
- Division into sub-circuits allows isolation for additions or maintenance.
- Emergency systems must be segregated from general sub-circuits.
- Mixed circuits are limited to 20 amps or lower.
- A common use for mixed circuits is to supply light and power to an outbuilding.

9.4 Electrical heating control devices

The selection of control devices for electric heating applications will usually be done by the manufacturer. You need to understand the operation of heating control devices and the importance of selecting the correct replacement part in the event of a fault occurring.

Manual heat control

FIGURE 9.13 Switched elements

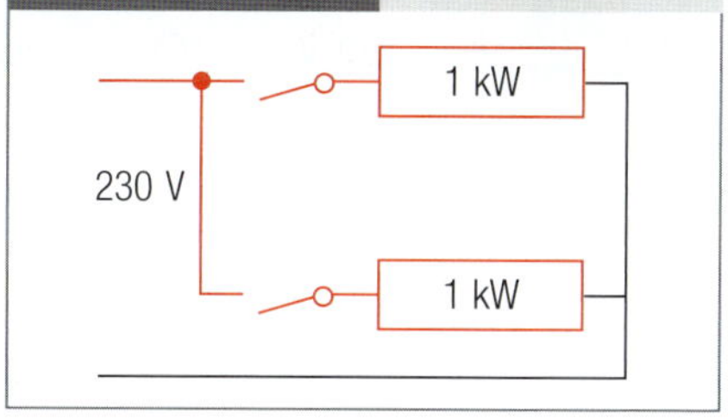

The simplest method of heat control is to provide switches so the user can switch elements on or off depending upon the level of comfort required, as shown in Figure 9.13. The possible selection would be:

1 Off – no elements on
2 Low – one element on
3 High – both elements on.

The three heat switch shown in Figure 9.14 allows the user to select one of four settings:

- OFF – no elements switched on
- Position 1 (Low) – one element switched on
- Position 2 (Medium) – two elements connected in series
- Position 3 (High) – two elements connected in parallel.

Note that as required in **AS/NZS 3000:2007** clause 2.3.1.1, the neutral is not switched.

The advantage of the three heat switch is that it provides for the selection of three heat levels from two elements using just one switch.

In Figure 9.14 on page 344 the circuit comprises two 1 kW elements, 52.9 ohms each. On the low setting they will total 105.8 ohms drawing 2.17 amps dissipating 500 watts. On the medium setting 1 kW is dissipated and on the high setting 2 kW are dissipated drawing 8.7 amps, and making it suitable to plug in to a 10 amp socket outlet without overloading the outlet.

Control of heating

The temperature setting using manual control will depend upon the preference of the user. To achieve more accurate temperature control needs automatic control.

FIGURE 9.14 Three heat switch

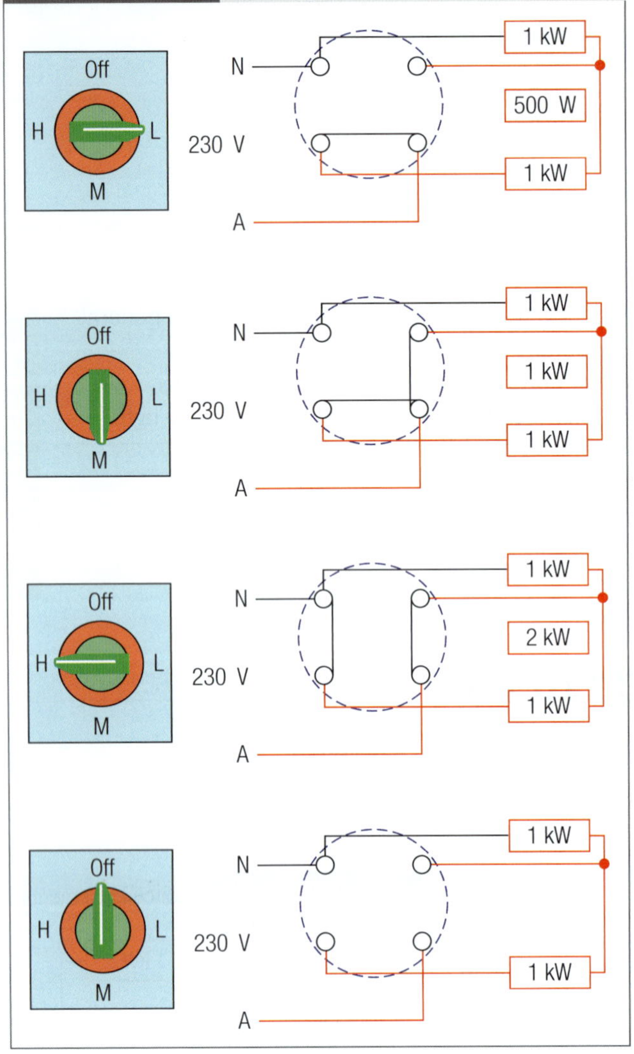

Automatic heat control

You will encounter two main types of automatic temperature control:

- a continuous open loop control known as a 'simmerstat'
- a feedback device known as a 'thermostat'.

Electronic and programmable thermostats are sophisticated versions of these devices.

Simmerstats

The simmerstat, known as an infinite control, is usually used to control the hotplates on a cooking range. Simmerstats do not sense the temperature. A time setting controls the average current supplied to the heating element by adjusting the period of time the element is switched on. When the control knob is turned it increases the pressure on a control spring that in turn restricts the movement of a bimetal strip, as shown in Figure 9.15.

Initially current passes through a normally closed contact attached to the bimetal strip heating the hotplate and the internal simmerstat element. When the bimetal strip is heated to a point where it bends and overcomes the pressure of the spring, the contact opens, stopping current to the internal heater as well as the hotplate. As the bimetal strip cools, it straightens, causing the contact to close again, repeating the cycle. The tighter the pressure of the spring, the hotter the hotplate becomes.

Thermostats

Unlike simmerstats, thermostats are designed to hold the temperature within a reasonable range.

Operation

The temperature set point is adjusted using a knob or a screwdriver adjustment that in turn adjusts the tension of a spring working against a heat-expandable system such as a strut and tube, bimetal strip or a filled thermal system, as shown in Figure 9.16. When the heat reaches the pre-set value the pressure from the expansion system overcomes the spring and operates the contacts. As the heat falls the expansion system releases the pressure on the spring and the contacts return to their original position.

FIGURE 9.15 Simmerstat operation

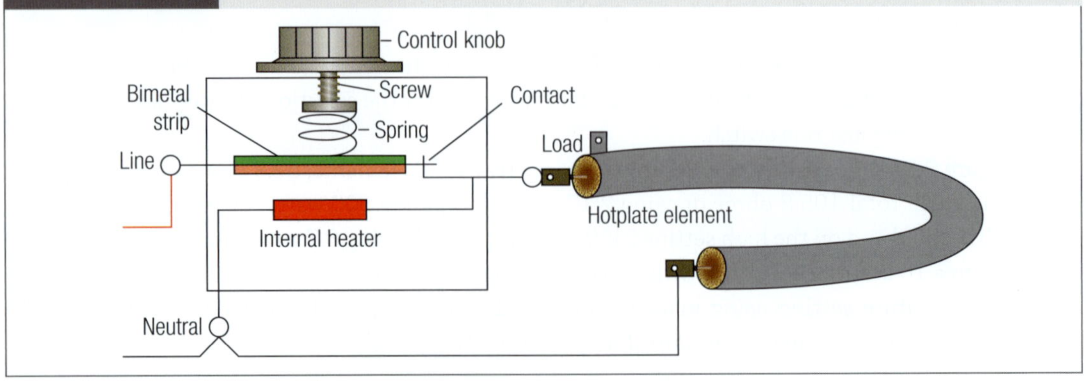

FIGURE 9.16 Filled system thermostat

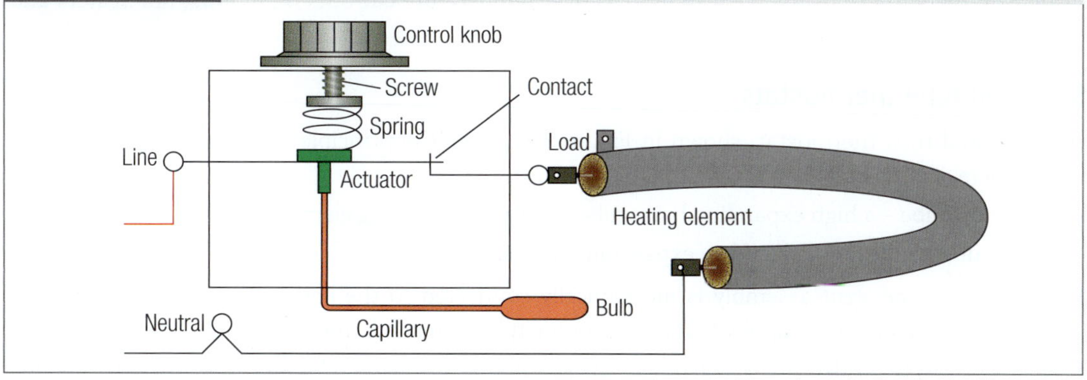

Sensitivity

The sensitivity of the control device refers to the delay time between the temperature reaching the set point and the operation of the contacts.

To prevent the contacts from switching continuously and causing arc damage a delay known as differential action is built into the control device. When the temperature rises to the set point the contacts open but do not close immediately and the temperature drops below the set point. When the temperature has dropped one or more degrees below the set point, the contacts will close, energising the element.

Differential

The temperature difference between the set point where the contacts open and the point where they close again is known as the differential range and is adjusted with a screw that tensions a spring restraining the contact.

Bimetal thermostats

The bimetal temperature control relies on two dissimilar metals fused together. When the fused metal (bimetal) strip is heated the two metals expand at different rates and the strip bends, operating the switch mechanism as described above. There is only one temperature at which the strip is straight – otherwise it is bent one way to close the switch and the other way to open it, as shown in Figure 9.17.

Some bimetals are in the form of a disc and produce a snap action. This device can maintain temperature control within ±5°C.

FIGURE 9.17 Bimetal disc thermostat operation

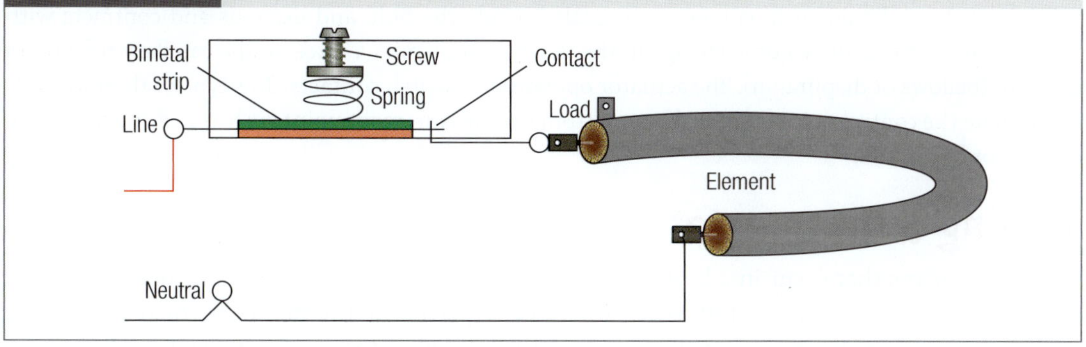

Poorly adjusted thermostats can cause considerable waste of electrical energy as rooms and processes may be heated to unnecessarily high temperatures.

Bimetal thermostats are used primarily in domestic appliances such as steam irons and hot water systems. Bimetal thermostats are also used as over temperature cut-outs on motors, fan heaters and hot water cylinders.

Strut and tube thermostats

The strut and tube thermostat, shown in Figure 9.18, uses the principle of thermal expansion of a metal. There are two basic parts:

- the outer tube – a high expanding brass tube forming the sensing element
- the strut assembly – made of low expansion metal alloy.

One end of the strut assembly is mechanically connected to the end of the tube. The principle of differential expansion is applied when the outer tube is heated. The outer tube expands, placing a mechanical strain on the strut which is not expanding at the same rate. The strut pulls the moving contact open.

Adjustment of the temperature is made by turning the knob to bring the fixed contacts closer or further away from the moving contacts. This type of thermostat has a high sensitivity with a switching differential of 0.1°C.

FIGURE 9.18 Strut and tube thermostat

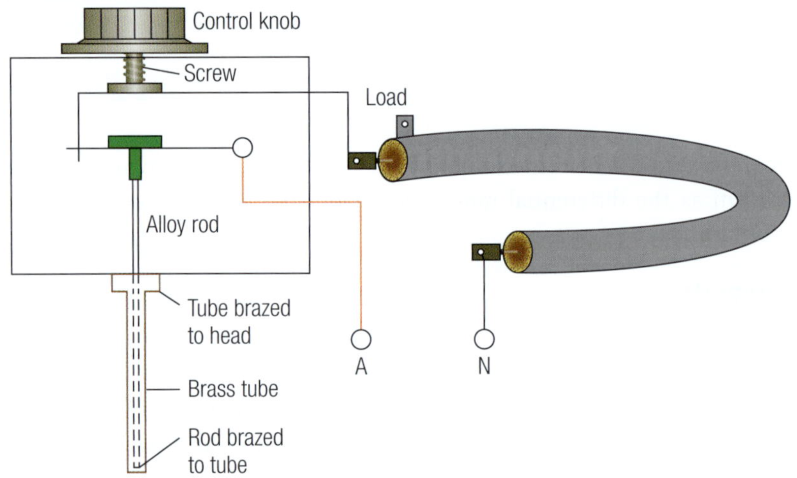

Filled system thermostats

Filled thermal system thermostats have a sensing bulb at the end of a small capillary tube, as shown in Figure 9.16. A liquid such as mercury is sealed inside the bulb and expands and contracts with temperature change. The volume change in the mercury transmits a force up the capillary tube to an actuator (bellows or diaphragm). The actuator operates the switch contacts. This type of thermostat is used where the control point is some distance from the measurement point such as in a cooking oven, air duct or liquid tank.

Testing a thermostat

Since most errors that occur in a heating process are due to the sensor itself, the operation must be checked against a known reference. This means placing a thermometer, mercury-in-glass or digital near where the probe for the thermostat is located so that both temperature devices are physically exposed to the desired temperature. The range of the thermostat can usually be adjusted by locating the 'range adjustment' screw through the front of the thermostat with the adjustment knob removed. If the temperature cannot be adjusted to follow closer than 1.5°C, the thermostat should be replaced.

Electronic heat control

The control devices mentioned above use the on/off position of a switch contact to maintain the controlled temperature. Even the most sensitive device will still take time to respond to a temperature change due to the differential inherent in the device.

Continuous control of temperature can be achieved using a solid state device known as a TRIAC. The TRIAC is an electronic switch that can control the portion of an a.c. cycle that is applied to the load, thus controlling the average current to within half an a.c. cycle (at 50 Hz this is 1/100th of a second).

Using an electronic sensor such as the resistive thermal device (RTD) shown in Figure 9.19, the controller can cause the TRIAC to be switched on later in the a.c. half cycle to reduce the current to the load if the temperature is higher than the set point, and switch the current on earlier if the temperature falls below the set point. As the a.c. cycle passes through zero it switches off and has to be switched on again during the next half cycle.

You should not try to make any adjustments with the electrical supply connected as it will usually involve exposure to live terminals. Do not work live unless there is no other possible option.

FIGURE 9.19 Electronic control of a heater

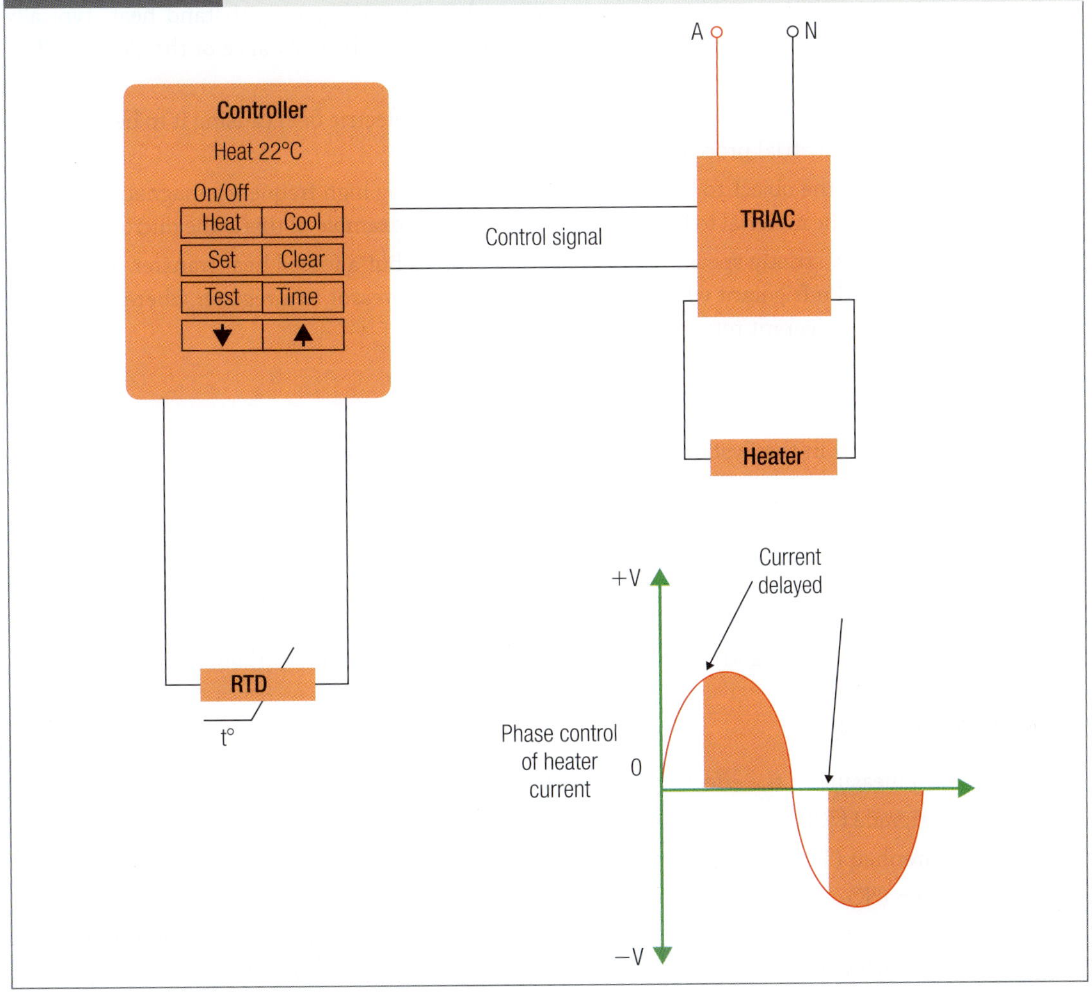

Another system using the thyristor control triggers the current on when the a.c. voltage passes through zero, letting a measured number of cycles through before switching off again as the voltage passes through zero. The average current is controlled by the length of each burst of pulses. This is known as zero switching control. Zero switching reduces disturbances in the supply voltage caused by switching large heating loads with the current still flowing.

SUMMING UP...

- Temperature is controlled by manual, continuous and closed loop control.
- A simmerstat is a continuous switch control.
- A thermostat is a feedback control device.
- Types of thermostat are bimetal, tube and strut, and filled thermal system.
- The set point is the required temperature setting.
- The operation of a thermostat can be checked against a thermometer reading.
- TRIACs are used to control the average current to the load.

9.5 Fixed electrical heating appliances

You will be expected to be able to install, service and repair a variety of electric heating systems in domestic and commercial premises.

Electrical energy is converted to heat using one of the following:

- Resistance heating – the heating element is selected for its ability to withstand heat, typically nichrome, and produces heat by the passage of current through the resistance of the element. This is used for most space heating and cooking appliances.
- Capacitive heating – the material to be heated is placed in an electric field causing it to heat up. This is usually used in industrial processes.
- Inductive heating – the object to be heated is placed in a strong high frequency magnetic field. The induced a.c. causes the material to heat up. The most common example of this is the microwave oven.
- Heat pump – this is, strictly speaking, not a heating process but a forced heat transfer. The heat is collected by a liquid refrigerant which is circulated via a compressor to a register where the heat is removed and the refrigerant returns to collect more heat.

Heat energy

The unit of heat energy in the Système International (SI) system of units is the joule. When no heat energy is present a substance will be solid and brittle. As the substance absorbs heat energy the molecules vibrate more rapidly, as shown in Figure 9.20, and the substance develops more elastic properties, becoming a liquid and finally a gas. On Earth certain materials occur naturally as solids, liquids and gases. To maintain suitable living spaces for human beings and animals, heat must be added or removed. To preserve food and prevent decay, heat must be removed to reduce bacterial action. To allow substances to be worked and manipulated, heat must be added to make them more malleable.

Temperature

Temperature is a measure of the effect of heat energy present in a substance and is measured in:

- degrees centigrade (°C)
- degrees Fahrenheit (°F)
- degrees Kelvin (°K).

FIGURE 9.20 Molecules move rapidly when heat is increased

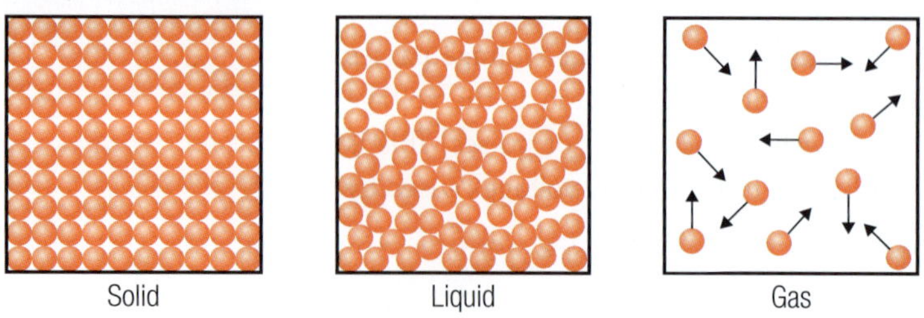

The centigrade scale is based on the freezing and boiling point of water. Zero (0) degrees centigrade is the temperature required to turn water from a liquid to a solid (ice). One hundred (100) degrees centigrade is the temperature required to turn water from a liquid to a gas (steam).

The Fahrenheit scale is based on different criteria so that 212 degrees Fahrenheit is equal to 100 degrees centigrade and 32 degrees Fahrenheit is equal to 0 degrees centigrade.

To convert between the two scales:

°F = (°C × 9/5) + 32 and °C = (°F − 32) × 5/9

The Kelvin is the fundamental unit of temperature in the SI system of units. Zero degrees Kelvin (K) is the point at which no heat activity occurs (absolute zero). The freezing point of water (0°C) is 273.15°K while the boiling point of water (100°C) is 373.15°K.

The Kelvin and Centigrade scales align with each other and conversion between them is a simple addition or subtraction of the conversion factor:

°K = °C + 273.15 and °C = °K − 273.15

EXAMPLE 9.1

Convert 30°C to Kelvin.

°K = °C + 273.15
= 30° + 273.15
= 303.15°K

Specific heat capacity

When a substance is heated the temperature of the substance increases. The quantity of increase depends upon the mass of the body. The specific heat of a substance is the amount of heat required per kilogram to raise the temperature through one degree K.

Specific heat (c) is measured in joules/kg°C

You can determine the heat gained or lost by a substance using the formula:

$Q = mc\Delta T$

Where:

Q = quantity of heat in joules
m = mass in kilograms
c = specific heat capacity in (J/kg°C) (see Table 9.1)
ΔT = change in temperature in Kelvin

Note: since 1 centigrade degree equals 1 Kelvin, changes can be measured in K OR °C.

Table 9.1 shows specific heat capacity for common materials.

EXAMPLE 9.2

Determine the amount of heat dissipated when an aluminium motor frame with a mass of 22 kg warms from an ambient temperature of 20°C to a working temperature of 100°C.

$Q = mc\Delta T$
$= 22 \times 897 \times (100 - 20)$
= 1.578 MJ

TABLE 9.1 Specific heat capacity for common materials

Substance	Specific heat capacity (J/kg°C)
Air, dry (sea level)	1005
Aluminium	897
Argon	520
Brass	375
Ice (0°C)	2093
Glass, crown	670
Glass, pyrex	753
Gold	129

(Continued)

TABLE 9.1 *(Continued)*

Substance	Specific heat capacity (J/kg°C)
Ice, water (−5°C)	2090
Quartz glass	700
Silver	235
Tungsten	132
Water, pure liquid (20°C)	4182
Water, vapour (27°C)	3985
Wood	1700
Zinc	388

Note: (J/kg°C = J/kg°K)

Source: Adapted from The Engineering ToolBox, 'Specific heat of some common substances', http://www.engineeringtoolbox.com/specific-heat-capacity-d_391.html. The content in The Engineering ToolBox is copyrighted but can be used with no warranty or liability. Important information should always be double checked with alternative sources. All applicable national and local regulations and practices concerning these aspects must be strictly followed and adhered to.

Thermal conductivity

Thermal conductivity is the flow of internal heat energy from a region of higher temperature to a region of lower temperature. The four factors that determine the rate of energy flow are:

- temperature difference (also known as the temperature gradient)
- length of the conductive path
- cross-sectional area
- type of material.

The rate of heat transfer is measured in watts (W; joules per second).

You can determine the rate of heat flow by applying the following formula:

Rate = kA T1 − T2 watts
l

Where:

k is the thermal conductivity of the substance in j/s.m.°C
l is the length of the heat conducting path
A is the cross-sectional area of the heat conducting material.

Conductivity (k) varies with the type of material and is greatest for metals, lower for non-metallic solids, lower again for liquids, and extremely low for gases (see Table 9.2). Most metals are good conductors of heat – silver is the highest at 420 j/s.m.°C followed by gold, aluminium and tungsten. Materials that have a relatively low conductivity, such as air at 0.023 j/s.m.°C, are good thermal insulators.

EXAMPLE 9.3

Heat energy transfer

Determine the heat loss from a room through a glass window 4.0 m × 2.00 m × 5 mm thick. The inner room temperature is 24°C and the outside temperature is 12°C, given that the k for glass is 0.84 j/s.m.°C.

$$\frac{\Delta Q}{\Delta t} = k \times A \times \frac{T1 - T2}{l}$$

$$= 0.84 \times 8 \times \frac{(24 - 12)}{0.005}$$

$$= 1612.8 \text{ j/s}$$
$$= 1612.8 \text{ watts}$$

If this situation was continued for 10 hours, the amount of electrical energy wasted in commercial units would be 16 kilowatt hours (Kwh). Check your latest electricity account to see what this would cost.

This example shows that a double-glazed window with a layer of still air in between would significantly reduce the heat energy lost from a building. Insulating materials are rated with an R (thermal resistance) value. The R value is defined by the thickness **l** and the thermal conductivity **k**.

$$R = \frac{l}{k}$$

You should always check that your customers understand the economic benefits of having thermal insulation and double glazing in their buildings.

TABLE 9.2 Thermal conductivity of common materials

THERMAL CONDUCTIVITY – k – (J/s.m.°C) AT 25°C	
Material/Substance	**k**
Vacuum	0
Air, atmosphere (gas)	0.024
Polystyrene, expanded Styrofoam™	0.03
Styrofoam™	0.033
Rock wool insulation	0.045
Paper	0.05
Cork	0.07
Straw slab insulation, compressed	0.09
Timber	0.14
Plaster	0.2
Water	0.58
Glass	0.84
Concrete, dense	1.0–1.8
Brick dense	1.31
Porcelain	1.5
Brass	109
Zinc	116
Aluminium	205
Copper	401

Source: Adapted from The Engineering ToolBox, 'Thermal conductivity of some common materials and gases', http://www.engineeringtoolbox.com/thermal-conductivity-d_429.html. The content in The Engineering ToolBox is copyrighted but can be used with no warranty or liability. Important information should always be double checked with alternative sources. All applicable national and local regulations and practices concerning these aspects must be strictly followed and adhered to.

Thermal stability

The ability of a material to resist changes in physical characteristics as its temperature changes is known as thermal stability.

Materials used in electrical equipment, particularly insulation, are classified with their maximum allowable temperature. This temperature limit is not the point at which the insulation fails but is the temperature above which the useful working life of the material will be significantly shortened. The thermal life expectancy of an insulation material, when operated at the maximum, continuous, total temperature rating, is generally designed to be approximately 20 000 to 25 000 hours under normal environmental conditions. If the operating temperature is increased by 10°C the working life of the insulation can be reduced by one half.

You should pay particular attention to the section of permanent wiring that runs into the terminal box on any heating equipment. It may require sleeving to be placed over it to increase its temperature rating to match the temperature at the location of the terminals. Any replacement part must be the same temperature class (or better) as the part that you replaced.

TABLE 9.3 Electrical insulation temperature classes

Class	Temperature
Y	90°C
A	105°C
E	120°C
B	130°C
F	155°C
H	180°C & over

Source: Three Bond Company Ltd, *Three Bond Technical News*, 1 December 1985, Table 1, http://www.threebond.co.uk/portals/0/tech13.pdf

Methods of heat transfer

Heat transfer takes place in the following ways:

- Convection – usually in fluids such as air or water, the heated fluid in contact with the heating element rises, making way for the cooler fluid to move in and take its place, as shown in Figure 9.21. When the heated fluid loses some of its heat it falls and is drawn in to the heating element, creating a circulation of the fluid.

FIGURE 9.21 Room heating using convection currents in the air

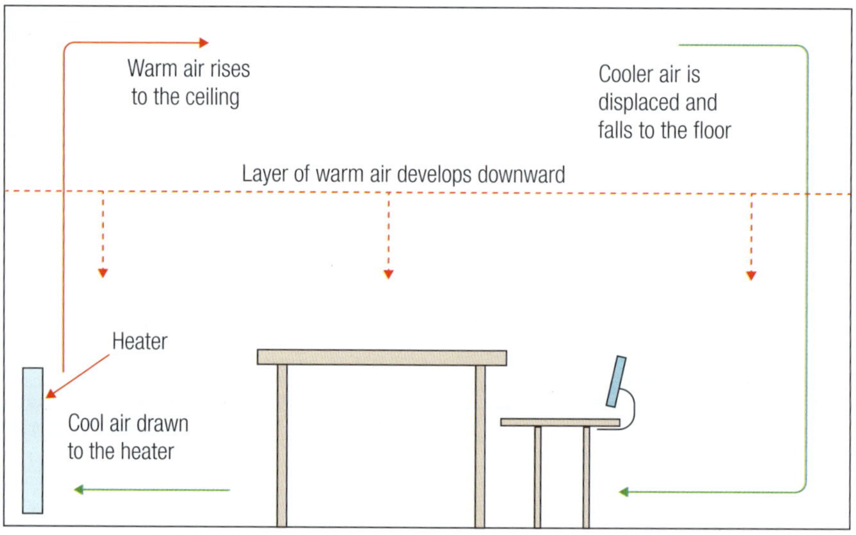

FIGURE 9.22 Thermal insulation improves heating efficiency

Source: Ed Hawkins

- Conduction – when a solid object is heated the heat is conducted through the material at a rate depending upon the type of material. Some solids such as metals will transfer heat readily and are known as thermal conductors. Solid materials that do not conduct heat well are known as thermal insulators. Figure 9.22 shows thermal insulation used to prevent heat escaping through the ceiling of a house.
- Radiation – heat is radiated in the infrared range of the electromagnetic spectrum and does not heat the air as it passes through. This means heat is not wasted by being carried away on convection currents. Since human skin is sensitive to radiated heat, radiation or infrared heating is used to provide comfort in large areas that are difficult to heat such as inside public halls.

Two phase stove

Manufacturers of cooking appliances provide links and terminals to allow their appliances to be connected to single phase, two phase or three phase supplies, as shown in Figure 9.23. The multiphase configuration requires only one neutral conductor. As the nature of the load involves switching resistance elements on or off, it is considered an unbalanced load. If a common neutral conductor is used:

- the active conductors must be controlled by linked circuit breakers or linked switches (AS/NZS 3000:2007 clause 2.2.1.2)
- the current carrying capacity of the common neutral conductor must be no smaller than that of the largest associated active conductor (AS/NZS 3000:2007 clause 3.5.2 (iv)).

FIGURE 9.23 Two phase hotplate connection

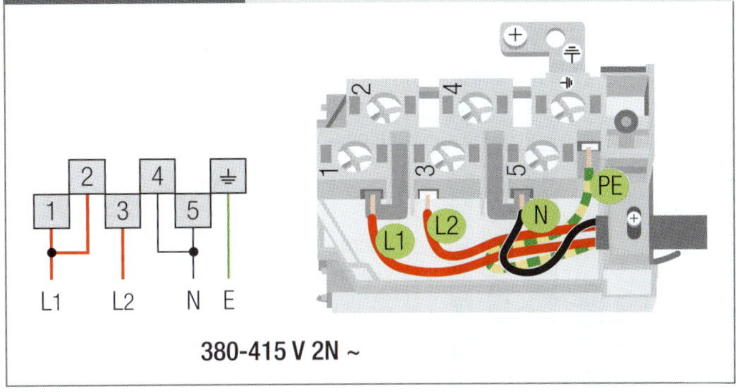

Reverse cycle air-conditioning

Reverse cycle air-conditioning is used to heat buildings during cold weather by making use of plant and equipment that would otherwise be idle. Normally the air-conditioning unit will remove heat from the air within a closed area such as a room or other closed space and transfer the heat to a compressed liquid refrigerant. The refrigerant is circulated through a pipe to the outside of the building where the heat is removed by a fan and the refrigerant returned to the compressor to repeat the cycle.

In cold weather the cycle is reversed by turning the unit to heating mode. A valve reverses the flow of the liquid refrigerant so that heat from outside the building is collected by the outside register (see Figure 9.24) and transferred to the refrigerant, which becomes a low pressure vapour. The refrigerant returns to the compressor where it becomes a high pressure vapour and gives off its heat, which is blown into the room by the fan (see Figure 9.25). With the heat removed the refrigerant becomes a liquid and is returned to the outside register to collect more heat.

FIGURE 9.24 Reverse cycle air-conditioner – outside register and heat pump

Source: Ed Hawkins

FIGURE 9.25 Reverse cycle air-conditioner – inside unit

Source: Ed Hawkins

SUMMING UP...

- Heat is a form of energy that affects the speed of vibration of the molecules in a substance.
- The unit of heat energy is the joule.
- Temperature is a measure of the effect of heat.
- The Kelvin scale is the basic temperature scale for all others.
- Specific heat capacity varies with each material.
- Thermal conductivity is high for metals, low for gases.
- Thermal insulating materials for buildings have an R-value allocated to them.
- Thermal stability is affected if the rated temperature is exceeded.
- Heat is transferred by:
  - conduction
  - convection
  - radiation.
- A common neutral conductor can be used for a two phase stove.
- The reverse cycle air-conditioner transfers heat from the external environment to the inside of a building.

FIGURE 9.26 Tankless water heater

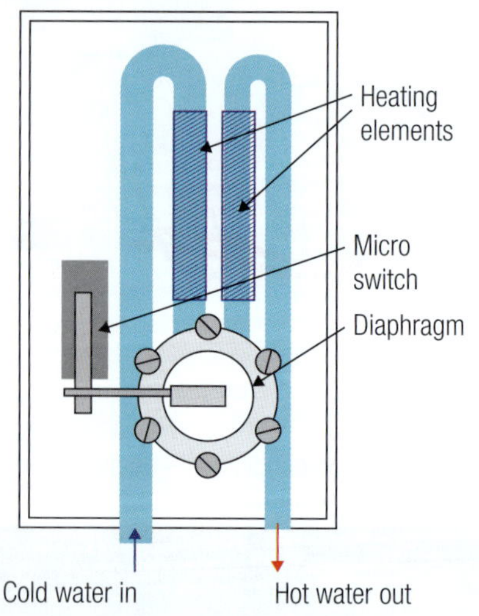

9.6 Electrical water heater operation

You will encounter water heaters in the majority of commercial, industrial and domestic installations. An electrical practitioner is expected to be able to install, service and repair a variety of water heater types.

Types of water heaters

Water heaters in domestic or industrial installations may be grouped as follows:

- Instantaneous heaters – are only switched on when hot water is required and are normally arranged for supply at the standard tariff rates. Included among these are jugs, kettles and urns. The elements are enclosed within a small copper container of about 1 litre capacity. When water is flowing through the unit it is switched on and the water is heated very quickly. The switch can be controlled automatically or manually. The automatic type has a diaphragm operated flow switch (see Figure 9.26) that comes on when water flow increases to a preset limit. If the water is turned off with the switch on, the water will quickly boil, creating a dangerous situation. Instantaneous water heaters are available for three phase, 400 volts, 12 to 18 amps rated between 9000 and 13 000 watts, and single phase 230 volts at 3000 watts.
- Continuous water heaters – are in the circuit all the time and have a storage capacity. They can be supplied on the standard tariff rates or on controlled tariffs such as off peak. The power rating of continuous water heaters is lower than that for instantaneous water heaters because of their storage capacity. Water heaters used on controlled tariffs are usually larger as they have to deliver hot water during periods when the supply is turned off during the day. The power rating of a storage heater is typically around 2000 watts and the capacity can vary from 60 to 300 litres.

Water heaters

In both types of water heater the water temperature is limited to 70°C by a thermostat built in to the storage tank. Early elements were a coiled nichrome wire mounted on a ceramic cylinder and inserted in a copper tube. These have mostly been replaced by an immersion type that comprises a copper tube with a coiled nichrome wire insulated and supported by a mineral insulating powder. Elements are fitted low in the tank and the top section is heated by convection. A booster element is sometimes fitted near the top to allow for periods of heavy use.

Never bridge out the contacts of the thermostat or cut-out for any reason.

Temperature control on water heaters is carried out by the use of bimetal disc thermostats (see Figure 9.17).

Intrinsic safety

AS/NZS 3000:2007 clause 4.8 specifies a number of safety features required for the installation of water heaters. Water heaters of the mains pressure type are not vented to the atmosphere so dangerous pressures can build up while the water is being heated. An automatic pressure relief valve is fitted with a manual override for testing purposes (clause 4.8.2.1).

The pressure relief valve should be manually exercised periodically.

Other safety features include:

- an over temperature cut-out, connected directly in the supply circuit, to disconnect the heater from the supply should the thermostat fail. The cut-out operates when the water temperature reaches 95°C and is not the auto resetting type (clause 4.8.1)
- a temperature operated pressure relief valve that opens before the water temperature reaches 99°C and allows the water to escape
- an isolating switch mounted adjacent to the water heater or on the switchboard (clause 4.8.2.3).

Wear appropriate PPE when dealing with hot water and hot metal containers. The thermostat body may be hot when you have completed your test, so let it cool down before touching it.

Testing of over temperature cut-out point of a thermostat

To test the over temperature cut-out:

1 Clamp the cut-out to the side of a small metal container.
2 Clip the leads of an ohm meter across the contacts and note they have continuity (zero ohms).
3 Fill the container with water and slowly heat it on a small hot plate while monitoring the temperature with a thermometer.
4 The cut-out should operate when the water temperature reaches around 95° centigrade.
5 The ohm meter should indicate an open circuit (infinity).

Switchboard requirements

The switchboard requirements for water heaters will depend upon whether they are to be supplied with:

- a controlled load tariff, off peak or similar
- a special heating tariff
- regular power and light tariff.

If a regular tariff is to apply then the circuit can be treated the same as any other load circuit connected to the switchboard and protected by a circuit breaker fed from the main switch. There is no requirement to connect through a RCD device.

If the water heater is to be supplied by a controlled or special tariff, a separate main switch for that tariff will need to be provided with its own meter position on the meter panel. This may also include a time switch or control centre provided by the utility supplier.

Twin element water heater

Twin element water heaters usually have a continuous standard tariff connected to the top element and a controlled tariff connected to the lower element. The top or booster element is set 10° below the bottom element so that it only operates when the bottom element is switched off or when the system is experiencing a period of high demand.

Although the circuit diagram for dual element hot water supplies in Figure 9.27 appears to contravene clause 2.3.2.1.1 regarding switching of neutrals, the rule applies to the installation wiring that terminates at the appliance only. The internal connections of the appliance are tested and approved and must not be altered.

Since the circuit has two actives and a common neutral, an isolating switch must be provided that opens both

FIGURE 9.27 Manufacturer's information for dual element connection

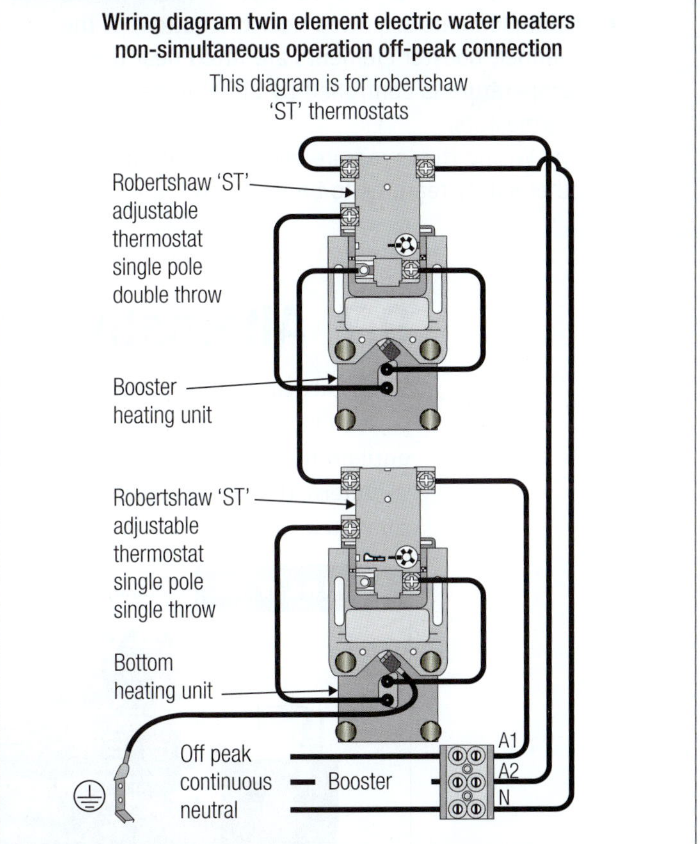

Electrical circuit for twin element models – Robertshaw 'ST' thermostats

Source: Based on Rheem Australia Pty Ltd, *Owners Guide and Installation Instructions*, p. 36, http://www.rheem.com.au

Heating water represents the biggest single source of greenhouse gas emissions in an average Australian household (not including cars).

active conductors at the same time and the circuit breakers for the active conductors must be linked mechanically (clause 2.2.1.2).

Solar heating

Solar heaters are not, strictly speaking, electric water heaters but rely on insolation from the sun during the day. At night or on cloudy days, however, they need an electric booster in the storage tank to maintain the water temperature.

SUMMING UP...

- Instantaneous water heaters heat the water on demand.
- Storage water heaters run continuously.
- Instantaneous water heaters run on standard tariffs.
- Continuous water heaters run on discount tariffs.
- Water temperature is limited to 70°C to prevent scalding from taps and outlets.
- Main elements are fitted near the bottom of the cylinder, booster elements are fitted near the top.
- Temperature is controlled by bimetal disc thermostats.
- The Wiring Rules require electrical cut-outs and other safety features to be installed.
- The cut-out temperature can be tested simply if it is suspect.
- Isolating switches are required either beside the heater or at the switchboard.
- If special tariffs are to be used then a separate main switch is required.
- The twin element water heater can have a different tariff connected to each element.
- Solar water heaters use the sun's energy to heat water but require an electric booster when the sun does not shine.

9.7 Alternative supplies

Electrical distributors in Australia and New Zealand maintain a very reliable and continuous supply of power. Occasionally breakdowns do occur and there is a need for an alternative supply to be provided until normal distribution is restored.

Alternative supplies such as the diesel generator shown in Figure 9.28 can be provided by electrical converters or engine-driven generator sets.

FIGURE 9.28 Stand-by generator set

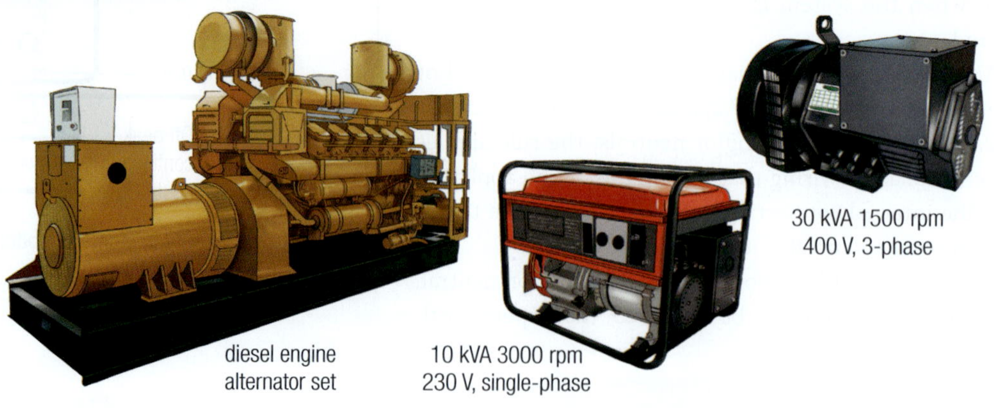

Source: Peter Phillips, *Electrical Principles*, 2nd edn, Cengage Learning, Melbourne, p. 570

Reasons for the installation of alternative supplies

In the past the main reasons to have an alternative supply of electricity were to ensure safety systems continued to operate and vital production was not interrupted, such as in operating theatres in hospitals. With the spread of electronic data storage a wider need for uninterruptible power supplies (UPS) has developed.

Alternative supplies include the flywheel UPS, which maintains a mechanical flywheel in the form of a spinning disc. The spinning disc provides drive for an alternator and the energy to start a standby generator if the main supply fails.

Other typical alternative supplies include:

- Standby UPS – typically consists of an AC/DC and DC/AC inverter, a battery, a static switch, a low pass filter to reduce switching frequency from the output voltage, and a surge suppressor [as shown in Figure 9.29]. The standby system operates with the switch setting the AC input as a primary power source, and alternating to the battery and inverter as backup sources in case the primary power fails. The inverter normally remains on standby, only activating when the power fails and the transfer switch automatically switches the electrical load to the inverter.
- Standby-Ferro UPS – relies on a specialised saturating transformer with multiple power connections. The current flows from the AC input, moving through the transformer and on to the output. If there is a power failure, the transfer switch activates the inverter to pick up the output load.
- Line Interactive UPS – the battery and AC power inverter is permanently connected to the UPS output, the battery can be charged by operating the inverter in reverse while AC supply is at normal levels. In case of power failure, the transfer switch diverts the current from the battery to the system output. Because the inverter is continuously connected to the output, the UPS provides additional filtering and lowers the risk of switching transients.
- Double Conversion – UPS systems are commonly used for higher voltage applications and they feature a configuration similar to those of standby units, but with a primary power path focused on the inverter rather than the AC main. This type of UPS system takes virtually no time to transfer between modes because input AC power failure does not trigger the transfer switch. Instead, the input AC charges the backup battery, which in turn powers the output inverter.
- Delta Conversion UPS – under power failure this UPS acts similarly to a double conversion unit, but provides more efficient energy performance by converting power from input to output rather than cycling between power and battery sources.

Source: Thomas.net, 'Types of uninterruptible power supply (ups) systems', http://www.thomasnet.com/articles/electrical-power-generation/ups-system-types.

Always dispose of used fuel and lubricants in the proper disposal system so they do not contaminate the soil and waterways.

Uninterruptible power supplies

FIGURE 9.29 Online UPS block diagram

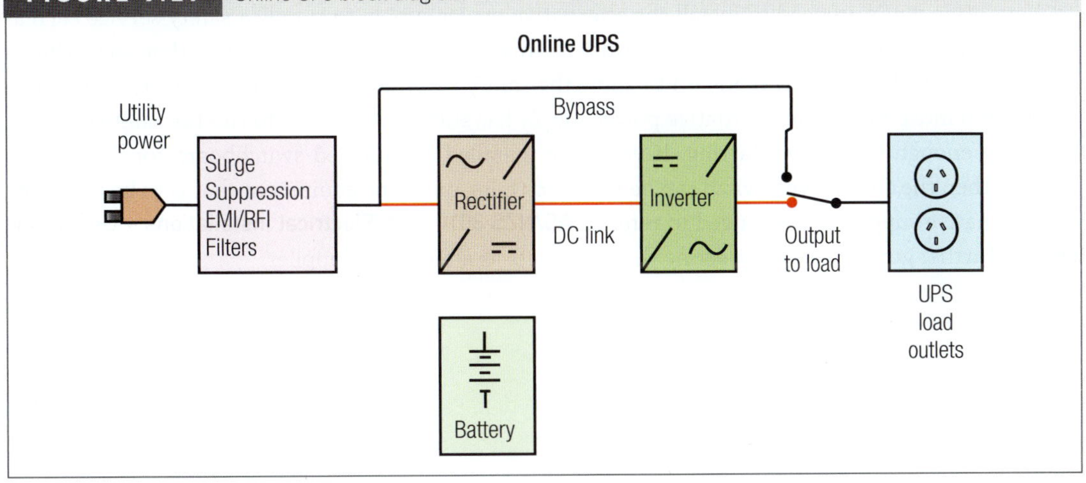

Always ensure the standby power supply is switched off and locked out before isolating the installation to carry out maintenance or additions. The power supply may automatically start when you power down the installation and liven it up again.

Characteristics and operation of UPSs

When recommending a UPS to your customer you should choose one based on the following:

- Regulated output voltage with low harmonic distortion unaffected by input voltage or load changes
- An input current with reduced harmonic distortion
- A low degree of electromagnetic interference and acoustic noise
- Minimal transition times between normal and backup operations
- High levels of reliability and efficiency
- Relatively low cost, weight, and size requirements.

Source: Thomas.net, 'Types of uninterruptible power supply (ups) systems', http://www.thomasnet.com/articles/electrical-power-generation/ups-system-types.

Australian Standards

Section 7.3 Electricity generating systems in **AS/NZS 3000:2007** specifies the requirements for electricity generating systems connected to electrical installations that are intended to supply part of the installation continuously or occasionally. Among those specifically mentioned are:

- engine driven generator sets
- stand alone power systems
- grid connected inverter systems
- battery systems (clause 7.3.2).

The alternative supply cables, protection and control equipment must be provided to the same standard as the rest of the installation using circuit breakers, RCDs, isolating switches and earthing systems where required (clauses 7.3.3 and 7.3.4). If the operating frequency is not 50Hz then the current carrying capacity tables in **AS/NZS 3008.1** may not apply (clause 7.3.7).

Connection of a standby system

Clause 7.3.8.1.1 of **AS/NZS 3000:2007** requires you to make the MEN connection at the switchboard and you are not to switch the normal installation and load neutral during the changeover to the standby supply. This may mean that you will have to disconnect any MEN system provided by the manufacturer of the generating system. You must connect all the earth connections to the main earth connection and not switch them anywhere. In **AS/NZS 3000:2007** Figure 7.2 you can see how the changeover switch disconnects the standby supply neutral but the other neutrals are not switched. You do not need to switch the neutral from the standby generator, but if you do it must comply with clause 2.3.2.1.1(b)(ii)(B).

Very often you may need to install the alternative power supply to supply a safety service such as fire fighting or emergency evacuation lighting. In this case, clause 7.2.2.3 requires that you install a changeover switch on the switchboard to allow the safety systems to be connected to the alternative supply. You must ensure the alternative power supply has sufficient capacity to run the safety systems, including enough for motor starting. If you have installed a dedicated switchboard for your safety systems then the changeover switch can be installed there, or on the equipment that is to be run using the alternative supply. You will need to refer to **AS/NZS 3010:2005 Electrical installations – Generating sets** for further guidance on connecting generator sets as alternative supplies.

SUMMING UP...

- Types of alternative supply:
  - engine driven generator sets
  - stand alone power systems
  - grid connected inverter systems
  - battery systems.
- A UPS is an uninterruptible power supply.
- Battery type UPS uses inverters to convert the battery d.c. to a.c.
- Standby-Ferro UPS relies on a saturating transformer.
- Permanently connected UPS provides better filtering and lowers the risk of voltage transients.
- Double conversion systems provide backup very quickly.
- The Wiring Rules apply to alternative supplies connected to an installation.
- UPSs should have a high level of reliability and efficiency.
- Neutrals in the installation wiring must not be switched during changeover.

9.8 Installation of batteries

As an electrical practitioner you will encounter many instances where misuse and poor maintenance of batteries has contributed to a system failure. For a full description of the principle of operation for the different types of cells discussed in this section refer to *Electrical Principles* by Peter Phillips.

Common types of batteries

Batteries are a widely used source of independent and portable electricity supply. Batteries are made up from combinations of cells that produce electricity by chemical reaction. There are two classes of battery:

- primary – made up of non-rechargeable primary cells
- secondary – made up from rechargeable secondary cells.

Common rechargeable batteries and their uses are shown in Table 9.4.

TABLE 9.4 Uses of common batteries

Battery type	Characteristics	Typical uses	Advantages
Sealed lead acid (SLA)	Can hold a charge for up to 3 years	Backup emergency power source	Inexpensive
Nickel-cadmium (Ni-Cad)	Fast, even energy discharge	Appliances, audio and video equipment, toys; most popular battery	Relatively inexpensive; widely available
Nickel-metal hydride (Ni-MH)	Typical power capacity 1.2 V – 1200 to 1500 mAh; extended life 2300 mAh; 2.5 to 4 hours battery life	Portable computers; cellular phones; same as for Ni-Cad batteries	No memory effect; unused capacity remains usable
Lithium ion (Li-Ion)	Stable and safe; highest energy capacity	Portable computers; cellular phones; same as for Ni-Cad batteries	Twice the charge capacity of Ni-Cad; slow self-discharge

Source: Adapted from Search Mobile Computing, 'Battery', http://searchmobilecomputing.techtarget.com/definition/battery

The primary cell is popular for portable lighting, radios and other applications where a source of electricity may not be available.

There are three main types in common use:

Uses of cells and batteries

- Carbon zinc dry cell – produces about 1.5 volts and has a shelf life of approximately one year and a low discharge current. It is available in many physical forms and sizes to suit user requirements. The main advantage of this cell is that it is cheap and convenient.
- Alkaline cell – a development of the zinc carbon cell with an output of 1.5 volts. It is more expensive but advantages include longer life, higher output current and more stable output voltage.
- Lithium cell – comes in a variety of types ranging from 2.9 volts to 3.7 volts, the 2.9 volt version being the most common. A widely used shape of lithium cell is the button, shown in Figure 9.30, designed to fit into small portable devices such as cameras and information systems to maintain memory settings during power down.

FIGURE 9.30 Common lithium cells

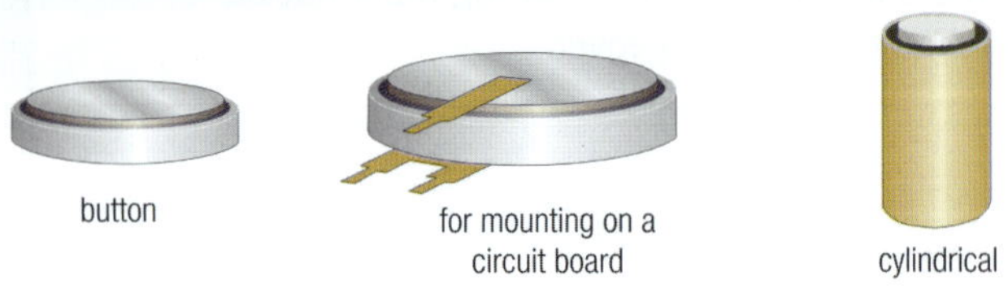

Source: Peter Phillips, *Electrical Principles*, 2nd edn, Cengage Learning, Melbourne, Figure 25.46, p. 624

Secondary cells

A primary cell is of no further use once it has been discharged and must be disposed of and replaced. Secondary cells can be recharged and reused throughout their useful life. The main types of secondary cells are:

> Do not try to recharge primary cells or dispose of them by burning as they will explode.

- Lead acid cell – has three main forms and is widely used in the transport industry to start engines and the electrical supply and telecommunication industries as backup power supplies. A standard lead acid cell is known as a 'vented wet cell' as it is ventilated to the atmosphere and has a liquid electrolyte. The electrolyte requires testing and topping up and it must be used and stored in an upright position. A variation on this is the valve regulated lead acid (VRLA) type, also known as maintenance free, where the electrolyte is absorbed within the frames of the cell plates and gases produced by the charging process are retained within the cell. A pressure relief valve allows for any pressure above 200 KPa caused by rapid charging to be vented off. The condition of this type can only be tested with a voltage tester. The gel cell, or gel battery shown in Figure 9.31, is another version of the sealed valve type. They are usually low capacity, around 10 ampere hours (Ah) or less, and are used widely for emergency lighting, backup power for alarm systems and portable equipment. The output voltage of a lead acid cell is 2.5 volts and the batteries are usually 6 volts (three cells in series) and 12 volts (six cells in series). Batteries made from lead acid cells have a high current capacity.
- Nickel cadmium (Ni-Cad) cell – has an output voltage of 1.2 volts but small versions are supplied in the same physical size and shape as the carbon zinc cell. The loss of 0.3 volts is made up for by the higher discharge current available and the fact that it can be recharged. The Ni-Cad battery will self-discharge when idle but can be stored for a long period when discharged. If not fully discharged before recharging the cell develops a 'memory' and will not deliver its full capacity.
- Nickel iron – has an output voltage of 1.3 volts and a low current capacity. This type of cell was used for backup power in telecommunication systems and applications where large currents were not required. The state of charge cannot be determined by testing the electrolyte. Due to its expense the nickel iron battery has been replaced by cheaper, more modern types.

FIGURE 9.31 Commonly used gel cell battery

Source: Peter Phillips, *Electrical Principles*, 2nd edn, Cengage Learning, Melbourne, Figure 25.49, p. 627

- Nickel metal hydride (Ni-MH) – has a terminal output of 1.2 volts and is used as a replacement for Ni-Cad cells as it does not tend to develop a memory. While this battery has a 40% higher capacity than a Ni-Cad it does have a high internal discharge rate. The automotive industry presently uses Ni-MH batteries for hybrid motor vehicles.
- Lithium ion (Li-ion) – has a terminal output of 3.6 volts and twice the capacity of a similar Ni-Cad with a much reduced internal discharge rate. Li-ion cells can rupture if overcharged or overheated

so a built-in electronic circuit disconnects the battery when the voltage falls to a preset value. The protection circuit continues to draw current, which will discharge the cell over time.

Storage and handling

The following should be observed when handling and storing batteries:

- Batteries should not be charged in a place where ventilation is poor. Lead acid batteries give off hydrogen and oxygen, forming an explosive mixture that can be easily ignited by a spark.
- If you need to dilute electrolytes you should always add the acid to water otherwise the acid will 'run' through the water and cause severe burns.
- Batteries can be heavy, so make sure you use correct lifting techniques when handling them.
- Lead acid batteries should be placed on a 'float charge' when stored as they will deteriorate rapidly if left in a discharged state.
- Nickel-cadmium cells should be stored in a discharged state and even in service should be regularly discharged completely.
- Disposal of used batteries should be done through a recycling service specialising in battery disposal; at no time should any battery be disposed of by burning as this will cause an explosion.

Charge/discharge cycle

To get the maximum life out of a battery you need to ensure that it always works well within its designed operating limits. You should make sure the correct charging system is used with the battery you are installing.

Several different charging methods are used depending on the type of battery and the user requirements:

- Constant voltage charger is basically a d.c. power supply which in its simplest form may consist of a step down transformer from the mains with a rectifier to provide the d.c. voltage to charge the battery. Such simple designs are often found in cheap car battery chargers.
- Constant current chargers vary the voltage they apply to the battery to maintain a constant current flow, switching off when the voltage reaches the level of a full charge. This design is usually used for nickel-cadmium and nickel-metal hydride cells or batteries.
- Taper current involves charging from a crude unregulated constant voltage source. It is not a controlled charge but the current diminishes as the cell voltage (back emf) builds up. Pulsed chargers feed the charge current to the battery in pulses. The charging rate (based on the average current) can be precisely controlled by varying the width of the pulses, typically about one second.

The life of the battery can be shortened by the rate of discharge and the depth of discharge (DOD), for example, continuously over-discharging Ni-MH cells can result in a 40% loss of cycle life; and over-discharge of lithium-ion batteries can result in 66% loss of capacity.

Sudden increases in discharge current must be avoided, especially rises that exceed the capacity of the battery. Protection and control systems should be installed to remove the battery from the circuit if its capacity is exceeded.

Internal resistance

The internal components of the cell have an effective electrical resistance. The internal resistance of the cell acts on a loaded circuit as if it was in series with the load. As the load current increases the voltage drop across the internal resistance increases, reducing the output voltage. When the voltage falls, less power is available for the load and more is lost within the cell. The discharge current from a cell is determined by the internal resistance.

When cells are connected in series the internal resistance of the resulting battery is determined by the sum of the resistance of the cells. When cells and batteries are connected in parallel the total internal resistance is reduced and the discharge current is increased. To calculate the effect of internal resistance, read Chapter 25 in *Electrical Principles* by Peter Phillips.

Make sure that you use the correct charger for each type of secondary cell or injury and damage will result.

Do not dispose of used secondary cells in the normal waste systems as they contain dangerous pollutants.

Always wear PPE such as a face shield and rubber gloves when handling electrolytes.

More batteries are damaged by bad charging techniques than all other causes combined.

FIGURE 9.32 Checking the state of charge using a hydrometer

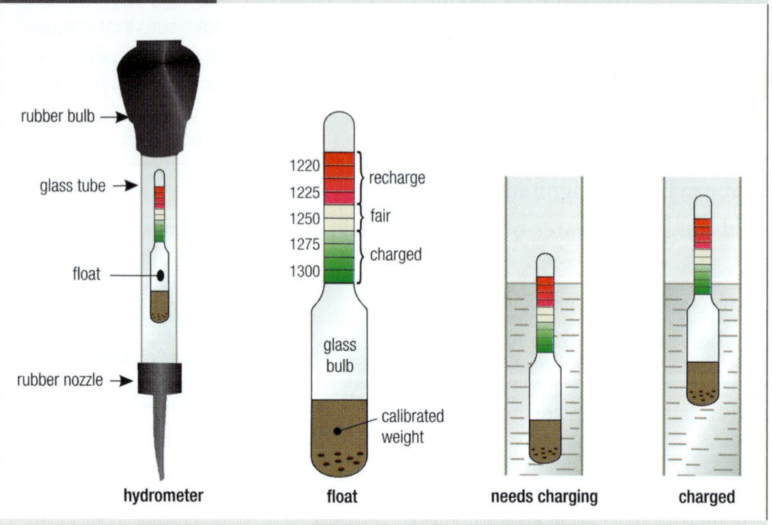

Source: Peter Phillips, *Electrical Principles*, 2nd edn, Cengage Learning, Melbourne, Figure 25.57, p. 634

State of charge

The state of charge (SOC) is a ratio expressed as a percentage of the available state of charge remaining in a cell or battery in relation to its rated capacity when fully charged, as shown in Figure 9.32. The SOC is not a valid indicator of the battery's runtime when taken alone. SOC readings will indicate 100%, even if a battery whose acceptance has dropped to 50% is fully charged.

Australian Standards

Some battery installations have terminal voltages that bring them into or close to the range of low voltage, and some installations have very high prospective fault current. To ensure the safety of battery installations in buildings you should consult the following AS documents:

- **AS 3011.1-1992 Electrical installations – Secondary batteries installed in buildings – Vented cells**
- **AS 3011.2-1992 Electrical installations – Secondary batteries installed in buildings – Sealed cells.**

These standards set out the requirements for battery rooms and short circuit protection with the purpose of preventing explosive gases from building up and arcs forming to ignite explosive atmospheres.

Commissioning procedures

Before and after you connect the battery to the charging system you will need to conduct the following checks and tests:

- Make sure the ventilation is effective before starting to commission the battery.
- Make sure you have adjusted the charge voltage to the manufacturer's instructions.
- If you are going to conduct a capacity test, make sure the battery is fully charged using the float charge method.
- Adjust the electrolyte level before connecting the battery for charging.
- While charging, the battery should be kept at a constant temperature of 25°C.

SUMMING UP...

- Batteries are used when other electrical supplies are unavailable or unsuitable.
- Primary cells are used once, then disposed of.
- Secondary cells can be recharged many hundreds of times.
- Correct charging and discharging is vital to the length of battery life.
- Battery installations in buildings must comply with AS 3011.
- Ventilation and voltage are important factors when commissioning a battery installation.
- PPE must be worn when handling electrolytes.
- Internal resistance of cells affects the load voltage.
- The state of charge cannot always be tested by measuring the electrolyte.

9.9 Fire protection – residential fire and smoke alarms

It is essential that you understand the types of fire detection systems available and their limitations.

Types of fire and smoke alarms

When a fire develops it produces a number of changes in the air surrounding it. By installing devices to detect the changes it is possible to cause an alarm to sound long before the fire is a danger to life and property. With the alarm to alert them, people can have time to evacuate the area. The alarm will also allow prompt arrival of fire-fighting teams to conduct fire containment and suppression.

The choice of fire alarm will be influenced by the:

- risk level where the fire is to be detected
- nature of the building or structure
- purpose and use of the building
- local legislation
- local fire and rescue recommendations.

Fire investigation statistics show that most domestic building fires originate in the living room, bedrooms, kitchen and other areas, in that order.

Since most fires that cause death occur between 7.30 p.m. and 7.30 a.m. and while people are asleep, it is important to have fire alarms installed in domestic premises.

Most fire detection devices are installed on or near the ceiling as this is the area where the first signs of fire become apparent:

- temperature rise
- smoke
- flames.

In some fire detection systems the detector also contains the alarm device and detects the changed conditions and sounds the alarm to alert people to the presence of fire.

In large buildings the detector is connected to a central monitoring system which can activate fire fighting systems using carbon dioxide, water or dry chemicals.

Regulations and standards

There are a number of authorities that oversee the protection and control of fire:

- Building Code of Australia (BCA) – this code is called in fire service acts throughout Australia and specifies the basic requirement for fire safety in buildings.
- State laws – each state has a fire service or fire safety act to provide fire-fighting forces and coordination of community resources in a fire emergency. The state acts usually are specific requirements for that state and address the types of fire exposure.
- Local government – controls types of building that can be located in an area including any fire access roads required.
- Local fire and rescue services – are responsible for the rescue of people, control and suppression of fires in buildings. They have extra requirements to suit local conditions and the need to consider the safety of their personnel. They have preferences for the location of fire alarms based upon their experience on the ground.

Installation and operation

Where you need to locate the fire alarm will depend upon the layout of the building and the type of detector that is being used.

Smoke detectors

Smoke detectors are designed to detect a developing fire while it is only smouldering material such as a cigarette left on a soft furnishing. Smoke alarms are best positioned on ceilings where smoke rises and gathers. Smoke alarms should be installed in hallways near bedrooms and on evacuation routes in multiple occupancy buildings (see Figure 9.33).

FIGURE 9.33 Location of household alarms

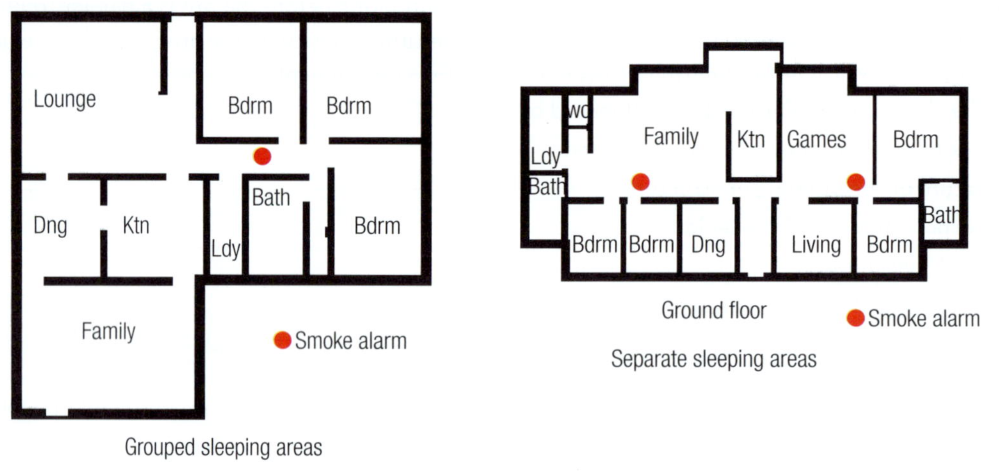

Source: Fire and Rescue NSW

> Dispose of old ionisation type detectors in the hazardous waste disposal system as they contain radioactive isotopes that will stay in the environment for a long time.

Typical smoke detectors are:

- Ionisation smoke detectors – use americium-241 that radiates active particles as it decays. A small current flows between the detector plate in the detector chamber caused by the ionisation of nitrogen in the atmosphere. When smoke from a fire or other source enters the detector chamber the ionised particles are neutralised, stopping the current flow. When the current flow falls below a set level the fire alarm is tripped (Australian Radiation Protection and Nuclear Safety Agency, 'Fact sheet 6 – radioactivity in domestic smoke alarms', http://www.arpansa.gov.au).
- Photoelectric smoke detectors – use the principle that smoke will scatter light beams. A small pulsed beam of light is transmitted across the detector chamber so that it cannot be detected by a small photocell close by. When smoke enters the detection chamber the light scatters and is detected by the photocell, causing a current to flow and the alarm is tripped (How Stuff Works, 'Photoelectric detectors', http://home.howstuffworks.com/home-improvement/household-safety/fire/smoke1.htm).

Heat detectors

Heat detectors respond to a rise in temperature (50°C to 60°C) caused by combustion. You will usually locate them on ceilings or high on walls, but you may need to place them in a storage area where fire might develop unnoticed. Typical heat detectors are:

- Fixed temperature heat detector – monitors the temperature and trips the alarm when it exceeds 50°C to 60°C. These detectors provide protection from slow-developing fires but have a temperature lag, meaning the temperature usually exceeds the setting by the time the alarm is sounded.
- Rate of rise detectors – are better to use where smoke or flames don't reach and detect a rapid rise in temperature. They operate when temperatures rise faster than 8°C per minute.
- Rate-compensation detectors – are used where a sudden rush of warm air might trip an alarm such as beside a heating chamber or the entrance to an air-conditioned building. The compensator allows them to ignore momentary rises in temperature.
- Combination detector – provides a higher degree of protection by combining the rate of rise detection and fixed temperature detector in one unit.

- Electronic thermal detector – uses the resistance change in a thermistor to detect a fall in current through the circuit caused by the increasing temperature of the thermistor. The electronic circuit can be connected to detect both fixed temperature and rate of rise. The reliability of electronic detectors is not high enough for them to be used where the lives of people are involved.

Wiring methods

AS/NZS 3000:2007 clause 7.2.7 specifies the type and location of wiring systems you must use for fire detection and control. The main requirement is that the wiring should be able to sustain the supply to the fire fighting and evacuation systems. You must use a wiring system that can withstand a fire due to its WS (wiring systems) classification, as explained in **AS/NZS 3000:2007** Appendix H in detail. If the wiring enclosure complies with the protection requirements then you need not use fire rated cables in your wiring system.

Cable types that meet the requirements are:

- Glass fibre covered cables – have a temperature rating of 15°C and are usually used for hot appliances such as cooking ranges and heaters.
- Radox® cable – can survive up to 280°C for short periods without compromising the insulation. It will not sustain a flame and is self-extinguishing (RADOX® Solar Cables, http://www.radoxsolar.com.au/products.php).
- Pyrolex™ Ceramifiable® cable – turns into a ceramic shield when exposed to fire, maintaining the electrical system to supply pumps and emergency equipment (CSIRO, 'Ceramifiable plastic cables', http://www.csiro.au/Organisation-Structure/Flagships/Future-Manufacturing-Flagship/Sustainable-High-Performance-Materials/High-performance-composites/Ceramifiable-plastic-cables.aspx).
- Mineral insulated metal sheathed (MIMS) – has a temperature rating of 250°C and can withstand a direct flame for a considerable time. This is used for mains in high-rise buildings and to supply essential services such as fire pumps.
- V105 PVC cable – has a temperature rating of 105°C but can only be considered as heat resistant as it will burn when subject to a naked flame. This grade of cable can be used for the internal wiring of hot appliances such as heaters.

SUMMING UP...

- Fire detectors detect the changes in the air caused by the combustion process.
- Detectors can sense either smoke or heat.
- Rate of rise detectors suit areas not reached by smoke or flame.
- Each jurisdiction has its own fire responsibilities under the law.
- On large buildings the fire detectors are connected to a central alarm system.
- Automatic fire-fighting systems involve carbon dioxide, dry chemical or water.
- The Building Code of Australia sets out fire control requirements in buildings.
- The correct wiring systems must be installed to maintain the fire-fighting systems.

9.10 Emergency and evacuation lighting and lighting control

You will need to be aware of the legal and occupational health and safety (OH&S) requirements that apply to the installation, control and servicing of emergency lighting as this is a vital safety system.

Factors and requirements of emergency and evacuation lighting

Emergency lighting is a term used in relation to the failure of the normal lighting system, while evacuation lighting refers to the lighting provided to illuminate and indicate the path to be followed when exiting a building or other closed space in the event of fire or other emergency. They are quite often associated with each other and, as with emergency lighting, evacuation lighting needs to continue to function even when the normal lighting system has failed.

The illumination levels required may vary considerably. For example, evacuation lighting usually only has to provide enough light to allow people to pass through an area without becoming trapped or tripping and falling. The spacing of evacuation lighting should be such that the minimum illuminance is 0.2 lux (see Section 9.11 below for an explanation of lighting terms) and should continue to provide that level of illuminance for 1.5 hours or any greater period required by local authorities. Emergency lighting on the other hand may be required to be bright enough for work activities to continue until normal lighting is restored, such as in hospital operating theatres and emergency service control centres.

The emergency lighting should be spaced so that the minimum illumination level is high enough for essential work activities to continue.

There are three main characteristics of emergency lighting:

- Maintained – fittings have a lamp with a battery connected and are illuminated all the time. When the power fails, the lamp is sustained by the battery until it is discharged.
- Non-maintained – do not normally illuminate until the power fails. The luminaire will switch on and run from the battery that has been kept in a charged state. The advantage of this type is that the globe will be less likely to fail due to age.
- Sustained – have a dual supply and two globes. One globe is illuminated during normal power conditions. When the power fails the second lamp illuminates, drawing its power from an emergency supply.

Emergency evacuation signage may be integrated in an existing emergency lighting system. The system of lighting must incorporate battery backup light fittings capable of illuminating the exit signage and providing clear direction on the safe means of getting out of the building (Safelincs, Fires Safety Products, 'Which emergency light: maintained or non-maintained?', http://www.safelincs.co.uk/blog/2012/06/30/which-emergency-light-maintained-or-non-maintained).

Point and central bank supply systems

Emergency lighting can be powered from either a central battery system or a point (local battery) system that comes on when the main power fails.

- Point system – each emergency light or sign has its own battery charged continuously from the main power (Figure 9.34). The advantages of the local battery system are:
  - lower installation cost
  - the installation is more reliable; a defect in one light does not affect the rest
  - the power is supplied over the local supply so in the case of a local failure the lights are switched on.
- The disadvantage is that more time is needed for maintenance.
- Central system – the power to the emergency lighting is supplied from a central battery or backup generator that switches on when the main supply fails. The advantage of a central system is that less time is needed to maintain and check batteries but you still need to check the lights. The use of LED technology should reduce this cost. The disadvantages of this system are:
- The installation can be vulnerable. If there is a battery fault or a short circuit the whole system or the section affected becomes unavailable.

FIGURE 9.34 Local battery emergency lighting

Source: Ed Hawkins

- The installation is expensive as the cable has to be run with fire resistance specifications.
- It is unaffected by a local power failure.

Lighting control

Emergency lighting can now be monitored and tested from a central location using a wi-fi linked monitoring system. Each light is fitted with a transmitter/receiver module with its own unique address. The main station can call up any light and test that it works, then time the period of battery discharge and record the date and time as well as the results of the test. A printout is available to the technician to go and carry out maintenance when needed.

SUMMING UP...

- Emergency lighting is required to maintain essential services and safe passage during an emergency or power outage.
- Illumination levels must be maintained to minimum levels.
- Three main emergency lighting modes are:
  - maintained
  - non-maintained
  - sustained.
- Point system uses a battery at each light.
- Point system responds to local power failure.
- Central battery distributes power to lights.
- Central battery does not respond to local power failure.
- Modern technology provides central monitoring and control of emergency lighting.

9.11 Lighting concepts and incandescent lighting

Selection of lighting

In addition to your knowledge of lighting control circuits you will need to be able to advise your customer on how comfort is affected by the choice and positioning of light fittings.

Basic concepts of lighting

Light comprises minute energy particles known as photons that travel through space in the form of electromagnetic waves similar to radio waves. Your eyes respond to the electromagnetic waves that occupy what is known as the visible light spectrum. The visible light spectrum exists between the higher ultraviolet radiation frequencies and the lower thermal radiation frequencies. Your perception of colour is based on the variation in frequencies from the highest, which appear blue, to the lowest, which appear red. Generally you see objects not because they emit their own light but because they reflect the light from an outside source. Your ability to see depends on the light emitted by a separate source such as the sun. The light falling on an object is all absorbed into the surface of the object except for the reflected frequency that is known as colour.

Terminology

As with any energy source utilised by human beings, light is measured and quantified by its useful volume and the amount of energy consumed to produce it.

Luminous flux (Φ_0)

Luminous flux is the quantity of light emitted by a light source or reflected from a surface regardless of the direction, as shown in Figure 9.35.

FIGURE 9.35 Total lumens produced

Luminous flux is measured in lumens (lm). All artificial lights are rated in lumens and their efficiency (luminous efficacy) is stated in lumens per watt (lm/W).

EXAMPLE 9.4

If you need to find the luminous efficacy of a 100 W incandescent lamp that produces 1750 lm:

$$\text{Efficacy} = \frac{\Phi_0}{P}\ \text{lm/W}$$

$$= \frac{1750}{100}\ \text{lm/W}$$

$$= 17.5\ \text{lm/W}$$

so it would have a luminous efficacy of 17.5 lm/W.

Luminous intensity (I)

The luminous intensity indicates the amount of light (lumens) emitted in a narrow angle in a specific direction. The unit of luminous intensity is the candela (cd) and is the measurement from which all other lighting units are derived. It is approximately equal to the old unit 'candlepower' and is generally taken to be equivalent. Typical values for luminous intensity are shown in Table 9.5.

TABLE 9.5 Luminous intensity of typical light sources

Light source	Candela
Light emitting diode (LED)	0.005
Candle	1
100 watt incandescent globe	150
Motor car headlamp	100 000
Lighthouse	300 000
Xenon flash tube	1 000 000

Source: Adapted from http://hyperphysics.phy-astr.gsu.edu/hbase/vision/photom.html

The relationship between the candela and lumen is: a point source of 1 cd in a given direction will deliver 1 lumen onto 1 square metre of surface at a distance of 1 m.

Light manufacturers provide candela distribution curves from measurements taken at angles to the vertically downward direction to help plan light distribution levels.

Illuminance (E)

This is a measurement of the distribution of light falling on a surface and is given in lux (lx) where 1 lux = 1 lumen per square metre ($lx = lm/m^2$).

The most important starting point when designing a lighting plan is the surface where a visual task is to be performed, known as the working plane. The Australian Standards specify the level of light in lux for each of a number of tasks to be performed.

Luminance (L)

Not to be confused with illuminance, luminance refers to the amount of light *reflected* in a given direction from an illuminated surface. Luminance is measured in candela per square metre (cd/m^2) and is an important consideration since it can cause glare.

Glare

Glare is the contrast between a bright object and a darker background and can result in visual discomfort. Glare does not only result from reflected light but can be caused directly from light sources placed at a low angle.

The inverse square law

As the distance from a light increases the illuminance decreases, making the light appear fainter. This is caused by the light rays spreading apart with distance and the same number of lumens must cover a larger surface area so the lumens per square metre, or lux level, is reduced, as shown in Figure 9.36.

FIGURE 9.36 How the inverse square law works

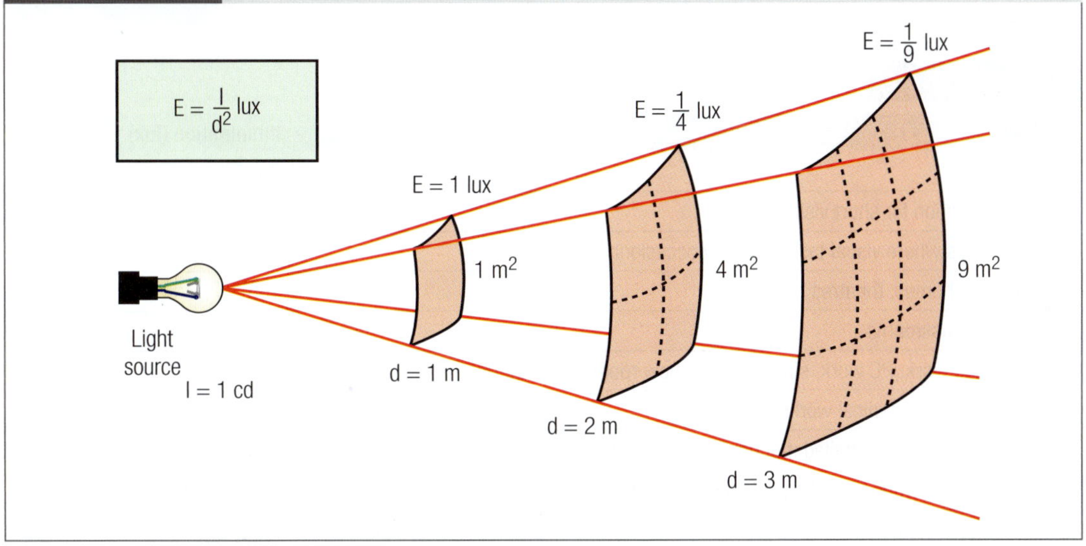

This is important since the further the working plane is away from the light source, the stronger the lamp needs to be.

However, the light does not always fall at right angles to the surface.

Cosine law

The amount of light falling on the working plane is not only determined by the height and output of the lamp but it is also affected by the incident angle. As the incident angle increases, the spread of light increases, therefore reducing the lumens per square metre.

The inverse square law is altered to take this into account by multiplying the luminous intensity by the cosine of the incident angle.

$$E = \frac{I}{D^2} \times \cos\theta$$

Where:

E = illuminance in lux (lx)

I = source intensity in candelas taken from the distribution curves (cd)

D = distance in metres (m)

θ = incident angle (degrees)

FIGURE 9.37 Thorn Cinqueline Distribution graph

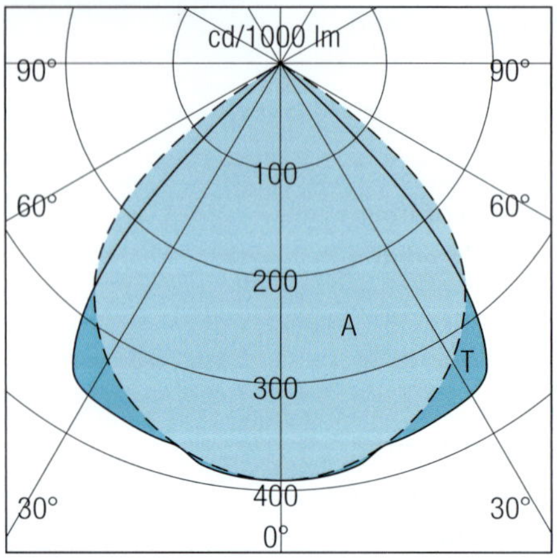

Cinqueline 600 x 600 with white infill panel
Lamp: 3 x 14W T16
DLOR: 0.70 LOR: 0.70
SHR max: 1.65 LMC: C

EXAMPLE 9.5

Your customer has chosen to install a light fitting with the luminous distribution curve shown in Figure 9.37. The light will be mounted flush in a 2.4 m high ceiling. Will it illuminate the work table 900 mm high to the level of 150 lux identified for the task to be carried out directly below it?

$D = 2.4 - 0.9 = 1.5\,\text{m}$

I (from graph at 0°) = 395 cd

$$E = \frac{I}{D^2}\ \text{lux (Inverse square law applies directly below)}$$

$$= \frac{395}{2.25}\ \text{lux}$$

$$= 175\ \text{lux}$$

The chosen light fitting will provide the required illumination.

The lux levels recommended for particular task areas are shown in Table 9.6.

TABLE 9.6 Illumination levels for typical tasks

Activity	Illumination (lux)
Public areas	20–50
Simple orientation for short visits	50–100
Working areas where visual tasks are only occasionally performed	100–150
Warehouses, homes, theatres	150
Offices and classrooms	250
Normal office work, PC work, study library, show rooms, laboratories	500
Supermarkets, mechanical workshops	750
Normal drawing work, detailed mechanical workshops, operating theatres	1000
Detailed drawing work, very detailed mechanical works	1500–2000
Performance of visual tasks of low contrast and very small size for prolonged periods of time	2000–5000
Performance of very prolonged and exacting visual tasks	5000–10 000

Source: The Engineering ToolBox, 'Efficiency requirements in buildings', http://www.engineeringtoolbox.com/light-level-rooms-d_708.html

Since the energy required to light buildings contributes 20% of greenhouse gas emissions the building industry has an important contribution to make towards reducing those emissions. The Building Code of Australia (BCA) contains energy efficiency measures for all building classifications. The maximum amount of power used by lighting is limited to:

BCA requirements for lighting efficacy

- 5 watts per square metre (5 W/m^2) of lighting indoors
- 4 watts per square metre (4 W/m^2) of lighting in outdoor areas (including verandahs)
- 3 watts per square metre (3 W/m^2) of lighting in garages.

This means that the more lumens you need for lighting the greater the luminous efficacy must be for the lamps you install to achieve it. Compact fluorescent lamps use 25% of the power required by older incandescent lamps.

AS/NZS3000:2007 clause 1.4.64 refers to light fittings as luminaires, meaning the complete assembly including lamp holders, reflectors and other components. It also must contain the means of connecting it to the supply. Some basic types of luminaire are shown in Figure 9.38.

FIGURE 9.38 Examples of luminaire types

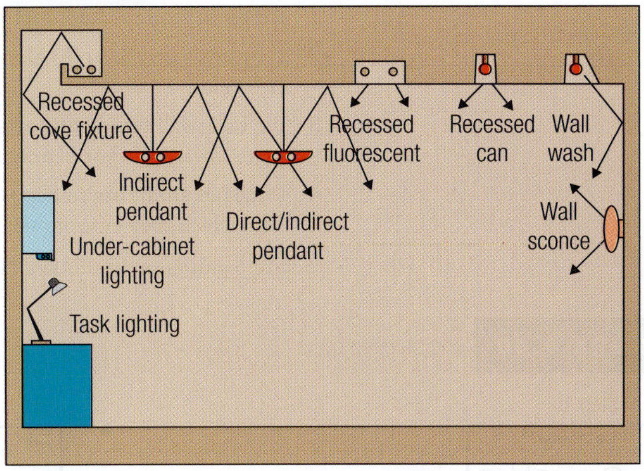

Source: Archtoolbox – Architect's Technical Reference, http://archtoolbox.com/images/materials/elect/lighting.gif

The incandescent lamp

The invention of a cheap, easily replaceable electric lamp during the latter part of the 19th century was the start of an industry that today is valued in the billions of dollars. Figure 9.39 shows an incandescent filament lamp. The filament of tungsten is connected across the supply voltage and a current flows causing the filament to heat up. The filament reaches a temperature where visible light is emitted, known as the range of incandescence. If the filament was exposed to oxygen at this temperature it would rapidly disintegrate. To avoid this, the glass bulb is evacuated and replaced with an inert gas such as argon. As well as preventing oxidisation the inert gas conducts the heat from the filament to the surrounding glass where it is dissipated, prolonging the life of the lamp. The filament will eventually erode away due to a process of evaporation. While cheap and convenient, 90% of the power required by this type of lamp is dissipated as heat.

Whenever an Edison screw lamp is installed, such as shown in Figure 9.39, the centre pin must always be connected to active and the outside thread must be connected to neutral.

Heat produced by incandescent lamps places a heavy load on air-conditioning systems so the industry has worked to produce more efficient versions of the incandescent lamp.

FIGURE 9.39 Incandescent lamp with Edison screw base

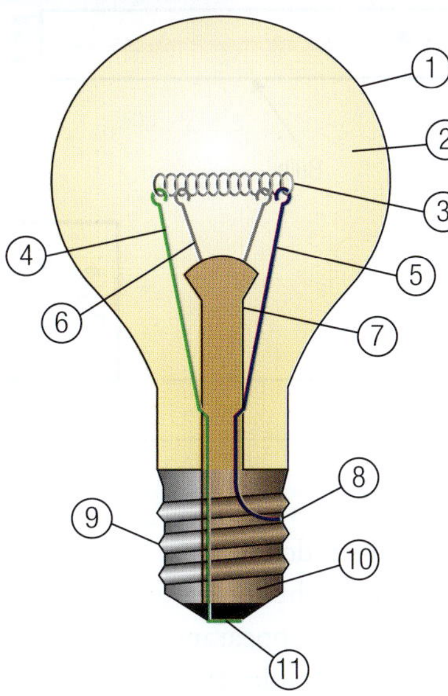

1. Glass bulb
2. Inert gas
3. Tungsten filament
4. Contact wire (goes to foot)
5. Contact wire (goes to base)
6. Support wires
7. Glass mount/support
8. Base contact wire
9. Screw threads
10. Insulation
11. Electrical foot contact

The Chicago Miniature Lighting Company gives this description:

> Tungsten-Halogen lamps are similar to conventional incandescent lamps in that they utilize a tungsten filament in a gas-filled, light transmitting envelope and emit the same type of light. The major differences are that in the tungsten-halogen lamp a trace of an element called a 'halogen' (such as iodine or bromine) is added as a vapour to the inert gas normally used in incandescent lamps, and the gas pressure and bulb temperature during operation are much higher than in non-halogen incandescent lamps. The higher gas pressure retards tungsten evaporation, allowing the filament to operate at a higher temperature. Therefore, the lamp operates at higher efficiencies than conventional incandescent lamps. To withstand those high operating pressures and temperatures, the lamp bulbs are made of fused quartz or high temperature 'hard' glass.
>
> Source: Chicago Miniature Lighting Company, 'Tungsten halogen lamps', p. 1, http://www.digikey.com/classic/documentredirector.aspx?doc=http://www.digikey.com/Web%20Export/Supplier%20Content/CML_28/PDF/CML_TechNotes_TungstenHal.pdf.

Incandescent lamps are often considered the least energy efficient type of electric lighting commonly found in residential buildings.

Figure 9.40 shows the cycle created by convection in the tungsten halogen lamp, as described by the Smithsonian Institution website:

1 Tungsten atoms evaporate from the hot filament and move toward the cooler wall of the bulb.
2 Tungsten, oxygen and halogen atoms combine at the bulb-wall to form tungsten oxyhalide molecules.
3 The bulb-wall temperature keeps the tungsten oxyhalide molecules in a vapor. The molecules move toward the hot filament where the higher temperature breaks them apart.
4 Tungsten atoms are re-deposited on the cooler regions of the filament – into the exact places from which they evaporated.

Source: Smithsonian Institution, 'Lighting a revolution', http://americanhistory.si.edu/lighting/tech/th.htm.

FIGURE 9.40 The halogen cycle

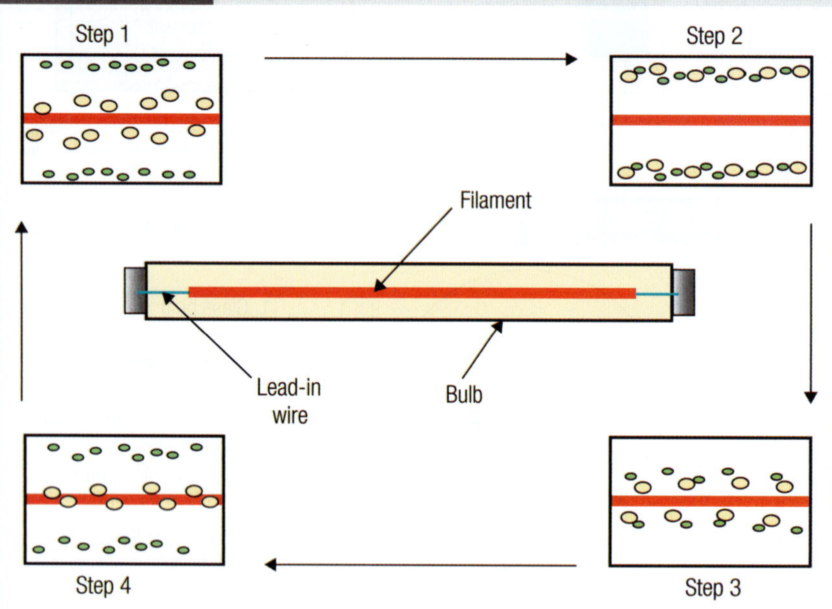

You must be careful not to touch the lamp envelope (bulb or tube) with your fingers as the oils from your skin will reduce the life of the lamp. Tungsten halogen lamps have a life expectancy of around 2500 to 3000 hours compared to that of a normal incandescent lamp of 1000 to1100 hours.

This type of lamp has a colour rendering that spans the middle range of ultraviolet down to the infrared ranges. Compared to the normal incandescent lamp which has most colour rendering at the yellow and red end of the colour spectrum.

The effect a light source has on the appearance of coloured objects is described by its colour rendering properties. These are grouped into grades based on the 'general colour rendering index' (CRI).

The colour rendering index indicates how closely the colour of an object matches its appearance under the relevant light source. To determine the CRI values of light sources, fifteen defined test colours commonly found in the environment are each illuminated under the reference light source (CRI = 100) and then under the source being evaluated. The greater the difference in the appearance of the test colours rendered, the poorer the colour rendering properties of the light source under examination.

> To prevent the possibility of fire when installing flush ceiling downlights, you must ensure that the metal 'chimney' extends above the thermal insulation and that no thermal insulation is blocking the escape of heat from the fitting.

Light and colour define the atmosphere of a room and influence mood and sense of well-being by their perceived 'warmth' or 'coldness'. Guaranteeing correct colour perception under artificial light forms is a very important part of your task as a lighting designer (adapted from OMS Lighting, 'Colour rendering index', http://www.omslighting.com/lqs/459/lqs-methodology/ergonomics/colour-rendering-index).

Although the theoretical luminous efficacy of the tungsten halogen lamp is 683 lm/W, in practice they achieve around 37 lm/W. This is still superior to 17 lm/W or less for a normal incandescent lamp.

Lighting layout

Visual comfort is determined by the nature of the space and the level of lighting. To achieve good lighting you should work to provide an environment where the visual task is conducted with accuracy and speed. In order to achieve this you will need to pay particular attention to:

- the size of the workspace
- the height of the working plane
- the height at which the luminaires are mounted

- reflective glare from ceiling and walls and sometimes the floor
- the output data for the fittings you will be using
- the combined lumen output of all the lamps.

You will need to obtain exact information about the workspace before deciding on the type and location of light fittings. Usually you will have a plan prepared by an architect or engineer and your work will involve placing everything according to the plan. If you are making alterations within a small space such as a house or shop then you will need to ensure the correct levels of illuminance are maintained with respect to **AS/NZS1680.0:2009 Interior lighting** and **AS/NZS1680.2.2:2008 Interior and workplace lighting**.

SUMMING UP...

- Light is an energy form that occupies the visible portion of the electromagnetic spectrum.
- The candela is the basic measure of light.
- Luminous efficacy is the measure of the efficiency of a lighting source.
- The inverse square law helps determine the reduction in illuminance over distance.
- The visual task is the activity that is to be illuminated.
- Glare is direct or reflected light at a low level that contributes to visual discomfort.
- The working plane is the area to be illuminated.
- A luminaire is another name for a light fitting.
- Distribution curves show the level of luminous intensity at various angles around a luminaire.
- The Building Code of Australia requires increased light efficiency to reduce greenhouse gas emissions.
- Incandescent filament lamps use heat to produce light output.
- Tungsten halogen lamps use the halogen convection cycle to produce gains in efficiency and lamp life.
- The colour rendition index (CRI) is used to determine the level of visual comfort provided by a lamp.
- AS/NZS 1680.0:2009 Interior lighting and AS/NZS 1680.2.2:2008 Interior and workplace lighting set out the requirements for lighting living and workspaces.

9.12 Fluorescent low intensity discharge lighting

Electrical practitioners are called on to install and maintain low intensity discharge lighting in commercial buildings where the light levels on work surfaces and in access areas are a legal OH&S requirement.

Types of low intensity discharge lamps

The basic fluorescent lighting circuit is shown in Figure 9.41. This circuit helps to illustrate the operation of a fluorescent light, although you will seldom be called upon to connect one of these in a new installation.

The tube is sealed and filled with an inert gas – typically – argon with a small amount of mercury. The inside of the glass tube is coated with phosphor powders.

FIGURE 9.41 Basic fluorescent lamp circuit

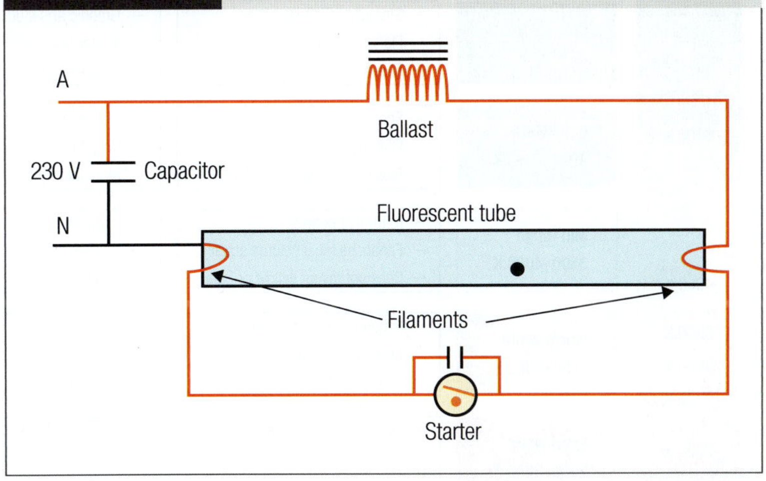

The starter is a small neon lamp with one electrode made from a bimetal strip. When the 230 volts a.c. supply is applied:

1. The neon lamp is illuminated by current flowing through the two filaments.
2. The heat from the neon lamp closes the thermal switch made from the bimetal strip.
3. Sufficient current flows through the filaments to heat the argon gas and vaporise the mercury.
4. Without the neon lamp keeping it heated the thermal switch opens.
5. The magnetic field in the ballast collapses causing a large inductive voltage to form across the ends of the tube.
6. The vaporised mercury ionises and the tube becomes conductive.
7. The current passing through the tube prevents the starter from operating again.
8. The electrons passing through the ionised tube produce an emission of ultraviolet light (UV).
9. The UV activates the phosphor powders on the inside of the tube producing visible light.
10. The ballast now acts to limit the current flow through the ionised tube.
11. The capacitor corrects the lagging power factor caused by the ballast.
12. The small capacitor in the starter reduces radio frequency interference caused by the opening switch contacts.

The fluorescent lamp in all its forms is widely used in industrial and commercial buildings and is gaining wide acceptance as a replacement for inefficient incandescent filament lamps (adapted from *Energy Efficiency Guide for Industry in Asia*, 'Features of fluorescent lamps', UNEP, p. 6, http://www.retscreen.net/fichier.php/903/Chapter-Lighting.pdf).

Operation of fluorescent lights

The expected service life of a fluorescent tube can be as much as 8000 hours. This can be affected by the number of times it is switched on since the starting process places the most wear on the lamp. If the lamp is located where it is subject to continuous starting, the life expectancy can be as low as 3000 hours.

A fluorescent lamp should run for at least 20 minutes after switch on before turning it off.

The light colour or colour rendering is an important consideration when choosing a fluorescent lamp (see Figure 9.42). Personal taste will largely influence the selection of colour temperature; for example, cool colours might be chosen for an office or factory, but warm colours preferred for areas for social gatherings such as hotels and clubs (adapted from *Energy Efficiency Guide for Industry in Asia*, 'Effect of temperature', UNEP, p. 8, http://www.retscreen.net/fichier.php/903/Chapter-Lighting.pdf).

FIGURE 9.42 Fluorescent lamp colour rendering applications

Colour temp.	Effects on colours	Typical applications
Daylight white 5000 K +	– Strongly enhances blues – Flattens reds – Bluish tint to whites and greens	– Graphic arts studios – Winter goods shops, e.g. furriers – Seasonal affective disorder
Cool white 4000–5000 K	– Enhances blues – Flattens reds – Bluish tint to whites and greens	– Offices – Hospitals – Manufacturing
Mid range 3500–4000 K	– Neutral appearance – Enhances most colours equally – Does not favour yellow or blue	– Retail stores – Supermarkets – Showrooms
Warm white 2700–3000 K	– Enhances red and orange – Blues appear darker – Yellow tint to whites and green	– Residential lighting – Restaurants – Hotel lobbies
Extra warm 2000–2500 K	– Strongly enhances red and orange – Blues appear almost black – Whites appear strongly orange	– Bread and meat displays – City beautification – Not for general lighting

Source: Adapted from Osram, 'Light colours: Fluorescent lamps and their selection criteria', p. 14, http://www.osram.com/media/resource/hires/333565/light-can-be-white-en.pdf

The luminous efficacy of a 40 watt fluorescent tube is around 80 lm/W compared to the most efficient tungsten halogen lamp at 37 lm/W.

The development of low-cost electronic ballasts has improved the efficiency and starting time for fluorescent lamps and has led to the availability of compact fluorescent lamps (CFLs) to directly replace tungsten filament lamps in domestic premises. Table 9.7 shows the replacement equivalent CFLs for standard tungsten filament lamps.

TABLE 9.7 CFL replacement ratings

CFL	Tungsten filament
8–12 W	40 W
13–18 W	60 W
18–22 W	75–100 W
23–30 W	100 W
25–28 W	150 W

Source: Adapted from eartheasy, 'LED light bulbs: Comparison charts', http://eartheasy.com/live_led_bulbs_comparison.html

While the potential exists for CFLs to have a life of 20 000 hours this may depend on the quality of the electronic ballast. One advantage of the electronic ballast is the reduction of the stroboscopic effect caused by fluorescent lamps when the a.c. cycle passes through zero and can cause moving machinery to appear stationary.

All imported equipment, including lighting control equipment such as ballasts and transformers, must comply with **AS/NZS 3820:2009 Essential safety requirements for electrical equipment** and must be labelled with its minimum energy performance rating.

When you install any type of low intensity discharge lighting you must ensure the fittings and control equipment do not create a heat hazard while the power factor must be kept equal to or above 0.8 by the addition of capacitors or other approved correction.

SUMMING UP...

- Fluorescent lights use mercury vapour and inert gas to produce ultraviolet light.
- Visible light is produced when UV activates fluorescent powders.
- The control equipment for a basic fluorescent lamp comprises:
  - ballast
  - capacitor
  - starter.
- Fluorescent lamps are more efficient than tungsten filament lamps.
- Colour selection depends on the application for the lighting.
- Compact fluorescent lamps (CFLs) can be used to reduce domestic power consumption.
- Power factor correction equipment must be provided when fluorescent lamps are used.
- Imported ballasts must comply with Australian MEPS requirements.

9.13 High intensity discharge lighting

You will find high intensity discharge (HID) lighting being used to illuminate large areas such as sports arenas, work areas and freight handling yards. Four commonly used HID lamps are:

- mercury vapour
- metal halide
- low pressure sodium
- high pressure sodium.

Mercury vapour

The modern mercury vapour lamp has a high luminous efficacy but is still the least efficient of the HID lamps (see Figure 9.43).

FIGURE 9.43 High pressure mercury vapour lamp

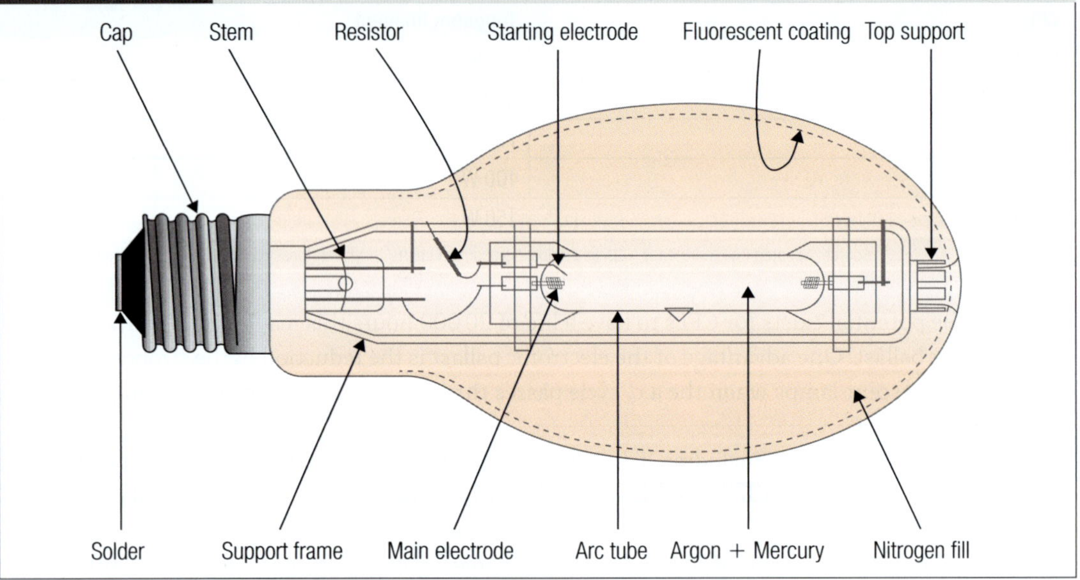

The Edison Tech Center describes how the mercury vapour lamp operates:

> There are several versions of the lamp but the basic principle is the same.
>
> **Low Pressure:**
>
> The first mercury vapor lamps were in a lower pressure tube. One would tip the lamp, and electrical contacts on each side of the lamp would send electricity through a liquid mercury which started the lamp. After that the lamp would heat fast and mercury became a vapor. The light would intensify as the arc grew stronger in the tube.
>
> **High Pressure:**
>
> Lamps of today are high pressure lamps with a fused quartz inner discharge tube. The high pressure helps increase efficiency and this was developed in 1936, 35 years after the low pressure lamps came out.
>
> This lamp starts with a small arc between the starting electrode and the main electrode. This arc goes through argon gas which easily strikes, even in cold weather. This little arc heats the tube, and over several minutes the tube gets hot enough to vaporize the solid mercury stuck to the sides. The mercury vaporized creates a strong light between the two main electrodes.
>
> To prevent the arc from infinitely getting stronger a ballast limits the current. Some lamps are 'Self Ballasted', they use an incandescent filament to act as a resistor, limiting current.
>
> Source: Edison Tech Center, 'Mercury vapor lamps', http://www.edisontechcenter.org/MercuryVaporLamps.html.

The capacitor shown in Figure 9.44 provides the required power factor correction. Advantages are:

- good efficiency (lamps after 1980s have a high lumen per watt rating)
- colour rendering is better than that of high pressure sodium street lights
- expected working life is 24 000 hours.

Metal halide

The metal halide (MH) lamp is a development of the mercury vapour lamp. Like the mercury vapour, this lamp consists of an arc tube within an outer envelope. The arc tube contains a starting gas (usually argon), mercury and MH salts. Traditional quartz MH arc tubes are similar in shape to mercury vapour (MV) arc tubes, but they operate at higher temperatures and pressures.

Connected as shown in Figure 9.45, the MH lamp starts when the ignitor supplies a high starting voltage, higher than those normally supplied to the lamp electrodes, through a gas mixture in the arc tube. The gas in the arc tube ionises, allowing current to flow and start the lamp. The ballast regulates the lamp starting current and lamp operating current. As pressure and temperature increase, the materials within the arc tube vaporise and emit light and UV radiation. As in the MV lamp an outer envelope, usually made of borosilicate glass, provides a stable thermal environment for the arc tube. Some MH lamps have a phosphor coat used to both diffuse the light and change the lamp's colour properties.

FIGURE 9.44 High pressure mercury vapour lamp circuit diagram

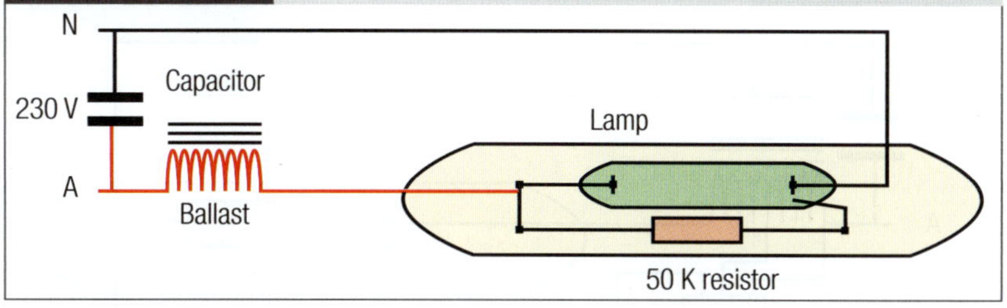

FIGURE 9.45 Circuit diagram for metal halide lamp

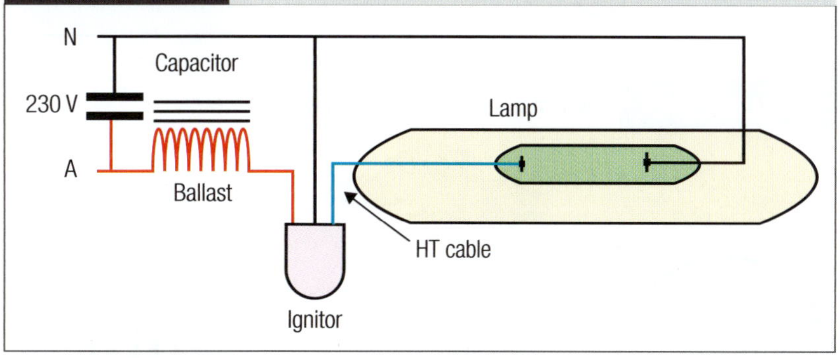

Low-pressure sodium lamps

Low-pressure sodium lamps (see Figure 9.46) are mercury-free. After ignition, the sodium vapour in the glass tube emits monochromatic yellow light with a wavelength of 590 nm. This enables a luminous efficacy of up to 178 lumen per watt to be achieved. The Edison Tech Center describes how the sodium lamp works:

> The lamp works by creating an electric arc through vaporized sodium metal. Other materials and gases are used to help start the lamp or control its color....
>
> The LPS lamp was the first sodium lamp to be developed. It is known by its signature monochromatic yellow color. It is mostly used in Europe since it did not appeal in other markets due to its poor CRI or color rendering. It is among the most efficient lamps in the world because it uses all the current it gets to create light at the most sensitive color (frequency) to the human eye. An incandescent lamp in contrast creates light at all frequencies from infrared (non-visible) to UV at the other end of the spectrum. The energy used to make non-visible light is a waste of energy since it does not help do the principle [sic] job of an electric light. The LPS lamp is also called a SOX lamp (SO for sodium).

Source: Edison Tech Center, 'Sodium lamp', http://www.edisontechcenter.org/SodiumLamps.html.

i A mercury vapour lamp cannot be replaced with a metal halide lamp in the same fixture. Metal halide lamps have power and voltage characteristics that are different from those for mercury vapour.

The remarkably high efficacy of the low pressure sodium lamp is due to the fact that the wavelength of light it radiates happens to be very close to the peak sensitivity of the human eye under normal viewing conditions. Only about 30% of the input power is converted into visible light. This percentage is comparable with other modern discharge lamps.

Advantages are:

- high-contrast visibility in mist and fog
- maximum luminous efficiency.

FIGURE 9.46 Low pressure sodium vapour lamp

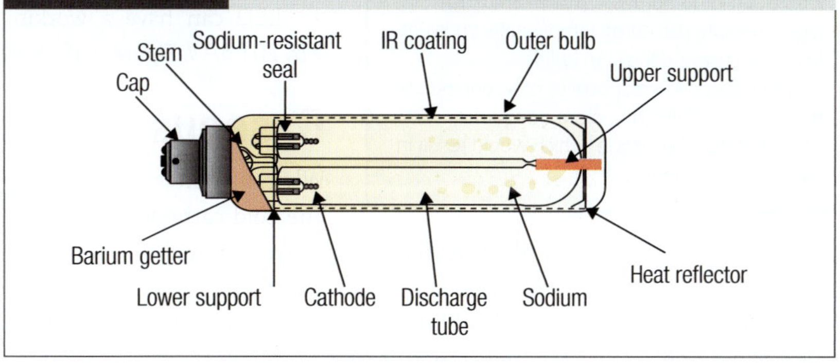

FIGURE 9.47 Circuit diagram for high pressure sodium vapour lamp

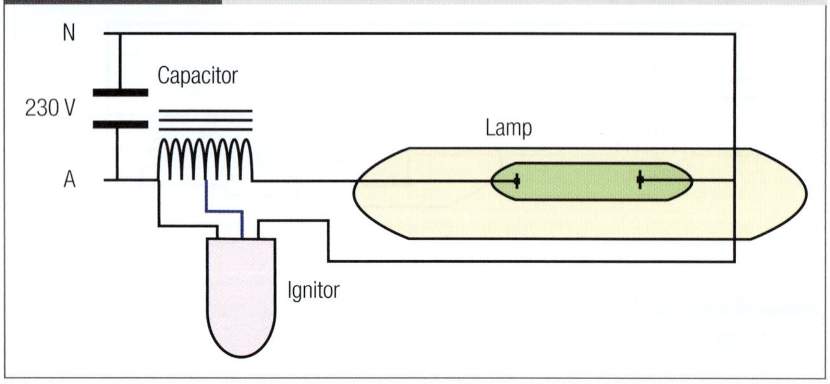

FIGURE 9.48 LED lamps to replace standard types

Source: iStockphoto/choness

High-pressure sodium

High-pressure sodium lamps provide the highest luminous efficacy of all high-pressure discharge lamps – as much as 150 lumens per watt! Emitting a high quality broad band light this lamp has an effective life of 25 000 hours. The discharge tube is made from sintered aluminium oxide to withstand the erosion from the sodium at high pressure. A high voltage starting pulse is delivered to the arc tube by the ignitor (see Figure 9.47) while the ballast limits the operating current. Common applications are outdoor lighting around factories and commercial buildings and street lighting.

Advantages are:

- very high luminous efficiency
- extremely long service life.

LED lighting and its applications

Light-emitting diodes (LED) are light sources utilising diodes that emit light when connected in a circuit. The effect is a form of electroluminescence where LEDs release a large number of photons outward. The most important part of an LED is the semiconductor chip located in the centre of the light source. When sufficient voltage is applied to the diode junction, electrons move across the junction and combine with positive charges on the other side. Each time an electron combines with a positive charge, electromagnetic energy in the form of light is released. Different semiconductor materials are used to create LEDs that emit different colours.

LEDs last much longer than conventional lamps as they don't have a filament to burn out. LEDs emit light immediately so they suit applications that are subject to frequent or potential on-off cycling such as traffic signals. Because of their small size LEDs are built into arrays to suit a range of different voltages. An LED can have a working life of 10 000 hours with a luminous efficacy of 2 to 20 lm/W. Figure 9.48 shows typical LED lamps.

High-pressure discharge lamps contain small quantities of mercury as an environment-relevant substance. Metal halide lamps can also contain thallium iodide as an additive. This is why discharge lamps must be disposed of separately from domestic waste and industrial waste similar to domestic waste. Breakage of high-pressure discharge lamps emits traces of toxic mercury and thallium halides.
During transport to disposal or collection points, make sure that the lamps are adequately protected from breakage, which would result in the emission of mercury.

Gaseous discharge display lighting

When current is passed through neon gas sealed in a glass container it glows orange. The cold-cathode tubes used in advertising signs and hoardings are commonly called 'neon tubes', although only the clear-glass red-light types actually contain neon. The most widely used gas is mercury and argon, the same as in ordinary fluorescent lamps. Mercury and argon produce a cyan (blue) colour which can be used directly or in tubes with internal phosphor coatings to produce a wide range of colours.

The voltages used in large display signs can be lethal. Isolate and lock out the supply before removing any covers.

Every 'neon' tube has a tricarbonate-emission coated electrode at each end. The length of the tubes used for signs and displays requires a high voltage transformer of around 4000 to 15 000 volts.

The transformer core is designed to provide a high voltage for starting but limits the build-up of current as the tube warms up and conducts. The arc from discharge lamps creates radio frequency interference (RFI) that must be suppressed by polyester capacitors to prevent them interfering with television and radio broadcasts.

FIGURE 9.49 Types of lighting compared by efficiency

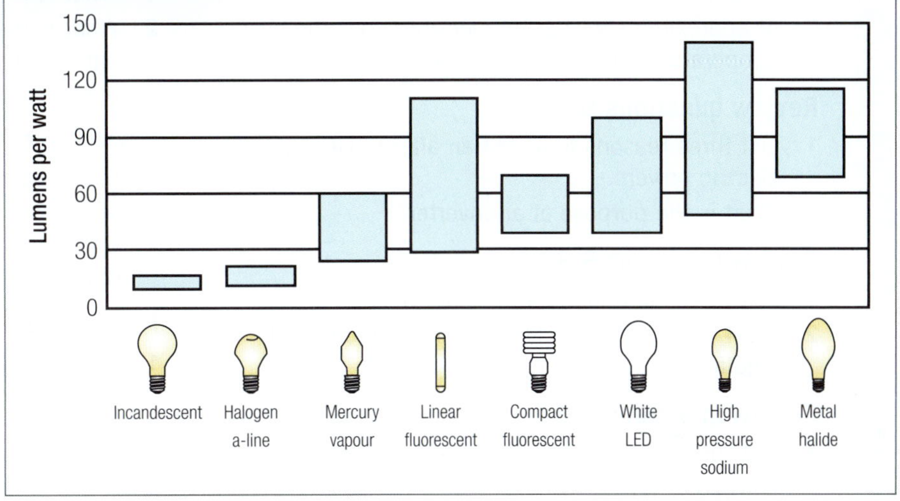

Lighting types compared for efficiency

The concern in the world for efficient use of energy to reduce greenhouse gases, in particular carbon dioxide, has led to a strong focus on the efficiency of lighting. Figure 9.49 shows a comparison of the luminous efficacy of the main types of lighting available. You can contribute to this effort by advising your customers on the most suitable lighting system for their application.

SUMMING UP...

- High intensity discharge (HID) lighting is cheap to run.
- Mercury vapour discharge lights are the least efficient of HID lights.
- Halide lamps are a development from mercury vapour lights.
- Mercury vapour and metal halide lamps will not work in each other's fittings.
- Low pressure sodium lamps are useful in fog due to their yellow light.
- Low pressure sodium lamps are highly efficient.
- High pressure sodium lamps are efficient and produce a wide band of light.
- Power supply authorities require power factor correction for HID lamps.
- LEDs emit light when electrons combine with positive charges.
- LEDs are built into arrays to make larger lamps.
- Neon display lighting uses high voltages to overcome the length of the tubes used for illuminated signs.
- Discharge lamps must be fitted with radio frequency interference (RFI) suppression.

REVIEW EXERCISES

Answer these questions and check your answers at the back of the book.

Review questions 9.1

1 Which TPS wiring configuration would most suit a building where the wiring had to be installed in the floor space?

2 What would be the switching circuit most suitable to switch the lights on three flights of stairs?

Review questions 9.2

1 Does the term socket outlet apply to cord extension socket?

2 When can you use one switch with two socket outlets?

Review questions 9.3

1 Does a load larger than 20 amps have to have its own circuit?

2 Under what circumstances may you switch the earth conductor?

Review questions 9.4

1 What is the purpose of the three heat switch?

2 Name a heat control device that adjusts the temperature to a preset value.

Review questions 9.5

1 List three electrical load types used for generating heat.
2 List three scales used for measuring temperature.

Review questions 9.6

1 Why are hot water cylinders referred to as continuous water heaters?
2 At approximately what temperature does the over temperature cut out trip on a hot water cylinder?

Review questions 9.7

1 List three reasons to install an alternative supply of electric power.
2 What is the purpose of an inverter?

Review questions 9.8

1 List three common types of batteries.
2 What is the class of cell that can be recharged regularly?

Review questions 9.9

1 List three effects of fire detected by fire alarm systems.
2 Where are fire alarms best positioned for early detection of fire?

Review questions 9.10

1 How much lighting does evacuation lighting have to supply?
2 What is the particular feature of non-maintained emergency lighting?

Review questions 9.11

1 What is the basic unit of light measurement?
2 What is the maximum power allowed for indoor lighting under Building Code of Australia (BCA) rules?

Review questions 9.12

1 In the basic fluorescent lighting circuit what is the purpose of the ballast?
2 Once a fluorescent tube is lit can the starter be removed without affecting the operation?

Review questions 9.13

1 What would be a typical application for a mercury vapour discharge lamp?
2 Is a metal halide lamp interchangeable with a mercury vapour lamp the same size?

ONLINE RESOURCES

Back-up power: Off-grid Energy Australia
http://www.offgridenergy.com.au/index.php/grid-battery-back-up.html

Batteries: Electropaedia
http://www.mpoweruk.com/life.htm

Central control of emergency lighting: Stanilite Nexus
http://www.tnb.ca/aus/pdfs/Nexus-LX-Brochure.pdf

Energy efficiency: Australian Building Codes Board (ABCB)
http://www.abcb.gov.au/en > Education events and resources > Publications > ABCB handbooks > Energy Efficiency Provisions for Electricians and Plumbers

Fire and rescue NSW
https://www.fire.nsw.gov.au/page.php?id=443

Heater controls: Appliance411
http://www.appliance411.com/parts/showcase.dryer-thermo.shtml

Heat and temperature: Zone Land Education
http://zonalandeducation.com/mstm/physics/mechanics/energy/heatAndTemperature/heatAndTemperature.html

Lighting measurements: Illuminating Engineering Society (IES)
http://www.ies.org/pdf/education/IES-Color-3-Webcast-Handout.pdf

Low voltage installation: Powercor Australia
http://www.powercor.com.au

Low voltage sub-circuits: Schneider Electric
http://www.schneider-electric.com.au

Mercury vapour lamps: Edison Tech Center
http://www.edisontechcenter.org/MercuryVaporLamps.html

Socket outlets: A1 Testing Tagging
http://www.a1testingandtagging.com.au/faqs/

Solar water heating: Solahart
http://www.solahart.com.au/solar-water-heating/solar-water-heating.aspx

10

ELECTRICAL INSTALLATIONS – ARRANGEMENT, CONTROL AND PROTECTION

This chapter examines the electrical installation safety principles that apply to the control and protection methods required for low voltage electrical installations. These are described in AS/NZS 3000:2007 Wiring Rules Section 1.5 Fundamental principles, and further sections describe methods employed to comply with those principles.

This chapter is not intended as an alternative to reading the Wiring Rules. The references stated within this text were correct at the time of writing but may change as the Australian Standards are updated regularly.

Chapter outline

10.1 Safety principles to which electrical systems in buildings and premises shall comply

Preparing to select and install control and protection equipment

As an electrical practitioner you will be required to design and execute a low voltage electrical installation so that no damage or injury will occur as a result of your work.

Safety principles

The scope of the **AS/NZS 3000:2007 Wiring Rules** is to set out installation standards to protect users of electricity from the adverse effects of a faulty low voltage electrical installation.

The dangerous effects are identified in **AS/NZS 3000:2007** clause 1.5.1 as:

- shock current – causing death to people or their livestock
- excessive temperatures – causing fire damage to buildings and structures
- explosive atmospheres – resulting in destruction and damage to buildings and structures as well as endangering the lives and safety of people and livestock.

The contributing factors to the listed dangers are identified in **AS/NZS 3000:2007** as:

- direct and indirect contact with conductors that are live in normal operation
- indirect contact or contact with conductors that have become live during fault conditions
- thermal effects caused by the heating of conductors due to the passage of current
- over-voltage – and other unwanted voltages caused by events external to the installation and faults generated within the installation
- over-current – causing excess heat in conductors and compromising the effectiveness of the insulation
- fault currents – causing voltage rises and possibility of indirect contact and thermal effects
- overload – especially motors causing heating that can ignite adjacent materials
- injury from mechanical movement – machinery such as shafts, belts and presses can be capable of dangerous unexpected movement with disastrous results (**AS/NZS 3000:2007 Wiring Rules**, pp. 40–50).

Compliant methods for providing protection

In keeping with the purpose of preventing fire and shock **AS/NZS 3000:2007** specifies a number of protection systems to be used in a low voltage installation.

Protection against electric shock

Each hazard has a compliant method set out to prevent injury or damage. It is also desirable to combine protection methods so that there is more than one system in operation at any one time.

Protection against direct contact

Prevention of electric shock due to direct contact requires one or a combination of the following methods:

- Insulation – conductors are covered with insulating material to prevent direct contact.
- Barriers or enclosures – these do not need to be insulating enclosures or barriers, provided all possibility of direct contact is eliminated.
- Obstacles – such as locked equipment rooms and cages to prevent access to live equipment.
- Placing out of reach – live conductors placed at a height where they cannot be touched by any person without the assistance of a portable ladder or other equipment (**AS/NZS 3000:2007 Wiring Rules**, pp. 40–3).

Protection against indirect contact

To prevent electric shock through indirect contact one or a combination of the following methods must be employed:

- Automatic disconnection of supply – faults within the installation can cause metal enclosures to become live. Devices are required to detect fault currents and disconnect the supply before voltages reach dangerous levels.
- Class II equipment or by equivalent isolation – enclosures may themselves be insulated, preventing the metal from becoming a shock hazard in the event of a fault.
- Residual current devices (RCDs) – used to detect the leakage of current to earth and interrupt the supply before a lethal current can develop.
- Electrical separation – isolation transformers can be used so that any fault developing in an appliance or other equipment will not have a reference to earth so any one fault will not cause a shock hazard.
- Extra low voltage – using either SELV (separate extra low voltage) or PELV (protected extra low voltage). In both these cases contact with a circuit conductor during a single fault condition will not constitute a shock hazard (**AS/NZS 3000:2007 Wiring Rules**, pp. 44–7).

Read AS/NZS 3000:2007 clause 1.4.12, which provides the definition for arm's reach.

Protection against thermal effects in normal service

Thermal effects can lead to burns on contact or damage due to fire. The following steps should be taken to reduce the risk of injury or damage:

- Location – equipment must not be installed where it will cause a fire hazard during normal operation. Equipment that reaches high temperatures should be installed beyond arm's reach (clauses 4.2.2.3 and 4.2.3).
- Separation of flammable materials – protection must be provided to prevent heat transfer through sparks or high temperatures (clause 4.2.2.3).
- Protection from emission of arcs or sparks – equipment must be totally enclosed in arc-resistant material, shielded by flame proof barriers or separated by distance (clause 4.2.2.4).

Equipment must not be installed in such a way that it contributes to the spread of fire or restricts escape from a fire-affected location (clause 1.5.12(d).

Protection against over-current

Equipment such as fuses and circuit breakers must be installed to protect sections of the installation from the effects of over-current. The equipment must limit the over-current to a safe value and duration by interrupting the circuit (**AS/NZS 3000:2007** clause 1.5.9).

Protection against earth fault currents

Where earth protection is provided the conductors must be of sufficient cross-sectional area to cause the over-current protection equipment to operate before dangerous voltages develop (**AS/NZS 3000:2007** clause 1.5.10).

Protection against abnormal voltages

Abnormal voltages have many causes, some within the installation and some from external causes. You must use the following methods to reduce or remove this hazard:

- Faults between live parts – high voltage, low voltage and extra low voltage equipment must be separated physically or by an insulating barrier rated for the highest voltage in the installation.
- Induced voltages – can be reduced by careful planning of circuit routes as well as use of screened cables in the adjacent circuits.
- Voltage disturbances and electromagnetic – suitable surge protection should be provided to prevent the build-up of over-voltages from external and internal sources.
- Voltage in unused conductors – these conductors must be removed from the installation or terminated both ends in junction boxes or other suitable enclosures (**AS/NZS 3000:2007** clause 1.5.11 and subclauses).

Protection against injury from abnormal mechanical movement

You will need to install devices to disconnect or isolate machinery that is likely to move unexpectedly during maintenance or while changing materials (clause 1.5.13).

Some of the equipment you could install are:

- light curtains
- lockouts
- limit switches
- electrically operated brakes.

Design and selection of equipment

You must ensure that your electrical installation:

- protects people and their livestock by the use of compliant protection devices and protection from thermal effects
- connects and functions correctly – by ensuring the compatibility with supply voltage for insulation selection and correct choice of cable sizes and accessories to suit the anticipated load conditions
- is provided with under- and over-voltage protection where required to prevent malfunctioning of equipment
- minimises inconvenience – by dividing circuits into smaller circuits each with its own protection to isolate a faulty circuit from the rest of the installation
- avoids over-design and unnecessary expense by correct estimation of maximum demand
- facilitates safe operation – the normal use of the installation must not involve the operator or others in exposure to any dangerous electrical or mechanical hazards (**AS/NZS 3000:2007** clause 1.6.1).

SUMMING UP...

- The main purpose of the Wiring Rules is to prevent fire and shock.
- Compliant installations will provide protection against overload and short circuit.
- Electric shock can be prevented by the use of residual current devices (RCDs).
- Protection against mechanical movement can be provided by limit switches and mechanical brakes.
- The installation must function correctly and be compatible with the available supply.
- Circuits must be arranged to take into account cable length and current carrying capacity.
- All equipment must be installed in accordance with manufacturer's instructions, operate in a safe and reliable manner and not cause danger from electric shock.

10.2 Circuit and control arrangements

You must plan your installation, determine the number of sub-circuits and the number and rating of control and protection devices.

FIGURE 10.1 Block diagram of an electrical installation

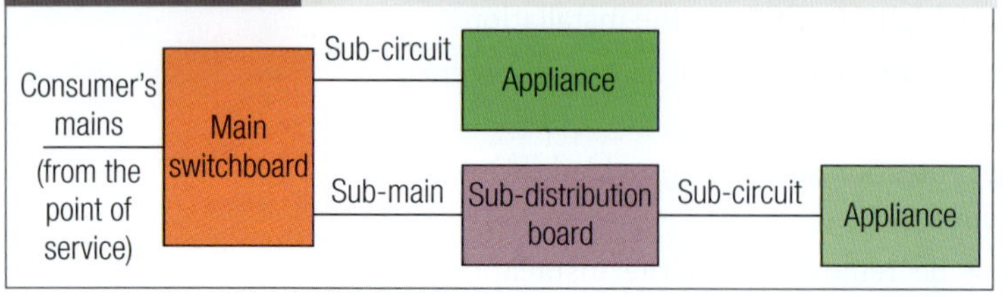

Division of electrical installations into circuits

AS/NZS 3000:2007 clause 1.6.5 requires that every electrical installation shall be divided into circuits as necessary to avoid danger and minimise inconvenience in the event of a fault and to facilitate safe operation and inspection, testing and maintenance (see Figure 10.1).

When designing and setting out the circuits you will need to take into account the requirements of clause 2.2.1.1, including the relationship of the equipment, any requirement for operation as a group and any special need identified by the user.

The load and operating characteristics of the equipment in relation to the rating of the circuit components must be considered along with limitation of the consequences of circuit failure and loss of supply to critical equipment due to short circuit and overload.

The layout of the equipment must simplify the process of locating a fault and carrying out maintenance work. Provision must be made to allow alterations and additions to be carried out without removing the supply from the rest of the installation. If safety services such as fire control systems are installed they must be separate from the rest of the electrical installation (**AS/NZS 3000:2007** clauses 2.1.2 and 2.2.1.1).

The advantages of this layout are:

- a fault will only trip off one circuit
- a fault is easier to find
- repairs or alterations can be carried out without removing the supply to the installation
- cable sizes can be smaller leading away from the protective devices installed on the switchboards. This allows for the decreasing demand towards the final sub-circuits.

This system does have the disadvantage that a fault occurring on one of the circuits from the main switchboard will shut down the supply to all circuits of related downstream distribution boards. Full guidance on circuit arrangements can be found in **AS/NZS 3000:2007** Appendix C.

Determining the number of circuits

The number of circuits required in an installation will be determined by:

- the size and nature of the installation
- the required current carrying capacity (**AS/NZS 3008.1.1:2009 Electrical installation – Selection of cables** clause 3.1)
- over-current requirements (**AS/NZS 3000:2007** clause 2.5 and paragraph B3)
- voltage drop requirements (**AS/NZS 3000:2007** clause 3.6 and **AS/NZS 3008.1.1:2009** clause 4.1)
- automatic disconnection of supply requirements (**AS/NZS 3000:2007** clause 5.7 and paragraphs B4 and B5).

As outlined in the Wiring Rules, you can install as many main switches on the main switchboard as you need to but usually you will put in one for each tariff to be provided by the utility distributor.

You will usually install at least one lighting circuit, but the size of the premises or user preference may dictate more. The number of power circuits you decide to install will depend on the loading. You can find out the number of socket outlets allowed on a circuit from **AS/NZS 3000:2007** paragraph C5 and Table C8.

AS/NZS 3000:2007 clause 2.5 specifies that you must restrict the current flowing in the circuit to the current rating of the protective device. In turn you must not use a device that exceeds the current carrying capacity of the cable. If a circuit supplies one item of equipment the current is the nominal load current of the equipment

EXAMPLE 10.1

The load current of a 10 kW 400/230 V three phase heater:

$$I = \frac{10000}{\sqrt{3} \times 400} \text{ A}$$

$$I = 14.5 \text{ A}$$

The circuit conductors and the protective device must have a current carrying capacity of not less than 16 A (nearest standard rating).

Where you have connected more than one item of equipment, the circuit current could be simply assessed as the sum of the individual load currents. This would provide a safe and conservative solution but it does not take into account the normal operating conditions. All equipment might not be operating simultaneously at full load. If a sub-main supplies a distribution board associated with lots of socket outlets you can use a system called 'after diversity maximum demand' to work out the circuit current.

Division of sub-circuits

Under **AS/NZS 3000:2007** Section C5.1 you must connect each item of equipment that has a current rating in excess of 20 A per phase to a separate and distinct circuit. When more than one item of equipment is to be connected to a circuit you will need to take into account:

- the number of lighting points, socket outlets or appliances that are to be supplied in combination
- the way different items of equipment are used, including seasonal or daily variations
- the circuit current under load and the matching of cable and protective device ratings to cut down the risk of overload faults
- the effects of an overload fault on the circuit, including loss of supply to security, emergency, medical or data and telecommunications services.

FIGURE 10.2 SELV circuit

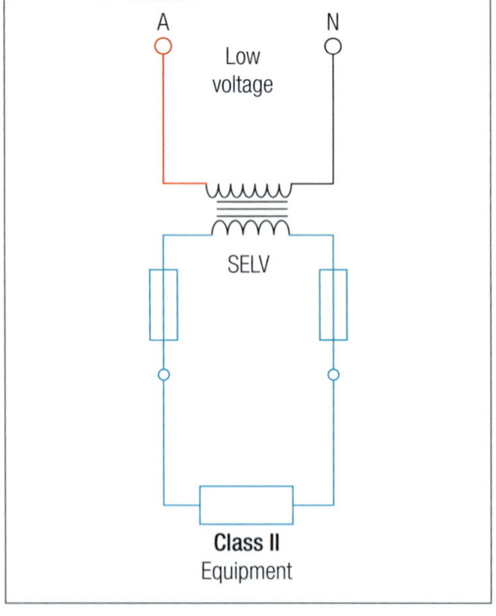

FIGURE 10.3 PELV circuit

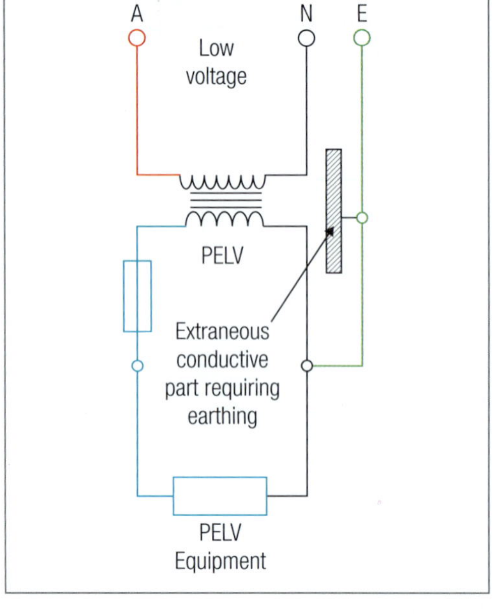

Number of points in final sub-circuits

If you are to use the values in **AS/NZS 3000:2007** Table C8 there must be two power circuits in the installation. Multiple socket outlets count as one socket outlet for each socket on the multiple combination. Where you know that the socket outlet will be used for a particular appliance such as a dishwasher, use the load of the appliance. If the installation has permanent heating or cooling systems installed then you can increase the number of socket outlets.

Application and arrangements of SELV and PELV circuits

Separated extra low voltage (SELV) and protected extra low voltage (PELV) provide protection against both direct and indirect contact by using extra low voltage (less than 50 V a.c. or 120 V ripple free d.c.) on the equipment conductors. SELV is described by **AS/NZS 3000:2007** clause 1.4.83 as separated from earth and the rest of the installation so that a fault cannot cause an electric shock. You can use a completely separate supply such as batteries or diesel-driven generators or an isolating transformer.

You will notice in the SELV circuit shown in Figure 10.2 that there are no earthed points. You are not allowed to earth either the exposed conductive parts or the live parts of a SELV circuit. You must separate it from other electrical systems as set out in clause 1.5.11.2. You must guarantee the separation of a SELV system from other circuits by containing the conductors in multi-conductor cables, providing them with an additional insulating sheath (clause 3.9.8.3).

PELV is described in **AS/NZS 3000:2007** clause 1.4.76. It has the same requirements as a SELV system, except PELV circuits have at least one point always earthed (see Figure 10.3).

Requirements for an isolated supply

When you use an isolated supply, you must make sure that all conductors are separated from earth and fully insulated from any earthed metal part of the installation. You must connect switching devices in all live conductors. Connecting a switch in the active conductor of the primary circuit of an isolated supply is all right for control and disconnection (**AS/NZS 3000:2007** clause 7.4).

Further information about isolated supplies is contained in Section 10.4 later in this chapter.

SUMMING UP...

- Every installation must be divided into sub-circuits.
- Any load greater than 20 A must be on a separate circuit.
- Each circuit must start at a protective device capable of opening the load current.
- The number of points on a circuit is specified in the Wiring Rules.
- The layout of an installation must aid fault finding.
- Safety systems must be given special treatment.
- SELV circuits must be electrically separated from the low voltage supply conductors and earth conductors.
- PELV circuits are earth referenced to the installation earth system.
- SELV and PELV circuits must be provided with at least one main switch but there can be more if required.

10.3 Hazards and risks in an electrical installation

You must pay strict attention to the risks involved in working with electricity as well as the hazards that you can create for others. You, the electrical practitioner, are legally responsible for the standard of your installation work.

Effects of current on your body

The severity of electric shock can range from a mild tingling sensation to electrocution and death, depending on the conditions that are present. In addition to electrocution, electricity can cause serious burns and side effects such as falling off a ladder or platform. See Table 10.1 for the effects of current on the body.

Factors that affect the severity of an electric shock are listed by the US Department of Labor Occupational Safety and Health Administration, and include:

- the voltage – this will have a large bearing on the amount of current passing through your body
- the path of current through your body – shock through your arm, across your chest and out your other arm or feet will allow the current to pass through your chest and heart, causing your muscles to cramp (suffocation) and your heart to go into spasm (fibrillation)
- the length of time the current flows through your body – the longer you are in contact, the more severe will be your injuries
- the frequency of the current – alternating current (a.c.) has the tendency to cause your muscles to spasm and stay in contact with the supply source. Direct current (d.c.) has the tendency to jolt your muscles, giving greater chance of breaking contact ('How electrical current affects the human body', https://www.osha.gov/SLTC/etools/construction/electrical_incidents/eleccurrent.html).

TABLE 10.1 Effect of current on the body

Current amps a.c.	Effect on the human body	Current mA a.c.
0.001 A	Mild sensation.	1 mA
0.005 A	Slight shock felt; not painful but disturbing. Average individual can let go.	5 mA
0.006 A to 0.016 A	Painful shock, loss of muscular control. You may not be able to let go.	6 mA to 16 mA
0.017 A to 0.099 A	Extreme pain, respiratory arrest, severe muscular reactions. You cannot let go. You could die if nobody helps you.	17 mA to 99 mA
0.1 A to 2 A	Ventricular fibrillation. Muscular contraction and nerve damage begins to occur. You are likely to die if no one helps you.	100 mA to 2000 mA
Greater than 2 A	Cardiac arrest, internal organ damage and severe burns. You will most likely die even if you are helped in time.	Greater than 2000 mA

Source: Adapted from US Department of Labor Occupational Safety and Health Administration, 'How electrical current affects the human body', https://www.osha.gov/SLTC/etools/construction/electrical_incidents/eleccurrent.htm

You must remember that time is very important as well as the size of current. The longer you are in contact the more severe your injuries. A 50 V touch voltage under dry conditions can be sustained indefinitely, while 100 V under the same conditions must be disconnected. Wet conditions alter the maximum duration considerably. See **AS/NZS 3000:2007** Figure B4.

Risk from thermal effects of current

When current flows through a conductor heat is generated due to the current and resistance of the conductors. The amount of heat (I^2R) generated within a conductor is dependent on the load, length, cross-sectional area and type of material the conductor is made from. You will be mostly using copper circuit conductors to carry current to a load so you must keep the heat produced by the cables as low as possible by selecting the correct size conductors for the task.

If the temperature rises above the temperature rating of the cable, e.g. V75, V90, etc. then the cable's insulation will break down. The heat generated may start a fire in adjacent structural material. If the installation has a high prospective fault current an arc flash will cause extensive damage.

Heating equipment has risks associated with it due to the danger of exposed elements. There is very little risk from hot water services and panel heaters that have sheathed or enclosed elements, but a cooking hotplate has an exposed coiled sheathed element just waiting to be accidentally touched. **AS/NZS 3000:2007** Table 4.1 shows the temperature limits for electrical equipment within arm's reach.

Risk of injury from mechanical movement

In an installation where there are electrically operated machines and equipment, you always face the possibility of injury from mechanical movement.

Some examples of the types of machine movements that you may be exposed to are:

- electrically operated guillotine without the appropriate guards
- large exposed flywheels on large presses such as in automotive plant
- unguarded belt and geared drives in coupled motors
- grinding machines, lathes, milling machines and pedestal radial drills in a workshop without the required guards
- someone incorrectly using portable electric equipment on building sites including:
  - rotary hammer drills
  - angle grinders
  - reciprocating saws.

Faced with the examples above, you can minimise or prevent risks by installing guards over moving parts. You can reduce the danger of injury from large machines by setting up safety zones and barriers.

You must always keep in mind the prevention of injury from mechanical movement of electrically operated equipment when designing electrical equipment control systems.

Examples of protection

You can prevent the operation of a press without the guard at the front pulled down by fitting a limit switch or light curtain. If the operator tries to work the press, the electrical control circuit is incomplete because the guard on the front of the press needs to be pulled right down to close the limit switch or restore the light curtain.

You can prevent operation of sheet-metal guillotine unless both hands of the operator are clear of any moving part and the cutting blade. This is achieved by placing two operating buttons so they must be operated by two hands. Both buttons need to be pressed for the guillotine to operate.

To ensure operators stand clear of a restricted area you can fit devices such as photoelectric switches and light beam sensors so that the machine will not operate while the sensors are interrupted by the presence of the operator.

A passenger lift automatically closes its doors some time after the last person has entered the lift. You can fit a light beam so that if a person enters the lift late while the doors are closing they will open again and prevent injury.

Protection against injury from mechanical movement is required by **AS/NZS 3000:2007** clause 4.2.1.2 which restricts the starting or reversal of a motor.

FIGURE 10.4 Direct contact

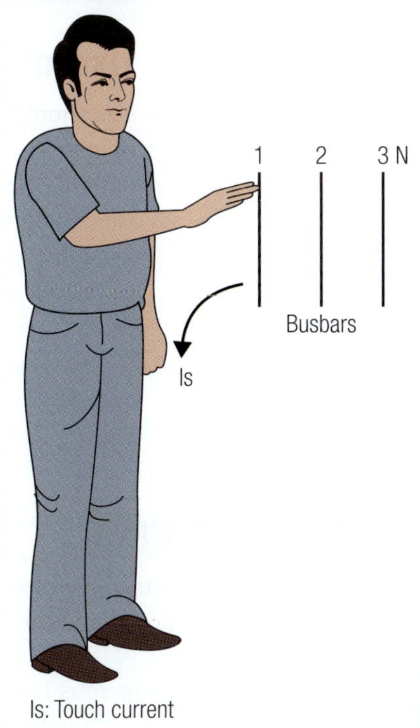

Protection against direct contact (basic protection)

Figure 10.4 shows direct contact as contact with electrical conductors that are live during normal operation of the installation as defined by **AS/NZS 3000:2007** clause 1.4.34.

Insulation

You must use non-conducting materials as the front-line protection from electric shock. Insulation has the following classifications under **AS/NZS 3000:2007** clause 1.4.60:

- Basic insulation – applied to live parts to provide protection against electric shock. It must be suitable for the temperature and voltage to which it is to be exposed. Direct contact usually occurs when the insulation fails, is damaged or removed.
- Supplementary insulation – is an independent insulation applied in addition to basic insulation. This ensures protection if the basic insulation should fail.
- Double insulation – comprises both basic and supplementary insulation. This applies particularly to sheathed cables and flexible cords.
- Reinforced insulation – is applied to live parts giving an insulation category equivalent to double insulation. Examples of this are bundled aerial cables.

Barriers or enclosures

You can provide a degree of protection by placing live parts behind barriers or enclosures with the following ratings:

- IP2X – a 12.5 mm diameter sphere shall not be able to enter (**AS/NZS 3000:2007** Figure G1a)
- IPXXB – standard test finger penetrating to 80 mm must not contact hazardous parts (**AS/NZS 3000:2007** Figure G1c)
- IP4X – the access probe of 1.0 mm will not penetrate from above horizontal top surfaces that are readily accessible.

You must fasten the barriers or enclosures in place firmly enough to withstand the stresses from normal operation. Anyone must only be able to remove barriers using a tool or key. And finally you should provide an interlocking device to switch off the source of supply if anyone has to remove the barrier or open the enclosure (**AS/NZS 3000:2007** clause 1.5.4.4).

Obstacles

You can only use this method of protection where access is limited to competent people. Obstacles should prevent unintentional bodily approach or unintentional contact with live parts. No one should be able to remove obstacles except with a key or tool. The purpose of this form of protection is to prevent accidental contact with live parts and not the deliberate circumvention of the obstacle (**AS/NZS 3000:2007** clause 1.5.4.5).

Placing out of reach

You should only use this method of protection under the same requirements as obstacles above. Parts that are at different voltages must not be located within arm's reach or less than 2.5 m apart.

The definition of arm's reach can be modified by the addition of barriers in the horizontal or vertical direction (**AS/NZS 3000:2007** clause 1.5.4.6).

Use of extra low voltage

You should use extra low voltage where the possibility of direct contact is unavoidable during normal operation of the installation (**AS/NZS 3000:2007** clause 1.5.7). Such situations are control circuits within a plant where limit switches or solenoids may be easily damaged, leaving exposed conductors. Garden lighting is a vulnerable system where digging implements may damage cables or fittings. See Section 10.2 earlier in this chapter for discussion of SELV and PELV.

Where no particular requirement is made for extra low voltage installations they must comply with other parts of **AS/NZS 3000:2007**.

You must ensure the conductors and insulation of cables are suitable for the installed conditions and don't need any extra protection unless the installation conditions such as the proximity of low voltage and high voltage circuits require it (**AS/NZS 3000:2007** clause 7.5.11.1).

SUMMING UP...

- The supply voltage, skin resistance and path through the body all determine the severity of an electric shock.
- a.c. is more dangerous than d.c. as it causes the heart muscles to fibrillate.
- The longer the duration of contact the greater the likelihood of death.
- Heating produces the possibility of contact burns and fire risk.
- Mechanical hazards can be mitigated by the addition of electrical protection.
- Insulation classes include basic, supplementary, double and reinforced insulation.
- Barriers or enclosures only provide protection if accessed by a key or tool.
- Obstacles usually only prevent accidental contact with live parts.
- Simultaneously accessible parts at different voltages must not be located within arm's reach.
- Extra low voltage is used where the possibility of direct contact is unavoidable during normal operation of the installation.

FIGURE 10.5 Indirect contact

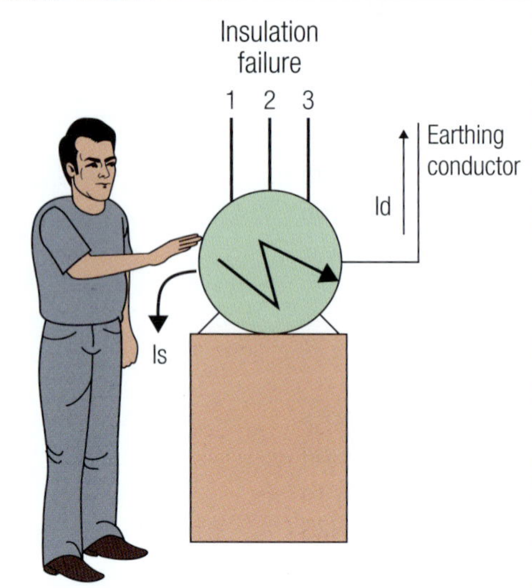

10.4 Protection against indirect contact

You must make sure that you prevent contact with electrical conductors that are not live during normal operation of the installation but may become live due to fault conditions (**AS/NZS 3000:2007** clause 1.4.34) – see Figure 10.5.

Metal frames and enclosures may become live for a variety of reasons:

- An electric motor may be overloaded, therefore overheating the windings to an extent where the insulation varnish of the windings melts, exposing the copper which makes contact with the frame of the motor.
- A cooking appliance may have such a low insulation resistance to earth that the leakage of current is high enough to liven the earthed metal frame but not operate the over-current protective device.
- An appliance such as an earthed portable tool, for example, a drill, may not have been effectively earthed and because it is continuously handled, one of the live conductors may loosen and contact the metal frame. The frame will become live and the operator becomes part of the return path of the fault current.

Selecting protection and control methods for an installation

EXAMPLE 10.2

Figure 10.6 shows how an electric shock can be received from indirect contact when an active conductor makes contact with the metal frame of an appliance and the earth has become loose from the earth terminal.

In this situation the over-current protective device can only operate if the rated current of the device is exceeded, e.g. 16 amperes.

Only the electrical resistance of the person's skin would limit the amount of current flowing through their body and determine the severity of the electric shock. The current would continue to rise to a lethal value.

If RCD protection is fitted, 30 mA of fault current flowing through the person will cause an imbalance in the RCD, tripping it and opening the circuit.

FIGURE 10.6 Earth protection failure

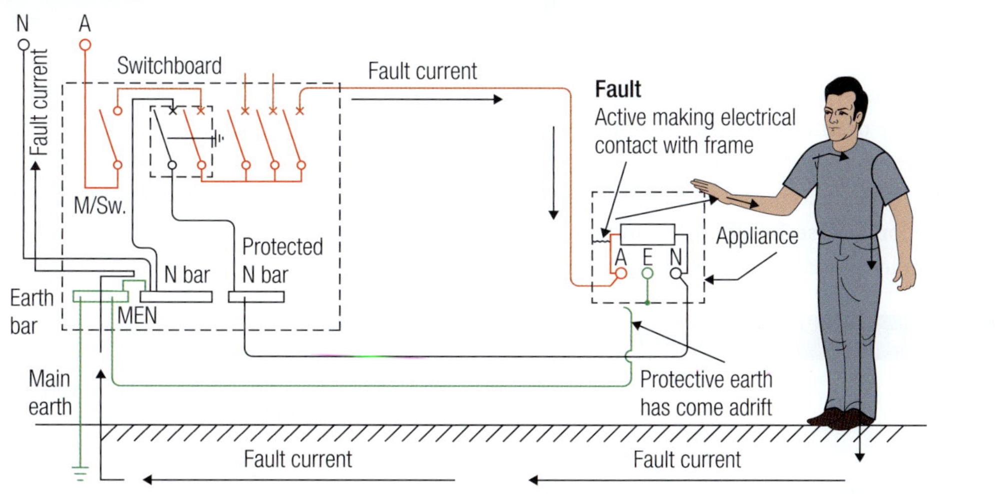

Methods of protection

You must provide all your electrical installations with protective equipment that suits the type of fault that is likely to develop. The most important is protection from electric shock followed by protection from over-currents. You will need to provide other forms of protection depending on the type of equipment and the conditions under which you have installed it.

Automatic disconnection of supply

You must take steps to limit the level of touch voltage arising in the event of a live conductor coming in contact with the conductive metal parts of an installation. You can achieve this by providing a system of earthing in which exposed conductive parts are connected to a protective earthing conductor. The fault will be disconnected by an over-current protective device or a residual current device (RCD) (**AS/NZS 3000:2007** clause 1.5.3).

Figure 10.7 shows an installation providing automatic disconnection utilising a system of earthing and disconnection of a fault by the use of an over-current device (a circuit breaker) as well as an RCD.

The use of class II equipment

Class II equipment has an extra layer of insulation (double insulation) to prevent contact with metal made live during a fault and does not rely on a protective earth (**AS/NZS 3000:2007** clause 1.4.28). Double insulated equipment has a durable and substantially continuous enclosure of insulating material which envelops all metal parts, except small parts such as nameplates, screws and rivets, which are isolated from live parts by insulation. Class II equipment may have parts operating at SELV.

FIGURE 10.7 Earth fault and RCD protection

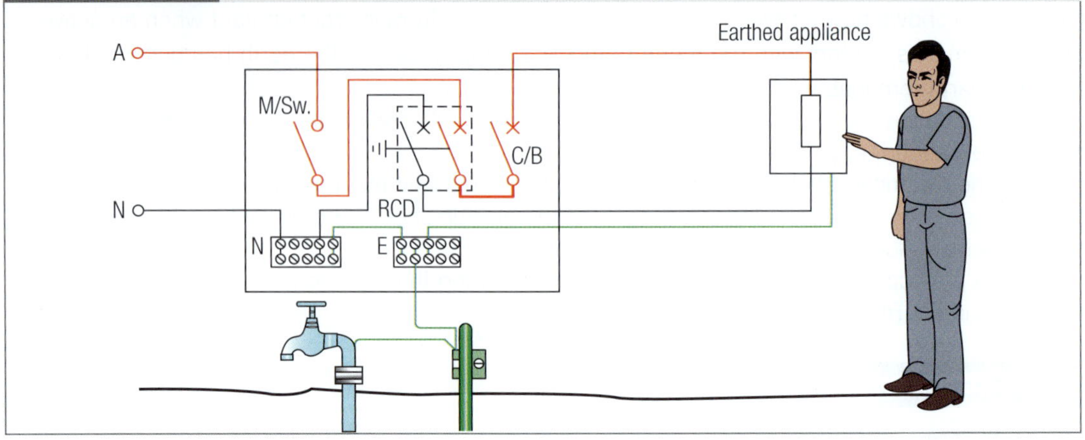

Metal-encased Class II equipment has a substantially continuous metal enclosure, in which double insulation is used throughout, except for those parts where reinforced insulation is used because the application of double insulation is impractical. If the equipment with double insulation or reinforced insulation throughout has an earthing terminal or earthing contact, it is considered to be of Class I construction (clause 1.4.28(a), (b) and (c), Note 2).

Examples of insulation-encased and metal-encased equipment are shown in Figure 10.8.

FIGURE 10.8 Class II double insulation

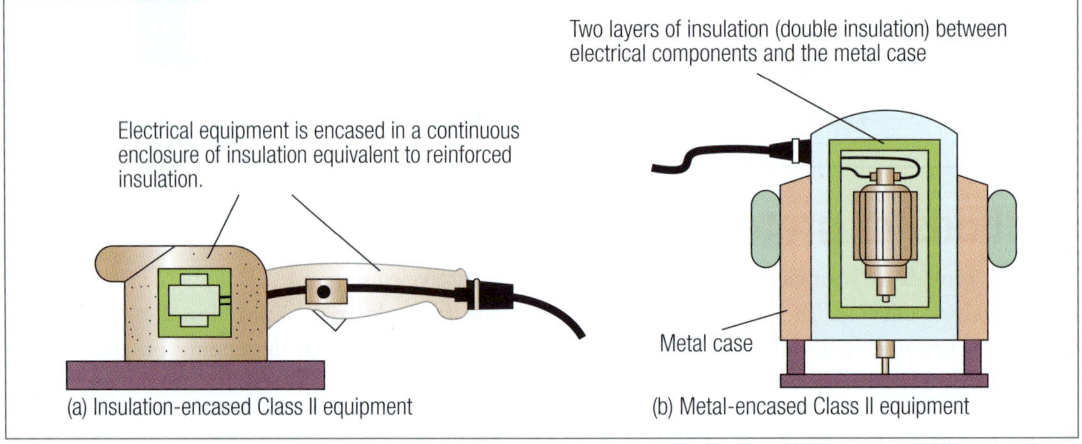

Electrical separation

You can prevent electric shock by isolating both poles of the circuit from the installation protective earth (Figure 10.9). Contact with one pole of the circuit will not cause a lethal current to flow. The transformer must be double wound with windings double insulated, as specified in **AS/NZS 3000:2007** clause 7.4.3.

You can use the supply from a generator or an uninterruptible power supply (UPS) but it must not exceed 500 V. You must separate the output from the input by double insulation or equivalent. All the rules for mechanical protection still apply to separated circuits (clause 1.5.5.5).

Touch current

When someone touches live conductors, a current flows depending upon the level of the voltage and the resistance of the skin surface of the person touching the conductor (see Figure 10.10).

Protection requirements for damp situations

You must check for the presence of moisture, and if it is permanently or intermittently present and likely to affect the safety of the installation extra protection methods are required. Such a situation would be the outer wall of a brick veneer house or similar structure with the weather wall made from hygroscopic (water absorbing) materials (**AS/NZS 3000:2007** clause 1.4.40).

The damp situation poses two threats:

- ingress of water – causing low resistance faults to develop
- exposure – of the damp material to contact with a live part.

You must provide similar protection from exposure as for direct contact, for example, insulated shrouds must be used where the back of a socket outlet is exposed to an external wall.

Any cable running down the inside of an external wall should be provided with a drip loop to prevent transferring moisture to the inside wall lining as well as preventing moisture from gathering on the terminals of the electrical fittings.

AS/NZS 3000:2007 Section 6.1 requires special treatment of the following damp situations:

- baths, showers and other fixed water containers
- swimming pools, paddling pools and spa pools or tubs
- fountains and water features
- saunas
- refrigeration rooms
- sanitisation and general hosing-down operations.

FIGURE 10.9 Electrical separation

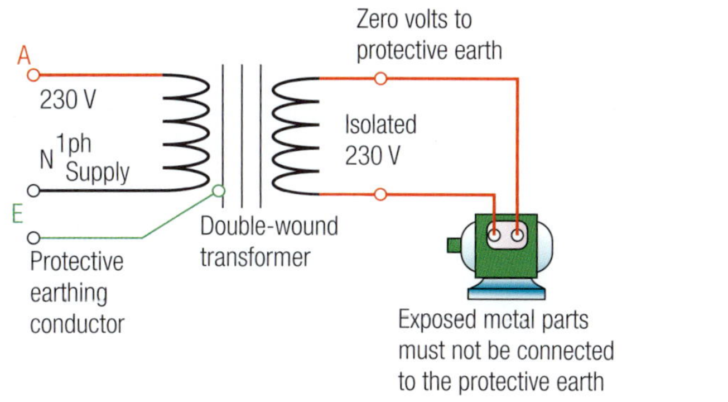

FIGURE 10.10 Touch current

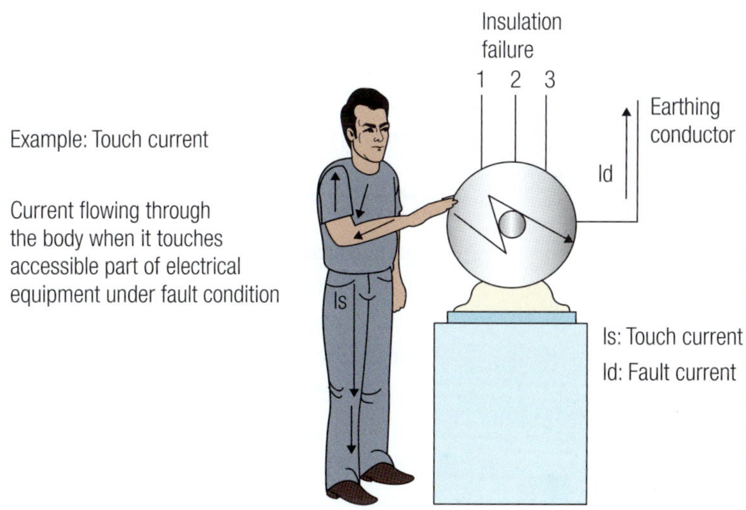

SUMMING UP...

- Indirect contact with live parts of an electrical installation may occur when fault currents flow, causing lethal touch voltages to be present on conductive metal parts.
- The components of the 'automatic disconnection of supply' method of protection against indirect contact comprise the protective device, the conductive path from the supply and the return path through the protective earth.
- Touch voltage is a voltage that is within arm's reach while in contact with earth.
- Touch current flows when a person or animal makes contact with a live part of an installation.
- The current path when a short circuit fault to exposed conductive parts of an appliance occurs is through the exposed metal part and back to the supply through the earth conductor, causing the protective device to open the circuit.
- Protection against indirect contact by the use of Class II equipment uses double insulation instead of earthing.
- Additional protection can be provided by use of residual current devices (RCDs).
- Protection against indirect contact can be achieved by the use of extra low voltage and electrical separation using isolating transformers.
- Damp situations require the same level of protection from live conductors as metal enclosures.
- Specific requirements are set out for particular damp situations.

10.5 Earthing

Earthing is a system of protection used in an electrical installation to protect against the risk of electric shock and fire hazard when a breakdown of insulation occurs.

The general mass of Earth is a good electrical conductor apart from some places of high-resistivity pumice or clay rock. With the neutral solidly connected to earth at the substation and at selected points along the service line, the system of supply has a common earth potential. Any fault to earth lets a large enough current flow to operate a protective device.

AS/NZS 3000:2007 clause 2.3.2.1.1 prohibits the switching of earth conductors. In the past, different systems of earthing have been used throughout Australia and New Zealand, but now the only earthing system that is acceptable is the multiple earthed neutral (MEN).

Earthing terms

The following terms are defined by **AS/NZS 3000:2007**. Refer to the clauses below for full details.

Earthed

Earthed means connected to the supply neutral and the general mass of earth in accordance with the appropriate requirements of this standard (clause 1.4.43). Figure 10.11 shows the connections for a multiple earthed neutral (MEN) earthing system.

Earthed situation

An earthed situation is where there is a reasonable chance of a person touching exposed conductive parts and at, the same time, coming into contact with earth or with any conducting medium which may be in electrical contact with the earth or through which a circuit may be completed to earth (clause 1.4.44).

The following situations are deemed to be earthed situations:

- Within 2.5 m in any direction from a conductive floor (such as earthen, concrete, tile or brickwork flooring), permanently damp surface, metallic conduit or pipe, metallic cable sheath or armour or any other conductive material on which a person may stand.
- External to a building, except for an isolated piece of equipment such as switchgear or a luminaire, which is mounted more than 2.5 m from the ground and from any exposed conductive part or other conductive material, which is in contact with earth.
- Within 2.5 m of the ground, floor or platform in rooms containing socket outlets, the earthing terminals of which are earthed, and where there is a reasonable chance of a person making contact with any exposed conductive part of electrical equipment and any exposed conductive part of an appliance connected to any of the socket outlets.
- All parts of a bathroom, laundry, lavatory, toilet or kitchen.

Equipotential bonding

Equipotential bonding is special electrical connections intended to bring exposed conductive parts or extraneous conductive parts to the same or approximately the same potential, but not intended to carry current in normal service (clause 1.4.52).

Main earthing conductor

This is the conductor connecting the main earthing connection or bar to the earth electrode (clause 1.4.65).

Protective earthing conductor

A protective earthing conductor is a conductor, except a main earthing conductor, connecting any portion of the earthing system to the portion of the electrical installation or electrical equipment required to be earthed, or to any other portion of the earthing system (clause 1.4.79).

FIGURE 10.11 MEN earthing system

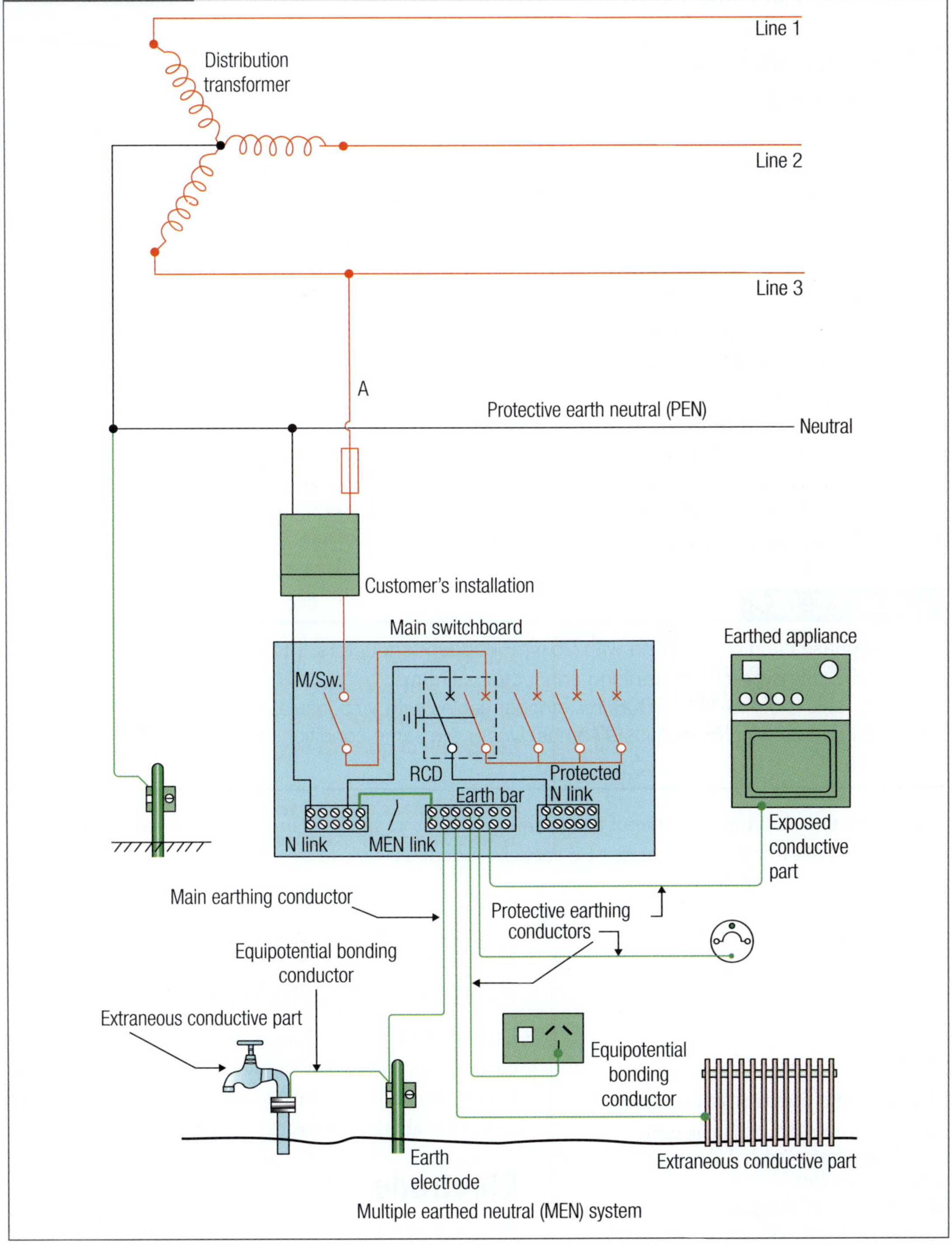

Functional earthing

Functional earthing is an arrangement provided to ensure correct operation of electrical equipment or to permit reliable and proper functioning of electrical installations (clause 1.4.56).

Multiple earthed neutral (MEN) system

The MEN is a system of earthing in which the parts of an electrical installation, required under this Standard to be earthed, are connected to the general mass of earth and, in addition, are connected within the electrical installation to the neutral conductor of the installation (**AS/NZS 3000:2007** clause 1.4.66).

MEN connection

In every electrical installation there shall be a MEN connection (also known as the MEN link) at the main switchboard. The function of the MEN connection is to connect the earthing system within the electrical installation to the supply neutral conductor by means of a connection from the main earthing terminal/connection or bar to the main neutral link (**AS/NZS 3000:2007** clause 5.3.5.1).

Selection of the minimum size earthing conductor

The size of the earthing conductor for any particular electrical installation is selected so that it meets the requirements of **AS/NZS 3000:2007** clause 5.2. In general, the minimum size copper earthing conductor that is required for the protection of any portion of an electrical installation is determined using clause 5.5.1 and Table 5.1 of **AS/NZS 3000:2007**.

In some circumstances, the determination of the minimum size earthing conductor may be different. When selecting an earthing conductor for a particular circuit, the earthing conductor is smaller than the associated active conductor. This means that the current-carrying capacity of the earthing conductor is also less.

EXAMPLE 10.3

A single-phase circuit is supplied with 25 mm^2 copper conductors. From Table 5.1 of AS/NZS 3000:2007, the correct size earthing conductor is 6 mm^2.

From Table 3 of AS/NZS 3008.1.1, copper cables with V75 insulated, installed enclosed in air, the current carrying capacity is 97 amps. Under the same conditions 6 mm^2 is only rated at 38 amps.

The current-carrying capacities of the earthing conductors specified in Table 5.1 of **AS/NZS 3000:2007** are intended to carry the fault current for very short periods while at the same time allowing the circuit protection device to operate.

FIGURE 10.12 Earth electrode

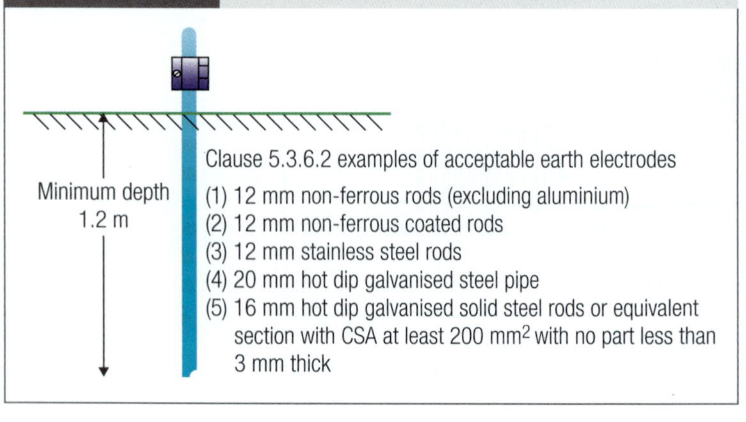

Parts of an earthing system and the purpose of each

Each part of the earthing system plays an important part in the prevention of shock in the case of a fault. Every connection and cable must be carefully checked as the installation proceeds.

Electrode

This is a rod, metal strip or cable embedded in the earth as means of connecting the parts of the electrical installation that are required to be earthed to the general mass of earth (see Figure 10.12). The most common type of earth electrode consists of a copper-sheathed steel rod driven into the ground to a depth of at least 1.2 m (**AS/NZS 3000:2007** clause 5.3.6.3). A number of alternatives are acceptable, such as 3 m of 35 mm^2 bare copper cable or a galvanised strip buried horizontally 0.5 m in the ground. The reinforcing steel of building foundations may also be used, provided they are in the earth (clause 5.3.6.2).

Main earthing conductor

This is the conductor that connects the main earthing terminal at the main switchboard to the earth electrode and so to the general mass of earth (**AS/NZS 3000:2007** clause 5.5.1). You must connect all parts required to be earthed to the main earthing terminal so that they connect to earth via the main earthing conductor and the earth electrode. You may use a soldered connection, a single terminal or an earth bar for the main earth terminal.

The earth bar

This is a brass bar with a number of tunnel terminals where you can connect the earthing conductors to the main earth. An earth bar is provided with most commercially available switchboards and enclosures.

MEN link

The MEN link, also known as the MEN connection, is a conductor (if it is insulated cable it is required to be green and yellow) that that you must use to connect the main earth terminal to the neutral link (**AS/NZS 3000:2007** clause 5.3.5). By doing so, you connect the main neutral and all protective earthing conductors to the supply neutral (see Figure 10.13).

FIGURE 10.13 MEN link

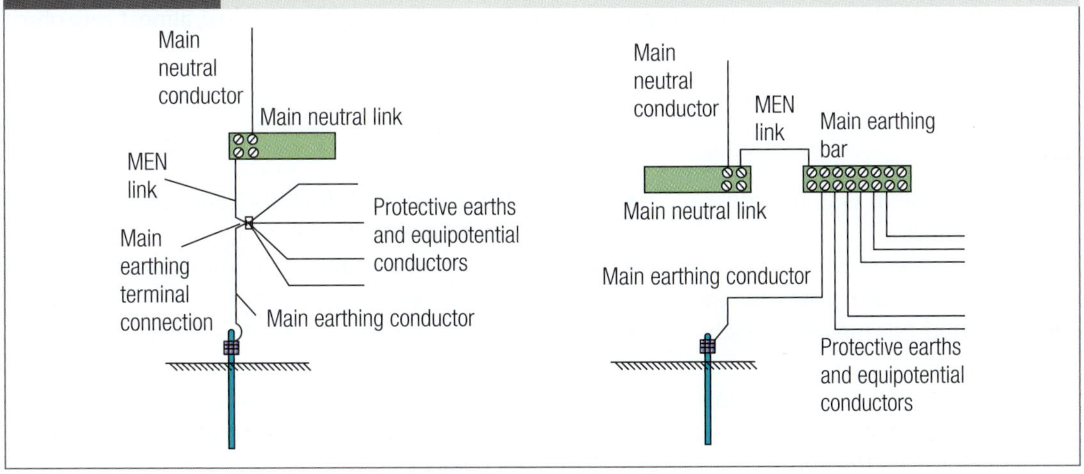

Protective earthing conductor

These conductors are commonly green and yellow insulated cable, which you must provide with each circuit to connect the associated exposed conductive parts to the main earth and subsequently to the main neutral and the general mass of earth. You must provide a protective earth at every socket outlet (**AS/NZS 3000:2007** clause 5.4.2) and at every lighting point (clause 5.4.3). You don't always have to use a cable as the protective earthing conductor; you may use a metallic wiring enclosure such as steel conduit, or metallic cable sheath or armouring (clause 5.3.2.2).

> It is vitally important to maintain the earth continuity of metal enclosures that protect single insulated cables.

Requirements for equipotential bonding

Equipotential bonding ensures all metal not associated with the installation is connected to the earth system to prevent dangerous touch voltages from developing under fault conditions.

Equipotential bonding conductor

This conductor connects conductive parts that are not part of the electrical installation but might be in contact with earth (known as extraneous conductive parts) to the earthing system of the installation.

This eliminates any potential difference that might occur between the extraneous and exposed conductive parts. Examples of extraneous conductive parts are:

- metallic water pipe
- metal pool fence (refer to **AS/NZS 3000:2007 Wiring Rules**, pp. 242–3).

Installation of a MEN earthing system for a single phase installation

The main earthing conductor and the main neutral conductor at each installation are connected together. You do this by connecting the main earthing conductor directly to the main neutral at the neutral bar or with the MEN link or connection. The MEN connection may be a cable or a solid copper bar in the case of an installation with a large main switchboard.

The purpose of the earth connection to the neutral is to stabilise the neutral voltage to zero at the start of the installation and provide the lowest possible impedance to fault currents. This is necessary so that during fault conditions sufficient current will flow to operate the circuit protective device without lethal touch voltages developing.

Under normal circumstances, the earth will not conduct current away from the neutral since the impedance of the neutral is much lower. The earth provides a parallel path for the return current to the substation. However, the impedance to the distribution transformer via the earth system will be generally from 20 ohms to 50 ohms, but via the neutral conductor it will be very low, almost zero ohms (**AS/NZS 3000:2007** clause 5.1.3).

SUMMING UP...

- The purpose of the earthing system is to provide protection against electric shock.
- Earth conductors must not be switched.
- An earthed situation exists if an earth potential and an active conductor are within 2.5 metres of each other.
- Equipotential bonding prevents lethal touch voltages developing on extraneous metal parts.
- The main earthing conductor joins the earth electrode to the earth bar or connection.
- The MEN connection (MEN link) connects the earth connection to the neutral link.
- Earth conductors are smaller than the protected circuit conductors.
- Earth conductors carry current briefly and only under fault conditions.
- The purpose of the MEN system is to provide a stable voltage between neutral conductors and the installation earth.

10.6 Protection against overload and short circuit current

An important part of your work as an electrical practitioner is to ensure that you select the correct equipment to rapidly remove overload and short circuit currents and ensure the safety of users of your installation.

Overload current or fault currents in an electrical installation

An overload is a current exceeding the rated current carrying capacity in a cable or the rated current value of a motor. Fault currents are excessive currents caused by short circuit faults and can cause extensive damage if not interrupted quickly.

Over-currents

Current overloads occur in a circuit that is electrically sound when, for example, the mechanical load on a motor is exceeded or if a power circuit has too many heaters plugged in.

EXAMPLE 10.4

A circuit supplying a number of 10A socket outlets in a domestic installation protected by a 20A fuse or a circuit breaker may have twenty socket outlets connected. Using three or four appliances simultaneously could easily exceed the 20 amperes as the drawing in Figure 10.14 shows.

FIGURE 10.14 Overloaded circuit that is electrically sound

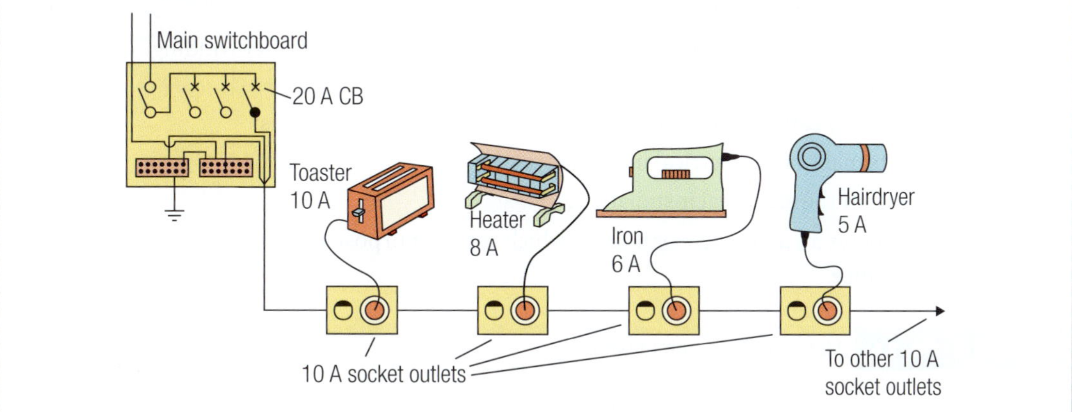

Over-current may occur when equipment is not operating correctly such as a motor with a number of shorted turns in the stator winding. The motor will appear to be working normally but it is drawing higher than its rated current. The damage may not be severe enough to operate the circuit protective device.

Fault current

Fault current results from an insulation failure bridging the two poles of the circuit so that the entire available supply capacity is available to flow into the fault. Insulation failure can occur by various means, for example, the misuse of portable equipment when the cords become damaged to an extent where the circuit conductors become exposed. The result can be a short circuit between earthing conductor and live conductors or a short between live conductors.

A fault current can occur in an earthed appliance, such as a wall oven if the terminals fail and the active conductor makes contact with the earthed frame of the appliance. If the earthing system is effective it will have a resistance low enough to allow the flow of fault current necessary to operate the circuit protective device. This will prevent the build-up of a dangerous touch voltage on the metal parts of the installation.

The aim of the earthing system is to limit the touch voltage in an installation to a value less than required to create a shock hazard. Figure 10.15 shows the path of the fault current. The supply to a circuit must be automatically disconnected where a fault could result in a touch voltage greater than 50 V a.c. or 120 V d.c. (**AS/NZS 3000:2007** clause 1.5.5.3(b)).

FIGURE 10.15 Limiting touch voltage during a fault

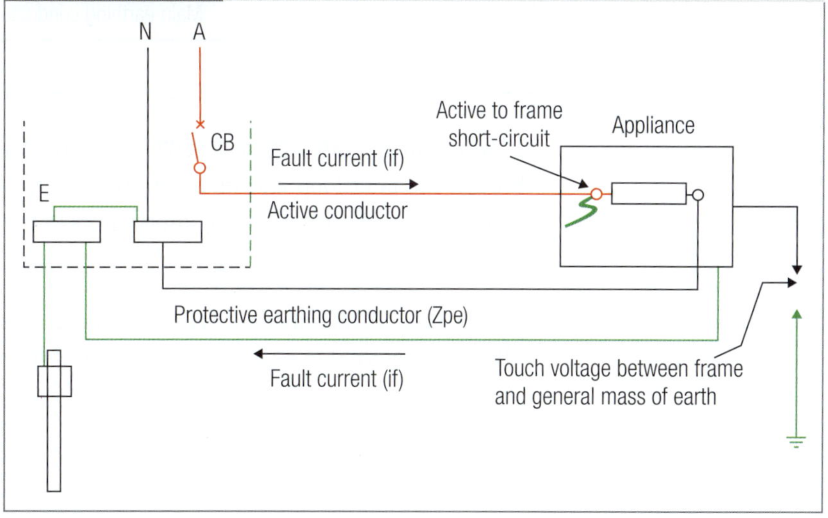

Equivalent circuit of an earth fault loop

An earth fault loop can be reduced to its component parts and treated as a series resistance circuit. Using the equivalent circuit allows the level of fault current to be determined.

Fault loop impedance

The fault loop in an MEN system comprises the following parts, starting and ending at the point of fault (**AS/NZS 3000:2007** Appendix B, clause B4.4):

- the protective earthing conductor (Pe), including the main earthing terminal/connection or bar and MEN link
- the neutral-return path, consisting of the neutral conductor (N), between the main neutral terminal or bar and the neutral point at the transformer (the earth return path, has a relatively high resistance and can be ignored in an MEN system)
- the path through the neutral point of the transformer and the transformer winding
- the active conductor as far as the point of the fault.

The fault loop normally can be looked at as consisting of two sections, that is:

- conductors external to the electrical installation
- conductors within the electrical installation.

Figure 10.16 shows an active to earth fault, which for the purpose of calculations, is deemed to be of negligible resistance.

FIGURE 10.16 MEN fault loop current path

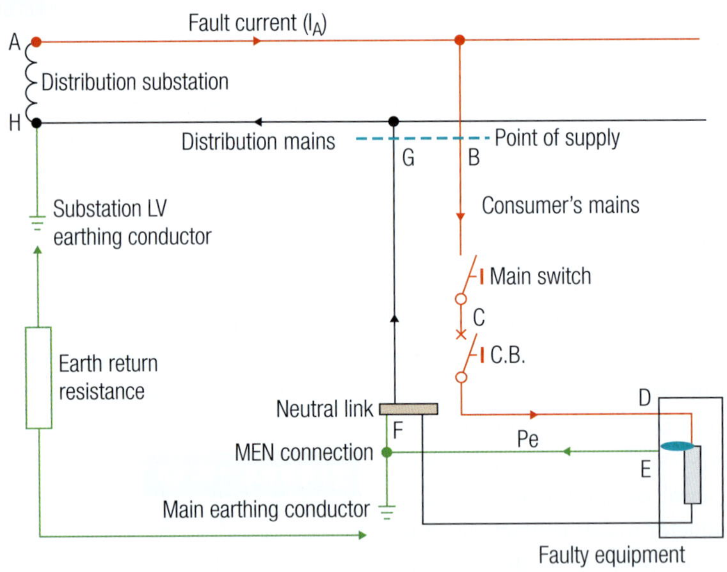

As outlined in **AS/NZS 3000:2007** clauses B4.4 and B4.5, at the instant of the fault, current will flow through the fault loop and its magnitude is only limited by the total system impedance Z_s which is obtained from all the individual impedances in the fault loop as follows:

$$Z_S = Z_{AB} + Z_{BC} + Z_{CD} + Z_{DE} + Z_{EF} + Z_{FA}$$

In Figure 10.16, impedances Z_{AB}, Z_{EF} and Z_{FA} are all external to the electrical installation and may be collectively referred to as Z_{ext}. The remainder, which are internal, may be referred to as Z_{int}. Therefore:

$$Z_S = Z_{ext} + Z_{int}$$

Protection against over-current

The protective device, that is, the circuit breaker or fuse, provides over-current protection. Protection against voltage rise due to an earth fault is not always provided by over-current devices, as this voltage is dependent on the current in the fault loop and the impedance of the circuit.

When earth fault current flows through an over-current protection device, it detects the current as an additional load, and if the current is high enough the protective device operates to open the circuit provided the following conditions are met:

- the resistance of the protective earthing conductor to the main earth is low enough to operate the circuit protective device
- the resistance of the main earthing conductor is no more than 0.5 ohms (represented by the symbol Ω), as specified by **AS/NZS 3000:2007** clause 5.5.1.4.

> The final impedance between the earth connection and the neutral at the supply source through the general mass of earth can vary considerably. High resistivity soil could result in a high resistance (50 ohms is common for a driven electrode).

EXAMPLE 10.5

Figure 10.17 shows the path of the fault current. The fault loop includes the series circuit of the distribution transformer, the supply lines, the active conductor and protective earthing conductor of the faulty circuit. It also includes the parallel path of the general mass of earth, which is relatively high impedance.

FIGURE 10.17 Example of MEN fault loop current path

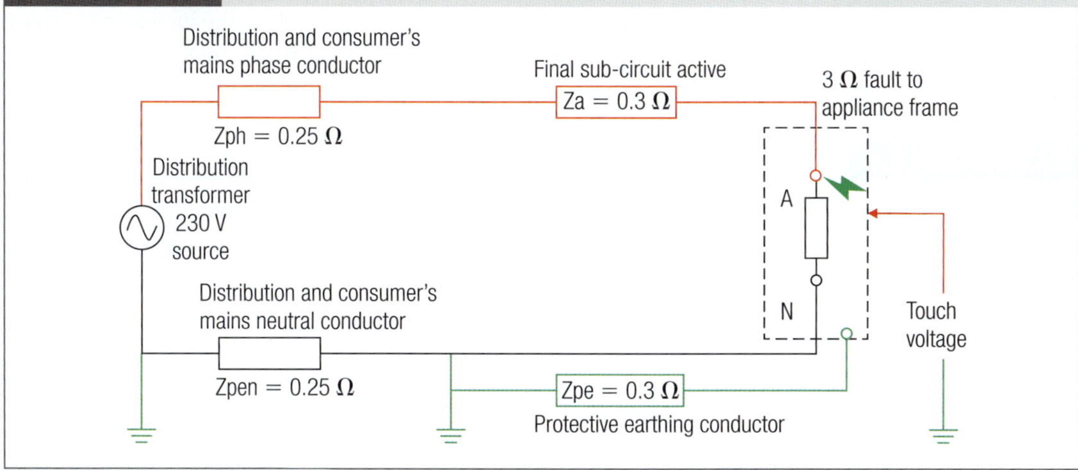

The impedance of the fault path shown is:

Z = Zph + Zpen + Za + Zf + Zpe

  = 0.25 + 0.25 + 0.3 + 3.0 + 0.3 = 4.1 ohms

Hence the magnitude of the fault is:

I = VZ

  = 230 × 4.1

  = 56.1 amperes

Therefore, the touch voltage (i.e. the voltage drop between the exposed conductive part of the appliance and earth) is:

Vt = If × Zpe

  = 56.1 × 0.3

  = 16.83 volts

This value of current is high enough to operate the 20 A protection device, but not necessarily in the prescribed time depending on the type. For example, a type C circuit breaker requires 7.5 times its rated current (i.e. 150 A) to trip in the prescribed time when a fault occurs. The circuit breaker will trip eventually and until it does the touch voltage will not rise above 16.83 volts — this is acceptable protection (**AS/NZS 3000:2007** clause 1.5.5.2(d)).

Level of fault current possible at a given point in an installation from the fault loop

As the circuit extends further from the point of supply the prospective fault current is reduced by the impedance of the circuit conductors. Basic electrical calculations are used to determine the possible fault current at any point.

Prospective fault current

During a short circuit the only limit to the value of current is the impedance of the faulty circuit, and the available short-circuit energy. The circuit impedance includes the impedance of the supply source, usually a substation transformer or group of transformers. If the short circuit occurs close to the source of supply, the only limit to the flow of current is the source impedance, leaving out terminal resistance and assuming a short circuit of zero impedance. A fault of zero impedance is often called a 'bolted fault', referring to tests where the short circuit is created by a bolted terminal. The fault levels at the consumer's terminals of your installation can be provided to you by the electricity supplier.

The impedance of a transformer is usually stated as the percentage of the primary-rated voltage that is necessary to cause full-load current in the secondary if the load terminals are short-circuited. This condition is approaching that of a short circuit between busbars on the low-voltage side of a supply transformer. A common transformer impedance value is 5%. This means the rated current is 5% of the total current that is available. When a secondary short circuit occurs with a normal primary supply voltage it will be 20 times the full load current, that is, full rated current x 100/5.The increasing current demand of modern installations results in an increase in the capability of the supply source to deliver high values of short-circuit current.

EXAMPLE 10.6

A three-phase 400 V, 500 kVA transformer with an impedance of 5% is to supply the 400 V busbars on a factory switchboard. The fault current can be calculated using the following steps:

1 Rated current $= \dfrac{500\,000}{400 \times \sqrt{3}} = 722\,\text{A}$

2 Fault current $= \dfrac{722 \times 100}{5}\,\text{kA} = 14\,433$ amps or 14.3 kA

At any point in the circuit, other than point of supply, the current would be reduced by the impedance of the circuit between the source and the fault. Some impedance is present in the fault itself. The most severe condition would be a three phase fault at the supply terminals, and calculations are usually based on this.

Using impedance and data from the electricity distributor

The fault-current level at a particular point in an installation is generally determined from information obtained from electricity distributors. The fault level is given in MVA.

Determining prospective fault-current levels

EXAMPLE 10.7

An electricity distribution company has stated that the fault level at a 400/230 volt substation transformer is 18 MVA. The substation feeds a main switchboard at the installation with four single X90 double insulated copper 120 mm^2 copper cables. The route length from the substation to the main switchboard is 40 metres having an impedance of 0.068 ohms.

To determine the fault current at the switchboard:

1 The substation fault current (If) $= \dfrac{18 \times 10^6}{400 \times \sqrt{3}} = 25\,981$ amps

2 The impedance (Z) of the transformer $= \dfrac{Vp}{If} = \dfrac{230}{25\,981} = 0.00885$ ohms

3 The total impedance at the switchboard $= 0.00885 + 0.0068 = 0.01565$ ohms

4 The fault current available at the switchboard $= \dfrac{230}{0.01565} = 14\,696$ amps

---

Methods and devices that comply with AS/NZS 3000:2007 Wiring Rules

If the prospective fault current at a switchboard is greater than 11 KA, full arc flash protection should be worn when you are carrying out voltage tests.

Fuses and circuit breakers are used to protect wiring and equipment from current overloads and short circuits. Residual current devices (RCDs) are used for protection against electric shocks and for protection of equipment.

Circuit excess-current protection

The damage to an installation caused by fault current is related to the size of excess current and the time during which it flows. To minimise the damage, protective devices should operate quickly in the event of excess current. Most acceptable protective devices operate within 4 hours at 1.5 times the designed load current of the circuit they protect. This applies to miniature circuit breakers (MCBs) and to most high rupturing capacity (HRC) fuses. Semi-enclosed rewireable fuses do not meet disconnecting time requirements and are no longer acceptable.

Rupture or breaking capacity

The maximum current that a protective device can safely interrupt during fault conditions is called the rupturing or breaking capacity. A short circuit may cause a current from 1000 amps to 50 000 amps to flow in the circuit. Protective devices must therefore be capable of interrupting or breaking this fault current. For example, if a fuse or MCB with a breaking capacity of 1500 amps was subjected to a short-circuit current of 33 000 amps, the device would be incapable of interrupting or breaking the current. This would result in serious damage to the device and the electrical equipment it was intended to protect.

Prospective short-circuit current

This is the highest value of current possible at the point at which the circuit is short-circuited. It is a calculation that is made to get some idea of the size of the current that would flow in a short circuit. It varies according to the capacity of the supply, its voltage and the resistance (d.c.) or impedance (a.c.) of the short circuit. This includes the size and proximity of the transformer, and the size, length and type of mains cable.

Cut-off characteristics

This is a measure of how quickly a protective device can interrupt a short-circuit current as it rises in the circuit. A device having a quick cut-off characteristic, such as a HRC fuse, limits the energy input to the fault, thereby reducing electromechanical and thermal stress in the electrical system to a minimum. AS/NZS 3000:2007 Table 8.1 shows maximum impedances to comply with disconnection time of 0.4 seconds for a range of protective devices.

Discrimination

The disconnection times of protective devices must be chosen so that a fault on a sub-circuit is contained within the faulty circuit. Upstream devices such as sub-main circuit breakers must not trip first and cause wider disruption than necessary. **AS/NZS 3000:2007** clause 2.5.7.2.3 sets out the levels of discrimination required.

SUMMING UP...

- Over-currents are those that exceed the rated current.
- Over-currents can occur in electrically sound circuits.
- Fault current results from a short circuit between two opposite electrical potentials in the installation.
- The fault loop comprises the active and neutral service conductors as well as the circuit active and the protective earth.
- The level of fault current is determined by the supply voltage and the impedance of the service and installation conductors.
- The electrical distributor will supply fault MVA at the point of supply to enable calculation of fault currents.
- Most HRC fuses and circuit breakers can be used as fault current protection.
- Fault current limiters must be able to break the prospective fault current at the point of installation.
- Circuit protection devices must break the circuit before lethal touch voltages have time to develop.
- Protective devices must discriminate to ensure that upstream protection devices do not trip first.

10.7 Devices for automatic disconnection of supply

When you choose a device to disconnect the supply in the event of a fault there are a number of possibilities to consider. Manufacturers provide excellent technical information about their products to enable you to make the correct choice quickly.

Circuit breakers

The primary function of a circuit breaker is to provide over-current or overload protection and short circuit protection. A circuit breaker can be reset with a switch action and cannot be held in against an overload. Where a fuse may require authorised personnel to carry out repairs, any person with access can operate the circuit breaker to restore supply.

Types of circuit breakers

Circuit breakers operate on electromechanical principles, but electronic circuit breakers are also available. Aside from the electronic type there are basically two electromechanical principles – thermal and magnetic or a combination of both.

Thermal type

This circuit breaker works on the principle of heat generated by excess current.

When an over-current flows, a bimetallic strip in the circuit breaker bends and this action operates the tripping mechanism. Figure 10.18 shows the basic operation in the closed operating state and the tripped state when the bimetal strip has bent due to an over-current.

The thermal type circuit breaker does not operate instantaneously. It requires time for the heat to build in the heater before the bimetal bends. This characteristic makes it suitable for loads that have a short-time overload condition, such as motors and welders.

Thermal circuit breakers are not satisfactory for short-circuit protection. It is common practice to have a set of HRC fuses, commonly referred to as fault current limiters, to provide the short-circuit protection.

FIGURE 10.18 Thermal circuit breaker operation

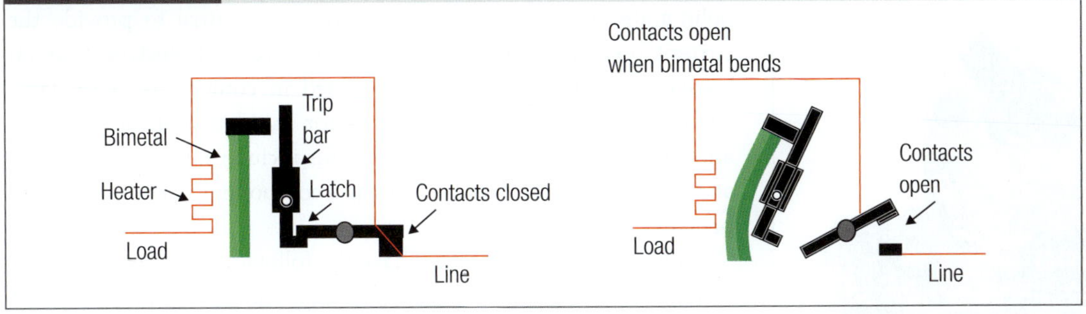

Magnetic types

The magnetic type circuit breaker works on the principle of current flowing through the coil of an electromagnet. The electromagnet is designed so that when an overload occurs, the magnetic field is of sufficient strength to attract an armature, which is attached to a tripping mechanism that opens the line contacts, as shown in Figure 10.19.

FIGURE 10.19 Magnetic circuit breaker components

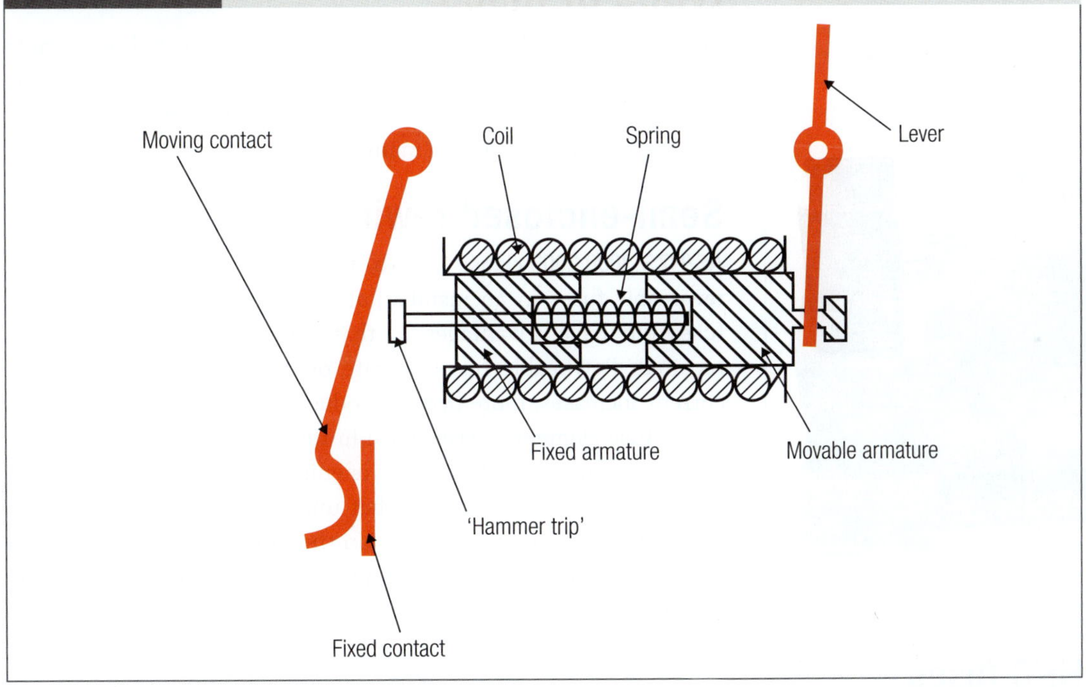

Figure 10.19 is a very basic explanation of the principle of operation. In actual practice, the circuit breaker has a restraining mechanism to prevent the tripping mechanism operating under normal operating conditions or just over the nominal current rating.

Generally a spring holds the tripping mechanism in the closed position and when an overload occurs the force of the spring needs to be overcome before the circuit breaker operates. Due to the instantaneous operation of the electromagnet, this type of circuit breaker is ideally suited to short-circuit protection.

Thermal magnetic type

The thermal magnetic type circuit breaker incorporates a thermal element to provide the time delay characteristic and magnetic action to provide short circuit protection.

Figure 10.20 shows how the two methods of operation are combined.

FIGURE 10.20 Thermal magnetic circuit breaker

Source: Ed Hawkins

FIGURE 10.21 Semi-enclosed rewireable fuse

Source: Ed Hawkins

Solid-state or electronic

Solid-state circuit breakers use electronic circuitry to provide the current breaking functions. Before the advent of electronic circuit breakers, solid-state devices were used in conjunction with large circuit breakers as adjustable relays for the detection of overloads and earth faults. Circuit breaker designs include solid-state overload and earth fault detection facilities as a component part of the circuit breaker.

Solid-state circuit breakers have the following features:

- readily interchangeable plugs to change the current ratings within the limits of the size (e.g. one frame size is suitable for ratings of 75, 100 and 150 amps with the appropriate plugs)
- tripping characteristics that are adjustable within the limits and may be tailored to provide discrimination
- earth fault protection that may be an optional feature.

Operating principles of common types of fuses

Fuses have been the main line of protection against short circuits and overloads for most types of installations. While this is rapidly changing, there are still many fuses in use and the HRC fuse is still a suitable protective device in many cases.

Semi-enclosed rewireable fuse

AS/NZS 3000:2007 clause 1.7.2(h) forbids the use of semi-enclosed rewireable fuses in an installation (see Figure 10.21). Rewireable fuses did have advantages in that they were cheap and easily repaired. Rewireable fuses operate on the principle that, as the current increases, more heat is produced, causing the temperature of the fuse element to rise which in turn will melt if its maximum temperature is reached, and thus opening the circuit.

The main disadvantage is that under short-circuit conditions the unsuppressed arc will maintain the fault current longer than is safe. It is also possible to replace the fuse element with the incorrect size.

HRC fuses

High rupturing fuses (HRC) are cartridge type fuses, which are more accurate and reliable than semi-enclosed fuses (see Figures 10.22 and 10.23). The HRC fuse comprises an element enclosed in an insulating tube filled with powdered quartz, to quench any arc that may develop. The HRC fuse cartridge (Figure 10.24) may be used in a porcelain or plastic carrier (Figure 10.25).

To ensure that a semi-enclosed rewireable fuse is not accidentally inserted, the HRC fuse holder has an extra moulding in the centre.

Operation of HRC fuses

The following sequence of events, shown also in Figure 10.26, occurs when a HRC fuse is subjected to a very high current:

- The silver fuse wire melts and vaporises.
- An arc is produced.

FIGURE 10.22 Semi-enclosed rewireable fuse carrier

Source: Ed Hawkins

FIGURE 10.23 HRC fuse

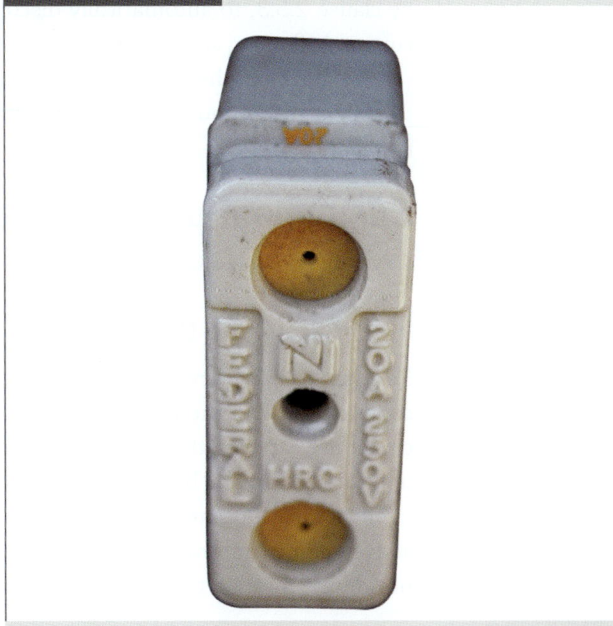

Source: Ed Hawkins

FIGURE 10.24 HRC cartridge

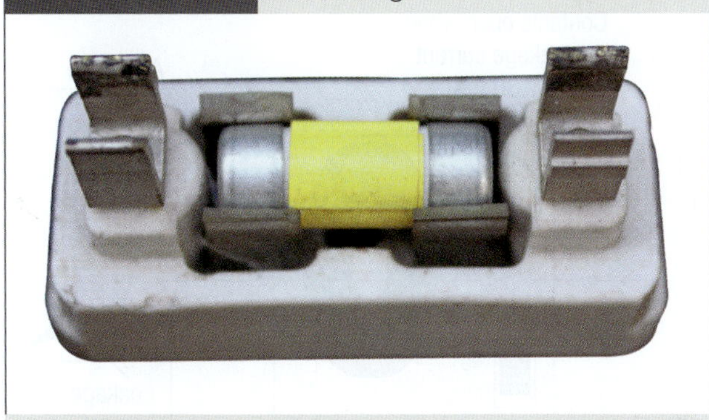

Source: Ed Hawkins

FIGURE 10.25 HRC fuse carrier

Source: Ed Hawkins

- Fusion is created between the silver vapour and quartz powder due to the heat produced.
- A highly insulating silver silicate compound is produced when the arc energy is absorbed.
- The arc is extinguished.
- Current stops flowing.
- The circuit is safely interrupted.

FIGURE 10.26 HRC fuse operation

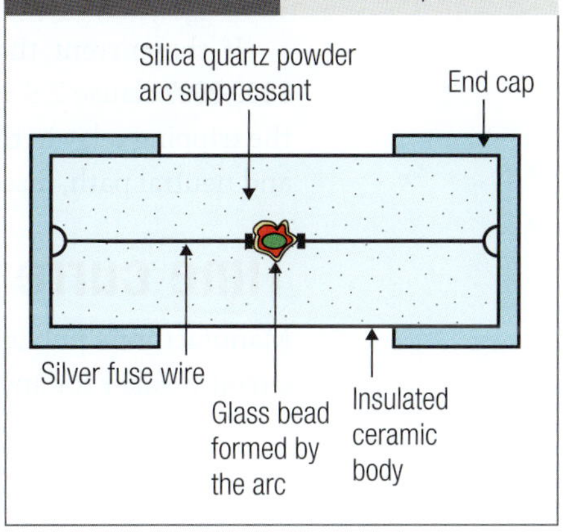

Operating principles of residual current devices

Residual current devices (RCDs) are earth leakage current operated detection devices. They are used to protect people against shocks and are also used to protect wiring and equipment against earth faults.

AS/NZS 3000:2007 clause 2.5.1 states that protective devices shall be installed, where required by clause 2.5.3, to automatically disconnect the active conductors of a circuit when an earth-leakage current is flowing above a predetermined value.

For details in the selection of RCDs, see AS/NZS 3000:2007 clause 2.5.2.

RCDs are compulsory in lighting final sub-circuits and socket outlet final sub-circuits in domestic installations.

Component parts of a residual current device

A typical two pole RCD and its internal component parts is shown in Figure 10.27.

Principle of operation

Residual current devices operate on the principle of an imbalance of current flow between active conductor(s) and the neutral conductor. When the active conductor and the neutral conductor of a circuit are carrying the same value of current, the magnetic field around each conductor is equal and opposite, producing balanced magnetic fields.

If a leakage occurs due to an earth fault, as shown in Figure 10.28, the current flow through the active and neutral conductors is no longer equal since some current flows through the earth and does not return through the neutral conductor.

FIGURE 10.27 Residual current device internal parts

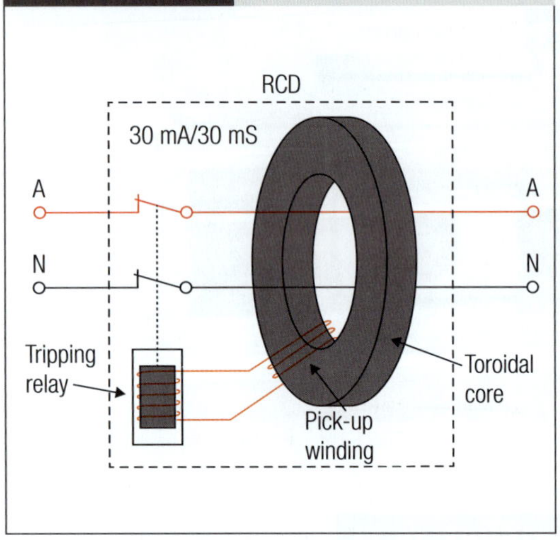

FIGURE 10.28 Residual current device operation

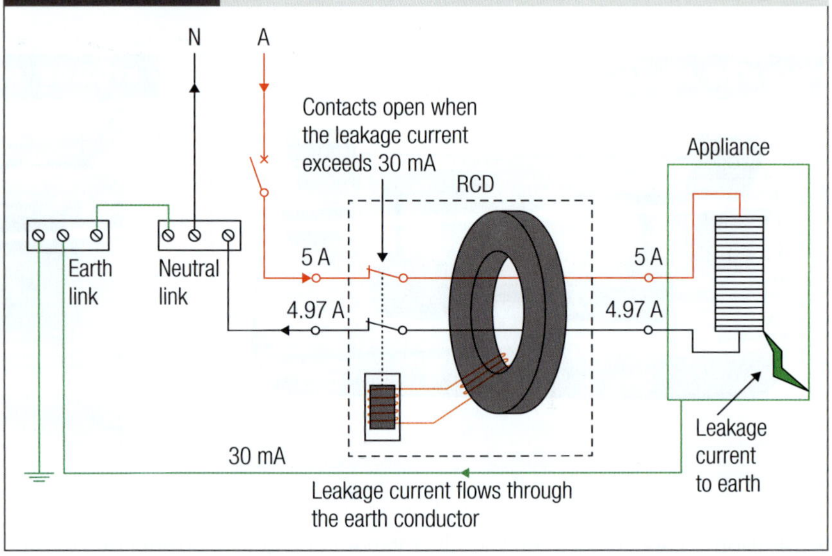

If the current that flows through the earth conductor exceeds 30 mA, as required by AS/NZS 3000:2007 clause 2.6.1, a voltage is induced into the pick-up, or secondary, coil which in turn energises the tripping relay in the RCD. When the tripping relay is energised, contacts in the RCD open the active and neutral path, disconnecting the circuit from the supply.

Time current curves

Manufacturers publish time/current curves for their protective devices allowing selection of a suitable circuit breaker for any given situation.

Tripping characteristics of circuit breakers

The most common type of circuit breaker that is used in circuits up to 125 A is the miniature circuit breaker (MCB). They are categorised according to the following criteria and their characteristics depend on the type:

- number of poles
- protection against external influences
- method of mounting
- values of rated voltage, frequency and current
- values of rated short-circuit capacity.

Range of instantaneous tripping current

Miniature circuit breakers are categorised as 'types' according to their tripping ranges as follows.

Type B

The tripping range of a type B circuit breaker is 3 to 5 times the rated current, that is, the mean is 4 times the rated current.

Type B miniature circuit breakers are only used in situations where a low mean tripping current is required, for example, where the rated temperature limit in a cable would be exceeded with a higher mean tripping current. See Figure 10.29.

FIGURE 10.29 Tripping characteristics of a type B circuit breaker

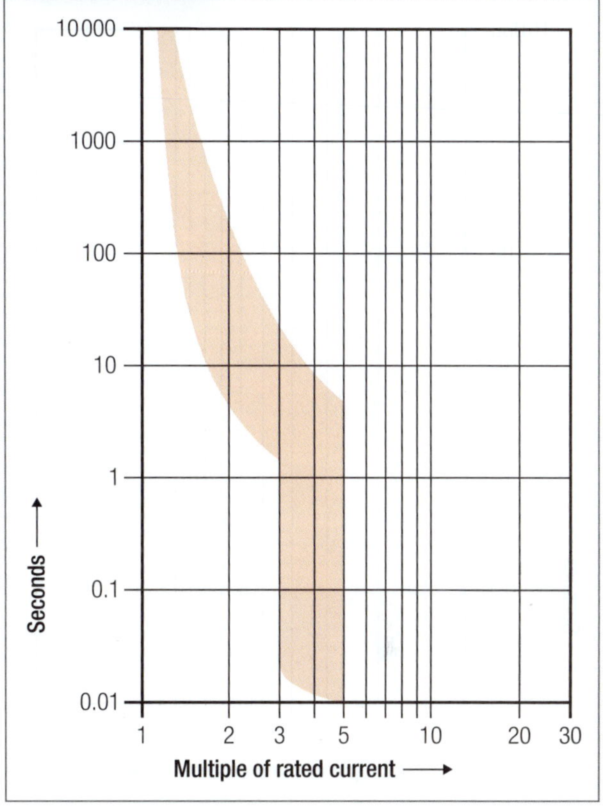

Type C

The tripping range of a type C circuit breaker is 5 to 10 times the rated current, that is, the mean is 7.5 times the rated current. This is the most commonly used miniature circuit breaker, and is used for general over current protection of sub-mains and final sub-circuits. See Figure 10.30.

The tripping of a type D circuit breaker is 10 to 50 times the rated current. With the type D circuit breaker, the mean tripping current needs to be specified for the particular application. Type D miniature circuit breakers are used in situations where transient over-current would cause nuisance tripping, such as circuits supplying motors or welders. See Figure 10.31.

FIGURE 10.30 Tripping characteristics of a type C circuit breaker

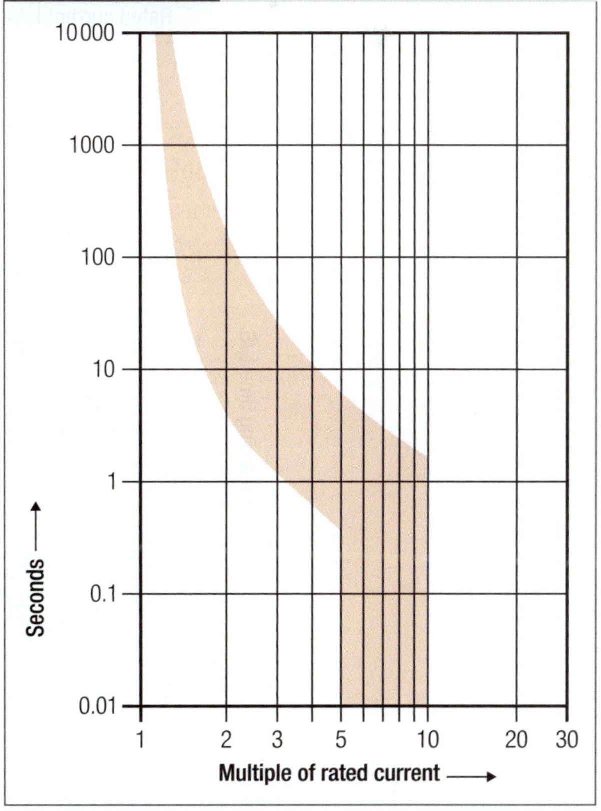

EXAMPLE 10.8

Determine the trip time for a type C 20 A circuit breaker when there is an over-current of 60 amps:

1 Over-current multiple drawn in the circuit $= \frac{Ia}{In} = \frac{60}{20} = 3$

Where Ia = over-current and In = rated current

2 On the characteristic curve at the 3 times factor, project a line vertically until it crosses the curve. Note that because the curve is wide, rather than just a line, this means that the operating time is approximately between 1.5 and 40 seconds.

FIGURE 10.31 Tripping characteristics of a type D circuit breaker

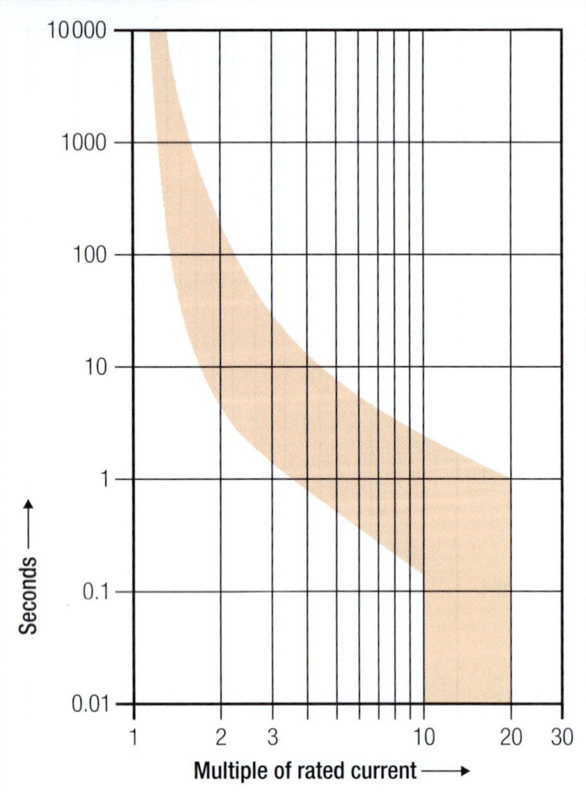

Fusing characteristics HRC fuses

The maximum current that a fuse can safely interrupt is determined by the inverse time characteristic. This means that the higher the current flowing, the shorter the disconnection time.

HRC fuses comply with the **AS/NZS 3000:2007** disconnection time requirements due to their special construction to reduce arcing time (see Section 10.7 earlier in the chapter).

The graphs in Figure 10.32 show the disconnection times of a range of HRC fuses. When the available fault current is high compared to the rating of the fuse, the fuse link operates in a very short time – less than half a cycle – preventing current from reaching the peak value.

Figure 10.33 shows the difference between the fault current available and the 'Let through' value determined by the fuse characteristic.

Tripping characteristics of RCDs

RCDs are classified according to their rated residual current at which tripping operation is ensured. This is also known as the sensitivity of the RCD. There is also a specific maximum time in which the RCD must operate.

The RCD must also be able to distinguish between true earth leakage current and other disturbances which may cause nuisance tripping.

FIGURE 10.32 Current/time characteristics of HRC fuses

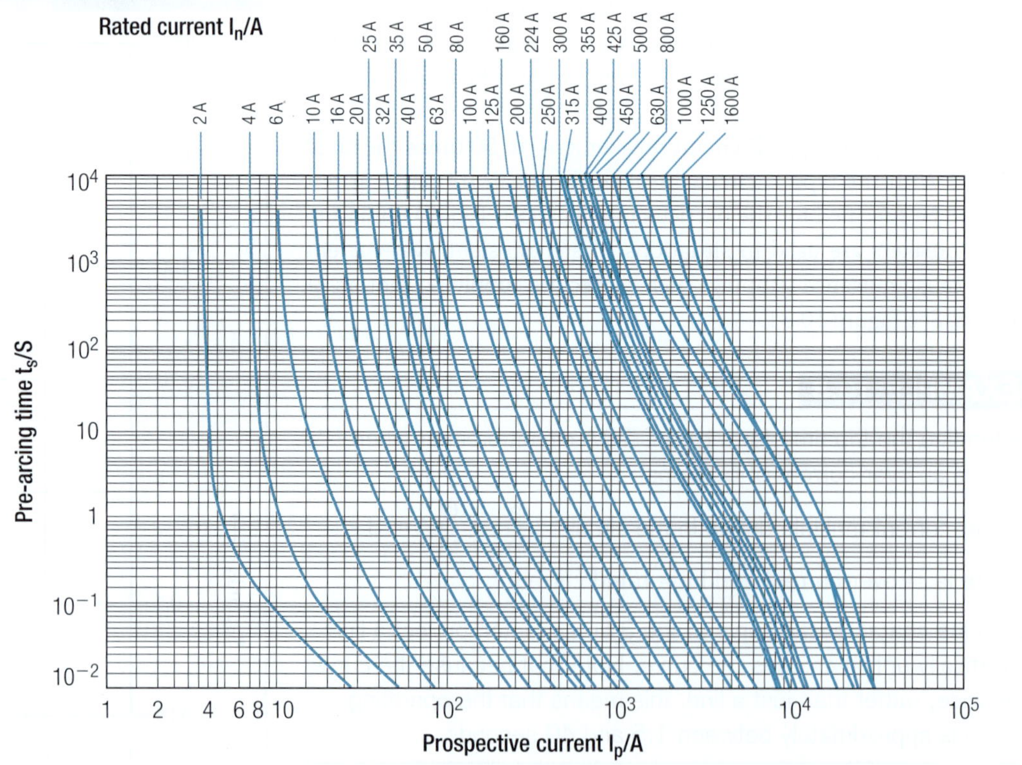

Source: ABB Inc., *Fuse Gear*, p. 19, http://www.abb.com/search.aspx?abbcontext=products&q=hrc%20fuse%20links%2C%20gg

RCDs are classified into four different types:

- type I – rated residual current ≤ 10 mA
- type II – rated residual current ≤ 30 mA
- type III – rated current ≤ 300 mA
- type IV – rated current >30 mA ≤ 2000 mA.

Type IV RCDs are not suitable for protection against electric shock but are used in certain industries to prevent arcing.

Portable RCD are divided into two classes:

- Class L
- Class H.

The graph in Figure 10.34 shows the operation times and leakage current levels for each type of RCD compared to the IEC 60479-1 Standard curves for protection of life.

Curve C1 shows that when a current greater than 30 mA passes through a human being from one hand to their feet, the person concerned is likely to be killed, unless the current is interrupted in a relatively short time. The point 500 mS/100 mA close to the curve C1 corresponds to a probability of heart fibrillation of the order of 0.14%.

The type I RCD operates at 5 to 10 mA within 40 mS in zone 2 of the graph; this zone has no harmful effect. The type II RCD operates within 300 mS at 30 mA leakage current and within 40 mS at 150 mA leakage current. In this zone, this operating characteristic is well below the level to cause ventricular fibrillation.

FIGURE 10.33 Fault current limiting

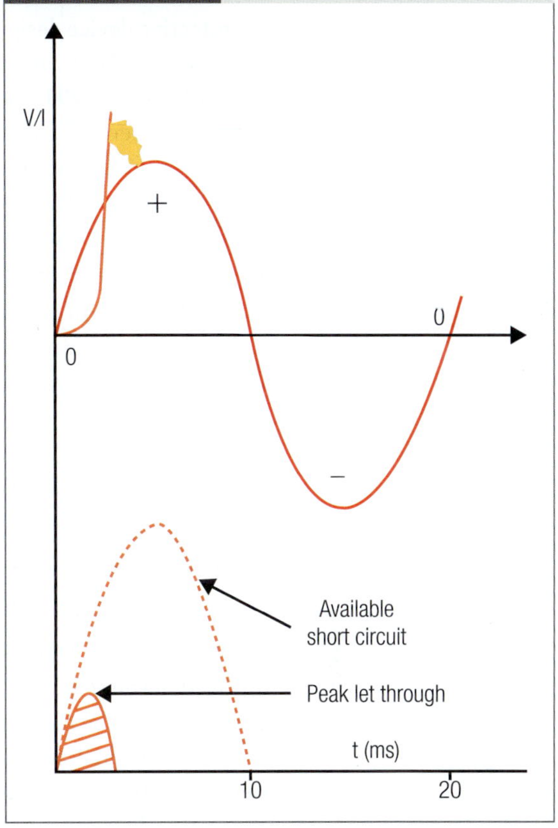

FIGURE 10.34 Response characteristics of RCDs

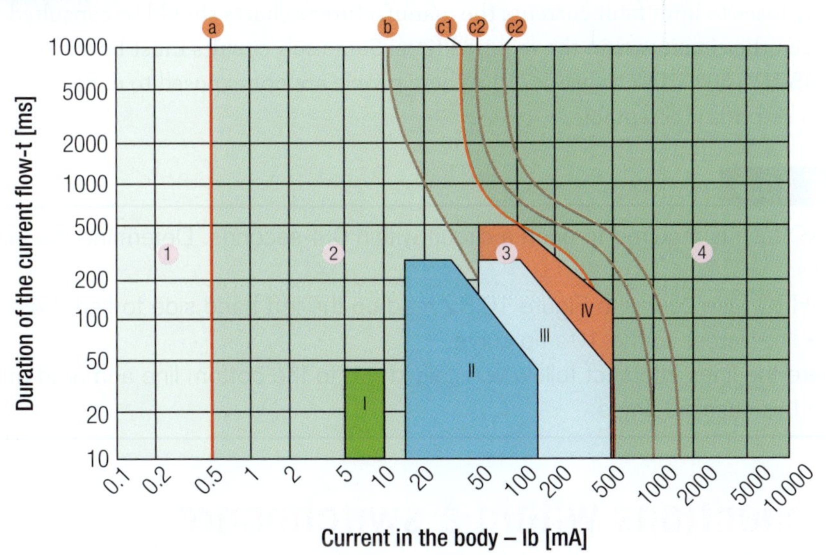

Source: Adapted from ABB Inc., *Protection against earth faults with Residual Current Devices, 2/2*, p. 19, http://www05.abb.com/global/scot/scot209.nsf/veritydisplay/c4e584f06cc6c4fbc1257ad800496193/$file/2CSC420004B0201_RCDs%20EN.pdf

Fault current limiter

A protective device designed to or selected to open the circuit and limit the instantaneous fault current is known as a fault current limiter.

AS/NZS 3000:2007 clause 2.5.1.1 sets out the requirements for current limit protection at the commencement of the consumer's mains. This may be provided by the distributor's service fuse. If devices provided for overload protection are not capable of opening the prospective fault current, it must be supplemented by a fault current limiter installed at another location (clause 2.5.4.3).

Clause 2.5.4.5 specifies the conditions that must be met by current limiters:

- Must be able to open the prospective fault current at the point of installation.
- All short-circuit currents anywhere in the installation must be interrupted before cable temperatures exceed their permissible limit.
- The nominal current of the current limiter can be greater than the current carrying capacity of the cable.

Maximum earth fault loop impedance

The maximum fault loop impedances are set out in AS/NZS 3000:2007 Table 8.1 and the following values must be considered when determining the allowable fault loop:

- protective device rating
- protective device type.

To determine whether or not a given circuit complies, Table 8.2 can be used, or if the power is available it can be measured using an earth fault loop tester.

For individual sub-circuits the conductor size, protective device rating and type of device will determine the acceptable combined active and earth conductor resistance, as well as the maximum resistance allowed for the protective earth conductor.

Current limiter fuse selection

When selecting fuses to limit fault currents the manufacturer's charts should be consulted. The chosen fuse should open the circuit within the required time. Some sub-circuits must be disconnected within 0.4 seconds (AS/NZS 3000:2007 clause 5.7.2). Where people are not exposed to danger a disconnection time of up to 5 seconds is acceptable.

EXAMPLE 10.9

A 10 amp HRC fuse is required to open a circuit within 0.4 seconds. Determine the fault current limit as follows:

1 From the HRC fuse curves in Figure 10.32 read up the left hand side to find 0.4 seconds.
2 Follow the line across to the 10 amp curve.
3 From where the lines intersect follow vertically down to the bottom line and read off the fault current, in this case 50 amps.

RCD connections within a switchboard

When connecting RCDs in a switchboard, the lighting and power circuits should each have their own earth leakage protection to prevent the total loss of all systems in the event of an earth fault occurring on any single plugged-in appliance.

The line and load connections must be connected on the correct side of the RCD. If the supply active and neutral are connected to opposite sides of the RCD, the RCD will remain in a continuously tripped state.

Combination RCD/MCBs help to save space. The use of these devices saves the need for a protected neutral link for each RCD. Figure 10.35 shows a typical connection to protect an installation using two-pole RCDs.

FIGURE 10.35 Switchboard connection for socket outlet and light circuits with RCD

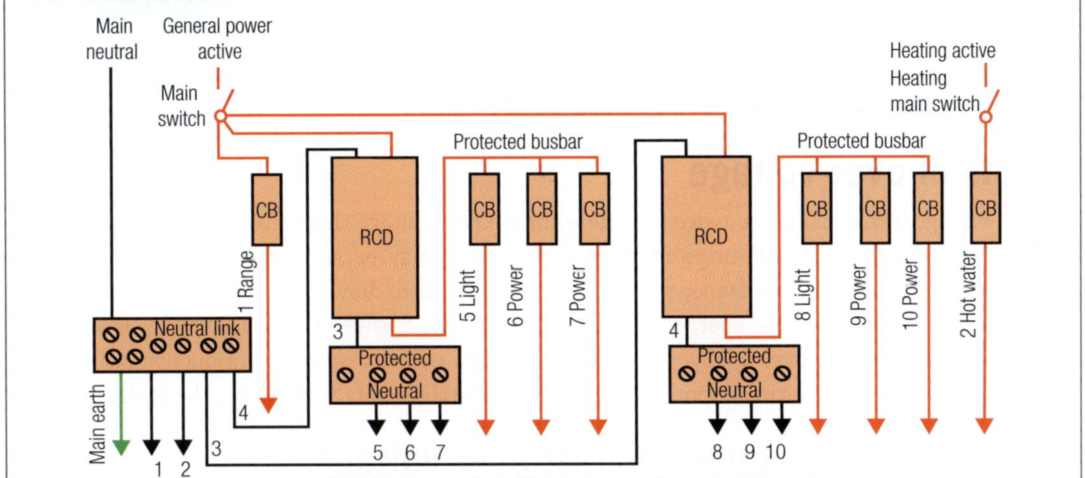

SUMMING UP...

- The thermal component of a thermal/magnetic circuit breaker protects against over-current while the magnetic component protects against short-circuit faults.
- Semi-enclosed rewireable fuses are not allowed in any new installation.
- HRC fuses can disconnect fault currents quickly.
- RCDs detect the difference between inflowing current and outflowing current and open the circuit if the difference is too high.
- Time/current curves can be used to select suitable fuses, circuit breakers or RCDs for given circuit conditions.
- The impedance of a fault loop is the sum of the following impedances in series:
  - source of supply
  - service line active and neutral
  - installation active conductors
  - protective earth conductor.
- The maximum impedance of an earth fault loop must be low enough to ensure operating of a protection device.
- Certain types of HRC fuse can be used as fault current limit protection.

10.8 Protection against over-voltage and under-voltage

You will need to pay strict attention to the voltage levels in your installation as large variations in voltage can cause damage to equipment or cause machinery to operate in an erratic and unsafe manner.

> The incorrect selection of protective devices can result in nuisance tripping. This will cause expense to the user through lost production and unnecessary maintenance calls.

Over-voltage

Often referred to as a voltage surge, voltage transient or voltage spike, over-voltage is a condition in an installation where a higher than the nominal operating voltage is inadvertently injected into an installation, thereby having adverse effects on the wiring and equipment.

Causes of over-voltage

Over-voltage may be generated externally by:

- lightning strikes – either directly or adjacent to the supply network lines produce large magnetic fields that induce momentary increases in supply line voltage
- sudden change in loads – in a.c. circuits the switching in and out of power factor correction capacitors can cause voltage transients

- operating switches that control inductive loads – such as motors, fluorescent light ballasts and welders, cause momentary over-voltage transients to be induced back into the supply lines
- when a fuse or circuit breaker operates – under fault conditions
- modern electronic equipment – capable of injecting voltage pulses into the system if the equipment's own filter is faulty or inadequate (an unsuppressed rectifier).

Effects of over-voltage

An over-voltage transient can be twice the line voltage. Over-voltage that is 10% above normal supply voltage can reduce the life of tungsten filament lamps. It may exceed the voltage rating of some equipment such as older TVs and video players and cause motors to draw more current and overheat. The over-voltage will register on the energy meter as increased use of power if the over-voltage is sustained.

Large, short duration over-voltages of some thousands of volts caused by inductive switching or lightning strikes may damage insulation, particularly in mineral insulated, metal sheathed (MIMS) cables.

Methods for protection against over voltage

AS/NZS 3000:2007 Section 2.7 and Appendix F provide requirements and information for the installation of surge protection devices (SPDs). Figures 10.36 and 10.37 show the method of connection for over-voltage and surge protection devices in a circuit.

FIGURE 10.36 Location of over-voltage protection devices

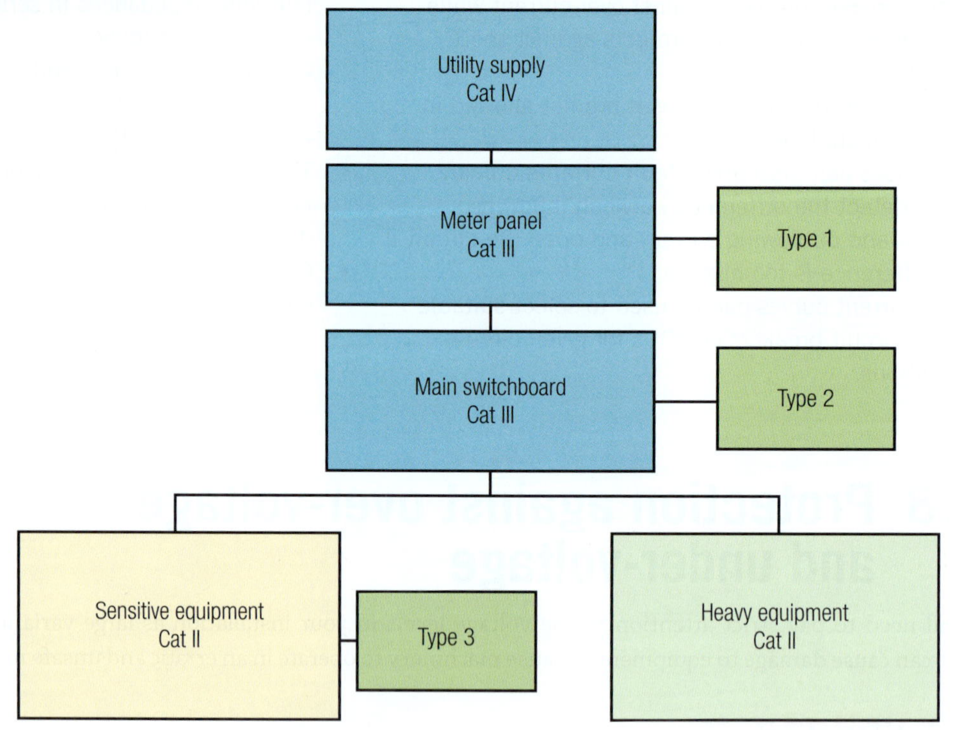

Classification of protection equipment

Surge protection devices are classified into types according to discharge capacity:

- Type 1:
  - Tested with a 10/350 μs waveform (Class I test), which simulates the current produced by a direct lightning strike.
  - Has the ability to discharge very high currents to earth, providing a high voltage protection (U_p) level.

  - Designed for use in incoming power supply panels where the risk of lightning strike is high, for example, in buildings with an external protection system.
  - Must be accompanied by downstream Type 2 protection.
- Type 2:
  - Tested with an 8/20 μs waveform (Class II test), which simulates the current produced in the event of a switching or lightning strike on the distribution line or its vicinity.
  - Has ability to discharge high currents to earth, providing a medium voltage protection (U_p) level.
  - Designed for use in distribution panels located downstream of Type 1 protection or in incoming power supply panels in areas with low exposure to lightning strikes.
- Type 3:
  - Tested with a combined 1.2/50 μs – 8/20 μs waveform (Class III test), which simulates the current and voltage that can reach the equipment to be protected.
  - Has ability to discharge medium currents to earth, providing a low voltage protection (U_p) level.
  - Always installed downstream of a Type 2 protection.
  - Designed to protect sensitive equipment or equipment located more than 20 m downstream of the Type 2 device.

FIGURE 10.37 Connection of surge protection devices

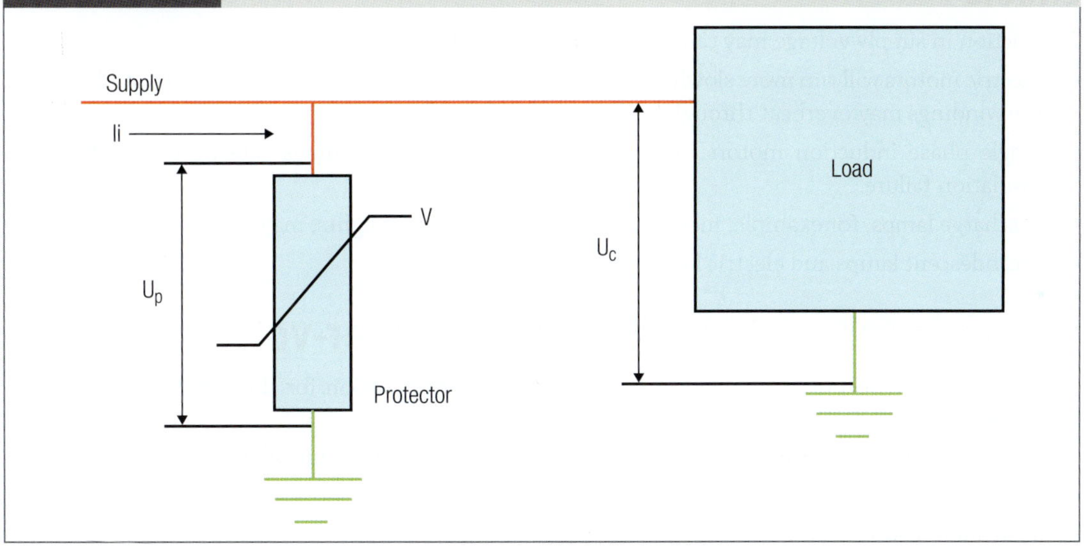

Protection parameters

The following circuit parameters are used to design a surge protection system (also see Figure 10.38):

- U_p – maximum residual voltage between the terminals of the protection device during the application of a peak current.
- I_n – peak current in 8/20 μs waveform means the protection device can withstand 20 times without reaching end of life.
- I_{max} – peak current with 8/20 μs waveform which the protection device can withstand without reaching end of life.
- U_c – maximum effective voltage that can be applied permanently to the terminals of the protection device.
- I_{imp} – peak current with 10/350 μs waveform which the protection device can withstand without reaching end of life.

FIGURE 10.38 DIN rail mounted surge protectors

Source: Ed Hawkins

For selecting an over-voltage surge protection device three main electrical parameters have to be evaluated:

- continuous operating voltage – U_c
- voltage protection level – U_p
- energy capability.

Under-voltage

Under-voltage is a condition that exists in an installation when the supply voltage falls below the nominal value, that is, the value required to make equipment operate correctly and safely.

Causes

Under-voltage has a variety of causes. Some of the causes are:

- the run of circuit conductors is long and there is high resistance in the conductors
- large loads such as electric motors being switched on
- the loss of power (no-volt condition).

Effects

A reduction in supply voltage may cause the following effects:

- Electric motors will run more slowly or stall, even though the load remains unchanged, and as result the windings may overheat through lack of circulating air.
- Single phase induction motors may cut back to their start windings causing overheating and insulation failure.
- Discharge lamps, for example, mercury vapour and fluorescent lights, may flicker or go out.
- Incandescent lamps and electric heating will be less effective.

Methods for protection against under-voltage

AS/NZS 3000:2007 Section 2.8 provides requirements and information for the installation of under-voltage protection.

Under-voltage protection may be provided by various means. These include:

- moving-iron attraction relays
- solid state relays
- no-volt protection.

Moving-iron attraction relays

A relay is an electromagnetic device that uses the principle of magnetism to attract an iron armature towards an electromagnet when energised and in doing so a set of contacts close (or open) to operate another circuit. They use small currents to operate a contactor (a larger current electromagnetic switching device) to directly control a motor circuit. Figure 10.39 shows an auxiliary relay.

The relay trips off when the voltage falls below the level required to provide enough holding current through the electromagnet. Standard electromagnet starters have relays that use an electromagnet to pull in the contacts of the starter. Under-voltage protection and no-volt protection is built in to the design.

Solid-state (electronic) relays

Solid-state relays perform the same function as electromagnetic relays, controlling a small current, which in turn controls a larger current to operate a load such as a motor. The principle of operation relies on electronic circuitry.

FIGURE 10.39 Small auxiliary relay

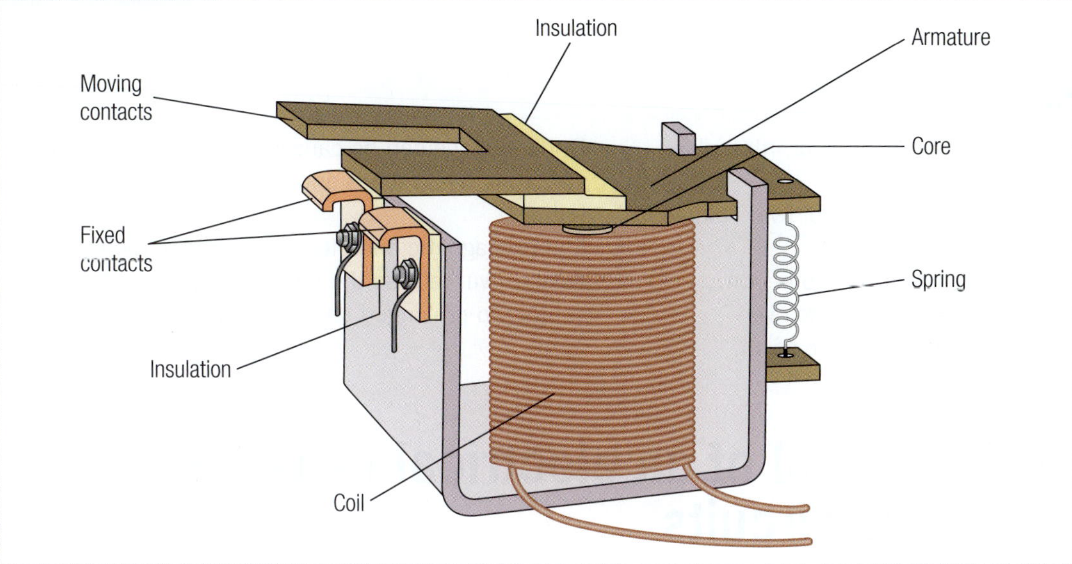

The solid-state relay switches off when the gate voltage (voltage value required to make them function correctly) falls below a predetermined value. The appropriate gate voltage must be supplied to the relay to switch it on again (see Figure 10.40).

FIGURE 10.40 Voltage and phase monitoring relay

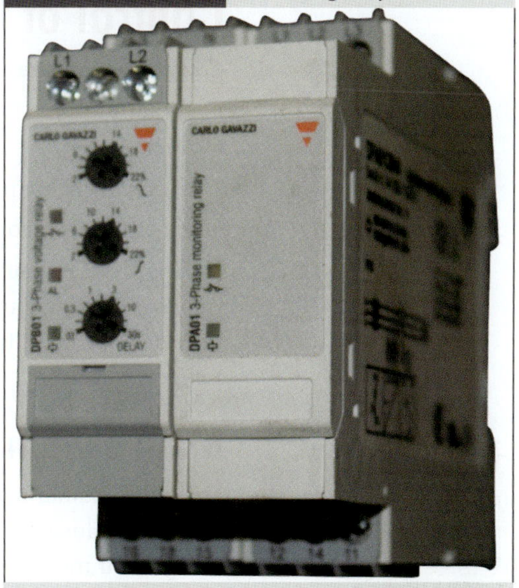

Source: Ed Hawkins

Under-voltage protection for motors

Every electric motor should have protection against under-voltage where the loss and subsequent restoration of voltage or a drop in voltage could cause danger to persons or property.

No-volt protection

No-volt protection is achieved when the voltage to a relay coil falls below that required to release it, and the relay de-energises and interrupts the supply to the equipment.

Figure 10.41 shows a typical arrangement for a no-volt release circuit. The contact **St** closes momentarily on start-up and the relay **Rl** energises and closes the circuit, holding the contact **Rl_1** closed. The circuit is maintained until the voltage is removed or falls below the hold-in voltage of the relay.

FIGURE 10.41 No-volt release relay circuit

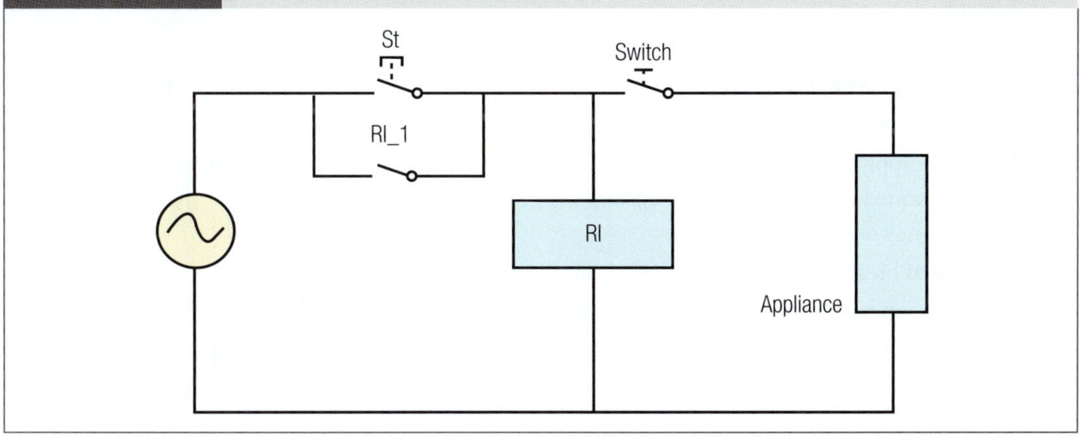

Once the relay is de-energised the equipment cannot be re-energised without manually resetting the no-volt release relay.

SUMMING UP...

- Over-voltage can be caused by external or internal incidents.
- Lightning strikes are a common cause of over-voltage in some locations.
- There are three levels of over-voltage protection.
- Surge diverters conduct when exposed to over-voltage, dissipating the excess energy to earth.
- Under-voltage can be caused by long service lines, switching heavy loads and a high voltage phase failure.
- Electromagnetic and electronic voltage monitoring devices will release the circuit if equipment is subject to prolonged under-voltage.

10.9 Control of an electrical installation and circuits

You must provide any electrical installation that you install with devices to prevent or remove hazards associated with the installation (**AS/NZS 3000:2007** Section 2.3).

Control of the installation

Control devices are classified according to their functions:

- isolation
- emergency
- mechanical maintenance
- function (control) (clause 2.3.2.2.1).

The devices you install for isolation must effectively isolate all active supply conductors. Any isolation devices must use mechanical contacts, so you cannot use semi-conductor devices. You must provide clear identification for all isolation devices. It must not be possible to mistakenly believe that a circuit is isolated.

Main switch

i Readily accessible means capable of being reached quickly and without climbing over or removing obstructions, etc. and in any case not more than 2.0 m above the ground, floor or platform (refer to AS/NZS 3000:2007 clause 1.4.2).

You must install a main switch on the main switchboard of every electrical installation (**AS/NZS 3000:2007** clause 2.3.3). Where multiple supplies are provided each supply must be controlled by a main switch. You must ensure the parts of the installation that supply a safety service have a main switch or switches separate from the main switches for the rest of the installation. All main switches must be readily accessible and clearly identified as 'MAIN SWITCH'.

Where an alternative supply is provided an isolating switch must be provided at the source of supply or at the switchboard.

Emergency switching

Under **AS/NZS 3000:2007** clause 2.3.5 you must provide emergency switching where it may be necessary to control the supply to remove an unexpected danger. In the case of electric motors the supply may need to be sustained to operate braking systems.

The emergency device must:

- be capable of breaking the full load of the circuit
- be manually operated directly interrupting the main circuit
- be provided with means of operation capable of latching in the 'OFF' or 'STOP' position

- not re-energise the relevant part of the circuit upon release
- only able to be reset manually where danger is likely to occur. An emergency stop switch is shown in Figure 10.42.

Appliances and accessories

Under AS/NZS 3000:2007 clause 2.3.7 you must provide appliances and accessories with devices for isolation and switching as follows:

- Socket outlets – must be provided with a switch with a current rating at its opening voltage not less than the current rating of the socket it controls.
- Cooking appliance – shall be provided with a switch acting in all active conductors mounted near the appliance.
- Water heaters – shall be provided with an independent isolating switch additional to any control incorporated in the heater, installed adjacent to the heater or on the switchboard where the circuit starts.
- Heating cables in floor or ceiling – cables or groups of cables that comprise the heating system must be provided with a functional switch installed in a readily accessible position.
- Motors – must be provided with a switch capable of starting and stopping the motor and isolating the motor for mechanical maintenance.

Your installation must include provision to enable isolation of electrical equipment and to prevent electrical equipment from being inadvertently energised. You must do this in such a way that a deliberate action is needed to energise the circuit.

Acceptable methods are:

- provision for the fitting of a padlock
- warning tags or notices
- location within a lockable space or enclosure.

If there is a possibility that an enclosure may contain more than one supply, a notice to this effect must be placed on the outside of the enclosure (see Figure 10.43). If an interlocking switch is fitted a sign will not be necessary.

You must provide suitable isolation devices to isolate each active conductor of an a.c. circuit (clause 2.3.2.2). You must not under any circumstances insert a switch or circuit breaker in the neutral conductor of a consumer's mains or where the neutral is used as combined protective earthing conductor and neutral. The active and neutral conductors may be switched together as in the case of an RCD where the active and neutral contacts are capable of making and breaking the load (clause 2.3.2.1.1(b)).

The control circuit of a fire pump motor must be directly connected between the active and neutral of the pump circuit with the active conductor connected directly to the coil of the operating device within the starter (clause 7.2.9.4). In this case a switch is required in the neutral conductor of the control circuit.

FIGURE 10.42 Emergency stop switch

Source: Dreamstime/Michael Beer

FIGURE 10.43 Dual power supply warning

SUMMING UP...

- All electrical circuits must have the capacity to be isolated.
- All installations must have a main switch located on the main switchboard.
- Safety services circuits must have their own separate main switch.
- Switching the earth conductor is prohibited.
- Switching the neutral conductor is only allowed if the active conductor is switched at the same time on a double pole switch.
- Socket outlets and permanently wired appliances must have an isolating switch.
- Alternative supplies must have an isolating switch either at the main switchboard or at the source of the supply.
- Emergency switching devices must not cause a hazard when reset for normal operation.

Installing control and protection equipment

10.10 Switchboards and distribution boards

You will be required to install a main switchboard in any electrical installation for the primary control and protective devices including the main switch.

Switchboards

The main switchboard in a small installation will be the starting point for all the sub-circuits forming part of the installation. For a larger installation you will need to locate sub-distribution boards closer to the loads to reduce the length of sub-circuit cables and associated voltage drop problems. If you are installing mechanical services such as air-conditioning and heating, you may need to provide a dedicated switchboard. All switchboards must be constructed in a way that provides basic protection.

In situations where the removal of covers exposes live parts you must identify the covers with a suitable label.

Construction suitability

The mechanical strength of the switchboard must be suitable to withstand the stresses both mechanical and thermal that are likely to be encountered during its service (clause 2.9.3.2).

FIGURE 10.44 Circuit breaker alignment

Source: Ed Hawkins

Where the switchboard is enclosed in a case or surround, you should pass any wiring systems entering the switchboard enclosure through openings that provide a close fit to restrict the spread of fire (clause 2.9.7). You must rigidly fix all bare conductors. Live parts of a switchboard must be separated by distance or a barrier that will prevent arcing between the conductors of opposite polarity and between live parts and earth. Where two or more circuit breakers are mounted in the same row, the operating mechanism of each must be aligned so that they all are 'OFF' or 'ON' in the same direction as shown in Figure 10.44.

Distributor equipment

The electrical distributor will provide certain equipment for the metering and connection of supply but there are some items that you must provide on behalf of your customer.

CT metering

The electrical distributor requires your customer to provide adequate mounting and installation facilities for metering equipment. Your customer is responsible for ensuring access is maintained. You must install meters in a location that is readily accessible for fixing, reading and testing. Electrical distributors usually require that you make provision for one location per property for metering equipment. You will usually need to install current transformers when a three phase load on any tariff exceeds 100 amps.

The distributor will provide:

- a current transformer test block
- current transformers
- potential fuses (including 10 amp fuse links)
- CT and potential cables.

You must mount all transformers and potential fuses on the load side of the main circuit breaker. If you install busbars, the section passing through the window of the transformer must be easily removable. You must not make any joins within the current transformer window (Aurora Energy Tasmania, *Services and Installation Rules*, p. 75, http://www.auroraenergy.com.au/electricity-network/electrical-contractors /service-installation-rules-manual). Figure 10.45 shows a current transformer.

You must keep the cable length between CT and metering equipment to 15 metres or shorter. You usually will use 4 mm² conductor or larger for metering (Aurora Energy Tasmania, *Services and Installation Rules*, p. 112). Figure 10.46 shows mounting CT in panel.

FIGURE 10.45 Current transformer

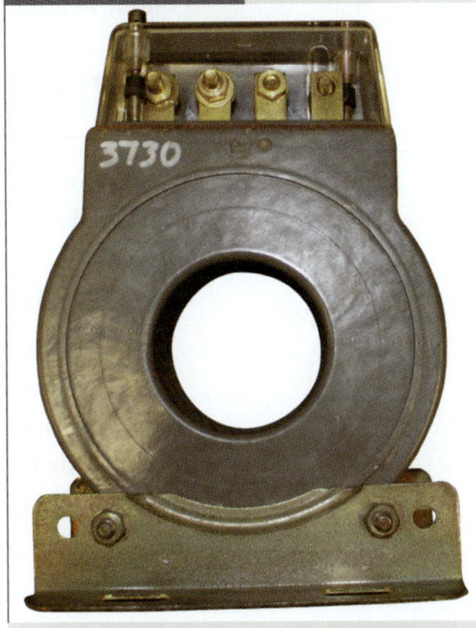

Source: Ed Hawkins

Equipment location and compliance

You must pay strict attention that switchboards and their associated equipment are installed in locations that allow easy access for isolation and escape in emergencies. You must identify all equipment by labelling.

> It is important that all equipment is correctly installed and prepared before applying for the supply connection as repeat visits by the distributor's workforce will be expensive and time-wasting.

Position of main switches and overload protective devices

You must install a device to provide protection of a conductor against overload at the origin of every circuit and at each point where there is a reduction in the current carrying capacity of the conductors (clause 2.5.1.2).

Identification

You must clearly mark the relationship of switches, circuit breaker fuses, RCDs and other similar equipment to enable ready identification of the circuits and equipment they supply (AS/NZS 3000:2007 clause 2.9.5).

Bars and links must be identified to show if they are earth or neutral, protected or unprotected, etc. This may not be needed if the colours of the conductors clearly identify the function of the link. All terminals must be marked to show the circuit with which they

FIGURE 10.46 Mounting CT in a panel

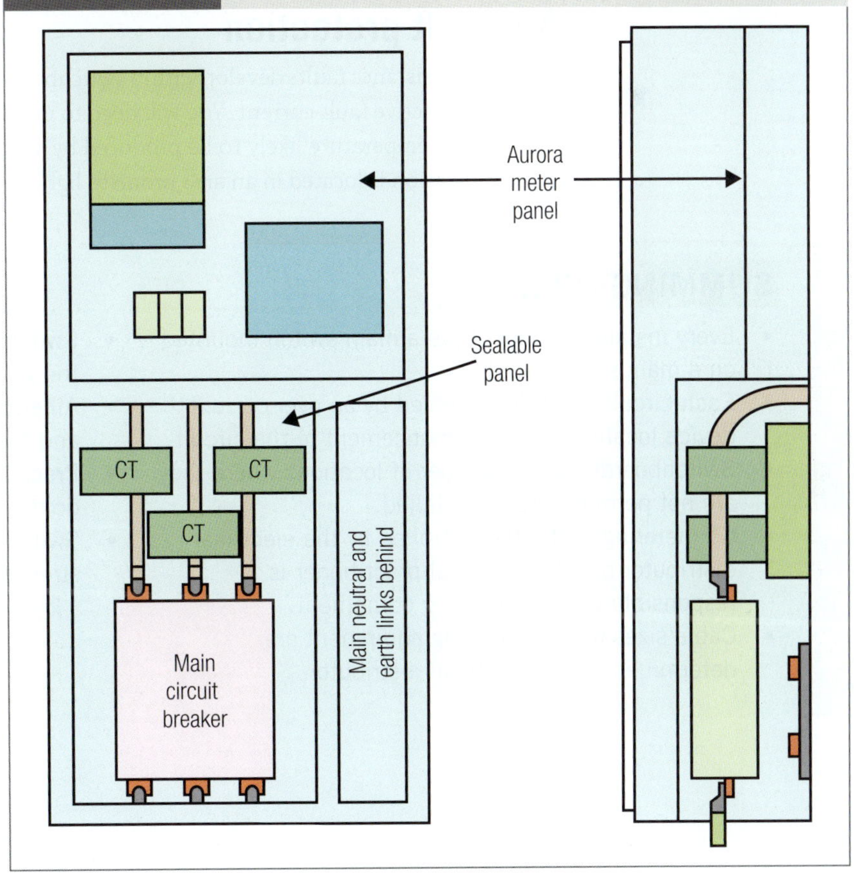

are associated. A lockable isolating switch is located under the meter panel. Ideally, this should totally remove all supply voltage from the installation.

Compliance requirements

You will find details of prohibited locations, height restrictions and suitable equipment in the distributor's service and installation rules. If they do not specify any special extra requirement then you will still need to comply with **AS/NZS 3000:2007 Wiring Rules**.

Location and access

You must not mount a switchboard or metering equipment any lower than 1.2 metres above a floor or platform (refer to **AS/NZS 3000:2007** clause 2.9.2.5). You can vary this if the access to live parts is only possible to authorised persons and access is locked against accidental entry. You do not need to apply this to domestic switchboards.

Switchboards may not be installed in certain areas such as:

- above water containers or fixed cooking appliances
- in cupboards unless set aside for the purpose
- near baths and showers
- near swimming pools, spas or saunas
- within a refrigeration room
- sanitisation area unless provided with IPX6 protection
- within a fire isolated stairway
- near fire-hose reels
- near automatic sprinklers
- within hazardous areas unless specifically designed for the area.

Arc fault protection

When low resistance faults develop within switchboards they may generate arc faults that are only limited by the prospective fault current. You will need to use materials and equipment that is rated to withstand the highest temperature likely to be produced by such a fault (**AS/NZS 3000:2007** clause 4.2.2.4). When your installation is located in an area prone to lightning strikes a surge diverter should be considered.

SUMMING UP...

- Every installation must have a main switch mounted on a main switchboard.
- Each circuit must be protected by an over-current device located at the commencement of the circuit.
- Switchboards have a number of locations where they are not permitted to be installed.
- CT metering is strictly controlled by the electrical distributor but the electrical practitioner is responsible for the mounting equipment.
- Cable sizes for the metering equipment are determined by the electrical distributor.
- Switchboards must be no lower than 1.2 m above the ground floor or platform.
- All equipment must be labelled to show its function and the circuits with which it is associated.
- Precautions must be taken to ensure that arc faults do not cause the spread of fire from the switchboard.
- Switchboards must be constructed to withstand the stresses of the environment where they are located.

REVIEW EXERCISES

Answer these questions and check your answers at the back of the book.

Review questions 10.1

1 Complete the list by writing a method of protection against each of the dangers listed below.

Danger or damage	Protection method
Direct contact	
Indirect contact	
Thermal effects	
Unwanted voltages	
Over-current	
Fault currents	
Overload	
Injury from mechanical movement	

2 List four requirements for the installation, design and selection of equipment.

Review questions 10.2

1 If an installation is provided to supply only one load such as a farm pump, does it have to have a main switch?

2 Are the metal parts of a SELV circuit allowed to be connected to the installation earth system?

Review questions 10.3

1 List three factors that determine the amount of heat generated within a conductor.

2 List three types of loads that are intended to produce heat.

Review questions 10.4

1 State the clause in AS/NZS 3000:2007 Wiring Rules that defines 'touch current'.

2 What is required in regards to any exposed conductive parts when a separated circuit supplies a single item of electrical equipment?

Review questions 10.5

1 What do the Wiring Rules regard as 'Extraneous conductive parts'?

2 When must other metallic piping systems associated with fire sprinklers, gas, hot water or flammable liquid have an equipotential bond?

Review questions 10.6

1 Briefly describe the cause of an overload current.

2 List the parts that comprise the complete fault loop impedance in the MEN system.

Review questions 10.7

1 List the devices that are used to protect wiring and equipment from current overloads and short circuits.

2 State two protective functions of circuit breakers.

Review questions 10.8

1 Name three other terms that are used to refer to over-voltage.

2 Describe two effects of under-voltage.

Review questions 10.9

1 When is a main switch required?

2 List four appliances that must have a functional switch located close to them.

Review questions 10.10

1 State the reason for the inclusion of a sub-distribution board within an installation.

2 If a neutral link has five black insulated conductors and a green/yellow MEN link connected to it, does it require a written label to identify it?

ONLINE RESOURCES

Circuit breakers: ABB
http://www.abb.com/search.aspx?abbcontext=products&q=minature%20circuit%20breaker%20index

Circuit protection: Bussmann by Eaton
http://www.cooperbussmann.com

Extra low voltage control of lighting systems: CentraLite
http://www.centralite.com > Products > Elegance

HRC fuses: ABB
http://www.abb.com/search.aspx?abbcontext=products&q=hrc%20fuse%20links%2C%20gg

Standards Australia, with link to frequently asked questions
http://www.wiringrules.standards.org.au/

11

ELECTRICAL INSTALLATIONS – CABLE SELECTION AND COORDINATION

Electricians need to be able to make design decisions as to the suitability of electrical cables and equipment in an installation. The selected cabling system must be suitable and cost-effective for the environment in which it is installed. An electrical installation must work as intended and be reliable. If a fault occurs, it must be possible to disconnect the fault without disrupting the rest of the installation. Provision must also be made to allow for safe maintenance and testing. This chapter outlines the selection and application of a range of commonly used cabling systems and protection devices.

Note: Parts of this chapter are adapted from the *UEE07 Electrotechnology Training Package. UEENEEGOO7B Select and Arrange Equipment for General Electrical Installations. Learner Workbook*, published by Industry Skills Unit Meadowbank, Tafe NSW, 2009, and referred to throughout the chapter as the Tafe NSW, *Learner Workbook*.

Chapter outline

11.1 Performance requirements – design and safety

AS/NZS 3000:2007 Wiring Rules Section 1.5 outlines what is required of an installation to comply with the standard. Effectively, the standard identifies risk sources associated with an electrical installation and provides measures to control them. Persons, livestock and property must be protected against dangers and damage that may arise during normal use. It is the responsibility of the installing electrician to ensure their work complies with the standard. **AS/NZS 3000:2007** clause 1.5.1 states that the three major risks associated with an electrical installation are:

- shock current
- excessive temperatures
- explosive atmospheres.

Factors to consider about the electrical supply

It is important that the electrical equipment and wiring systems installed are compatible with the characteristics of the local supply. Most installations are supplied by large electrical distributors. However, it is not uncommon for private supply sources to be in use. Larger factories may have their own steam turbines as a source of supply. In remote areas, solar, wind and small internal combustion generators are common.

Overseas, countries such as the United States use a 110 V 60 Hz system; American equipment will therefore not be compatible with Australia's 230 V 50 Hz systems. **AS/NZS 3000:2007** clause 1.6.2 lists nine characteristics of typical Australian and New Zealand installations, which include:

1 Nature of current, a.c. or d.c. – generally the supply is a.c.

2 Nature and number of conductors, that is number of phases, neutral and protective earthing conductors. The number of supply phases will depend on the maximum demand and load types. Small installations up to 100 amps will be connected only to one phase and a neutral. If the load exceeds 100 amps or a three phase motor is installed, three phases and a neutral will be connected. This is called a four-wire three phase supply, shown in Figure 11.1.

   In some areas a three phase supply is not always available. Many rural distributors use a three-wire two phase supply as shown in Figure 11.1. Heavy loads are wired across A and B phases but are connected to 460 volts single phase, not 400 volts. A three phase 400 V motor would not be compatible with the two phase 460 volt supply

3 Voltage and voltage tolerances. The nominal supply voltage is 230 volts for single phase supply and 400 volts for three phase supply (not 240/415 volts). All calculations are performed at 230/400 V. It is important to ensure the supply voltage is within tolerance. If the voltage is either too high or too low damage to electrical wiring and equipment will occur. Clause 1.6.2 of **AS/NZS 3000:2007** specifies a tolerance of +10% to −6% of the nominal voltage in Australia and +6% to −6% of the nominal voltage in New Zealand.

FIGURE 11.1 Number of phases

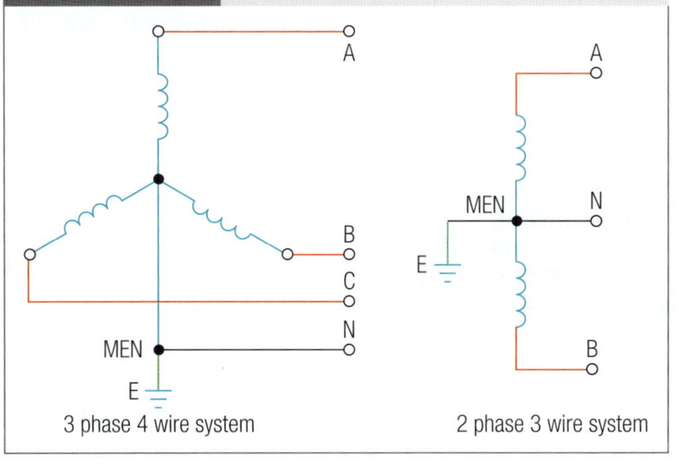

EXAMPLE 11.1

Determine the maximum and minimum voltages to comply with permissible voltage tolerances for an Australian single phase supply:

$V_{max} = 230 + (230 \times 10/100)$
$= 246$ V

$V_{min} = 230 - (230 \times 6/100)$
$= 216.2$ V

EXAMPLE 11.2

Determine the maximum and minimum voltages to comply with permissible voltage tolerances for an Australian three phase supply:

$$V_{max} = 400 + (400 \times 10/100)$$
$$= 440\ V$$
$$V_{min} = 400 - (400 \times 6/100)$$
$$= 376\ V$$

This voltage tolerance of a supply should not be confused with the voltage drop along the length of a cable route, which is discussed later in this chapter.

4 Frequency and frequency tolerances – generally 50 Hz for Australia and New Zealand. Large aircraft and ships use a higher onboard frequency of 400 Hz.
5 Maximum current that can be supplied – if an installation's maximum demand is greater than that of the supply available it is normal for the supply distributor to ask for a financial contribution to any upgrades in infrastructure.
6 Prospective short-circuit current – the maximum possible current that could flow under short-circuit conditions. Protection devices at or near the point of supply must be capable of interrupting this current without damage.
7 Protective measures inherent in the supply – the MEN system is used in Australia and New Zealand. Many countries in Europe and Asia do not ground the neutral conductor.
8 Limits on the use of equipment – devices such as motors can draw large currents when starting; this in turn will cause disruptions such as low voltage and distortion to the supply current.
9 Harmonic current or other limitations. Modern electronic equipment such as computers, inverters, electronic ballasts and variable speed drives cause harmonics. Harmonics can cause transformers and neutral conductors to overheat (Tafe NSW, *Learner Workbook*, pp. 24–6).

Maximum demand in consumer's mains and sub-mains

Maximum demand is the highest current expected to flow in a conductor at any given time. It is not necessarily the total of the connected load on the conductor. 'Diversity' is the term used to describe the difference between the actual full connected load and the current assigned for a circuit's maximum demand. Take as an example a power circuit with eight by 10 amp outlets. The maximum expected current to be drawn by this circuit will not be 80 amperes. It is very unlikely that all eight socket outlets will operate at the 10 amps maximum current all of the time, as some will be lightly loaded or have no load.

Methods of determining maximum demand

The same logic can also be applied to consumer's and sub-mains. It is extremely rare for all of the load to be energised and drawing current at the same time because loads cycle on and off over time. Installing a consumer's or sub-main cable rated for a capacity based on the full connected load in most cases would be overkill. Maximum demand of installations will be covered in detail later in this chapter.

AS/NZS 3000:2007 clause 2.2.2 lists four methods that may be used to determine the maximum demand of a cable:

- *Calculation* – most commonly used to determine maximum demand of consumer's and sub-mains in smaller installations up to 400 A. Tables C1 to C3 are used depending on the type of installation to make an estimation of the maximum demand based on previous similar installations.
- *Assessment* – used in large and complex installations.
- *Measurement* – the current is measured over a 15-minute period when current is expected to be at its highest using a maximum demand recorder or indicator such as that shown in Figure 2.21 in Chapter 2. The black pointer rises and falls with the current just like any other ammeter. The red pointer only rises and does not fall when current decreases, therefore recording the highest value of current reached over the test period. The black knob to the bottom right of the meter is used to reset the red pointer.

- *Limitation* – most commonly used to determine maximum demand of final sub-circuits. The maximum current that can flow is restricted by the current rating of a circuit protection device (fuse or circuit breaker). This is not recommended for consumer's or sub-mains as operation of the protective device will result in loss of supply to all of the installation (Tafe NSW, *Learner Workbook*, p. 27).

Voltage drop limits

AS/NZS 3000:2007 clause 3.6.5 requires the voltage drop within an installation to be restricted to 5% of the nominal supply voltage, as shown in Figure 11.2. Installations which are supplied directly from a supply authority distribution transformer can increase the value to 7%. This is not to be confused with the previous values of +10% and −6% value which related to the characteristics of the supply.

This percentage of voltage drop is spread across the whole installation, from the point of supply to the load. It is not applied separately to consumer's main, sub-main and final sub-circuit. It is a general rule of thumb that approximately a 3%/2% split is made between the consumer's mains and the final sub-circuit.

FIGURE 11.2 Limitation of voltage drop

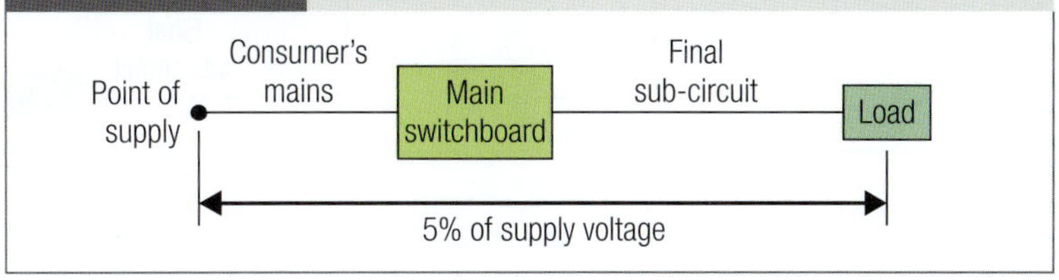

If an installation contains both consumer's and sub-mains, the consumer's and sub-mains are combined. The 3%/2% split is then made between the combination of consumer's/sub-mains and the final sub-circuit as shown in Figure 11.3.

FIGURE 11.3 Limitation of voltage drop

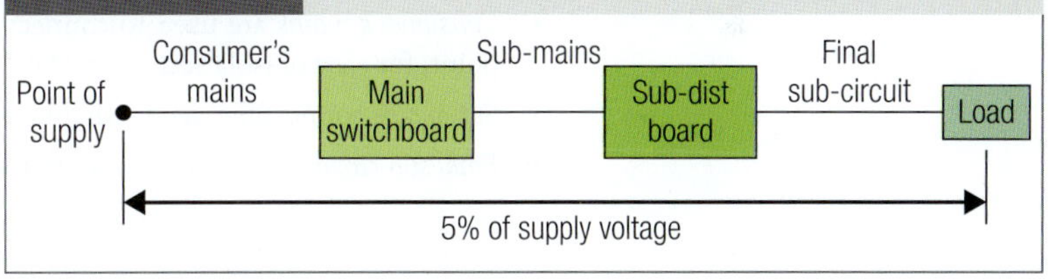

Long runs of cable anywhere in the installation will cause the voltage drop to increase, possibly by more than the permissible amount. This can cause electric motors to overheat and fail. A simple solution to the problem is to increase cable size. The permissible voltage drop (V_p) is calculated as:

For 230 volt circuits: $V_p = 230 \times 5/100$
$= 11.5$ V

For 400 volt circuits: $V_p = 400 \times 5/100$
$= 20$ V

FIGURE 11.4 Compliant installation concerning voltage drop

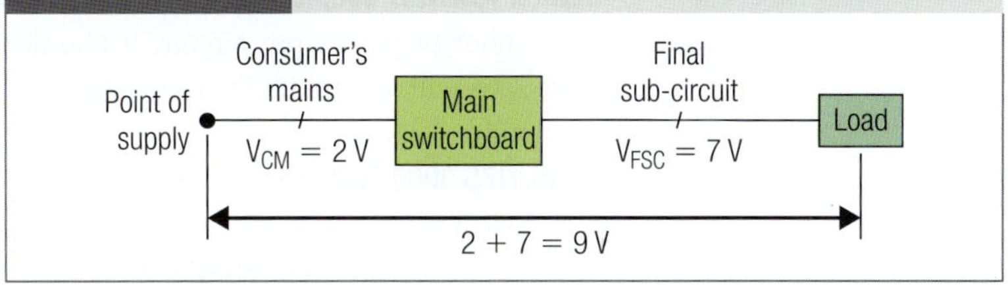

The voltage drop on each section of cable is series connected, therefore the total voltage drop is the sum of each individual voltage drop across each cable. Figure 11.4 shows a single phase 230 volt example. The installation is compliant as the total voltage drop is 9 volts, which does not exceed the maximum permitted voltage drop for a 230 volt circuit of 11.5 volts.

FIGURE 11.5 Non-compliant installation concerning voltage drop

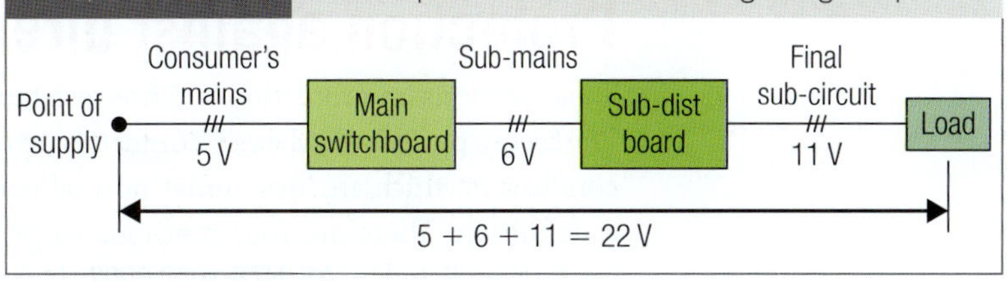

Figure 11.5 shows an example of a three phase 400 volt installation. The sum of the voltage drops is 22 volts which exceeds the maximum permitted voltage drop of 20 volts, making the installation non-compliant.

FIGURE 11.6 Voltage drop across final sub-circuit of 6 volts

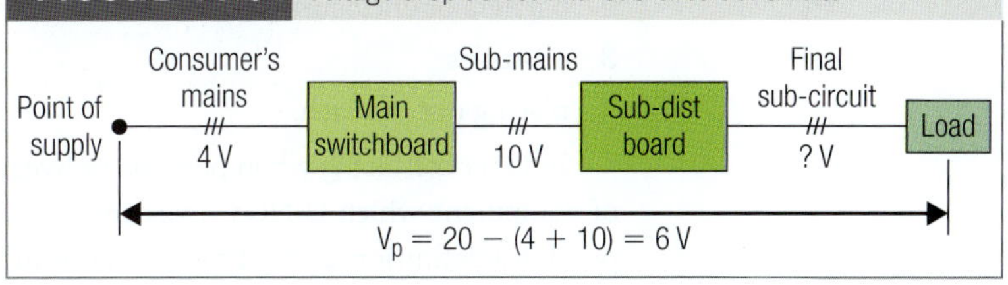

Figure 11.6 shows an example of a three phase 400 volt installation. The sum of the voltage drops on the consumer's and sub-mains is 14 volts. So that the installation does not exceed the maximum permitted total voltage drop of 20 volts on an installation, the permitted voltage drop on any final sub-circuit originating at the distribution board is 6 volts.

Voltage drop is discussed in greater detail later in this chapter (Tafe NSW, *Learner Workbook*, pp. 28–9).

FIGURE 11.7 Circuit arrangements

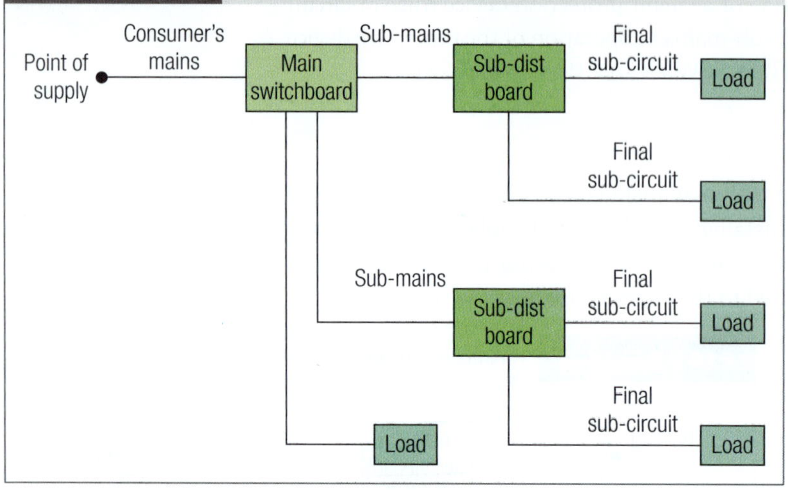

Dividing electrical installations into a number of circuits

In most cases, it is not practical to place the entire electrical load on just one circuit. The load would require large conductor sizes which would not be suitable for the terminals of most equipment. Dividing an installation into a number of circuits allows smaller, more practical and cost-effective conductors to be used.

Consider **AS/NZS 3000:2007** clause 1.6.1. Parts (d) and (e) require that the installation is designed so that inconvenience is minimised in the event of a fault. It also provides safer operation, and facilitates inspection, testing and maintenance. Splitting an installation into appropriately grouped circuits is a means of compliance with this clause. Figure 11.7 shows typical arrangements for dividing an installation into a number of different types of circuits.

As outlined in the Tafe NSW, *Learner Workbook*:

- *Consumer's mains* are used to connect the entire installation to the street supply at only one point, for ease of isolation.
- *Sub-mains* connect between switchboards; no load is connected directly to a sub-main.
- *Final sub-circuits* connect the load to the switchboard.

Source: *Select and Arrange Equipment for General Electrical Installations. Learner Workbook*, Industry Skills Unit Meadowbank, Tafe NSW, 2009, p. 33.

AS/NZS 3000:2007 Section 2.2 provides examples of typical loads that are divided into various circuits. **AS/NZS 3000:2007** Appendix C5 provides guidance to the number of circuits required and the number of points per circuit. It should be noted that electrical equipment with a current rating above 20 A will require its own separate final sub-circuit, meaning there will be only one point per final sub-circuit.

AS/NZS 3000:2007 clause 1.5.14 lists 16 external factors that need to be considered in the design of an electrical installation. Electrical equipment and the selected wiring system must be compatible with the environment in which they are installed. For example, electrical equipment exposed to the weather must have a suitable international protection (IP) rating.

Protection against direct and indirect contact

These sections on protection and fire-resistant construction are based in part on the Tafe NSW, *Learner Workbook*, pp. 33–46. Direct contact is when persons or livestock come in contact with electrical conductors which are 'live' under normal use. **AS/NZS 3000:2007** clause 1.4.34 and Figure 1.2 provide a definition. There are four methods to provide protection against direct contact, also called basic protection, listed by **AS/NZS 3000:2007** clause 1.5.4.2, which are:

1 insulation
2 barriers or enclosures
3 obstacles
4 placing out of reach.

Indirect contact is when persons or livestock come in contact with an electrical conductor or piece of equipment which is NOT 'live' under normal use. **AS/NZS 3000:2007** clause 1.4.35 and Figure 1.3 provide a definition. Some type of failure such as damage to insulation or a conductor breaking free of a termination must have occurred for this to happen.

Methods used to protect against indirect contact are called *fault protection*. They reduce the risk of electric shock to an acceptable level by reducing the value and time of that touch current. DO NOT CONFUSE WITH DIRECT CONTACT! There are four methods to provide protection against indirect contact listed by **AS/NZS 3000:2007** clause 1.5.5.2. These are protection by:

1 automatic disconnection of supply (most commonly used)
2 the use of class II (double insulated equipment)
3 the use of class III (electrically separated equipment)
4 the use of ELV to limit the fault current that can pass through a body to a value lower than the shock current (ELV).

Protection against thermal effects of current

Heat is produced by the flow of electric current. **AS/NZS 3000:2007** clause 1.5.8 requires that precautions must be taken to prevent the risk of ignition of flammable materials by the heat or arcing resulting from the normal use of electrical equipment. Adequate ventilation equipment is required to be provided to maintain normal operating temperatures.

AS/NZS 3000:2007 Section 4.2 provides guidance to acceptable methods of protection against the risks of ignition of flammable materials and injury by burns from the thermal effects of current, in normal service. They include not installing equipment in a position where it might cause a fire hazard. Areas used to store flammable materials may be considered as hazardous locations. Specialist equipment that is not capable of creating sparks or high temperature sources is installed.

Electrical equipment such as wall ovens and other cooking appliances that can produce high surface temperatures are required to be installed in such a way as to prevent ignition of the adjacent combustible materials, such as timber constructional components. These include insulating, enclosing or mounting on non-combustible material, screening and mounting at a sufficient distance from any combustible material to prevent ignition.

Sparks produced by electrical equipment such as the brushes of an electric motor are required to be totally enclosed in or screened by an arc resistant material, or mounted at a safe distance from the combustible material.

To prevent burns, **AS/NZS 3000:2007** clause 4.2.3 requires that the surface temperature of accessible parts of electrical equipment installed within arm's reach be limited to the values listed in **AS/NZS 3000:2007** Table 4.1. Metallic materials generally conduct heat better than non-metallic materials. Metal will transfer more heat in less time to the skin than a non-metal. For this reason the surface temperature of metallic materials is restricted more than a non-metallic material such as plastic or glass. For example, the metallic door of a wall oven must not exceed 70°C whereas the glass front may not exceed 80°C.

Protection against faults causing unwanted voltages

AS/NZS 3000:2007 Section 1.5.11 defines three sources of unwanted voltages:

- a fault between live parts of circuits supplied at different voltages, also known as an intermix of circuits. In other words, accidental contact between a 230 V 50 Hz a.c. low voltage power circuit and a 12 V d.c. extra low voltage control circuit of an air-conditioner
- electromagnetic mutual induction in unused conductors (refer to Chapter 12.5 of *Electrical Principles*, 3rd edition, by Peter Phillips)
- any harmful influence between different circuits and installations.

AS/NZS 3000:2007 clause 1.5.11.4 requires that any redundant or disconnected cables be suitably terminated in the same way as for live terminals. Best practice is to remove and recycle these conductors.

Segregation puts a physical barrier between cabling systems supplied at different voltages. This guards against faults such as insulation failure or a wire breaking away from a terminal strip and making contact with another terminal supplied at a different voltage.

Terminations of conductors operating at different voltages should be made in separate compartments. Conductors should be tied in place to prevent conductors defeating these barriers.

Protection against the effects of over-current

There are three types over-currents:

- *Overload* – current flowing due to additional connection of load in an undamaged circuit. The conductors will eventually overheat and be damaged without operation of the circuit protection device.
- *Fault current* – current flowing due to insulation failure/damage from active to earth. This current is the cause of touch voltage.
- *Short circuit* – this is the largest possible current flow due to a fault of almost zero impedance. The circuit protection must operate very quickly as cable damage and fire is otherwise certain (Tafe NSW, *Learner Workbook*, p. 43).

When higher than normal current flow occurs, there is a danger from excessive heat and electromagnetic effects. Under short-circuit conditions, cables produce not only heat but extremely high mechanical forces due to electromagnetic effects, which may cause the conductors to move and create additional faults and increased damage.

Refer to **AS/NZS 3000:2007** clause 1.5.9 for the full details of requirements for protection against over-currents by automatic disconnection of supply and limiting the maximum over-current.

Refer to **AS/NZS 3000:2007** clause 2.5.2 which specifies four devices deemed suitable to protect against all types of over-currents. These include fuse-combination units (CFS units).

Semi-enclosed rewireable fuses are not suitable for protection against over-currents. Over-currents are discussed in greater detail later in this chapter.

AS/NZS 3000:2007 clause 1.5.10 requires that protective earthing conductors must be capable of carrying an earth fault current without attaining excessive temperature. To achieve this, the protective earthing conductor needs to be of sufficient c.s.a. (cross-sectional area) and the circuit protection device will need to operate quickly to prevent the cable overheating.

AS/NZS 3000:2007 clauses 1.5.13 and 2.3.6 specify the requirement for electrical and mechanical isolation of equipment. Section 1.8 in Chapter 1 provides detailed information on isolation devices and their appropriate use.

Fire-resistant construction

The aim of fire-resistant construction under the National Building Code of Australia is not to make the building fireproof. Instead, it aims to slow the spread of fire so that occupants of the building can be evacuated in a safe manner and emergency service personnel (the fire brigade) have time to respond and extinguish the blaze before it gets out of control. In very large buildings, fire-resistant construction breaks the building into compartments to stop the spread of fire. In multi-occupancy buildings the intention is to avoid the fire spreading from one occupancy to the next, or to an adjacent building, and to protect other property from physical damage caused by structural failure of a building as a result of fire.

AS/NZS 3000:2007 clause 1.5.12 requires that electrical equipment that is selected and installed in an installation should not reach temperatures high enough to start a fire or be placed in a location that would prevent people escaping the fire.

As part of normal installation, electrical services may need to pass through fire-resistant construction.

AS/NZS 3000:2007 clause 3.9.9.3 requires that where a wiring system passes through elements of building construction that is required to be fire-rated, any holes shall be close fitting to the wiring system and at least 50 mm away from any other services. Refer to clause 3.9.9.3 for required sizes of the cross-sectional area of openings. Any penetration made must have the fire-resistant rating restored. Fire barrier pillows and sealant compounds are commonly used to seal cable penetrations.

SUMMING UP...

- AS/NZS 3000:2007 clause 1.6.2(c) specifies tolerances of the nominal voltage for Australia and for New Zealand.
- AS/NZS 3000:2007 clause 2.2.2 lists the four methods that may be used to determine the maximum demand of a cable: calculation, assessment, measurement and limitation.
- AS/NZS 3000:2007 clause 3.6.2 requires the voltage drop within a typical installation to be restricted to 5% of the nominal supply voltage.
- Installations are split into various circuits to minimise inconvenience in the event of a fault and to facilitate safe operation, inspection, testing and maintenance (see AS/NZS 3000:2007 clause 1.6.1).
- Protections against direct contact (basic protection) are listed in AS/NZS 3000:2007 clause 1.5.4.2.
- Protections against indirect contact (fault protection) are listed in AS/NZS 3000:2007 clause 1.5.5.2.
- Over-currents are caused by an overload, a fault or a short circuit.
- AS/NZS 3000:2007 clause 1.5.9 requires protection against over-currents.
- AS/NZS 3000:2007 clause 2.5.2 specifies the four devices deemed suitable to protect against all types of over-currents.
- AS/NZS 3000:2007 clause 3.9.9.3 requires that where a wiring system passes through elements of a building construction that is required to be fire-rated, any penetration made must have the fire-resistant rating restored.

11.2 Final sub-circuit arrangements

The total load of an installation is split into a number of final sub-circuits to connect the loads to the supply at main and distribution switchboards. The number of points connected to each circuit is restricted to allow smaller and more practically sized conductors to be installed. Circuit protection is selected to match the load while at the same time ensuring the conductor is protected against over-currents appropriate to the size and current rating for the given installation method. This topic explains the selection process.

Daily and seasonal demand

Maximum demand is the maximum expected current that may be drawn by a particular circuit at any given time. For single items of equipment, it is set by assessment of the connected load. In light and power circuits, or other circuits with more than one point per final sub-circuit, the maximum demand of a final sub-circuit is set by limitation (**AS/NZS 3000:2007** clause C2.5.1 in Appendix C). The purpose of the final sub-circuit will determine the number of points that may be connected to that final sub-circuit.

In Section 11.1 earlier in the chapter we showed that the load in an installation is divided into a number of circuits that are logically arranged to group loads of a similar type onto the same circuit. The more loads connected to a single circuit the higher the demand will be, for example, light and power circuits tend to have a large number of points connected to a single circuit to reduce the cost of the installation. If too much load is connected at the same time the circuit protection device will operate, automatically disconnecting the circuit. Circuits for appliances such as air-conditioning, hotplates and hot water systems are arranged so that each appliance is on its own separate final sub-circuit, for maintenance and testing purposes.

The key to circuit design is to limit the number of lights or socket outlets on the one circuit so the circuit protection device does not trip, while still using the minimum number of circuits to keep installation costs to a minimum. Table C8 of **AS/NZS 3000:2007** gives guidance to the recommended number of points per final sub-circuit.

- *Daily demand* – different types of installations will use power in different ways. For example, socket outlets installed in a domestic installation may rarely ever be used, but in a non-domestic installation such as a factory, the socket outlet is probably installed for a purpose and will most likely be in frequent use. For the circuit to function correctly as intended, the number of points per final sub-circuit will be less in the factory.

- *Seasonal demand* – the demand of circuits that supply appliances such as air-conditioners will vary throughout the year. Most air-conditioning units use more current to heat than to cool. The higher current must be taken as the maximum demand.
- *Connected load* – this is the actual current drawn by the circuit with no 'diversity' applied. It can be found on the compliance plate of an appliance or calculated using the power equation transposed to find current.

Single phase appliances use:

$P = V \times I \times \lambda$

where

P = power in watts (W)

V = voltage in volts (V)

I = current in amperes (A)

λ = power factor (0-1)

Three phase appliances use:

$P = \sqrt{3} \times V_L \times I_L \times \lambda$

where

P = power in watts (W)

V_L = line voltage in volts (V)

I_L = line current in amperes (A)

λ = power factor (0-1)

For an electrician, it is important to know the current drawn by the load in order to determine the required cross-sectional area of the cable. The power equation is transposed to determine the current flow (Tafe NSW, *Learner Workbook*, pp. 60–1).

$$I = \frac{P}{V \times \lambda} \qquad I_L = \frac{P}{\sqrt{3} \times V_L \times \lambda}$$

EXAMPLE 11.3

Determine the current drawn by a 230 volt, 3.6 kW single phase air-conditioner operating at a rated power factor of 0.85 (adapted from Tafe NSW, *Learner Workbook*, pp. 60–1).

Solution

$$I = \frac{P}{V \times \lambda}$$

$$I = \frac{3600}{230 \times 0.85}$$

$$I = 18.4 \text{ A}$$

EXAMPLE 11.4

Determine the line current drawn by a 400 volt, 50 kW three-phase kiln (resistive load) (adapted from Tafe NSW, *Learner Workbook*, pp. 60–1).

Solution

$$I_L = \frac{P}{\sqrt{3} \times V_L \times \lambda}$$

$$I_L = \frac{50\,000}{400 \times 1.0}$$

$$I_L = 125 \text{ A}$$

Number and types of circuits required

This section is based in part on the Tafe NSW, *Learner Workbook*, pp. 60–1.

AS/NZS 3000:2007 clause C5.2 and Table C8 provide guidance on the recommended number of points per final sub-circuit. To the left of the table the first two columns show the c.s.a. of the conductor and a number of circuit protection device ratings. The ratings from Tables C5 and C8 align with different methods of installation of the conductor (covered later in this topic). The remaining columns show the contribution to the total current caused by each point. The total current should not exceed the rating of the circuit protection device. Conductor sizes of less than 2.5 mm^2 are not permitted for use in a circuit that supplies socket outlets – this is indicated by NP. The footnotes that follow Table C8 provide further explanation of the correct use of Table C8. Key points include:

- footnote c – a double socket outlet is treated as two single points
- footnote f – socket-outlet (surface sockets or plug bases) installed more than 2.3 m above a floor for the connection of a luminaire may be included as a lighting point. An appliance (such as a fan) rated at not more than 150 W, which is permanently connected, or connected by means of a socket-outlet installed more than 2.3 m above a floor, may be included as a lighting point.

EXAMPLE 11.5

Determine the minimum number of circuits and points per final sub-circuit for the following domestic installation.

- A house wired with 2C+E TPS cable has the following load installed:
  - 36 lights (10 A CB)
  - 26 double 10 A socket outlets (20 A CB). Note that double socket outlets (SO) count as two singles.
  - 1 22 A A/C (25 A CB)
  - 1 4.4 kW hot water system (HWS; 20 A CB) (adapted from Tafe NSW, *Learner Workbook*, pp. 60–1).

Solution

TABLE 11.1 Final sub-circuit arrangement

Circuit number	Purpose	Protection device rating (A)	Working calculations	Points per circuit
1	Light 1	10 A	$18 \times 0.5 = 9$ A	18
2	Light 2	10 A	$18 \times 0.5 = 9$ A	18
3	Power 1	20 A	$9 \times 2 = 18$	9
4	Power 2	20 A	$9 \times 2 = 18$	9
5	Power 3	20 A	$8 \times 2 = 16$	8
6	A/C	25 A	$1 \times 22 = 22$ A	1
7	HWS	20 A	$I = \frac{P}{V} = \frac{4400}{230} = 19.1$ A	1

Source: *Select and Arrange Equipment for General Electrical Installations. Learner Workbook*, Industry Skills Unit Meadowbank, Tafe NSW, 2009, p. 62

EXAMPLE 11.6

Determine the minimum number of circuits and points per final sub-circuit for the following non-domestic installation.

- A shop is wired using TPS cable and has the following load installed:
  - 44 lights (16 A CB)
  - 14 10 A double socket outlets (20 A CB) (adapted from Tafe NSW, *Learner Workbook*, pp. 60–1).

Solution

TABLE 11.2 Final sub-circuit arrangement

Circuit number	Purpose	Protection device rating (A)	Working calculations	Points per circuit
1	Light 1	16 A	$22 \times 0.5 = 11$ A	22
2	Light 2	16 A	$22 \times 0.5 = 11$ A	22
3	Power 1	20 A	$5 \times 2 \times 2 = 20$	5
4	Power 2	20 A	$5 \times 2 \times 2 = 20$	5
5	Power 3	20 A	$4 \times 2 \times 2 = 16$	4

Source: *Select and Arrange Equipment for General Electrical Installations. Learner Workbook*, Industry Skills Unit Meadowbank, Tafe NSW, 2009, p. 63

Because a shop is considered a non-domestic installation and the load does not include any mention of air-conditioning, 20 amperes is allocated to each socket outlet which reduces the number of points per final sub-circuit and increases the number of required circuits.

EXAMPLE 11.7

Determine the minimum number of circuits and points per final sub-circuit for the following non-domestic installation:

- A three phase factory unit is wired using TPS cable and has the following load installed:
  - 10 hi-bay MV lights rated at 1.85 A each (16 A CB)
  - 17 10 A double socket outlets (20 A CB)
  - 3 three phase 32 A socket outlets (32 A CB) (adapted from Tafe NSW, *Learner Workbook*, pp. 60–1).

Solution

TABLE 11.3 Final sub-circuit arrangement

Circuit number	Purpose	Protection device rating (A)	Working calculations	Points per circuit
1	Light 1	16 A	$5 \times 1.85 = 9.25$ A	5
2	Light 2	16 A	$5 \times 1.85 = 9.25$ A	5
3	Power 1	20 A	$5 \times 2 \times 2 = 20$	5
4	Power 2	20 A	$4 \times 2 \times 2 = 16$	4
5	Power 3	20 A	$4 \times 2 \times 2 = 16$	4
6	Power 4	20 A	$4 \times 2 \times 2 = 16$	4
7	Power 5	32 A	$1 \times 32 = 32$	32
8	Power 6	32 A	$1 \times 32 = 32$	32
9	Power 7	32 A	$1 \times 32 = 32$	32

Source: *Select and Arrange Equipment for General Electrical Installations. Learner Workbook*, Industry Skills Unit Meadowbank, Tafe NSW, 2009, p. 64

The current drawn by each light is considerably higher than 0.5 A per point so the actual connected load is used for the calculation. The 32 A three phase outlets are rated at more than 20 A, so each outlet requires an individual circuit.

Cook tops, stoves, ovens and hotplates in domestic installations are allowed some diversity in calculation of their maximum demand. When the elements are energised they heat up, and within a couple of minutes the temperature controllers operate and begin to cycle the elements on and off. Not all of the elements will be on or off at the same time, so the current drawn by the appliance will be less than the rating on the name plate. A lower current is given for the maximum demand to reduce installation cost. Even if all elements are switched on at the same time the circuit protection device will take time to operate. The current flowing is only a small overload in relation to the nominal rating of the protection device and the cable it protects. In other words, by the time the circuit breaker is about to trip the temperature controllers will have started to cycle the elements, and the current drops to a value that will not operate the protection device or damage the conductors.

AS/NZS 3000:2007 clause C2.5.3 and Table C4 provide guidance on the recommended assessed maximum demand on final sub-circuits that supply domestic cooking appliances such as cook tops, ovens and ranges. A cook top and oven may be connected on the same circuit, effectively creating a range. The maximum demand is based on the combined wattage. In the case of non-domestic cooking appliances, maximum demand should be based upon the full connected load as per Table C8.

EXAMPLE 11.8

List the assessed maximum demand from Table C4 and calculate the current from the connected load for each cooking appliance name plate rating below (adapted from Tafe NSW, *Learner Workbook*, pp. 60–1).

Solution

TABLE 11.4 Domestic cooking final sub-circuit maximum demand

Cooking appliance name plate rating	Table C4 assessed maximum demand	Calculated full connected load
Not greater than 5000 W	16 A	$I = \frac{P}{V} = \frac{5000}{230} = 21.74$ A
10000 W to 13000 W	32 A	$I = \frac{P}{V} = \frac{13000}{230} = 56.53$ A

Source: *Select and Arrange Equipment for General Electrical Installations. Learner Workbook*, Industry Skills Unit Meadowbank, Tafe NSW, 2009, p. 65

Current requirements for final sub-circuits

This section is based in part on the Tafe NSW, *Learner Workbook*, pp. 66–71.

To protect conductors from overload and short-circuit faults, protection devices such as circuit breakers or HRC fuses are used. They must be able to carry the maximum demand continuously without operation, and still be able to detect an overload or short circuit and disconnect the circuit before the circuit conductors are damaged.

Protection from overload – coordination between conductors and protective devices

When selecting the circuit protection device (circuit breaker or fuse) it is important to remember the protection device protects the cable, not the load. **AS/NZS 3000:2007** clause 2.5.3.1 states:

$I_B \leq I_N \leq I_Z$

where

I_B = the maximum demand current in amperes

I_N = the nominal current of the protective device

I_z = the current capacity of the conductor

If the statement in the above equation is true, the circuit protection rating has been selected so that the load will not operate the protection and the protection device will operate before the cable overheats. The circuit will operate as intended.

EXAMPLE 11.9

For the circuits listed in Table 11.5 indicate 'yes' or 'no' as to which circuits comply with AS/NZS 3000:2007 clause 2.5.3.1 and the statement $I_B \leq I_N \leq I_Z$ (adapted from Tafe NSW, *Learner Workbook*, p. 67).

Solution

TABLE 11.5 Final sub-circuit coordination between conductors and protection device circuit breaker

Load maximum demand I_B	Protective device rating I_N	Conductor current carrying capacity I_Z	Complies Yes/No
10 A	10 A	10 A	Yes
18 A	10 A	13 A	No
32 A	25 A	33 A	no
40 A	50 A	56 A	Yes

Source: *Select and Arrange Equipment for General Electrical Installations. Learner Workbook*, Industry Skills Unit Meadowbank, Tafe NSW, 2009, p. 67

Figure 11.8 shows a typical circuit breaker time current characteristic curve. I_N is shown at the bottom left of the graph. The line rising above the 1 × I_N graduation does not cross the blue curve

FIGURE 11.8 Inverse time current characteristic curve

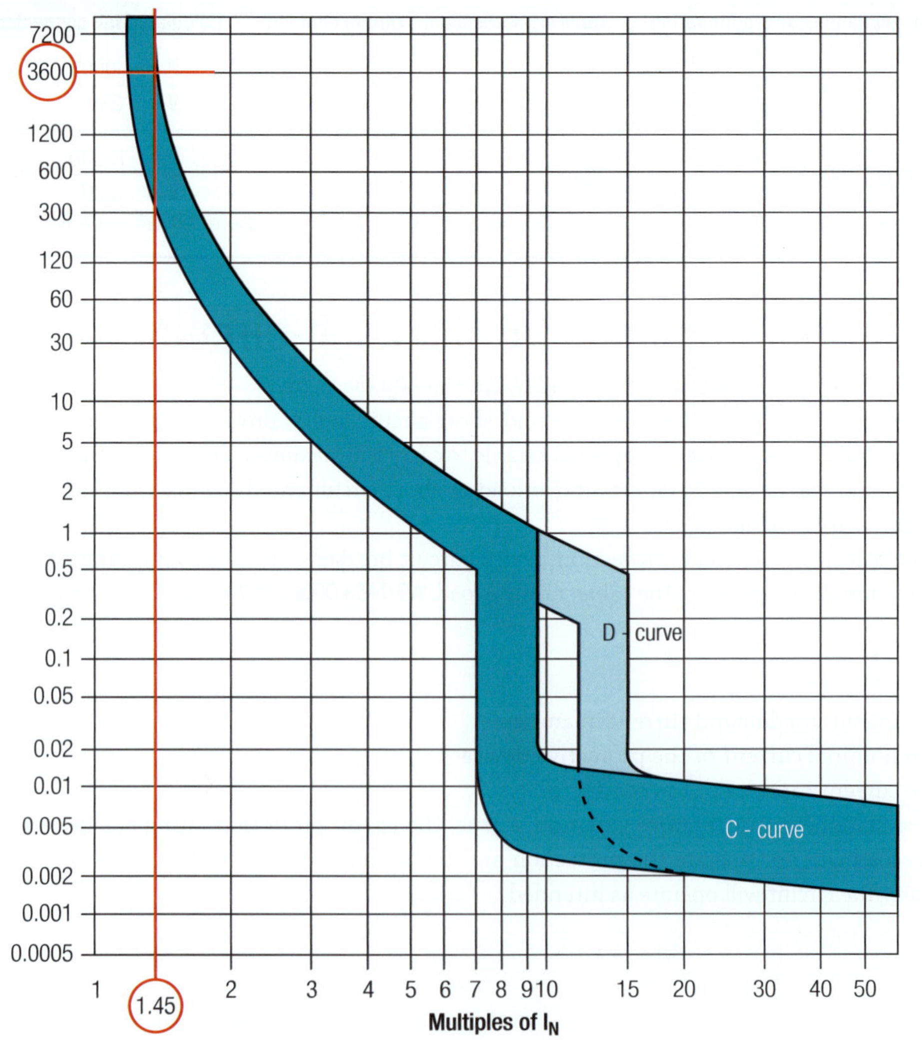

that indicates the tripping characteristic of the circuit breaker. Because the two lines do not cross, this indicates the circuit breaker will not operate. If the load current I_B exceeds the value of I_N the breaker will operate. The higher the load current is above the nominal rating of the circuit breaker, the shorter the time the breaker will take to operate.

As outlined in **AS/NZS 3000:2007** clause 2.5.3.1, to ensure the circuit protection will operate in the case of an over-current before the cable overheats we use the equation:

$I_2 \leq 1.45 \times I_Z$

where

I_2 = the current ensuring effective operation of the protective device (operates in 1 hour or 3600 seconds) as shown in Figure 11.8 (CBs $1.45 \times I_N$, fuses $1.6 \times I_N$)

I_z = the current capacity of the conductor

EXAMPLE 11.10

A circuit has a maximum demand (I_B) of 18 A. It is protected by a 20 A (I_N) circuit breaker and is wired in 2.5 mm² TPS cable rated at 22 A (I_Z). Does the circuit comply with the requirements of AS/NZS 3000:2007 clause 2.5.3.1 (adapted from Tafe NSW, *Learner Workbook*, pp. 257–8)?

Solution

$I_B \leq I_N \leq I_Z$, is 18 A ≤ 20 A ≤ 22 A? Yes.

$I_2 \leq 1.45 \times I_Z$,

$I_2 = 1.45 \times I_N$

$= 1.45 \times 20 = 29$ A

is $29 \leq 1.45 \times 22 = 31.9$ A? Yes.

Where the protection is a circuit breaker, if the equation $I_B \leq I_N \leq I_Z$ is true, $I_2 \leq 1.45 \times I_Z$ will also be true in every case. A true statement in both equations means the circuit will not trip under normal use but operate under overload conditions before the cable is damaged by heat.

If an HRC fuse is used as protection $I_2 = I_N \times 1$. It is quite often the case that the cable size and current rating will require upgrading to comply with **AS/NZS 3000:2007** clause 2.5.3.1 as shown in Example 11.11.

EXAMPLE 11.11

A circuit has a maximum demand (I_B) of 18 A. It is protected by a 20 A (I_N) HRC fuse and is wired in 2.5 mm² TPS cable rated at 22 A (I_Z). Does the circuit comply with the requirements of AS/NZS 3000:2007 clause 2.5.3.1 (adapted from Tafe NSW, *Learner Workbook*, pp. 257–8)?

Solution

$I_B \leq I_N \leq I_Z$, is 18 A ≤ 20 A ≤ 22 A? Yes.

$I_2 \leq 1.45 \times I_Z$

$I_2 = 1.6 \times I_N$

$= 1.6 \times 20 = 32$ A

is $29 \leq 1.45 \times 22 = 31.9$ A? No.

Upgrade cable size – 4.0 mm² TPS cable rated at 27 A

$I_2 \leq 1.45 \times I_Z$

$I_2 = 1.6 \times I_N$

$= 1.6 \times 20 = 32$ A

is $29 \leq 1.45 \times 27 = 39.15$ A? Yes.

Circuit breakers have a bimetallic tripping mechanism which closely mimics the heating effect a current has on a conductor. It requires an overload current of 1.45 times the rated current of a circuit breaker to cause it to operate in one hour. If 1.45 times the rated current of a conductor flows through it for one hour, it will be approaching its maximum rated operating temperature. If this current flows for longer than one hour the maximum cable operating temperature will be exceeded. It is expected that the circuit breaker operating in one hour will prevent the cable from overheating.

HRC fuses do not have a bimetallic strip mechanism. Melting of the fuse element requires a current flow of 1.6 times the rating of the HRC fuse to cause it to operate in one hour. This slightly higher current will cause the cable to reach a higher temperature in one hour. Cables protected by HRC fuses must be de-rated by the ratio of the heating effect of the cable (1.45) to the conventional tripping current of the fuse (1.6). Therefore 1.45/1.6 is equal to a de-rating factor of 0.9.

If an HRC fuse is the protection device, the cable rating (I_Z) must be de-rated to 90% of its original capacity. For example, if the cable was originally rated at 20 amperes the cable would only have a current carrying capacity of 18 amperes ($I_Z \times 0.9$) after it had been de-rated. Once de-rated, I_N is not less than I_Z. To ensure the fuse will operate in the case of an over-current we use the equation:

$$I_B \leq I_N \leq I_Z \times 0.9$$

where

I_B = the maximum demand current in amperes

I_N = the nominal current of the protective device

I_z = the current capacity of the conductor

EXAMPLE 11.12

For the circuits listed in Table 11.6, indicate 'yes' or 'no' as to which circuits comply with AS/NZS 3000:2007 clause 2.5.3.1 and the statement $I_B \leq I_N \leq I_Z \times 0.9$ (adapted from Tafe NSW, *Learner Workbook*, p. 67).

Solution

TABLE 11.6 Final sub-circuit arrangement between conductors and protection device HRC fuse

Load maximum demand I_B	Protective device rating I_N	Conductor current carrying capacity I_Z	Complies Yes/No
10 A	10 A	10 A	No
18 A	10 A	13 A	No
32 A	25 A	33 A	No
40 A	50 A	56 A	Yes

Source: *Select and Arrange Equipment for General Electrical Installations. Learner Workbook*, Industry Skills Unit Meadowbank, Tafe NSW, 2009, p. 69

Cables protected by HRC fuses will generally need to be increased in size to accommodate the 90% de-rating. To determine the minimum current carrying capacity of a conductor protected by an HRC fuse, use the equation:

$$I_Z \geq \frac{I_N}{DR}$$

where

I_Z = the current capacity of the conductor.

I_N = the nominal current of the protective device

DR = the de-rating factor that applies to fuse protection (0.9)

Note: the de-rating factor for a CB is 1

EXAMPLE 11.13

List the required current carrying capacity of conductors protected by the protection devices in Table 11.7 (adapted from Tafe NSW, *Learner Workbook*, p. 70).

Solution

TABLE 11.7 Circuit de-ratings

Protection (I_N)	De-rating	Current carrying capacity (I_Z)
20 A CB	1	$I_z = \frac{20}{1} \geq 20$ A
20 A HRC fuse	0.9	$I_z = \frac{20}{0.9} \geq 22.2$ A
32 A CB	1	$I_z = \frac{32}{1} \geq 32$ A
32 A HRC fuse	0.9	$I_z = \frac{32}{0.9} \geq 35.55$ A
40 A HRC fuse	0.9	$I_z = \frac{40}{0.9} \geq 44.44$ A

Source: *Select and Arrange Equipment for General Electrical Installations. Learner Workbook*, Industry Skills Unit Meadowbank, Tafe NSW, 2009, p. 70

Cable selection

AS/NZS 3000:2007 clause 3.4.1 requires that selected cables have a current carrying capacity not less than that specified by the **AS/NZS 3008.1.1:2009 – Electrical installations – Selection of cables** series. (Section 11.5 later in this chapter discusses the use of AS/NZS 3008.1.)

When a conductor carries an electric current, work is done to overcome the resistance of the conductor, and as a result heat is produced. If the temperature rise in the cable exceeds safe limits the insulation of the cable will be permanently damaged. **AS/NZS 3000:2007** clause 3.4.2 requires the maximum operating temperature of insulated conductors not to exceed values listed in **AS/NZS 3000:2007** Table 3.2. Maximum normal operating temperatures of commonly used conductors include:

- V-75 and V-90 polyvinyl chloride (PVC) insulated 75°C
- X-90 cross length polyethylene (XLPE) insulated 90°C.

A simplified cable and protection device selection process is shown in **AS/NZS 3000:2007** Table C5 for single phase and Table C6 for three phase circuits. These tables show how to select cables and circuit breakers to suit a number of installation conditions for cables ranging from 1 mm^2 to 25 mm^2. **AS/NZS 3008.1.1:2009** provides more detailed current ratings for a larger range of cables and cross-sectional areas.

Installation conditions

The environment into which a cable is installed will affect the current carrying capacity of that cable. Cables installed in thermal insulation will be unable to transfer heat into the air, and as a result the cable will retain heat and have a much lower current rating than that of cable installed in air or underground. In domestic installations, **AS/NZS 3000:2007** clause 3.4.1 requires that in New Zealand it should be assumed that thermal insulation is installed in ceilings, walls and floors. In Australia national and local building codes should be consulted regarding thermal insulation in walls and under floor. Thermal insulation is generally installed in ceiling spaces. Cables clipped to structural members or laid flat on plasterboard sheeting should be considered partially surrounded by thermal insulation.

EXAMPLE 11.14

What is the maximum rating of a circuit breaker protecting the circuits listed in Table 11.8 (adapted from Tafe NSW, *Learner Workbook*, p. 71)?

Solution

TABLE 11.8 Circuit protection device ratings

Circuit	Maximum circuit breaker rating	AS/NZS 3000:2007 reference
2.5 mm^2 T + E cable installed in air	25 A	Table C5 unenclosed in air
2.5 mm^2 T + E cable installed partially surrounded by thermal insulation	20 A	Table C5 partially surrounded
6.0 mm^2 4C + E, circular cable installed enclosed in the ground	32 A	Table C6 enclosed in ground
4.0 mm^2 3C + E cable installed enclosed in air	32 A	Table C6 enclosed in air
10.0 mm^2 4C + E cable installed completely surrounded by thermal insulation	25 A	Table C6 completely surrounded

Source: *Select and Arrange Equipment for General Electrical Installations. Learner Workbook*, Industry Skills Unit Meadowbank, Tafe NSW, 2009, p. 72

When selecting cable cross-sectional areas (c.s.a.) the size must be both cost-effective and practical. A power circuit wired in 6.0 mm^2 will be impossible to terminate. **AS/NZS 3000:2007** Table C5 is used to select the cable c.s.a. and circuit protection. Table C8 is then used to determine the number of points per final sub-circuit (Tafe NSW, *Learner Workbook*, p. 71).

EXAMPLE 11.15

Record in Table 11.9 the minimum cable c.s.a., maximum protection device rating and maximum number of points per final sub-circuit for the load of a single phase domestic installation (house) described below:

- 22 light points, wiring installed clipped to timbers in the roof
- 24 double 10 A socket outlets, wiring installed clipped to timbers in the roof
- 1 15 A socket outlet for a split system A/C, wiring installed clipped to timbers in the roof
- 1 6.0 kW cook top, wiring installed clipped to timbers under the floor
- 1 4.6 kW wall oven, wiring installed clipped to timbers under the floor
- 1 4.6 kW storage HWS, wiring installed enclosed in conduit in air (adapted from Tafe NSW, *Learner Workbook*, p. 74).

Solution

TABLE 11.9 Final sub-circuit design

Circuit number	Purpose	Cable c.s.a. (mm^2)	Protection device rating (A)	Installation method	Number of points
1	Light 1	1.0	10 A	Partially surrounded	11
2	Light 2	1.0	10 A	Partially surrounded	11
3	Power 1	2.5	20 A	Partially surrounded	8
4	Power 2	2.5	20 A	Partially surrounded	8
5	Power 3	2.5	20 A	Partially surrounded	8
6	A/C	2.5	20 A	Partially surrounded	1
7	CT oven	4.0	32 A	Unenclosed in air	2
8	HWS	2.5	20 A	Enclosed in air	1

Source: *Select and Arrange Equipment for General Electrical Installations. Learner Workbook*, Industry Skills Unit Meadowbank, Tafe NSW, 2009, p. 74

In Example 11.15, for cost reasons it is desirable to use the smallest conductor size. According to Table C8 each lighting point contributes 0.5 A to the circuit current. This means the 22 lights would take 11A, which would overload a 10 A circuit breaker that is protecting a 1.0 mm^2 cable. The lights must therefore be split over two circuits of 11 points.

The 24 socket outlets are considered as 48 points at 1 A each (by Table C8). Table C5 shows 2.5 mm^2 cable is rated at 20 A when installed partially surrounded by thermal insulation. Obviously there is a need for more than one circuit, so the 24 socket outlets are split over three circuits of 8 points each. The 8 points will contribute 16 A of load which will not trip the 20 A circuit breaker.

The air-conditioner circuit is rated at 20 A and requires its own separate circuit as per **AS/NZS 3000:2007** clause C5.1.

The cook top and oven are connected on one circuit in the same way a range (combined cook top and oven appliance) would be. The wattage ratings of the cook top and oven are combined and given an assessed maximum demand of 32 A by Table C4. Table C5 shows 4.0 mm^2 cable is rated at 32 A when installed unenclosed in air. A 32 A circuit breaker is used to protect the circuit. An alternative arrangement is to install two separate circuits for the cook top and oven based on wattage ratings of each individual appliance.

The HWS is assessed as full connected load by Table C8 which is approximately 20 A, which will require a separate final sub-circuit. Table C5 indicates a 2.5 mm^2 cable when installed enclosed in air and protected by a 20 A circuit breaker.

EXAMPLE 11.16

Record in Table 11.10 the minimum cable c.s.a., the maximum protection device rating and maximum number of points per final sub-circuit for the load of a three phase domestic installation (house) described below (note that the Greek letter phi (Φ) denotes the phase angle between phasors in an AC circuit):

- 48 light points, wiring installed clipped to timbers in the roof
- 1 three-in-one fan/heat lamps (4 × 275 W), wiring clipped to timbers in the roof
- 30 double 10 A socket outlets, wiring installed clipped to timbers in the roof
- 1 18 A three phase (3 Φ) ducted A/C, wiring installed enclosed in conduit in air
- 1 8.4 kW 1 Φ range, wiring installed clipped to timbers under the floor
- 1 22.0 kW 3 Φ spa heater, wiring installed enclosed in conduit in air
- 1 3.9 kW 1 Φ sauna, wiring installed clipped to timbers in the roof
- 1 4.6 kW 1 Φ storage HWS, wiring installed enclosed in conduit in air (adapted from Tafe NSW, *Learner Workbook*, p. 75).

Solution

TABLE 11.10 Final sub-circuit design

Circuit number	Purpose	Cable c.s.a. (mm^2)	Protection device rating (A)	Installation method	Number of points
1	Light 1	1.5	16 A	Partially surrounded	28
2	Mixed circuit light 2 + fan 3 in 1	1.5	16 A	Partially surrounded	20 + 3
3	Power 1	2.5	20 A	Partially surrounded	10
4	Power 2	2.5	20 A	Partially surrounded	10
5	Power 3	2.5	20 A	Partially surrounded	10
6	A/C	2.5	20 A	Enclosed in air	1
7	Range	2.5	25 A	Unenclosed in air	1
8	Spa	6.0	32 A	Enclosed in air	1
9	Sauna	2.5	20 A	Partially surrounded	1
8	HWS	2.5	20 A	Enclosed in air	1

Source: *Select and Arrange Equipment for General Electrical Installations. Learner Workbook*, Industry Skills Unit Meadowbank, Tafe NSW, 2009, p. 75

In Example 11.16, 1.5 mm^2 cable protected by a 16 A breaker is used for the lighting and mixed circuits. The mixed circuit is made up of 24 lighting points plus the 3-in-1 fan heater. Table C8 'footnote f' says that a fan less than 150 watts is counted as a light point depending on the number of sections in which it is switched. The fan and the light of the 3-in-1 are switched separately and are considered two lighting points equalling 1 A.

The heat lamps are treated as a permanently connected fixed or stationary appliance. Full connected load of the heat lamps is equal to (4 × 275) ÷ 230, which equals 4.8 A. If the lamps are split evenly across the two circuits, the 24 light points, two light/fan points and heat lamps would cause a load current of 12 + 1 + 4.8 A which equals 17.8 A. Over time, this current is likely to overload 1.5 mm^2 cable and trip the 16 A circuit breaker. The solution is to move a number of lights from the mixed to lighting circuit. If 28 of the 48 lighting points are connected to light circuit 1, both the light and mixed circuits will operate correctly.

An alternative arrangement is to run three circuits of 1.0 mm^2 protected by a 10 A circuit breaker. It is also common practice, but not a requirement, to provide the 3-in-1 unit with a separate circuit.

The 30 socket outlets are considered 60 points at 1 A each (by Table C8). Table C5 shows 2.5 mm^2 cable is rated at 20 A when installed partially surrounded by thermal insulation. The 30 socket outlets are split over three circuits of 10 points each. The 10 points will contribute 20 A of load which will not trip the 20 A circuit breaker.

The three phase air-conditioner circuit is rated at 18 A per phase, and for testing and maintenance reasons it requires its own separate circuit. Table C6 shows 2.5 mm^2 cable is rated at 20 A when installed enclosed in conduit in air.

The 8.4 kW range is given an assessed maximum demand of 25 A by Table C4. Table C5 shows 2.5 mm^2 cable is rated at 25 A when installed unenclosed in air. A 25 A circuit breaker is used to protect the circuit.

The spa is considered as a permanently connected fixed or stationary appliance by Table C8 and has a connected load of $I_L = \dfrac{P}{\sqrt{3} \times V_L} = \dfrac{22\,000}{\sqrt{3} \times 400} = 31.8\text{ A}$. Table C6 shows 6.0 mm^2 cable is rated at 32 A when installed enclosed in air.

The 3.9 kW sauna is considered as a permanently connected fixed or stationary appliance by Table C8 and has a connected load of $I = \dfrac{P}{V} = \dfrac{3900}{230} = 17\text{ A}$. Table C5 shows 2.5 mm^2 cable is rated at 20 A when installed partially surrounded by thermal insulation.

The HWS is assessed as a full connected load by Table C8 which is approximately 20 A and will require a separate final sub-circuit. Table C5 indicates a 2.5 mm^2 cable when installed enclosed in conduit in air and protected by a 20 A circuit breaker.

EXAMPLE 11.17

A three phase non-domestic installation (factory unit) is wired using TPS and orange circular cable. Record in Table 11.11 the minimum cable c.s.a., maximum protection device rating and maximum number of points per final sub-circuit for the load described below:

- 16 MH hi-bay lights (2 A each) split over two circuits unenclosed in air
- 24 twin 36 W fluorescent lights (0.333 A each) unenclosed in air
- 15 10 A double socket outlets wiring installed enclosed in conduit in air
- 3 32 A 3 Φ socket outlets wiring installed enclosed in conduit in air
- 1 hard-wired machine 54 A/phase installed enclosed in conduit in air
- 2 hard-wired machines 34 A/phase enclosed in conduit in air
- 1 4.6 kW HWS wiring installed enclosed in conduit in air (adapted from Tafe NSW, *Learner Workbook*, p. 77).

Solution

TABLE 11.11 Final sub-circuit design

Circuit number	Purpose	Cable c.s.a. (mm^2)	Protection device dating (A)	Installation method	Number of points
1	Light 1	1.5	20 A	Unenclosed in air	8
2	Light 2	1.5	20 A	Unenclosed in air	8
3	Light 3	1.5	20 A	Unenclosed in air	24
4	Power 1	2.5	20 A	Enclosed in air	5
5	Power 2	2.5	20 A	Enclosed in air	5
6	Power 3	2.5	20 A	Enclosed in air	5
7	32 A O/L 1	6.0	32 A	Enclosed in air	1
8	32 A O/L 2	6.0	32 A	Enclosed in air	1
9	32 A O/L 3	6.0	32 A	Enclosed in air	1
10	Machine 1	16.0	63 A	Enclosed in air	1
11	Machine 2	10.0	40 A	Enclosed in air	1
12	Machine 3	10.0	40 A	Enclosed in air	1
13	HWS	2.5	20 A	Enclosed in air	1

Source: *Select and Arrange Equipment for General Electrical Installations. Learner Workbook*, Industry Skills Unit Meadowbank, Tafe NSW, 2009, p. 77

In non-domestic installations such as factory units, less diversity is applied to circuits. Therefore fewer points are allowed per final sub-circuit and more often it is wise to use the connected load as a basis to select cable sizes. When industrial lights such as used in lighting circuits 1 and 2 are installed with current ratings higher than the 0.5 A recommended by Table C8, the actual full connected load should be used. Eight lights per circuit will cause a current of 16 A to flow, so a 1.5 mm^2 cable and 20 A circuit breaker is selected from Table C5. An over-rating of cable and protection is selected to prevent nuisance tripping during lamp warm-up.

Lighting circuit 3 has a full connected load of approximately 8 A. Table C8 recommends 0.5 A be allocated for each of the 24 lights, equalling 12 A. Both currents are less than the current rating from Table C5 for 1.0 mm^2 cable protected by a 16 A circuit breaker. It should be noted that in non-domestic installations lighting circuits are significantly longer than in domestic installations. Lighting circuits are often upgraded to 2.5 mm^2 cables to reduce the effect of voltage drop (this is discussed later in the chapter in Section 11.5).

In power circuits 1 to 3, the 15 double socket outlets are considered 30 points at 2 A each by Table C8. Table C5 shows 2.5 mm^2 cable is rated at 20 A when installed enclosed in conduit in air. The socket outlets are split over three circuits of five points each. The five points will contribute 20 A of load which will not trip the 20 A circuit breaker.

Each of the three 32 A socket outlet circuits will require a separate circuit. Table C6 indicates the use of 6.0 mm^2 cable protected by a 32 A circuit breaker for the installation method of enclosed in conduit in air.

The three machine circuits will each require a separate circuit. Table C6 requires machine circuit 1 to be protected by a 63 A circuit breaker and supplied by 16 mm^2 cable. Machine circuits 2 and 3 will be protected by two 40 A circuit breakers and supplied by 10 mm^2 cables.

The HWS is assessed as a full connected load by Table C8, which is approximately 20 A, so it will require a separate final sub-circuit. Table C5 indicates a 2.5 mm^2 cable when installed enclosed in air and protected by a 20 A circuit breaker.

SUMMING UP...

- The maximum number of points and maximum demand of a final sub-circuit can be determined by AS/NZS 3000:2007 clause C5.2 and Table C8.
- Circuit protection devices for a variety of cable sizes and installation methods can be selected from:
  - AS/NZS 3000:2007 Table C5 for single phase cables
  - AS/NZS 3000:2007 Table C6 for three phase cables.
- Correct coordination of load, circuit protection and cable current carrying capacity is achieved when the statements $I_B \leq I_N \leq I_Z$ and $I_2 \leq 1.45 \times I_Z$ are true.
- Cables must be de-rated to 90% of the current carrying capacity when protected by HRC fuses.

11.3 Maximum demand on consumer's mains and sub-mains

The following text on maximum demand is based in part on the Tafe NSW, *Learner Workbook*, pp. 112–17.

When designing a new electrical installation it is necessary to predict the performance requirements for all parts of the installation. This is especially so for consumer's mains and sub-mains in an installation, which will have some diversity as all loads may not be connected at the same time. The maximum demand in the consumer's mains or sub-mains is the largest current expected to be carried by a conductor. In most installations within the scope of this book (installations up to 400 A) only one set of consumer's mains are permitted under local supply authority service rules. There can be as many sub-mains as required, as shown in Figure 11.7.

Electrically protected consumer's mains have either a fuse or a circuit breaker fitted at the connection point, or point of supply. In installations with a maximum demand of 400 A or greater, the consumer's mains are generally run directly to a distribution transformer. The consumer's mains are then electrically protected with fuses or a circuit breaker at the transformer. Electrically unprotected consumer's mains do not have either a fuse or circuit breaker fitted at the connection point, or point of supply. These are the vast majority of installations.

Determining maximum demand on consumer's mains and sub-mains

The maximum demand of consumer's mains and sub-mains is the subject of clause 2.2.2 of **AS/NZS 3000:2007** and, as for final sub-circuits, can be determined by:

- calculation, using **AS/NZS 3000:2007** Appendix C
- assessment (used for unusual loads, or large and complex installations where the number and type of loads is not known; see **AS/NZS 3000:2007** Table C3)
- measurement
- limitation by fixed current circuit breaker (inconvenient if the CB operates) – refer also to clause 2.2.2.

Once the maximum demand has been determined it is used to determine the minimum required cable size (c.s.a.) of the consumer's and/or sub-mains. The supply authority may set minimum cable sizes and types, and installation conditions for consumer's mains.

Calculation of maximum demand in consumer's mains and sub-mains is the simplest method of ensuring that minimum adequate cable sizes can be used. Many loads in an installation are not always turned on at the same time, or only for short periods. This means that some 'diversity' can be applied in determining the maximum demand in consumer's mains and sub-mains, allowing mains and sub-mains to have a maximum demand much less than the total current of all the individual loads, and cables smaller than that required to carry the total current for all loads.

Individual loads are allocated to load *groups* such as lighting, socket outlets, cooking ranges, etc. Based on results from similar previously constructed installations, each load group is given a contributing value of current to the maximum demand, which may be less than the total load ('diversity allowance').

The contribution to the maximum demand of each load group is given in **AS/NZS 3000:2007** Table C1 (domestic installations), Table C2 (non-domestic installations) and clause C2.5.2 (welding machines; in Appendix C). The maximum demand per phase is the total of the contributions by all the load groups on that phase supplied by that active conductor.

Determining the maximum demand in consumer's mains and sub-mains has some differences compared to determining it for the final sub-circuit. Be careful not to get the two confused. *Do not use Tables C1 or C2 for final sub-circuits*. Instead, use **AS/NZS 3000:2007** Tables C4, C5, C6 and C8.

The following questions should be considered when calculating maximum demand in consumer's mains and sub-mains:

- Is the installation domestic or non-domestic?
  - Use Table C1 for domestic installations.
  - Use Table C2 for non-domestic installations.
- What loads are to be supplied? This information is obtained from job specifications and or job plans.
- What equipment is in each load group?
  - Equipment in each load group is specified in column 1 of Tables C1 and C2.
  - Footnotes to the tables give additional information on the equipment in the various load groups.
- Is the installation to be supplied with single, two or three phase?

Local supply authorities may stipulate the maximum total load for single and two phase supplies. For example, single phase for total loads up to 100 A, two phase for total loads greater than 100 A and up to 200 A (maximum of 100 A per phase), and three phase for total loads over 200 A (split over three phases). An installation is also supplied with a three phase supply where there are three phase loads, such as three phase motors.

Follow the instructions in column 2 of Table C1 for each load group in single domestic installations, that is, individual houses (consumer's mains) or individual home units in a block (sub-mains). Calculations must be done separately for each conductor and only for the loads connected to it. You do *not* calculate the contribution for all phases once, then divide by three. Instead, add together the contribution of each load group supplied through the same active conductor.

If the load is across all active conductors, it must be balanced to satisfy the local supply authorities. If there is a large difference in the maximum demand between any two active conductors of a multi-phase supply, rearrange single phase loads to balance the load across all active conductors.

Single domestic

The maximum demand of consumer's mains in single domestic premises, or of individual units (townhouses or villas) in blocks of home units (townhouses or villas), is calculated using column 2 of **AS/NZS 3000:2007** Table C1 (the following text on maximum demand is based in part on the Tafe NSW, *Learner Workbook*, pp. 114–17).

EXAMPLE 11.18

Use Table 11.12 to calculate the maximum demand of the single phase consumer's mains for a single domestic dwelling (house) with the following loads:

- 15 lighting points
- 16 double 10 A sockets outlets (doubles count as two single points)
- 4 single 10 A socket outlets
- 1 4.6 kW storage type hot water system
- 1 11.4 kW cooking range (adapted from Tafe NSW, *Learner Workbook*, p. 117).

Solution

From AS/NZS 3000:2007 Table C1 column 2.

TABLE 11.12 Maximum demand calculation for a single phase domestic installation

Load group	Load	Calculation	Demand (A)
Ai	15 × lights	3 A	3.0
Bi	36 × 10 A socket outlets	10 + 5 = 15 A	15.0
C	11.4 kW range	11 400/230 × 0.5 = 24.8 A	24.8
F	4.6 kW hot water	4400/230 = 19.1 A	19.1
		Maximum demand	61.9

Source: *Select and Arrange Equipment for General Electrical Installations. Learner Workbook*, Industry Skills Unit Meadowbank, Tafe NSW, 2009, p. 117

EXAMPLE 11.19

Using Table 11.13 calculate the maximum demand of the single phase consumer's mains for a single domestic dwelling (house) with the following loads:

- 22 light points
- 3 500 W floodlights
- 24 double 10 A socket outlets
- 1 15 A socket outlet
- 1 6.0 kW cook top
- 1 3.9 kW wall oven
- 1 4.6 kW storage HWS (adapted from Tafe NSW, *Learner Workbook*, p. 118).

Solution

From AS/NZS 3000:2007 Table C1 column 2.

TABLE 11.13 Maximum demand calculation for a single phase domestic installation

Load Group	Load	Calculation	Demand (A)
Ai	22 × lights	3 + 2 = 5 A	5.0
Aii	3 × 500 W floodlights	$\frac{3 \times 500}{230} \times 0.75 = 4.9$ A	4.9
Bi	48 × 10 A socket outlets	10 + 5 + 5 = 20 A	20.0
Bii	1 × 15 A socket outlet	10 A	10.0
C	6 kW cook top	$\frac{6000}{230} \times 0.5 = 13$ A	13.0
C	3.9 kW wall oven	$\frac{3900}{230} \times 0.5 = 8.47$ A	8.47
F	4.6 kW HWS	$\frac{4400}{230} = 20$ A	20
		Maximum demand	81.37 A

Source: *Select and Arrange Equipment for General Electrical Installations. Learner Workbook*, Industry Skills Unit Meadowbank, Tafe NSW, 2009, p. 118

EXAMPLE 11.20

Using Table 11.14 calculate the maximum demand of the three phase consumer's mains for a single domestic dwelling (house) with the following loads:

- 48 light points (two circuits)
- 1 3-in-1 fan/heat lamps, 4 × 275 W heat lamps (connected to light circuit)
- 30 double 10 A socket outlets (three circuits)
- 1 18 A 3 Φ ducted A/C
- 1 7.8 kW 1 Φ range
- 1 22.0 kW 3 Φ spa heater
- 1 3.6 kW sauna
- 1 4.4 kW storage HWS (adapted from Tafe NSW, *Learner Workbook*, p. 121).

Solution

From AS/NZS 3000:2007 Table C1 column 2.

TABLE 11.14 Maximum demand calculation for a three phase domestic installation

Load Group	Load	Calculation	A (A)	B (A)	C (A)
Ai	24 lights + fan	3 + 2 = 5 A		5	5
Bi	20 socket outlets	10 A	10		10
Bi	20 socket outlets	10 + 5 = 15 A		15	
D	Air-conditioning	18 × 0.75 = 13.5 A	13.5	13.5	13.5
C	7.8 kW range	$\frac{7800}{230} \times 0.5 = 16.9$ A			16.9
G	Spa	$\frac{22\,000}{\sqrt{3} \times 400} \times 0.75 = 23.8$ A	23.8	23.8	23.8
D	Sauna	$\frac{3600}{230} \times 0.75 = 11.8$ A		11.8	
F	4.6 kW HWS	$\frac{4400}{230} = 19.1$ A	19.1		
			66.4	69.1	69.2

Source: *Select and Arrange Equipment for General Electrical Installations. Learner Workbook*, Industry Skills Unit Meadowbank, Tafe NSW, 2009, p. 121

In Example 11.20, **AS/NZS 3000:2007** Table C1 column 2 is used to determine the maximum demand of the installation. The 3-in-1 fan heater is wired to the lighting circuit on B phase. Table C3 'footnote e' says that a fan less than 150 watts is counted as a light point. The fan and the light of the 3-in-1 fan heater will add two extra points to B phase, increasing the total number of points to 26. The additional points do not take the total above another group of 20, so no additional current is added. The heat lamps, even though they are wired to a lighting circuit, are included in load group Bi as a permanently connected load because they are greater than 150 W but less than 10 A, so they are treated as an additional socket outlet. The heater pushes the number of socket outlets above 20 so another 5 amperes is added to B phase.

If there is a difference of more than 25 amperes between the total and the highest and lowest phase, a single phase load will need to be moved to another phase until the calculation balances. The air-conditioner and spa are three phase loads so they contribute current to all three phases and cannot be shifted to balance the load.

Multiple domestic

The majority of duplexes, villas, townhouses and units are only connected to a single phase supply, for cost purposes. A three phase consumer's main is run to the main switchboard and a single phase sub-main to each unit. So if there are 30 units in total, only 10 units would be connected to each phase. The maximum demand of consumer's mains in multiple domestic premises (blocks of home units) is calculated using column 3, 4 or 5 of **AS/NZS 3000:2007** Table C1. Column 3 is used where there are between two and five units per phase in the installation. Column 4 is used for between six and 20 units per phase. Column 5 is used for 21 or more units per phase (Tafe NSW, *Learner Workbook*, p. 122).

Community loads

Loads in common areas (foyers, stairwells, community laundries, garages, common recreational or outdoor areas, etc.) are covered in load groups H to M. In large unit blocks the community loads may be connected over three phases (lifts, fire pumps, air-conditioning in common areas, etc.) but in smaller blocks, where the common area loads are only single phase, the supply authority may allow connection of these loads to single phase only, to reduce the metering required. If this is the case, the number of units per phase may need to be uneven to allow for community loads and to make the calculation balance (Tafe NSW, *Learner Workbook*, p. 122).

EXAMPLE 11.21

Using Table 11.15 calculate the maximum demand of the two phase consumer's mains for a multiple domestic dwelling of a group of four townhouses that are to be connected to a two phase supply, two units per phase. Each unit contains the following loads:

- 18 light points
- 6 metres track lighting
- 11 double 10 A socket outlets
- 1 6.0 kW 1 Φ range
- 1 15 A socket for a clothes dryer
- 1 4.6 kW storage HWS (adapted from Tafe NSW, *Learner Workbook*, p. 124).

Solution

From AS/NZS 3000:2007 Table C1 column 3.

TABLE 11.15 Calculation of multiple installation maximum demand, two phase supply

No. of units A phase = 2, B phase = 2, C phase = 0

Load group	Load	Calculation	A (A)	B (A)	C (A)
Ai	Lights		6	6	0
Bi	Socket outlets	10 + (5 × 2) = 20 A	20	20	0
C	Range		15	15	0
D	Clothes dryer	2 × 15 × 0.75 = 22.5 A	22.5	22.5	0
F	HWS	2 × 6 = 12 A	12	12	0
		Total	75.5	75.5	0

Source: *Select and Arrange Equipment for General Electrical Installations. Learner Workbook*, Industry Skills Unit Meadowbank, Tafe NSW, 2009, p. 124

In Example 11.21, each unit contains both lights and track light. **AS/NZS 3000:2007** Table C1 'footnote d' regards each metre of lighting track as the equivalent of two lighting points. This, however, is irrelevant to the calculation of consumer's mains maximum demand. The number of units per phase is the basis of the calculation, not the load contained within each unit. As there is only a two phase supply no load is connected to C phase.

EXAMPLE 11.22

Using Table 11.16 calculate the maximum demand of the three phase consumer's mains for a multiple domestic dwelling of a group of 24 townhouses that are to be connected to a three phase supply, eight units per phase. Each unit contains the following loads:

- 11 lighting points
- 7 double socket outlets
- 3 single socket outlets
- 1 15 A socket outlet
- 1 9.2 kW range
- 1 4.6 kW storage water heater.
- There is no communal load (adapted from Tafe NSW, *Learner Workbook*, p. 124).

Solution

From AS/NZS 3000:2007 Table C1 column 4.

TABLE 11.16 Calculation of multiple installation maximum demand, three phase supply, no communal load

No of units A phase = 8, B phase = 8, C phase = 8

Load group	Load	Calculation	A (A)	B (A)	C (A)
Ai	Lights	5 + (0.25 × 8)	7	7	7
Bi	Socket outlets	15 + (3.75 × 8)	45	45	45
Bii	15 A outlet	10 A	10	10	10
C	Ranges	2.8 × 8	22.2	22.2	22.2
F	Hot water heater	6 × 8	48	48	48
		Total	132.2	132.2	132.2

Source: *Select and Arrange Equipment for General Electrical Installations. Learner Workbook*, Industry Skills Unit Meadowbank, Tafe NSW, 2009, p. 122

EXAMPLE 11.23

Using Table 11.17 calculate the maximum demand of the three phase consumer's mains for a block of 22 units that are to be connected to a three phase supply. The communal load is greater than 25 A.

Each unit contains the following load:

- 16 light points
- 16 double 10 A socket outlets
- 1 6.0 kW cook top
- 1 3.0 kW oven
- 1 4.6 kW storage HWS.

Communal load:

- 20 fluorescent (0.33 A each) car park
- 50 60 W lamps house lights
- 15 single 10 A socket outlets house power
- 2 15 A socket outlets clothes dryers (adapted from Tafe NSW, *Learner Workbook*, p. 125).

Solution

From AS/NZS 3000:2007 Table C1 column 4.

TABLE 11.17 Calculation of multiple installation maximum demand, three phase supply, with communal load

No of units A phase = 8, B phase = 8, C phase = 6 + communal

Load group	Load	Calculation	A (A)	B (A)	C (A)
Ai	Lights	5 + (0.25 × 8) 5 + (0.25 × 6)	7	7	6.5
Bi	Socket outlets	15 + (3.75 × 8) 15 + (3.75 × 6)	45	45	37.5
C	Cook top	2.8 × 8 2.8 × 6	22.4	22.4	16.8
C	Oven	Included above	–	–	–
F	Hot water	6 × 8 = 48 A 6 × 6 = 36 A	48	48	36
H	Communal lights	$(20 \times 0.33) + \left(\frac{50 \times 60}{230}\right) = 19.65$ A	–	–	19.65
I	Communal power	15 × 2 = 30 A Maximum 15 A	–	–	15
J	Dryers	2 × 15 × 0.5 = 15 A	–	–	15
		Total	122.4	122.4	146.45

Source: *Select and Arrange Equipment for General Electrical Installations. Learner Workbook*, Industry Skills Unit Meadowbank, Tafe NSW, 2009, p. 125

In Example 11.23, there is 49.65 A of communal load which would prevent the installation from balancing. Units are moved from C phase to A and B phases to accommodate the additional communal load on C phase.

Non-domestic maximum demand

When calculating maximum demand for non-domestic installations such as factories, shops and offices, **AS/NZS 3000:2007** Table C2 is used. The load that is installed is likely to be used more frequently than in a domestic installation, so therefore less diversity is applied. The contribution to the maximum demand of each load group will be closer to the connected load.

EXAMPLE 11.24

Using Table 11.18 calculate the maximum demand of a three phase industrial installation comprising:

- 6 twin × 36 W fluorescent lights rated at 0.46 A each
- 12 mercury vapour high bay lights rated at 1.8 A each
- 21 single phase double 10 A socket outlets
- 2 three phase 10 A socket outlets
- 2 15 A single phase socket outlets
- 2 three phase 20 A socket outlets
- 1 single phase 2.2 kW instantaneous water heater
- 1 single phase 3.6 kW storage water heater
- 1 three phase 5 kW, 9 A compressor
- 1 three phase 4.1 kW, 8 A milling machine
- 1 three phase 2.2 kW, 5 A lathe

- 1 three phase 370 W, 1 A pedestal drill
- 1 three phase 560 W, 3 A grinder
- 1 single phase, 400 V electric arc welder rated at 14 A (adapted from Tafe NSW, *Learner Workbook*, p. 132).

Solution

From AS/NZS 3000:2007 Table C2 column 3.

TABLE 11.18 Calculation of non-domestic maximum demand

Load Group	Load	Calculation	A (A)	B (A)	C (A)
A	6 × 0.46 A fluorescent lights (C Φ)	6 × 0.46 = 2.8 A			2.8
A	12 × MV lights (A & C Φ)	6 × 1.8 = 10.8 A	10.8		10.8
Bi	21 × 2 × 1 Φ 10 A socket outlets (14/Φ) + 2 × 3 Φ 10 A socket outlets 16 points per phase	1000 + (750 × 15) = 53.3 A 230	53.3	53.3	53.3
Bi	2 × 3 Φ 10 A socket outlets	2 extra points included above	–	–	–
Biii	2 × 3 Φ 20 A S/0s	20 + (0.75 × 20) = 35 A	35.0	35.0	35.0
Biii	2 × 1 Φ 15 A socket outlets (A & B Φ)	(0.75 × 15) = 11.25 A	11.25	11.25	
D	1 × 9 A motor	9 A	9	9	9
D	1 × 8 A motor	8 × 0.75 = 6 A	6	6	6
D	1 × 5 A motor	5 × 0.5 = 4 A	2.5	2.5	2.5
D	1 × 3 A motor	3 × 0.5 = 1.5 A	1.5	1.5	1.5
D	1 × 1 A motor	1 × 0.5 = 0.5 A	0.5	0.5	0.5
G	1 × 1 Φ 3.6 kW storage HWS	$\frac{3300}{230} = 14.3$ A			14.3
H	1 Φ × 400 V 14 A arc welder	Full connected load	14	14	
		Total	143.85	133.05	135.7

Source: *Select and Arrange Equipment for General Electrical Installations. Learner Workbook*, Industry Skills Unit Meadowbank, Tafe NSW, 2009, p. 125

In Example 11.24, the lighting circuits have been split into three circuits. The fluorescents and one group of mercury vapour lamps are connected to C phase to balance the load.

As there is no mention of heating or cooling equipment such as air-conditioning in the load details or description, it can be assumed there is none. The socket outlets therefore fall into load group Bi not Bii. Load group B in Table C2 is only for socket outlets, not for permanently connected load. It differs from load group B in Table C1, as permanently connected load less than 10 A is not included in load group B. Permanently connected loads (regardless of the current rating) are included in one of the other load groups. The 21 double socket outlets are split evenly across all three phases, seven doubles per phase.

It is important to remember we are not dealing with the final sub-circuits in this case. This is how many socket outlets are indirectly connected to each phase via the consumer's mains. Note that the three phase 10 A socket outlets are included in the count of points in load group Bi. The load contribution for socket outlets is first calculated in watts and then converted to current by dividing by a phase voltage of 230 volts. Both the 20 and 15 A socket outlets fall into load group Biii as they are greater than 10 A and are calculated accordingly.

There are a number of motors in this installation. The highest is rated at full load current, the next at 75%, and all of the remaining motors are rated at 50% each. If all motors are of the same rating, the same rule would apply to each individual motor. For example, if the installation contained four motors each rated at 12 A, the motor current contribution would equal 12 + (12 × 0.75) + (2 × 12 × 0.5) = 33 A.

The hot water system is a full connected load and is connected to C phase to balance the load.

Single phase welders, as shown in Figure 11.9, are only connected between two phases so that they are supplied at 400 volts rather than 230 volts. Because each welder is a single load they are considered single phase. Groups of welders will be connected so that the load is balanced across the three phases. That is, the first welder is between A and B phases, the second between B and C phases, and the third between C and A phases. Welder loads are calculated in accordance with **AS/NZS 3000:2007** clause C2.5.2, in this case as a full connected load.

FIGURE 11.9 Single phase welders connected to a three phase supply to balance the load

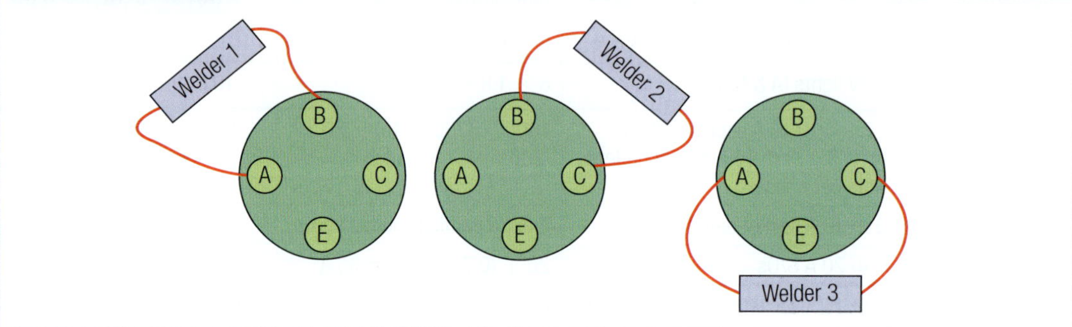

EXAMPLE 11.25

Using Table 11.19 calculate the maximum demand of a three phase industrial installation comprising:

- 36 high bay MV luminaires rated at 2.6 A each (12/phase)
- 21 10 A double single phase socket outlets (7/phase)
- 4 20 A three phase socket outlets
- 2 lathes with three phase induction motors rated at 28 A
- 3 400 V single phase spot welders with a rated primary current of 40 A and a duty cycle of 40% (adapted from Tafe NSW, *Learner Workbook*, p. 134).

Solution

From AS/NZS 3000:2007 Table C2 column 3.

TABLE 11.19 Calculation of non-domestic maximum demand

Load group	Load	Calculation	A (A)	B (A)	C (A)
A	Lights (12 per Φ)	12 × 2.6 = 2.8 A	31.2	31.2	31.2
Bi	10 A socket outlets	$\frac{1000 + 13 \times 750}{230} = 46.7$ A	46.7	46.7	46.7
Biii	4 × 3 Φ 20 A socket outlets	20 + (3 × 0.75 × 20) = 65	65	65	65
D	Lathe 1	28 A	28	28	28
D	Lathe 2	28 × 0.75 = 21 A	21	21	21
D	Welder 1	1st welder 40 × 0.63 = 25.2 A	25.2	25.2	–
D	Welder 2	1st welder 40 × 0.63 = 25.2 A 2nd welder 40 × 0.63 × 0.6 = 15.2 A		15.2	25.2 –
D	Welder 3	2nd welder 40 × 0.63 × 0.6 = 15.2 A	15.2	–	15.2
		Total	232.3	232.3	232.3

Source: *Select and Arrange Equipment for General Electrical Installations. Learner Workbook*, Industry Skills Unit Meadowbank, Tafe NSW, 2009, p. 134

In Example 11.25, single phase resistance (spot) welders are installed. The machines installed have a 40% duty cycle (ratio of time on to time off during weld). **AS/NZS 3000:2007** clause C2.5.2.3(a)(ii) allows for a multiplication factor of 0.63 of the rated primary current of each resistance welder. This reduces the resistance welder's current contribution to the consumer's mains from 40 to 25.2 A. The second welder on each phase has its contribution further reduced by another 60% according to **AS/NZS 3000:2007** clause C2.5.2.3(b).

Non-domestic energy demand method

In some cases the exact load details of equipment being installed in a non-domestic installation will not be known at the time of construction. Installations such as factory units, shops and offices are mostly rental properties. The electrical demand will vary from tenant to tenant. In these cases an estimate based on experience gained from other similar installations can be made using **AS/NZS 3000:2007** Table C3 (Tafe NSW, *Learner Workbook*, p. 138).

The maximum demand is assessed depending on the area, use and climate control (air-conditioning) of the installation. The demand is given in VA and is converted to current using the equation:

$$I_L = \frac{S}{\sqrt{3} \times V_L}$$

where

I_L = the line current/maximum demand current in amperes
S = the energy demand in VA
V_L = the line voltage of the supply in volts

EXAMPLE 11.26

Using Table 11.20 calculate the maximum demand of a three phase small retail complex consisting of three shops at street level (280 m² each) and three offices (250 m² each) on the first floor. All shops and offices have reverse cycle air-conditioning, light and power. Determine the maximum demand of this commercial installation (adapted from Tafe NSW, *Learner Workbook*, p. 138).

Solution

From AS/NZS 3000:2007 Table C3 column 4.

TABLE 11.20 Calculation of maximum demand for a small retail complex

Type of occupancy		Energy demand		
	Load	**Area**	**Average VA/m²**	**Demand**
Shops	Light and power	840 m²	70	58800 VA
	Air-conditioning	840 m²	30	25200 VA
Offices	Light and power	750 m²	50	37500 VA
	Air-conditioning	750 m²	25	18750 VA
			Total	140250 VA

Source: *Select and Arrange Equipment for General Electrical Installations. Learner Workbook*, Industry Skills Unit Meadowbank, Tafe NSW, 2009, p. 138

$$\text{Maximum demand } I_L = \frac{S}{\sqrt{3} \times V_L} = \frac{140250}{\sqrt{3} \times 400} = 203 \text{ A per phase}$$

EXAMPLE 11.27

Using Table 11.21 calculate the maximum demand of the consumer's mains of a three phase factory unit complex with the following load:

- 1200 m^2 of light industrial units fitted with light and power and ventilation
- 850 m^2 of offices fitted with light and power and reverse cycle air-conditioning
- 500 m^2 of open-air car park (adapted from Tafe NSW, *Learner Workbook*, p. 141).

Solution

From AS/NZS 3000:2007 Table C3 column 4.

TABLE 11.21 Calculation of non-domestic maximum demand for a factory unit complex

Type of occupancy		Energy demand		
	Load	**Area**	**Average VA/m^2**	**Demand**
Factory units	Light and power	1200 m^2	15	18000 VA
	Ventilation	1200 m^2	15	18000 VA
Offices	Light and power	850 m^2	50	42500 VA
	Air-conditioning	850 m^2	25	21250 VA
Car park		500 m^2	5	2500 VA
			Total	102250 VA

Source: *Select and Arrange Equipment for General Electrical Installations. Learner Workbook*, Industry Skills Unit Meadowbank, Tafe NSW, 2009, p. 141

$$\text{Maximum demand } I_L = \frac{S}{\sqrt{3} \times V_L} = \frac{102250}{\sqrt{3} \times 400} = 147.6 \text{ A per phase}$$

Maximum demand of sub-mains

The methods used to calculate the maximum demand of sub-mains are the same as for consumer's mains, as stated in **AS/NZS 3000:2007** clause 2.2.2. They are calculation, assessment, measurement and limitation. The most commonly used method is calculation. The one variation is that only the load connected to the sub-main is included in the maximum demand calculation of the sub-main.

In modern domestic installations the space available for the customer's equipment on a switchboard has decreased, because supply authorities are requiring larger 'footprints' for their metering equipment. There has also been an increase in the number and types of protection devices. Two pole combination RCD/MCB and voltage surge diverters fitted in main switchboards require more space. It is now common practice in larger installations to run a sub-main to a location such as the garage or kitchen. A distribution board placed in the kitchen shortens the runs of a majority of the final sub-circuits, reducing cost, and provides the convenience to the customer of not having to go outside to reset a tripped circuit breaker. Sub-mains are also used to supply out-buildings such as granny flats or garages.

Unlike consumer's mains, sub-mains are electrically protected at their origin. The nominal rating of the protection device and from there the rating of the cable is determined by the sub-main's maximum demand (Tafe NSW, *Learner Workbook*, pp. 144–5).

EXAMPLE 11.28

Using Table 11.22 calculate the maximum demand of a single phase sub-main supplying a single domestic installation (granny flat) comprising:

- 8 lighting points
- 5 double 10 A socket outlets
- 1 single 10 A socket outlet
- 1 3.3 kW heat pump storage HWS
- 1 4.4 kW range
- 1 15 A socket outlet for a split system air-conditioner (adapted from Tafe NSW, *Learner Workbook*, p. 144).

Solution

From AS/NZS 3000:2007 Table C1 column 2.

TABLE 11.22 Calculation of sub-main maximum demand for an outbuilding

Load group	Load	Calculation	Demand (A)
Ai	8 lights	3 A	3.0
Bi	11 × 10 A socket outlets	10 A	10.0
F	Hot water system	$\frac{3300}{230} = 14.4$ A	14.4
C	4.4 kW range	$\frac{4400}{230} \times 0.5 = 9.6$ A	9.6
D	Air-conditioner	15 × 0.75 = 11.25 A	11.25
		Maximum demand	48.25

Note: For Bi check against AS/NZS 3000:2007 Table C1 footnote h.

Source: *Select and Arrange Equipment for General Electrical Installations. Learner Workbook*, Industry Skills Unit Meadowbank, Tafe NSW, 2009, p. 143

EXAMPLE 11.29

Using Table 11.23 calculate the maximum demand for the sub-main of a single unit in a multiple domestic installation. A group of four townhouses are to be connected to a two phase supply, two units per phase. Each contains the following load:

- 18 light points
- 6 m track lighting
- 11 double 10 A socket outlets
- 1 6.0 kW 1 Φ range
- 1 15 A socket for a clothes dryer
- 1 4.6 kW storage HWS (adapted from Tafe NSW, *Learner Workbook*, p. 145).

Solution

From AS/NZS 3000:2007 Table C1 column 2.

TABLE 11.23 Calculation of sub-main maximum demand for a single unit in a four-unit complex

Load group	Load	Calculation	Demand (A)
Ai	18 lights + 6 m track light (30 points total)	3 + 2 = 5 A	5.0
Bi	11 × 10 A double socket outlets (22 points)	10 + 5 = 15 A	15.0
C	6 kW range	$\frac{6000}{230} \times 0.5 = 13.1$ A	13.1
C	Clothes dryer	15 × 0.5 = 7.5 A	7.5
F	Hot water system	$\frac{4400}{230} = 19.1$ A	19.1
		Maximum demand	59.7

Source: *Select and Arrange Equipment for General Electrical Installations. Learner Workbook*, Industry Skills Unit Meadowbank, Tafe NSW, 2009, p. 145

Table 11.23 shows the calculation of a sub-main for one of the units of the installation given in Example 11.29. In the calculation for consumer's mains the number of units connected to each phase and column 3 of **AS/NZS 3000:2007** Table C1 is used to determine the maximum demand. When calculating the sub-main maximum demand for an individual unit, column 2 of **AS/NZS 3000:2007** Table C1 and the actual load within the individual unit is considered to calculate the maximum demand of the sub-main.

EXAMPLE 11.30

Using Table 11.24 calculate the maximum demand of a single phase sub-main for a single unit in a block of 22 units where the communal load is greater than 25 A.

Each unit contains the following load:

- 16 light points
- 16 double 10 A socket outlets
- 1 6.0 kW cook top
- 1 3.0 kW oven.

Communal load:

- 20 fluorescent (0.33 A each) car park
- 50 60 W lamps house lights
- 15 single 10 A socket outlets house power
- 1 4.4 kW storage HWS
- 2 15 A socket outlets clothes dryers (adapted from Tafe NSW, *Learner Workbook*, p. 146).

Solution

From AS/NZS 3000:2007 Table C1 column 2.

TABLE 11.24 Calculation of domestic sub-main maximum demand for one unit in a block of units

Load group	Load	Calculation	Demand (A)
Ai	16 lights	3 A	3.0
Bi	16 × 10 A double socket outlets (32 points)	10 + 5 = 15 A	15.0
C	6 kW range	$\frac{6000}{230} \times 0.5 = 13.1$ A	13.1
C	3.0 kW oven	$\frac{3000}{230} \times 0.5 = 13.1$ A	6.52
F	Hot water system	$\frac{4400}{230} = 19.1$ A	19.1
		Maximum demand	56.72

Source: *Select and Arrange Equipment for General Electrical Installations. Learner Workbook*, Industry Skills Unit Meadowbank, Tafe NSW, 2009, p. 146

EXAMPLE 11.31

Using Table 11.25 calculate the maximum demand of a single phase sub-main for the communal load of a block of 22 units, where the communal load is greater than 25 A. Each unit contains the following load:

- 16 light points
- 16 double 10 A socket outlets
- 1 6.0 kW cook top
- 1 3.0 kW oven
- 1 4.4 kW storage HWS.

Communal load:

- 20 fluorescent (0.33 A each) car park
- 50 60 W lamps house lights
- 15 single 10 A socket outlets house power
- 2 15 A socket outlets clothes dryers (adapted from Tafe NSW, *Learner Workbook*, p. 146).

Solution

From AS/NZS 3000:2007 Table C1 column 2.

TABLE 11.25 Calculation of domestic sub-main for communal load

Load group	Load	Calculation	Demand (A)
H	20 × fluorescent lights	20 × 0.33 = 6.6 A	6.6
H	50 × 60 W house lights	$\frac{50 \times 60}{230} = 1.3$ A	1.3
I	15 × 10 A single socket outlets (15 points)	2 × 15 = 30 A 15 A maximum	15.0
J	2 × clothes dryers	2 × 15 × 0.5 = 15 A	15.0
		Maximum demand	37.9

Source: *Select and Arrange Equipment for General Electrical Installations. Learner Workbook*, Industry Skills Unit Meadowbank, Tafe NSW, 2009, p. 146

Sub-mains in non-domestic installations

Once again the procedure to determine the maximum demand of sub-mains in a non-domestic installation is the same as for consumer's mains, but only the load in that section of the building is included (Tafe NSW, *Learner Workbook*, p. 147).

EXAMPLE 11.32

Using Table 11.26 calculate the maximum demand of a sub-main for a factory unit with the following load:

- 6 high bay MV luminaires rated at 2.6 A each (one circuit)
- 30 twin 36 W fluorescent lights 0.333 A each (one circuit)
- 20 10 A double single phase socket outlets (five circuits, four points per circuit)
- 4 20 A three phase socket outlets
- 4 three phase induction motors rated at 32 A
- 1 400 V single phase arc welding machine with a rated primary current of 40 A
- 1 4.4 kW 230 V HWS (adapted from Tafe NSW, *Learner Workbook*, p. 147).

Solution

From AS/NZS 3000:2007 Table C21 column 3.

TABLE 11.26 Calculation of maximum demand of a sub-main for a factory unit

Load group	Load	Calculation	A (A)	B (A)	C (A)
A	6 × high bay lights	6 × 2.6 = 15.6 A	15.6		
A	30 × twin 36 W fluorescent lights	30 × 0.33 = 10 A		10	
Bi	8 × 10 A double socket outlets 16 points per phase	$\frac{1000 + (750 \times 15)}{230} = 53.3$ A		53.3	53.3
Bi	4 × 10 A double socket outlets 8 points per phase	$\frac{1000 + (750 \times 7)}{230} = 53.3$ A	27.2		

(Continued)

TABLE 11.26 *(Continued)*

Biii	4 × 3 Φ 20 A socket outlets	20 + (3 × 20 × 0.75) = 65 A	65	65	65
D	motor 1 32 A		32	32	32
D	motor 2 32 A	32 × 0.75 = 24 A	24	24	24
D	motor 3 32 A	32 × 0.5 = 16 A	16	16	16
D	motor 4 32 A	32 × 0.5 = 16 A	16	16	16
H	1 Φ × 400 V 40 A arc welder	Full connected load	40		40
G	1 × 1 Φ 3.6 kW storage HWS	$\frac{4400}{230} = 19.1$ A		19.1	
		Total	235.8	235.4	246.3

Source: *Select and Arrange Equipment for General Electrical Installations. Learner Workbook*, Industry Skills Unit Meadowbank, Tafe NSW, 2009, p. 147

EXAMPLE 11.33

Using Tables 11.27 and 11.28 calculate the maximum demand for the sub-mains for each shop and office in a three phase small retail complex consisting of three shops at street level (280 m^2 each) and three offices (250 m^2 each) on the first floor. All shops and offices have reverse cycle air-conditioning, light and power (adapted from Tafe NSW, *Learner Workbook*, p. 148).

Solution

From AS/NZS 3000:2007 Table C3 column 4.

TABLE 11.27 Calculation of maximum demand for sub-mains for each shop in a three phase small retail complex

Type of occupancy		**Energy demand**		
	Load	**Area**	**Average VA/m²**	**Demand**
Shops	Light and power	280 m²	70	19 600 VA
	Air-conditioning	280 m²	30	8 400 VA
			Total	28 000 VA

Source: *Select and Arrange Equipment for General Electrical Installations. Learner Workbook*, Industry Skills Unit Meadowbank, Tafe NSW, 2009, p. 148

$$\text{Maximum demand } I_L = \frac{S}{\sqrt{3} \times V_L} = \frac{28000}{\sqrt{3} \times 400} = 40.5 \text{ A per phase}$$

TABLE 11.28 Calculation of maximum demand for sub-mains for each office in a three phase small retail complex

Type of occupancy		**Energy demand**		
	Load	**Area**	**Average VA/m²**	**Demand**
offices	Light and power	250 m²	50	12 500 VA
	Air conditioning	250 m²	25	6 250 VA
			Total	18 750 VA

Source: *Select and Arrange Equipment for General Electrical Installations. Learner Workbook*, Industry Skills Unit Meadowbank, Tafe NSW, 2009, p. 148

$$\text{Maximum demand } I_L = \frac{S}{\sqrt{3} \times V_L} = \frac{18750}{\sqrt{3} \times 400} = 28 \text{ A per phase}$$

SUMMING UP...

- Maximum demand is the maximum expected current that a conductor must carry.
- Maximum demand can be determined by calculation, assessment, measurement and limitation.
- Calculation is the most commonly used method to determine the maximum demand of consumer's mains and sub-mains.
- Limitation is the most commonly used method to determine the maximum demand of final sub-circuits.
- AS/NZS 3000:2007 Table C1 is used to calculate the maximum demand of consumer's mains and sub-mains of single and multiple domestic installations.
- AS/NZS 3000:2007 Table C2 is used to calculate the maximum demand of consumer's mains and sub-mains of non-domestic installations.
- AS/NZS 3000:2007 Table C3 is an alternative energy demand method used to determine the maximum demand of consumer's mains and sub-mains of non-domestic installations.
- AS/NZS 3000:2007 Table C4 is used to determine the maximum demand of final sub-circuits for domestic cooking appliances only.
- Appendix C of AS/NZS 3000:2007 provides examples of maximum demand calculations, starting at C2.3.2.

11.4 Cable selection based on current carrying capacity requirements

Cable sizes are selected by their current carrying capacity to achieve the smallest cross-sectional area of the conductor, to minimise the cost of installation. At the same time, the current flow must not cause the cable's insulation to exceed its rated temperature. **AS/NZS 3000:2007** Tables C5 and C6 provide guidance to the current carrying capacities of copper conductors ranging in sizes from 1 mm^2 to 25 mm^2. This topic introduces you to Section 3 of **AS/NZS 3008.1**, a series of standards which specifies current carrying capacities of both copper and aluminium conductors ranging in sizes of 1 mm^2 to 650 mm^2. It should be noted that **AS/NZS 3008.1.1:2009** is for Australian conditions and **AS/NZS 3008.1.2:2009** is for New Zealand conditions.

Installation conditions of wiring systems

The following section on installation conditions is based in part on the Tafe NSW, *Learner Workbook*, pp. 156–8. The installation conditions of a conductor will affect its current carrying capacity, and thus the size of cable needed for a circuit. **AS/NZS 3008.1.1:2009** clause 3.4 specifies a number of insulation conditions:

- ambient temperature
- cables installed in:
  - air
  - thermal insulation
  - conduits, ducts and trunking
  - underground.

Table 3 of **AS/NZS 3008.1.1:2009** provides guidance to different installation methods:

- Table 3(1) unenclosed in air
- Table 3(2) enclosed
- Table 3(3) buried direct in the ground
- Table 3(4) in underground wiring enclosures.

Generally tables in **AS/NZS 3008.1.1:2009** only show the current carrying conductors. In other words, in single phase circuits only the active and neutral conductors are shown. In three phase circuits the load is assumed to be balanced, so the neutral current is zero. Therefore, to simplify the diagram the neutral conductor is not shown. Single core cables can be configured in trefoil, laid flat touching and laid flat separated as shown in Figure 11.10.

FIGURE 11.10 Single core cable configurations

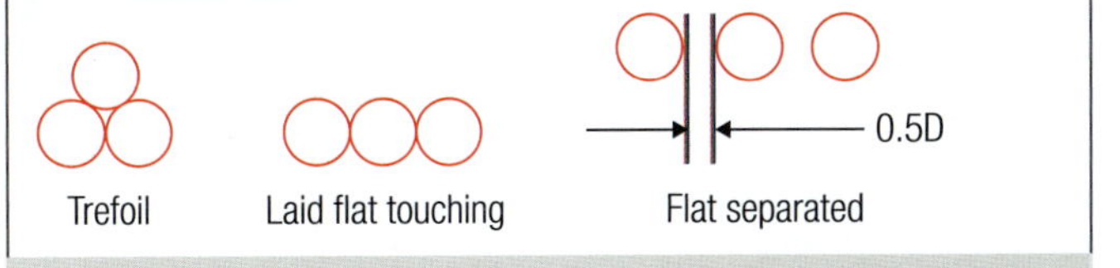

Source: AS/NZS 3008.1.1:2009 Electrical installations – Selection of cables, Figure 1A, p. 21

Separation between single cables improves the heat dissipation of the conductors and improves current carrying capability. Laying the cables in trefoil reduces magnetic effects. It is important not to confuse separation between conductors with cable support systems that space the cable from surfaces such as walls and ceilings, as shown in Table 3(1) of **AS/NZS 3008.1.1:2009**.

If cables are installed so that they are in contact with cables of another circuit they are said to be *grouped*, as shown in Table 1 of **AS/NZS 3008.1.1:2009**.

External influences that require a de-rating factor

External influences that can affect the current rating of a cable are specified in clause 3.5 of **AS/NZS 3008.1.1:2009**. Factors include grouping of cables, ambient temperature, depth of laying, different soil types for underground cables, varying loads, thermal insulation, direct sunlight, harmonic currents, paralleling of cables and electromagnetic interference. If any of these influences are present where a cable is installed, a de-rating (DR) factor is applied to the cable.

TABLE 11.29 Cable installation conditions that avoid de-rating

Condition	Value
Ambient air temperature	40°C
Ambient soil temperature	25°C
Depth of laying cable underground	0.5 m
Soil thermal resistivity	1.2°C.m/W
Cable grouping	single circuit
Harmonic distortion	<15% THD AS/NZS 3008.1.1:2009 Table 2
Circuit protection	Circuit breaker

Source: AS/NZS 3008.1.1:2009 Electrical installations – Selection of cables, Figure 1B, p. 21

The circuit protection device selected to protect the cable will also affect the cable current carrying capacity:

- circuit breaker – 100% of current carrying capacity of cable (DR = × 1)
- HRC fuse – 90% of current carrying capacity of cable (DR = × 0.9)
- semi-enclosed rewireable fuses, existing installations only, 80% of current carrying capacity of cable (DR = × 0.8).

AS/NZS 3008.1.1:2009 clauses 3.4.2 to 3.5.2 and 3.5.2.2 specifies values for installation conditions and external influences on cables that do not require de-rating. You should make yourself familiar with these values. A summary is shown in Table 11.29.

If cables are installed as described in Table 11.29, the cable is not required to be 'de-rated'. If any variation to these installation conditions occurs the cable will have to be de-rated. This means its current carrying capacity will be lower, and a larger cable c.s.a. may be required. Footnotes at the bottom of **AS/NZS 3008.1.1:2009** Tables 4 to 20 provide guidance to applicable de-rating factors.

EXAMPLE 11.34

Determine the de-rating factors for the cable installation methods and conditions shown in Table 11.30 (adapted from Tafe NSW, *Learner Workbook*, p. 161).

Solution

TABLE 11.30 Cable installation conditions de-ratings

Cable installation methods and conditions	De-rating factor	Table column
A single phase V75 T + E cable installed on perforated cable tray in a factory with an ambient air temperature of 55°C.	0.72	T27(1) C10
A three phase V75 4 core + E cable installed in an underground conduit with an ambient soil temperature of 35°C.	0.89	T27(2) C7
A three phase V75 4 core + E cable installed in an underground conduit at a depth of 0.6 m.	0.99	T28(2) C3

Source: AS/NZS 3008.1.1:2009 Electrical installations – Selection of cables, Tables 27 and 28, pp. 83 and 84

The minimum distances that cables of different circuits must be spaced by to avoid a de-rating for grouping is shown in **AS/NZS 3008.1.1:2009** Figure 1.

EXAMPLE 11.35

Determine the minimum spacing required to avoid de-rating of grouped cables installed as listed in Table 11.31 (adapted from Tafe NSW, *Learner Workbook*, p. 162).

Solution

TABLE 11.31 Minimum spacing between different circuits to avoid de-rating for grouping

Multiple circuit installation method	Horizontal spacing ↔	Vertical spacing ↕
Single core cables in trefoil or laid flat supported on perforated or unperforated cable trays	2 × cable diameter	4 × cable diameter
Single core cables in trefoil or laid flat fixed directly to a wall, floor, ceiling or similar surface	2 × cable diameter	6 × cable diameter
Multi-core cables supported on perforated or unperforated cable trays	2 × cable diameter	4 × cable diameter
Multi-core cables fixed directly to a wall, floor, ceiling or similar surface	2 × cable diameter	6 × cable diameter

Source: AS/NZS 3008.1.1:2009 Electrical installations – Selection of cables, Figure 1, p. 21

Coordination of cables and protection devices

As discussed in Section 11.2 of this chapter, coordination of cables and protection devices is specified by **AS/NZS 3000:2007** clause 2.5.3.1 and achieved by ensuring the following statement is true.

$I_B \leq I_N \leq I_Z$

where
I_B = the maximum demand current in amperes
I_N = the nominal current of the protective device
I_z = the current carrying capacity of the conductor

AS/NZS 3000:2007 clause 3.4.1 requires that a conductor must have a current carrying capacity not less than determined by the **AS/NZS 3008.1.1:2009** series. Cables are not permitted under normal use to exceed temperatures specified by either **AS/NZS 3000:2007** Table 3.2 or **AS/NZS 3008.1.1:2009** Table 3.1. In the case of V75 and V90 insulated cables this temperature is 75°C; for cross length polyethylene (XLPE) insulated cables the maximum permitted temperature is 90°C.

Conductor size based on current carrying capacity (I_Z)

This section is based in part on the Tafe NSW, *Learner Workbook*, pp. 163–7.

Determine the minimum current carrying capacity (I_Z) by:

1 Determining the current requirements, maximum demand (I_B) for the circuit.

2 Determine the current rating of the protective device (I_N) to be used. **AS/NZS 3000:2007** Tables 8.1, 8.2 and B1 show details of standard protection device ratings up to 200 A.

3 Decide which cable type and installation method to use.

4 Apply a de-rating/rating factor from tables of **AS/NZS 3008.1.1:2009** for the installation environment conditions where applicable.

If a de-rating factor is to be used you will need to calculate a 'look up current rating'. In other words, a minimum current rating for the required cable to take to the tables as a reference to find cable size:

$$I_Z = \frac{I_N}{\text{DR factors}}$$

where

I_N = the nominal current of the protective device

I_Z = the required current rating of the conductor

Select a minimum conductor size for the look up current rating (or next largest) from tables of **AS/NZS 3008.1.1:2009**. The actual current rating of the cable under these conditions will be the current rating from the table times de-rating factor/s. If there is more than one de-rating factor the overall de-rating factor is the product of all de-rating factors that apply.

EXAMPLE 11.36

A three phase load has a maximum demand of 180 A. The circuit will be protected by a 200 A HRC fuse. Determine the minimum current carrying capacity of the cable (adapted from Tafe NSW, *Learner Workbook*, pp. 163–7).

Solution

$$I_Z = \frac{I_N}{\text{DR factors}} = \frac{200}{0.9} = 223\text{ A}$$

Example 11.36 shows that a cable with a current rating of 233 A is required rather than 200 A because of the de-rating.

The current carrying capacities for various types of commonly used cables and installation methods are given in **AS/NZS 3008.1.1:2009** Tables 4 to 21, as shown in Figure 11.11.

FIGURE 11.11 Tables for determining current carrying capacities for commonly used cables and installation methods

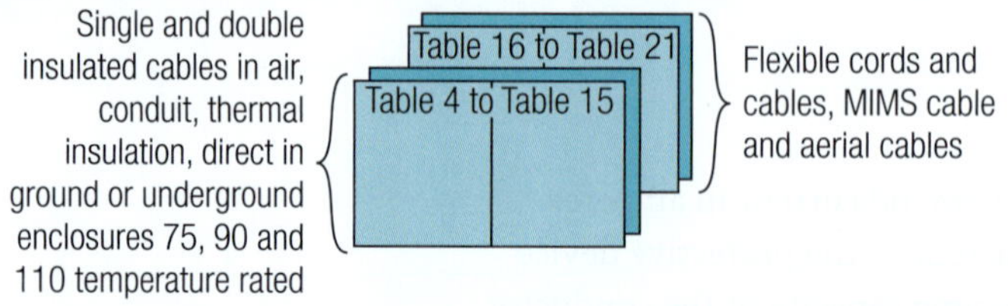

Source: *Select and Arrange Equipment for General Electrical Installations. Learner Workbook*, Industry Skills Unit Meadowbank, Tafe NSW, 2009, Figure 4, p. 164

The current carrying capacity of a cable may be reduced or increased when particular external influences are present. In these cases a de-rating factor or rating factor must be applied before the correct minimum size conductor can be determined. **AS/NZS 3008.1.1:2009** Tables 22 to 29 provide de-rating/rating factors for various external influences, as shown in Figure 11.12.

AS/NZS 3008.1.1:2009 Tables 4 to 15 are the most frequently used current carrying capacity tables, as they cover the most commonly used cables types and cable grouping arrangements. Tables 16 to 21 cover the more unusual cables such as flexible, MIMS and aerial cables.

AS/NZS 3008.1.1:2009 Table 3 gives directions on which of these tables to use for various cable configurations and installation methods. Table 3 is arranged in four parts, as shown in Figure 11.13.

FIGURE 11.12 De-rating factor tables

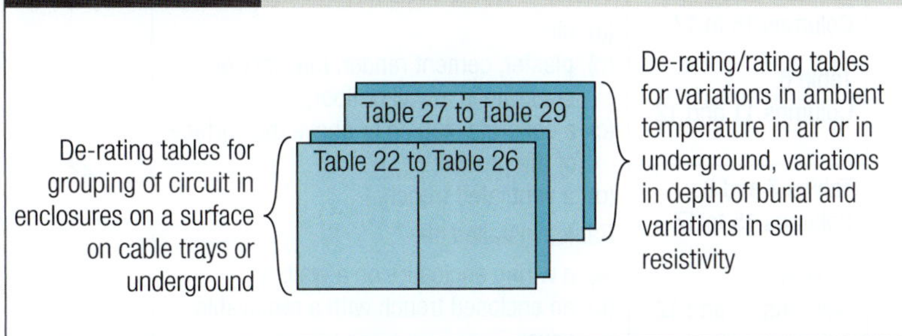

Source: *TAFE Select and Arrange Equipment for General Electrical Installations. Learner Workbook*, Industry Skills Unit Meadowbank, Tafe NSW, 2009, Figure 5, p. 164

FIGURE 11.13 AS/NZS 3008.1.1:2009 Table 3

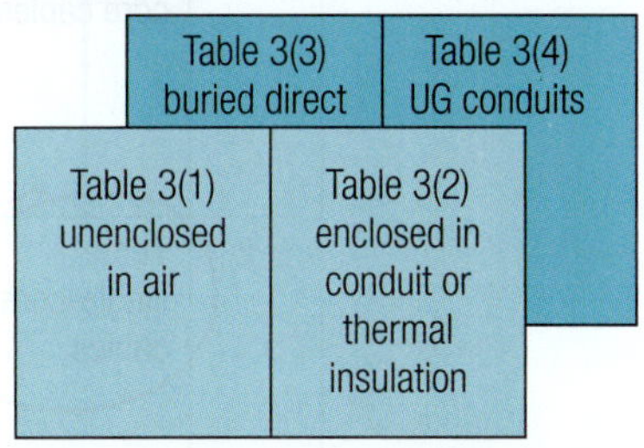

Source: *Select and Arrange Equipment for General Electrical Installations. Learner Workbook*, Industry Skills Unit Meadowbank, Tafe NSW, 2009, Figure 7, p. 165

Each Table 3(1) to 3(4) has the same format, as shown in Figure 11.14.

FIGURE 11.14 AS/NZS 3008.1.1:2009 Table 3 format

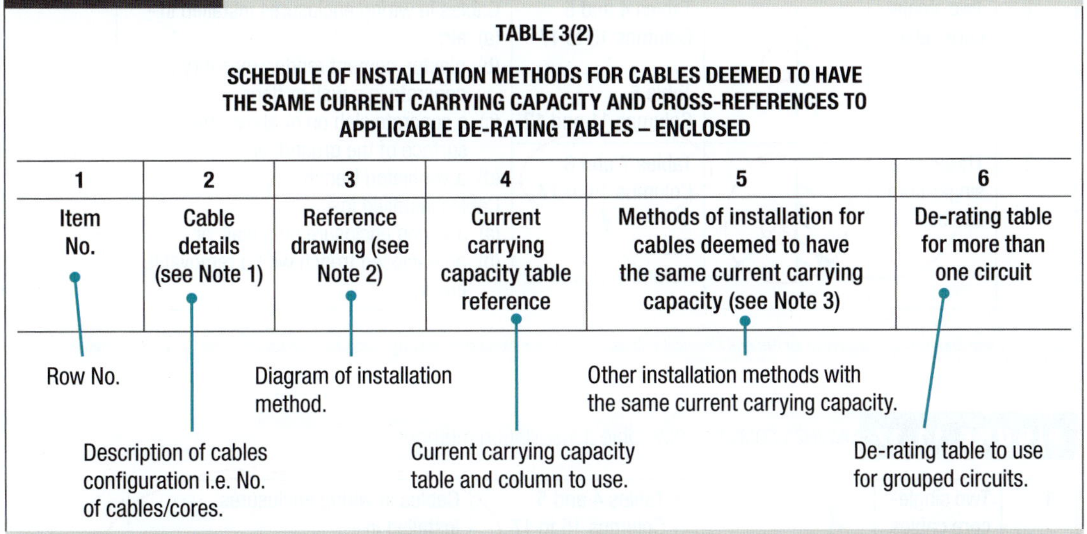

Source: *Select and Arrange Equipment for General Electrical Installations. Learner Workbook*, Industry Skills Unit Meadowbank, Tafe NSW, 2009, Figure 7, p. 165

The following is a guide to using **AS/NZS 3008.1.1:2009** Table 3.

1 Look up in Table 3 how the cable is to be installed. For example, you would use Table 3(2) enclosed for a cable enclosed in conduit.
2 Go down column 2 and match a description of the cable configuration you are to use against a reference drawing in column 3 that shows you how the cable is to be installed, as shown in Figure 11.15.
3 Column 4 on the same row gives the current carrying capacity tables and columns to use to select the minimum conductor size, as shown in Figure 11.16.
4 The reference drawing in column 3 may not fully show how you intend to install the cable. In this case check column 5 for a description of installation methods deemed to be the same, as shown in Figure 11.17.

FIGURE 11.15 AS/NZS 3008.1.1:2009 Table 3 cable configuration

TABLE 3(2)

SCHEDULE OF INSTALLATION METHODS FOR CABLES DEEMED TO HAVE THE SAME CURRENT CARRYING CAPACITY AND CROSS-REFERENCES TO APPLICABLE DE-RATING TABLES – ENCLOSED

1	2	3	4	5	6
Item No.	**Cable details (see Note 1)**	**Reference drawing (see Note 2)**	**Current carrying capacity table reference**	**Methods of installation for cables deemed to have the same current carrying capacity (see Note 3)**	**De-rating table for more than one circuit**
1	Two single-core cables		Tables 4 and 5 Columns 15 to 17 Table 6 Columns 11 and 12	Cables in wiring enclosures installed in– (a) air; (b) plaster, cement render, masonry or concrete in a wall or floor; (c) a concrete slab on or above the surface of the ground; or (d) a ventilated trench. Cables installed in– (a) a wiring enclosure on a wall; or (b) an enclosed trench with a removable cover.	22
2	Three single-core cables		Tables 7 and 8 Columns 15 to 17 Table 9 Columns 11 and 12		

Source: *Select and Arrange Equipment for General Electrical Installations. Learner Workbook*, Industry Skills Unit Meadowbank, Tafe NSW, 2009, Figure 8, p. 166

FIGURE 11.16 AS/NZS 3008.1.1:2009 Table 3 current carrying capacity tables

1	Two single-core cables		Tables 4 and 5 Columns 15 to 17 Table 6 Columns 11 and 12	Cables in wiring enclosures installed in– (a) air; (b) plaster, cement render, masonry or concrete in a wall or floor; (c) a concrete slab on or above the surface of the ground; or (d) a ventilated trench. Cables installed in– (a) a wiring enclosure on a wall; or (b) an enclosed trench with a removable cover.	22
2	Three single-core cables		Tables 7 and 8 Columns 15 to 17 Table 9 Columns 11 and 12		

Source: *Select and Arrange Equipment for General Electrical Installations. Learner Workbook*, Industry Skills Unit Meadowbank, Tafe NSW, 2009, Figure 9, p 166.

FIGURE 11.17 AS/NZS 3008.1.1:2009 Table 3 installation methods

1	Two single-core cables		Tables 4 and 5 Columns 15 to 17 Table 6 Columns 11 and 12	Cables in wiring enclosures installed in– (a) air; (b) plaster, cement render, masonry or concrete in a wall or floor; (c) a concrete slab on or above the surface of the ground; or (d) a ventilated trench. Cables installed in– (a) a wiring enclosure on a wall; or (b) an enclosed trench with a removable cover.	22
2	Three single-core cables		Tables 7 and 8 Columns 15 to 17 Table 9 Columns 11 and 12		

Source: *Select and Arrange Equipment for General Electrical Installations. Learner Workbook*, Industry Skills Unit Meadowbank, Tafe NSW, 2009, Figure 10, p. 166

5 When cables are to be installed with cables of other circuits, a de-rating factor must be applied. Column 6 gives the table to use to find the correct de-rating factor for groups of cables, as shown in Figure 11.18. For example, a three-core cable installed unenclosed in air on a cable tray with two other circuits.

FIGURE 11.18 AS/NZS 3008.1.1:2009 Table 3 de-rating tables

1	2	3	4	5	6
Item No.	**Cable details (see Note 2)**	**Reference drawing (see Note 3)**	**Current carrying capacity table reference**	**Methods of installation for cables deemed to have the same current carrying capacity (see Notes 4, 5 and 6)**	**De-rating table for more than one circuit**
9	Two-core cables	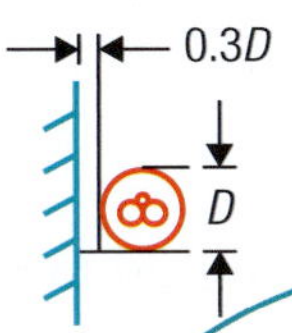 	Tables 10 and 11 (see Note 5) Columns 2 to 4 Table 12 Columns 2 and 3	Cables with minimum spacings in air as shown and installed– (a) spaced from a wall or vertical surface: (b) supported on ladders, racks, perforated or unperforated trays, cleats or hangers; (c) in a switchboard or similar enclosure; or	24
10	Three-core cables	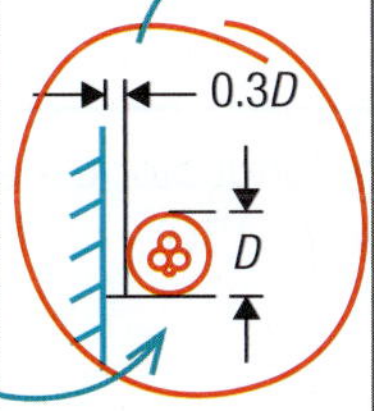 	Tables 13 and 14 (see Note 5) Columns 2 to 4		
11			Table 15 Columns 2 and 3	(d) suspended from a catenary or as a self-supported overhead cable.	22

Source: *Select and Arrange Equipment for General Electrical Installations. Learner Workbook*, Industry Skills Unit Meadowbank, Tafe NSW, 2009, Figure 11, p. 167

How de-rating (and rating) factors are applied

The minimum current carrying capacity in a cable (I_Z) when installed with cables of other circuits is decreased according to the de-rating factor that applies. The result can mean a larger cable is required for the circuit.

- De-rating factor applied to a particular cable results in a reduction in current carrying capacity for that cable.
- De-rating can be avoided by installing cables with minimum spacing as shown in Figure 1 of **AS/NZS 3008.1.1:2009**.
- When using HRC fuses as the circuit protective device, a de-rating factor of 0.9 will automatically apply to the current carrying capacity of the cable.
- Rating factors (Tables 27 to 29) are applied to cables in ambient temperatures (other than 40°C air temperature or 25°C soil temperature) in the same way as de-rating factors. Rating tables are not given in the Table 3 schedule.
- I_Z = current carrying capacity from **AS/NZS 3008.1.1:2009** × de-rating factor for installation conditions (Tafe NSW, *Learner Workbook*, p. 163).

AS/NZS 3008.1.1:2009 Tables 4 to 21

Table 3 can be used as an index to find which table from Tables 4 to 14 has the equivalent current carrying capacity for that cable configuration and installation conditions. Tables 15 to 21 are not mentioned in Table 3, as they are for different cable types (flexible cords, MIMS cables, aerial conductors, etc.). When selecting conductor size for these cable types, consult the appropriate table directly. De-rating and rating factor from the appropriate Tables 22 to 29 must be applied where necessary (Tafe NSW, *Learner Workbook*, p. 163).

Neutral conductor size

AS/NZS 3000:2007 clause 3.5.2 requires that in single phase consumer's mains, sub-mains or final sub-circuits the neutral conductor/s have a current carrying capacity of not less than the largest associated active conductor/s. This means that if an active conductor of a circuit is rated at 63 A, the neutral conductor must also be rated at not less than 63 A. Selecting a neutral cable the same size as the active cable achieves this.

In multi-phase consumer's mains, sub-mains or final sub-circuits the same applies; that is, the neutral conductor/s must have a current carrying capacity of not less than the largest associated active conductor/s. If the circuit supplies harmonic generating loads, the neutral conductor current rating is determined by adding 100% of the largest active harmonic current to the maximum out of balance load current. In this case, the neutral conductor will be larger than the associated active conductors (Tafe NSW, *Learner Workbook*, pp. 169–70).

Protective earthing conductor size

AS/NZS 3000:2007 clause 5.3.3.1.2 requires protective earthing c.s.a. sizes to be in accordance with **AS/NZS 3000:2007** Table 5.1.

EXAMPLE 11.37

Complete Table 11.32 to show the required protective earthing conductor c.s.a. for the given active conductors (adapted from Tafe NSW, *Learner Workbook*, p. 170).

Solution

TABLE 11.32 Protective earthing conductor sizes

Active conductor	Protective earthing conductor
2.5 mm^2 TPI copper active conductors enclosed in LD PVC conduit.	2.5 mm^2
1.0 mm^2 TPI copper active conductors enclosed in LD PVC conduit.	2.5 mm^2
1.0 mm^2 TPI copper active conductors in a V90 T + E cable.	1.0 mm^2
Sub-mains are 3-phase 95 mm^2 copper XLPE single core cables installed in underground enclosures.	25 mm^2
Sub-mains are 3-phase 95 mm^2 aluminium XLPE single core cables installed in underground enclosures.	16 mm^2

Source: *Select and Arrange Equipment for General Electrical Installations. Learner Workbook*, Industry Skills Unit Meadowbank, Tafe NSW, 2009, p. 170

EXAMPLE 11.38

Complete Table 11.33 to show the required protection device and conductor c.s.a. for a sub-main cable.

- Cable selection based on current carrying capacity maximum demand current – 170 amperes.
- Cable – four core, non-armoured, V75 insulated and sheathed circular cable with copper conductors.
- Installation method – clipped directly to a vertical surface, open to the air.
- Circuit protection – circuit breaker with fixed current setting (adapted from Tafe NSW, *Learner Workbook*, p. 171).

Solution

TABLE 11.33 Cable selection based on current carrying capacity with no de-rating

Minimum circuit breaker rating	200 A
Table number 3(?)/Item number	T3(1) Item 13
Table number/Column number	T13 C5
Cable size	95 mm^2 (213 A)
Protective earthing conductor cable size	25 mm^2
Coordination between conductors and protective devices (AS/NZS 3000:2007 clause 2.5.3.1)	
I_B = Maximum demand current	170 A
I_N = Nominal current rating of protective device	200 A
I_z = Current carrying capacity of the selected cable after any de-rating has been applied	213 A
Is $I_B \leq I_N \leq I_z$?	Yes

Source: *Select and Arrange Equipment for General Electrical Installations. Learner Workbook*, Industry Skills Unit Meadowbank, Tafe NSW, 2009, p. 171

EXAMPLE 11.39

Complete Table 11.34 to show the required protection device and conductor c.s.a. for a sub-main cable.

- Maximum demand current – 110 amperes.
- Cable – four single core, non-armoured, XLPE insulated and sheathed circular cable with aluminium conductors.
- Installation method – laid in trefoil on a perforated cable tray as a single circuit.
- Circuit protection – circuit breaker with fixed current setting (adapted from Tafe NSW, *Learner Workbook*, p. 172).

Solution

TABLE 11.34 Cable selection based on current carrying capacity with no de-rating

Minimum circuit breaker rating	125 A
Table number 3(?)/Item number	T3(1) Item 5
Table number/Column number	T8 C7
Cable size	35 mm^2 (140 A)
Protective earthing conductor cable size	6 mm^2
Coordination between conductors and protective devices (AS/NZS 3000:2007 clause 2.5.3.1)	
I_B = Maximum demand current	110 A
I_N = Nominal current rating of protective device	125 A
I_z = Current carrying capacity of the selected cable after any de-rating has been applied	140 A
Is $I_B \leq I_N \leq I_z$?	Yes

Source: *Select and Arrange Equipment for General Electrical Installations. Learner Workbook*, Industry Skills Unit Meadowbank, Tafe NSW, 2009, p. 171

EXAMPLE 11.40

Complete Table 11.35 to show the required protection device and conductor c.s.a. of a single phase final sub-circuit.

- Maximum demand current – 30 amperes.
- Cable – V75 thermoplastic insulated (TPI) copper conductors.
- Installation method – cables run in medium duty conduit with two other circuits.
- Circuit protection – circuit breaker with fixed current setting (adapted from Tafe NSW, *Learner Workbook*, p. 172).

Solution

TABLE 11.35 Cable selection based on current carrying capacity with de-rating

Minimum circuit breaker rating	32 A
Table number 3(?)/Item number	T3(2) Item 1
Table number/Column number	T4 C15
De-rating table numbers/Column numbers	T22C6
De-rating factors	0.7
Required current carrying capacity	$I = \frac{I_N}{DR} = \frac{32}{0.7} = 45.7\text{ A}$
Cable size	6 mm² (52 A)
Protective earthing conductor cable size	2.5 mm²
Coordination between conductors and protective devices (AS/NZS 3000:2007 clause 2.5.3.1)	
I_B = Maximum demand current	30 A
I_N = Nominal current rating of protective device	32 A
I_z = Current carrying capacity of the selected cable after any de-rating has been applied	52 × 0.7 = 36.4 A
Is $I_B \leq I_N \leq I_z$?	Yes

Source: *Select and Arrange Equipment for General Electrical Installations. Learner Workbook*, Industry Skills Unit Meadowbank, Tafe NSW, 2009, p. 171

EXAMPLE 11.41

Complete Table 11.36 to show the required protection device and conductor c.s.a. for a sub-main cable.

- Maximum demand current – 100 amperes.
- Cable – two unserved 1 kV/1 kV three-core MIMS copper cables.
- Installation method – run touching on cable ladder.
- Circuit protection – HRC fuses (adapted from Tafe NSW, *Learner Workbook*, p. 174).

Solution

TABLE 11.36 Cable selection based on current carrying capacity with de-rating

Minimum HRC fuse rating	100 A
Table number 3(?)/Item number	N/A MIMS cable
Table number/Column number	T19 C4
De-rating table numbers/Column numbers	HRC fuse
De-rating factors	0.9
Required current carrying capacity	$I = \frac{I_N}{DR} = \frac{100}{0.9} = 111.1\text{ A}$
Cable size	25 mm² (125 A)
Protective earthing conductor cable size	Cable sheath
Coordination between conductors and protective devices (AS/NZS 3000:2007 clause 2.5.3.1)	
I_B = Maximum demand current	100 A
I_N = Nominal current rating of protective device	100 A
I_z = Current carrying capacity of the selected cable after any de-rating has been applied	125 × 0.9 = 112.5 A
Is $I_B \leq I_N \leq I_z$?	Yes

Source: *Select and Arrange Equipment for General Electrical Installations. Learner Workbook*, Industry Skills Unit Meadowbank, Tafe NSW, 2009, p. 175

AS/NZS 3008.1.1:2009 clause 3.5.2.2(a) exempts unserved MIMS cables from de-rating for grouping of cables. The only de-rating factor applicable is 0.9 for protection by an HRC fuse.

Cable current carrying capacity

For a specified cable c.s.a. the current carrying capacity must be determined to select a suitable rating of the protection device. The method is the reverse of selecting a cable:

1. Consult **AS/NZS 3008.1.1:2009** Table 3, depending on the installation method, to find current rating table and de-rating table reference.
2. Determine initial cable current carrying capacity from Tables 4 to 20.
3. Determine appropriate de-rating from Tables 21 to 29.
4. Apply, if required, any de-rating to initial cable current carrying capacity.
5. Select rating of circuit protection device so that it is less than the rating of the cable after de-rating has been applied.

EXAMPLE 11.42

Complete Table 11.37 to show the current carrying capacity and a suitable rating of the circuit protection device for a final sub-circuit with the following specifications:

- Cable – 10 mm² copper 4-core + E, V75 orange circular cable.
- Installation method – laid touching two other circuits on perforated cable tray.
- Circuit protection – miniature circuit breaker (adapted from Tafe NSW, *Learner Workbook*, p. 175).

Solution

TABLE 11.37 Cable current carrying capacity and protection selection

Cable size	10 mm²
Protective earthing conductor cable size	6 mm²
Table number 3(?)/Item number	T3(1) Item 10
Table number/Column number	T13 C2
Initial current carrying capacity	54 A
Circuit breaker rating	50 A
Coordination between conductors and protective devices (AS/NZS 3000:2007 clause 2.5.3.1)	
I_B = Maximum demand current	50 A by limitation
I_N = Nominal current rating of protective device	50 A
I_z = Current carrying capacity of the selected cable after any de-rating has been applied	54 A
Is $I_B \le I_N \le I_z$?	Yes

Source: *Select and Arrange Equipment for General Electrical Installations. Learner Workbook*, Industry Skills Unit Meadowbank, Tafe NSW, 2009, p. 175

EXAMPLE 11.43

Complete Table 11.38 to show the current carrying capacity and a suitable rating of the circuit protection device for two single core, TPI conductors, circuit breaker protected for a final sub-circuit with the following specifications:

- Cable – 4 mm² V75 two single core TPI copper conductors.
- Installation method – enclosed in medium duty conduit in a concrete slab above the ground with three other circuits.
- Circuit protection – miniature circuit breaker (adapted from Tafe NSW, *Learner Workbook*, p. 177).

Solution

TABLE 11.38 Cable current carrying capacity and protection selection

Cable size	4 mm²
Protective earthing conductor cable size	2.5 mm²
Table number 3(?)/Item number	T3(2) Item 1
Table number/Column number	T4 C15
Initial current carrying capacity	32 A
De-rating factors	0.65
Table numbers/Column numbers	T22 C7
Current carrying capacity after de-rating	$= 32 \times 0.65 = 20.8$ A
Maximum circuit breaker rating	20 A
Coordination between conductors and protective devices (AS/NZS 3000:2007 clause 2.5.3.1)	
I_B = Maximum demand current	20 A by limitation
I_N = Nominal current rating of protective device	20 A
I_z = Current carrying capacity of the selected cable after any de-rating has been applied	20.8 A
Is $I_B \leq I_N \leq I_z$?	Yes

Source: *Select and Arrange Equipment for General Electrical Installations. Learner Workbook*, Industry Skills Unit Meadowbank, Tafe NSW, 2009, p. 177

EXAMPLE 11.44

Complete Table 11.39 to show the current carrying capacity and a suitable rating of the protection device for a three phase sub-main with the following specifications:

- Cable – single core XLPE 120 mm² aluminium.
- Installation method – direct in the ground, cables laid in trefoil spaced from another circuit by 300 mm at a depth of 0.5 m.
- Circuit protection – HRC fuse (adapted from Tafe NSW, *Learner Workbook*, p. 178).

Solution

TABLE 11.39 Cable current carrying capacity and protection selection

Cable size	120 mm²
Protective earthing conductor cable size	25 mm²
Table number 3(?)/Item number	T3(3) Item 2
Table number/Column number	T8 C23
Initial current carrying capacity	277 A
De-rating factor (grouping)	0.88
Table numbers/Column numbers (grouping)	T25(1) C5
De-rating factor (HRC fuse)	0.9
De-rating factor (depth)	N/A for 0.5 m
Current carrying capacity after de-rating	$= 277 \times 0.88 \times 0.9 = 219$ A
Maximum circuit HRC fuse rating	200 A
Coordination between conductors and protective devices (AS/NZS 3000:2007 clause 2.5.3.1)	
I_B = Maximum demand current	200 A by limitation
I_N = Nominal current rating of protective device	200 A
I_z = Current carrying capacity of the selected cable after any de-rating has been applied	219 A
Is $I_B \leq I_N \leq I_z$?	Yes

Source: *Select and Arrange Equipment for General Electrical Installations. Learner Workbook*, Industry Skills Unit Meadowbank, Tafe NSW, 2009, p. 178

Adjustable circuit breakers

An adjustable circuit breaker may be used to better match the setting of the breaker to the capacity of the cable. This way no capacity between the preset size of the breaker and the cable is lost.

The current rating of the device can be adjusted between 100% to 63% of its rated nominal value (I_N). Nominal ratings of adjustable circuit breakers are typically 250 A and 1600 A (Tafe NSW, *Learner Workbook*, p. 180). The current carrying capacity of a conductor is determined from **AS/NZS 3008.1.1:2009**, then the circuit breaker percentage setting is calculated for the circuit breaker by the equation:

$$\text{Setting } \% = \frac{I_Z}{I_N} \times \frac{100}{1}$$

where

I_N = the nominal current of the protective device in amperes

I_Z = the current capacity of the conductor in amperes

EXAMPLE 11.45

Four single core 70 mm² XLPE copper cables are installed laid flat touching as a single circuit on a cable ladder. What size and current setting of an adjustable circuit breaker will allow the full optimisation of the cable current carrying capacity (CCC)?

Table 3(1) Item 5 Tables 8 Columns 5 to 7 CCC = 240 A

$$\text{Setting } \% = \frac{I_Z}{I_N} \times \frac{100}{1} = \frac{240}{250} \times \frac{100}{1} = 96\%$$

A 250 A circuit breaker is now set to 96% of its rated value to match the protection device to the cable rating. Had a preset circuit breaker been used, a 95 mm² cable would have been required (adapted from Tafe NSW, *Learner Workbook*, p. 180).

Paralleling cables

To increase the current carrying capacity of a circuit, the conductors may be run in parallel. Smaller conductors are used, allowing a tighter radius when bending and easier installation than one single larger conductor. **AS/NZS 3000:2007** clause 3.4.3 prohibits the paralleling of conductors smaller than 4 mm². The cables used in parallel must be identical in material, c.s.a. and route length. The current rating of the parallel group is the single cable current rating multiplied by the number of parallel conductors, multiplied by any de-rating applicable, including grouping (Tafe NSW, *Learner Workbook*, p. 183).

EXAMPLE 11.46

Complete Table 11.40 to show the current carrying capacity and a suitable rating of protection device for a three phase sub-main.

- Cable – 70 mm² copper single core XLPE cables.
- Installation method – two circuits touching laid in trefoil on cable ladder.
- Circuit protection – HRC fuse (adapted from Tafe NSW, *Learner Workbook*, p. 183).

Solution

TABLE 11.40 Current carrying capacity of parallel cables

Cable size	70 mm² × 2
Protective earthing conductor cable size	25 mm²
Table number 3(?)/Item number	T3(1) Item 5
Table number/Column number	T8 C5

(Continued)

TABLE 11.40 *(Continued)*

Initial current carrying capacity	240 A
De-rating factor (grouping)	0.95
Table numbers/Column numbers (grouping 2 circuits)	T23 C7
De-rating factor (HRC fuse)	0.9
Current carrying capacity after de-rating	$= 240 \times 0.95 \times 0.9 = 410$ A
Maximum circuit HRC fuse rating	400 A
Coordination between conductors and protective devices (AS/NZS 3000:2007 clause 2.5.3.1)	
I_B = Maximum demand current	400 A by limitation
I_N = Nominal current rating of protective device	400 A
I_z = Current carrying capacity of the selected cable after any de-rating has been applied	410 A
Is $I_B \leq I_N \leq I_z$?	Yes

Source: *Select and Arrange Equipment for General Electrical Installations. Learner Workbook*, Industry Skills Unit Meadowbank, Tafe NSW, 2009, p. 183

SUMMING UP...

- The current carrying capacity of a conductor is affected by ambient temperature and installation conditions.
- The four methods of cable installation shown in AS/NZS 3008.1.1:2009 Table 3 are:
  - installed in air
  - enclosed
  - buried direct
  - in underground enclosure.
- AS/NZS 3008.1.1:2009 cable configuration diagrams only show the current carrying conductors.
- Single core cable configurations include:
  - trefoil
  - laid flat touching
  - laid flat separated.
- De-rating factors are assigned to cables depending on the installation method and conditions.
- AS/NZS 3008.1.1:2009 Table 3 provides guidance for the use of Tables 4 to 15 which specifies current ratings of copper and aluminium cables with insulation rated at 75°C, 90°C and 110°C.
- AS/NZS 3008.1.1:2009 Tables 16 to 21 provides guidance for the current rating of specialist cables.

11.5 Cable selection based on voltage drop requirements

Electrical cables made from either copper (Cu) or aluminium (Al) have resistance. When an electric current flows a voltage drop across the length of the cable will occur. This voltage drop will reduce the supply voltage available at the terminals of the load supplied by the cables.

As stated in Section 11.1 earlier in the chapter, **AS/NZS 3000:2007** clause 3.6.5 requires that the voltage drop within an installation be restricted to 5% of the nominal supply. A total voltage drop of a maximum of 11.5 V across the installation is permitted for single phase circuits and 20 V for three phase circuits. Installations that are supplied directly from a supply authority substation or distribution transformer are allowed a 7% drop. **AS/NZS 3000:2007** clause 7.5.7 requires the voltage drop within extra low voltage portions of an installation to be restricted to 10% of the nominal supply.

Excessive voltage drop in an installation can cause:

- a reduction in the effective operation of appliances and lighting
- overloading of cables if a fault occurs by delaying the operating time of circuit protection devices
- overheating of motors, noticeable when the voltage reduction is more than 5%.

You must consider the effect of voltage drop when selecting cables, especially for circuits which have long route lengths (that is, length of the cable) and circuits with relatively high currents. Voltage drop (V_d) in the cables of a circuit is caused by the current in the circuits (I) and the resistance (R) of the circuit.

$V_d = I \times Z_c$

where

V_d = the voltage drop on the cable in volts

I = the current in the cable in amperes

Z_c = the impedance of the section of cable in ohms

In smaller cables the resistive component is the most significant component of the impedance (Tafe NSW, *Learner Workbook*, p. 196).

Relevant tables in AS/NZS 3008.1.1:2009

Factors that determine the voltage drop in a cable are the:

- length of the cable
- c.s.a. of the cable
- current flowing in the cable
- type of material the cable is made of (copper or aluminium)
- operating temperature of the cable and its ability to dissipate heat
- installation method of the cable (trefoil, laid flat or in a multi-core cable).

The voltage drop of any given combination of the above can be predicted before the cable is selected and installed by using Tables 40 to 51 of Section 4 of **AS/NZS 3008.1.1:2009**.

Tables 40 to 51 show values of V_c in millivolts per ampere metre for a number of copper and aluminium cable configurations. As shown in Figure 11.19, the value of V_c was determined by passing a current of one ampere through a 1 metre length of conductor, in sizes from 1 mm² to 630 mm² for both copper and aluminium conductors at operating temperatures ranging from 45°C to 110°C. The temperature used is the maximum normal use operating temperature as given in **AS/NZS 3008.1.1:2009** Table 1. The voltage drop across the 1 metre length of conductor is then recorded as V_c.

All of the values listed in Tables 40 to 51 are three phase values of V_c. When performing a voltage drop calculation, the value of V_c is obtained directly from the tables.

FIGURE 11.19 AS/NZS 3008.1:2009 mV/A.m

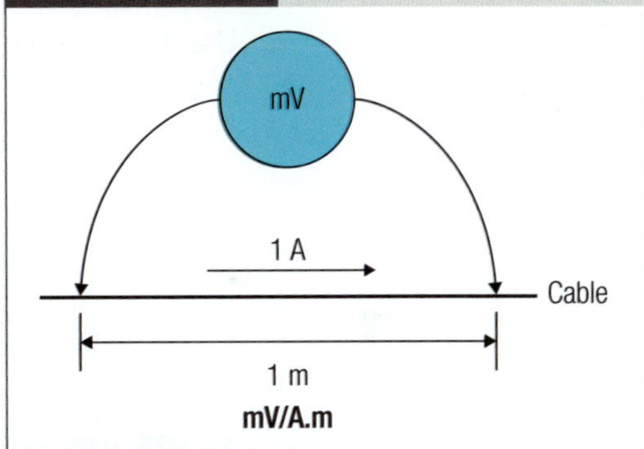

Source: *Select and Arrange Equipment for General Electrical Installations. Learner Workbook*, Industry Skills Unit Meadowbank, Tafe NSW, 2009, Figure 1, p. 198

The three phase values of V_c were obtained using the line current (I_L) of a balanced three phase circuit. The three phase values of V_c require adjustment when applied to a single phase circuit to allow for current flowing in the neutral. If the voltage drop of a single phase cable needs to be calculated, the three phase value of V_c for a given cable size (from tables), must be converted to a single phase value of V_c with this equation:

Single phase V_c = Table $V_c \times 1.155$

EXAMPLE 11.47

Calculate the single phase value of V_c for a 16 mm² V75 cable operating at 75°C (adapted from Tafe NSW, *Learner Workbook*, pp. 197–8).

Solution

From AS/NZS 3008.1.1:2009 V_c of 16 mm² = 2.43 mV/A.m (Table 42 column 6).

Single phase $V_c = 2.43 \times 1.155$

$= 2.80665$ mV/A.m

When a single phase value of V_c has been calculated, the single phase value of V_c must be converted to a three phase value of V_c. An appropriate table of **AS/NZS 3008.1.1:2009** is then used to determine the minimum size of conductor with a value of V_c lower than the one calculated.

Table V_c = Single phase $V_c \div 1.155$

A characteristic of a three phase circuit is that it has lower losses than a single phase circuit of the same material, length and c.s.a. Voltage drop is a loss. The three phase V_c for an equivalent conductor is always smaller than the V_c of a single phase circuit (Tafe NSW, *Learner Workbook*, pp. 197–8).

EXAMPLE 11.48

Calculate the three phase value of V_c for a single phase value of 0.68295 mV/A.m and determine the minimum required c.s.a. of a two core XLPE aluminium cable operating at 90°C when laid flat touching (adapted from Tafe NSW, *Learner Workbook*, pp. 197–8).

Solution

Three phase $V_c = 0.68295 \div 1.155$
$= 0.5912987$ mV/A.m

From AS/NZS 3008.1.1:2009 (Table 44 column 8) V_c of 120 mm² = 0.589 mV/A.m.

Calculation of voltage drop

It is possible to predict the voltage drop across a cable using the equations supplied by **AS/NZS 3008.1.1:2009** clause 4.2. To determine the actual voltage drop for a given cable size:

1 Determine the value of cable V_c from Tables 40 to 51 of **AS/NZS 3008.1.1:2009**

2 Use the equation:

$$V_d = \frac{L \times I \times V_c}{1000}$$

where

V_d = the voltage drop on the cable in volts
L = the length of the cable run in metres
I = the current in the cable in amperes
V_c = the value of V_c from AS/NZS 3008.1.1:2009 Tables 40 to 51

AS/NZS 3000:2007 clause 3.6.2 provides guidance on the current to be used in the voltage drop calculation. When calculating voltage drop for light and power circuits where the load is distributed throughout the length of the cable run, the current is taken to be half the rating of the protective device being used. For example, if a power circuit is protected by a 20 A circuit breaker, 10 A is used to determine the cable voltage drop.

EXAMPLE 11.49

Calculate the voltage drop on a 6 mm² V90 three phase multi-core copper cable with a length of 30 m if protected by a 32 A circuit breaker cable (adapted from Tafe NSW, *Learner Workbook*, p. 201).

Solution

From AS/NZS 3008.1.1:2009 (Table 42 column 6) V_c of 6 mm² = 6.49 mV/A.m.

$$V_d = \frac{L \times I \times V_c}{1000} = \frac{30 \times 32 \times 6.49}{1000} = 6.23 \text{ V}$$

To find the total voltage drop for an entire installation, the voltage drops of the consumer's mains and final sub-circuits are added together.

EXAMPLE 11.50

Calculate the voltage drop for the three phase installation shown in Figure 11.20 cable (adapted from Tafe NSW, *Learner Workbook*, p. 201).

FIGURE 11.20 Total voltage drop in a three phase installation

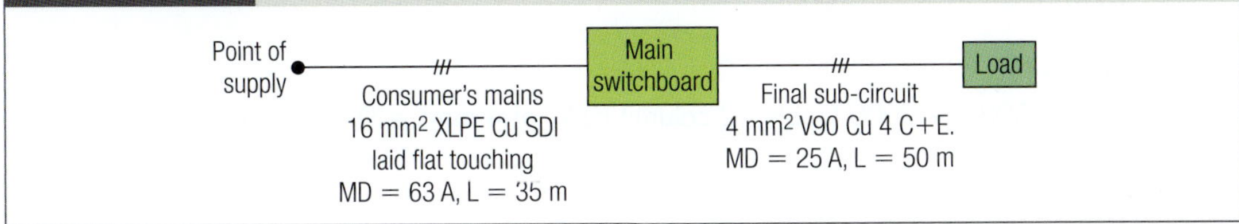

Solution

Consumer's mains ($V_{d\ cm}$)
From AS/NZS 3008.1.1: 2009 (Table 42 column 6) V_C of 16 mm² cable = 2.55 mV/A.m.

$$V_d = \frac{L \times I \times V_c}{1000} = \frac{35 \times 63 \times 2.55}{1000} = 5.622\ \text{V}$$

Final sub-circuit ($V_{d\ fsc}$)
From AS/NZS 3008.1.1: 2009 (Table 42 column 6) V_c of 4 mm² = 6.49 mV/A.m.

$$V_d = \frac{L \times I \times V_c}{1000} = \frac{50 \times 25 \times 9.71}{1000} = 12.137\ \text{V}$$

Total voltage drop (V_d total)

$$V_{d\ total} = V_{d\ cm} + V_{d\ fsc} = 5.622 + 12.137 = 17.759\ \text{V}$$

The installation complies with AS/NZS 3000:2007 clause 5.6.2 as the total voltage drop does not exceed 5% of 400 V (or 20 V).

Single phase installations

When an installation contains single phase circuits, the values of V_c must be converted to single phase values by multiplying the table value by 1.155 before use in the voltage drop equation (Tafe NSW, *Learner Workbook*, p. 202).

EXAMPLE 11.51

Calculate the voltage drop for the single phase installation shown in Figure 11.21 (adapted from Tafe NSW, *Learner Workbook*, pp. 202–4).

FIGURE 11.21 Total voltage drop in a single phase installation

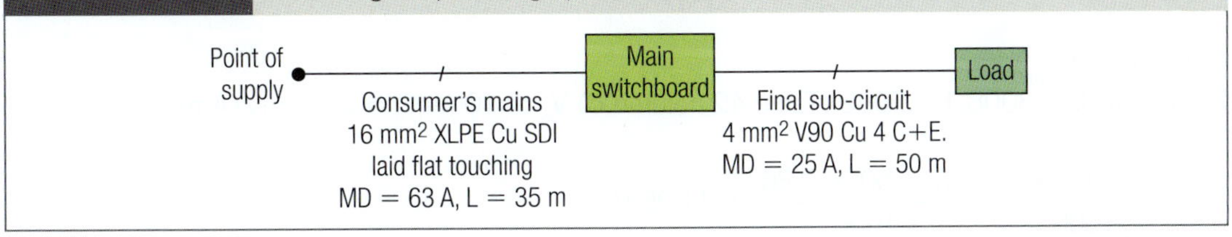

Solution

Consumer's mains ($V_{d\ cm}$)

From AS/NZS 3008.1.1:2009 (Table 41 column 6) V_c of 16 mm² = 2.55 mV/A.m.

$$V_d = \frac{L \times I \times V_c}{1000} = \frac{35 \times 63 \times 2.55 \times 1.155}{1000} = 6.499\ V$$

Final sub-circuit ($V_{d\ fsc}$)

From AS/NZS 3008.1.1:2009 (Table 42 column 6) V_c of 4 mm² = 9.71 mV/A.m.

$$V_d = \frac{L \times I \times V_c}{1000} = \frac{50 \times 25 \times 9.71 \times 1.155}{1000} = 14\ V$$

Total voltage drop ($V_{d\ total}$)

$$V_{d\ total} = V_{d\ cm} + V_{d\ fsc} = 5.622 + 12.137 = 20.5\ V$$

The installation does not comply with AS/NZS 3000:2007 clause 5.6.2 as the total voltage drop exceeds 5% of 230 V (or 11.5 V).

When an installation contains both single and three phase sections, the voltage drops of the three phase sections are added together and then converted to single phase values by dividing them by √3. The total single phase voltage drop is the sum of the converted three phase voltage drop and the total of the single phase section.

EXAMPLE 11.52

Calculate the voltage drop for the installation shown in Figure 11.22 (adapted from Tafe NSW, *Learner Workbook*, pp. 202–4).

FIGURE 11.22 Total voltage drop in an installation

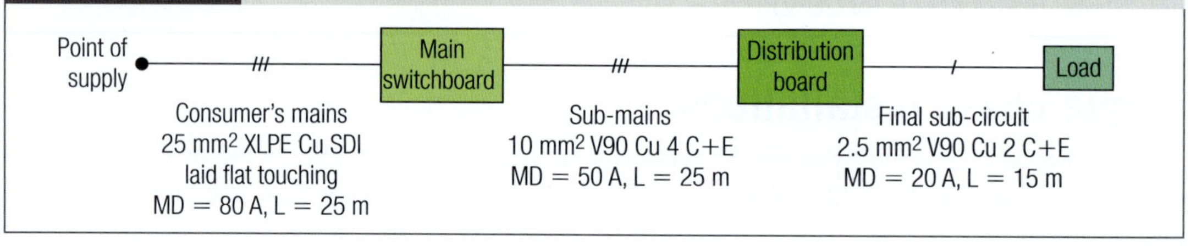

Solution

Consumer's mains ($V_{d\ cm}$)

From AS/NZS 3008.1.1:2009 (Table 41 column 8) V_c of 25 mm² = 1.62 mV/A.m.

$$V_d = \frac{L \times I \times V_c}{1000} = \frac{25 \times 80 \times 1.62}{1000} = 3.24\ V$$

Sub mains ($V_{d\ sm}$)

From AS/NZS 3008.1.1:2009 (Table 42 column 6) V_c of 10 mm² = 3.86 mV/A.m.

$$V_d = \frac{L \times I \times V_c}{1000} = \frac{25 \times 50 \times 3.86}{1000} = 4.825\ V$$

Final sub-circuit ($V_{d\ fsc}$)
From AS/NZS 3008.1.1:2009 (Table 42 column 6) V_c of 2.5 mm² = 15.6 mV/A.m.

$$V_d = \frac{L \times I \times V_c}{1000} = \frac{15 \times 20 \times 15.6 \times 1.155}{1000} = 5.4\ V$$

Total voltage drop ($V_{d\ total}$)

$$V_{d\ total} = \frac{V_{d\ cm} + V_{d\ sm}}{\sqrt{3}} + V_{d\ fsc} = \frac{3.24 + 4.825}{\sqrt{3}} + 5.4 = 10\ V$$

The installation complies with AS/NZS 3000:2007 clause 5.6.2 as the total voltage drop does not exceed 5% of 230 V (or 11.5 V).

An alternative method is to convert each section's voltage drop to a percentage of the nominal supply voltage. To comply with **AS/NZS 3000:2007** clause 5.6.2, the total percentage must not exceed 5%.

EXAMPLE 11.53

Using the percentage method, determine if the installation shown in Example 11.52 complies with AS/NZS 3000:2007 clause 5.6.2 (adapted from Tafe NSW, *Learner Workbook*, pp. 202–4).

Solution

Consumer's mains

$$\%_{cm} = \frac{V_{d\ cm}}{V_{nom}} \times \frac{100}{1} = \frac{3.24}{400} \times \frac{100}{1} = 0.82\%$$

Sub-mains

$$\%_{sm} = \frac{V_{d\ sm}}{V_{nom}} \times \frac{100}{1} = \frac{4.825}{400} \times \frac{100}{1} = 1.2\%$$

Final sub-circuit

$$\%_{f\ sc} = \frac{V_{d\ fsc}}{V_{nom}} \times \frac{100}{1} = \frac{5.4}{230} \times \frac{100}{1} = 2.347\%$$

Voltage drop total

$$\%_{total} = \%_{cm} + \%_{sm} + \%_{fsc} = 0.82 + 1.2 + 2.347 = 4.367\%$$

The installation complies with AS/NZS 3000:2007 clause 5.6.2 as the total voltage drop does not exceed 5% of 230 V (or 11.5V).

Parallel conductors

When conductors are connected in parallel, the voltage drop is calculated using a current value of the total circuit current divided by the number of conductors in the parallel group, in accordance with **AS/NZS 3000:2007** clause 3.6.3. The same procedure is adopted for determining minimum c.s.a. and maximum length of cables in parallel.

EXAMPLE 11.54

Calculate the voltage drop for the consumer's mains shown in Figure 11.23 (based on Tafe NSW, *Learner Workbook*, p. 224).

FIGURE 11.23 Total voltage drop on parallel cables

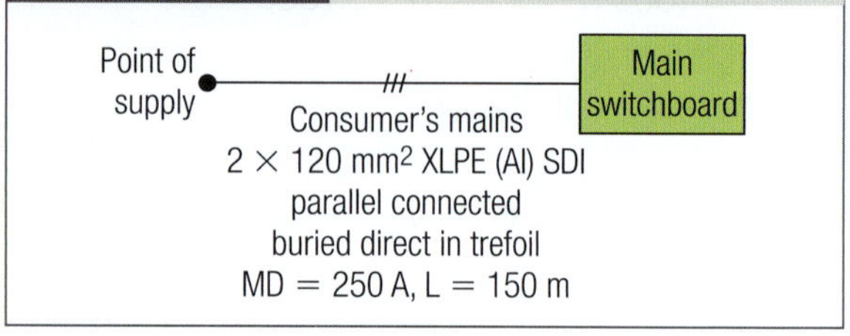

Solution

Consumer's mains ($V_{d\ cm}$)

From AS/NZS 3008.1.1:2009 (Table 41 column 8) V_c of 120 mm² = 0.385 mV/A.m. Note: current is reduced to 250/2 = 125 A because of the parallel grouping.

$$V_d = \frac{L \times I \times V_c}{1000} = \frac{150 \times 125 \times 0.385}{1000} = 7.219\ V$$

Selecting cables based on voltage drop requirements

In the previous examples, the equation used determines the actual voltage drop on a given section of cable. The following examples demonstrate how to determine the minimum cable size based on voltage drop as per **AS/NZS 3008.1.1:2009** clause 4.2. To determine the minimum required cable size:

1 Calculate the maximum permissible voltage drop (V_p) for the final sub-circuit.

2 Calculate the maximum permissible value of V_c using the equation:

$$V_c = \frac{1000 \times V_d}{L \times I}$$

where

V_c = the value found from AS3008.1.1 tables in mV/A.m
V_p = the permissible voltage drop on the circuit run in volts
L = the route length of circuit in metres
I = the current to be carried by the cable in amperes

3 Determine the minimum cable size from Tables 40 to 51 of **AS/NZS 3008.1.1:2009**.
The selected conductor size must also be large enough to satisfy the current carrying requirements.

EXAMPLE 11.55

Calculate the maximum permissible value of V_c for a three phase V90 copper multi-core cable, if the permissible voltage drop is 14 V, the length of the cable run is 45 m and the maximum demand is 25 A cable (adapted from Tafe NSW, *Learner Workbook*, p. 206).

Solution

$$V_c = \frac{1000 \times V_p}{L \times I} = \frac{1000 \times 14}{45 \times 25} = 12.4\ mV/A.m$$

From AS/NZS 3008.1.1:2009 (Table 42 column 6) 4.0 mm² has a V_c of 9.71 mV/A.m which is less than the calculated value of 12.4 mV/A.m.

EXAMPLE 11.56

Calculate the maximum value of V_c and the minimum c.s.a. of cable for the installation shown in Figure 11.24 (adapted from Tafe NSW, *Learner Workbook*, p. 206).

FIGURE 11.24 Cable selection based on voltage drop requirements three phase

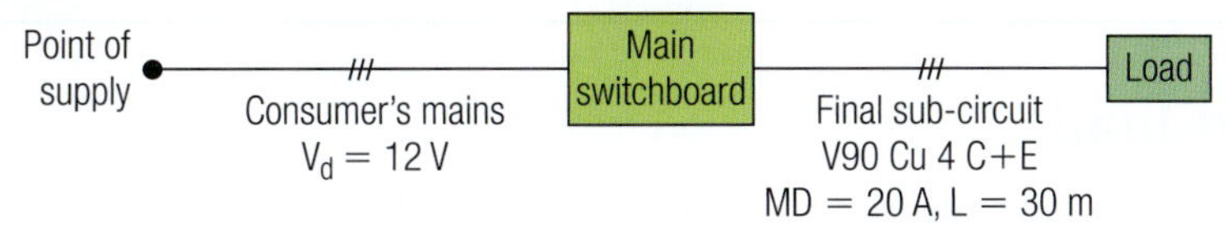

Solution

Permissible voltage drop on final sub-circuit (fsc):

$V_p = 5\%$ of $400\text{ V} = 20\text{ V}$

$V_{p\ fsc} = V_p - V_{d\ cm} = 20 - 12 = 8\text{ V}$

Maximum value of V_c

$$V_c = \frac{1000 \times V_p}{L \times I} = \frac{1000 \times 8}{30 \times 20} = 13.333\text{ mV/A.m}$$

From AS/NZS 3008.1.1:2009 (Table 42 column 6) 4.0 mm² has a V_c of 9.71 mV/A.m which is less than the calculated value of 13.333 mV/A.m.

Single phase installations

When an installation contains single phase circuits, values of V_c must be calculated using the single phase permissible voltage drop (V_p). The calculated single phase V_c is then converted to a three phase value by dividing it by 1.155. The cable size can then be selected using the appropriate operating temperature for the cable from Tables 40 to 51 of **AS/NZS 3008.1.1:2009** (Tafe NSW, *Learner Workbook*, p. 202).

EXAMPLE 11.57

Calculate the maximum value of V_c and the minimum c.s.a. of the cable for the installation shown in Figure 11.25 (adapted from Tafe NSW, *Learner Workbook*, p. 208).

FIGURE 11.25 Cable selection based on voltage drop requirements single phase

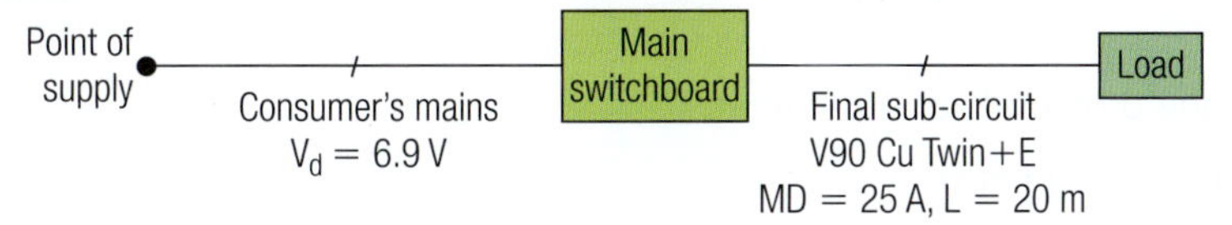

Solution

Permissible voltage drop on final sub-circuit:

$V_p = 5\%$ of $230\text{ V} = 11.5\text{ V}$

$V_{p\ fsc} = V_p - V_{d\ cm} = 11.5 - 6.9 = 4.6\text{ V}$

Maximum value of V_c:

$$V_c = \frac{1000 \times V_p}{L \times I} = \frac{1000 \times 4.6}{20 \times 25} = 9.2\text{ mV/A.m}$$

Convert single V_c to three phase value:

$$V_{c\,3\Phi} = \frac{V_{c1\Phi}}{1.155} = \frac{9.2}{1.155} = 7.965\text{ mV/A.m}$$

From AS/NZS 3008.1.1:2009 (Table 42 column 6) 6.0 mm² has a V_c of 6.49 mV/A.m which is less than the calculated value of 7.965 mV/A.m.

Distributed load

In circuits which supply only one point, such as air-conditioners, hot water systems and motors, the entire current must flow the entire length of the circuit. This current is then used when calculating voltage drop. In circuits that supply more than one point such as lighting points and socket outlets the load is spread across the length of the cable. The switchboard end or start of the run will typically carry more current than the cable towards the end of the run. The maximum demand of circuits with distributed loads is generally set by limitation, meaning the current rating of the protective device (Tafe NSW, *Learner Workbook*, p. 203). **AS/NZS 3000:2007** clause 3.6.2 exception 1 allows the current used to calculate voltage drop to be reduced to half of the current rating of the protective device to average the current over the length of the run.

EXAMPLE 11.58

Calculate the maximum value of V_c and the minimum c.s.a. of the final sub-circuit shown in Figure 11.26 (adapted from Tafe NSW, *Learner Workbook*, p. 210).

FIGURE 11.26 Cable selection based on voltage drop distributed load

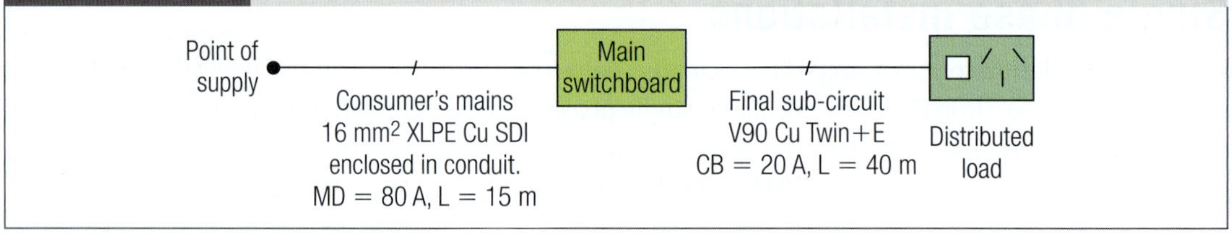

Solution

Consumer's mains ($V_{d\,cm}$)

From AS/NZS 3008.1.1:2009 (Table 41 column 8) V_c of 16 mm² = 2.55 mV/A.m.

$$V_d = \frac{L \times I \times V_c}{1000} = \frac{15 \times 80 \times 2.55 \times 1.155}{1000} = 3.534\,\text{V}$$

Permissible voltage drop on final sub-circuit:

$$V_p = 5\% \text{ of } 230\text{ V} = 11.5\text{ V}$$

$$V_{p\,fsc} = V_p - V_{d\,cm} = 11.5 - 3.534 = 7.8\text{ V}$$

Maximum value of V_c

$$V_c = \frac{1000 \times V_p}{L \times I} = \frac{1000 \times 7.8}{40 \times 10} = 19.5\text{ mV/A.m}$$

Note that half the rating of the protection device is used as the value of current (I).
Convert single V_c to three phase value

$$V_{c\,3\Phi} = \frac{V_{c\,1\Phi}}{1.155} = \frac{19.5}{1.155} = 16.883\text{ mV/A.m}$$

From AS/NZS 3008.1.1:2009 (Table 42 column 6) 2.5 mm² has a V_c of 15.6 mV/A.m which is less than the calculated value of 16.883 mV/A.m.

Three phase installations with single phase circuits

As discussed earlier in this section, the value of the three phase V_d is calculated and then converted to a single phase V_d. This is done by either converting both three phase and single phase voltage drops to a percentage, or by dividing the three phase value of V_d by √3. A single phase value for V_p is then found and used to calculate a single phase value for V_c. This is then converted back to a three phase value of V_c so that the cable size can be determined by looking up the table (Tafe NSW, *Learner Workbook*, p. 203).

EXAMPLE 11.59

Calculate the maximum value of V_c and the minimum c.s.a. of the final sub-circuit shown in Figure 11.27 (adapted from Tafe NSW, *Learner Workbook*, p. 211).

FIGURE 11.27 Cable selection based on voltage drop three phase to single phase

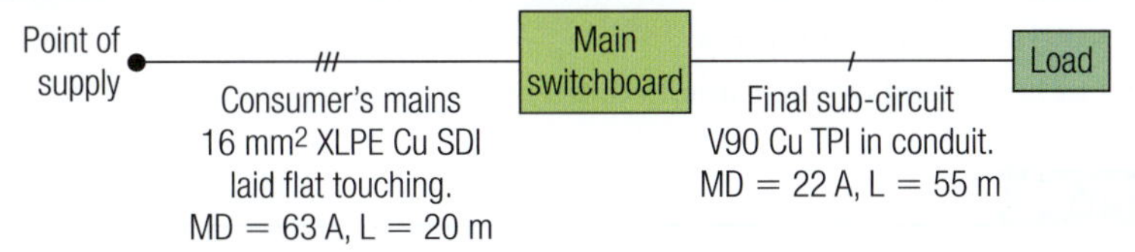

Solution

Consumer's mains ($V_{d\ cm}$)

From AS/NZS 3008.1.1:2009 (Table 41 column 8) V_c of 16 mm² = 2.55 mV/A.m.

$$V_d = \frac{L \times I \times V_c}{1000} = \frac{20 \times 63 \times 2.55}{1000} = 3.213\ \text{V}$$

Three phase permissible voltage drop on final sub-circuit

$$V_p = 5\%\ \text{of}\ 400\ \text{V} = 20\ \text{V}$$

$$V_{p\ fsc\ 3\Phi} = V_p - V_{d\ cm} = 20 - 3.213 = 16.787\ \text{V}$$

Single phase permissible voltage drop on final sub-circuit

$$V_{p\ fsc\ 1\Phi} = \frac{V_{p\ fsc\ 3\Phi}}{\sqrt{3}} = \frac{16.787}{\sqrt{3}} = 9.7\ \text{V}$$

Maximum value of V_c

$$V_c = \frac{1000 \times V_p}{L \times I} = \frac{1000 \times 9.7}{55 \times 22} = 8\ \text{mV/A.m}$$

Convert single V_c to three phase value

$$V_{c3\Phi} = \frac{V_{c1\Phi}}{1.155} = \frac{8}{1.155} = 6.94\ \text{mV/A.m}$$

From AS/NZS 3008.1.1:2009 (Table 41 column 6) 6.0 mm² has a V_c of 6.49 mV/A.m which is less than the calculated value of 6.94 mV/A.m.

Maximum length of cable based on voltage drop

Under normal operating conditions, the voltage drop across a given length of cable effectively limits the length of the run. The equations shown in **AS/NZS 3008.1.1:2009** clause 4.2 are transposed to determine the maximum permissible length. To determine the maximum length for a given cable size:

1 Calculate the maximum permissible voltage drop (V_p) for the final sub-circuit.

2 Determine the value of V_c for the given cable size from Tables 40 to 51 of **AS/NZS 3008.1.1:2009**.

3 Calculate the maximum permissible length of the cable using the equation:

$$L = \frac{1000 \times V_p}{V_c \times I}$$

where

V_p = the permissible voltage drop on the circuit run in volts
V_c = the value found from AS/NZS 3008.1.1 tables in mV/A.m
L = the route length of circuit in metres
I = the current to be carried by the cable in amperes

EXAMPLE 11.60

Calculate the maximum permissible length of the final sub-circuit shown in Figure 11.28 (adapted from Tafe NSW, *Learner Workbook*, p. 215).

FIGURE 11.28 Maximum cable length based on voltage drop

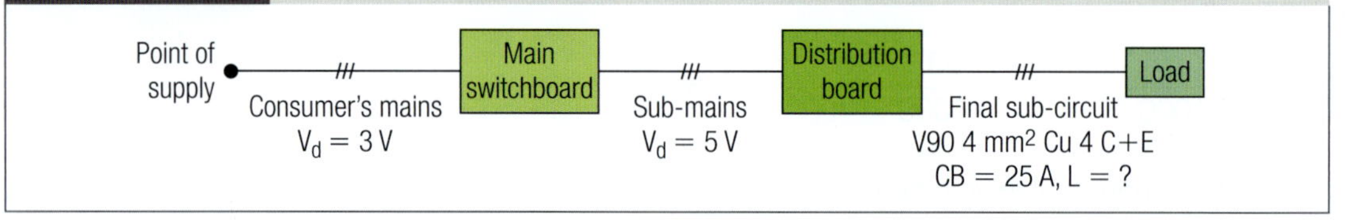

Solution

Three phase permissible voltage drop on final sub-circuit:

$$V_p = 5\% \text{ of } 400\text{ V} = 20\text{ V}$$
$$V_{p\ fsc\ 3\Phi} = V_p - (V_{d\ cm} + V_{d\ sm}) = 20 - (3 + 5) = 12\text{ V}$$

Single phase permissible voltage drop on final sub-circuit:

$$V_{p\ fsc\ 1\phi} = \frac{V_{p\ fsc\ 3\phi}}{\sqrt{3}} = \frac{12}{\sqrt{3}} = 6.936\text{ V}$$

From AS/NZS 3008.1.1:2009 (Table 42 column 6) V_c of 4.0 mm² = 9.71 mV/A.m.
Maximum permissible length:

$$L = \frac{1000 \times V_p}{V_c \times I} = \frac{1000 \times 6.936}{9.71 \times 1.155 \times 25} = 24.73\text{ m}$$

EXAMPLE 11.61

Calculate the maximum permissible length of the final sub-circuit shown in Figure 11.29 (adapted from Tafe NSW, *Learner Workbook*, p. 217).

FIGURE 11.29 Maximum cable length based on voltage drop

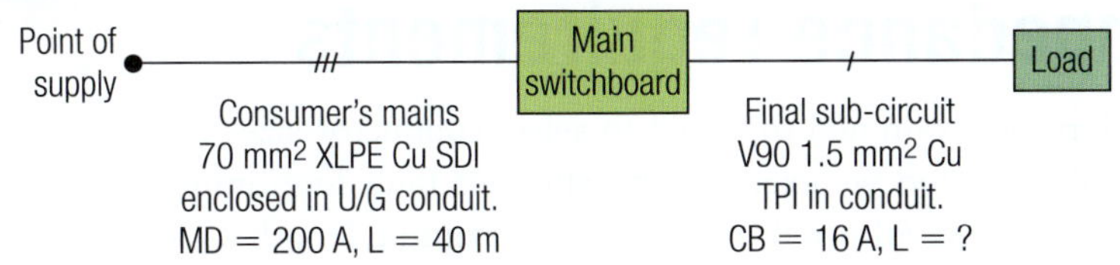

Solution

Consumer's mains ($V_{d\,cm}$)

From AS/NZS 3008.1.1:2009 (Table 41 column 8) V_c of 70 mm² = 0.623 mV/A.m.

$$V_d = \frac{L \times I \times V_c}{1000} = \frac{40 \times 200 \times 0.623}{1000} = 4.984 \text{ V}$$

Three phase permissible voltage drop on final sub-circuit:

$$V_p = 5\% \text{ of } 400 \text{ V} = 20 \text{ V}$$
$$V_{p\,fsc\,3\phi} = V_p - V_{d\,cm} = 20 - 4.984 = 15.016 \text{ V}$$

Single phase permissible voltage drop on final sub-circuit:

$$V_{p\,fsc\,1\phi} = \frac{V_{p\,fsc\,3\phi}}{\sqrt{3}} = \frac{15.016}{\sqrt{3}} = 8.679 \text{ V}$$

From AS/NZS 3008.1.1:2009 (Table 42 column 6) V_c of 1.5 mm² = 28.6 mV/A.m.

Maximum permissible length:

$$L = \frac{1000 \times V_p}{V_c \times I} = \frac{1000 \times 8.679}{28.6 \times 1.155 \times 16} = 16.4\text{m}$$

SUMMING UP...

- AS/NZS 3000:2007 clause 3.6.5 requires the voltage drop within an installation to be restricted to 5% of the nominal supply; that is, for a 230 V supply, maximum voltage drop = 11.5 V, for a 400 V supply = 20 V.
- Installations supplied directly from a supply authority substation or distribution transformer are allowed a voltage drop up to 7% of the nominal supply voltage.
- AS/NZS 3000:2007 clause 7.5.7 requires the voltage drop within extra low voltage portions of an installation to be restricted to 10% of the nominal supply.
- You can determine the value of V_c for the given cable size from AS/NZS 3008.1.1:2009 Tables 40 to 51.
- To determine the actual voltage drop for a given cable size, use the equation:

$$V_d = \frac{L \times I \times V_c}{1000}$$

- To determine the minimum permissible cable c.s.a. for a given circuit, use the equation:

$$V_c = \frac{1000 \times V_d}{L \times I}$$

- Calculate the maximum permissible length of the cable using the equation:

$$L = \frac{1000 \times V_P}{V_c \times I}$$

11.6 Cable selection based on fault loop impedance requirements

In previous sections we examined the need to select conductor sizes based on current rating and voltage drop. In this section we examine the effect of fault loop impedance on the selection of conductor sizes in circuits. Voltage drop limits the maximum length of conductors when current is flowing under normal operating conditions, that is, from phase to phase or phase to neutral. Earth fault loop impedance limits the maximum length of conductors when current is flowing under earth fault conditions, that is, from phase to earth. The protective earthing conductor is usually smaller than the active or neutral conductors so its impedance will be higher than that of the active or neutral conductors.

Under earth fault conditions the combined impedance of the fault path from active to protective earthing conductor will be higher than the impedance of a fault path from active to active, or active to neutral. In long cable runs, because of the higher impedance from active to protective earthing conductor, the fault current will be lower than that of a fault on a cable of a shorter length. The lower the earth fault current, the longer the circuit protection device will take to operate. In the time that it takes the circuit protection device to operate, a touch voltage will be present on the exposed conductive parts of the apparatus under fault. If a person is in simultaneous contact with the exposed conductive parts and earth they are said to be in indirect contact with live parts (Tafe NSW, *Learner Workbook*, p. 226).

AS/NZS 3000:2007 – maximum fault loop impedance

One of the fundamental safety principles required by the standards is fault protection. **AS/NZS 3000:2007** clause 1.5.5.1 specifies protection methods from indirect contact with live parts. The most commonly used method for providing this protection is automatic disconnection of supply. See **AS/NZS 3000:2007** clauses 1.5.5.3 and 2.4.2 on achieving automatic disconnection of the supply.

AS/NZS 3000:2007 clause 1.5.5.3 requires each circuit in an electrical installation to be protected such that automatic disconnection of supply will occur within the specified disconnection time when a fault of negligible impedance occurs between an active conductor and a protective earthing conductor, or an exposed conductive part anywhere in the electrical installation. This condition is met when the impedance of the path taken by the fault current, known as the earth fault loop, is low enough to allow sufficient current to flow to cause the protective device to operate within the specified time.

The earth fault loop in a MEN system is shown in Figure B5 of **AS/NZS 3000:2007**. The fault current flows from the supply transformer to the fault through the active conductors. It returns on the fault side of the installation MEN connection through the *protective earthing* conductor. From the MEN connection, the fault current returns along the *neutral* conductors to the supply transformer. **AS/NZS 3000:2007** clause 5.7.4, Tables 8.1 and 8.2, and Appendix B4 provide further guidance regarding compliance with fault loop impedance requirements (Tafe NSW, *Learner Workbook*, p. 227).

Disconnection times

When an earth fault occurs, a touch voltage appears on exposed conductive parts. This touch voltage will disappear quickly if the earth fault loop impedance is low enough to ensure a large current flow which quickly operates the circuit protection device. This is done in miniature circuit breakers by a magnetic trip mechanism. If the fault current is too low, the breaker will trip by its thermal mechanism. The longer time a person is in contact with a touch voltage the greater the risk of injury to the person. **AS/NZS 3000:2007** clauses 1.5.5.3(d) and 5.7.2 specify the time in which automatic disconnection of supply must occur (Tafe NSW, *Learner Workbook*, p. 230).

Using AS/NZS 3008:2009 to determine cable impedances

The total earth fault loop impedance will be the sum of the supply transformer impedance and the impedance of all cables in the path between the supply and the fault.

For the purpose of simplification in cable ratings up to 400 A, the reactance will be treated as negligible, and calculations will be done using only the a.c. resistance. Cables less than, say, 120 mm² only have to consider resistance but larger cables should take reactance into account. The a.c. resistance of conductors in any given combination of cables can be predicted before the cable is selected and installed by using Tables 34 to 39 of Section 4 of **AS/NZS 3008:2009** (Tafe NSW, *Learner Workbook*, p. 227).

The unit values in Tables 34 to 39 are given in ohms per kilometre (Ω/km). To calculate the a.c. resistance of a given conductor use the equation:

$$R = \frac{R_c \times L}{1000}$$

where

R = the resistance of the cable in ohms (Ω)

R_c = the table value in ohms per km (Ω/km)

L = the length of the conductor in metres (m)

EXAMPLE 11.62

Calculate the a.c. resistance of a 16 mm² SDI XLPE (Cu) cable, 45 m long (use normal operating temperature) (adapted from Tafe NSW, *Learner Workbook*, p. 228).

Solution

From AS/NZS 3008.1.1:2009 (Table 34 column 5) R_c of 16 mm² = 1.47 Ω/km.

$$R = \frac{R_c \times L}{1000} = \frac{1.47 \times 45}{1000} = 0.06615\ \Omega$$

EXAMPLE 11.63

Using AS/NZS 3008.1.1:2009 Tables 34 to 39, determine for the diagram shown in Figure 11.30 the impedance of the distribution mains, impedance of the consumer's mains, impedance of the final sub-circuit and total earth fault loop impedance. Use normal maximum operating temperature of the cable (adapted from Tafe NSW, *Learner Workbook*, p. 229).

FIGURE 11.30 Total fault loop impedance Z_s

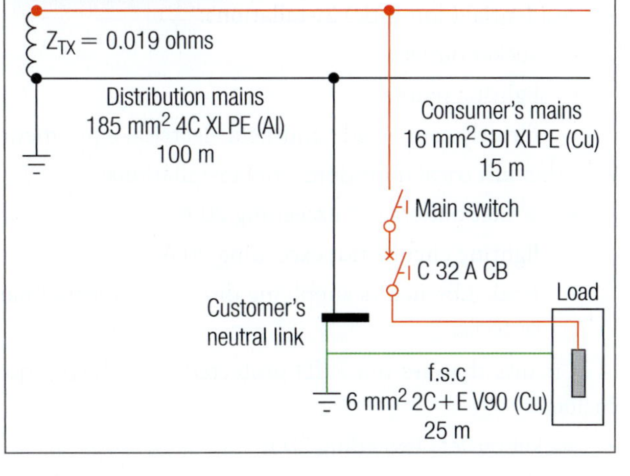

Solution

Distribution mains (Z_{Dist}) active + neutral

From AS/NZS 3008.1.1:2009 (Table 35 column 10) R_c of 185 mm² = 0.212 Ω/km.

$$R = \frac{R_c \times L}{1000} = \frac{0.212 \times 100}{1000} \times 2 = 0.0414\ \Omega$$

Consumer's mains (Z_{cm}) active + neutral

From AS/NZS 3008.1.1:2009 (Table 34 column 5) R_c of 16 mm² = 1.47 Ω/km.

$$R = \frac{R_c \times L}{1000} = \frac{1.47 \times 15}{1000} \times 2 = 0.0441\ \Omega$$

Final sub-circuit active ($Z_{A\ fsc}$)
From AS/NZS 3008.1.1:2009 (Table 35 column 4) R_c of 6 mm² = 3.75 Ω/km.

$$R = \frac{R_c \times L}{1000} = \frac{3.75 \times 25}{1000} = 0.09375\ \Omega$$

Final sub-circuit protective earth (Z_{PEfsc})
From AS/NZS 3008.1.1:2009 (Table 35 column 4) R_c of 2.5 mm² = 9.01 Ω/km.

$$R = \frac{R_c \times L}{1000} = \frac{9.01 \times 25}{1000} = 0.22525\ \Omega$$

Total fault loop impedance

$$\begin{aligned} Z_s &= Z_{Tx} + Z_{Dist} + Z_{cm} + Z_{fsc} \\ &= 0.019 + 0.0414 + 0.0441 + 0.09375 + 0.22525 \\ &= 0.4235\ \Omega \end{aligned}$$

Calculating expected fault loop impedance

Any circuit protected by an RCD satisfies the requirements of earth fault loop impedance. RCDs that protect light and power circuits must have a rated residual current of not greater than 30 mA (**AS/NZS 3000:2007** Section 2.6.3). Calculation of earth fault loop impedance in these circuits is pointless. The low current (< 30 mA) and extremely fast operation (< 300 mS) ensure automatic disconnection of supply within the required time. Type 'S' RCDs that have a rated residual current in the range of 100 mA to 300 mA (**AS/NZS 3000:2007** clause 2.6.2.3), which are used to protect against the initiation of fire, will also satisfy earth fault loop impedance requirements. **AS/NZS 3000:2007** Section 2.6.3 requires the following circuits to have additional protection by RCDs:

- residential (domestic) installations:
  - socket outlets
  - lighting points
  - directly connected hand-held electrical equipment
- other electrical (non-domestic) installations:
  - socket outlets not exceeding 20 A
  - lighting circuits not exceeding 20 A
  - final sub-circuits supplying directly connected hand-held electrical equipment, e.g. hair dryers or tools.

Circuits that are not RCD protected and which require earth fault loop impedance to be applied include:

- socket outlets exceeding 20 A
- fixed or stationary equipment (mass exceeds 18 kg)
- sub-mains (Tafe NSW, *Learner Workbook*, p. 230).

The total earth fault loop impedance (Z_S) is calculated by a variation of Ohm's law as per **AS/NZS 3000:2007** Section B4.5:

$$Z_s = \frac{U_o}{I_a}$$

where
Z_s = the total earth fault loop impedance in ohms (Ω)
U_o = the nominal phase voltage in volts (V)
I_a = current causing automatic operation of the protective device in amperes (A)

Figure 11.31 shows values I_a for 0.4 and 5.0 second disconnection times.

The value of I_a is the current required to operate the protection device within the maximum required time. For circuit breakers, the value used is the mean of the maximum and minimum operating currents.

FIGURE 11.31 Current causing automatic operation of the protective device I_a

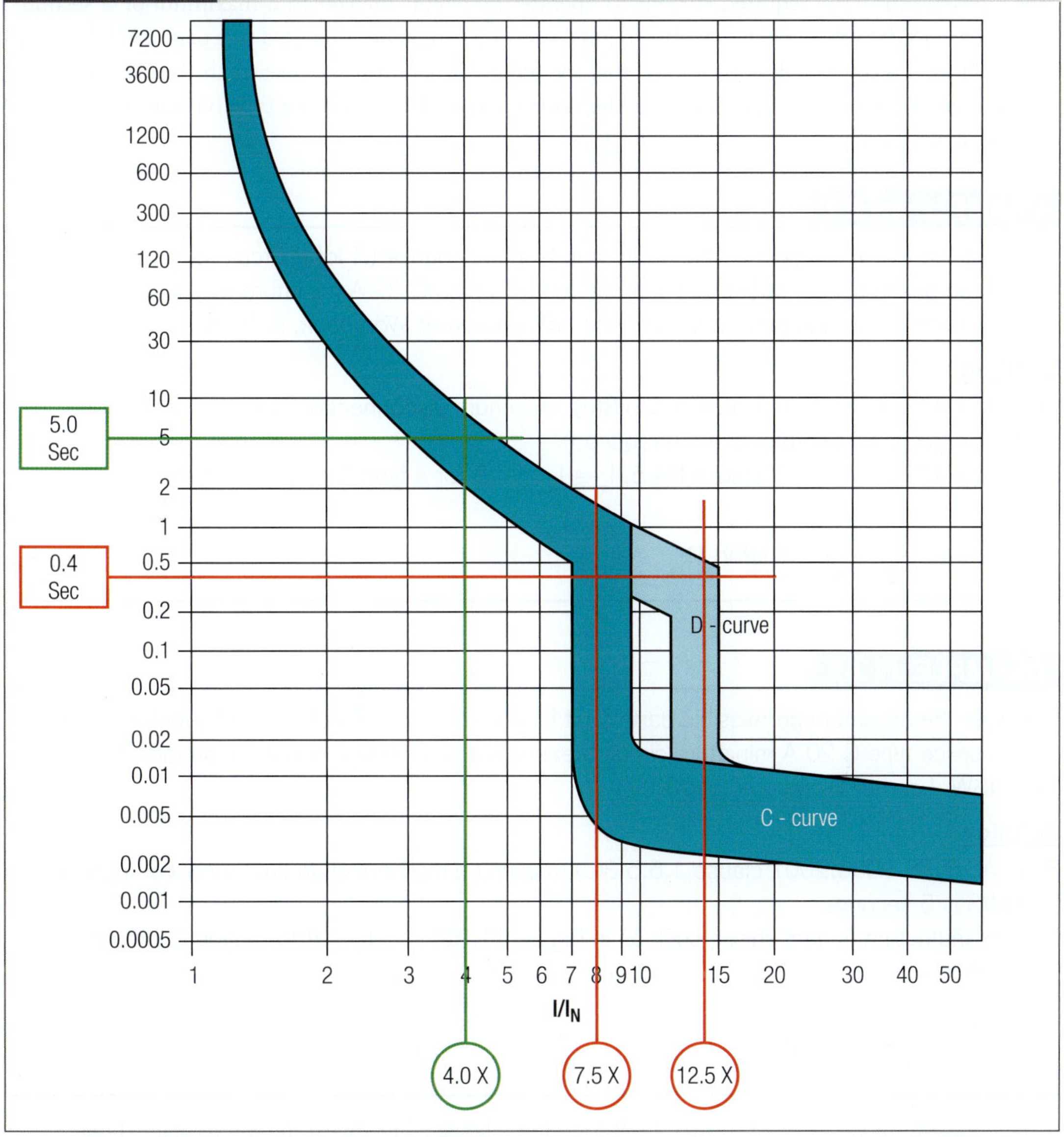

In circuits that require a 0.4 second disconnection time and that are protected by a circuit breaker, the required mean current as per **AS/NZS 3000:2007** Section B4.5 will be:

- $4 \times I_N$ for a type B circuit breaker
- $7.5 \times I_N$ for a type C circuit breaker
- $12.5 \times I_N$ for a type D circuit breaker.

Figure 11.31 shows the time current characteristics of a miniature circuit breaker. The current required to operate the circuit breaker in 0.4 seconds for socket outlets, class I and hand-held equipment is $7.5 \times I_N$ (nominal current rating) for type C circuit breakers, and $12.5 \times I_N$ for type D circuit breakers is specified by **AS/NZS 3000:2007** Section B4.5. For fixed or stationary devices with a maximum disconnection time of 5 seconds, the required current is $5 \times I_N$ for type B, C and D circuit breakers. In this case the protection device is operated by the thermal tripping mechanism rather than magnetically.

A different brand of breaker may have a different required current to operate the breaker in 5 seconds. For this reason, Tables 8.1, 8.2 and B1 of **AS/NZS 3000:2007** only show a 0.4 second disconnection time. To determine the required current to operate the circuit breaker in a maximum of 5 seconds, manufacturer's data sheets like that shown in Figure 11.32 must be consulted (see page 492). It should be noted that it is completely coincidental that the required current (I_a) to operate the breaker thermally in 5 seconds, as shown in Figure 11.32, is the same value of $4 \times I_N$ as for a type B circuit breaker that operates magnetically.

EXAMPLE 11.64

Calculate the maximum permissible earth fault loop impedance (Z_s)of a circuit supplying a 32 A three phase socket outlet that is protected by a type C 32 A miniature circuit breaker in a 230/400 volt installation (adapted from Tafe NSW, *Learner Workbook*, p. 233).

Solution

From AS/NZS 3000:2007 clause 1.5.5.3(d), maximum disconnection time for socket outlet is 0.4 seconds (RCD protection not required).

From AS/NZS 3000:2007 clause B4.5, $I_a = I_N \times 7.5$ for a type C circuit breaker.

$$Z_S = \frac{U_o}{I_a} = \frac{230}{7.5 \times 32} = 0.958\ \Omega$$

EXAMPLE 11.65

Calculate the maximum permissible earth fault loop impedance (Z_s) of a circuit supplying a hot water service type C 20 A miniature circuit breaker in a 230/400 volt installation (adapted from Tafe NSW, *Learner Workbook*, p. 233).

Solution

From AS/NZS 3000:2007 clause 1.5.5.3(d), maximum disconnection time for fixed appliance HWS is 5 seconds.

From manufacturer's data sheet, such as in Figure 11.32, $I_a = I_N \times 4$ for a type C circuit breaker.

$$Z_S = \frac{U_o}{I_a} = \frac{230}{4 \times 20} = 2.875\ \Omega$$

The maximum value of total earth fault loop impedance (Z_s) can also be found using Table 8.1 of **AS/NZS 3000:2007**. Only 0.4 second disconnection times are shown for circuit breakers. If an earth fault loop impedance is required for a fixed or stationary appliance with a 5 second disconnection time, it will have to be calculated.

Internal earth fault loop impedance (Z_{int})

The total earth fault loop impedance is made up of two parts known as the *external* and *internal* earth fault loop impedances (see Figure B5 **AS/NZS 3000:2007**). In the vast majority of cases the impedance of the external section will be unknown. To simplify calculations it is assumed that at the circuit protection device (reference point), as shown in Figure B5 (**AS/NZS 3000:2007**), at least 80% (0.8 x) of the nominal supply voltage (230 V) is available under earth fault conditions; that is, when calculating the internal earth fault loop impedance, 80% of the nominal supply voltage is used. If a larger value of voltage is present, a higher earth fault current will flow. The operating time of the protection device will be shorter and will disconnect the circuit automatically in less time than is required.

The internal earth fault loop impedance (Z_{int}) is calculated by a variation of Ohm's law (**AS/NZS 3000:2007** Section B5.2.1):

$$Z_{int} = \frac{U_o \times 0.8}{I_a}$$

where

Z_{int} = the total earth fault loop impedance in ohms (Ω)
U_o = the nominal phase voltage in volts (V)
I_a = current causing automatic operation of the protective device in amperes (A)

Once the maximum permissible internal earth loop impedance is known, the maximum length of the cable can be determined so that the impedance of the cable is less than or equal to the maximum permissible internal earth fault impedance (Tafe NSW, *Learner Workbook*, p. 233).

EXAMPLE 11.66

Calculate the maximum permissible earth internal fault loop impedance (Z_{int}) of a circuit supplying a 32 A three phase socket outlet that is protected by a type C 32 A miniature circuit breaker in a 230/400 volt installation (adapted from Tafe NSW, *Learner Workbook*, p. 235).

Solution

From AS/NZS 3000:2007 clause 1.5.5.3(d), maximum disconnection time for socket outlet is 0.4 seconds (circuit not RCD protected).
From AS/NZS 3000:2007 clause B4.5, $I_a = I_N \times 7.5$ for a type C circuit breaker.

$$Z_S = \frac{U_o}{I_a} = \frac{230 \times 0.8}{7.5 \times 32} = 0.7664\ \Omega$$

EXAMPLE 11.67

Calculate the maximum permissible earth fault loop impedance (Z_s) of a circuit supplying a hot water service type C 20 A miniature circuit breaker in a 230/400 volt installation (adapted from Tafe NSW, *Learner Workbook*, p. 235).

Solution

From AS/NZS 3000:2007 clause 1.5.5.3(d), maximum disconnection time for fixed appliance HWS is 5 seconds.

From manufacturer's data sheet, such as Figure 11.32, $I_a = I_N \times 4$ for a type C circuit breaker.

$$Z_S = \frac{U_o}{I_a} = \frac{230}{4 \times 20} = 2.3\ \Omega$$

Selecting cables to satisfy fault loop impedance and current

The major impact that earth fault loop impedance has on a circuit is to limit its length for a given c.s.a. As shown earlier, in the design stage **AS/NZS 3008.1.1:2009** Tables 34 to 39 can be used to predict the earth fault loop impedance of a cable for a given length. In most cases, if the cable has been selected correctly based on current carrying capacity and voltage drop, the earth fault loop impedance will not be an issue.

Normally voltage drop is the most significant factor that limits the length of a cable. Long cables which are lightly loaded however can be an issue. When cable sizes are increased to compensate for long runs, the rating of the circuit protection should *not* be increased to suit the current rating of the larger cable. A higher protection device rating will increase the required current (I_a) to operate the protection device, so this will decrease the maximum permissible fault loop impedance.

Table B1 of **AS/NZS 3000:2007** specifies maximum route lengths for a number of standard circuit protection devices and cable size combinations. Only lengths relating to 0.4 seconds disconnection times are shown. Circuits supplying socket outlets and lighting points which are RCD protected are not restricted in length by earth fault loop impedance as the RCD will operate under active to earth fault conditions despite excessive earth fault loop impedance. Voltage drop in the circuit must still be considered.

The circuits in the following examples are of a type that are typically not RCD protected and will have a 0.4 second disconnection time (Tafe NSW, *Learner Workbook*, p. 236).

EXAMPLE 11.68

Determine the maximum route length based on earth fault loop impedance for a 4 mm² copper 4C+e V90 cable supplying a 32 A three phase socket outlet that is protected by a type C 32 A miniature circuit breaker in a 230/400 volt installation (adapted from Tafe NSW, *Learner Workbook*, p. 238).

Solution

From AS/NZS 3000:2007 Table B1, maximum length = 67 m.

EXAMPLE 11.69

Determine the maximum route length based on earth fault loop impedance for a three phase 63 A socket outlet for a welder protected by a 63 A type D circuit breaker wired in 16 mm² copper 4C+E in a 230/400 volt installation (adapted from Tafe NSW, *Learner Workbook*, p. 238).

Solution

From AS/NZS 3000:2007 Table B1, maximum length = 45 m.

If a cable/circuit breaker combination is not listed in **AS/NZS 3000:2007** Table B1 or the circuit supplied requires a 5 second disconnection time, the earth fault loop impedance must be calculated.

AS/NZS 3000:2007 Section B5.2.2 specifies a method to determine maximum route length of a conductor based on earth fault loop impedance (L_{max}) using the equation:

$$L_{max} = \frac{0.8 \times U_o \times S_{ph} \times S_{pe}}{I_a \times \rho \times (S_{ph} + S_{pe})}$$

where, as stated in **AS/NSZ 3000:2007**:

L_{max} = maximum route length in metres
U_o = the nominal phase voltage in volts (V)
ρ = resistivity at normal working temperature in Ω-mm²/m
= 22.5×10^{-3} for copper
= 36×10^{-3} for aluminium
I_a = current causing instantaneous operation of the protective device in amperes (A)
also = the current that assures operation of the protective fuse concerned, in the specified time
S_{ph} = cross-sectional area of the active conductor of the circuit concerned in mm²
S_{pe} = cross-sectional area of the protective earthing conductor concerned in mm²

This calculation was used to determine the maximum lengths of cables in Table B1 of **AS/NZS 3000:2007**. If a circuit has a disconnection time of 5 seconds the mean tripping current of the protection device must be found from manufacturer's data. The multiples of the nominal current are then applied to the nominal current rating to find I_a.

The characteristic curves of circuit breakers in the 5 second range vary widely as demonstrated by Examples 11.69 and 11.70. The earth fault loop impedance must be calculated to suit the characteristic of the brand of circuit breaker actually used. The 5 second trip characteristics of circuit breakers in the lower current ranges (10–63 A) are usually different to the characteristics of circuit breakers in the range of 250–400 A.

EXAMPLE 11.70

Using Section B5.2.2 of AS/NZS 3000:2007, calculate the maximum route length based on earth fault loop impedance for a 4 mm² copper 2C+e V90 circuit supplying a cook top protected by a C 25 A Clipsal miniature circuit breaker in a 230/400 volt installation (adapted from Tafe NSW, *Learner Workbook*, p. 245).

Solution

From AS/NZS 3000:2007 clause 1.5.5.3(d), maximum disconnection time for fixed appliance (cook top) is 5 seconds.

From manufacturer's data sheet shown in Figure 11.32, $I_a = I_N \times 4$ for a circuit breaker.

$$L_{max} = \frac{0.8 \times U_o \times S_{ph} \times S_{pe}}{I_a \times \rho \times (S_{ph} + S_{pe})} = \frac{0.8 \times 230 \times 4.0 \times 2.5}{4 \times 25 \times 22.5 \times 10.3 \times (4.0 + 2.5)} = 81.78 \text{ m}$$

EXAMPLE 11.71

Using Section B5.2.2 of AS/NZS 3000:2007 and Figure 11.33 below, calculate the maximum route length based on earth fault loop impedance of a 240 mm² 4 × single core aluminium XLPE conductors, supplying a three phase sub-main protected by a 400 A type C miniature circuit breaker in a 230/400 volt installation (adapted from Tafe NSW, *Learner Workbook*, p. 245).

Solution

From AS/NZS 3000:2007 clause 1.5.5.3(d), maximum disconnection time for a sub-main is 5 seconds.

From the manufacturer's data sheet shown in Figure 11.32, $I_a = I_N \times 8$ for a circuit breaker.

FIGURE 11.32 Current causing automatic operation of the protective device I_a

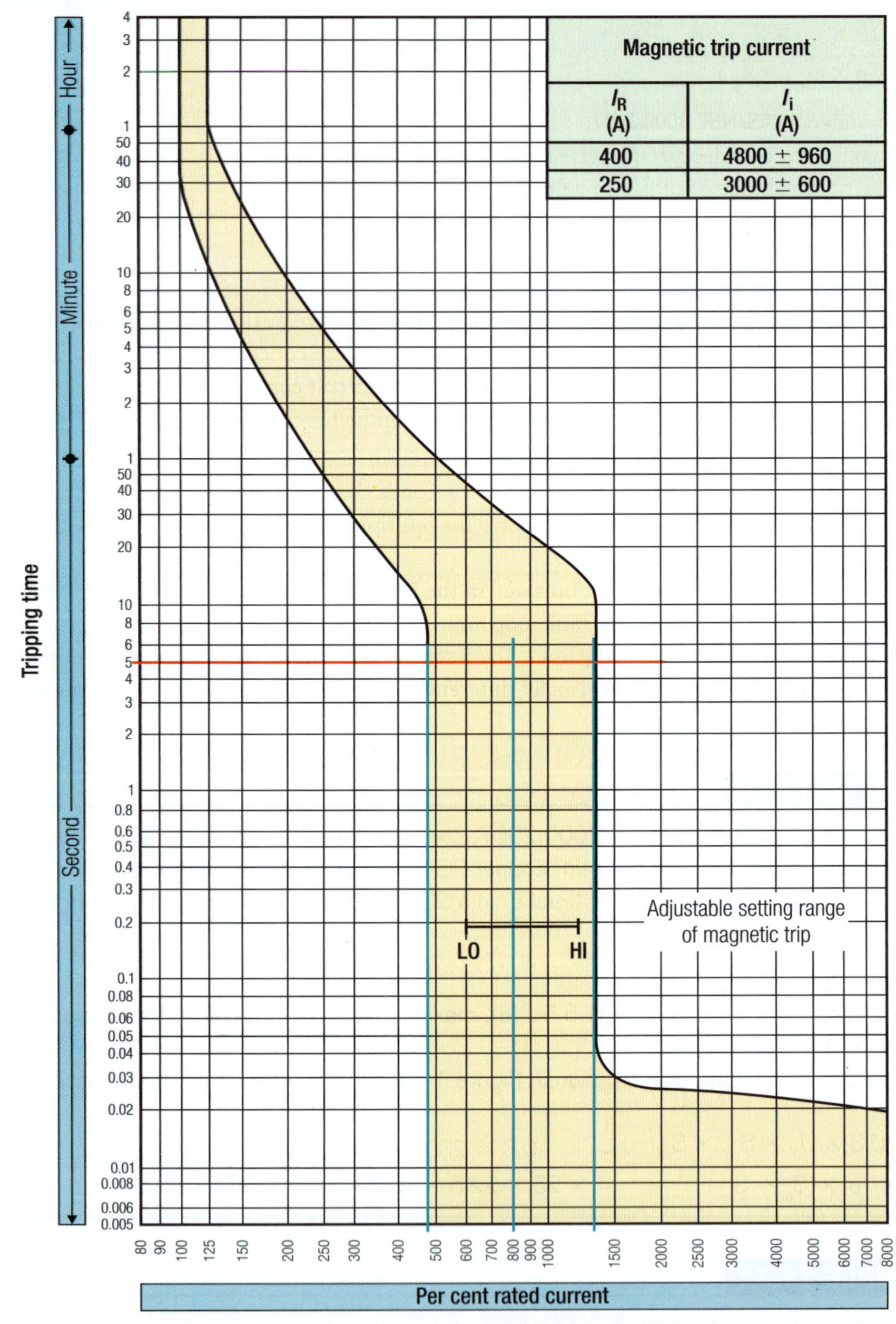

I_R (A)	I_i (A)
400	4800 ± 960
250	3000 ± 600

Source: NHP Electrical Engineering Products, *ELR-C 06-010 Moulded Case Circuit Breakers Technical Catalogue*, 2008, p. 24, http://www.nhp.com.au

On a long run of cable the situation can arise where cable size has been increased to compensate for voltage drop; that is, the maximum demand (I_B) will be much lower than the current carrying capacity of the cable (I_Z). Figure 11.33 shows a scenario where both current carrying capacity and voltage drop comply with **AS/NZS 3000:2007** requirements, but the earth fault loop impedance does not.

AS/NZS 3000:2007 clause 2.5.3.1 covers protection against overload $I_B \leq I_N \leq I_Z$. In this case I_B (150 A) is less than I_N (400 A) and I_N is less than I_Z (403 A), so protection against overload is provided.

FIGURE 11.33 Sub-main fault loop impedance

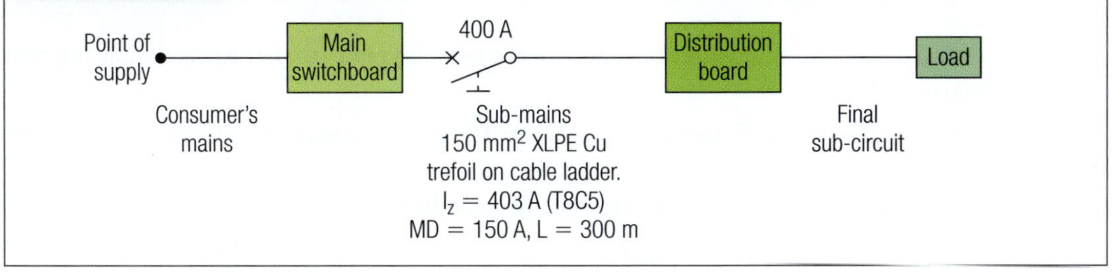

Under **AS/NZS 3000:2007** clause 3.6.2 voltage drop should not exceed 5% of nominal supply voltage. When calculating voltage drop it is permitted to use the lower maximum demand current rather than the rating of the circuit breaker.

$$V_d = \frac{L \times I \times V_c}{1000} = \frac{300 \times 150 \times 0.316}{1000} = 14.22\ \text{V}$$

The voltage drop on the circuit is not excessive and complies with **AS/NZS 3000:2007** requirements. While 150 A would be a more suitable rating of a circuit breaker than the 400 A device, it still complies with requirements for overload protection and voltage drop.

Under **AS/NZS 3000:2007** clause 1.5.5.3, circuits under fault conditions between active and earth must be automatically disconnected from the supply within the specified time. In this case for a sub-main, the required disconnection time is 5 seconds (Tafe NSW, *Learner Workbook*, pp. 246–7).

Under **AS/NZS** 3000:2007 clause 5.2.1, determination of maximum length is made with the equation:

$$Z_{int} = \frac{0.8 \times U_o}{I_a}$$

From manufacturer's data, such as the graph shown in Figure 11.33, the current required to operate the 400 A circuit breaker in 5 seconds would be 8 × 400 A which equals 3200 A:

$$Z_{int} = \frac{0.8 \times U_o}{I_a} = \frac{0.8 \times 230}{8 \times 400} = 0.0575\ \Omega$$

Following is an explanation of the solution to achieve compliance for Example 11.70.

From Table 34 of **AS/NZ 3008.1.1:2009**:

Active 150 mm^2

$$Z_{PH} = \frac{0.160 \times 300}{1000} = 0.048\ \Omega$$

Protective earth 50 mm^2

$$Z_{PE} = \frac{0.494 \times 300}{1000} = 0.1482\ \Omega$$

Total cable impedance

$$Z_{Cable} = Z_{PH} + Z_{PE} = 0.048 + 0.1482 = 0.1962\ \Omega$$

The combined impedance of the cables (active and protective earthing conductor) is larger than the permitted internal earth fault loop impedance (Z_{int}). The 400 A circuit breaker would take longer than 5 seconds to operate. The circuit *does not* comply.

Solution 1 – Decrease the rating of the circuit breaker to 160 A. From the manufacturer's data the current required to operate the 160 A circuit breaker in 5 seconds would be 7.5 × 160 A which equals 1200 A.

$$Z_{int} = \frac{0.8 \times U_o}{I_a} = \frac{0.8 \times 230}{7.5 \times 160} = 0.153\ \Omega$$

The circuit still does not comply. The impedance of the cable is greater than the permissible internal earth fault loop impedance.

Solution 2 – Decrease the rating of the circuit breaker to 160 A and increase the size of the protective earthing conductor.

Required maximum impedance of protective earthing conductor:

$$Z_{PE} = Z_{int} - Z_{PH} = 0.153 - 0.048 = 0.105\ \Omega$$

Increase the protective earthing conductor to70 mm^2.

$$Z_{PE} = \frac{0.342 \times 300}{1000} = 0.1026\ \Omega \text{ (Table 34 column 5 of AS/NZS 3008.1.1:2009)}$$

$$Z_{Cable} = Z_{PH} + Z_{PE} = 0.048 + 0.1026 = 0.15062\ \Omega$$

The cable impedance (0.15062 Ω) is now less than the permissible internal fault loop impedance (0.153 Ω). The circuit does comply.

SUMMING UP...

- RCD protection devices have a lower I_a and offer an alternative method of earth fault loop impedance compliance.
- Above the MEN connection, the path of the fault loop is active to neutral and below the installation's MEN connection the path is active to earth.
- AS/NZS 3000:2007 clauses 1.5.5.3(d) and 5.7.2 specify the time in which automatic disconnection of supply must occur; that is, 0.4 seconds for circuits supplying socket outlets and 5 seconds for fixed or stationary appliances or sub-mains.
- To calculate the a.c. resistance of a given conductor use the equation:

$$R = \frac{R_c \times L}{1000}$$

- The total earth fault loop impedance (Z_S) is calculated as required by AS/NZS 3000:2007 Section B4.5 with the equation:

$$Z_S = \frac{U_o}{I_a}$$

- The internal earth fault loop impedance (Z_{int}) is calculated as required by AS/NZS 3000:2007 Section B5.2.1 with the equation:

$$Z_s = \frac{U_o \times 0.8}{I_a}$$

- AS/NZS 3000:2007 Table 8.1 provides guidance to the maximum permissible total fault loop impedance.
- AS/NZS 3000:2007 Section B5.2.2 specifies a method for determining maximum route length of a conductor based on earth fault loop impedance (L_{max}) with the equation:

$$L_{max} = \frac{0.8 \times U_o \times S_{ph} \times S_{pe}}{I_a \times \rho \times (S_{ph} + S_{pe})}$$

- AS/NZS 3000:2007 Table B1 provides guidance to the maximum permissible length of a conductor based on fault loop impedance.
- Voltage drop is more likely to restrict maximum permissible length of a conductor than fault loop impedance.

11.7 Selecting protection devices

This section covers three types of electrical faults that must be protected against:

- *Overload* – too much load is applied to an otherwise undamaged circuit, causing a larger current to flow than under normal operation.

- *Short circuit* – a fault of negligible impedance which occurs between live conductors, usually due to damage to insulation or failure of equipment. Known as the prospective short circuit current, current will only be limited by the impedance of the cables and will be of an extremely high value.
- *Earth fault* – a live conductor comes in contact with conductive parts which are 'earthed', usually due to insulation failure. The current that flows will be of a lower value than the prospective short circuit current as the earth fault loop impedance will be higher than the cable impedance of a phase-to-phase fault.

AS/NZS 3000:2007 clause 1.5.5.1 requires that one or more methods are used to prevent a touch voltage occurring on the conductive parts under fault. Automatic disconnection of supply is the most commonly used method. A combination of circuit protection and a protective earthing conductor is used to provide a low impedance path so that a sufficient current flows to operate the protection device in the required time (Tafe NSW, *Learner Workbook*, p. 254).

Coordination between conductors and protection devices

To understand what occurs in a circuit under fault conditions it is necessary to understand the normal operation of a circuit. Figure 11.34 shows a healthy circuit. The current is limited by the impedance of the load. Under normal conditions the impedance of the cable is so low it is considered negligible. The line current carried by the cable is calculated using Ohm's law. As the circuit is a balanced three phase load, no current flows in the neutral. For single point loads like that shown in Figure 11.34, this is also the maximum demand current (I_B) of the circuit. From **AS/NZS 3000:2007** Table C6, the current rating (I_Z) of the cable is 32 A.

FIGURE 11.34 Normal circuit operation

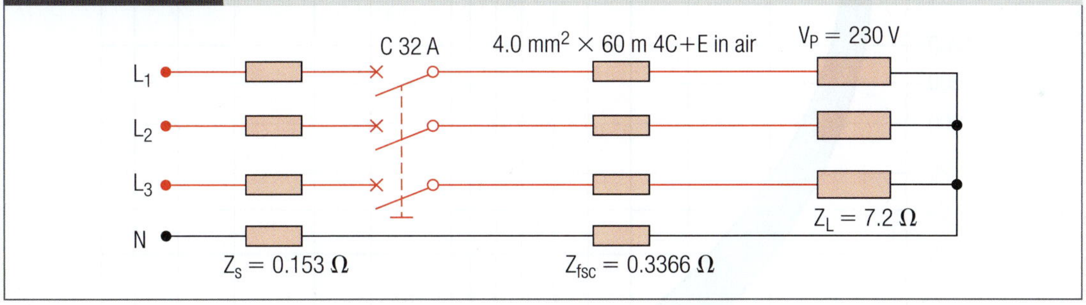

The circuit line current can be calculated using the equation:

$$I_L = I_P = \frac{V_P}{Z_P}$$

where

I_L = the line current in amperes (A)
I_P = the phase current in amperes (A)
V_P = the nominal phase voltage in volts (V)
Z_P = the phase impedance in ohms (Ω)

EXAMPLE 11.72

Calculate the line and phase currents of the circuit shown in Figure 11.34 (adapted from Tafe NSW, *Learner Workbook*, p. 254).

Solution

$$I_L = I_P = \frac{V_P}{Z_P} = \frac{230}{7.2} = 31.95\text{ A}$$

To protect the cable from an overload, the circuit protection device must operate in a suitable time so the cable does not operate above its maximum (normal) operating temperature for a prolonged period of time. In Section 11.2, it was shown that for cables to be protected against overload current **AS/NZS 3000:2007** clause 2.5.3.1 must be met; that is:

$$I_B \leq I_N \leq I_Z$$

In this case the maximum demand of 31.95 A (I_B) is less than the 32 A rating of the protection device (I_N) and is also less than the 32 A (I_Z) current rating of the cable.

The ratio of the current actually flowing through the circuit (I) to the nominal rating of the circuit protection device (I_N) gives the multiples of rated current (I_M); that is:

$$I_M = \frac{I_N}{I}$$

This ratio is used to predict the operating time of the protection device. Figure 11.35 shows the time-current characteristic curve of the circuit protection device in Figure 11.34. In this case the ratio

FIGURE 11.35 Normal circuit operation time current characteristic curve

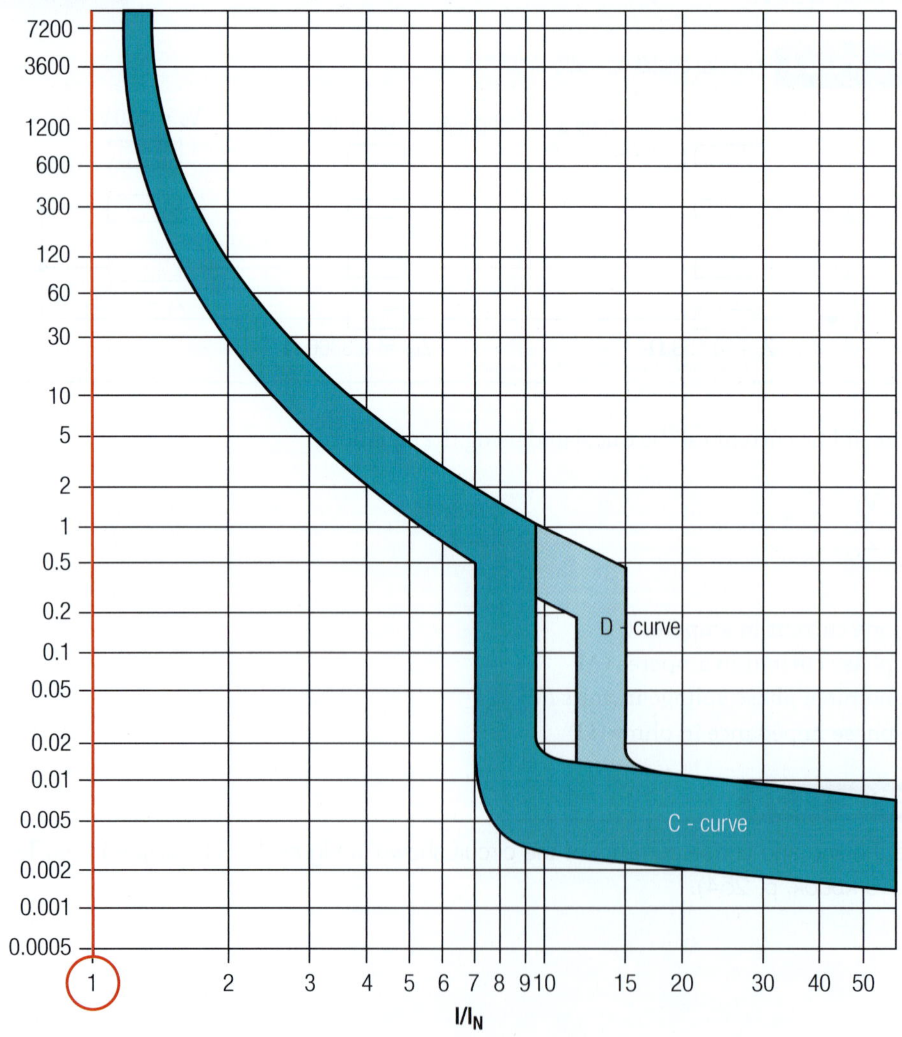

is equal to 1. When plotted on the graph of Figure 11.35 (in red) the line never crosses the blue curve of the circuit breaker's operating characteristic, indicating the breaker will not trip.

Overload condition

Figure 11.36 shows the same circuit in an overload condition. By adding more load in parallel to each phase, the circuit impedance has decreased from 7.2 Ω to 3.6 Ω. The cable impedance is still considered negligible. The circuit line current will increase to a value higher than the rated current of the cable (I_Z) (Tafe NSW, *Learner Workbook*, p. 256).

FIGURE 11.36 Overload circuit condition

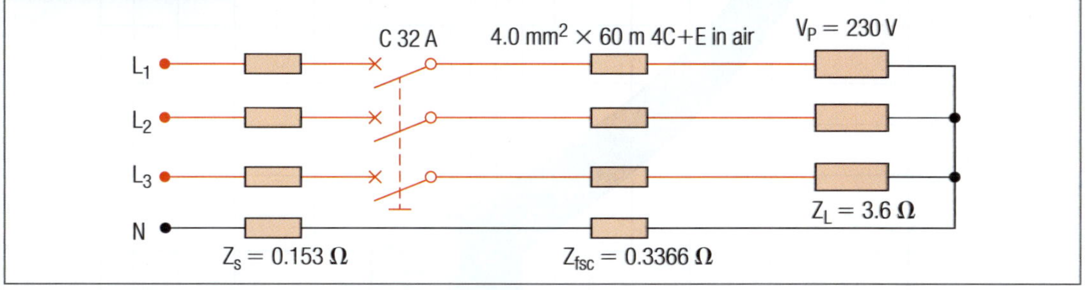

EXAMPLE 11.73

Calculate the line and phase currents of the circuit shown in Figure 11.36 and determine the operating time of the circuit breaker (adapted from Tafe NSW, *Learner Workbook*, p. 258).

Solution

$$I_L = I_P = \frac{V_P}{Z_P} = \frac{230}{3.6} \approx 64 \text{ A}$$

$$I_M = \frac{I}{I_N} = \frac{64}{32} = 2 \text{ x}$$

From Figure 11.37 the maximum operating time is 120 seconds and the minimum operating time is approximately 30 seconds.

Short circuit condition

Figure 11.38 shows a circuit condition known as a bolted symmetrical fault. The impedance of the fault is considered to be negligible. The only impedance remaining to limit the circuit current is the impedance of the cable. The fault forms a star point on the supply side of the load, so the cable impedances form a balanced star connected circuit. As the circuit is balanced, no current will flow in the neutral; therefore the neutral conductor impedance is not included when calculating the short circuit current.

The circuit protection device, either circuit breaker or HRC fuse, must be capable of breaking this prospective short circuit current without damage to either the cables or the protection device itself (Tafe NSW, *Learner Workbook*, p. 262).

FIGURE 11.37 Overload time current characteristic curve

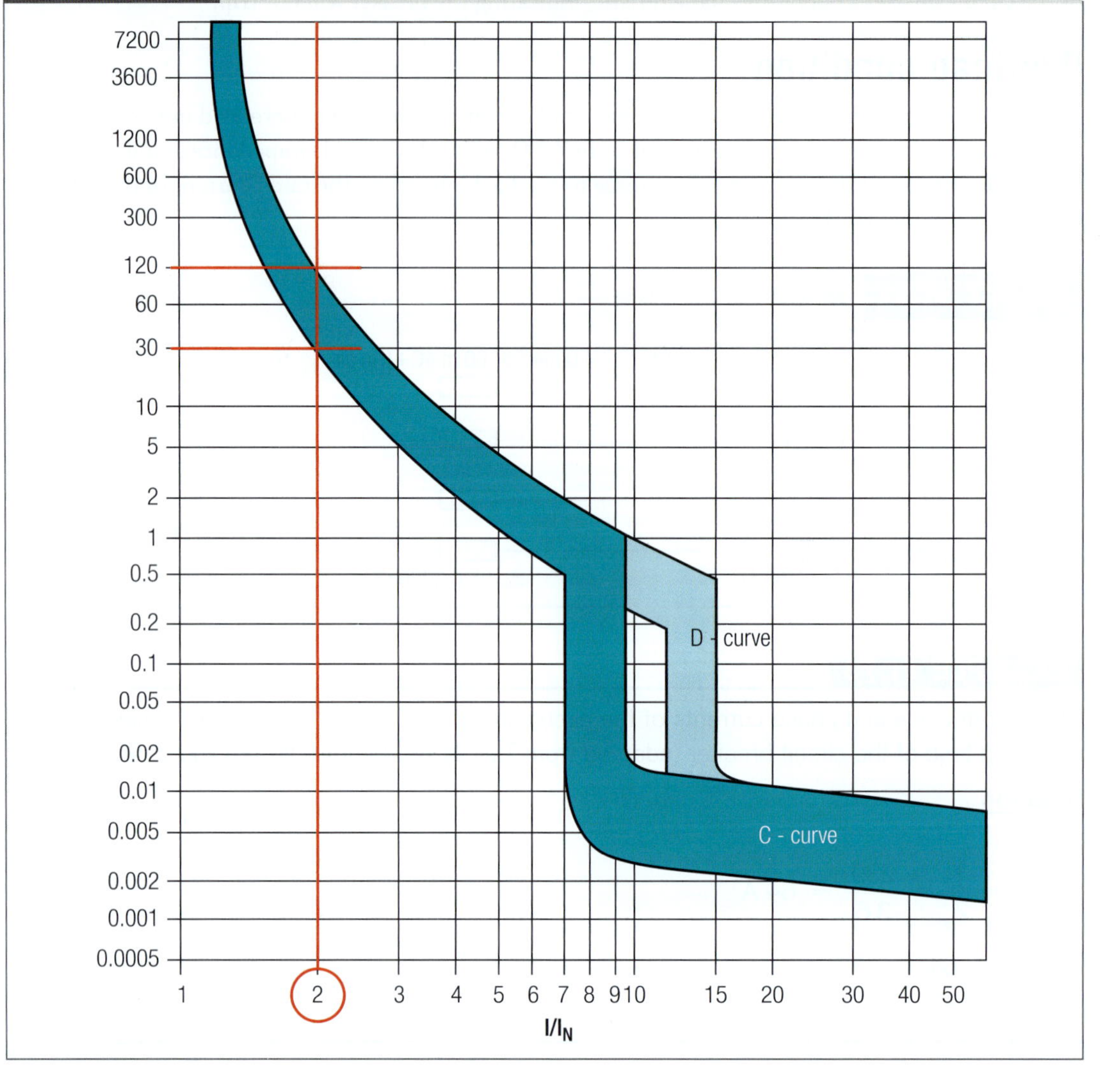

FIGURE 11.38 Short circuit condition

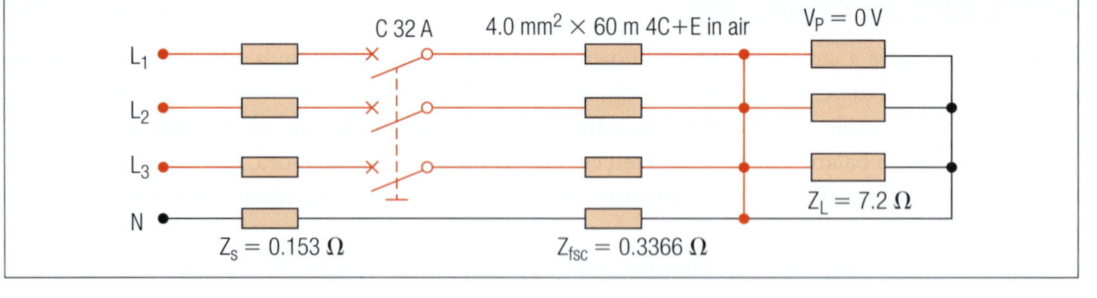

EXAMPLE 11.74

Calculate the line and phase currents of the circuit shown in Figure 11.38 and determine the operating time of the circuit breaker (adapted from Tafe NSW, *Learner Workbook*, p. 263).

Solution

$$I_L = I_p = \frac{V_p}{Z_p} = \frac{230}{0.153 + 0.3366} \approx 470\text{ A}$$

$$I_M = \frac{I}{I_N} = \frac{470}{32} \approx 15\text{ x}$$

From Figure 11.39, the maximum operating time is approximately 0.015 seconds and the minimum operating time is approximately 0.002 seconds.

FIGURE 11.39 Short circuit time current characteristic curve

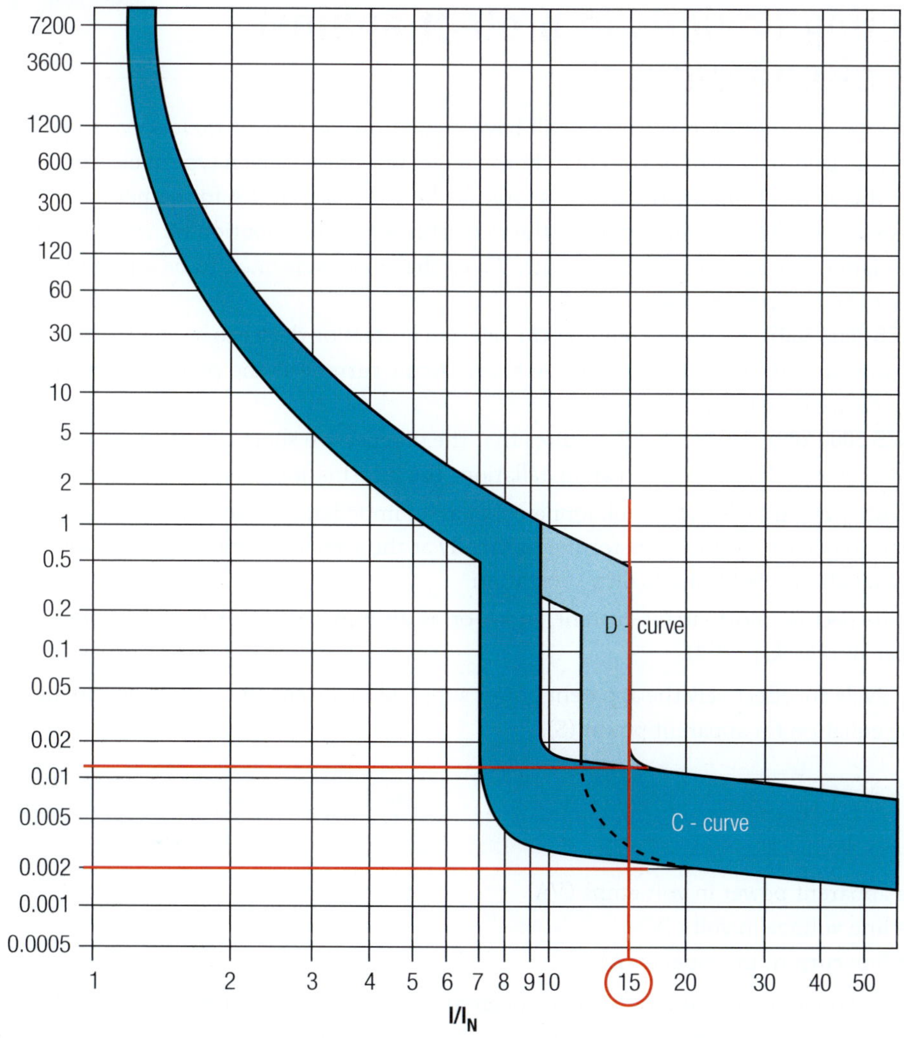

Hazards caused by a short circuit

Hazards caused by a short circuit

An uninterrupted prospective short circuit current will result in catastrophic failure and destruction of the installation, usually resulting in fire.

Hazards associated with short circuit conditions include:

- extremely large currents
- cables reaching very high temperatures
- insulation failure
- destruction of protection devices
- ionisation of enclosures
- severe magnetic stresses on conductors and supports
- fire and explosion (arc blasts)
- fatal injury to persons and livestock.

Source: *UEE07 Electrotechnology Training Package. Learner Workbook. Industry Skills Unit Meadowbank*, Tafe NSW, 2009, pp. 262–3.

Selecting devices to protect against short circuit current

The local supply authority in your area will have regulations as part of their service and installation rules that will deem a value of prospective short circuit current for a given type of electrical installation, usually at the point of supply. This value is typically between 6 to 10 kA for domestic installations. In non-domestic installations, depending on the kVA rating of the distribution transformer, the value can be much higher. For example, in New South Wales the value is deemed to be a minimum of 25 kVA regardless of transformer rating.

AS/NZS 3000:2007 clause 2.5.2 requires the selected protection device to be able to interrupt, without damage, any current up to the prospective short circuit current at the point within the installation where the device is installed.

AS/NZS 3000:2007 clause 2.5.4.1 requires that the prospective short circuit current be calculated or measured at every relevant point in an installation. The information used will either be:

- as a fault level, in MVA (mega volt amperes) or as an impedance percentage (Z%), similar to voltage regulation percentage in a transformer, normally at the secondary winding of a transformer where the installation is fed from its own substation
- as a prospective short circuit current, in kA or as an impedance per phase, in ohms (Tafe NSW, *Learner Workbook*, p. 264).

Fault levels are effectively the apparent power supplied to the circuit under short circuit conditions. Recall the equation for apparent power (S).

$$S = \sqrt{3} \times V_L \times I_L$$

where

S = the apparent power in volt amps (VA)

V_L = the line voltage in volts (V)

I_L = the line current in amperes (A)

The equation above is used to calculate apparent power supplied by devices such as alternators and transformers. It can be transposed to find the line current supplied by such devices under normal operating conditions. The apparent power output of supply authority transformers is usually given in kVA.

$$I_L = \frac{S}{\sqrt{3} \times V_L}$$

EXAMPLE 11.75

Calculate the full load current available from a 500 kVA 400/230 V supply authority distribution transformer in Figure 11.38 and determine the operating time of the circuit breaker (adapted from Tafe NSW, *Learner Workbook*, p. 266).

Solution

$$I_L = \frac{S}{\sqrt{3} \times V_L} \frac{500 \times 10^3}{\sqrt{3} \times 400} = 721.7\text{ A}$$

The same equation is used to express prospective short circuit values as volt amps (VA) rather than amperes. As the line current is extremely high in short circuit faults, the fault level is also extremely high and usually specified in mega volt amps (MVA) rather than kVA.

$$I_{sc} = \frac{MVA \times 10^6}{\sqrt{3} \times V_L}$$

where

MVA = the fault level (apparent power) in mega volt amps (MVA)

V_L = the line voltage in volts (V)

I_{sc} = the prospective short circuit current (line current) in amperes (A)

EXAMPLE 11.76

Calculate the prospective short circuit current for a 500 kVA 400 V supply authority distribution transformer with a fault level of 10 MVA breaker (adapted from Tafe NSW, *Learner Workbook*, p. 267).

Solution

$$I_{sc} = \frac{MVA \times 10^6}{\sqrt{3} \times V_L} = \frac{10 \times 10^6}{\sqrt{3} \times 400} = 14{,}433\text{ A (14.433 kA)}$$

Supply authority transformers have a percentage impedance rating. To determine the percentage impedance, the primary winding is short circuited and the secondary winding is connected to a variable supply. The voltage on the variable supply is increased from zero volts until the current in the primary reaches the maximum rated current for normal operation. The voltage on the secondary is recorded and converted to a percentage of the secondary nominal voltage. This gives the percentage of the nominal supply required to cause the nominal full load current to flow under short circuit conditions. It can now be predicted what current will flow under short circuit conditions if the full nominal supply voltage was connected. This method is much safer and less destructive than shorting the secondary terminals of a 400 kVA transformer and trying to measure a real short circuit current (Tafe NSW, *Learner Workbook*, p. 267).

The prospective short circuit current can be calculated using the equation:

$$I_{sc} = \frac{100 \times kVA \times 10^3}{Z\% \times \sqrt{3} \times V_L}$$

where

I_{sc} = the prospective short circuit current (line current) in amperes (A)

kVA = the maximum rated output (apparent) power in kilo volt amps (kVA)

V_L = the line voltage in volts (V)

EXAMPLE 11.77

Calculate the prospective short circuit current for a 500 kVA 400 V supply authority distribution transformer with a percentage impedance of 5% (adapted from Tafe NSW, *Learner Workbook*, p. 268).

Solution

$$I_{sc} = \frac{100 \times kVA \times 10^3}{Z\% \times \sqrt{3} \times V_L} = \frac{100 \times 500 \times 10^3}{5 \times \sqrt{3} \times 400} = 14\,433.75 \text{ A } (14.433 \text{ kA})$$

The prospective short circuit current can now be converted to an impedance per phase value at the point of supply (Z_S).

$$Z_p = \frac{V_p}{I_p}$$

where

Z_p = the phase impedance in ohms (Ω)
V_p = the phase voltage in volts (V)
I_p = the phase current in amperes (A)

Once the impedances of each part of the supply system are known, the prospective fault current can be calculated at each point, as shown in the following examples (Tafe NSW, *Learner Workbook*, p. 265).

EXAMPLE 11.78

Calculate the nominal impedance per phase of the supply system at the point of supply for services with a prospective short circuit current of 6 kA (adapted from Tafe NSW, *Learner Workbook*, pp. 265–6).

Solution

$$Z_p = \frac{V_p}{I_p} = \frac{230}{6000} = 0.0383\ \Omega$$

EXAMPLE 11.79

For the installation shown in Figure 11.40 calculate the prospective short circuit current at the:

a transformer
b main switchboard
c sub-distribution board
d end of the final sub-circuit (adapted from Tafe NSW, *Learner Workbook*, p. 269).

FIGURE 11.40 Prospective short circuit calculation

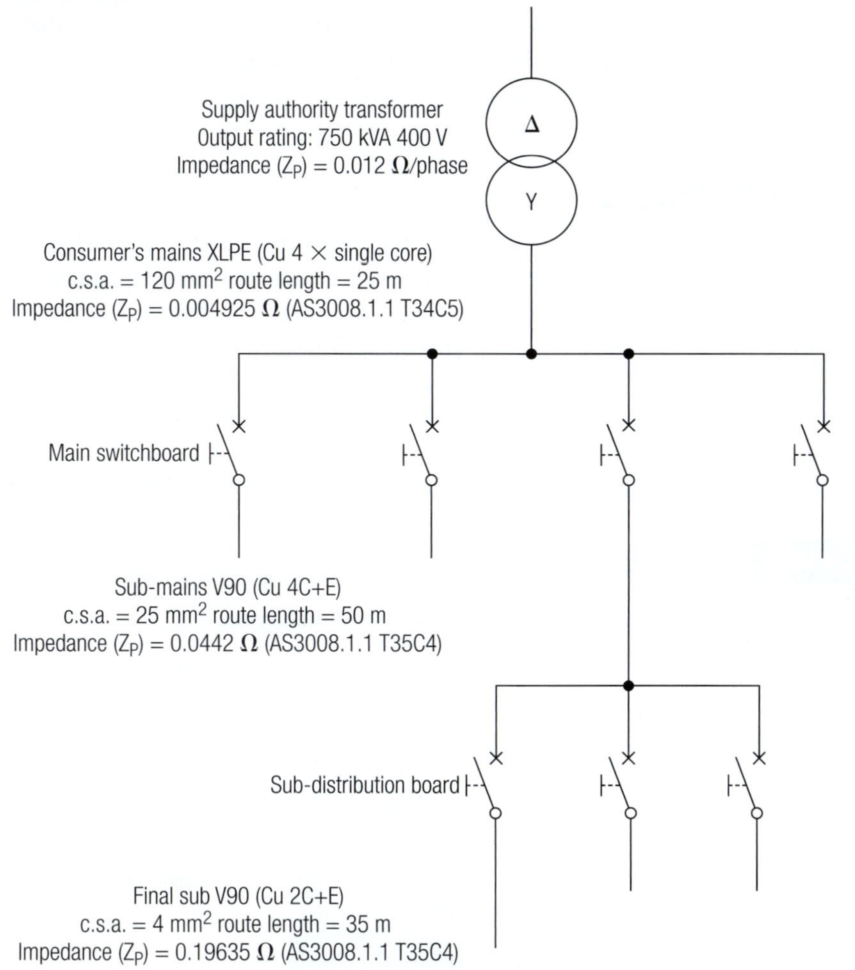

Source: *Select and Arrange Equipment for General Electrical Installations. Learner Workbook*, Industry Skills Unit Meadowbank, Tafe NSW, 2009, Figure 8, p. 269

Solution

a transformer

$$I_{sc} = \frac{V_p}{Z_p} = \frac{230}{0.012} = 19\,166 \text{ A} = 19.166 \text{ kA}$$

b main switchboard

$$= \frac{230}{0.012 + 0.004925} = 13\,589.3 \text{ A} = 13.589 \text{ kA}$$

c sub-distribution board

$$I_{sc} = \frac{V_p}{Z_p} = \frac{230}{0.012 + 0.004925 + 0.0442} = 5196.6\text{A} = 5.196 \text{ kA}$$

d end of the final sub-circuit

$$I_{sc} = \frac{V_p}{Z_p} = \frac{230}{0.012 + 0.004925 + 0.0442 + 0.19635} = 893 \text{ A}$$

Example 11.79 shows that the prospective short circuit (PSC) decreases as the location of the fault moves away from the supply transformer. The greater the distance from the supply transformer, the higher the impedance of the cable and the lower the prospective short circuit current that can flow.

When selecting short circuit protection devices, the device's breaking capacity, or kA rating, must be higher than the prospective short circuit current. Otherwise the device may not be able to interrupt the fault or may be damaged in the process.

Fault current limiters

A fault current limiter (FCL) is a device which limits the prospective fault current when a fault occurs. A suitably rated HRC fuse or fault current limiting circuit breaker is placed 'upstream' in series with protection devices which have a lower fault current breaking capacity than the FCL. This allows the downstream protection devices to be physically smaller and cheaper.

The fault current limiter protects the downstream protection devices by operating before the fault current rises above a value that may damage the downstream device. The fault current that flows until the FCL operates is known as 'let through', shown in the green shaded area of Figure 11.41. The downstream devices only have to be capable of interrupting the much smaller let through current of the upstream device.

FIGURE 11.41 Let through energy

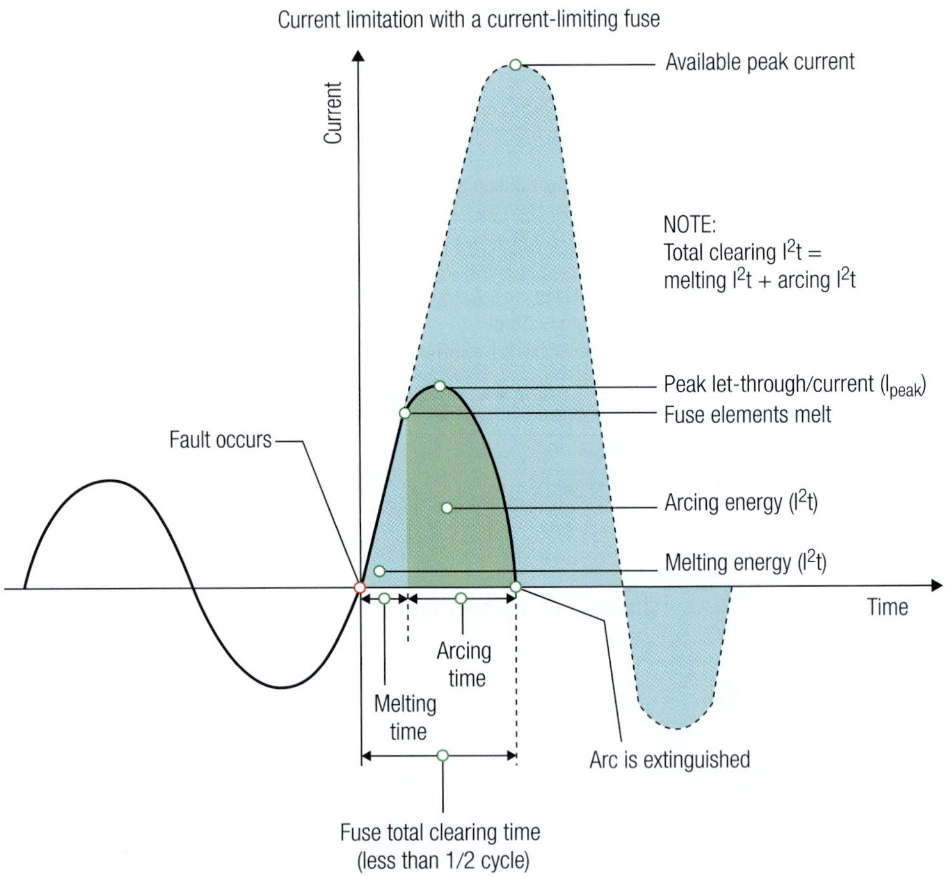

Source: Kenneth Cybart, Senior Engineer, Littelfuse, Inc., *Industrial Equipment News*, 2013, http://www.ien.com/article/getting-selective-about/139083

This is known as *co-ordination* or *back-up protection*. Figure 11.42 shows how the 100 A service protective device (service fuse) is coordinated with the 20 A 3 kA circuit breaker to limit the current the circuit breaker is exposed to. In a domestic installation the service protective device (SPD), typically a 100 A HRC fuse with a breaking capacity of between 50 A to 33 kA, is also used as a fault current limiter. The 100 A nominal current rating is large enough to cover the maximum demand of most domestic installations (Tafe NSW, *Learner Workbook*, p. 273).

FIGURE 11.42 Back-up protection

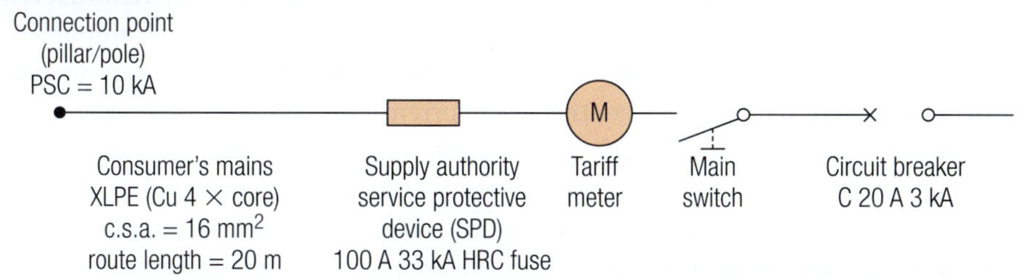

Source: *Select and Arrange Equipment for General Electrical Installations. Learner Workbook*, Industry Skills Unit Meadowbank, Tafe NSW, 2009, Figure 10, p. 272

If a short circuit occurs in the final sub-circuit, close to the main switchboard a high fault current will flow in the final sub-circuit. The 20 A 3 kA breaker will operate first because the fault current will be a greater multiple of its rating than the 100 A fuse. If the circuit breaker is unable to interrupt the fault current, the 100 A fuse will operate to clear the fault, limiting the let through current to a value that will not cause damage to the downstream 20 A circuit breaker. If the SPD operates, supply to the entire installation will be lost.

Figure 11.43 shows a let through energy chart for a BS fuse which is typically used in a service protection device. It shows how a prospective short circuit current of 10 kA is limited to approximately 2.5 kA. The chart is read by drawing a line from the prospective short circuit current value of 10 kA

FIGURE 11.43 Let through energy chart

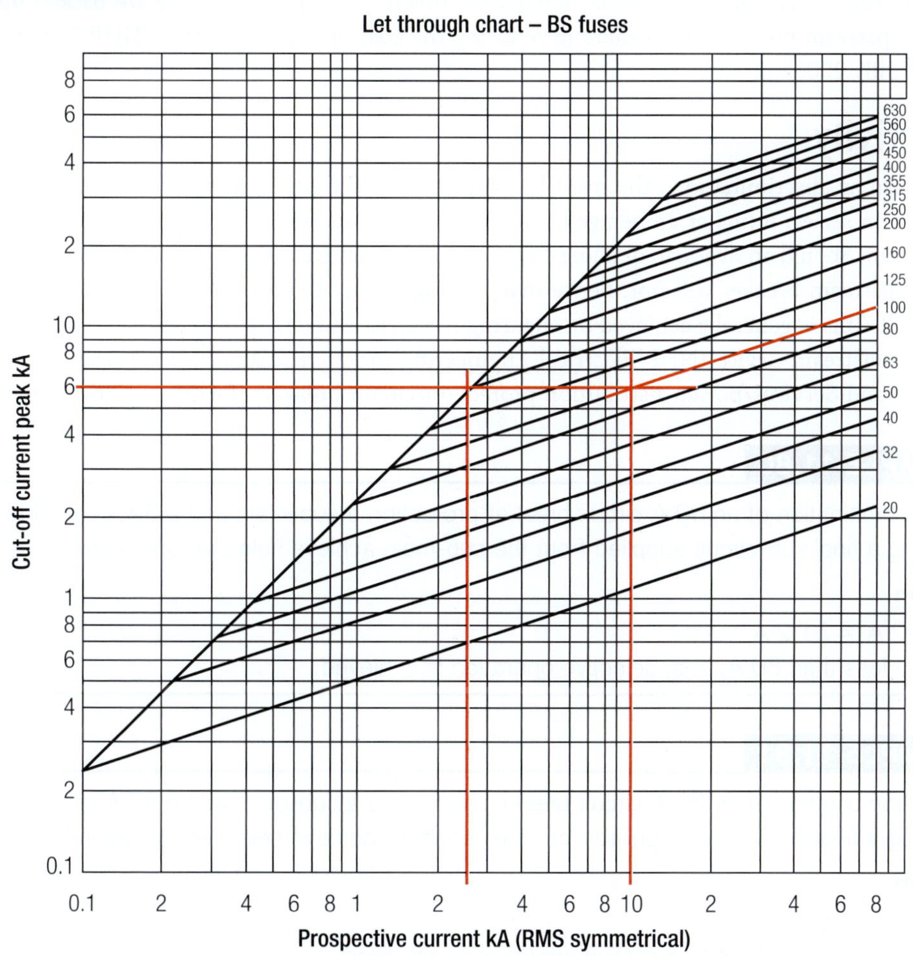

(shown in red) up until it crosses the line representing the current rating of the fuse, shown on the right-hand side of the chart – 100 A in this case. A horizontal line is then drawn to the left until it crosses the diagonal line. A line is then dropped to the bottom of the graph to determine the let through current.

Most faults occur away from the main switchboard. The increased distance from the main board decreases the fault current to a point where the lower current rated circuit breaker operates first. The fault is disconnected without loss of supply to the entire installation. This is known as 'selectivity' (discrimination). Fault current limiters are not expected to operate under overload conditions.

Protective device manufacturers provide coordination tables which demonstrate the combination of protective devices that may be connected in series to ensure that the short circuit ratings are achieved. Protection devices used should all be from the same manufacturer. If different brands of protection devices are used, the individual characteristic curves of the upstream and downstream devices must be compared to ensure effective protection.

Cascading (back-up)

As well as using HRC fuses, circuit breakers which are rated as fault current limiters are also used to protect against prospective short circuit current. The nominal current ratings (I_N) of both the upstream and downstream devices must be carefully considered.

Cascading is achieved by using an upstream device to assist (back up) a downstream device in clearing a fault current that happens to be greater than the breaking capacity of the downstream device.

In cascading applications, the upstream device may have to trip (unlatch) in order to give sufficient protection to the downstream device, thus interrupting supply of power to all devices downstream. Therefore, cascading is generally used in applications involving the supply of non-essential loads, such as basic lighting. The main benefit of cascading is that circuit breakers with breaking capacities lower than the prospective fault level can be safely used downstream provided they are backed up by the relevant upstream breaker. This results in reducing the cost of the installation (NHP *Technical News*, Issue 39, July 2003).

Selectivity

Also known as 'discrimination', the most basic form of selectivity is where two circuit breakers are connected in series. A higher amperage breaker is installed upstream, and a lower amperage breaker downstream. Should an overload or short circuit occur downstream, the downstream breaker will trip, but the upstream breaker will not, hence maintaining supply to parts of the system which are not under fault. This is generally achieved by ensuring the upstream device is at least 1.6 times the rating of the downstream device (NHP Electrical Engineering Products, http://www.nhp.com.au/Products/Products-and-Services/power-distribution-protection/circuit-breakers/Technical-Information/FAQs).

EXAMPLE 11.80

Will the combination of an 80 A circuit breaker protecting a sub-main and a 32 A circuit breaker protecting a final sub-circuit supplied from the sub-main achieve selective operation?

Solution

$1.6 \times 32\text{ A} = 51.2\text{ A}$

51.2 A is less than 80 A. Yes, selective operation is achieved.

EXAMPLE 11.81

Will the combination of an 80 A circuit breaker protecting a sub-main and a 40 A circuit breaker protecting a final sub-circuit supplied from the sub-main achieve selective operation?

Solution

$1.6 \times 40\text{ A} = 64\text{ A}$

64 A is less than 80 A. Yes, selective operation is achieved.

EXAMPLE 11.82

Will the combination of an 80 A circuit breaker protecting a sub-main and a 50 A circuit breaker protecting a final sub-circuit supplied from the sub-main achieve selective operation?

Solution

$1.6 \times 50\text{ A} = 80\text{A}$

80 A is not less than 80 A. No, selective operation is not achieved.

In Example 11.82 the ratings of the two breakers are so close that the sub-main breaker could operate at the same time as the downstream breaker.

The principle of selectivity (discrimination) is based upon an analysis of several types of circuit breaker characteristics. These include tripping characteristics (time-current curves), peak let through current (I_{peak}) and energy let through (I_2T).

SUMMING UP...

- Three types of electrical faults that must be protected against are:
  - overload
  - short circuit
  - earth fault.
- The ratio of the current actually flowing through the circuit (I) to the nominal rating of the circuit protection device (I_N) gives the multiples of rated current (I_M)

$$I_M = \frac{I_N}{I}$$

- The higher the multiples of current, the shorter the operating time.
- The normal full load current of a distribution transformer is determined by the equation:

$$I_L = \frac{S}{\sqrt{3} \times V_L}$$

- The prospective short circuit current (PSC) of a distribution transformer is determined from the fault level by the equation:

$$I_{sc} = \frac{MVA \times 10^6}{\sqrt{3} \times V_L}$$

- The PSC of a distribution transformer is determined from its percentage impedance by the equation:

$$I_{sc} = \frac{100 \times kVA \times 10^3}{Z\% \times \sqrt{3} \times V_L}$$

- The impedance of a distribution transformer is determined by the equation:

$$Z_p = \frac{V_p}{I_p}$$

- The cable impedance decreases the prospective fault current as the distance from the supply transformer increases.
- A fault current limiter (FCL) is a device which limits the prospective fault current when a fault occurs. This allows lower fault current rated protection devices to be used, which reduces cost.

REVIEW EXERCISES

Answer these questions and check your answers at the back of the book.

Review questions 11.1

1 What are the maximum and minimum supply voltages for your country?

2 What is the most likely method for determining the maximum demand of a final sub-circuit?

3 What is required if a cable is required to pass through a wall or floor that is of a fire-resistant construction?

Review questions 11.2

1 Under normal operating conditions, what is the maximum permissible temperature of a V90 insulated, 2.5 mm² TPS cable?

2 What is the recommended number of double 10 A socket outlets that can be connected to a circuit wired in 2.5 mm² TPS cable and protected by a 20 A type C circuit breaker in a factory without air-conditioning?

3 A house contains:
- 22 lighting points protected by a 10 A CB
- 16 double 10 A socket outlets protected by a 20 A CB
- 1 15 A socket outlet for a clothes dryer protected by a 20 A CB
- 1 13 A A/C unit protected by a 20 A CB
- 1 4.6 kW HWS protected by a 20 A CB.

Record in Table 11.41 the number of circuits and the number of points per final sub-circuit required.

TABLE 11.41 Final sub-circuit arrangements

Circuit number	Purpose/load	Protection device/ rating (A)	Number of points per circuit
1			
2			
3			
4			
5			
6			
7			

Review questions 11.3

1 A block of 33 townhouses is to be connected to a three phase supply. Using AS/NZS 3000:2007 calculate the maximum demand for the consumer's mains of the multiple domestic installation with the following load:

Each unit contains:
- 22 light points
- 18 double 10 A socket outlets
- 1 8.0 kW range
- 1 4.4 kW storage HWS
- 1 14 A air-conditioner

Community load:
- 24 compact fluorescent (0.22 A each) for car park lighting
- 10 bollard lights (0.15 A each)
- 10 single 10 A socket outlets

2 Using AS/NZS 3000:2007 calculate the maximum demand for a sub-main of an individual unit in the previous question.

3 Using AS/NZS 3000:2007 calculate the maximum demand in the consumer's mains for a factory with the following load:
- 18 high bay MH luminaires rated at 2.6 A each (six per phase)
- 33 10 A double single phase socket outlets (11 per phase)
- 4 32 A three phase socket outlets
- 3 20 A three phase socket outlets
- 2 three phase induction motors rated at 42 A
- 1 three phase induction motor rated at 28 A
- 2 400 V single phase arc welders at rated current of 18 A.

4 A retail complex consists of five shops at street level (330 m² each) and three offices (300 m² each) on the first floor. All shops and offices have reverse cycle air-conditioning. Determine the maximum demand of the three phase consumer's and sub-mains for this commercial installation.

Review questions 11.4

1 The maximum demand current of a single phase domestic distribution board has been calculated to be 70 amperes. To select a cable size for the sub-mains, consider the following:
- type of cable to be used – SDI, non-armoured, XLPE cable with copper conductors
- method of installation – the cable is to be enclosed in heavy duty PVC conduit in ceiling and partially surrounded by thermal insulation
- circuit protection – circuit breaker.

Table No. ____ Column No. ____ Cable Size ____ PE Size
I_B = ____ I_N = ____ I_Z = ____ $I_B \leq I_N \leq I_Z$ (Y/N)

2 A final sub-circuit is to be run in a non-domestic installation for the purpose of supplying power to a reverse cycle air-conditioner. The full load current of the appliance has been calculated to be 35 A. The type of cable to be used is two single core, V75 insulated and unsheathed copper cables with earthing conductor and will be protected by circuit breakers. The cable is to be run in a non-metallic conduit and saddled to an external wall.

Table No. ____ Column No. ____ Cable Size ____ PE Size
I_B = ____ I_N = ____ I_Z = ____ $I_B \leq I_N \leq I_Z$ (Y/N)

3 Determine the current rating of a cable, protection device and protective earthing conductor size for a final sub-circuit wired in 6 mm² V90 TPI copper cables. The circuit is to be installed in an HD underground conduit at a depth of 0.5 m. The conduit is laid in a trench touching one other circuit in separate conduits.

De-rating Table ____ Column No. ____ DR ____
Table No. ____ Column No. ____ Cable Size ____ PE Size ____
I_B = ____ I_N = ____ I_Z = ____ $I_B \leq I_N \leq I_Z$ (Y/N)

Review questions 11.5

1 Determine the total voltage drop for the circuit shown in Figure 11.44.

FIGURE 11.44 Cable voltage drop

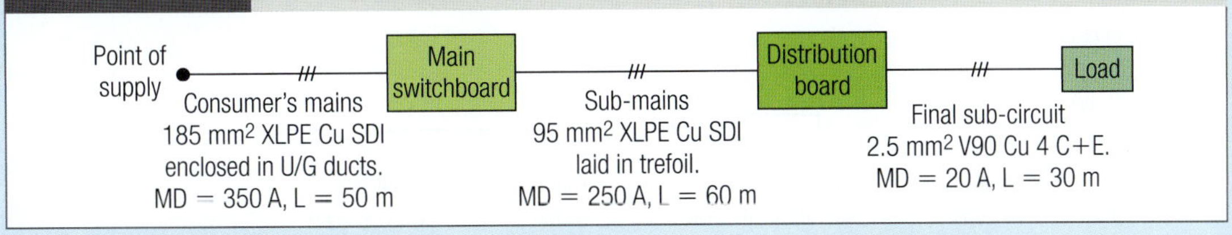

2 Determine the minimum c.s.a. of the final sub-circuit shown in Figure 11.45.

FIGURE 11.45 Cable selection based voltage drop

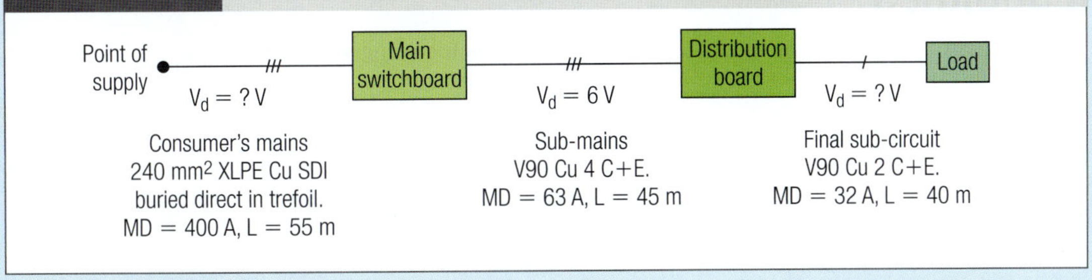

3 Determine the maximum length of the final sub-circuit cable shown in Figure 11.46.

FIGURE 11.46 Maximum cable length based voltage drop

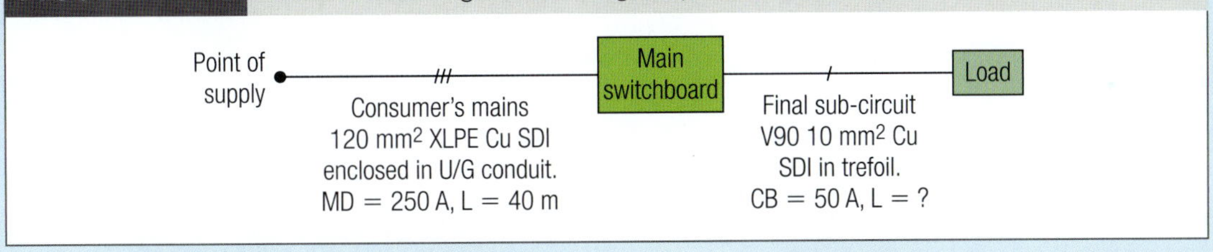

Review questions 11.6

1 Calculate the internal earth fault loop impedance for a circuit supplying a range protected by a Clipsal 32 A type C circuit breaker.

2 A 50 mm^2 copper sub-main is protected by a 125 A HRC fuse with a route length of 80 m.
  - a Determine from Table 8.1 AS/NZS 3000:2007 the maximum permissible internal earth fault loop impedance.
  - b Using Table B1 of AS/NZS 3000:2007, determine if the route length is compliant.

3 An underground 25 mm^2 XLPE copper sub-main is protected by a 100 A Clipsal circuit breaker.
  - a What is the maximum permissible operating time?
  - b Using Figure 11.32 determine the current I_a.
  - c Calculate the maximum permissible length to satisfy earth fault loop requirements.

Review questions 11.7

1 List four harmful effects that can be caused by a short circuit current.

2 Calculate the prospective short circuit current for a 750 kVA 400 V supply authority distribution transformer with a fault level of 15 MVA.

3 Calculate the impedance of a 750 kVA 400 V supply authority distribution transformer.

4 For the installation shown in Figure 11.47 determine:
  - a the impedance of the supply
  - b the impedance of the consumer's mains (using AS/NZS 3008.1 Table 35)
  - c the prospective short circuit current at the SPD

d the let through current to the 20 A circuit breaker (using Figure 11.44)

e if the kA rating of the 20 A circuit breaker is sufficient.

FIGURE 11.47 Prospective fault current

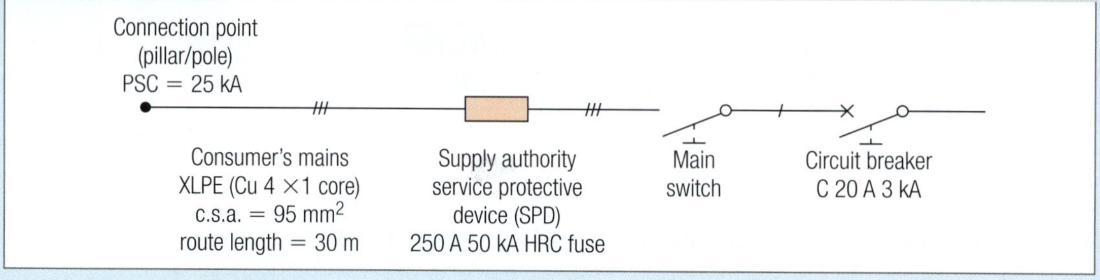

ONLINE RESOURCES

Clipsal
http://www.clipsal.com.au

NHP Electrical Engineering Products
http://www.nhp.com.au

12

INSTALLATION OF WIRING SYSTEMS

This chapter provides information to help you understand the reasons for selecting particular wiring systems. Your choices will be based on the fundamental requirement for protection and the environment surrounding the installation. Since it is impossible to address every conceivable situation, your task as an electrical practitioner will be to use your knowledge, your understanding of the Australian Standards and information circulated by the electrotechnology industry to develop safe and functional installations for your customers.

As your training progresses you will be able to continue to develop your skills to apply the AS/NZS 3000:2007 Wiring Rules to different situations.

Chapter outline

12.1 Standards, codes and requirements applicable to the installation of wiring systems

Installation of wiring

Strict codes and standards apply to the installation of electric wiring and equipment. The aim of the standards is to prevent the incidence of fire and shock for users of electrical installations.

Protection and support of cables

You will remember from Chapter 8 that you can expect to encounter four general classes of cables while performing installation work:

1 *thermoplastic sheathed* (TPS) cables, including single double insulated (SDI) either flat or circular
2 *thermoplastic insulated* (TPI) cables, which are sometimes also referred to as building wire
3 *armoured* cables, including neutral screened cables
4 *mineral insulated metal sheathed* (MIMS) cables which you probably know by the brand name 'Pyrotanex'.

You can provide mechanical protection and support for the cables in your installation in many ways: refer to **AS/NZS 3000:2007** clause 3.10.2.1.

Once you have brought all this together and selected the correct cable, you will need to select the most appropriate wiring system. The wiring system can be one or a combination of the following:

- buried within a wiring enclosure in the ground
- buried directly in the soil
- catenary supported system
- cable tray or cable ladder
- in a wiring enclosure like conduit, ducting, trunking or other approved means (clauses 3.10.2 and 3.10.3)
- resting on or secured to a surface
- resting on, in or under thermal insulation
- fixed intermittently, supported on timbers/steel or suspended in free air.

You will need to ensure your wiring is installed in accordance with clauses 3.1.2(a)–(f) and all of clause 3.9, to comply with the minimum installation standards and requirements.

> **i** The cable, the means of mechanical protection and their respective supports will be subjected to external influences. Refer to clauses 1.5.14 and 3.3 of AS/NZS 3000:2007 for information regarding these influences. In some cases this will require a de-rating or re-rating of the current carrying capacity of the cable.

> **i** All references to AS/NZS 3000:2007 including Amendment 2 were correct at time of publishing. Standards are updated from time to time with new amendments and rewrites, so please check for currency.

Segregation of low voltage cables from other services

You may often encounter other services that are planned or already installed that have the potential to harm your installation, be harmed by your installation or cause injury to users. Among the more common services that you need to keep separate are:

- Different electrical installations – you will encounter these in the same tray, trunking, conduits, pits, pillars and the like that supply different electrical installations (**AS/NZS 3000:2007** clause 3.9.8.2).
- Different voltage levels (**AS/NZS 3000:2007** clause 3.9.8.3) – extra low voltage (ELV) systems – your low voltage installation may cause harm to other services such as telecommunications, data networks and TV aerials.
- High voltage (HV) supplies – your installation will be adversely affected by HV sources in the same area.
- Other services such as:
  - gas (**AS/NZS 3000:2007** clause 3.9.8.4 and **AS/NZS 5601 Gas installations**)
  - water (**AS/NZS 3000:2007** clause 3.9.8.4 and **AS/NZS 3500 Plumbing and drainage set**).

AS/NZS 3000:2007 clauses 1.5.14 and 3.3 show how external influences play a part in the selection and installation of your wiring system.

Your choice of wiring system will be influenced by the following:

- Type of building construction – what spaces are available to run cables, difficulty of fixing and terminations.
- Flexibility of the system – how often will equipment change location, such as in machine shops and temporary buildings?
- Installation conditions – do you need to provide protection against mechanical damage; will your installation require you to work at heights?
- Ambient temperatures – you will need to consider the basic temperature such as in boiler houses and heat treatment installations.
- Appearance of the finished installation – can you run cables and conduits on the surface or must they be concealed?
- Durability – how long does your customer expect to use the installation? Domestic could be a lifetime, industrial could be until they switch to a new product.
- Economics – how much money is your customer prepared to pay? Check before committing to expensive fittings and systems. Where possible you must aim to be cost-effective, but you cannot compromise safety or standards (see BSE Class, 'Electrical installation systems – wiring systems', p. 1, http://bseclass.files.wordpress.com/2012/09/2-wiring-systems.pdf).

Prohibited cable locations

You must not install wiring between roofing material and the battens that support the roofing material. Nor must you install wiring between wall cladding and the noggins and plates that support the cladding material.

This space is a 'no go area' for wiring between the ridges of the cladding and its support, as it is usually left clear to allow the securing of the cladding with screws and or nails. The fastenings would penetrate any wiring installed in these areas.

Refer to **AS/NZS 3000:2007** clause 3.9.4.3.1.

Effect of building codes on cable installation

In addition to **AS/NZS3000:2007 Wiring Rules**, you will remember from Chapter 7 the other major code you must take into account, that is, the National Building Code, or Building Code of Australia (BCA).

Penetration of structural elements

You must not alter or damage the structural integrity of a building by penetrating any structural members. You must not go around drilling or cutting holes in structural members such as bearers and joists. If you have no other choice, then drilling or cutting holes in this way has to be allowed for, engineered and then certified to comply. This could be as simple as adding strengthening timber to joists where holes have been drilled through to having an engineer certify the drilling of holes through steel beams.

AS/NZS 3000:2007 clause 3.9.3.1 covers this in detail.

Fire protection integrity

From Chapter 7 you will remember that all utility installers must pay strict attention to preventing the start and spread of fire within a building or structure. You are responsible for the fire integrity of your installation by ensuring you comply with dimensions for the penetration of fire barriers or fire-rated walls as well as the requirement to reinstate any protection by the use of fire-rated materials. You can find the dimensions for clearance holes to prevent stress on your cables during and after installation by referring to **AS/NZS 3000:2007** clause 3.9.3.3.

Once you have made the appropriate clearance through the barrier and installed the wiring, you will need to use a suitably rated material to reinstate the barrier. The materials available to you include the following:

- fire check expanding foam for small to medium size penetrations
- fire rated putty for very small penetrations or finishing prior to painting or rendering
- fire rated pillows or bags and similar devices for use on cable trays, ladders and other such wiring systems or enclosures.

Above suspended ceilings

Do not rely on suspended ceilings to carry your weight; always use separate work platforms and scaffolds.

The BCA and Wiring Rules are very specific on wiring installed above suspended ceilings as this is an area that is easily accessible and likely to be disturbed. You will need to take the following fundamental requirements into consideration:

- Does the design of the suspended ceiling allow for the weight of cables?
- Is there adequate support available to fasten the cable to?
- Can trunking be mounted on the suspended ceiling or must it be supported separately?

Heritage buildings and premises

You may need to consult more than one authority for approval to do any work on heritage-listed buildings and structures.

The authorities range from local councils or government bodies to state and federal government bodies, which all have the common goal of preserving our history and architectural heritage.

Penetration of structures and finishes

As with any building you must not simply drill or cut holes in structural members or elements, and this is especially the case for older heritage listed buildings. Due to the age of the building or premises any penetration of these structural elements could lead to irreparable damage to the building or even total collapse of the premises. This will expose you to the possibility of fines being imposed by the authorities. Should you need to drill through or attach items to finished elements you may encounter problems with the integrity of the finish, which might be compromised or damaged. Another precaution you must take is to seek advice about the existence of dangerous old building materials such as asbestos.

Accessing cable routes

You will need to be extremely careful when selecting cable routes as you may not be allowed to simply fix a conduit to a wall or chase out a sandstone wall or structure. You will need to consider under-carpet wiring, hollow skirting or similar methods of concealment.

Types and colour of accessories

Where possible you will need to consider the colour and types of finished products or exposed accessories. For example, light fittings should match the period of the building. If the owner wants to simply install a fluorescent lamp fitting you might suggest a range of possible options such as a coach light or other appropriate fitting matching the era. Another example would be a power point or socket outlet mounted on or in a wall, which perhaps could be mounted in a cupboard or some other disguised location. In the electrical industry some manufacturers have a wide range of heritage accessories.

If the building where you are working is accessible to the general public, or is a tourist attraction, you may need to install exit and emergency lighting. This is generally the only time that you may deviate from the intent of preserving heritage values. If you must do any work on finished elements then you may have to develop different ways of using fastenings and fittings. You will need to consult with specialist advisers or obtain the services of those who have done this sort of work before. If you are unsure at all then you must contact the appropriate authorities for assistance and approval.

SUMMING UP...

- Four classes of cables used for installation work are:
  - thermoplastic sheathed cables
  - thermoplastic insulated cables
  - armoured cables
  - mineral insulated metal sheathed cables.
- Mechanical protection and support is provided by:
  - conduits
  - cable trunking
  - other constructed enclosures
  - aerial conductors
  - catenary support.
- Segregation required of installations in common:
  - tray
  - trunking
  - conduits
  - pits
  - pillars.
- Segregation of different voltage levels required:
  - telecommunications
  - data networks
  - TV aerials
  - extra low voltage
  - high voltage.
- Protection from other services are gas and water.
- Compliance is required with two main codes:
  - Building Code of Australia
  - AS/NZS 3000:2007 Wiring Rules.
- When working on heritage-listed buildings you must:
  - preserve building integrity
  - consult with authorities
  - consider health issues with old materials, e.g. asbestos.

12.2 Use of other installation standards called up by the Wiring Rules for special situations

As an electrical practitioner you must be able to consult widely to ensure your installation complies with other standards that may have more stringent requirements for particular situations.

Standards for special situations

Electromedical treatment areas

AS/NZS 3003:2011 Electrical installations – Patient areas is another installation standard that is called up by **AS/NZS 3000:2007**, and applies to electromedical areas. The layout of this standard is shown in Table 12.1. These are additional to the requirements of **AS/NZS 3000:2007**.

TABLE 12.1 Layout of AS/NZS 3003:2011 Electrical installations – Patient areas

Section	Subject	Description
Section 1	Scope and general	Specifies the area classifications, specifies that Wiring Rules apply except as varied by this standard, referenced documents and additional definitions
Section 2	General requirements for body protected and cardiac protected electrical areas	Classifies the patient areas, types of supply, controls, indicators, leakage protection, marking and testing
Section 3	Additional requirements for body protected electrical areas	Earthing as per AS/NZS 3000:2007
Section 4	Additional requirements for cardiac protected electrical areas	Earth free ELV, number and location of points controlled by RCDs, equipotential bonding
Section 5	Special patient areas	Home care and self harm patients
Section 6	Alterations, additions and repairs	Requirement for updating compliance
Section 7	Magnetic fields	Limits on total flux density and harmonics

(Continued)

TABLE 12.1 *(Continued)*

Section	Subject	Description
Section 8	Marking of patient areas	Marking and signage
Section 9	Routine inspection and testing	Inspection and test requirements
Appendices		Planning, selection, commissioning, measuring potential differences in earths, calculations, documentation and reporting

Some of the main points covered are:

- leakage protection of circuits supplying outlets and equipment in the area
- indication of supply status, monitoring of hazard currents and monitoring of fault currents (includes alarms and resetting)
- location of monitoring and protection equipment (including access)
- number of outlets per circuit (connection of medical equipment)
- earthing and equipotential bonding requirements (more stringent than that of the requirement in **AS/NZS 3000:2007**, for example ≤0.1 ohms for equipotential bonds).

Construction and demolition sites

Another installation standard that is called up by **AS/NZS 3000:2007** is **AS/NZS 3012:2010 Electrical installations – construction and demolition sites**. This mandatory requirement is specified by clause 7.8.2.1 of **AS/NZS 3000:2007**. The layout of **AS/NZS 3012:2010** is shown in Table 12.2.

Ensure that all scrap copper is collected and recycled.

TABLE 12.2 Layout of AS/NZS 3012:2010 Electrical installations – construction and demolition sites

Section	Subject	Description
Section 1	Scope, application, referenced documents and definitions	Outlines what is covered in the standard, verifies that the Wiring Rules apply except where varied and added to by this standard
Section 2	Installation	Supply, maximum demand, specific switchboards, control protection, wiring, leads, lights, lifts and structures
Section 3	Verification or inspection and testing	Application, frequency, personnel, construction wiring, RCDs, other equipment, generators, inverter sets, documentation
Appendices		Referenced documents, classification of structures, application for construction and demolition sites, colours of tags, verification of inverters, alternative supplies

Some points to remember:

- Maximum demand calculations are per **AS/NZS 3000:2007**.
- Socket outlets are given a rating in **AS/NZS 3012** and this rating needs to be taken into consideration when calculations are done.
- Each circuit must be protected by an over-current circuit breaker device.
- Each circuit must also be protected by a 30 mA RCD.
- Socket outlets must be of the double pole switched type with the active and neutral switched simultaneously.
- Each switchboard must comply with **AS/NZS 3000:2007** and must have at least one 15 amp socket outlet.
- Testing must also be carried out routinely by a licensed electrician as per these requirements and in accordance with **AS/NZS 3760:2010 Electrical testing and tagging**.

Relocatable installations

AS/NZS 3000:2007 calls up **AS/NZS 3001:2008 Electrical installations – Transportable structures and vehicles including their site supplies** which applies to relocatable installations. Relocatable or transportable structures refers to the following:

- vehicles offering accommodation, caravans, camper trailers, motor homes, camper vans, livestock transporters with accommodation, car transporters with accommodation
- transportable structures, such as 'granny flats', 'atco' transportable huts, relocatable homes, annexes rigid and non-rigid, tents
- temporary site huts, offices, 'dongers' accommodation, toilets, 'cribrooms' or 'smoko' huts provided on construction and demolition sites
- vehicles for vending vans or vending trailers, display vans/trailers, outside broadcast vans for TV
- mobile scoreboards, mobile traffic signs, mobile advertising signs, trailer mounted 'big screen TV'.

The layout of **AS/NZS 3001:2008** is shown in Table 12.3.

When working on relocatable and mobile structures make sure the wheels or supports are chocked or otherwise stable before starting work.

TABLE 12.3 Layout of AS/NZS 3001:2008 Electrical installations – Transportable structures and vehicles including their site supplies

Section	Subject	Description
Section 1	Scope, application, referenced documents, definitions, alterations additions and repairs	Describes the structures and types, verifies that the Wiring Rules apply except where varied and added to by this standard, definitions not covered in the Wiring Rules
Section 2	Supply arrangements	Socket outlets in caravan parks, socket outlets in other locations, generator or inverter, multiple structures on one site, low consumption structures
Section 3	Permanent installations in transportable structures	Control and protection, earthing, wiring systems, accessories, appliances, ELV transformers, damp areas
Section 4	Electrical installations in tents and annexes	Supply to equipment, connections, outlet boxes, lighting
Section 5	Connection to the site supply	Supply lead, installation of supply lead
Appendices		Maximum demand, connection instructions, NZ only verification of the structures and supply arrangement

Caravan parks

Caravan parks accommodate mobile domestic installations that may have non-compliant sections that can cause dangerous situations to develop. To ensure safety for all users of the caravan park the following must be addressed:

- length of supply leads
- service pillars can be treated as outbuildings MEN earthing only in Australia
- voltage drop on long runs
- install 16 mm^2 earth for low impedance only required in New Zealand
- socket outlets and RCD protection
- double pole switching for socket outlets in caravans.

Marinas and pleasure craft

AS/NZS 3000 calls up **AS/NZS 3004.1:2008 Electrical installations – Marinas and recreational boats** which cover marinas and pleasure craft. The additional requirements in this standard apply due to the wet and corrosive environment. Some special considerations are listed below:

- Metallic conduits and enclosures are normally not suitable. Non-metallic conduits should always be used.
- Aluminium conductors are unsuitable.

When working over water use a safety harness and wear a buoyancy aid.

- TPS or elastomer cables in conduits should always be used.
- Service pillars are distribution boards and their supplies can be considered sub-mains. Note these pillars cannot be considered outbuildings as per other standards like those applicable to caravan parks.
- Isolation transformers or RCDs are to be used on service pillars.
- You must remember tidal changes on pontoons so you need to allow for some flex and remove stress on supply conductors for the raising and lowering of the pontoons at different tide heights.

The layout of **AS/NZS 3004.1:2008** is shown in Table 12.4.

Do not throw off-cuts, cleaning materials or other waste into waterways.

TABLE 12.4 Layout of AS/NZS 3004.1:2008 Electrical installations – Marinas and recreational boats

Section	Subject	Description
Section 1	Scope, application, referenced documents, definitions, alterations, additions, marine environment considerations, galvanic corrosion and connections	Describes the structures and types, verifies that the Wiring Rules apply except where varied and added to by this standard, definitions not covered in the Wiring Rules
Section 2	Marina electrical installation	Wiring details, maximum demand, isolation transformers, switchboards and socket outlets
Section 3	Connection of recreational boats	Supply leads, installation of leads, earthing
Section 4	New installations or additions	Periodic verification
Appendices		Berthing and connection, low voltage configuration, calculating maximum demand

Shows and carnivals

Another installation standard that is called up by **AS/NZS 3000** is **AS/NZS 3002:2008 Shows and carnivals**, and a summary of the layout is shown in Table 12.5.

TABLE 12.5 Layout of AS/NZS 3002:2008 Shows and carnivals

Section	Subject	Description
Section 1	Scope, application, referenced documents, definitions and connections	Describes the structures and types, verifies that the Wiring Rules apply except where varied and added to by this standard, definitions not covered in the Wiring Rules
Section 2	Permanent site low voltage installation	Method of supply, connection details, location and mounting
Section 3	Supply by generating set or inverter assembly	Generating sets, inverters, RCD types
Section 4	Temporary distribution systems and electrical equipment	Cable, cascading reticulation, outlet boxes, portable outlet devices, festoon lighting, rides
Appendices		Testing, low voltage installation arrangements

Due to the need to connect nearly everything via leads, special considerations need to be applied. Some of these considerations are listed below:

- Maximum length of conductor size is limited to a voltage drop of no more than 5%.
- 'Cascadable reticulation units' are the same as 'assemblies for construction units' which can be daisy chained for supply requirements.
- Note this standard applies to the concessions, tents, living quarters and other structures as applicable to shows and carnivals.

SUMMING UP...

- Additional standards are provided for special installations.
- Medical treatment areas require protection for vulnerable people who may have increased sensitivity to electric shock.
- Demolition sites require extra mechanical protection as equipment is used in an area where heavy machinery may be working.
- Relocatable and transportable structures require strict attention to supply polarity and protection from overloading of circuits.
- Marinas and pleasure craft installations require protection from ingress of water and protection against electrolytic damage.
- Show and carnival installations must have additional standards to protect members of the public from dangers caused by temporary installations and structures.

12.3 Hazardous areas

An area that contains highly flammable or combustible materials is known as a hazardous area. Since the electrical installation has the potential to be an ignition source your installation must meet certain requirements to minimise the risk by taking precautions in such areas.

AS/NZS 3000:2007 clause 1.4.11 provides the definition of a hazardous area.

Installations in hazardous areas

Classification of hazardous areas

The classification of hazardous areas is broken into zones, based on the frequency of occurrence and the duration of the explosive atmosphere being present (*Hazardous Area Motors*, 'Hazardous area – zone classification by hazardous substance', p. 2, http://www.ferret.com.au/ODIN/PDF/Showcases/104868.pdf).

There are two standards that specify these requirements:

1 **AS/NZS 60079.10.1:2009 Explosive atmospheres – Classification of areas – Explosive gas atmospheres** which covers explosive gas atmospheres – see Table 12.6.

2 **AS/NZS 60079.10.2:2011 Explosive atmospheres – Classification of areas – Combustible dust atmospheres** which covers combustible dust atmospheres – see Table 12.7.

You will need to consult **AS/NZS 4761.1:2008 Competencies for working with electrical equipment for hazardous areas (EEHA) – Competency Standards** which covers competencies for persons working with electrical equipment in hazardous areas. This tells you how to ensure you carry out only the highest quality work in such areas and don't create hazards and associated risks. A summary of hazardous zones is shown in Tables 12.6 and 12.7.

TABLE 12.6 Zones for explosive gas atmospheres

Zone 0	An explosive gas atmosphere is present continuously or for long periods of time or frequently for short periods
Zone 1	An explosive gas atmosphere is likely to occur in normal operation occasionally
Zone 2	An explosive gas atmosphere is not likely to occur in normal operation but, if it does occur, it will only exist for a short duration only

Sources: AS/NZS 60079.10.1:2009 Explosive atmospheres – Classification of areas – Explosive gas atmospheres; *Hazardous Area Motors*, 'Hazardous area – zone classification by hazardous substance', p. 3, http://www.ferret.com.au/ODIN/PDF/Showcases/104868.pdf

TABLE 12.7 Zones for combustible dust atmospheres

Zone 20	An explosive atmosphere in the form of a cloud of combustible dust in air is present continuously or for long periods of time or frequently for short periods
Zone 21	An explosive atmosphere in the form of a cloud of combustible dust in air is likely to occur in normal operation occasionally
Zone 22	An explosive atmosphere in the form of a cloud of combustible dust in air is not likely to occur in normal operation but, if it does occur, it will only exist for a short duration only

Sources: AS/NZS 60079.10.2:2011 Explosive atmospheres – Classification of areas – Combustible dust atmospheres; *Hazardous Area Motors*, 'Hazardous area – zone classification by hazardous substance', p. 3, http://www.ferret.com.au/ODIN/PDF/Showcases/104868.pdf

When commencing work in hazardous areas make sure that you follow all the safety procedures set down for that area.

Classifying a hazardous area

The responsibility for classifying the hazardous area and the subsequent zones for explosive gas and combustible dust atmospheres rests with the parties in control of the installation. To assist in this an expert in the field should be appointed or employed to undertake the classification process.

Records must be kept of the classification process, and these records must be made available on demand.

AS/NZS requirements for hazardous areas

Designers of installations keep electrical equipment out of hazardous areas where possible. Where this is not possible, explosive-protected equipment will be stipulated along with an approved explosive-protected technique. Explosion protection is broken into two areas:

- equipment grouping
- temperature limitations.

Equipment grouping is broken down into three further groups as shown in Table 12.8.

TABLE 12.8 Equipment grouping table

Group	Description	Detail
Group I	Equipment for use in mines susceptible to firedamp	
Group II	Equipment for use in places with an explosive gas atmosphere and subdivided into the type of gas	**IIA** Propane **IIB** Ethylene **IIC** Hydrogen
Group III	Equipment for use in places with an explosive dust atmosphere and subdivided into the type of dust	**IIIA** Combustible flyings **IIIB** Non-conductive dust **IIIC** Conductive dust

Temperature limitations are broken down into three further groups as shown in Table 12.9.

TABLE 12.9 Equipment grouping table temperature limits

Group	Description	Detail
Group I	The surface temperature of equipment is not permitted to exceed:	**150°C** where coal dust can form a layer **450°C** if coal dust is prevented from forming a layer
Group II	A temperature class designated 'T' is assigned to the equipment:	**T1** 450°C **T2** 300°C **T3** 200°C **T4** 135°C **T5** 100°C **T6** 85°C
Group III	Maximum surface temperature shall not exceed:	• Maximum surface temperature assigned to equipment • Layer or cloud ignition temperature of the specific combustible dust for which it is intended

Methods of explosion protection are listed in Table 12.10. These methods separate ignition source from the explosive atmosphere and don't have to be exclusive but may have more than one method employed.

TABLE 12.10 Protection methods and equipment types

Methods of explosion protection		Equipment types/rating/methods
Exclusion method	The hazardous material is prevented from entering the electrical equipment so that arcing and heat cannot produce ignition suitable for gas or dust hazards	**Dust ignition protection type 'tD' or 't'** An enclosure providing dust ingress protection and a means to limit surface temperatures **Pressurisation 'p'** Enclosures for electrical equipment and machinery in which air or non-flammable gas is maintained at a pressure to exclude the entry of an explosive atmosphere external to the enclosure **Protection type 'n'** Non-sparking devices/components Including those that are encapsulated; hermetically sealed; enclosed break; sealed or energy limited **Encapsulation 'm'** Parts that are capable of igniting by heat or spark are enclosed in a compound such that the explosive atmosphere cannot be ignited **Oil immersion 'o'** Equipment or parts are immersed in a protective liquid such that the explosive atmosphere above the liquid or outside the enclosure cannot be ignited **Powder filling 'q'** Parts cable of igniting the explosive atmosphere are fixed in position and completely surrounded by filling material to prevent the ignition of external explosive atmosphere
Explosion containment method	If an explosion does occur within the electrical equipment, it will not ignite the hazardous material outside the electrical equipment	**Flameproof enclosure 'd'** An explosion within the enclosure will not damage the enclosure, but will prevent transmission of flame or heat to the explosive atmosphere external to the enclosure
Energy limitation method	The energy level in the electrical installation is kept below the minimum energy required to ignite the hazardous material; suitable for gas and dust hazards	**Intrinsic safety 'i'** Electrical circuits and apparatus in which the energy from sparks and thermal effects are insufficient to ignite a given explosive atmosphere under normal and fault conditions **Protection type 'n'** Non-sparking enclosed break devices
Avoidance of ignition source method	Equipment use and design are selected to prevent the occurrence of an ignition source; suitable for gas and dust hazards	**Increased safety 'e'** A higher standard of design parameters, such as of insulation, creepage, connections, sealing against foreign materials and temperature is applied to equipment that does not normally produce arcing or sparks **Protection type 'n'** Non-sparking devices
Dilution method	The level of an explosive gas in the air is eliminated or reduced below the level needed for an explosion	**Ventilation** Protection by movement of air and its replacement with fresh air by: • the effects of wind and temperature gradients • artificial means such as fans or extractors

The majority of the information required is called up in the **AS/NZS 60079** series of standards and covers inspection and maintenance requirements along with the design, selection and installation of equipment requirements.

Installation design factors

Installation and design factors that apply to hazardous areas relate to:

- defining and classifying the hazardous area
- methods and type of explosion protection required
- selection of wiring systems to meet the requirements for the type of explosive protection required
- selection of equipment to meet the requirements for the type of protection employed
- reducing the need for electrical equipment to be installed in the hazardous areas
- general installation, maintenance requirements, and associated techniques to be used.

SUMMING UP...

- Hazardous areas are classified by the nature and duration of the ignition hazard.
- Protection is provided by preventing arcing or by physical isolation of electrical terminals.
- Suitable wiring systems are chosen by their ability to prevent ingress of gas and vapours.
- Electrical equipment must be chosen to suit the level of hazard identified in the area.
- Electrical equipment must be relocated outside the hazardous area where possible.
- Design of installations in hazardous areas must take into account the nature of the hazard and the possible duration of exposure in order to select suitable equipment and wiring systems.
- Electrical installations in hazardous areas must have increased standards of protection to prevent ignition of flammable gases and vapours.

12.4 Requirements for the installation of cables and accessories in damp situations and ELV installations

AS/NZS 3000:2007 clause 1.4.40 defines a damp situation while Section 6 covers the full requirements.

Installing equipment and wiring in damp situations

You will often be called upon to install electrical equipment and wiring around swimming pools, bathrooms, wash down areas and other wet areas. The mixture of electricity, water and unprotected human bodies would be a lethal combination if special rules were not in force for protection of the public.

Restricted zones

You must install electrical equipment enclosures that are suitable and appropriate for the environmental conditions they are to work in. To assist you in the classification or rating of the equipment IP Ratings are provided. IP Ratings are based on an International Standard and are known as the International Protection (Egress) Ratings, IEC 60529. In Australia and New Zealand **AS 60529 Degrees of protection provided by enclosures (IP Code)** specifies the installation requirements. Based on the IEC 60529, it covers the degrees of protection provided by enclosures for electrical equipment.

To assist in understanding the zones and their classification please refer to Table 12.11.

TABLE 12.11 Zone classification ('zones of risk')

Zone number	Description	Application
Zone 0	Highest risk of electric shock; areas within and above a vessel containing water	Bath, shower, tub, pool, pond, fountain, other fixed water container, paddling pool, spa pool, water feature
Zone 1	High risk of electric shock; adjacent to Zone 0 where wetting is likely or within a source of water	Bath, shower, tub, pool, pond, fountain, other fixed water container, paddling pool, spa pool, water feature, sauna
Zone 2	Still a risk of electric shock; areas adjacent to Zone 1 where wetting of body can occur	Bath, shower, tub, pool, pond, other fixed water container, paddling pool, spa pool, sauna
Zone 3	Still a risk of electric shock; areas adjacent to Zone 2 in bath or shower	Bath, shower, tub, pond, fountain, other fixed water container, sauna
Classified Zones	1 The source of dampness in refrigerated rooms is condensation, therefore the whole room is classified as a damp zone 2 The portion of an area subject to sanitising and hosing is classified as a damp zone	1 Refrigerated rooms 2 Sanitisation and general hosing down operations

Selecting equipment for damp situations

With the popularity of spas and the increase in the variety of electrical appliances that people want to use in bathrooms you will need to observe the requirements listed in Table 12.12.

TABLE 12.12 Wiring Rules requirements for showers, baths and other fixed water containers

Equipment and installation	Zone 0	Zone 1	Zone 2	Zone 3	Wiring Rules clause number
IP rating	IPX7	IPX4, IPX5 in communal areas	IPX5 in communal areas		6.2.4.1
Socket outlets	Not permitted	Shaver unit or RCD protected and in cupboard	RCD protected or separated circuit or SELV or PELV		6.2.4.2
Switches and other accessories	Not permitted	Restricted location and required degree of protection	Restricted location		6.2.4.3
Luminaires	IPX7, specific for purpose, at specified voltage from SELV or PELV outside zone	Required degree of protection	Required degree of protection. Some exceptions	Required degree of protection	6.2.4.4
Other electrical equipment	IPX7, specific for purpose, at specified voltage from SELV or PELV outside zone	Required degree of protection. Floor heating in accordance with Clause 4.10.1 permitted			6.2.4.5
Switchboards	Not permitted	Not permitted	Not permitted	Not permitted	6.2.4.6

Many properties and suburbs in the warmer parts of Australia and New Zealand have electrical appliances and water treatment systems installed within pool areas. Table 12.13 lists the requirements for safe installation around swimming pools.

TABLE 12.13 Wiring Rules requirements for swimming pools, paddling pools, spa pools or tubs

Equipment and installation	Zone 0	Zone 1	Zone 2	Wiring Rules clause number
IP Rating	IPX8	IPX5	IPX4	6.3.4.1
Wiring system	Prevent entry of moisture or water siphoning/cable			6.3.4.2
Socket outlets	Not permitted	Restricted locations Required degree of protection and RCD protected or separated circuit or SELV or PELV		6.3.4.3
Switches and other accessories	Not permitted	Required degree of protection	Not permitted	6.3.4.4
Luminaires, appliance and other electrical equipment	IPX8, specific for purpose, at specified voltage from SELV or PELV outside of zone. Not earthed	IPX5 Supplied as SELV or PELV systems or Class II or RCD protected Class I	IPX4 Separated supply or supplied as SELV or PELV systems or Class II or RCD protected Class I	6.3.4.5
Switchboards	Not permitted			6.3.4.6

Water features and fountains have long been a feature in public spaces and larger properties. Electric pumps and display lighting are often required. Installation requirements for fountains and water features are shown in Table 12.14.

TABLE 12.14 Wiring Rules requirements for fountains and water features

Equipment and installation	Zone 0	Zone 1	Wiring rules clause number
IP rating	IPX8	IPX5	6.4.4.1
Wiring system	Sheathed copper cable or flexible cords suitable for immersion in water installed in wiring enclosure if subjected to damage. Prevent entry of moisture or water siphoning/cable		6.4.4.2
Socket outlets	Not permitted		6.4.4.3
Switches and other accessories	Not permitted	IPX5	6.4.4.4
Luminaires, appliance and other electrical equipment	IPX8, specific for purpose, at specified voltage from SELV or PELV outside of zone. Not earthed	IPX5, supplied as SELV or PELV systems or Class II or RCD protected Class I	6.4.4.5
Switchboards	Not permitted		6.4.4.6

Saunas are a fashionable installation in many houses and hotels requiring special attention to lighting, pump controls and heating. Requirements for saunas are summarised in Table 12.15.

TABLE 12.15 Wiring Rules requirements for saunas

Equipment and installation	Zone 1	Zone 2	Zone 3	Wiring Rules clause number
IP rating	IPX4B or IP24			6.5.4.1
Requirements for classified zones	Equipment belonging to sauna only	No requirement	Suitable for minimum temperatures; Equipment 125°C Wiring 175°C	6.5.4.2
Wiring system	Heat resistant, metallic sheathed or metal conduit not accessible		Heat resistant, metallic sheathed or metal conduit not accessible	6.5.4.3
Socket outlets, switches and other accessories	Not permitted except switches and accessories that are part of the heater			6.5.4.4
Other electrical equipment	Sauna heaters installed to manufacturer's instructions			6.5.4.5
Switchboards	Not permitted			6.5.4.6

Refrigeration rooms for preservation of food and other perishable goods need the special requirements summarised in Tables 12.16 and 12.17.

TABLE 12.16 Wiring Rules requirements for refrigerated rooms

Equipment and installation	Classified zone	Wiring Rules clause number
IP rating	IPX4B or IP24	6.6.4.1
Wiring system	Not affected by temperature Does not accumulate or channel moisture into electrical equipment	6.6.4.2
Types	Unenclosed sheathed cables, unsheathed insulated or sheathed cables in wiring enclosures with drainage facilities or served MIMS cable	
Sealing	Non setting sealing compound at point of entry of cables into refrigerated room, motors, luminaires, switches and other electrical equipment	
Socket outlets, switches and other accessories	Required degree of protection and sealed to prevent retention of moisture	6.6.4.3
Luminaires	Designed and constructed to prevent entry of moisture to live parts and prevent retention of moisture	6.6.4.4
Lamp holders	All insulated or type where external metal portion cannot become live Flexible pendant not suspended below 2.5 m from floor	
Fixed appliances and motors	Designed and constructed for location and conditions including suitable degree of protection	6.6.4.5
Door seal heating	Additional protection by 30 mA RCD	6.6.4.6
Switchboards	Not permitted	6.6.4.7

TABLE 12.17 Wiring Rules requirements for refrigerated rooms

Equipment and installation	Classified zone	Wiring Rules clause number
General	Equipment suitable for: • the temperature and pressure of the fluids used in the process • exposure to the relevant chemicals used in that process in accordance with clause 3.3	6.7.4.1
IP Rating	Minimum IPX5 where subject to low-pressure hosing; IPX6 where subject to high pressure hosing	6.7.4.2
Electrical equipment	Not affected by hosing, materials, temperature or pressure Protected against accumulated moisture Does not accumulate or channel moisture into electrical equipment	6.7.4.3
Switchboards	With IPX6 degree of protection are permitted	6.7.4.4

Extra-low voltage

AS/NZS 3000:2007 clause 1.4.98 defines extra low voltage (ELV) as any voltage not exceeding 50 volts a.c. or 120 volts ripple free d.c. There are two possible ELV systems allowed:

- separated extra-low voltage (SELV) system – defined by clause 1.4.83; the layout is shown in Figure 12.1
- protected extra-low voltage (PELV) system – defined by clause 1.4.76; the layout is shown in Figure 12.2.

FIGURE 12.1 SELV system

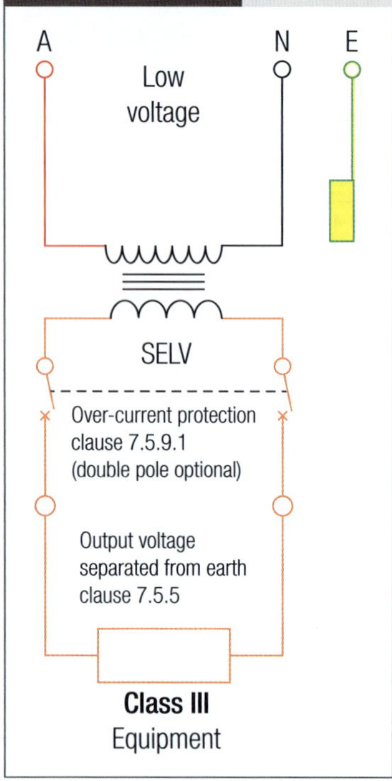

FIGURE 12.2 PELV system

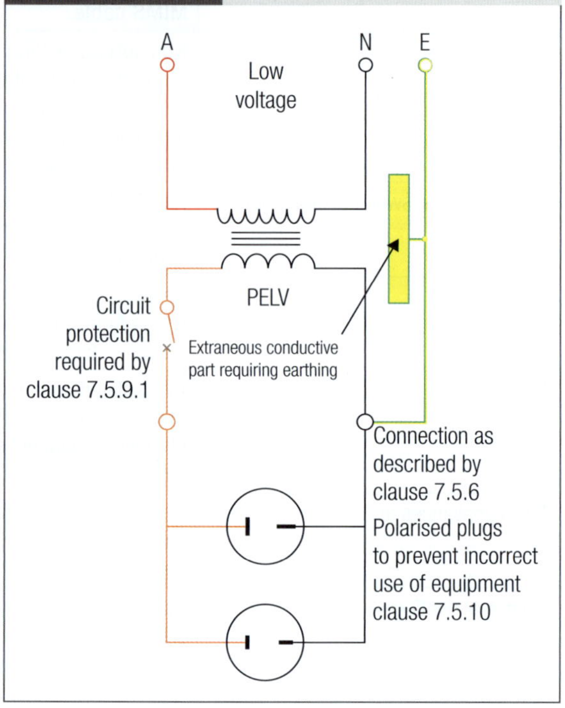

AS/NZS 3000 requirements

Generally Sections 2, 3, 4 and 5 of AS/NZS 3000:2007 are adopted to help you decide on the best way forward when selecting the devices and practices you need to use for a particular installation:

- Section 2 General arrangement, control and protection
- Section 3 Selection and installation of wiring systems
- Section 4 Selection and installation of appliances and accessories
- Section 5 Earthing arrangements and earthing conductors.

Other circumstances will arise, leading you to different sections of the Wiring Rules, depending on the installation and in particular damp situations and other special electrical installations.

These are comprehensively covered in:

- Section 6 Damp situations
- Section 7 Special electrical installations.

All the sections in Part 2 of the Wiring Rules (Installation practices) are your guide on how to comply with the fundamental requirements or principles as laid out in Part 1 (Scope, application and fundamental principles) of the Wiring Rules.

Remember, this is a minimum standard that you must achieve using the Wiring Rules. In some cases to comply, or meet the requirements, you may have to select an extra low voltage system. Some of this will be required to be done at the planning stage, by finding out what the customer needs and their intended use for the installation.

SUMMING UP...

- When working in restricted zones:
  - equipment enclosures must be suitable and appropriate
  - IP ratings provide classification of equipment
  - IP ratings are based on International Protection (Egress) Ratings, IEC 60529
  - AS 60529 covers the degrees of protection for enclosures.
- Damp situations are:
  - baths and spas
  - swimming pools
  - saunas
  - refrigeration rooms.
- Extra low voltage (ELV) is voltage not exceeding 50 volts a.c. or 120 volts d.c. ripple free.
- PELV is protected extra low voltage.
- SELV is separated extra low voltage.

12.5 Aerial cabling

Aerial cable types and requirements

Aerials are a convenient and cheap method of installing electrical cables between buildings or connecting to the reticulated service.

Aerial types

When installing aerials be sure to follow all the safety procedures for working at height.

The following aerial types are commonly used:

- Bare aerial conductors – have no insulation on them and are often used by the electrical distributor or local supply authority for their low, medium and high voltage distribution system. This type of cable is usually made from aluminium outer conductors with galvanised steel reinforcing in the centre. Aluminium is used as it is very light and relatively strong while still being one of the best conductors. Copper is a better conductor but is relatively heavy for the same current rating.
- Bundled aerial cables – are insulated and bunched in a two, three or four core configuration. These cables are often used from the electrical distributor or supply authority's distribution network in the street to the consumer's point of attachment. You may come across single and multicore aerial cabling in flat and twisted sections, as well as single and multicore neutral screened aerial cable. The conductors may be identified by ridges on the insulation.

Regulatory requirements

The **AS/NZS 3000:2007** requirements for aerial cabling are covered under clause 3.12 as well as Appendix D. There are other clauses which may also apply, for example, parts of clauses 3.2, 3.3, 3.5 and 3.8, and clause 3.7.2.9.

The local supply authority arrangements can be found in the particular state's Service Installation Rules. In Queensland these rules are also known as the *Queensland Electricity Connection and Metering Manual* and the particular requirement for aerial cabling is found in Section 5.4.

Refer to Chapter 7 for more details on these requirements.

You will find these documents cover conditions for installation where you can and cannot run your overhead lines. Some of the conditions include:

- clearance heights above roads
- forbidden routes such as over a pool
- distance from vegetation, particularly trees
- selection of UV resistant cabling.
- type and size for the aerial conductors.

You need to use the correct 'line taps' to stop the possible galvanic reaction between the aluminium and copper when you attach to consumer's mains at the point of attachment.

Termination of aerial cables

Refer to **AS/NZS 3000:2007** clauses 3.7.2.9 and 3.12.7 for specific requirements for termination of aerial cables.

The local supply authority arrangements can be found in the particular state's Service Installation Rules. In Queensland the particular requirement for aerial cabling is found in Section 5.2 of the *Queensland Electricity Connection and Metering Manual*.

Installation of consumer's mains

Refer to **AS/NZS 3000:2007** clauses 3.7.2.9 and 3.12 as well as Appendix D for specific requirements for the installation of consumer's mains via aerial cables.

The local supply authority arrangements can be found in the particular state's Service Installation Rules. In Queensland the particular requirement for aerial cabling is found in sections 5.1, 5.2 and 5.4 of the *Queensland Electricity Connection and Metering Manual*.

Testing installed cables

All testing, including cables, conductors and their respective fittings and supports must meet the requirements of **AS/NZS 3000:2007** Section 8 and in particular the results contained in:

- clause 8.3.6.2 – results for insulation resistance
- clause 8.3.7.2 – polarity
- clause 8.3.8.2 – correct circuit connections
- clause 8.3.9.2 – fault loop impedance.

Remember that the test voltage is applied at twice the rating of the voltage of the installation, that is, if the installation is a 230 V installation then you need to test at 500 V d.c. and if the installation is a 400 V installation then you need to test at 1000 V d.c.

Depending on manufacturer's specifications there may be additional requirements for testing.

SUMMING UP...

- Aerial cable types can be:
  - single and multicore
  - flat and twisted sections
  - single and multicore neutral screened.
- Aerial cable is commonly used from the electrical distributor's network in the street to the consumer's point of attachment.
- Code and standard requirements include:
  - UV resistance
  - restricted routes
  - clearances.
- Special connection devices are required between aluminium and copper conductors.
- Testing must meet the requirements of AS/NZS 3000:2007 Section 8.
- Special installation rules exist for utility providers in each state.

When working in trenches follow the safety procedures to prevent cave-ins and entrapment.

12.6 Underground cabling

As an electrical practitioner you will need to be able to plan a suitable route for underground cables and determine the depth at which they must be buried to afford safety and protection from damage.

Underground cabling systems

Underground cable systems are based on their category rating. This rating has been developed by taking into consideration the cable type and the mechanical protection afforded to the cable. Depth of laying the system or cable is also governed by these categories, along with the types of covering.

Category A can apply to one of the following systems (as covered in **AS/NZS 3000:2007** clause 3.11.3):

1 Cables enclosed in heavy-duty insulating conduit without further mechanical protection.
2 Cables enclosed in insulating wiring enclosures encased in concrete.
3 Sheathed cables enclosed in galvanised steel pipe without further mechanical protection. NOTE: Metal conduits are not suitable for this purpose.
4 Armoured sheathed cables or neutral-screened cables buried direct in the ground without mechanical protection.

Category B can apply where:

1 Cables are enclosed in medium-duty insulating conduit with additional mechanical protection.
2 Sheathed cables are buried direct in the ground with mechanical protection.

Category C applies where cables are chased in rock and covered with concrete.

You will find valuable information regarding the categories and the depths of laying and associated covering in Table 3.6 of the Wiring Rules.

Underground cable systems

A Category A underground wiring system laid in a channel chased in rock, in accordance with Category C system requirements, may be deemed to be Category C.

Wash work boots and digging equipment thoroughly before leaving the work site so that you do not spread weeds.

Dealing with other underground utilities

The underground environment can be a crowded place and you must contact other utility providers to find out where you can safely place your cables. The following utility providers should be contacted and their requirements for depth and separation complied with:

- water supply
- stormwater drainage
- sewage drainage
- gas supply
- telecommunications.

Even though Table 3.7 of the Wiring Rules specifies separation distances, individual distributors or authorities may require greater distances.

Regulatory requirements

You will find the **AS/NZS 3000:2007** requirements for underground cabling covered under clause 3.11. There are other clauses which may also apply, for example, parts of clauses 3.3, 3.8 and 3.10 and clause 3.7.2.10.

You must remember to protect against detrimental effects between different services as discussed in Section 12.1 in this chapter. Check again the requirements in **AS/NZS 3000:2007** clause 3.9.8.

The local supply authority arrangements can be found in the particular state's Service Installation Rules. In Queensland the particular requirement for underground cabling is found in Section 5.5 of the *Queensland Electricity Connection and Metering Manual*.

Advantages and disadvantages

Advantages of underground systems are:

- there are no overhead structures to snag kites or tall equipment
- they are usually not susceptible to bushfire threats, but if so they are easier to rectify than overheads and associated poles

- they are not an eye-sore as all reticulation is underground, and there are no wires, poles or steel towers cluttering the landscape
- they are not susceptible to the forces of nature, sun, wind and the like
- they are not often susceptible to fauna related issues such as rats and bandicoots
- they are not often susceptible to flora related issues such as roots breaking into conduits or pipes.

Disadvantages of underground systems are:

- the need to dig or chase access trenches
- the need to know where the services are if you are going to dig
- some faults are not obvious, whereas faults on overheads can often be seen.

SUMMING UP...

- Underground cabling systems are classed A, B and C depending upon the method of enclosure and protection.
- Other underground services may be endangered by your installation. Ensure that you know the location of each of the following:
  - water supply pipes
  - storm drains
  - sewerage drains
  - gas supply pipes
  - telecommunications cables.
- Before proceeding with your underground cable installation be sure to consult AS/NZS 3000:2007 clause 3.3. as well as the local state and utility requirements.
- While underground cables are concealed from view, require no further support, avoid damage from fauna, flora and UV, you will need to dig trenches, and avoid damage to other underground utilities.

Before making any wall penetrations carry out checks to ensure there are no services that you can damage within the wall.

12.7 Techniques for installing cables and wiring systems

As an electrical practitioner, the cables and systems you select will be critical to the safety and comfort of the users of your installation. Besides selecting the correct system to suit the location you will need to conduct tests to ensure the safety of the installation before handing it over to the customer.

Installation planning and testing

Cable routes through buildings and structures

This depends mainly on the type of structure.

If you look at a *domestic premises* you will find spaces such as the cavity between brick and the internal wall, or you may drill holes in timber if it is a timber frame clad house. You may have to use conduits for concrete poured or concrete filled cement blocks.

Other places that you may be able to run cables are along the beams, bearers, rafters, neatly clipped for timber or secured with clips should it be metal framed.

A lot of mains cables are now provided through underground conduits.

However, if you look at *commercial premises* you may find there are a lot of systems run in false ceilings secured to catenaries or tray work.

Some *industrial premises* use trays, either covered or uncovered, with the wiring neatly secured to these trays and coming down the walls in trunking or covered trays as shown in Figure 12.3.

Depending on the job some buildings may even have trunking or conduits embedded in concrete slabs and walls. This is usually done in the construction phase as it becomes harder to do after the event.

High-rise buildings provide their own challenges but often have a services riser for the supply of services and then employ any number of the above systems or a combination of these.

Wiring accessories

Wiring accessories are purpose built to suit the particular cable and or wiring system chosen. Some of the more common items and their application are:

- plastic saddles to suit plastic conduits
- galvanised saddles to suit steel conduits
- copper saddles to suit copper pipes and copper clad MIMS cables
- cable ties for securing wiring to trays and/or bunching on catenaries or bunching in trunking
- girder clips to secure cables in/on girders and purlins
- plastic or brass clips to secure TPS to timber and brick or concrete
- elbows, bends and boxes for conduit, both metal and plastic
- lock nuts for both plastic and metal conduit threads and adaptors
- 'unistrut' brackets, accessories for the construction support for particular wiring systems. Figure 12.4 shows a variety of cable enclosures.

FIGURE 12.3 TPS cable run on cable tray

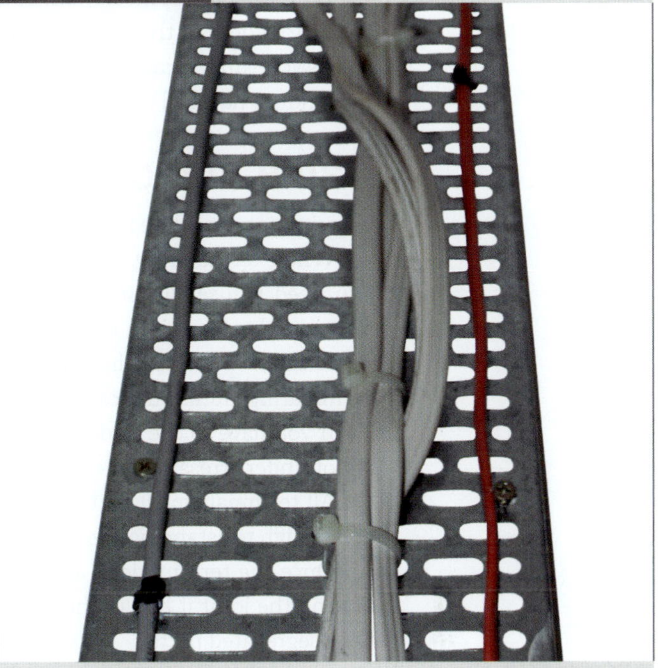

Source: Ed Hawkins

FIGURE 12.4 An example of a cable enclosures

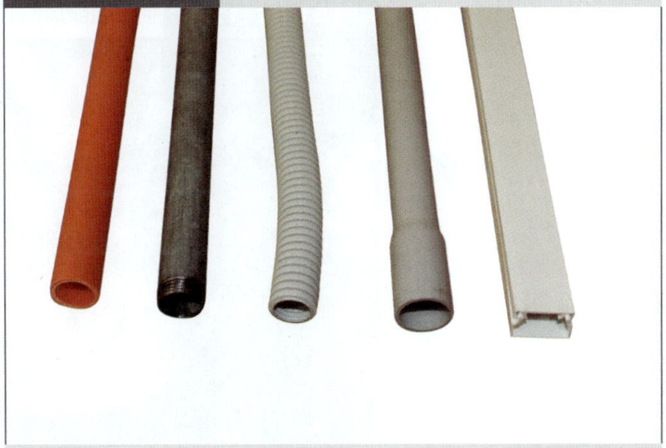

Source: Ed Hawkins

Drawing-in, placing and fixing of cables

Most of the this process is covered in **AS/NZS 3000:2007** Section 3 which stipulates what you need to know as good practice.

When drawing in cables it is imperative not to damage the insulation of the wiring. To assist you with this process you need to ensure you don't have sharp edges that will cut the insulation as it passes through or over items.

A steel conduit is usually put in place or constructed and secured first. At the ends of the conduit run you can also use bushing to ensure that you draw cable over a rounded surface rather than a sharp edge. Draw wires are then used to attach the building wires to assist with drawing the building wires into the conduits. Pull-through springs (snakes) can also be used. You should exercise care as this metal spring may touch exposed live parts. The fibreglass tape (snake) will be used to access the conduits, then pull in the wiring. Fibreglass has insulating properties, making it the ideal tool of choice.

A 'slippery lube' may also be employed when pulling wires through conduits as this lessens the friction on the cables and the lube does not affect the insulation properties of the wire.

For larger cables, due to length and weight of the cable, you may also need to use cable rollers, cable guides, blower kits for larger conduits or ducts and even a high strength glass rod (large fibreglass snake). The blower kits use a balloon or parachute that is pushed via air or gas through larger conduits.

Remember when you are pulling in cables to be careful not to cut the insulation and ensure that you aren't stressing the cable, as this could affect its current carrying capacity. In conduit, strain is not generally a problem once the cable is installed since it is well supported. If it is TPS clipped to timber you need to use enough clips to prevent the cable from sagging and becoming strained.

Personal safety is of paramount importance as there may well be sharp edges or pinch points that could damage fingers and hands. Gloves may be an option to assist you in this process. Put yourself in a good ergonomic position and do not work from ladders when pulling cables.

When you pull cables on tray and up risers you must secure the cable so as you do not put stress on the cable and/or the insulation. You do this by using cable ties at suitable intervals. Remember to avoid the sharp edges of the folded-up section of the tray when pulling cables along the tray.

As you run the cable, ensure that you don't run too near other services or across other services that may induce harmful effects.

Cable and conductor terminations

Terminations must be made in a suitable manner to prevent stress on the cable or the termination.

When terminating copper conductors use only a copper compatible terminal; with aluminium you need to use aluminium compatible terminals. To terminate aluminium to copper you need to use suitable aluminium to copper terminals which are readily available from most wholesalers. This is especially so when terminating aluminium overhead cables to consumer's copper mains cables or when terminating aluminium cable to copper busbar or terminals in switchboards.

If you have to use soldered connections then you must ensure that you prepare the joint so that it is not under stress even before the solder has been applied.

All terminals must be tensioned accordingly or to manufacturer's specifications so that you don't create high resistance joints which could then lead to sparking, arcing and eventually connection failure.

Once the terminations are complete you must ensure that suitable insulation is replaced on and around the joint if it is in the vicinity of other connections that are not in a suitable terminating box with segregated parts.

Check the terminations again before testing and/or sealing up in the case of motors, terminals of breakers and the like.

FIGURE 12.5 Fire retardant filler installed

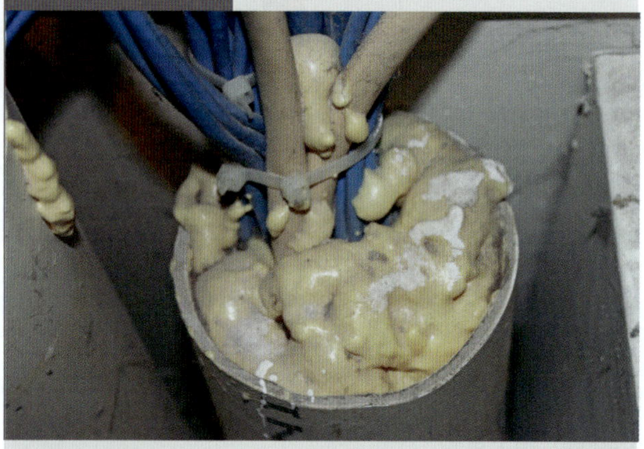

Source: Ed Hawkins

Maintaining fire integrity

Where wiring goes through a fire barrier or fire-rated wall you must ensure suitable clearance around the cable and then you must reinstate the fire rating by using fire stop putty or sealer (as shown in Figure 12.5), fire pillows or other suitable means. This is the same for tray and duct work that also goes through walls or penetrations.

This must meet AS/NZS 3000:2007 clause 3.9.9.3. See Chapter 10 for information.

Make sure that you do not touch the test probes while using the insulation tester; it is not likely to kill you, but you may lose your balance and fall.

Inspecting and testing for compliance

All testing including cables, conductors and their respective fittings, and supports must meet the requirements of AS/NZS 3000:2007 Section 8 and in particular:

- clause 8.3.6.2 – results for insulation resistance
- clause 8.3.7.2 – polarity
- clause 8.3.8.2 – correct circuit connections
- clause 8.3.9.2 – fault loop impedance.

Using an insulation tester similar to that shown in Figure 12.6, set the test voltage to twice the rating of the voltage of the installation, so that if the installation is a 230 V installation then you need to test at 500 V d.c. and if the installation is a 400 V installation then you need to test at 1000 V d.c.

FIGURE 12.6 insulation and resistance tester

Source: Ed Hawkins

Depending on manufacturer's specifications there may be other requirements for testing also. The minimum requirement you are aiming for is a reading of 1 MΩ.

SUMMING UP...

- Choose the wiring systems to suit the type of building.
- Plan the cable routes to suit the available structural system.
- Select accessories to suit the wiring system and the environment.
- Remove sharp edges and pinch points, so they don't pierce the insulation.
- Don't stress the cable when drawing in – use assistance.
- Maintain fire integrity by keeping required clearance around conductors using fire retarding putty or foam to seal penetrations; and using fire retarding pillows for tray ways.
- Carry out testing to AS/NZS 3000:20007 Section 8 requirements such as:
  - minimum value of insulation resistance must be 1 MΩ
  - use a test voltage that is double the rated voltage
  - check that connections are 'tight and bright'
  - check the correct connectors have been used
  - check that fault loop impedance complies with AS/NZS 3000:20007 Table 8.2.

REVIEW EXERCISES

Review questions 12.1

1 What methods of mechanical protection can be employed for cables?

2 Why is it important to maintain fire integrity when laying cables through walls or penetrating fire barriers?

Review questions 12.2

1 What standard, other than AS/NZS 3000:2007, applies to electromedical treatment areas?

2 What standard, other than AS/NZS 3000:2007, applies to construction and demolition sites?

Review questions 12.3

1 What standards are called up for hazardous areas?

2 Apart from explosive gas what else can be found in a hazardous area? State an example.

Review questions 12.4

1 List the restricted zones found in damp situations.

2 State the voltage range that defines an ELV installation.

Review questions 12.5

1 Describe two types of aerial cables.

2 State the minimum height for an aerial conductor that crosses a driveway.

Review questions 12.6

1 Describe two permissible underground wiring systems.

2 What are the advantages of an underground wiring system?

Review questions 12.7

1 How do you decide on the wiring system for a particular building or customer?

2 List the tests you must carry out on installed cables.

ONLINE RESOURCES

Building industry standards and codes: SAI store
http://infostore.saiglobal.com/store/getresource.aspx?path=/publishing/shop/productguides/productguides/guide_to_standards-building_and_construction.pdf

Cabling systems design: Cabling Installation and Maintenance
http://www.cablinginstall.com/articles/print/volume-13/issue-5/contents/technology/cable-trays-evolving-with-building-design.html

Fire pillows: Wormald
http://www.wormald.com.au

Hazardous areas: ICEWeb Cool Engineering
http://www.iceweb.com.au/> Australian Electrical Equipment in Hazardous Areas Standards (Gas) pages

Low voltage aerial cables
http://cablesaustralia.com.au > Products > Low voltage cables

Medical areas
http://www.ihea.org.au/files/vic/presentation_as_3003.pptx

Safety near swimming pools: PowerCor
http://www.powercor.com.au > Electrical safety > Safety around the home

Underground cables: City of West Torrens
http://www.westtorrens.sa.gov.au/files/b6571c61-fa6d-439d-9759-a196010ea067/Guidelines_for_consumer_mains_or_underground_electrical_work.pdf

13

ELECTRICAL INSTALLATIONS – VERIFICATION AND TESTING

This chapter provides an outline of the final tests and checks required by legislation and good work practice to ensure your completed installation complies with the AS/NZS 3000:2007 Wiring Rules as amended. These procedures are designed to prevent the occurrence of fault current voltage rise, electric shock from exposed metal and incorrect polarity, fire from arc faults and any other hazards caused by omissions and oversight during installation work. You know that you have done a good job with your wiring; here is how you prove it.

Chapter outline

Preparing to verify and test electrical installations

13.1 Electrical safety

Electrical testing involves possible exposure to hazards from faulty circuit conditions as well as access to confined spaces, working at heights and mobile machinery movements. It is the responsibility of the competent person carrying out the tests to make sure that safe practices are followed for each test.

Isolation

Before commencing any resistance tests on an installation you will need to ensure that:

- a job safety analysis (JSA) has been carried out to mitigate general workplace hazards
- electrical isolation procedures have been carried out on the circuits to be tested; this may, in the case of new consumer mains, require the removal of the service fuses.

Some utility suppliers allow the removal of service fuses by a qualified person but a utility employee is usually required to replace them. If multiple service fuses are present, double-check to make sure the correct fuse has been removed.

Carry out voltage tests with a Category III or IV 600 volt (or higher) rated meter to ensure the removal of all voltage sources from the installation, paying particular attention to the possible presence of dual supplies (see Figure 13.1).

While working on the switchboard you will need to erect safety barriers to prevent interference from passing traffic. To conduct tests once the installation has been energised you will need to carry out a 'Live Work' safety assessment. Appropriate personal protective equipment (PPE) must be worn to suit the live exposure. The very minimum required will be safety glasses and gloves. If high fault currents are possible then a face mask and flameproof vest will be needed.

Equipment

All test equipment used must have an up-to-date calibration check. You will need the following test equipment:

- an insulation resistance tester capable of testing at 1000 volts d.c. as well as having accurate low and medium range resistance scales (see Figure 13.2)
- a pair of resistors to assist with polarity testing 5 ohms and 10 ohms
- a trailing lead long enough to reach from the switchboard to the furthest point in your installation (see Figure 13.3)
- a fault loop tester
- an RCD tester
- a polarity tester.

FIGURE 13.1 Beware of dual supplies

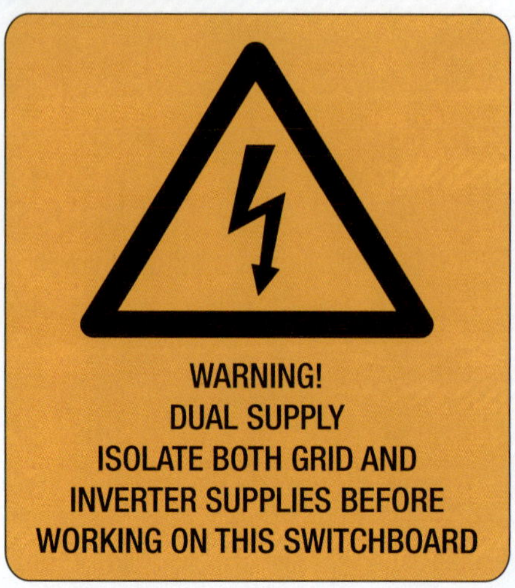

FIGURE 13.2 Insulation resistance tester

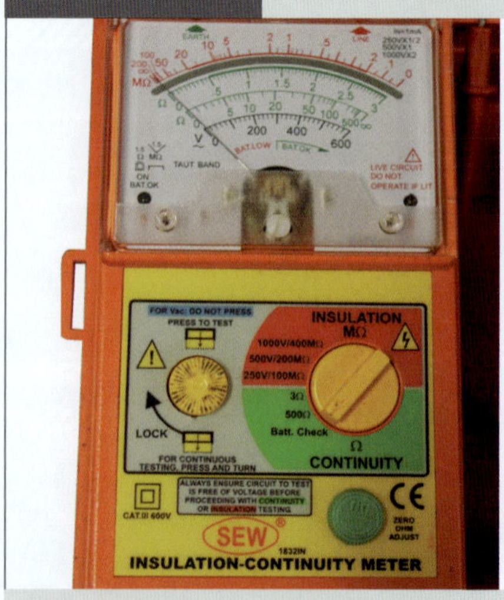

Source: Ed Hawkins

FIGURE 13.3 Resistance box and leads

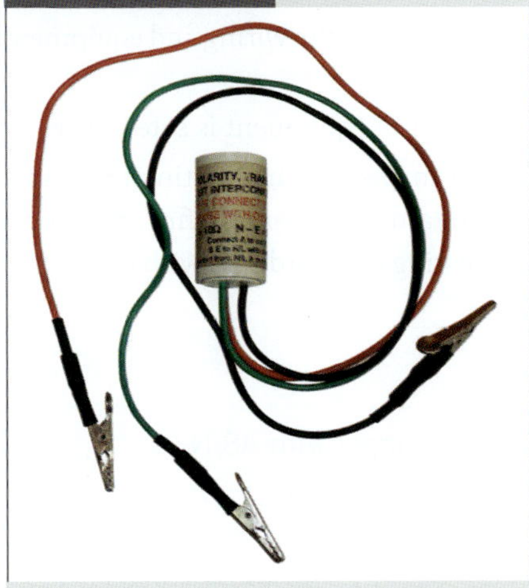

Source: Ed Hawkins

FIGURE 13.4 Safety barrier

Source: Ed Hawkins

Safety barriers are required when working on switchboards (see Figure 13.4).

SUMMING UP...

- Safety barriers are required when working on switchboards.
- Conduct lockout and isolation checks.
- Beware of dual supplies: check with the utility supplier before removing service fuses.
- Once a service fuse is removed, supply can only be restored with the authority of the utility supplier.

13.2 Legislated regulations

Every state and territory in Australia has legislation to control the installation and commissioning of medium low installations. All require testing to be carried out in compliance with **AS/NZS 3000:2007** as amended. Additional restrictions may be required to suit local conditions.

Each state has a Workplace Health and Safety Act to enforce safety in the workplace with associated regulations (see also Chapter 1).

Technical standards

Common to all of these state regulations is the requirement that all low voltage installations must comply with **AS/NZS 3000:2007 Wiring Rules**.

Other standards that relate to installation and equipment testing are:

- **AS/NZS 3017:2001 Electrical installations – Testing and inspection guidelines**
- **AS/NZS 3760:2010 In service safety inspection and testing electrical equipment**.

These standards call up other standards depending upon the location of the installation or type of equipment.

Requirements

Typically, legislation and regulations require:

- installations and equipment to be inspected and tested to ensure they are safe

- a person or agency to be responsible for the various aspects of ensuring electrical installations are safe
- records to be kept of test results that show an electrical installation is safe for connection to the supply
- records to be kept of periodic inspection and tests that show construction site wiring and equipment is safe to use
- records to be kept of periodic inspection and tests that show electrical equipment is safe to use.

AS/NZS 3000:2007 Section 8 provides the minimum standard of inspection and testing that must be achieved to satisfy the fundamental safety principles of the standard in preventing fire and shock. This requires a visual inspection in accordance with clause 8.2 and testing in accordance with clause 8.3.

SUMMING UP ...

- All states in Australia have mandatory electrical testing requirements.
- All installations must comply with AS/NZS 3000:2007 Wiring Rules.

13.3 Visual inspection of installations for compliance with the Wiring Rules

Verifying and testing electrical installations

Many non-compliance issues can be found by a rigorous visual check.

Such things as cables without adequate support or protection, missing earth bonding, damaged main earth connection and missing labels are all easily identified by a visual check.

Initial visual checks

The general condition of the installation should show:

1. Cables correctly supported.
2. Aerial cables at the correct height
3. Equipment mounted securely and at the correct height.
4. No exposed terminals or missing covers.
5. Underground cables at the correct depth (this should be checked before allowing trenches to be backfilled; clause 8.2.2) – see Figure 13.5.
6. Main earth electrode is installed as required by **AS/NZS 3000:2007** clause 5.3.6.4 in open air and not surrounded by concrete (clause 8.2.2) – see Figure 13.6.

FIGURE 13.5 Underground cable marker

Source: Ed Hawkins

FIGURE 13.6 Main earth electrode

Source: Ed Hawkins

You will need to determine the maximum demand and voltage drops in the installation and then check that the consumer's mains and sub-mains are the correct size to suit both maximum demand and voltage drop requirements as required by **AS/NZS 3000:2007** clauses 2.2.2 and 3.6.2.

Check the circuit protection to ensure that double insulation, conduits and other physical protection is continuous and complies with **AS/NZS 3000:2007** clauses 1.5.13 and 1.5.14.

The circuit over-current protection devices must be inspected to see if they have sufficient fault current rating to safely carry all possible fault currents in the installation and sufficient current rating to protect the circuit wiring against over-current as required by **AS/NZS 3000:2007** clauses 2.5.1.2 and 2.5.2.

Circuit protective devices must include RCDs in power and lighting circuits to comply with **AS/NZS 3000:2007** clause 2.6.3. Carry out the following inspections to ensure compliance:

1 Carefully examine the switchboard and meter panels to ensure they are mounted at the correct height with safe entry and exit provided in case of an emergency.
2 Check that fire proofing has not been compromised by any alterations and replace it if it does not comply.
3 Check to see that the metal frame has the correct size earth wire connected to suit the size of the active conductor in the consumer's mains.
4 Make sure the location of the earth electrode is noted on the switchboard along with the location of any underground mains (clause 8.2.2).

Final visual check

A final visual check will need to be carried out after all tests are complete to ensure that no equipment disturbed during testing has been left in an unsafe state.

SUMMING UP...

- Visual checks of equipment and wiring should show up any missing covers or connections.
- Visual checks should be made on trenches before backfilling.
- Visual checks should be made of cable sizes and protection equipment.
- A final visual check should be conducted before energising the installation.

13.4 Testing installations

AS/NZS 3000:2007 clause 8.3.4 sets out a sequence of tests:

1 Continuity of the earthing system
2 Insulation resistance
3 Polarity
4 Correct circuit connections
5 Verification of impedance required for automatic disconnection of supply
6 Operation of RCDs.

AS/NZS 3000:2007 Table 8.2 is provided so that a high impedance fault loop can be detected and corrected before supply is connected.

There are several provisions that are made for certain types of equipment such as mineral insulated heating elements. **AS/NZS 3000:2007** clause 8.3.6.2 allows certain specific equipment to be acceptable down to 0.01 megohms.

It also mentions that short sub-mains or consumer's mains 50 m or less could be 50 megohms or higher; note that this is not a requirement, but a guidance.

The set-out procedure in **AS/NZS 3000:2007** clause 8.3.4 will reveal most faults in an installation if followed in order from top to bottom. It may be more appropriate, depending on the layout of the installation, to carry out some of the tests out of order; for example, the MEN link continuity test

FIGURE 13.7 IR tester range settings

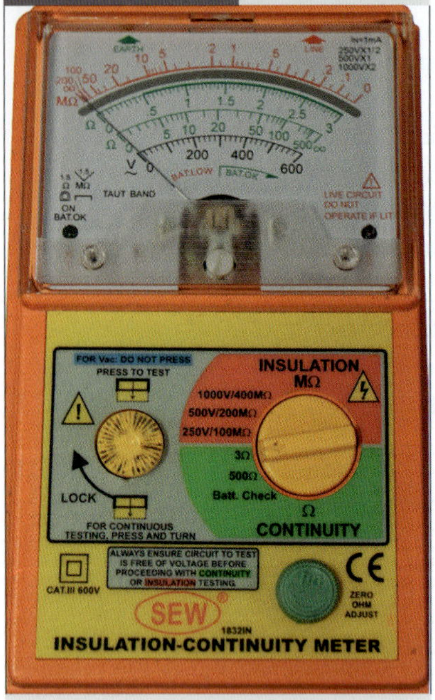

Source: Ed Hawkins

should be conducted last when all other tests are completed and the link does not have to be removed any more.

Continuity of the earthing system

Earth resistance tests are required to ensure that the earthing system will conduct sufficient current to operate the protective devices if there is a fault between live parts, other than the neutral, and the general mass of earth. If an earth fault occurs, the fault current must be able to return to the supply without causing a touch voltage to develop anywhere in the installation (clause 8.3.5.2). Figure 13.7 shows tester range settings.

AS/NZS 3000:2007 clause 8.3.5 requires that the following tests must be carried out:

- resistance of the main earthing conductor
- resistance of the protective earthing conductors
- resistance of the bonding conductors.

Main earth

To test the resistance of the main earth conductor:

1. Disconnect the MEN link from the neutral bar.
2. Prove the continuity tester to ensure it is operating correctly and that the battery is charged.
3. Switch the selector switch to the 3 Ω (ohm) range.
4. Ensure the pointer is reading zero (0) ohms when the two leads are touched together by adjusting the zero adjustment knob.
5. Connect one lead of the continuity tester to the MEN link which has been left disconnected from the neutral bar.
6. Connect the remaining lead of the continuity tester onto the earth electrode that is being tested. The reading *must be no greater than 0.5 Ω*.

> **i** AS/NZS 3000:2007 clause 8.3.5.2 states the requirements for the resistance of earthing conductors. The main earth must be tested from the earth electrode (not just the clip) to the earth bar at the main switchboard. The absolute maximum permitted for this is 0.5 ohms.

> **i** Make sure the test lead connection is to the actual earth electrode and not just to the earth clip.

> **i** Make sure the test lead connection is to the water pipe and not just to the earth clip.

Earth bonding

All metal entering the ground in the vicinity of the earth electrode must be earth bonded using the same size conductor as the main earth. Any exposed metal within the main switchboard and meter panel enclosures must be earthed using the same size conductor as the main earth. If the service is overhead to a raiser the raiser may need to be earthed.

FIGURE 13.8 Setting continuity zero on the IR tester

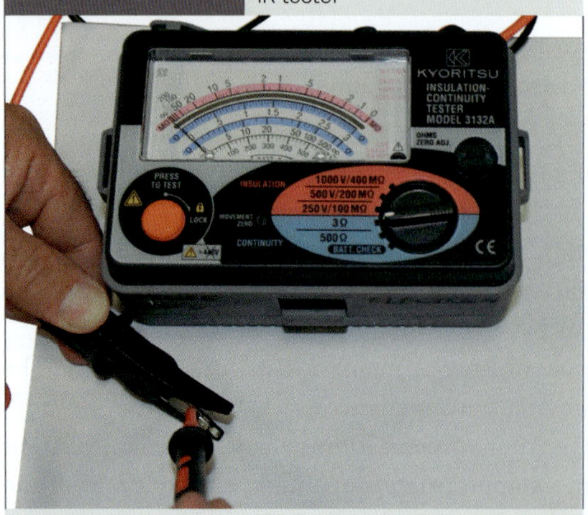

Source: Ed Hawkins

To check the equipotential bonding conductors:

1. Remove earth connections from any sub-circuits such as water heaters that may cause a false reading.
2. Prove the continuity tester to ensure it is operating correctly and that the battery is charged.
3. Switch the selector switch to the 3 Ω range.
4. Ensure the pointer is reading zero (0) ohms when the two leads are touched together by adjusting the zero adjustment knob (see Figure 13.8).
5. Connect one lead of the continuity tester to the MEN link which has been left disconnected from the neutral bar.

6 Connect the remaining lead of the continuity tester onto the water pipe that is being tested.

7 The meter reading *must be no greater than 0.5 Ω.*
Replace earth connections removed from sub-circuits.

AS/NZS 3000:2007 clause 8.3.5.2 states that the resistance shall be low enough to permit the passage of current necessary to operate the over-current protective device.

Protective earthing conductors

To test the earth continuity to the sub-circuits:

1 Disconnect the MEN link.

2 Disconnect the water pipe equipotential bond to prevent parallel earth circuits.

3 Prove the continuity tester to ensure it is operating correctly and that the battery is charged.

4 Switch the selector switch to the 3 Ω range.

5 Connect the trailing lead to one of the continuity tester terminals. Ensure the pointer is reading zero (0) ohms when the two leads are touched together by adjusting the zero adjustment knob (make sure the trailing lead is included to allow for its resistance).

6 Connect one lead of the continuity tester to the MEN link which has been left disconnected from the neutral bar.

7 Using the trailing lead connect the other lead of the continuity tester to the point of the circuit that is required to be earthed – for example, the socket outlet, lighting point, etc., as shown in Figure 13.9.

8 The maximum resistance reading should be no greater than determined using the Wiring Rules Table 8.2 as shown in Example 13.1.

Table 8.1 of AS/NZS 3000:2007 enables you to determine the maximum allowable values for the fault loop impedance of a circuit using an earth fault loop tester with power available.

EXAMPLE 13.1

To determine the maximum fault loop resistance of a power sub-circuit wired in 2.5 mm^2 cable with a 2.5 mm^2 earth conductor protected by a 16 amp type C circuit breaker:

- Open your AS/NZS:3000: 2007 to Table 8.2.
- Locate the row that shows the required active (2.5 mm^2) and earth (2.5 mm^2) conductor sizes, and circuit breaker rating (16 amps).
- Next select the column Re under the circuit breaker type C.
- Where the row and column intersect read the maximum allowable earth resistance for the sub-circuit (0.16 ohms).

R_e – the resistance, in ohms, of the protective earthing conductor.
R_{ph} – the resistance, in ohms, of the active phase (ph) conductor.
R_{phe} – the combined series resistance, in ohms, of both the active phase (ph) conductor and the protective earthing (e) conductor.

Table 8.2 of AS/NZS 3000:2007 lists the allowable resistance for individual sub-circuits taking into account the conductor size and type of protective device for testing with power unavailable.

Insulation resistance

Despite all your care the insulation of the cable may be accidentally damaged while you are installing it. Insulation may have been stripped back to expose the bare active copper conductor. In the case where the exposed active conductor happened to come in contact with some form of metal that wasn't earthed, the metal would become live.

To ensure the installation remains safe, insulation tests must be carried out every time some form of wiring on an installation or equipment is completed. The resistance must be high enough to stop electric current leaking from the conductors and on to conductive parts where it can cause damage due to short-circuit faults, arcing or lethal touch voltages. A minimum value of 1 million ohms (1M Ω) is required

FIGURE 13.9 Trailing lead in power outlet

Source: Ed Hawkins

AS/NZS 3000:2007 clause 8.3.6.2 states the minimum permissible insulation resistance values for a low voltage installation.

The insulation that surrounds or covers the copper conductor within a cable is primarily there to restrict contact with the LIVE conductor which is carrying the electric current. This is intended to prevent the conductor coming into contact with a metal structure, frame, equipment, etc.

Be sure you have tested and confirmed that the installation is de-energised.

Make sure that nobody is working where they might make contact with conductors under test.

The current from an IR tester is not sufficient to kill but if anyone is up a ladder and felt the test voltage they might fall and injure themselves.

Make sure the consumer's mains active and neutral conductors are completely disconnected from the supply.

The MEN link connects the neutral link to the earth system. If not removed the MEN link will cause a zero reading to earth on all circuits when insulation to earth tests are carried out.

If fluorescent fittings are present allow several seconds for the capacitors to charge as they initially give a low active to neutral resistance reading. This rises to infinity as the capacitors charge up.

for the resistance of the insulation covering conductors within an installation or equipment. Safety from electric shock and short-circuit damage are assured by such a large value of resistance.

EXAMPLE 13.2

The following application of Ohm's law shows that the current that can flow is well below any value known to cause damage or injury.

Using a supply voltage of 230 volts and the minimum resistance of 1 MΩ:

$$I = \frac{V}{R}$$

$$= \frac{230}{1000\,000}$$

$= 0.00023\,A$ or
$0.23\,mA$ (not even 1 milliamp)

To ensure that the insulation is tested to the highest degree, it is stressed using a d.c. voltage at twice the a.c. RMS (root mean squared) value. This means 500 volts d.c. for 230 volts a.c. and 1000 volts d.c. for 400 volts a.c.

Order of testing

Testing should be completed in the following order:

1 whole installation
2 each final sub-circuit if necessary (only if a low reading is obtained on the whole installation)
3 equipment.

The whole installation

To carry out insulation tests on the whole installation, you should complete the following steps:

1 Connect the un-metered consumer's mains active and neutral conductor together at the point of supply.
2 Bridge out the kWh meter tails at the consumer's meter panel with a bridging wire fitted with alligator clips.
3 Disconnect the MEN link if it is not already disconnected.
4 At the switchboard, switch ON the main switch as well as all circuit breakers and RCDs.
5 Throughout the entire installation, switch all switches to the 'on' position, e.g. socket outlets, lights, exhaust fans, etc. This ensures the switch wires are also tested.
6 Switch 'off' or disconnect all current consuming appliances from the installation, e.g. hot water units, electric stoves, etc. that contain metal sheathed heating elements. Remove all lamps and fluorescent tubes.

Taking readings

To take a reading, you should:

1 Connect one lead of the insulation resistance tester to the joined consumer's mains active and neutral conductors.

2 Connect the other lead of the insulation resistance tester to the MEN link which has been disconnected from the neutral bar.
3 Carry out the insulation resistance test. See also Figure 13.10.
Except where special provisions are made the resistance must not be less than 1 MΩ.

FIGURE 13.10 Tester setting for 230 volt a.c. insulation

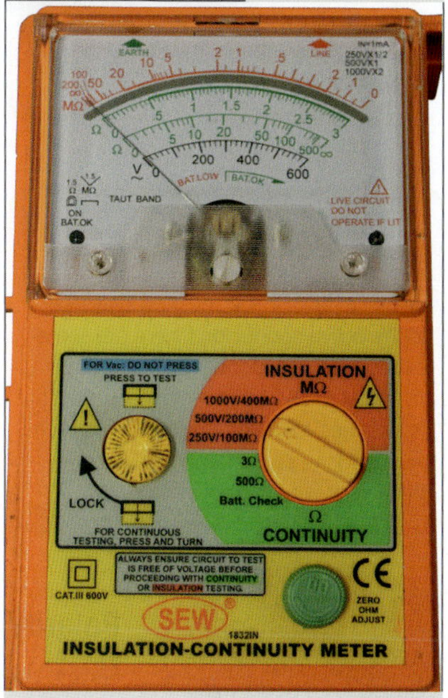

Source: Ed Hawkins

Isolating a low reading

If the insulation resistance of the whole installation was less than 1 MΩ, then further insulation resistance tests must be carried out on each final sub-circuit to determine which circuit contains the fault:

1 Leave the MEN link disconnected from the neutral bar.
2 Switch the off circuit breaker of final sub-circuit 1.
3 Disconnect neutral of final sub-circuit 1 from neutral bar.
4 Connect a bridging wire between the final sub-circuit number 1 active and neutral conductors.
5 Leave all switches within the installation switched to the 'on' position.
6 Connect one lead of the insulation resistance tester to the disconnected end of the MEN link.
7 Connect the other lead of the insulation resistance tester to the bridged active and neutral conductors of the final sub-circuit number 1.
8 Carry out insulation resistance test. Resistance must *not be less than 1 MΩ*.
Continue to test the other final sub-circuits using the same procedure.

Polarity tests

AS/NZS 3000:2007 clause 8.3.7.1 states that polarity testing is necessary to ensure no shock hazard arises from the incorrect connection of active, neutral and earthing conductors.

Tests must be carried out to ensure that:

- no transposition of active and neutral conductors of the consumer's mains or sub-mains has occurred
- no incorrect combinations have occurred between active, neutral and earthing conductors, thereby resulting in exposed conductive parts of the electrical installation becoming LIVE
- switches are connected in the neutral conductor, resulting in parts of appliances, e.g. heating elements and lamp holders, remaining energised when the switches are in the 'OFF' position.

> Make sure the insulation resistance tester is still operating correctly by shorting out and then open-circuiting the test leads:
> Leads together – (0 Ω)
> Leads apart – (∞ Ω).

> Once you have finished testing make sure you discharge any capacitors by briefly shorting them out with a suitably insulated lead.

Testing the consumer's mains polarity

When preparing to test the polarity of the consumer's mains to ensure that the active is connected to the main switch and that the neutral is connected to the neutral bar, the insulation resistance tester must have the selector switch switched to the 3 Ω range.

To test the consumer's mains active for correct polarity, connect the trailing lead from the tester to the active conductor tail at the point of supply and the other lead of the tester to the main switch at the main switchboard.

The outcome of this test should show the meter pointer move from infinity ohms (∞ Ω) to approximately zero ohms (0 Ω). This will indicate that the consumer's mains active is in fact continuous from the point of supply to the main switch at the main switchboard.

When testing the consumer's mains neutral for correct polarity, connect the trailing lead from the tester to the neutral conductor tail at the point of supply and the other lead of the tester to the neutral bar at the main switchboard.

> If the point of service is in an elevated position review your job safety analysis and ladder safety procedures before starting these tests. If the point of supply is in a pit then you will need to consider a gas test before commencing work. Check to make sure the service is completely disconnected from the consumer's mains before starting this test.

> **i**
> You will need to bridge out the consumer's mains neutral at the meter panel if it hasn't been fitted off at the meter neutral link.

> **i**
> AS/NZS 3000:2007 clause 2.3.2.1.1 requires that all switches within an electrical installation must be switched in the active conductor. (There are some exceptions, e.g. double pole switches.)
> This is to ensure that when the switch is switched to the off position, the switch wire and the equipment connected to the switch wire are not energised.

> **i**
> AS/NZS 3000:2007 clause 4.4.5 deals with polarisation and phase sequence. Commencing at the earth socket, the polarity for all socket outlets is earth, active and neutral in a clockwise direction when viewed from the front of the outlet. This applies for all 10 A, 15 A and 20 A socket outlets.
> In general, when viewed from the front of the outlet, whether it is a single, double or multiple socket outlet, the active socket is on the left, the neutral socket is on the right and the earth socket is perpendicular on the bottom.

The outcome of this test should show the meter pointer move from infinity ohms (∞ Ω) to approximately zero ohms (0 Ω). This will indicate that the consumer's mains neutral is in fact continuous from the point of supply to the neutral bar at the main switchboard.

Polarity and connection of sub-circuits

The following tests will ensure the sub-circuits do not interconnect and each protective device numbered is correctly associated with its neutral link terminal number. The polarity test will ensure that no neutrals are switched and that active, neutral and earth are correctly connected at the points or appliances.

- At the switchboard remove circuit number 1 from its circuit breaker and the corresponding neutral number 1 from the neutral link.
- Measure between neutral number 1 and the neutral link to ensure there is no connection between neutrals of different circuits.
- Measure between circuit active number 1 and the other sub-circuit actives to ensure there are no crossover connections.
- Connect the resistance box to the neutral, active and earth as shown in Figure 13.11.

Polarity of socket outlets

Using the 500 ohm resistance range on the insulation resistance tester as shown in Figure 13.12, measure the following resistances at each outlet on the circuit first with the switch on, then with the switch off:

- earth (E) to active (A)
- earth (E) to neutral (N)
- active (A) to neutral (N).

  Test results:
- N to E = 5 ohms

FIGURE 13.11 Resistance box connection

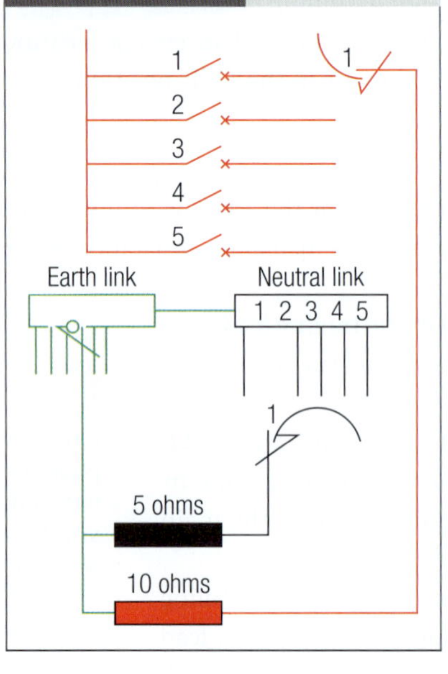

FIGURE 13.12 Meter reading power outlet polarity

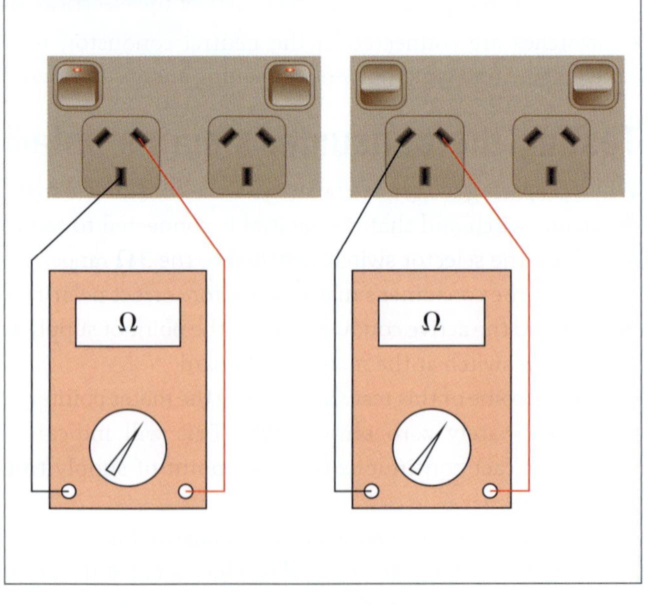

FIGURE 13.13 Meter reading for lighting outlet polarity

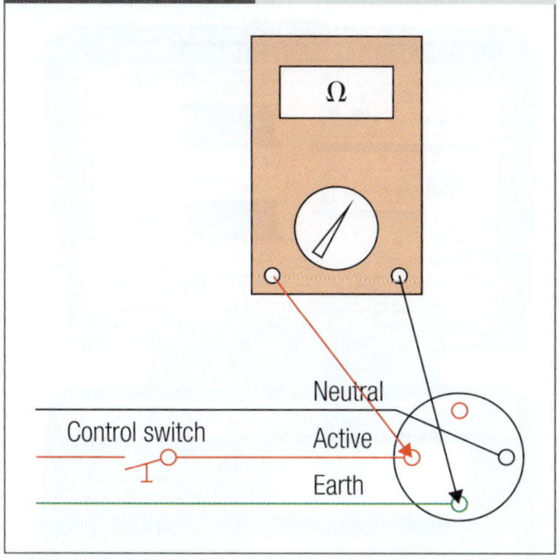

- A to E = switch on 10 ohms
- A to E switch off = infinity
- A to N switch on = 15 ohms
- A to N switch off = infinity.

Polarity of light sockets

With the resistor network connected to each circuit in turn, test the lighting points as shown in Figure 13.13 to ensure that the switch is in the active conductor and the earth is in the earth terminal.

Test results:

- E to N = 5 ohms
- E to A = 10 ohms
- A to N with control switch on = 15 ohms
- A to N with control switch off = infinity.

AS/NZS 3000:2007 clause 8.3.7.2 requires all Edison screw lamp holders to have the outer contact connected to neutral.

Polarity of connected equipment

Remove the resistance box from the switchboard and restore all circuits including the MEN link. For each item of equipment in the circuit that is permanently connected test the resistance, as shown in Figure 13.14, between:

- active (A) and neutral (N)
- active (A) and earth (E)
- neutral (N) and earth (E).

If the equipment is fitted with a functional switch such as a heater or range then test the active terminal to the supply side of the switch to ensure the switch is connected correctly.

The synthetic fibre material used in some brands of ranges and water heaters could be glass fibre. This is a skin and eye irritant requiring PPE and a respirator.

FIGURE 13.14 Meter reading for connected equipment polarity

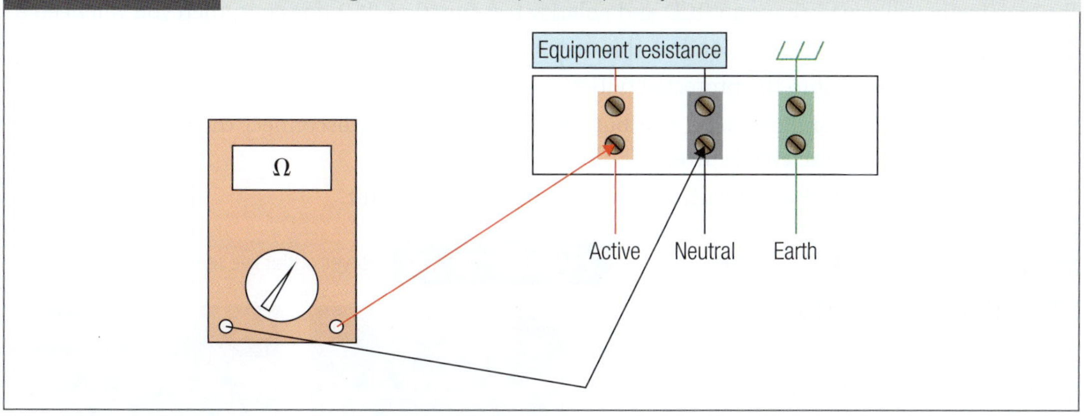

On completion of the resistance and polarity tests, be sure to reconnect the MEN link to the neutral bar and retest its continuity, remove all temporary short circuits and reconnect all conductors removed for testing. Replace all covers.

Test results:

- E to N = 0 ohms
- E to A = infinity
- A to N with control switch on = equipment resistance
- A to N with control switch off = infinity.

Power on tests

When the installation has been energised, the final polarity checks can be carried out along with the testing of the fault loop impedance and the function of the RCDs.

FIGURE 13.15 The fault loop impedance tester

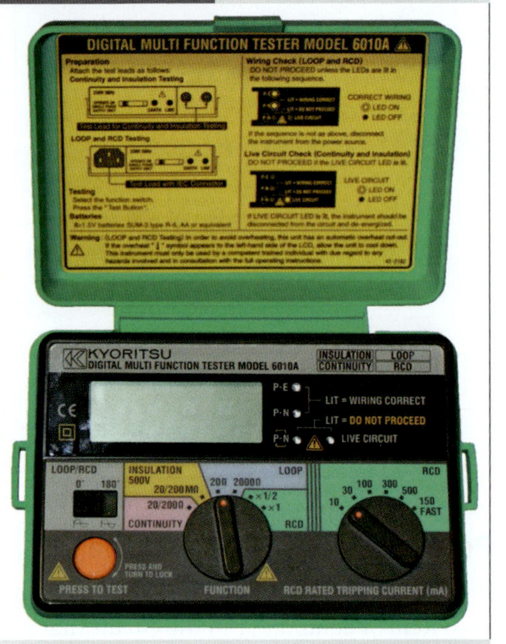

Source: Ed Hawkins

Power on earth fault loop impedance test

Operate your fault loop impedance tester, as shown in Figure 13.15, according to the manufacturer's instructions.

EXAMPLE 13.3

Use AS/NZS 3000:2007 Table 8.1 to find the maximum earth fault loop impedance of a circuit protected by a 16 amp type C circuit breaker:

1 Locate the row containing the current rating (16 A).
2 Next find the column to match the type of breaker (type C).
3 Finally read the value in the box where the row and column meet (1.92 ohms).

The reading on your tester should not exceed this value. If it does, there is a fault in the system that will prevent the circuit breaker operating in case of an earth fault.

Power on verification of operation of RCD devices

The loop tester will usually have an RCD and polarity test function. If it does not, then you will need to use a purpose-made RCD tester. Conduct this test as follows:

1 Plug the RCD tester into the protected outlet.
2 Turn on the switch and press the test button. The RCD should disconnect the circuit within the required time for the selected current.
3 Reset the RCD and repeat for each outlet on each protected circuit.

Power on verification of outlet polarity

A final check of the polarity of each socket outlet is with the power on (see Figure 13.16). Usually this will be done in conjunction with the fault loop and RCD tests, making a separate test unnecessary.

Using an a.c. voltage tester set to read up to 450 volts, measure the voltage between:

- earth (E) and active (A)
- earth (E) and neutral (N)
- active (A) and neutral (N)
- first with the switch on then with the switch off.

FIGURE 13.16 Power outlet live tests

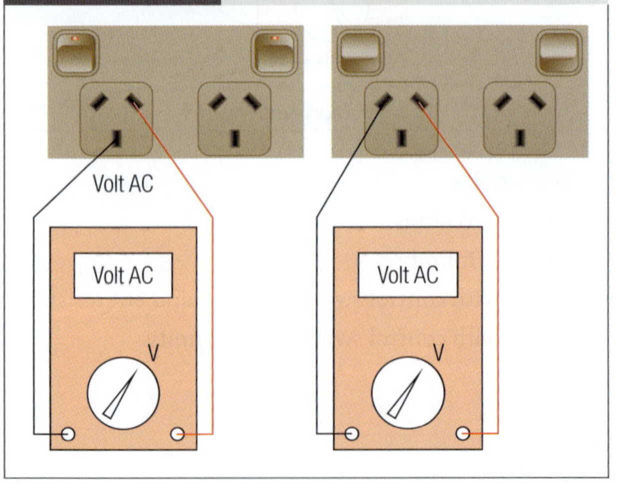

When the installation has been energised a sign should be placed at the switchboard to the effect the equipment is live. This can be removed when handing the premises over to the customer. It is quite easy to become used to working on the unpowered switchboard and mistakes can be made.

AS/ NZS 3000:2007 clause 8.3.9.2 requires the earth fault loop impedance at a socket outlet to be no higher than the value shown in Table 8.1. If an RCD trips during the test it is satisfactory but does not indicate the effectiveness of the MEN connection and upstream conductors. These will need to be tested separately at the switchboard.

When conducting any tests on an energised installation you must use appropriate PPE to mitigate the hazards of electric shock and arc flash.

AS/NZS 3000:2007 clause 8.3.10 requires the testing of RCDs using a tripping device connected to the protected circuit. The test button on the RCD does not show that it can be tripped by a circuit fault.

Test results:

- E to N = 0 volts
- E to A control switch on = 230 volts
- E to A control switch off = 0 volts
- A to N with control switch on = 230 volts
- A to N with control switch off = 0 volts.

Finally, test the voltage between all metal parts of the installation against a known good earth to make sure that no live metal is present.

Only use Cat III 500 volt or higher rated voltmeter to protect against the possibility of an arc flash.

Testing of equipment

AS/NZS 3012:2003 Electrical installations – Construction and demolition sites sets out conditions for testing electrical equipment that include:

- earth continuity
- insulation resistance
- leakage current
- the operation of RCDs.

Testing instruments discussed previously in this chapter can be used to carry out these tests, but the volume of equipment on any single work site would make this expensive and inefficient.

To speed up the process of compliance testing, a portable appliance tester (PAT) is used (see Figure 13.17).

PATs can be sophisticated enough to carry out all the tests required and produce a report as well as retaining the result of the test in memory so that deterioration over time can be detected and reported automatically.

Records must be kept of the results of all tests, including the date of the test and the date the next test is due.

FIGURE 13.17 Portable appliance tester (PAT)

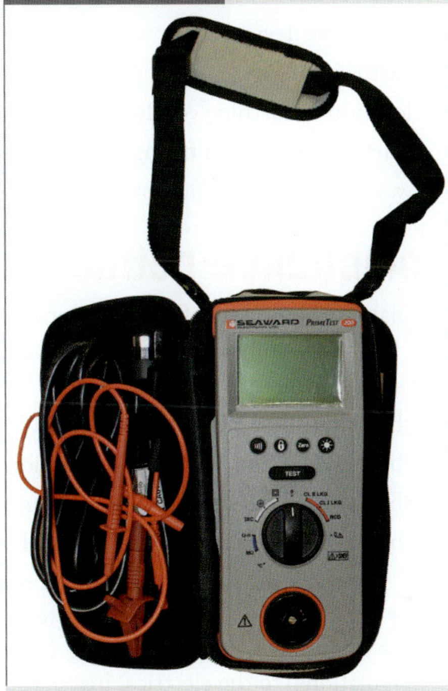

Source: Ed Hawkins

Collect all copper off-cuts for recycling.

Use correct disposal techniques for plastic sheathing and damaged fluorescent fittings.

Turn off all lights and unnecessary appliances after completing your energised testing.

SUMMING UP...

- Testing on installations includes:
  - continuity tests on the earthing system
  - insulation resistance tests
  - polarity tests on mains, sub-mains and sub-circuits.
- Visual check for correct circuit connections to circuit breakers and neutral link.
- Verification of impedance is required for automatic disconnection of supply.
- Test RCDs to ensure correct time and current sensitivity.
- Test building site equipment to ensure correct connection and polarity and insulation integrity.
- Records must be kept of test results.
- Use suitably rated test equipment when carrying out tests with the system energised.
- Observe manufacturer's safety precautions when using a portable appliance tester (PAT).
- Ensure that you work in an environmentally sustainable manner.

Recording and reporting on the results of verifying and testing electrical installations

13.5 Documentation

Required documentation will vary slightly depending upon the jurisdiction and utility supplier.

Variations due to climate mean that some requirements in the Northern Territory and Queensland may not be appropriate in Victoria or Tasmania. However, all regulators and utility suppliers have the same basic requirements:

- labelling of the switchboard and installation to comply with **AS/NZ 3000:2007**
- the provision of notices of compliance.

Labelling

AS/NZS 3000:2007 requires the following labelling:

- protective devices
- earth conductor
- location of earthing conductor
- warning of recessed fittings, etc.
- warnings regarding dual supply such as with solar grid connections.

Compliance notifications

Regulators require a certificate of compliance after work has been carried out on an electrical installation that has been connected to the supply. This must be provided by the person who carried out the work and should include:

- the name and address of the person for whom the work was performed
- the details of the electrical installation tested
- the day the electrical installation was tested
- the electrical contractor licence number under which the electrical installation was tested
- a statement on the certificate that certifies the electrical installation (to the extent it is affected by the electrical work) has been tested to ensure it is electrically safe and is in accordance with the requirements of the Wiring Rules (and any other standard applying under the appropriate regulations) to the electrical installation.

Compliance forms are available online from the electrical regulator in each state or territory, or obtainable in book form with pages for providing duplicate copies.

The electrical contractor is usually required to keep a copy of the certificate for a period of time specified by the electrical regulator (five years in Queensland, 10 years in Tasmania) after the certificate is given to the person for whom the work was performed.

SUMMING UP...

- AS/NZS 3000:2007 specifies certain mandatory labelling at switchboards and other parts of the installation.
- Not all jurisdictions have the same documentation requirements.
- Utility suppliers have their own supplementary documentation requirements.
- All tests on installations and equipment must be recorded.
- Customers must be given a copy of all test reports.
- Copies of all certificates of electrical compliance must be kept for a specific period of time.

REVIEW EXERCISES

Answer these questions and check your answers at the back of the book.

Review questions 13.1

1 What is the minimum category of voltage tester that you should use on mains and switchboards?

2 What is the purpose of a 'trailing lead'?

Review questions 13.2

1 List the mandatory tests required by AS/NZS 3000:2007 Wiring Rules.

2 Why do you need to test the continuity of the earthing system?

3 Does AS/NZS 3000:2007 specify any type of visual inspection?

Review questions 13.3

1 List four faults that might be found by visual checks.

2 If a switchboard is supplied with 80 amp mains, what size earth conductor should be connected to the frame of the switchboard?

Review questions 13.4

1 Describe how protective earthing conductors should be tested for continuity.

2 State the minimum resistance for protective earthing conductors.

3 Describe how the required fault loop impedance should be determined.

4 Describe how the fault loop impedance of a circuit should be tested.

5 Describe how you should measure the resistance of conductors of the final sub-circuits.

6 Determine if the following circuit complies with AS/NZS 3000:2007, and include an explanation of your answer:

- A final sub-circuit of 2.5 mm^2 V75 TPS Red, Black and Earth has been installed in a domestic installation to supply a circuit of 10 A socket outlets.
- The circuit is protected by a 16 A type C circuit breaker.
- The resistance (Re) of the 2.5 mm^2 protective earthing conductor has been measured from the disconnected MEN link to the earth socket of the outlet and found to be 0.55 Ω.
- The resistance in the 2.5 mm^2 final sub-circuit active conductor (Rph) has been measured to be 0.6 Ω.

Review questions 13.5

1 What documentation should be left with the owner of the premises?

2 List four items that should be included on a Certificate of Electrical Compliance.

3 List three information or warning labels that you might be required to place in a main switchboard.

ONLINE RESOURCES

For amendments and new requirements you will need to access the state or territory websites shown below:

Australian Capital Territory: ACT Government
http://www.legislation.act.gov.au/a/2004-12/default.asp

New South Wales
http://www.fairtrading.nsw.gov.au > Tradespeople > Electricians > Electrical compliance requirements

Northern Territory: NT Government
http://notes.nt.gov.au/dcm/legislat/legislat.nsf/linkreference/electricity%20reform%20%28safety%20and%20technical%29%20regulations?opendocument

Queensland: Queensland Government Department of Justice and Attorney-General
http://www.justice.qld.gov.au > Electrical safety > Electrical contractors and workers > Issuing certificates of compliance

South Australia
http://www.sa.gov.au > Water, energy & environment > Electrical, gas and plumbing safety and technical regulation > Electricity trades > Electrical installations > Testing and verification of electrical installations

Tasmania
http://workplacestandards.tas.gov.au > Industry and safety > Industries > Electrical work

Victoria
http://www.esv.vic.gov.au > Legislation and regulations > Legislation administered by ESV

Western Australia: Government of Western Australia Department of Commerce
http://www.commerce.wa.gov.au > Energy safety
http://www.commerce.wa.gov.au > WorkSafe

14

FAULTS OF ELECTRICAL CIRCUITS AND EQUIPMENT, AND FAULT-FINDING TECHNIQUES

This chapter introduces the concepts of electrical circuit and equipment faults and fault-finding techniques, and guidance to determine if a repair can be carried out economically. Consideration is given to safe disposal of unserviceable equipment and safe practices when testing and repairing electrical equipment.

Chapter outline

14.1 Troubleshooting concepts

All electrical faults can be grouped into open-circuit or short-circuit faults. Variations of insulation and conductor failure can be traced by choosing appropriate test equipment and procedures.

Preparing to troubleshoot and repair faults in electrical equipment

Circuit operation

Despite the complexity that may exist within any given electrical system, for troubleshooting purposes all electrical systems can be reduced to the following essential parts (see Figure 14.1):

1 A supply of electromotive force – battery, generator, mains supply, etc.
2 A current path made from conductors of copper, aluminium, carbon or other materials to suit the operating conditions.
3 An electrical load utilising one of the main electrical effects that convert electricity into another energy form such as heating, ionisation, magnetism or electrolysis.
4 A return path to the supply.
5 A means to control the flow of current such as a switch, relay, variable resistor or one of the modern solid-state equivalent devices.
6 Insulation material to prevent unwanted flow of current other than through the required circuit devices.

FIGURE 14.1 Essential circuit operation

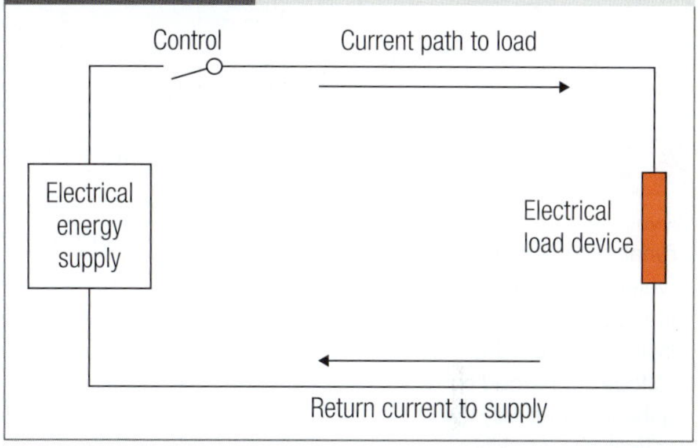

Common faults

Based on the essential electric circuit, faults will develop in a predictable manner:

1 Failure of the voltage supply. This could have many causes but can be quickly established using a voltmeter to measure the level of voltage available.
2 Failure of the current path to or from the load. This is known as an open circuit fault and is the most common cause of electrical system failure to such an extent that all fail safe requirements specify an open circuit condition must stop the system and leave it in a safe state (see Figure 14.2).
3 Failure of the insulation between the conductors. This is known as a short circuit fault and can cause severe damage to wiring and equipment (see Figure 14.3). Circuit breakers and fuses are required to remove the supply from the circuit quickly when a short circuit occurs.
4 Failure within the load section of the circuit. In the case of a purely electrical load this will usually be a variation of the open circuit or short circuit fault, for example, an open circuit element or insulation failure. In the case of a motor the mechanical load may contribute to the problem in the form of a misaligned coupling, seized bearing, clogged fan duct or similar restriction to free movement of the shaft. A solenoid controlling flow of fluids may have a seized valve or linkage.

When testing for live on a low voltage system make sure that all safety precautions for live testing have been put in place before you start.

Basic causes of circuit failure

FIGURE 14.2 Open circuit fault

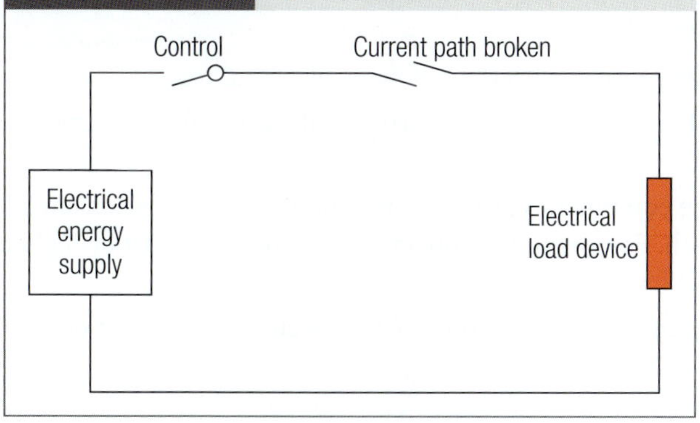

FIGURE 14.3 Short circuit fault

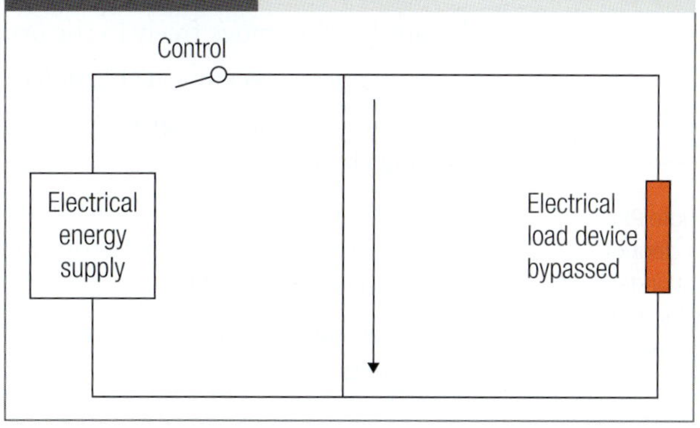

5 Failure of the control device which either fails to open, leaving the load permanently connected, or fails to close, preventing the load from receiving any current.

6 Latent faults introduced during installation or servicing, such as high resistance joints, where terminals have been over-tightened and cut strands of conductor; or not correctly tightened, allowing the contacting surfaces to oxidise and causing a high resistance joint that can overheat.

7 Faults introduced by inexperienced or careless operators who overload motors or ignore operating procedures or leave waste materials to cover ventilation holes. Misusing plugs such as filing down a 15 amp plug to fit a 10 amp socket, causing the switch on the socket outlet to fail.

8 Damage by rodents and other pests. Mice and rats can gain access to the backs of cooking ranges and refrigeration equipment. Insects can build in light fittings causing arcing and short circuit faults.

Symptoms and causes

If you see green water running from the hot water outlets you should immediately isolate the installation from the voltage supply and carry out earth continuity tests to verify the integrity of the earth system. Follow up with checks for insulation failure in the mains, sub-mains and sub-circuits.

Most electrical equipment faults produce symptoms that give clues to where the problem may lie.

- Worn bearings, for example, will produce noisy running and if not replaced will finally collapse or cause 'poling' where the rotor side movement becomes so large it collides with the stator pole faces, causing costly damage to the machine.
- Short circuited stator windings will lead to tripping of the motor overloads. Single phase induction motors may fail to produce starting torque due to failed start windings or starting switch. Three phase motors may seem to have a power supply but may be running on 'single phase', meaning that one line is missing and causing the motor to fail on start-up.
- Rusty water coming through the hot-water pipes can indicate the lining of the water heater is damaged while green water will indicate a failed earth bond with fault current running through the water system – a very dangerous situation.

Fault diagnosis

Fault diagnosis

The initial report on the fault by the owner or operator is very important. The user knows what their equipment should be doing and can tell you a lot about the symptoms and history of the fault. You should obtain as much information as possible:

- Did the equipment stop suddenly?
- Did it run down over a period of time?
- Was there any damage done to it such as water spills or dropping?
- Was smoke or a burning smell associated with the fault?
- Was there a flash or sparking of any sort?

Once you have gathered as much information as you can from the user you will need to carry out an inspection of your own.

- Is there any visible damage?
- Can shafts be turned by hand?
- Do switches sound as if they are operating?
- Do any brushes move freely in the brush boxes?

If you cannot detect any visible defect, then proceed with the following electrical tests:

1 Use your continuity tester to make sure there is a continuous current path through to the motor, heating element or solenoid.

2 Carry out an insulation check to ensure the equipment has no low resistance faults between circuit conductors or between conductors and earth. For low voltage operation this reading must be higher than 1 megohm (1 MΩ).

3 Once you have determined that the equipment is safe to connect to the supply, switch it on and observe its operation.

Wear appropriate PPE and keep well clear of any equipment under test, as it may fail in a dangerous manner.

4 From any observed symptoms you should make an assumption as to the likely cause of the fault.
5 Conduct further tests to prove or disprove your assumption.

It is easy to make false assumptions, but as your experience with a particular piece of equipment develops your assumptions become more accurately focused and the cause of the fault located quickly.

Assumptions from observed effects

Your observations may lead you to make false assumptions that may waste time and divert you away from the true cause of the fault. It is important to follow up each assumption to establish its soundness.

Table 14.1 is a list of diagnostic checks that might be applied while checking an item of equipment. Each is accompanied by a false assumption that can delay the location of a fault.

TABLE 14.1 Diagnostic checks and possible false assumptions that can be made

Check	False assumption
Switch is in the 'ON' position	Switch is making contact
Solenoid indicator lamp is on	Current is flowing through the solenoid
Replacement part substituted	The replacement part is in good working condition
The overload trip indicator is displayed	An overload has occurred
Replacement part has the same part number	The parts are identical
The consol voltmeter indicates the presence of the supply	The voltmeter is functioning correctly
Switch is in the down position	The switch is closed

Verifying assumptions

Troubleshooting and repairing faults in electrical equipment

Methods of approach to fault-finding vary with the nature and type of equipment being tested. The three main approaches are:

- work forward
- work backwards
- half split.

1 Work forward

The work forward system involves starting at the beginning and following the circuit through until the fault is located. This procedure is very good for small circuits such as hot water systems or single appliances with one load and a limited number of control devices. To do this you should:

Diagnostic techniques

1 Isolate the supply.
2 Ensure that all control switches are closed.
3 Connect your ohm meter across the circuit terminals. The load should appear as a resistance, typically 22 to 23 ohms for a 2.4 kW element.
4 If the load resistance is not present then check that the neutral or return conductor is connected.
5 Work along the circuit testing until the load resistance reappears.
6 Do not restore the supply until your insulation and continuity tests show that it is safe to do so.

2 Work backwards

The work backwards system involves starting at the end and working back until the fault is found (see Figure 14.4). This system works well for control circuits where a contactor coil must be energised. To do this you should proceed as follows:

1 Isolate the supply.
2 Use your ohm meter set to the 1000 ohms range to establish the existence of the coil (this will be around 600 ohms for a 230 volt coil, higher for a 400 volt coil and lower for a 110 volt coil).

FIGURE 14.4 Start at the end

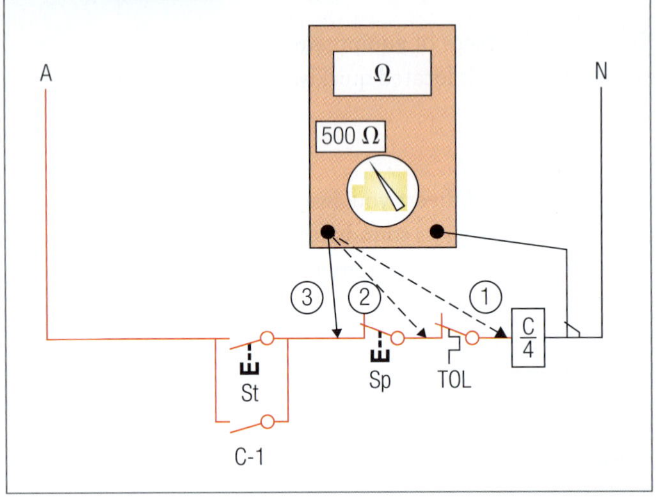

3 Test the continuity of the return conductor or neutral back to the supply terminals.
4 With one end of the ohm meter connected to the return or neutral side of the coil work back along the supply side of the circuit, operating the contacts and switches until the continuity fails.

3 Half split

For larger or more complex equipment the 'half split' method is very good. In this case you start taking measurements and examining the circuit from the middle and move first one way and then the other through the circuit until the fault is located. A system controlled by a programmable logic controller (PLC) or programmable relay lends itself ideally to this method. The PLC is a natural start point since the operation of the input devices can be monitored on the input module, an illuminated LED indicating a closed contact.

The output LEDs will indicate the output devices that have been energised, simply requiring you to carry out a physical check to see if the device (contactor, solenoid, etc.) has operated.

Preventing recurring faults

When making a repair you should, if possible, try to ensure the fault is less likely to occur again. If the failure was due to wear over time then the frequency of preventative maintenance checks will need to be increased.

Bearings may need to be changed more regularly to prevent failure in service. Resistance checks on solenoids, contactor coils and motor windings need to be done more regularly with coils and motors being changed out if the resistance is falling rapidly over a short time. Some design faults may be easily removed such as replacing equipment that uses operating springs to carry load current with a better quality product.

Intermittent faults

Areas of high vibration, such as ore processing crushers and hammer mills, require regular checking of terminals as nuts and screws often seem to vibrate undone under these circumstances. Metal fatigue in aluminium conductors and busbars must be monitored and vibration-absorbing mounting used. Examine studs and nuts during preventative maintenance shutdowns and ensure the clamped surface is protected by a washer on both sides. If the washer has worn or corroded then replace it with a new one.

Lighting circuits may be subject to occasional ingress of water from a leaking roof. By the time you arrive to investigate, the weather is dry and the fault has cleared. The nature of the fault may only become apparent after extensive investigation.

Recommissioning the repaired system

Once equipment has been repaired it should be checked for earth continuity, where necessary, and insulation resistance to earth.

Carry out a visual check to ensure that all covers and screws have been replaced and turn shafts by hand to ensure they move freely. Before restoring the supply voltage to any machine, ensure that everyone is clear, that any PPE required is worn and guards are in place. The equipment may start in a dangerous state so stand clear and monitor the operation as the machine starts up. If necessary use a clamp meter to monitor the current to ensure the machine is running correctly.

SUMMING UP...

- All electrical circuits have basic requirements.
- Most faults can be divided between conductor failure and insulation failure. Faults can be introduced by carelessness or misuse.
- Diagnosis starts with enquiries about the history of the fault.
- Assumptions can be made from the visible evidence.
- Assumptions must be tested before proceeding.
- The three main approaches for fault-finding are:
  - work forward from the supply
  - work backwards from the load
  - use the half split method.
- More frequent checks will help reduce breakdowns.
- Intermittent faults require experience and knowledge to solve.
- Recommissioning requires care and thorough testing of continuity and insulation.
- Strict care must be taken to isolate circuits from the supply.

14.2 Troubleshooting water heater equipment and circuits

Water heaters use the heating effects of electric current to heat or boil water. Faults may involve contact between live electrical terminals and damp materials, requiring caution when troubleshooting.

Hot water heater circuits

The most common hot water heater is the single phase storage heater that may sometimes be fitted with a booster element (see Figure 14.5). It is a basic heating circuit with a thermostat to control the temperature and a thermal cut out to override the thermostat in case of failure. It is modern practice to fit a fixed thermostat in order to prevent tampering that might cause burns if set too high or bacteria growth if set too low.

Where quick recovery of water temperature is required with high demand, three phase supplies are provided in order to provide extra power (see Figure 14.6). The elements are usually single phase elements mounted in banks of three, with each element in the bank controlled by a thermostat. The diagram below shows one possible circuit configuration using a neutral return. The system may

Faults in water heaters

FIGURE 14.5 Single phase HWC circuit

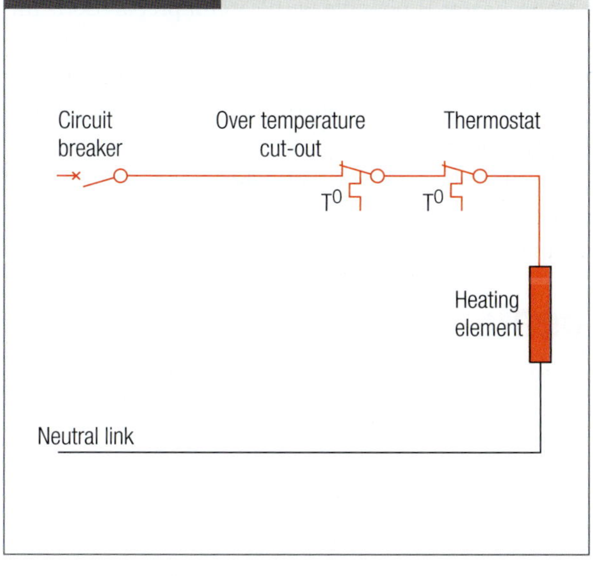

FIGURE 14.6 Three phase HWC circuit

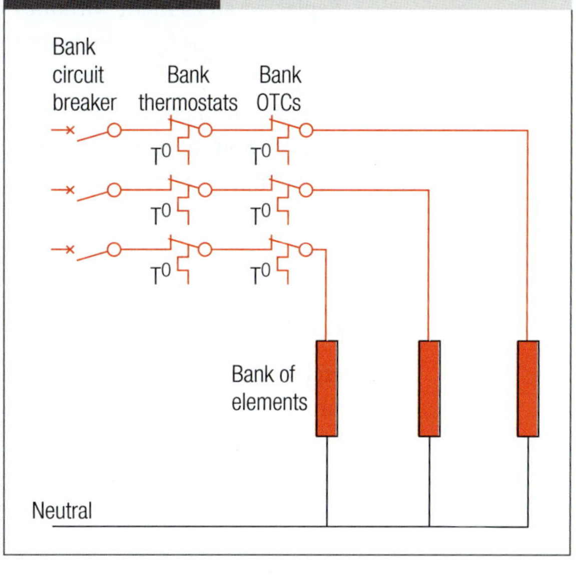

comprise more than one bank of heating elements depending on the peak demand, with each element and bank of elements installed at different heights.

In both single and three phase systems the power rating of the heating element will determine the current drawn. In both examples the single phase voltage across each element will be 230 volts.

In this case a 1 kilowatt element will draw:

$$I = \frac{W}{V} = \frac{1000}{230} = 4.4 \text{ amps}$$

The resistance of the element will be:

$$R = \frac{V}{I} = \frac{230}{4.4} = 52\,\Omega$$

Once the resistance is known the existing element can be tested to see if it is the correct size or even if it is still an effective element at all.

Replacing heating elements

When replacing heating elements, you should follow these steps:

1 Isolate the electrical supply to the hot-water cylinder at the switchboard. Test to make sure there is no voltage present on any terminals. If there is a booster element in the top of the tank, make sure that it is not supplied separately.

2 Turn off the cold water supply to the water heater. Have the replacement element close to hand with all the tools required to install it.

3 Disconnect the element leads from the terminal block and thermostat.

4 Drain the water heater – follow the draining instructions in the 'Maintenance Section' of the manufacturer's 'Instruction Manual' that came with the water heater. If the 'Instruction Manual' is not available, copies can be obtained from the manufacturer's website.

5 Remove a screw-in type heating element by turning the element anti-clockwise with a 38 mm socket wrench. Remove the existing gasket.

6 Remove a flange type heating element by removing the four screws holding the element in place. Remove the existing gasket.

7 Remember that many elements are curved and may not pull straight out.

8 Once the new element is in place, reconnect the leads inside the terminal cover.

9 Restore the cold water supply and check for leaks around the gasket.

10 Carry out continuity and insulation tests before restoring the voltage supply and checking the function of the new element.

> i
>
> While the water supply is turned off it is a good idea to inspect the state of the sacrificial anode, if it is easily accessible, and advise your customer to replace it if it has corroded (anodes slowly dissolve to prevent the inside of the hot water cylinder from rusting). New anodes can be obtained from a plumbing supply warehouse.

Replacing an element without draining the water heater is possible with careful preparation. This should not be done by an inexperienced person.

The thermostat and thermal cut-out are normally closed contacts and should give circuit continuity after the replacement of the element and before restoring the supply. Fixed thermostats and cut-outs are mounted on the same plate and are a single replacement part.

Before finally restoring the supply voltage, operate the pressure relief valve and make sure that water flows through it and stops when released. If this does not work properly you will need to have it attended to by a licensed plumber. In most cases it is unlawful to tamper with the water supply.

Locating faults in water heaters

When locating a fault in a water heating system, you should first look out for the symptoms. These can be varied, but once you find the symptom, you will be able to ascertain the cause of the failure.

Symptom: The water fails to heat

Possible causes:

- Lack of supply to the heating elements – caused by a tripped circuit breaker.
  - Check the resistance of the element, continuity of the earth connection and the insulation resistance to earth. If no obvious fault is found then reset the circuit breaker and monitor the operation of the hot water cylinder through several reheating cycles. The thermostat can be made to operate by draining some hot water off the system.
- Faulty electric thermostat – may not be controlling at the correct temperature or is stuck in the closed position, causing the over temperature cut-out to open.
  - If the over temperature cut-out trips, check the thermostat to see if it has opened and replace it if it proves faulty. If the over temperature cut-out cannot be reset then replace it with an equivalent type.
- Faulty element.
  - Test the element with an ohm meter to ensure that it has the correct resistance for the current and voltage rating shown on the name plate.
- Faulty booster element or thermostat (if fitted).
  - Carry out the same checks on any additional elements and replace them if they are faulty.

Before taking the covers off to access the supply and element terminals, isolate the supply to the water heater at the circuit breaker or fuse. Do not conduct any live testing until you have conducted earth continuity and insulation resistance tests.

Symptom: Water goes cold too soon

Possible cause:

- Water heater may be undersized for water heating demands.
  - Check the water heater is the correct size to meet the demand for hot water. As a rule of thumb the hot water cylinder should only be required to supply about 75% of its capacity so a 150 litre cylinder should only be required to supply 114 litres of hot water in use. Get a plumber to check the hot water demand to make sure it is not too high for the cylinder.
- Faulty main or booster element or thermostat.
  - The element may have a break that still arcs across but does not allow the water to reach the required temperature. If a booster element is fitted make sure that it is working; if it is, then check the main element and make sure that is working. On a three phase water heater make sure that each of the elements is working.

Symptom: The water is hotter than it should be

Possible cause:

- Heating unit thermostat setting is too high.
  - Check main and booster heating element thermostat settings and reduce if set too high.

In no circumstances should the temperature be set lower than 60°C, otherwise it will allow the growth of Legionella bacteria. If the hot water outlets in bathrooms need to be at reduced temperatures then temperature reduction valves will need to be fitted by a plumber.

Symptom: There is water running from the overflow tray

Possible cause:

- Leak from loose heating element bolt(s) or bad gasket.
  - Tighten the bolts to see if this cures the leak; if not, then a new gasket will need to be fitted. Depending upon the time since the element was last replaced you might advise the owner that it could be replaced to save expense at a later date.
- Faulty temperature and pressure relief valve.
- T&P (temperature and pressure) valve leak due to excessive pressure, overheating or is stuck.
- Leak from overhead or nearby plumbing connection.
- Leaking hot water cylinder (corrosion likely).
  - The last four possibilities will need the attention of a qualified plumber to rectify them.

Symptom: Rust-coloured water delivered to hot water outlets

Possible cause:

- Glass lined cylinder is corroding.
- Failure of sacrificial anode.
  - Replace sacrificial anode rod.

Symptom: Green-coloured water delivered to hot water outlets

Possible cause:

- Earth system failure causes electrical fault currents to corrode copper water pipes. This fault used to be most noticeable in old copper hot water cylinders and is symptomatic of a potentially lethal combination of a failed earthing system and electric fault currents finding their way to earth through the hot water system.
  - Immediately isolate the supply from the installation and conduct a full earth continuity check and insulation resistance check on the entire installation. Carry out any repairs before restoring the supply.

SUMMING UP...

- Water heater elements are controlled by thermostats.
- Over temperature cut-outs prevent the temperature reaching dangerous levels.
- Elements must be tested for the correct resistance and insulation values.
- Symptoms of hot water cylinder failure:
  - water fails to heat
  - water goes cold too soon
  - water is hotter that it should be
  - water is running from the overflow tray
  - rust-coloured water is delivered to hot water outlets
  - green-coloured water is delivered to hot water outlets.
- Use proper procedure to change elements.
- Use resistance tests to isolate faulty components.

14.3 Troubleshooting electrical appliance equipment and circuits

Most of the techniques that have been discussed so far can be applied equally to any electrical appliance such as a cooking range or heater.

Appliance circuit diagrams

Single and three phase heaters have similar circuits to the hot water heater with variations to suit their use. Some circuits use control devices that work differently from the thermostat to control temperature.

Fault-finding appliances

Simmerstat

The simmerstat is usually used to control the hotplates on a cooking range (see Figure 14.7). Simmerstats do not control the temperature. A time setting controls the average current supplied to the heating element by adjusting the period of time the element is switched on. When the control knob is turned it increases the pressure on a control spring that in turn restricts the movement of a bimetal strip.

Initially current passes through a normally closed contact attached to the bimetal strip heating the hotplate and the internal simmerstat element. When the bimetal strip is heated to a point where it bends and overcomes the pressure of the spring, the contact opens stopping current to the internal

FIGURE 14.7 Simmerstat

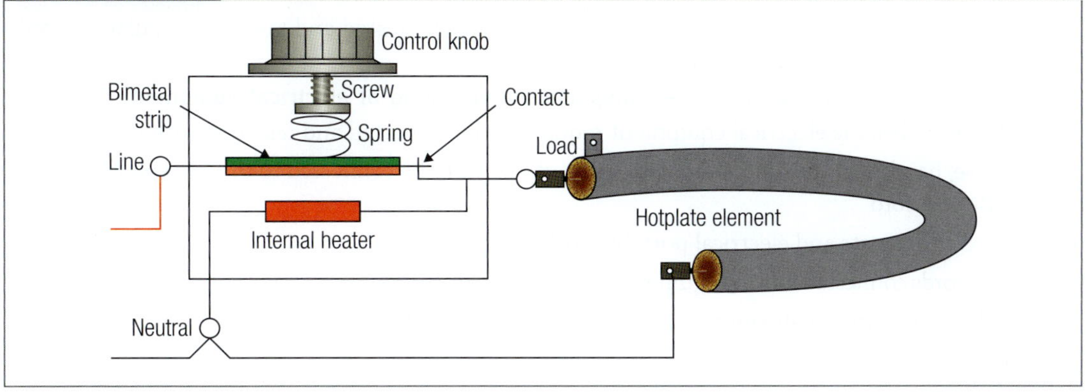

heater as well as the hotplate. As the bimetal strip cools, it straightens causing the contact to close again, repeating the cycle. The tighter the pressure of the spring the hotter the hotplate becomes.

Three heat switch

The three heat switch is used to switch two equal sized elements in the following combinations:

- no connection (off)
- two elements in series (low)
- one single element (medium)
- two elements in parallel (high).

In Figure 14.8 the circuit comprises two 1 kW elements 52.9 ohms each. On the low setting they will total 105.8 ohms drawing 2.17 amps dissipating 500 watts. On the medium setting 1 kW is dissipated and on the high setting 2 kW are dissipated, drawing 8.7 amps, making it suitable to plug in to a 10 amp socket outlet without overloading the outlet.

FIGURE 14.8 Three heat switch

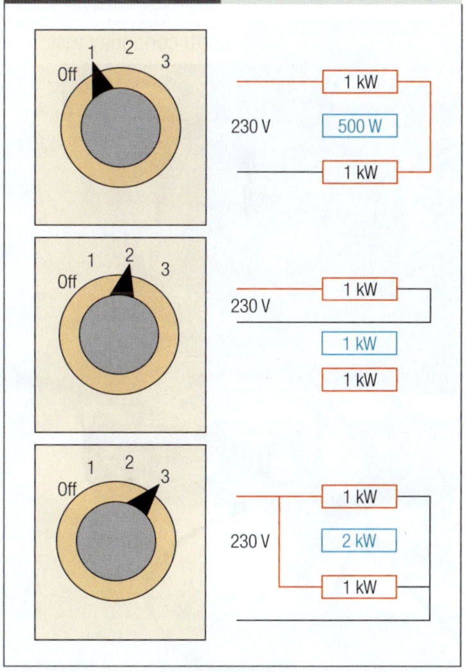

Causes of RCD operation

The RCD (residual current device) protection is designed to trip when a human body makes contact with an active conductor and at the same time makes contact with earth. The current setting is very small and the trip time short. This is to prevent death from electric shock. The RCD does not provide protection against over-current except in the case of a combination device that includes a circuit breaker and an RCD.

Other causes of RCD tripping are:

- the RCD is faulty
- the circuit is faulty
- something attached to the circuit is faulty.

To locate the cause of RCD tripping you should:

1. Unplug all appliances from the socket outlets protected by the RCD and reset it.
2. If the RCD does not trip then replace the appliances one at a time until the RCD trips. This will confirm which appliance needs further attention.
3. If the RCD continues to trip then disconnect the load circuits from it completely. A faulty RCD will continue to trip even with nothing connected.
4. If the RCD appears all right, carry out an insulation test on the circuits that you have removed (including the neutral conductors) – this should locate the faulty conductor.
5. Replace each circuit one at a time, making sure the RCD does not trip at each stage. Remember a neutral to earth fault can cause an RCD to trip but will not trip the over-current protection.

Do not disconnect circuit conductors unless they have been isolated from the supply.

Never connect the supply voltage to any appliance that cannot pass a visual inspection.

Equipment with mineral insulated metal sheathed heating elements may show a lower resistance; 0.01 MΩ is permissible.

Testing appliances

All Australian jurisdictions require that portable appliances be tested and tagged at regular intervals, depending upon the frequency of use.

AS/NZS 3760:2010 In-service safety inspection and testing of electrical equipment Section 1, clause 1.1 describes the electrical equipment required to be inspected and tested. It includes:

- portable, hand-held and stationary appliances, designed to be connected to the low voltage supplied by a flexible cord
- cord extension sets and electrical portable outlet devices
- flexible cords connected to fixed equipment in hostile environments
- portable isolation transformers, such as power adaptors and plug packs
- RCDs
- commercial and industrial battery chargers
- portable and transportable 415 V heavy duty tools such as high-pressure washers and concrete grinders.

FIGURE 14.9 Earth continuity test

Source: Ed Hawkins

Visual and physical inspection

Ninety per cent of all faults can be detected with a visual and physical inspection. Equipment failing a visual test is unsafe for use and no further testing is required.

Earth continuity test

The earth continuity test is to ensure the earth wire is connected from the plug to the equipment or outlet socket with sufficiently low resistance to provide a good fault current path, ensuring circuit protection devices operate (see Figure 14.9).

Using an ohm meter set to the 3 ohm range, measure the resistance of the earth wire. The resistance must not exceed 1 ohm. Painted or coated surfaces may prevent earth continuity to the casing or metal. Don't assume earthing to one point guarantees earthing at all points. The appliance shown below tests out at 0.6 ohm, well below 1 ohm.

Insulation resistance test

Insulation must have a high enough resistance to prevent voltage from the energised load conductors contacting any exposed metal and creating a touch voltage. The viability of the insulation is tested by applying a higher but safer d.c voltage using the following procedure (see also Figure 14.10):

Remember to turn off your tester when you have finished.

FIGURE 14.10 Testing insulation of mineral insulated element

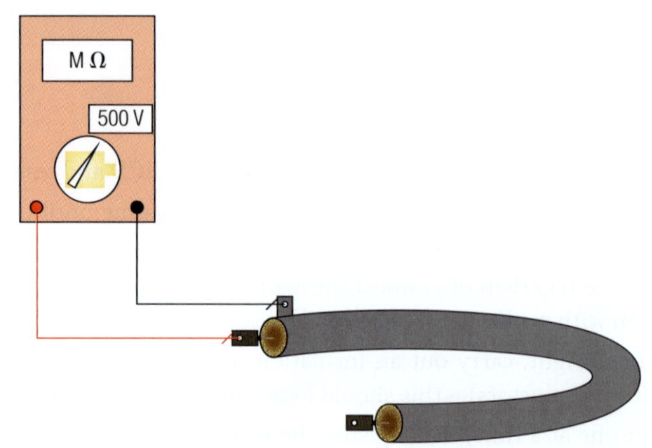

1 Connect one lead to both active and neutral conductors at the same time and connect the other lead to the earth.
2 Select the 500 volt range and switch on the tester.
3 Record the resistance measured in megohms. The insulation resistance must be higher than 1 MΩ.

Appliance controls

While it can be irritating if the switch controlling an appliance does not switch on, it can be very dangerous if it will not switch off. Before connecting any appliance to the low voltage supply, you must make sure the control devices work properly. The circuit through the load should be open when the control device is supposed to be 'Off'. With some switches this is straightforward and the switch can be operated and the resistance of the load seen to be present when in the 'On' position and open circuit when in the 'Off' position.

Some appliances such as a toaster have a switch that operates when a slice of toast is placed in it and times out, releasing the contacts when a selected time has elapsed. The only test for this switch is to test for an open circuit before operating it, as shown in Figure 14.11. If it proves satisfactory then press the operating plate down with a screwdriver and check that it latches and the element resistance can be read on the ohm meter. A 900 W 230 V toaster should show a reading of 59 ohms. If all other tests on the appliance prove it is safe to do so you will need to plug it in and see if the switch times out correctly.

FIGURE 14.11 Checking for open circuit with switch off

Source: Ed Hawkins

Locating faults in electrical appliances

Electrical appliances sold in Australia must comply with strict standards, but as they age materials and safety features suffer damage from heavy use or careless handling.

Completely failed appliance

Often you will be faced with a fault of an unknown nature or fault symptoms that have no obvious cause. The only answer is to work very carefully through the problem, making sure that you do not miss important signs or tests.

Very useful information can be gained from the user of the failed equipment.

The following information can be easily obtained:

- Has the appliance completely failed?
- Is the plug in and the power outlet switched on?
- Are the switches, if any, on the appliance switched on?
- Has the circuit breaker or RCD in the switchboard 'tripped'?
- Are there any other appliances or lights that are not working?

When you arrive on the site you will need to verify the user's answers for yourself. When the whole appliance is inoperative and you have checked the above points, you will need to start testing the appliance itself.

Conduct the following tests on the appliance:

1 Test the supply terminals and establish the presence of a voltage.
2 Start at the motor, element or other device and see that the return or neutral conductor is connected.
3 Test the active conductor and see that voltage is present.
4 If no voltage is present then work back through the control contacts and switches until you come to a point with voltage present.
5 Remove the supply from the appliance and check the control devices for open circuit faults.

Follow your live testing procedures and prevent access by untrained people to your work area while testing.

If voltage is present at the motor or element then the fault will be in that part of the circuit. Again you will need to isolate the supply and carry out resistance checks to determine the cause of the failure.

Intermittent faults

If the fault isn't there when you arrive, it cannot be traced. This makes intermittent faults a constant source of frustration to customer and service technician alike. The following list will help locate intermittent faults:

- Does the described fault appear to be limited to one or two inexpensive components? If so then it will be more economical to change them out immediately and eliminate them.
- Does logic, experience or the service manual suggest a particular cause? Eliminate these possibilities by cleaning contacts, tightening terminals or similar checks.
- Does experience or knowledge indicate this particular product has a weakness such as a poorly fitted door switch? Examine each of the known or suspected causes of failure and eliminate them as possibilities.

EXAMPLE 14.1

Single phase induction motors may start sometimes and not others; a likely suspect would be the centrifugal switch that may need servicing.

A likely cause of intermittent tripping of the RCD could be dampness forming in the windings of a motor during weekend or holiday shutdowns. This can be traced by using the insulation tester; if insulation is damaged, a low reading will often be detected even though the appliance appears to be working.

Repairing faulty appliances

One of the most difficult decisions to make is between following through with a repair or replacing the faulty appliance entirely. For example, your customer is not going to be very pleased to receive an invoice for $60 to repair an appliance that could be bought new for $70. On the other hand, a small business may lose many hundreds of dollars while a new piece of equipment is being purchased and installed, so a timely repair would be more desirable even if a new item was eventually obtained on your recommendation.

Repairs must be carried out in good time with materials at least the same quality as those that have failed. A customer may see the point in purchasing a better quality replacement part, but if the appliance is more than halfway through its useful life then it would be an unnecessary expense.

If you cannot obtain an exact replacement then you will need to act on the recommendation of the equipment manufacturer before using an alternative. 'Bodging in' a motor or switch might sound like a professional challenge, but usually it adds expense and delay where a complete new appliance would be a better alternative.

Once an appliance has been repaired it will need to be tested for earth continuity, unless it has double insulation, and a full insulation test carried out. With the resistance checks completed the supply can be restored. All exposed metal must be checked with a voltmeter against a known good earth to ensure there are no touch voltages present. A full functional check should now be conducted to ensure the appliance is working correctly and safely.

SUMMING UP...

- Appliances are consuming devices such as motors and heaters.
- Failure of the supply voltage should be checked first.
- Repairs may be more expensive than replacing the appliance.
- Repairs should be timely and thorough.
- Only use approved replacement parts.
- Carry out earth and insulation checks before restoring the supply.

14.4 Troubleshooting lighting circuits

Lighting circuits are one of the most commonly encountered systems for the electrical practitioner. They can range from a simple single light controlled by one switch through to complex electronic control systems. In many cases, the failure of the lighting circuit can have serious implications for safety and security.

Common lighting circuits

Figure 14.12 shows a set of lights with switch S1 controlling Lamp L1 and switch S2 controlling Lamps L2 and L3.

FIGURE 14.12 Loop at the light circuit

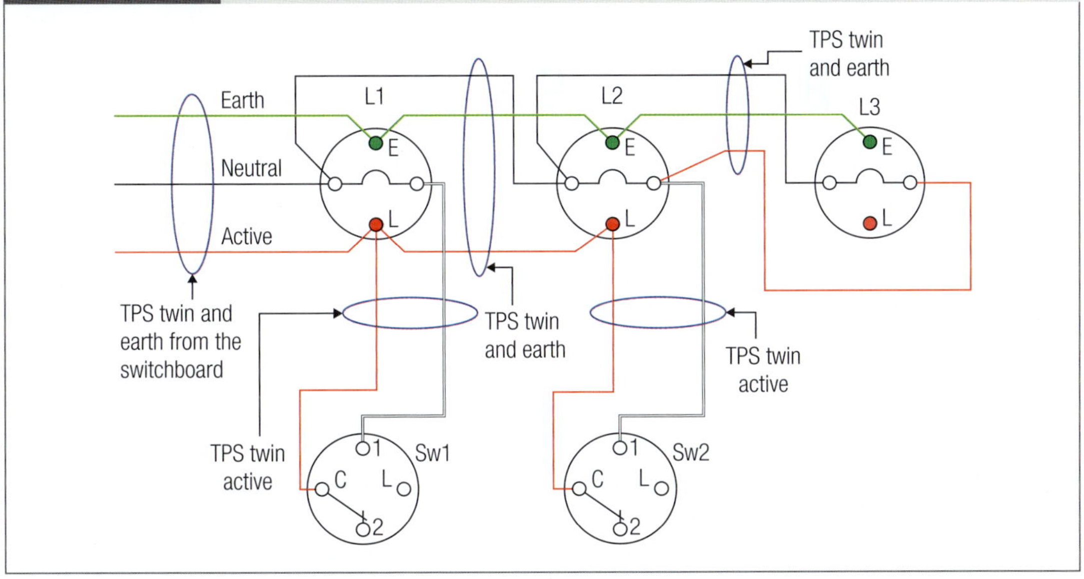

This method of using TPS (thermoplastic sheathed double insulated cable) provides a live supply feed to each light fitting and the twin active is run to the switch. All connections are made in the light fittings themselves, reducing the cost of junction boxes and installation time. Lamp L3 is on the end of the circuit and has the active conductor supplied from the switch wire at L2.

Faults in this system are not always easy to trace, especially if there is no access to the roof or floor space to allow you to trace the wiring. An open circuit fault in the switch circuit of one lamp can be quickly identified by checking that the other lights are working. This should limit your search to the wiring directly associated with the particular light.

If the circuit is tripping the over-current protection, and the fault cannot be isolated to the operation of one light, then you will need to work your way around the circuit, disconnecting each light fitting and testing each supply feed until you locate the fault. The half split method of fault location works well in this situation.

To identify the cause of RCD tripping remember to carry out earth resistance checks on the neutral conductor as well as the active. Careful visual inspection can be a considerable help, and black marks on the ceiling or wall around a light fitting can be a helpful indicator of trouble. If the weather has been wet recently then check the exterior light fittings for water ingress.

The wiring diagram in Figure 14.13 shows the same three lamps controlled in the same way from the same two switches. This time the circuit is connected using the loop at the switch system. Once again this technique saves on additional work and materials but has the added advantage that only one type of cable is needed. The supply feed is run to each switch in the circuit and one TPS twin and earth cable is run to the lamp.

Fault-finding lighting circuits

Fault-finding in this system is much easier as each light has no live connection when the switch is off. Any suspect lamp can be disconnected at the switch and the circuit tested. This saves a lot of time climbing a ladder to each light on the circuit.

FIGURE 14.13 Loop at the switch circuit

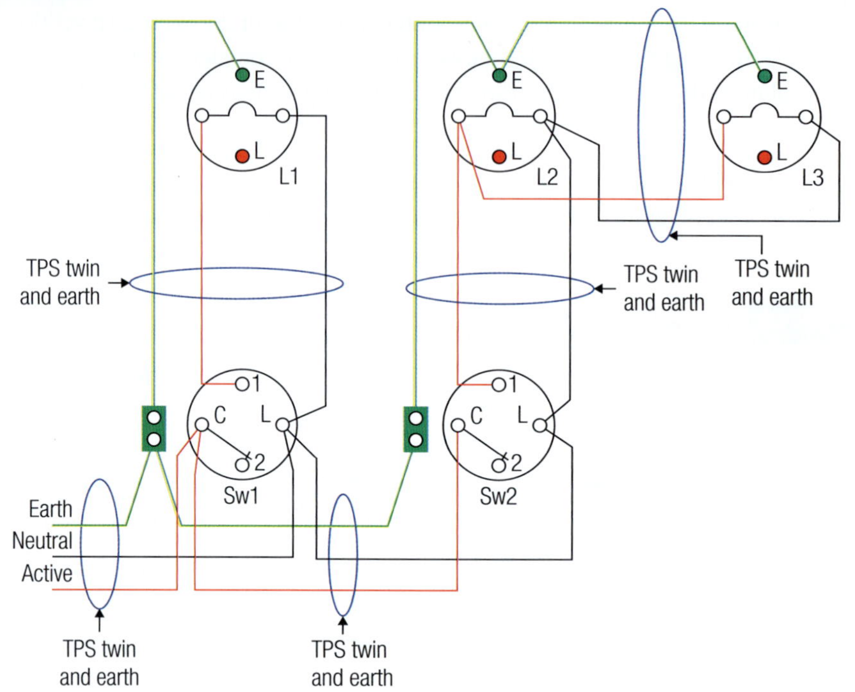

Lighting circuit faults

When fault-finding lighting circuits you should always start with a known good lamp or tube, and in the case of a fluorescent lamp a new starter.

The following faults and symptoms are commonly encountered in lighting circuits:

- The lamp fails to work even with a new lamp or tube. This means there is an open circuit fault in the fitting or the wiring associated with it. If an examination of the fitting and switch prove them satisfactory then a series of continuity tests from the switch to the light fitting and from the switchboard to the light fitting will help locate the fault.
- Every time a lamp is replaced it blows very quickly. The first thing to check is to make sure the power rating for the lamp is the correct size for the fitting. An oversize lamp will not be able to dissipate heat quickly enough and will have a shortened working life.
- The supply voltage ranges between 230–240 volts a.c. The installation might be close to a transformer on the street which means the nominal voltage could possibly be up around 240 volts or slightly higher. If this is the case, try a 250 volt rated lamp and see if that fixes the problem.
- If it is an ELV (extra low voltage) light fitting it could have a faulty transformer that is supplying a higher voltage than the lamp is designed for.
- Another cause of blown lamps in ELV fittings can be poor ventilation. Check around the fitting and make sure there is plenty of room for air to circulate.

Faults in fluorescent lighting circuits

Fluorescent lighting provides a large part of the lighting in modern commercial buildings due to its high efficiency and low maintenance costs.

The circuits providing the supply to the fittings are controlled in the same way as any other light so the techniques described previously can be applied to locating wiring faults (see Figure 14.14). The fittings have components that are subject to failure with age and require some time to locate and repair faults.

The tubes and starters have an economic life of about 3000 hours and should be changed on a routine basis to keep light levels on working surfaces and walkways at a safe value. Tubes and starters should be changed together since a faulty starter can damage a new tube by keeping the heaters on too long during the starting cycle.

Failed capacitors can cause short circuits that trip the circuit protection. In large installations each row of light fittings has its own fuse and one large capacitor. This means that a faulty capacitor can be quickly located and replaced.

The ballast should be checked with an insulation tester to ensure there are no earth faults between the windings and the iron core. Modern fittings now have electronic ballasts and it is not advisable to use an insulation tester on them.

FIGURE 14.14 Fluorescent light circuit

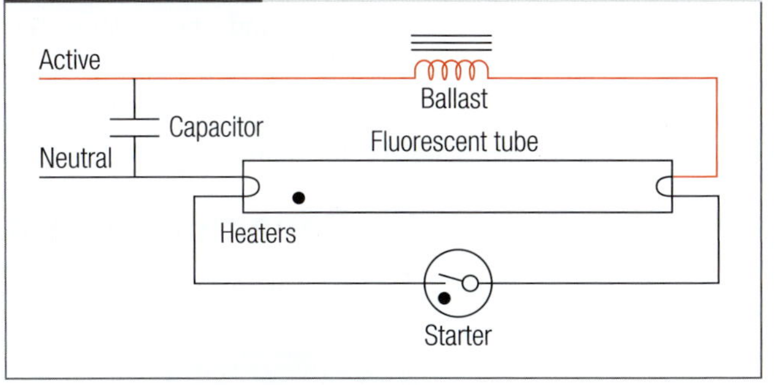

Some old fittings contain oil filled capacitors that may be leaking. These should be handled with disposable gloves and placed in a suitable container for disposal.

Lighting control devices

It might not be possible to switch the light on in the normal way, as time clocks, daylight switches and PIR (photo infrared) devices are used to prevent unnecessary power use when it is not needed. Some of these devices develop their own faults and may need to be bypassed before the light can be checked for correct operation.

SUMMING UP...

- TPS systems can be loop at the light or loop at the switch layouts.
- Loop at the switch systems don't require light fittings to be disconnected to locate faults.
- Over voltage can cause shorter lamp life.
- Fluorescent lights require scheduled tube replacement.
- Capacitors are the most likely cause of over-current tripping in fluorescent light circuits.
- Ballasts are the most likely cause of earth leakage faults in fluorescent fittings.
- Check transformers on ELV lighting systems.
- Time clocks, PIRs and other devices need to be checked if lights fail.
- Use caution when using the insulation tester around electronic equipment.

14.5 Troubleshooting single phase motor and control circuits

Single phase motors are widely used in domestic and commercial installations. The failure of a motor, for example, in a refrigeration system, can cause many hundreds of dollars worth of spoilt merchandise.

Diagrams of single phase motors

Electrical appliances use a range of single phase motors. They each have different ways of producing torque which can give benefits and disadvantages. Single phase motors used in electrical appliances can be one of the following:

- split phase motor
- capacitor start motor
- capacitor start capacitor run motor
- permanently split phase motor
- shaded pole motor
- universal series motor.

1 Split phase motor

Single phase induction motors (see Figure 14.15) rely on an extra winding called the start or auxiliary winding to provide starting torque. As the start winding is made from much finer conductor than the main winding a centrifugal switch is used to disconnect the starting winding at about 75% of full-load speed. A thermal cut-out switch may be included with the motor to protect the motor from high internal temperatures.

2 Capacitor start motor

To give improved starting torque a capacitor is added in series with the start winding (see Figure 14.16).

FIGURE 14.15 Single phase motor split phase

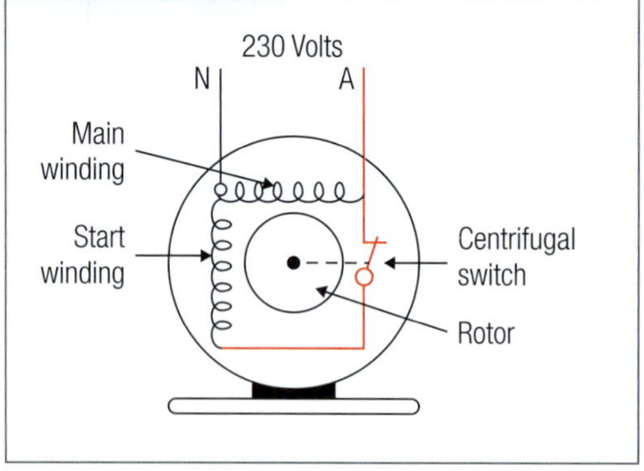

FIGURE 14.16 Single phase motor capacitor start

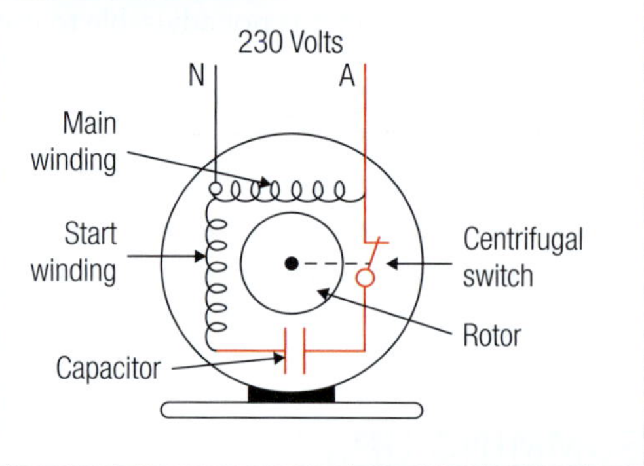

3 Capacitor start capacitor run motor

To give smoother running under varying loads the run capacitor helps with the starting torque, then stays connected to provide smoother running with varying loads (see Figure 14.17).

4 Permanently split phase

This motor (see Figure 14.18) has the auxiliary winding and the main winding made from the same size conductor. It does not have very high starting torque but is very easy to reverse.

FIGURE 14.17 Single phase motor capacitor start capacitor run

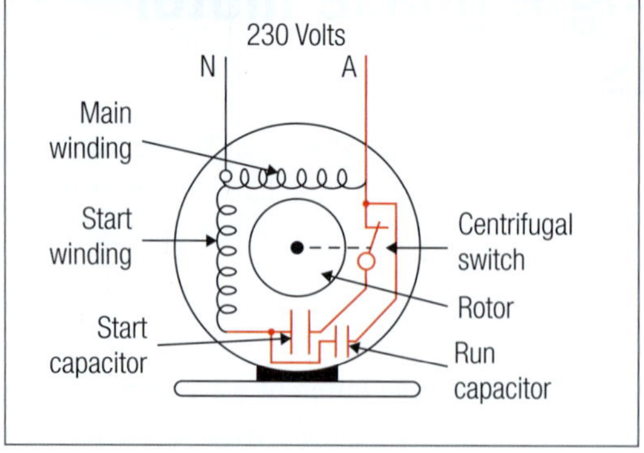

FIGURE 14.18 Single phase motor permanently split

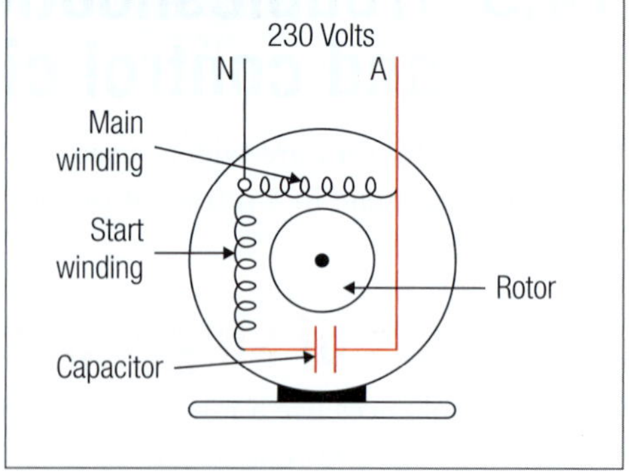

5 Shaded pole motor

This motor uses Lenz's law to cause a 'sliding' field along the pole faces to provide torque (see Figure 14.19). It does not have very high starting torque but is cheap and reliable. Its main enemies are dust and dirt that damage the bearings and stall the driven load. Repairs to this type of motor are not economically justified and it should be replaced if the windings are damaged or the bearings worn.

6 Universal series motor

The universal series motor (see Figure 14.20) can run at speeds higher than the synchronous speed of the mains supply and has very good starting torque. It is used mostly for appliances where speed is important such as vacuum cleaners and those requiring high torque at low speed such as power drills and other tools.

Faults in single phase motors

Faults in single phase motors usually involve:

- overload damage, especially to the start windings
- open circuit faults
- failure of the starting circuit to operate
- failure of the capacitors where fitted
- short circuit in the windings
- open circuit rotor
- insulation failure to the frame
- in the case of the universal motor, brush failure
- the constant for all motors – bearing wear or failure.

The tests that have been previously discussed such as insulation and continuity tests must be done on motors. Additionally, the shaft needs to be checked to ensure side movement is not present and that endplay is minimal.

Testing a low voltage capacitor

Capacitors are completely different sorts of devices from resistors and coils. An insulation tester will give an indication of their state if you follow these steps (see also Figure 14.21):

1 With the appliance isolated remove and discharge the capacitor using a 20 K 5 watt resistor.
2 Visually inspect the capacitor to see if there is any deterioration evidenced by rust-coloured streaks and cracked insulation.
3 Set your insulation tester to the 250 volt range to suit 400 volt rated capacitors and connect the test leads to the capacitor terminals and read the displayed resistance value:
   - reading changes from zero to infinity – good capacitor
   - reading changes to zero – short circuit capacitor
   - reading does not change from infinity – open circuit capacitor.
4 Discharge the capacitor again using the 20 K 5 watt resistor.
5 Charge the capacitor to about half scale before removing the test leads.
6 After about 30 seconds reconnect the leads and read the value displayed:
   - reading starts at about half scale and charges – good capacitor
   - reading starts at zero and charges – bad capacitor.

It is possible to test the rotor of an induction motor without removing it from the motor (see Figure 14.22). Connect an extra low voltage a.c. supply to the input terminals and monitor the current with an ammeter. Rotate the shaft of the motor slowly

FIGURE 14.19 Shaded pole motor

Source: Ed Hawkins

FIGURE 14.20 Universal motor

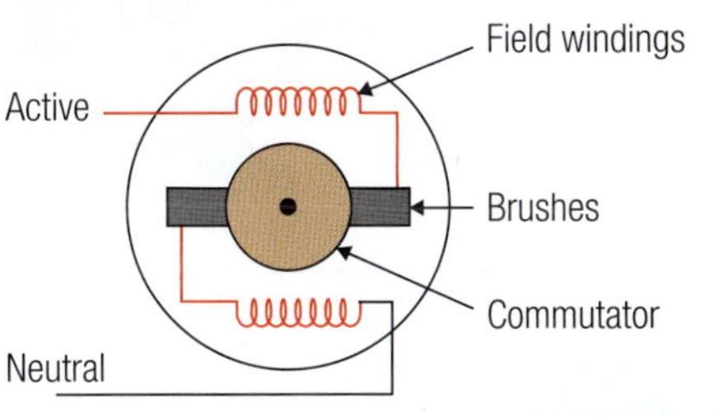

Fault-finding motors

Avoid making contact with capacitor terminals as the voltage stored may be able to sustain a lethal current.

FIGURE 14.21 Testing a capacitor

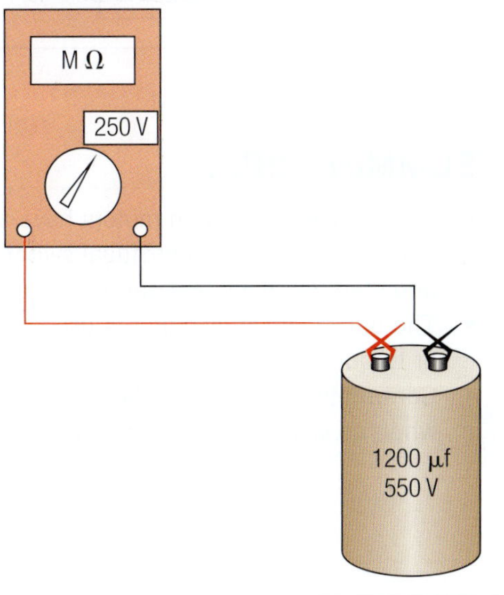

FIGURE 14.22 Testing a rotor

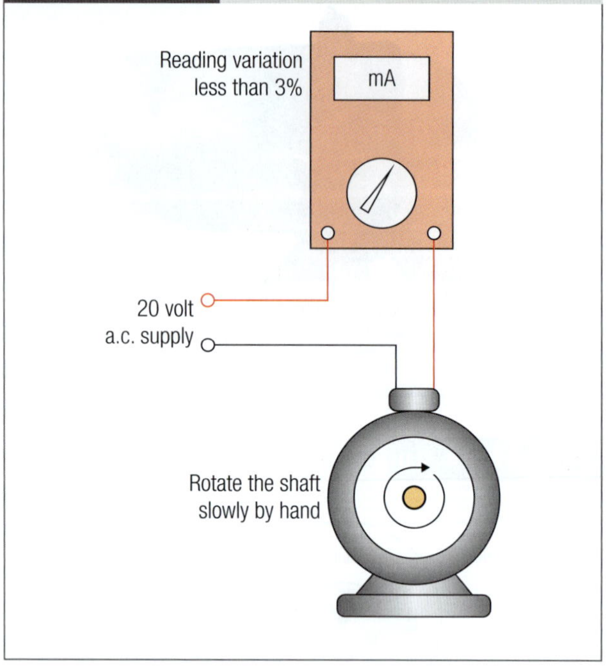

FIGURE 14.23 Testing the universal motor

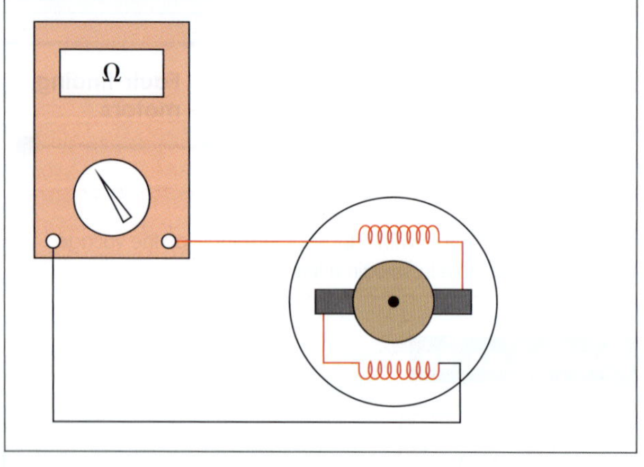

and see if the current changes. If the variations in current are more than 3% then the rotor should be removed and checked for cracks in the end rings and bars. It is not economically viable to repair rotor bars on small motors so a faulty rotor needs to be replaced with a new one.

Universal motors often fail due to repeated mechanical overload. The least troublesome effect is damage to the commutator which can be cleaned and returned to service. The resistance of the field coils can be measured separately by removing the brushes and comparing the resistance of each winding; if they are the same then it is very unlikely they are faulty. The armature can be given a preliminary test by connecting an ohm meter to the supply leads and measuring the resistance of the armature while rotating the shaft slowly. If the commutator is clean and the brushes are functioning correctly the resistance should not vary greatly in one revolution of the shaft. If the resistance varies considerably as the shaft is rotated then the motor will need to be dismantled and further investigation and testing carried out. See also Figure 14.23.

Thermal overloads

Thermal overloads are designed to protect the motor from damage caused by long-term over-current that would cause the windings to overheat and fail.

Some causes of thermal overload tripping would be:

- internal mechanical faults such as centrifugal switches, bearings, fan or a bent shaft
- external faults such as mechanical overload caused by jammed machinery, clogged fan ducts or misaligned conveyor belts.

Frequent tripping of the overload when running at rated speed and torque would indicate excessive current flow due to short circuits in the windings. It could also indicate a malfunctioning overload device.

It is not acceptable to keep resetting the thermal overload unless you have clearly established and removed the cause.

SUMMING UP...

- Single phase induction motors have additional equipment such as centrifugal switches, capacitors and start windings that can fail.
- The universal series motor is used where high speed or torque is required.
- Continuity and insulation tests need to be conducted when no visible fault is present.
- Preliminary tests on armatures and rotors can be carried out without dismantling the motor.
- Thermal overloads protect the motor against prolonged high currents.
- Thermal overloads should not be reset until a fault is identified.

14.6 Troubleshooting three phase induction motor and control circuits

The two main connections for a three phase motor are star or delta.

Care must be taken to check the delta rating of the motor as some 400 volt motors are only suitable for star connection at that voltage and will only run on 230 volts when connected in delta. Figure 14.24 shows a six lead terminal block provided with links to connect the motor as a star or delta circuit.

If the motor is connected for delta, the links will need to be removed in order to test each winding. With the motor connected in star the resistance of each winding can be tested but the links will need to be removed to identify any coil that has an insulation resistance fault. The motor shown in Figure 14.25 has three windings with the same resistance so it is very unlikely that the windings are damaged. Unequal measurements would indicate damage to one or the other of the windings.

Three phase motors do not have the auxiliary equipment that single phase motors need. The torque is produced by the rotating three phase stator field interacting with the induced rotor field. Three phase electric motors are highly effective and reliable machines, but they do fail.

Many of the general faults associated with single phase motors apply to three phase motors, such as bearing wear, mechanical load problems and winding failure. One symptom that is particular to a three phase motor is that the motor appears to have power but does not pull away.

This indicates a loss of one of the three active feeds to the motor and is termed 'single phasing' to describe the fact that only one phase is applied to the motor.

A three phase motor that is single phasing can continue to run unevenly and might eventually overheat and suffer winding damage. Three phase control circuits are designed to fail in a safe condition.

The circuit shown in Figure 14.25 is the standard magnetic starter with thermal overloads designed to stop the motor if the control circuit opens for any reason. The motor cannot be started until an operator manually restarts it. This starter does not provide phase failure or over temperature protection.

Faults in three phase motors

FIGURE 14.24 Six lead terminal block

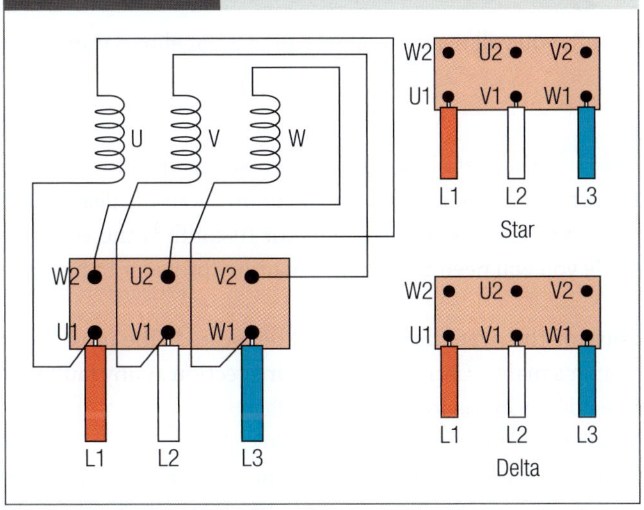

FIGURE 14.25 Direct online starter circuit

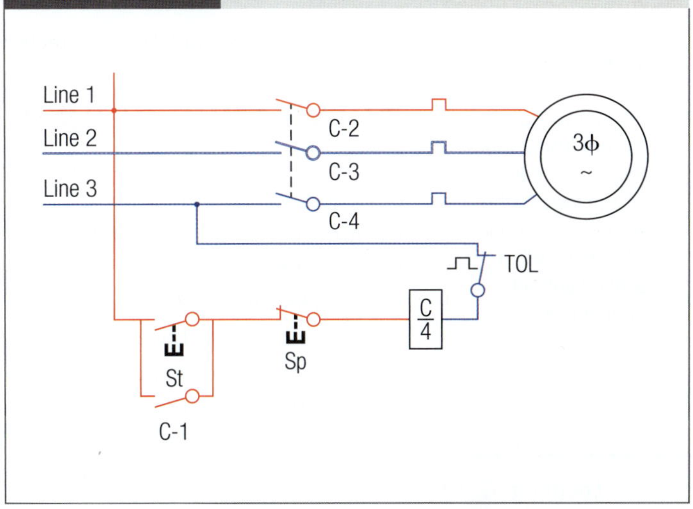

FIGURE 14.26 Thermal imaging camera

Source: shutterstock.com/Ivan Smuk

Modern starters provide over-current protection, over temperature protection, phase failure protection as well as 'ramp up' starting and 'ramp down' stopping. Indicator LEDs on the front indicate the nature of the failure if the starter trips out. Connections are provided for thermal sensors in the motor windings to be connected. The starter can be programmed to suit the characteristics of the motor and the nature of the load.

Taking electrical and temperature measurements can help identify and reduce 44% of premature motor failure. Modern thermal imaging cameras allow temperature monitoring and recording (see Figure 14.26). Any abnormal rise in temperature of windings or bearings should prompt attention before the motor fails in service.

SUMMING UP...

- An overload is the application of too much load to a motor.
- The larger the load, the more current is required to drive the load.
- High current causes additional motor heating and insulation breakdown.
- When one of the three lines no longer delivers voltage to the motor, the motor will single phase.
- Although a three-phase motor will probably never start (or restart) when single phased, most running motors will continue to run if single phasing occurs.
- Taking current measurements and comparing them to the motor's nameplate-rated current determines if the motor is overloaded.
- Taking voltage measurements determines loss of phase voltage or improper supply voltages.
- Taking resistance measurements with an insulation tester determines the condition of motor insulation.
- Taking regular temperature readings prevents unscheduled downtime.

14.7 Troubleshooting electrical installations

When an entire installation is affected by a fault, you will need to follow a process of elimination.

> Do NOT work with energised wires or circuits unless carrying out tests for voltage, in which case follow your live testing procedure. All voltage tests on the mains and switchboard must be done with a Category III 600 volt meter or higher.

Blown service fuse

The most common fault that affects all or part of the installation is a blown service fuse. First carry out a voltage test using a known good earth as reference to confirm the incoming active is dead. If the point of service is accessible remove the cover and test the active conductor to confirm the supply is missing. If the point of service is not accessible you will need to contact the utility provider to have the fuse checked and replaced.

Before replacing the service fuse the cause of the over-current should be located. The most likely place is in the switchboard itself or the consumer's mains. Carry out a visual inspection as any fault that can blow the service fuse might leave visual evidence in the form of arc damage.

If no visible damage is found then the next step is to remove the MEN link.

Locating faults in electrical installations

Switch off all the circuit breakers and carry out an insulation test on the consumer's mains and the busbars in the switchboard. If no earth fault is found then remove the lamps from the light fittings, unplug any portable appliances and check the active conductor leading away from each circuit breaker to ensure no earth faults exist. If no short circuits are found and no earth faults exist then carry out earth continuity checks on the main earth and sub-circuit earths (see Example 14.2 on page 571 for acceptable resistance values), replace the MEN link, replace the lamps and plug the appliances back in. When you are sure the installation is safe, restore the supply. Monitor the current through the consumer's mains with a clamp-tester while you switch on each circuit in turn. If the consumer's mains current rises to the rating of the service fuse do not turn on any more circuits. In this case the maximum demand has been exceeded and new mains or a current limiter will need to be installed.

Do not under any circumstances touch the neutral conductors if it is possible the main neutral is disconnected as a dangerous touch voltage may be present.

Missing neutral

A missing neutral from the supply is a possible fault that affects the whole installation. In this case the return current will be through the main earth and through the ground back to the transformer. This is a high impedance circuit that will cause considerable voltage drop.

The hot water circuit may appear to function normally but the cooking range will seem 'slow' and the light appear dim.

Use a clamp meter to check the current in the consumer's mains active conductor, then transfer the clamp meter to the consumer's mains neutral – the current should be the same.

If there is no neutral current then the neutral is open circuit before it comes to the switchboard. Switch off the main switch so this dangerous situation is removed. Check for damage to the consumer's mains, such as underground cables disturbed or cables in the roof space cut or pinched, before contacting the utility supplier.

Shocks from metal surfaces

Reports of shocks from metal parts of the installation indicate a failure of the earthing system. Since this is a dangerous situation, initial tests should not be conducted with supply connected until the integrity of the earthing system has been confirmed by the following procedure:

1 Isolate and lock out the supply to the whole installation.
2 Remove the neutral link and test the continuity of the main earth conductor to the earth electrode; this should be 0.5 ohms or less.
3 Remove the earth conductor from the hot water cylinder (to prevent back-feeds through the water pipes) and test the continuity of the earth bonding to water pipes and other metal parts of the installation from the main earth. The resistance should be 0.5 ohms or less.
4 Reconnect the hot water earth.
5 Carry out an earth continuity test for the earth conductors associated with each circuit. The resistance reading must comply with Table 8.2 of **AS/NZS 3000:2007 Wiring Rules**.

Table 8.2 of AS/NZS 3000:2007 Wiring Rules lists the allowable resistance for individual sub-circuits taking into account the conductor size and type of protective device for testing with *power unavailable*.

EXAMPLE 14.2

Determine the maximum fault loop resistance of a power sub-circuit wired in 2.5 mm^2 cable with a 2.5 mm^2 earth conductor protected by a 16 amp type C circuit breaker.

Open your AS/NZS 3000:2007 and turn to Table 8.2. Locate the row that shows the required active (2.5 mm^2) and earth (2.5 mm^2) conductor sizes, and circuit breaker rating (16 amps). Next select the column R_e under the circuit breaker type C.

Where the row and column intersect, read the maximum allowable earth resistance for the sub-circuit (0.16 ohms).

Re – the resistance, in ohms, of the protective earthing conductor.
Rph – the resistance, in ohms, of the active phase (ph) conductor.
Rphe – the combined series resistance, in ohms, of both the active phase (ph) conductor and the protective earthing (e) conductor.

Make sure the power circuit is isolated and locked off before removing any socket outlet from the wall for inspection.

With all appliances and all circuit breakers switched on, conduct an insulation test between earth and the active conductors. The insulation resistance should be at least 1 megohm. If the insulation test fails, then progressively switch off circuit breakers until the fault clears. Follow the faulty circuit until the faulty wiring or appliance is identified.

Poorly maintained electrical equipment will result in power wasted in worn bearings, excess power loss in motors that have damaged windings, as well as leakage of oil and dangerous gases from aged and damaged seals.

If no insulation failure is found then connect the MEN link back to the neutral bar and check that it has a maximum resistance to the main earth conductor of less than 0.5 ohms.

Restore the supply and switch each circuit breaker back on one at a time, checking between a known good earth and metal in the installation to make sure that no touch voltages exist.

It is important to remember to check equipment that operates using timer controls or thermostats that may not be connected in the circuit all the time.

SUMMING UP...

- Installation faults can include:
  - total loss of supply
  - loss of neutral conductor
  - faults in the earth system.
- Insulation and continuity checks must be carried out to establish the cause of the fault.
- The service fuse must be removed before testing the consumer's mains.
- Earth conductors must be tested for low resistance that complies with Section 8 of AS/NZS 3000:2007 Wiring Rules.
- Full live polarity checks must be made on the installation after restoring the supply.

REVIEW EXERCISES

Answer these questions and check your answers at the back of the book.

Review questions 14.1

1 List the four essential parts of an electric circuit.

2 State the purpose of control devices in an electric circuit.

3 List three main causes of electric circuit failure.

Review questions 14.2

1 Calculate the resistance of a 230 volt 1500 watt water heater element.

2 Calculate the maximum 230 volt resistance element that could be connected to a circuit protected by a 16 amp circuit breaker.

Review questions 14.3

1 If a heater controlled by a three heat switch was fitted with two 1500 watt elements would it be suitable to plug into a 10 amp socket outlet?

2 List four reasons that an RCD might be caused to trip.

3 Is it necessary to carry out an earth continuity test on a double insulated appliance?

Review questions 14.4

1 In a lighting circuit using the 'loop at the light' system, where would you need to look for a short circuit fault between neutral and active conductors?

2 In a lighting circuit using the 'loop at the light' system with all the light switches turned off an earth fault persists. Describe how you would proceed to locate the fault.

3 If a circuit is tripping the RCD, would there be any purpose in carrying out an insulation test on the neutral conductor?

4 Describe any visual indication that might help locate a short circuit fault in a light circuit.

Review questions 14.5

1 State the motor that is most used where low speed torque is a constant requirement.

2 State the name of the single phase induction motor that is the most easily reversed.

3 State why the shaded pole motor is only used on easy to start loads such as fans.

Review questions 14.6

1 While conducting continuity tests on the coils of a three phase motor you observe that two windings have 4 ohms each but the third is 500 ohms. Would this motor be suitable to return to service?

2 While conducting an ELV rotor test on a three phase motor you observe a difference of 10% change among several of the current readings. Would this motor be suitable to return to service?

Review questions 14.7

1 A 2.5 mm^2 sub-circuit protected by a 16 amp type C circuit breaker has an earth conductor resistance reading of 0.5 ohms. Is it safe to return to service?

2 A circuit supplying a cooking range fitted with MIMS elements has an insulation resistance reading of 0.8 MΩ. Is it safe to return to service?

ONLINE RESOURCES

Hot water temperature: Origin energy
http://www.originenergy.com.au/files/SMEfs_HotWater.pdf

Appliance advice: Siemens
http://www.siemens-home.com/ae/customer-services/fault-and-error-codes.html

Lighting circuits: Thorn Lighting
http://www.thornlighting.com.au/download/TechnicalHandbook_Australia_COMPLETE_2.pdf

Thermal overload relays: Global Spec Electronics
http://www.globalspec.com/learnmore/electromechanical components/relays_timers/thermal_overload_relays

Professional equipment: Grainger Industrial Supply
http://www.professionalequipment.com

Electrical safety guidelines: Australian Constructors Association
http://www.constructors.com.au > Safety > Electrical Safety Guidelines – Working on Fixed Electrical Installations.pdf

Appendix

SI base and derived measurement units

The International System of Units (SI) defines seven units of measurement as a basic set from which all other SI units are derived. The base units used in electrical measurement are the metre for length, kilogram for mass, second for time, ampere for electric current, Kelvin for temperature and candela for luminous intensity. All other units in electrotechnology are derived from these six base units. Those you are likely to use or need to know are listed below according to where they apply. The list includes a summary of various terms and laws you may find useful.

AC fundamentals

Impedance is the phasor addition of the resistance and reactance in an a.c. circuit, symbol Z, unit is the ohm. In any circuit, Z = supply voltage divided by total current.

Phase angle is the phase difference in electrical degrees between two electrical values.

Phasor diagram is a drawing with lines drawn to scale at the correct phase angle to represent currents or voltages in an AC circuit.

Power factor is the cosine of the angle (θ) between the supply current and supply voltage in an a.c. circuit.

Reactance is the opposition of inductance or capacitance in an a.c. circuit, symbol X, unit is the ohm.

Resonance is when the inductive reactance equals the capacitance reactance in an a.c. circuit.

RMS (root mean square), an a.c. current with the same heating effect as that value of d.c. current.

Capacitance

Capacitance is the property of a circuit or component that stores an electric charge.

Coulomb is the unit for electrical charge and equals 6.25×10^{18} electrons.

Farad is the unit of capacitance. One farad is the capacity of a capacitor to store one coulomb of charge if one volt is applied.

Permittivity (symbol ε) is the ability of a material to support an electric field. Relative permittivity is a comparison of the permittivity of a material to that of air. Also called dielectric constant (K).

Electrical units

Ampere is a base unit and is the unit for current flow. One ampere can be defined as equal to 6.24×10^{18} electrons flowing past a point in one second. One ampere is also the current produced by one volt acting across a resistance of one ohm. Another definition is the current which will produce an attractive force of 2×10^{-7} newton per metre of length between two straight, parallel conductors of infinite length and negligible circular cross-section placed one metre apart in a vacuum.

Resistance is measured in ohms. One ohm is defined as the resistance between two points of a conductor when a potential difference of one volt applied across these points causes a current of one ampere to flow in the conductor.

Volt is the unit of electromotive force. One volt is defined as the potential difference across a conductor that is carrying a current of one ampere and dissipating one watt of power in the conductor.

Watt is the unit for power. One watt of power is equal to one joule of energy expended per second. It is also the power dissipated by a current of one ampere flowing through a resistance of one ohm.

Electromagnetism

Absolute permeability (μ) of a material is its relative permeability multiplied by the permeability of air. That is $\mu = \mu_r \times \mu_0$.

Magnetic flux is the total number of lines of force in a magnetic field. The unit is the weber (Wb) and one weber equals 10^8 lines of force. Symbol is ϕ (phi).

Flux density is flux (ϕ) divided by area. The symbol is B and the unit is the tesla (T).

Magnetising force is magnetomotive force divided by the length of the magnetic circuit. The symbol is H and the unit is the ampere-turn per metre (At/m).

Magnetomotive force is the 'pressure' that produces a magnetic field. It equals current (I) multiplied by the number of turns (n) of a coil. The symbol is mmf and the unit is the ampere-turn (At).

Permeability is the ability of a material to conduct magnetic lines of force. The permeability of air (μ_0) is 1.26×10^{26}. Relative permeability (μ_r) is a comparison of the permeability of a material to air (which has a μ_r of 1).

Reluctance is magnetic resistance. A high permeability material has a low reluctance. Symbol is S, unit is At/Wb.

Electromagnetic induction

Fleming's right-hand rule (for generators): Thumb is direction of travel, first finger is direction of lines of force, second finger is direction of current.

Henry is the unit of inductance. One henry is the amount of inductance that causes an induced voltage of 1 V when the current is changing at 1 A per second.

Induced EMF is the voltage produced in a length of wire or a coil when it moves relative to a magnetic field. If the wire (coil) is connected to a circuit the EMF will cause current flow (induced current).

Inductance is the property of a circuit or component that opposes a change in the value of an electric current.

Lenz's law: The current induced in a conductor will set up a magnetic field that opposes the direction of the magnetic field causing the current.

Mutual inductance is the effect of the magnetic field of a coil inducing an EMF in another nearby coil.

Self-inductance is the same as inductance.

Time constants

A time constant is the time taken for a current or voltage to reach 63.2% of its final value. Symbol τ, measured in seconds. In RC circuit, $\tau = RC$, for LR circuit, $\tau = L/R$.

Universal time constant curve is an exponential curve that shows how current or voltage changes in an RC or an LC circuit. It has the same shape regardless of the component values in the circuit.

Greek letters

The Greek letters used in this book are identified below

Letter	Name	What it represents
Δ	delta	change in (for example Δ means change in time)
ε	epsilon	permittivity of an insulator, value only
η	eta	efficiency
λ	lamda	power factor (= cos θ), also wavelength in electronics
μ	mu	permeability of a ferromagnetic material, value only, also micro (10^{-6})
Ω	omega	resistance, unit is the ohm
Φ	phi	magnetic flux, unit is the weber (Wb)
Φ	phi	phase angle between phasors in an AC circuit
π	pi	constant used in circular measurement = 3.1416 (to 4 decimal places) also = 355/113
ρ	rho	resistivity, unit is the ohmmetre (Ωm)
τ	tau	time constant, unit is the second (s)
θ	theta	angle of rotation or angle in a triangle

Metric prefixes (or multipliers)

The four most commonly used metric prefixes in electrotechnology are mega, kilo, milli and micro. These and other prefixes are given in the following table. A prefix is expressed in mathematical form using scientific notation, in which each prefix is treated as a power of 10. Engineering notation only uses powers of 10 that are a multiple of 3.

Term	Symbol	Multiply by	Engineering notation	Example
tera	T	1 000 000 000 000	10^{12}	2 THz = 2 × 10^{12} Hz
giga	G	1 000 000 000	10^{9}	4.62 GHz = 4.62 × 10^{9} Hz
mega	M	1 000 000	10^{6}	2.2 MΩ = 2.2 × 10^{6} Ω
kilo	k	1 000	10^{3}	5 kV = 5 × 10^{3} V
milli	m	0.001	10^{-3}	3.4 mA = 3.4 × 10^{23} A
micro	μ	0.000 001	10^{-6}	89 μA = 89 × 10^{26} A
nano	n	0.000 000 001	10^{-9}	68 nF = 68 × 10^{29} F
pico	p	0.000 000 000 001	10^{-12}	120 pF = 120 × 10^{212} F

Temperature conversion

Key: C = 5/9 (F−32) F = (9/5C) + 32

Temperature conversion table

°C	°F	°C	°F	°C	°F	°C	°F	°C	°F	°C	°F	°C	°F
−40	−40	−10	14	20	68	50	122	80	176	110	230	140	284
−39	−38.2	−9	15.8	21	69.8	51	123.8	81	177.8	111	231.8	141	285.8
−38	−36.4	−8	17.6	22	71.6	52	125.6	82	179.6	112	233.6	142	287.6
−37	−34.6	−7	19.4	23	73.4	53	127.4	83	181.4	113	235.4	143	289.4
−36	−32.8	−6	21.2	24	75.2	54	129.2	84	183.2	114	237.2	144	291.2
−35	−31	−5	23	25	77	55	131	85	185	115	239	145	293
−34	29.2	−4	24.8	26	78.8	56	132.8	86	186.8	116	240.8	146	294.8
−33	−27.4	−3	26.6	27	80.6	57	134.6	87	188.6	117	242.6	147	296.6
−32	−25.6	−2	28.4	28	82.4	58	136.4	88	190.4	118	244.4	148	298.4
−31	−23.8	−1	30.2	29	84.2	59	138.2	89	192.2	119	246.2	149	300.2
−30	−22	0	32	30	86	60	140	90	194	120	248	150	302
−29	−20.2	1	33.8	31	87.8	61	141.8	91	195.8	121	249.8	151	303.8
−28	−18.4	2	35.6	32	89.6	62	143.6	92	197.6	122	251.6	152	305.6
−27	−16.6	3	37.4	33	91.4	63	145.4	93	199.4	123	253.4	153	307.4
−26	−14.8	4	39.2	34	93.2	64	147.2	94	201.2	124	255.2	154	309.2
−25	−13	5	41	35	95	65	149	95	203	125	257	155	311
−24	−11.2	6	42.8	36	96.8	66	150.8	96	204.8	126	258.8	156	312.8
−23	−9.4	7	44.6	37	98.6	67	152.6	97	206.6	127	260.6	157	314.6
−22	−7.6	8	46.4	38	100.4	68	154.4	98	208.4	128	262.4	158	316.4
−21	−5.8	9	48.2	39	102.2	69	156.2	99	210.2	129	264.2	159	318.2
−20	−4	10	50	40	104	70	158	100	212	130	266	160	320
−19	−2.2	11	51.8	41	105.8	71	159.8	101	213.8	131	267.8	161	321.8
−18	−0.4	12	53.6	42	107.6	72	161.6	102	215.6	132	269.6	162	323.6
−17	1.4	13	55.4	43	109.4	73	163.4	103	217.4	133	271.4	163	325.4
−16	3.2	14	57.2	44	111.2	74	165.2	104	219.2	134	273.2	164	327.2
−15	5	15	59	45	113	75	167	105	221	135	275	165	329
−14	6.8	16	60.8	46	114.8	76	168.8	106	222.8	136	276.8	166	330.8
−13	8.6	17	62.6	47	116.6	77	170.6	107	224.6	137	278.6	167	332.6
−12	10.4	18	64.4	48	118.4	78	172.4	108	226.4	138	280.4	168	334.4
−11	12.2	19	66.2	49	120.2	79	174.2	109	228.2	139	282.2	169	336.2

Example: 80 degrees C = 176 degrees F & 14 degrees F = −10 degrees C

Source: © 1998–2009 Alan Parker

LED watt conversion chart

Incandescent/ Halogens	Mercury vapour	Metal halide	High pressure sodium	Compact fluorescent (CFLs)	Light emitting diodes (LEDs)
40–60	15–25	5–15	5–15	12–15	5–8
60–75	25–35	15–25	15–25	15–18	7–10
75–100	35–45	20–35	20–35	18–23	10–15
100–150	50–60	25–40	25–40	23–35	15–20
150–200	70–85	35–45	35–45	30–45	20–25
200–250	90–110	40–55	40–55	45–60	25–30

Source: http://www.optimalled.com/resources-watt-conversion-chart.php

Bulb type efficacy rating chart

Poor / Average / Efficient

Incandescent/ Halogens	Mercury vapour	Metal halide	High pressure sodium	Compact fluorescent (CFLs)	Light emitting diodes (LEDs)
Efficacy: 15–25 Lm/W	Efficacy: 30–60 Lm/W	Efficacy: 80–105 Lm/W	Efficacy: 70–110 Lm/W	Efficacy: 40–70 Lm/W	Efficacy: 60–140 Lm/W
CRI: 98–100	CRI: 40–50	CRI: 60–70	CRI: 20–30	CRI: 60–90	CRI: 70–95
Lifespan: 3000 hours	Lifespan: 15 000 hours	Lifespan: 10 000 hours	Lifespan: 15 000 hours	Lifespan: 15 000 hours	Lifespan: 40 000 hours

Source: http://www.optimalled.com/resources-watt-conversion-chart.php

Solutions to review exercises

Chapter 1 solutions

Review questions 1.1

1 Australian Capital Territory: WorkSafe ACT (http://www.worksafe.act.gov.au), New South Wales: WorkCover NSW (http://www.workcover.nsw.gov.au), Northern Territory: Northern Territory Work Safe (http://www.worksafe.nt.gov.au), Queensland: Work Place Health & Safety Queensland (http://www.deir.qld.gov.au/workplace/index.htm), South Australia: Safe Work SA (http://www.safework.sa.gov.au), Victoria: WorkSafe Victoria (http://www.worksafe.vic.gov.au), Western Australia: WorkSafe WA (http://www.safetyline.wa.gov.au). New Zealand: WorkSafe New Zealand (http://www.business.govt.nz/worksafe).
2 An apprentice is classified as a 'worker'.
3 A work health and safety inspector can issue various notices, such as (a) a site non-disturbance notice (to prevent an accident site being changed before investigation is carried out), (b) an improvement notice (where the inspector believes that a person is contravening the model WHS Act), (c) a prohibition notice (if the inspector believes there is a serious risk to the health or safety of a person at a workplace resulting from an exposure to a hazard), and (d) a remedial action injunction (where a person issued with a prohibition notice fails to comply with the notice). An inspector can also enter any workplace without prior notice, particularly if notification would defeat the purpose of entry or delay an investigation of the suspected contravention.

Review questions 1.2

1 A hazard is defined by AS/NZS 4360:2004 as 'potential for harm, or adverse effect on an employee's health'. A risk is defined as 'likelihood that the hazard will cause injury or ill health'.
2 Physical hazards include: floors, slippery surfaces, stairs, steps, work platforms, ladders, falling objects, manual handling (lifting, pushing, pulling), excessively loud and prolonged noise, vibration, heat and cold, radiation, poor lighting, ventilation, air quality and fire.
3 Class A fire. Suitable extinguishers to fight a class A fire include: water, wet chemical, foam, dry powder and vaporising liquid types.

Review questions 1.3

1 Muscle sprains and strains, hernia, injuries to ligaments, inter-vertebral discs and other structures in the back, injuries to nerves, ligaments and tendons in the wrists, arms, shoulders, neck or legs.
2 The SMART lifting procedure is S (size up the load), M (move in close to the load), A (always bend your knees), R (raise the object with your knees), T (turn by moving your feet).
3 The symbol shows the package's centre of gravity.

Review questions 1.4

1 A hazardous substance contains ingredients that may be solids, liquids or gases that can harm people, other living organisms, property or the environment in the immediate and long term. Dangerous goods are classified on the basis of immediate or short-term physical or chemical hazards, such as fire, explosion, corrosion and toxicity.
2 One 9 kg bottle, two 4 kg bottles or no greater than 22 litres water capacity.
3 Packaging groups show the degree of danger in each group, with restrictions placed on incompatible chemicals being grouped together.
4 A safety data sheet is a document that provides information on the properties of hazardous chemicals and how they affect health and safety in the workplace. They provide information on a hazardous substance and dangerous goods additional to that on a label.

Review questions 1.5

1 Use only industrial rated ladders. Ladder should only be erected on flat solid surfaces and adequately supported at the base, set at a slope of 4 in 1. The ladder should extend at least 1 metre above the access point, it should be firmly secured or tied off, and when working near doorways or pedestrian traffic areas, lock the door shut or erect a barrier.
2 Precautions include keeping your body within the frame of the ladder when ascending or descending while keeping at least three points of contact with the ladder. Other precautions are never stand on a rung closer than 900 mm to the top of a single or extension ladder, or any higher than the second tread below the top plate of any stepladder. Do not work over other people, or allow anyone else to be on the ladder at the same time.
3 Light duty scaffolding.

Review questions 1.6

1 Risks include lack of oxygen, inhalation of fumes and dust, fire and explosions from flammable liquids, gases and vapours, crush injuries from machinery.

2 Controlling risks associated with working in confined spaces:
   a elimination (if it is practicable to do so, work should be done outside of the confined space using relevant equipment such as remote controlled devices or special tools)
   b isolation (risk sources should be disconnected before anyone enters the confined space)
   c engineering (purging the air from the space, setting up communication system, confirming air quality)
   d administration (providing training for those entering a confined space, having a standby person, setting up a rescue plan, signage, securing against accidental start-up of machinery)
   e PPE (using the correct respiratory protective equipment, which should be based on the level and type of contaminants and the work to be done).

Review questions 1.7

1 Around 120 dB, exposure time about 2 minutes.
2 Hand-arm vibration can be caused by using power tools that produce vibration, such as pneumatic and electric jackhammers, impact drills and sanders. Whole-body vibration occurs when the operator is seated or standing on equipment such as a tractor, forklift or any vehicle producing vibrations.
3 Symptoms of heat stress include dizziness, fatigue, headache, nausea, breathlessness, clammy skin or difficulty remaining alert.

Review questions 1.8

1 Factors are: current path through the body, amount of current, length of time the current flows, presence of moisture, phase of the heart cycle when the shock occurs, general health of the person prior to the shock.
2 No.
3 Six basic steps ensuring safe isolation are: consultation, isolation, securing the isolation, tagging, testing, re-testing as necessary.
4 Four basic steps of a general lockout procedure are: lock the isolation device so it can't be re-energised, tag the isolation device, try to reclose the isolation device to confirm it cannot be reclosed, repeating these steps if there is a change in conditions.

Chapter 2 solutions

Review questions 2.1

1 The three pillars of sustainability are economic development, social development and environmental protection.
2 The purpose of ISO 14001 is to provide a framework for organisations wishing to develop an environmental management scheme that can then be certified if required.

Review questions 2.2

1 Greenhouse effect refers to heat energy from the sun being absorbed by gases in the atmosphere, with some of the heat being re-radiated back to Earth.
2 Human activities that are causing an increase in the concentration of carbon dioxide in the Earth's atmosphere are burning fossil fuels such as coal, gas, oil and its derivatives such as petrol and diesel, removing vegetation and forests through land clearing, agriculture.

Review questions 2.3

1 There is a duty to notify relevant authorities of pollution incidents where material harm to the environment is caused or threatened. Material harm includes actual or potential harm to the health or safety of human beings or to ecosystems that results in actual or potential loss or property damage of a specified amount.
2 Individuals, like companies, have a responsibility to comply with the EPBC Act and state/territory environmental legislation. Specifically, individuals must not cause water pollution, air pollution and land pollution as a result of spills or illegal dumping. Also, individuals on a work site must dump waste only at a licensed facility, not harm protected animals and plants, manage dangerous chemicals responsibly, and not start major construction works without appropriate approval.

Review questions 2.4

1 Typical sustainable work practices associated with an outdoors work site include ensuring that contractors and workers know their environmental obligations, separating different wastes before disposal, reusing and recycling materials where possible, reducing water wastage, covering stockpiles and materials to reduce windblown dust, general good housekeeping practices, preventing paint and other chemicals from soaking into the ground or getting into stormwater drains, installation and maintenance of sediment and erosion control devices on the site.
2 Batteries and electronic circuit boards contain materials classified as hazardous. These should be disposed of correctly, which can include through relevant recycling depots or special disposal depots. Local councils often operate these facilities.

Review questions 2.5

1 Efficient use of energy, and use of renewable energy sources.
2 A more efficient appliance uses less energy to achieve a given task compared to a less efficient

appliance. Examples include appliances of the same type with different MEPS (star) ratings (e.g. refrigerators, washing machines), different types of water heaters (heat pumps), different types of lights (CFL, LED are more efficient than halogen lights).

Chapter 3 solutions

Review questions 3.1

1 Coach screw.
2 A self tapping screw needs a hole to be drilled before insertion, a self drilling screw cuts its own hole. Both types cut their own thread during fastening.
3 Fixing ratings (in newtons) are: single shear strength, axial tensile strength, pullout load and torsional strength (torque) in newton metres.

Review questions 3.2

1 Types include: nylon frame anchor, nylon drive anchor, nylon anchor, metal drive anchor, hammer-in anchor, concrete nails, plastic wall plugs.
2 Direct acting high velocity, where the charge acts directly on the fixing, indirect acting low velocity.
3 Impact drills apply the rotational force in short pulses, hammer drills apply a momentary forward driving force on the drill bit at the same time as rotating it.

Review questions 3.3

1 The following three threads are commonly used in Australia: British Standard (from England), Unified (from USA) and metric (from Europe).
2 Uses for washers include: spread the clamp load over a larger surface area, cover an oversize or elongated hole, reduce the friction of the rotated component (nut), protect the work piece surface from damage by the rotated component, provide a locking or vibration resistant function.
3 Spanner size is 10 mm. Bolts are tightened gradually, moving from one bolt to another until all are equally tight.
4 An adjustable spanner cannot grip a bolt head or nut as tightly as a fixed spanner. Unless used in low torque applications, the bolt head or nut can be damaged, and the spanner can easily slip off the bolt or nut unexpectedly.
5 Risk sources (hazards) associated with drilling steel and masonry include noise, vibration, dust and sharp objects such as metal filings and swarf. Drill bits can break, causing flying debris. If a larger sized drill bit jams, the torque from the power drill can cause a sudden, unexpected rotation of the power tool which can injure the operator. In locations close to walls, the operator's hands can be crushed between the wall and the drill handle. PPE can include ear muffs, safety glasses, gloves, dust mask, plus hard hat and high visibility vest.
6 Twist drill.
7 Auger drill bit, spade bit.
8 Masonry drill. It has a hardened tungsten tip that can withstand drilling into concrete.

Review questions 3.4

1 Advantages include: use in premises where other fixing options involving drilling into walls are not permitted, strength of the bond, use in close to edge fixing.
2 A hollow block wall may not be able to support the intended load.
3 From Table 3.14, load rating is 8.6 kN.
4 It should be cleaned with isopropyl alcohol and water mixture.
5 Generally takes about three days.
6 Distance, as per Figure 3.86, is no less than 20 mm.

Chapter 4 solutions

Review questions 4.1

1 An ISO standard is an international standard that other countries refer to when developing their own standards. The differences between an Australian or New Zealand standard to an ISO standard depend on local requirements.
2 AS 1100.101-1992.
3 To draw hidden outlines and hidden edges.

Review questions 4.2

1 Pure metals include aluminium, iron, copper, gold and silver.
2 Toughness refers to a metal that is able to resist fracture under shock loads, for example, a hammer head.

Review questions 4.3

1 By covering the metal surface with a coloured die, such as marking blue or whitewash.
2 To find the centre of a piece of round stock with jenny calipers, open the calipers to an amount approximately equal to half the diameter of the stock. Then scribe arcs from four positions equally spaced around the circumference of the stock. The centre of the four arcs gives the centre of the stock.

Review questions 4.4

1 A pipe vice, which can be either a hinged or chain type vice.
2 The teeth should point away from the handle so cutting occurs on the forward stroke.

Review questions 4.5

1 To carry away the cut material (swarf) from the drill point, and to allow lubricant or coolant to reach the drill point during drilling.

2 The rake faces of the cutting lips are ground parallel with the axis of the drill, giving a zero rake angle.

Review questions 4.6

1 Taper tap, used to start a thread; intermediate tap to complete tapping through an open-ended hole or as the second stage in tapping a blind hole; plug tap to complete a thread in a blind hole.

2 A die nut is used to clean a damaged or poorly cut thread. It is not used to cut a new thread.

Review questions 4.7

1 600 grams.

2 Defects making a hammer unsafe include loose hammer head (missing wedge), chipped, burred or damaged hammer head, splintered or split wooden handle, oily or greasy handle.

Review questions 4.8

1 Welding processes are either fusion or non-fusion. Fusion welding occurs when both parts being welded are melted so the parts when cooled form a welded bond. Examples are arc welding, spot welding and oxyacetylene welding. Hard soldering, brazing and soft soldering are forms of non-fusion welding, in which the parts are not heated above melting point, instead they are joined by a filler such as solder.

2 The gas associated with gas metal arc welding is to protect the molten weld pool from contamination by nitrogen or oxygen. Gases used include argon, helium, carbon dioxide and oxygen.

Review questions 4.9

1 The different directions of motion in a random orbital sander reduces swirl marks compared to just orbital motion.

2 A residual current detector (RCD). If the supply being used is not protected with an RCD, use an RCD protected portable power outlet unit.

Review questions 4.10

1 Metals formed into sheet metal include mild steel, brass, aluminium, copper, phosphor bronze, muntz metal and silicon steel.

2 A guillotine shear is used to cut sheet metal.

Review questions 4.11

1 A micrometer is more accurate.

2 $39 + (46 \times 0.02) = 39.92$mm

Review questions 4.12

1 A mechanical torque wrench will click when the torque setting is reached. A digital wrench produces an audible tone and lights an indicating LED.

2 By marking the shields and stator with witness marks using a punch. One shield is marked with a single punch, and a corresponding single punch is marked on the stator opposite the punch on the shield. Two punches are used on the other shield and corresponding punches are applied to the stator.

Chapter 5 solutions

Review questions 5.1

1 A site plan. This drawing is important as it shows the orientation of the plan to the real world by including a reference to north.

2 Architectural floor plan location symbols:

a single light point

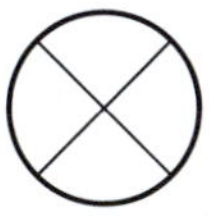

b twin fluorescent

c double socket outlet

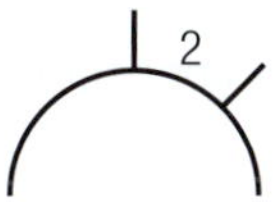

d hot water service

3 $55 \times 20 = 1100$ mm or 1.1 metre

Review questions 5.2

1 Types of electrical diagrams include block, circuit, wiring, architectural, ladder (for PLCs) and single line.

2 A circuit diagram shows component connections as an aid to determining how the circuit works. A wiring diagram shows the location of all components and wiring, including the location of every terminal and the wires that connect to each terminal.

3 AS/NZS standard symbols for:

a circuit breaker

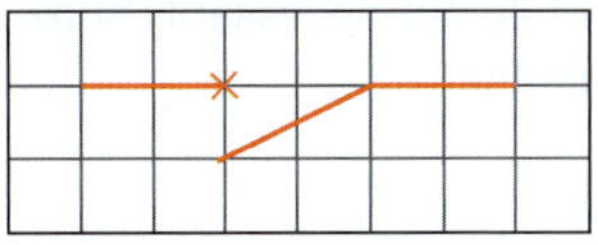

Circuit breaker

b normally open momentary push button

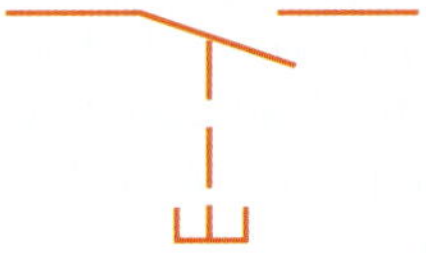

c normally closed emergency stop button

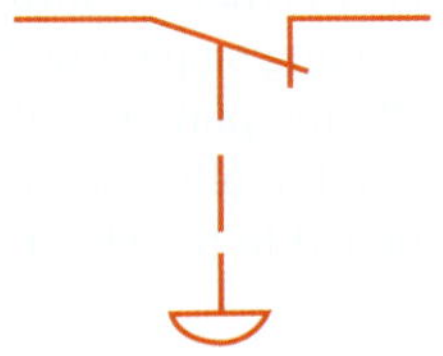

Review questions 5.3

1. Loads are shown at the bottom of a vertically orientated circuit diagram.
2. DPDT means double pole double throw.
3. AS/NZS standard symbols for:
   a. triple pole single throw N.O. switch

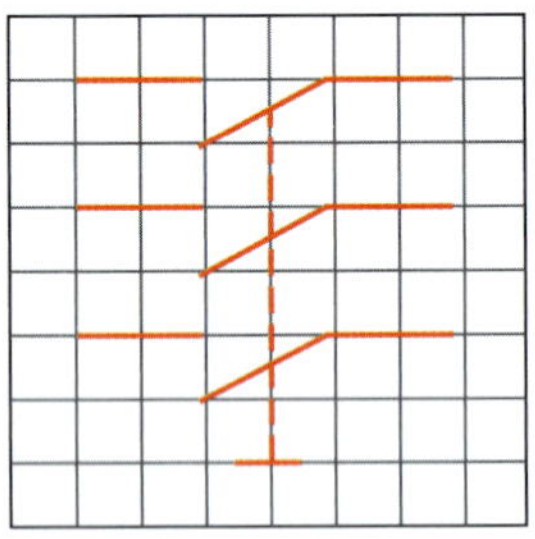

   b. two conductors crossing but not connecting

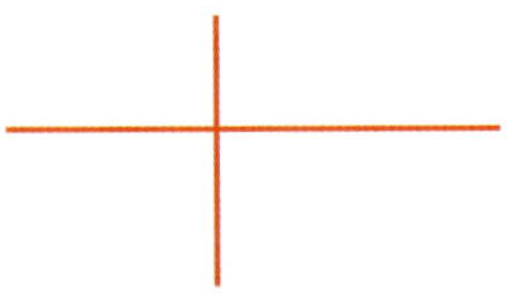

   c. N.O. time delay contact, delay close instant open

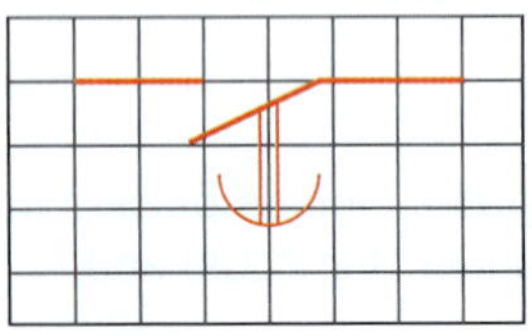

Review questions 5.4

1. Components should be in proportion and positioned true to how they are in the installation.
2. Cable terminations in a wiring diagram are shown as they would be in the installation. In a circuit diagram, connections are shown to make the circuit easy to interpret, and do not necessarily show where they are in the installation.

Review questions 5.5

1. Brick veneer has a non-load-bearing outer brick 'skin' and a load-bearing timber or steel frame, separated by a cavity. Double brick construction has two brick walls separated by a cavity with the inner skin of the brickwork being the load-bearing part.
2. No, cables must be protected as per AS/NZS 3000:2007 clause 3.9.4.4.
3. Cables installed likely to be disturbed are required to be fixed to prevent undue sagging.
4. Typical stages of construction at which electrical and communication services should be installed:
   a. base stage of construction – temporary supply, laying conduits
   b. frame stage of construction – 1st electrical fixing or 'rough in'
   c. fixing stage of construction – 2nd electrical fixing or 'fit out'.

Chapter 6 solutions

Review questions 6.1

1. In the planning stage of a job.
2. A risk assessment.

Review questions 6.2

1. The forms are hazard identification (risk source), risk assessment and SWMS.
2. A job safety analysis (JSA) is developed by all workers on the site and covers low-risk tasks. For high-risk tasks a SWMS is provided to workers.

Review questions 6.3

1. a. ELV – a voltage not exceeding 50 V a.c. or 120 V ripple-free d.c.
   b. LV – a voltage exceeding extra-low voltage, but not exceeding 1000 V a.c. or 1500 V d.c.
2. Separated extra low voltage (SELV), which is an extra-low voltage system that is electrically separated from earth and from other systems in such a way that a single fault cannot give rise to the risk of electric shock. Protected extra-low voltage (PELV), which is an extra-low voltage system that is not electrically separated from earth, but that otherwise satisfies all the requirements for SELV.
3. A distribution transformer provides voltages at a safe level for use in consumer installations.
4. High current can be present in the consumer's main and switchboard. Large electric motors take high current when starting. The main earth conductor carries high current under fault conditions.

Review questions 6.4

1. A high voltage is defined as a voltage that exceeds 1000 V a.c. or 1500 V d.c.
2. Clearance is the shortest distance between two conductive parts measured through air. Creepage is the shortest path between two conductive parts measured along the surface of the insulation separating the parts.

Review questions 6.5

1 Working live is only permitted if isolation creates a greater risk than working live.
2 A safe working method statement (SWMS).
3 The employer is responsible but everyone must contribute.

Chapter 7 solutions

Review questions 7.1

1 To prevent injury, fire and shock.
2 They must not leave an installation in a dangerous state.
3 To provide a guide to the application of the Act.
4 To describe actions for compliance with the regulations.

Review questions 7.2

1 1997.
2 Volume 1 Section c.
3 Electrical supply utility.
4 Referenced by regulation so it is a legal compliance.

Review questions 7.3

1 Electrical general requirements, Design & installation, Cable support & duct systems.
2 Under common law an agreement to supply and receive payment.
3 What is to be done, who will do it, how much is to be paid.
4 As soon as it is accepted.

Chapter 8 solutions

Review questions 8.1

1 Single insulated copper cable.
2 Current carrying capacity; operating voltage; operating temperature.
3 1000 V a.c.

Review questions 8.2

1 Green/yellow.
2 The earth pin is slightly longer than the other pins, is vertically aligned and sometimes is labelled 'EARTH'.

Review questions 8.3

1 Thermoplastic double insulated cable system.
2 They must be afforded mechanical protection.

Review questions 8.4

1 It is better suited to multicore cables.
2 To seal against moisture and dust as well as provide strain relief for terminations.

Review questions 8.5

1 Clause 1.4.98 defines all cable trunking.
2 Cables must be separately supported and not rely on the lid for support.

Review questions 8.6

1 The continuity of mechanical and earth protection must be maintained.
2 A stock bend or elbow.

Review questions 8.7

1 Alarm system and fire pumps.
2 Continuity test between earth and the metal sheath. Insulation test between conductors and insulation test between conductors and earth.

Review questions 8.8

1 Safety glasses and leather gloves.
2 To prevent the steel strands from springing out of shape.

Review questions 8.9

1 A trailing cable is suspended on a system that allows it to extend along a track or guide wire. A reeling cable is held on a reel that allows the cable to extend then winds it back.
2 To prevent moisture from running down the wire into the building and provide strain relief for conductor terminals.

Chapter 9 solutions

Review questions 9.1

1 Loop at the switch method.
2 Intermediate switching.

Review questions 9.2

1 Yes.
2 When the current rating of the switch is equal or greater than the combined load on the socket outlets.

Review questions 9.3

1 Yes.
2 Under no circumstances.

Review questions 9.4

1 To give three possible heat settings using two heating elements.
2 A thermostat.

Review questions 9.5

1 Resistive, inductive and capacitive.
2 Centigrade, Fahrenheit and Kelvin.

Review questions 9.6

1 Because they maintain a continuous supply of hot water.
2 95°C.

Review questions 9.7

1 To maintain vital production, provide a safe exit from a building and retain data storage systems.
2 To convert d.c. to a.c.

Review questions 9.8

1 Zinc carbon, lead acid and lithium ion.
2 Secondary cell.

Review questions 9.9

1 Smoke, temperature rise and flames.
2 On the ceiling.

Review questions 9.10

1 Enough to allow a safe exit from a building.
2 It does not come on until the power fails.

Review questions 9.11

1 Candela.
2 5 watts per square metre.

Review questions 9.12

1 To provide a voltage spike to start the tube then restrict the current flow through the tube.
2 Yes, but if it is turned off it will not restart.

Review questions 9.13

1 Lighting sports arenas, streets and freight yards.
2 Definitely not.

Chapter 10 solutions

Review questions 10.1

1

Danger or damage	Protection method (acceptable answers)
Direct contact	One of the following: • Provide insulation. • Separate by using barriers. • Place out of arm's reach.
Indirect contact	One of the following: • Earth all exposed metal parts. • Fit residual current devices. • Use an isolated supply. • Use extra low voltage.
Thermal effects	Temperature characteristics of selected electrical equipment must be suitable for the installed conditions.
Unwanted voltages	One of the following: • Insulation or separation of supplies. • Protection by devices that detect under- or over-voltage and separate the circuit from the supply.
Over-current	Conductors must be protected from over-current by devices that interrupt the supply in the event of an over-current (AS/NZS 3000:2007 clause 2.5.1).
Fault currents	One of the following: • Fit a circuit breaker with a short circuit release. • Fit an HRC fuse.
Overload	Fit a circuit breaker or fuse with a current rating that matches the current carrying capacity of the circuit conductors.
Injury from mechanical movement	Motors to be provided with switching devices to: • Start and stop the motor. • Provide emergency stopping. • Isolate the motor for mechanical maintenance.

2 Must comply with the following:
- operate in a safe and reliable manner
- not cause danger from electric shock
- not cause fire or high temperatures
- installed according to the manufacturer's instructions. (AS/NZS 3000:2007 clause 2.5.1)

Review questions 10.2

1 Yes, the supply to every electrical installation must be controlled by a main switch. (AS/NZS 3000:2007 clause 2.3.3.1)

2 No, SELV is a system that is electrically separated from earth and from any other systems. (AS/NZS 3000:2007 clause 1.4.83)

Review questions 10.3

1 The current flowing, the length of the cable and the cross-sectional area of the conductors.

2 Resistance elements, electric arc welders and induction heaters.

Review questions 10.4

1 AS/NZS 3000:2007 clause 1.4.94 defines touch current.

2 They must not be connected to the protective earthing conductor or any exposed conductive parts of the source of supply. (AS/NZS 3000:2007 clause 7.4.3(c))

Review questions 10.5

1 Any part that does not form part of the electrical installation but may be at the electrical potential of a local earth.

2 Bonding must be provided to avoid any potential differences that may occur between electrical equipment connected to the earthing system and any conductive piping (and including taps, etc.) that may independently be in contact with the general mass of earth. (AS/NZS 3000:2007 clause 5.6.2.1)

Review questions 10.6

1 Overload current is caused by a malfunction in the electrical equipment such as a fault in the mechanical load of a motor, a decrease in resistance of an electrical load due to age or damage, or addition of appliances to a circuit that exceeds the designed current rating for the circuit.

2 Fault loop from the point of fault comprises:
- protective earthing conductor
- return neutral path
- the path through the neutral point of the transformer and the transformer winding
- the active conductor as far as the point of the fault. (AS/NZS 3000:2007 Appendix B clause B4.4)

Review questions 10.7

1 Circuit breakers and fuses.

2 Overload protection and short circuit protection.

Review questions 10.8

1 Voltage surge, voltage spike and voltage transient.

2 Electrical control equipment may not function correctly and motors may run slowly and overheat.

Review questions 10.9

1 At the start of each separate installation and one for each supply where there are multiple supplies. (AS/NZS 3000:2007 clause 2.3.3.1)

2 Functional switches must be provided for:
- cooking appliances
- water heaters
- room heaters
- cables or groups of cables that comprise a heating system
- electricity converters.

Review questions 10.10

1 To reduce voltage drop when appliances are located a long distance from the main switchboard. To provide a dedicated distribution board for special equipment and services.

2 No, in this case the cable colours clearly indicate the purpose of the link. (AS/NZS 3000:2007 clause 2.9.5.3)

Chapter 11 solutions

Review questions 11.1

1 Single phase

$V_{max} = 230 + (230 \times 10/100)$
$= 253\ V$

$V_{min} = 230 - (230 \times 6/100)$
$= 216.2\ V$

Three phase

$V_{max} = 400 + (400 \times 10/100)$
$= 440\ V$

$V_{min} = 400 - (400 \times 6/100)$
$= 376\ V$

(AS/NZS 3000:2007 clause 1.6.2)

2 Limitation by protection device current rating. (AS/NZS 3000:2007 clause 2.2.2)

3 Where a wiring system passes through an element of building construction that is required to be fire-rated any holes shall be close fitting to the wiring system and at least 50 mm away from any other services. The cross-sectional area of the opening shall not be greater than 25 mm diameter (500 mm^2), or up to 50 mm diameter (2000 mm^2) for a single cable that leaves a gap of not more than 15 mm between the

cable and the opening. Any penetration made must have the fire resistant rating restored with devices such as fire pillows. (AS/NZS 3000:2007 clause 3.3.9.3)

Review questions 11.2

1 75°C. (AS/NZS 3000:2007 clause 3.4.2 Table 3.2 and/or AS/NZS 3008.1:2009 clause 3.2.2 Table 3.1)

2 1 double socket outlet is counted as 2 singles. Each socket outlet contributes 2 A. Load per socket outlet = 2 × 2 = 4 A. Maximum recommended number of points per circuit is the circuit protection device divided by the load per point.

Points = 20/4 = 5 double points per circuit (AS/NZS 3000:2007 clause C5.2 Table C8)

3 Points per final sub-circuit and protection device ratings.

Circuit number	Purpose/load	Protection device/Rating (A)	Number of points per circuit
1	Light 1	10 A	11
2	Light 2	10 A	11
3	Power 1 – 10 A S/O	20 A	16
4	Power 2 – 10 A S/O	20 A	16
5	Power 3 – 15 A S/O	16/20 A	1
6	A/C	20 A	1
7	HWS	20 A	1

Review questions 11.3

1 Maximum demand multiple domestic. (AS/NZS3000:2007 Table C1 Colum 4)

No. of Units A phase ___11___ B phase __11____ C phase ___11___

Load group	Load	Calculation	A	B	C
A(i)	Lights	5 + (11 × 0.25) =	7.75	7.75	7.75
B(i)	Socket outlets	15 + (11 × 3.75) =	56.25	56.25	56.25
C	Range	11 × 2.8 =	30.8	30.8	30.8
F	HWS	11 × 6 =	66	66	66
D	A/C	11 × 14 × 0.75 =	115.5	115.5	115.5
H	Community lighting	(24 × 0.22) + (10 × 0.15) =		6.78	
	Community power	10 × 2 = 20 A (15 A max)		15	
		Total	276.3	**298**	276.3

2 Maximum demand single domestic individual unit. (AS/NZS3000:2007 Table C1 Colum 4)

Load group	Load	Calculation	Demand
A(i)	Lights (22 points)	3 + 2 =	5
B(i)	Socket outlets (36 points)	10 + 5 =	15
C	Range	8000/230 × 0.5 =	17.4
F	HWS	4400/230	19.1
D	A/C	14 × 0.75 =	10.5
		Maximum demand	67 A

3 Non-domestic maximum demand unit. (AS/NZS 3000:2007 Table C2 Colum 3)

Load group	Load	Calculation	A	B	C
A	Lights – high bays	6 × 2.6 =	15.6	15.6	15.6
B(i)	10 A socket outlets	1000 + (21 × 750) /230 =	72.8	72.8	72.8
B(iii)	4 × 32 A socket outlets	32 + (3 × 32 × 0.75) =	104	104	104
B(iii)	3 × 20 A socket outlets	3 × 20 × 0.75 =	45	45	45
D	motor 1	42 × 100% =	42	42	42
D	motor 2	42 × 75% =	31.5	31.5	31.5
D	motor 3	28 × 50% =	14	14	14
H	welder 1	full rated current	18	18	
H	welder 2			18	18
		Total	342.9	**360.9**	342.9

4 Maximum demand of consumer's mains in multiple non-domestic installation. (AS/NZS 3000:2007 Table C3 Colum 4)

Type of occupancy		Energy demand		
	Load	**Area**	**Average VA/m²**	**Demand**
Shops	Light and power	1650 m²	70	115500 VA
	Air-conditioning	1650 m²	30	49500 VA
Offices	Light and power	900 m²	50	45000 VA
	Air-conditioning	900 m²	25	22500 VA
			Total	232500 VA

Review questions 11.4

1 Starting at Table 3(2) Item 1 No _**3**____ Then Table 5 column 18 ____Cable Size **25 mm²** _PE Size **6 mm²**
I_B = _**70A**____ I_N = ____ **80A**____ I_Z = _**90A**____
$I_B \leq I_N \leq I_Z$ (Y)

2 Starting at Table 3(2) Item 1 Table No ___ **4**_ Column No ___**15**_____ Cable Size _**6 mm²**_ PE Size _**2.5 mm²**
___ I_B = _**35 A**_ I_N = _**40 A**_ I_Z = ___ **40A**_
$I_B \leq I_N \leq I_Z$ (Y)

3 Starting at AS/NZ3008.1:2009 Table 3(4) Item 2 De-rating Table ___ **26(2)**______ Column No ____ **2**____ DR _____ **0.9**_____ Table No _**7**_____ Column No ___ **24**___ Cable Size _____ **6 mm²**_ PE Size _ **2.5 mm²**_ I_B = **40 A**___ I_N = _ **40 A**_____ I_Z = **45 × 0.9 = 40.5 A** $I_B \leq I_N \leq I_Z$ (Y)
The 6 mm² TPI cable has a rated current of 45 A from AS/NZ 3008.1:2009 Table 7 column 24. The cable is de-rated by 0.9 from AS/NZ 3008.1:2009 Table 26(2) column 2 to 40.5 A. A circuit breaker rating 40 A is selected to protect the cable. The maximum demand of the cable is set by limitation at 40 A.

Review questions 11.5

1 a Consumer's mains
V_c = 0.285 mV/A.m (Table 41 column 8)

$$V_d = \frac{L \times I \times V_c}{1000} = \frac{50 \times 350 \times 0.285}{1000} = 4.98\ \text{V}$$

b Sub-mains
V_c = 0.457 mV/A.m (Table 40 column 8)

$$V_d = \frac{L \times I \times V_c}{1000} = \frac{60 \times 250 \times 0.457}{1000} = 6.86\ \text{V}$$

c Final sub-circuit
V_c = 15.6 mV/A.m (Table 42 column 6)

$$V_d = \frac{L \times I \times V_c}{1000} = \frac{30 \times 20 \times 15.6}{1000} = 9.36\ \text{V}$$

d $V_T = V_{CM} + V_{SM} + V_{fsc} = 6.86 + 9.36 + 9.36 = 21.2\ \text{V}$
Installation does not comply $V_T > 5\%$ of 400 V (20 V)

2 a Consumer's mains
V_c = 0.227 mV/A.m (Table 40 column 8)

$$V_d = \frac{L \times I \times V_c}{1000} = \frac{55 \times 400 \times 0.227}{1000} = 4.99\ \text{V}$$

b Sub-mains

$$V_c = \frac{1000 \times V_d}{L \times I} = \frac{1000 \times 6}{45 \times 63} = 2.116\ \text{mV/A.m}$$

c.s.a. = 25 mm² (1.54 mV/A.m) (Table 42 column 6)

c Final sub-circuit

$$V_p = \frac{5\% \times 400 - (V_{CM} + V_{SM})}{\sqrt{3}} = \frac{20 - (4.99 + 6)}{\sqrt{3}} = 5.2\,V$$

$$V_c = \frac{1000 \times V_d}{L \times I} = \frac{1000 \times 5.2}{40 \times 32} = 4.06\,mV/A.m$$

c.s.a. = 10 mm² (3.86 mV/A.m) (Table 42 column 6)

3 a Consumer's mains

$V_c = 0.385$ mV/A.m (Table 41 column 8)

$$V_d = \frac{L \times I \times V_c}{1000} = \frac{55 \times 400 \times 0.385}{1000} = 3.85\,V$$

b Final sub-circuit length

$V_c = 3.86 \times 1.155 = 4.4583$ mV/A.m (Table 40 column 6)

$$V_p = \frac{5\% \times 400 - V_{CM}}{\sqrt{3}} = \frac{20 - 3.85}{\sqrt{3}} = 9.32\ V$$

$$V_d = \frac{L \times I \times V_c}{1000} \therefore V_p \times 1000 = L \times I \times V_c$$

$$\therefore L = \frac{V_p \times 1000}{I \times V_c} = \frac{9.32 \times 1000}{50 \times 4.4583} = 41.8\,m$$

Review questions 11.6

1 From AS/NZS3000:2007 clause B4.5.

$I_a = 7.5 \times IN$

$$Z_S = \frac{U_o \times 0.8}{I_a} = \frac{230 \times 0.8}{7.5 \times 32} = 0.766\ \Omega$$

2 a 5 second disconnection time AS/NZS 3000:2007 clause 1.5.5.3(d)

$Z_S = 0.43\,\Omega$ AS/NZS 3000:2007 Table 8.1

$Z_{int} = ZS \times 0.8 = 0.43 \times 0.8 = 0.344\,\Omega$

b $L_{max} = 90$ m AS/NZS 3000:2007 Table B1

80 m complies with AS/NZS 3000:2007

3 a 5 second disconnection time AS/NZS 3000:2007 clause 1.5.5.3(d)

b $I_a \approx 4 \times I_N = 4 \times 100 = 400\,A$

c $$L_{max} = \frac{U_o \times 0.8 \times S_{ph} \times S_{pe}}{I_a \times \rho \times (S_{ph} + S_{pe})}$$

$$= \frac{230 \times 0.8 \times 25 \times 6}{400 \times 22.5 \times 10^{-3} \times (25 + 6)} = 98.9\ m$$

Review questions 11.7

1 Harmful effects of short circuit current include any 4 of the following; extremely large currents, cables reaching very high temperatures, insulation failure, destruction of protection devices, ionisation of enclosures, severe magnetic stresses on conductors and supports, fatal injury to persons and livestock, fire and explosion (arc blasts).

2 $$I_{S/C} = \frac{MVA}{\sqrt{3} \times V_L} = \frac{15 \times 10^6}{\sqrt{3} \times 400} = 21650\,A$$

3 a $$I_{S/C} = \frac{MVA}{\sqrt{3} \times V_L} = \frac{750000}{\sqrt{3} \times 400} = 1090\,A$$

b $$Z_{Tx} = \frac{V_P}{I_{S/C}} = \frac{230}{1090} = 0.211\,\Omega$$

4 a $$Z_{Supply} = \frac{V}{I_{S/C}} = \frac{230}{25000} = 0.0092\,\Omega$$

b $R_c = 0.248\,\Omega$/km (AS3008.1 Table 35 column 5) @90°C

$$Z_{CM} = \frac{R_c}{1000} \times L = \frac{0248}{1000} \times 30 = 0.00744\ \Omega$$

c $Z_T = Z_{Supply} + Z_{CM} = 0.0092 + 0.00744 = 0.01664\ \Omega$

d $$I_{S/C} = \frac{V}{Z_T} = \frac{230}{0.01664} = 13822\,A\ (13.822\,kA)$$

e No, the 20 A 3 kA CB does not have the breaking capacity to interrupt 13.8 kA on its own. The let through energy of the 250 A fault current limiter is greater than 3 kA. A larger kA rating of the circuit breaker is required.

Chapter 12 solutions

Review questions 12.1

1 Sheathing, conduit, trunking and constructed enclosures.

2 The installation methods must not contribute to the spread of fire.

Review questions 12.2

1 AS/NZS 3003:2011 Electrical installations – Patient areas.

2 AS/NZS 3012:2010 Electrical installations – Construction and demolition sites.

Review questions 12.3

1 AS/NZS 60079.10.1 Gases and AS/NZS 60079.10.2 Dust.

2 Combustible dust such as flour in a flour mill.

Review questions 12.4

1 Zone 0; Zone 1; Zone 2 and Zone 3 as defined in AS/NZS 3000:2007 Figure 6.1.

2 No greater than 50 volts a.c. or 120 volts d.c. ripple free.

Review questions 12.5

1 Bundled and webbed.
2 4.6 m for insulated cable, 5 m for uninsulated. (AS/NZS 3000:2007 Table 3.8)

Review questions 12.6

1 Cables enclosed in insulating conduit. Cables enclosed in GI pipe.
2 There are no vulnerable overhead structures and no visual intrusion.

Review questions 12.7

1 Choose to suit the structural system and the environment.
2 Insulation resistance between active conductors and earth. Insulation resistance between circuit poles. Polarity of connections. Continuity of earth conductors.

Chapter 13 solutions

Review questions 13.1

1 Cat III 500 volt.
2 To allow continuity testing of sub-circuits.

Review questions 13.2

1 Earth resistance tests, insulation resistance tests, polarity tests, earth fault loop impedance tests, and operation of RCDs.
2 To ensure that a fault current is high enough to operate the circuit protection equipment.
3 Yes.

Review questions 13.3

1 Incorrect labelling, incorrect size circuit breaker, missing junction box covers and damaged cable insulation.
2 6 mm^2.

Review questions 13.4

1 With the continuity tester set to the lowest range.
2 As determined with reference to Table 8.2 in AS/NZS 3000:2007 Wiring Rules.
3 With reference to Table 8.1 in AS/NZS 3000:2007 Wiring Rules.
4 Using a fault loop impedance tester.
5 Using a trailing lead and the low resistance continuity tester.
6 The circuit complies as each conductor resistance is lower than required from Table 8.2.

Review questions 13.5

1 Certificate of Electrical Compliance.
2 The date of initial energisation, the name and licence number of the electrical practitioner, the results of continuity tests, and the results of insulation tests.
3 Location of the main earth, the presence of alternative power supplies, plan of underground cables.

Chapter 14 solutions

Review questions 14.1

1 Voltage source, current path, load and control device.
2 To remove the voltage source safely.
3 Loss of supply, open circuit and short circuit.

Review questions 14.2

1 35 ohms.
2 0.14 ohms.

Review questions 14.3

1 No, the parallel circuit would be 3000 watts and would draw 13 amps.
2 Faulty RCD, earth leakage from neutral conductor, earth leakage from active conductor, or earth leakage within the appliance.
3 No.

Review questions 14.4

1 At each light fitting.
2 Inspect each light fitting for a short circuit between active feed and earth.
3 Yes, the neutral conductor is part of the protected sub-circuit.
4 Soot marks around a ceiling fitting.

Review questions 14.5

1 Universal motor.
2 Permanently split motor.
3 It has a low starting torque.

Review questions 14.6

1 No.
2 No.

Review questions 14.7

1 Yes, it is less than the 0.61 ohms maximum required.
2 Yes, this is greater than the special allowance for this type of element.

Index